T0317639

LASERS AND OPTOELECTRONICS

LASERS AND OPTOELECTRONICS

FUNDAMENTALS, DEVICES AND APPLICATIONS

Anil K. Maini

Laser Science and Technology Centre (LASTEC), Delhi, India

This edition first published 2013

Registered office
John Wiley & Sons Ltd, The Atrium, Southern Gate, Chichester, West Sussex, PO19 8SQ, United Kingdom

For details of our global editorial offices, for customer services and for information about how to apply for permission to reuse the copyright material in this book please see our website at www.wiley.com.

Library of Congress Cataloging-in-Publication Data

Maini, Anil Kumar.
Lasers and optoelectronics : fundamentals, devices, and applications / Dr Anil K. Maini.
1 online resource.
Includes bibliographical references and index.
Description based on print version record and CIP data provided by publisher; resource not viewed.
ISBN 978-1-118-68894-6 – ISBN 978-1-118-68895-3 – ISBN 978-1-118-68896-0 – ISBN 978-1-118-45887-7 (cloth) 1. Lasers. 2. Optoelectronic devices. I. Title.
TA1675
621.36′6–dc23

2013023492

A catalogue record for this book is available from the British Library.

ISBN: 978-1-118-45887-7

Set in 9/11pt Times by Thomson Digital, Noida, India

1 2013

Affectionately dedicated to:
The loving memory of my parents
Shri Sukhdev Raj Maini
and
Smt Vimla Maini

Contents

Preface **xix**

Part I LASER FUNDAMENTALS **1**

1 Laser Basics **3**
1.1 Introduction 3
1.2 Laser Operation 3
1.3 Rules of Quantum Mechanics 3
1.4 Absorption, Spontaneous Emission and Stimulated Emission 4
1.5 Population Inversion 10
1.5.1 Producing Population Inversion 11
1.6 Two-, Three- and Four-Level Laser Systems 11
1.6.1 Two-Level Laser System 11
1.6.2 Three-Level Laser System 12
1.6.3 Four-Level Laser System 14
1.6.4 Energy Level Structures of Practical Lasers 15
1.7 Gain of Laser Medium 16
1.8 Laser Resonator 17
1.9 Longitudinal and Transverse Modes 18
1.10 Types of Laser Resonators 21
1.11 Pumping Mechanisms 23
1.11.1 Optical Pumping 24
1.11.2 Electrical Pumping 28
1.11.3 Other Methods of Pumping 29
1.12 Summary 29
Review Questions 30
Problems 30
Self-evaluation Exercise 31
Bibliography 33

2 Laser Characteristics **34**
2.1 Introduction 34
2.2 Laser Characteristics 34
2.2.1 Monochromaticity 34
2.2.1.1 Line-broadening Mechanisms 34
2.2.2 Coherence 36
2.2.2.1 Temporal Coherence 36
2.2.2.2 Spatial Coherence 38
2.2.3 Directionality 39

2.3 Important Laser Parameters 41
2.3.1 Wavelength 41
2.3.2 CW Power 42
2.3.3 Peak Power 42
2.3.4 Average Power 43
2.3.5 Pulse Energy 43
2.3.6 Repetition Rate 43
2.3.7 Pulse Width 44
2.3.8 Duty Cycle 44
2.3.9 Rise and Fall Times 45
2.3.10 Irradiance 45
2.3.11 Radiance 45
2.3.12 Beam Divergence 46
2.3.13 Spot Size 47
2.3.14 M^2 Value 48
2.3.15 Wall-plug Efficiency 48
2.4 Measurement of Laser Parameters 49
2.4.1 Measurement of Power, Energy and Repetition Rate 49
2.4.1.1 Choosing the Appropriate Sensor 49
2.4.1.2 Choosing the Appropriate Meter 50
2.4.2 Measurement of Spot Size 50
2.4.3 Measurement of Divergence 50
2.4.4 Measurement of M^2 Value 52
2.4.5 Measurement of Line Width 53
2.5 Laser Beam Diagnostic Equipment 56
2.5.1 Wavelength Meter 56
2.5.2 Laser Spectrum Analyzer 56
2.5.3 Laser Beam Profiler 57
2.5.4 Beam Propagation Analyzer 58
2.6 Summary 59
Review Questions 60
Problems 61
Self-evaluation Exercise 62
Bibliography 63

Part II TYPES OF LASERS **65**

3 Solid-state Lasers **67**
3.1 Introduction: Types of Lasers 67
3.2 Importance of Host Material 67
3.2.1 Lasing Species 68
3.3 Operational Modes 68
3.3.1 CW Output 69
3.3.2 Free-running Output 69
3.3.3 Q-switched Output 69
3.3.4 Cavity-dumped Output 72
3.3.5 Mode-locked Output 72
3.4 Ruby Lasers 76
3.5 Neodymium-doped Lasers 78
3.5.1 Nd:YAG Lasers 79
3.5.2 Nd:YLF Lasers 79

3.5.3 Nd:YVO$_4$ Lasers 82
3.5.4 Nd:Cr:GSGG Lasers 82
3.5.5 Nd:Glass Lasers 84
3.6 Erbium-doped Lasers 85
3.6.1 Er:YAG Laser 85
3.6.2 Er:Glass Laser 85
3.7 Vibronic Lasers 88
3.7.1 Alexandrite Laser 88
3.7.2 Titanium-sapphire Laser 90
3.8 Colour Centre Lasers 90
3.9 Fibre Lasers 91
3.9.1 Basic Fibre Laser 92
3.9.2 Fibre Lasers versus Bulk Solid-state Lasers 93
3.9.3 Operational Regimes 95
3.9.4 Photonic Crystal Fibre Lasers 96
3.9.4.1 Guiding Mechanisms in PCF 97
3.9.4.2 Subclasses of PCFs 97
3.9.4.3 PCF Lasers 98
3.9.5 Applications 98
3.10 Summary 101
Review Questions 102
Problems 102
Self-evaluation Exercise 102
Bibliography 104

4 Gas Lasers **105**
4.1 Introduction to Gas Lasers 105
4.1.1 The Active Media 105
4.1.2 Inter-level Transitions 106
4.1.3 Pumping Mechanism 106
4.2 Helium-neon Lasers 107
4.3 Carbon Dioxide Lasers 111
4.4 Metal Vapour Lasers 115
4.4.1 Helium-cadmium Laser 115
4.4.2 Copper Vapour and Gold Vapour Lasers 116
4.5 Rare Gas Ion Lasers 118
4.6 Excimer Lasers 120
4.7 Chemical Lasers 121
4.7.1 Hydrogen Fluoride/Deuterium Fluoride (HF/DF) Lasers 121
4.7.2 Chemical Oxygen Iodine Laser (COIL) 123
4.7.3 All Gas-Phase Iodine Laser (AGIL) 124
4.8 Carbon Dioxide Gas Dynamic Lasers 125
4.9 Dye Laser 125
4.9.1 Active Medium 126
4.9.2 Pump Mechanisms 127
4.9.3 Wavelength Selection 127
4.10 Free-electron Lasers 127
4.11 X-Ray Lasers 129
4.12 Summary 129
Review Questions 129
Self-evaluation Exercise 130
Bibliography 131

5 Semiconductor Lasers **132**
5.1 Introduction 132
5.2 Operational Basics 132
5.3 Semiconductor Laser Materials 135
5.4 Types of Semiconductor Lasers 136
5.4.1 Homojunction and Heterojunction Lasers 136
5.4.2 Quantum Well Diode Lasers 136
5.4.3 Distributed-feedback (DFB) Lasers 138
5.4.4 Vertical-cavity Surface-emitting Laser (VCSEL) 140
5.4.5 Vertical External-cavity Surface-emitting Lasers (VECSEL) 140
5.4.6 External-cavity Semiconductor Diode Lasers 141
5.4.7 Optically Pumped Semiconductor Lasers 143
5.4.8 Quantum Cascade Lasers 145
5.4.9 Lead Salt Lasers 147
5.5 Characteristic Parameters 148
5.5.1 Threshold Current 148
5.5.2 Slope Efficiency 148
5.5.3 Beam Divergence 149
5.5.4 Line Width 151
5.5.5 Beam Polarization 152
5.6 Gain- and Index-guided Diode Lasers 152
5.7 Handling Semiconductor Diode Lasers 152
5.8 Semiconductor Diode Lasers: Application Areas 153
5.8.1 Directed Energy 153
5.8.2 Coherence 153
5.8.3 Monochromaticity 153
5.9 Summary 154
Review Questions 155
Problems 155
Self-evaluation Exercise 156
Bibliography 157

Part III LASER ELECTRONICS AND OPTOELECTRONICS **159**

6 Building Blocks of Laser Electronics **161**
6.1 Introduction 161
6.2 Linear Power Supplies 161
6.2.1 Constituents of a Linear Power Supply 161
6.2.2 Rectifier Circuits 162
6.2.3 Filters 164
6.2.4 Linear Regulators 166
6.2.4.1 Emitter-follower Regulator 166
6.2.4.2 Series-pass Regulator 167
6.2.4.3 Shunt Regulator 170
6.2.4.4 Linear IC Voltage Regulators 171
6.3 Switched-mode Power Supplies 173
6.3.1 Linear versus Switched-mode Power Supplies 173
6.3.2 Different Types of Switched-mode Power Supplies 174
6.3.2.1 Flyback Converters 174
6.3.2.2 Forward Converter 178
6.3.2.3 Push-pull Converter 178

6.3.2.4 Switching Regulators 181
6.3.2.5 Three-terminal Switching Regulators 183
6.3.3 Connecting Power Converters in Series 184
6.3.4 Connecting Power Converters in Parallel 184
6.4 Constant Current Sources 186
6.4.1 Junction Field-effect-transistor-based Constant Current Source 186
6.4.2 Transistor-based Constant Current Source 187
6.4.3 Opamp-controlled Constant Current Source 189
6.4.4 Constant Current Source Using Three-terminal Regulators 189
6.4.5 Current Mirror Configurations 190
6.4.5.1 Basic Current Mirror 190
6.4.5.2 Widlar Current Source 191
6.4.5.3 Wilson Current Source 191
6.5 Integrated-circuit Timer Circuits 191
6.5.1 Digital IC-based Timer Circuits 191
6.5.2 IC Timer-based Multivibrators 193
6.5.2.1 Astable Multivibrator Using Timer IC 555 194
6.5.2.2 Monostable Multivibrator Using Timer IC 555 195
6.6 Current-to-voltage Converter 197
6.7 Peak Detector 199
6.8 High-voltage Trigger Circuit 200
6.9 Summary 202
Review Questions 203
Problems 204
Self-evaluation Exercise 206
Bibliography 207

7 Solid-state Laser Electronics 208
7.1 Introduction 208
7.2 Spectrum of Laser Electronics 208
7.2.1 Solid-state Lasers 208
7.2.2 Semiconductor Diode Lasers 209
7.2.3 Gas Lasers 211
7.2.4 Testing and Evaluation of Lasers 212
7.2.5 Laser Sensor Systems 213
7.3 Electronics for Solid-state Lasers 213
7.4 Electronics for Pulsed Solid-state Lasers 214
7.4.1 Electronics for Q-switched Solid-state Lasers 214
7.4.2 Capacitor-charging Power Supply 216
7.4.3 Simmer Power Supply 222
7.4.4 Pseudo-simmer Mode 224
7.4.5 Pulse-forming Network 225
7.4.6 Flashlamp Trigger Circuit 231
7.5 Electronics for CW Solid-state Lasers 233
7.5.1 Arc Lamps 233
7.5.2 Electrical Characteristics 234
7.5.3 Arc Lamp Power Supply 235
7.5.4 Modulated CW and Quasi-CW Operation of Arc Lamp 236
7.6 Solid-state Laser Designators and Rangefinders 237
7.7 Summary 238
Review Questions 239
Problems 240

Self-evaluation Exercise 240
Bibliography 241

8 Gas Laser Electronics 242
8.1 Introduction 242
8.2 Gas Discharge Characteristics 242
8.3 Gas Laser Power Supplies 242
8.4 Helium-Neon Laser Power Supply 244
8.4.1 Power Supply Design 247
8.4.2 Switched-mode Power Supply Configurations 250
8.4.3 Other Possible Configurations 253
8.4.4 Configurations for Special Applications 254
8.4.5 Ballast Resistance 257
8.5 Carbon Dioxide Laser Power Supplies 257
8.5.1 DC-excited CW CO_2 Laser 257
8.5.2 DC-excited Pulsed CO_2 Laser 257
8.5.3 RF-excited CO_2 Lasers 259
8.6 Power Supplies for Metal Vapour Lasers 260
8.7 Power Supplies for Excimer Lasers 261
8.8 Power Supplies for Ion Lasers 262
8.9 Frequency Stabilization of Gas Lasers 263
8.9.1 Dither Stabilization 264
8.9.2 Stark-cell Stabilization 265
8.9.3 Optogalvanic Stabilization 265
8.9.4 Stabilization using Saturation Absorption Dip 266
8.10 Summary 267
Review Questions 268
Problems 268
Self-evaluation Exercise 269
Bibliography 270

9 Laser Diode Electronics 271
9.1 Introduction 271
9.2 Laser Diode Protection 271
9.2.1 Laser Diode Drive and Control 272
9.2.2 Interconnection Cables and Grounding 274
9.2.3 Transient Suppression 275
9.2.4 Electrostatic Discharge 275
9.3 Operational Modes 276
9.3.1 Constant-current Mode 276
9.3.2 Constant-power Mode 277
9.4 Laser Diode Driver Circuits 278
9.4.1 Basic Constant-current Source 278
9.4.2 Laser Diode Driver with Feedback Control 279
9.4.3 Laser Diode Driver with Modulation Input 282
9.4.4 Laser Diode Driver with Protection Features 284
9.4.5 Laser Diode Driver with Automatic Power Control 286
9.4.6 Quasi-CW Laser Diode Driver 289
9.5 Laser Diode Temperature Control 291
9.5.1 Thermoelectric Cooling Fundamentals 292
9.5.2 Thermoelectric Cooler: Performance Characteristics 295

9.5.3 TE Module Selection 297
9.5.4 Heat Sink Selection 299
9.5.5 Thermoelectric Cooler Drive and Control Circuits 301
9.5.5.1 Temperature Sensing Circuits 301
9.5.5.2 Error Amplifier 303
9.5.5.3 Error Signal Processor 303
9.5.5.4 Output Stage 306
9.6 Summary 308
Review Questions 310
Problems 310
Self-evaluation Exercise 312
Bibliography 314

10 Optoelectronic Devices and Circuits **315**
10.1 Introduction 315
10.2 Classification of Photosensors 315
10.2.1 Photoelectric Sensors 315
10.2.2 Thermal Sensors 316
10.3 Radiometry and Photometry 316
10.3.1 Radiometric and Photometric Flux 316
10.3.2 Radiometric and Photometric Intensity 316
10.3.3 Radiant Incidence (Irradiance) and Illuminance 318
10.3.4 Radiant Sterance (Radiance) and Luminance 318
10.4 Characteristic Parameters 318
10.4.1 Responsivity 318
10.4.2 Noise Equivalent Power (NEP) 321
10.4.3 Detectivity and D-star 321
10.4.4 Quantum Efficiency 321
10.4.5 Response Time 322
10.4.6 Noise 323
10.5 Photoconductors 324
10.5.1 Application Circuits 326
10.6 Photodiodes 329
10.6.1 Types of Photodiodes 330
10.6.1.1 PN Photodiodes 330
10.6.1.2 PIN Photodiodes 331
10.6.1.3 Schottky Photodiodes 331
10.6.1.4 Avalanche Photodiodes 331
10.6.2 Equivalent Circuit 331
10.6.3 I–V Characteristics 333
10.6.4 Application Circuits 334
10.6.5 Solar Cells 336
10.7 Phototransistors 340
10.7.1 Application Circuits 341
10.8 Photo- FET, SCR and TRIAC 343
10.8.1 Photo-FET 343
10.8.2 Photo-SCR 343
10.8.3 Photo-TRIAC 344
10.9 Photoemissive Sensors 345
10.9.1 Vacuum Photodiodes 345
10.9.2 Photomultiplier Tubes 345
10.9.3 Image Intensifiers 346

10.10 Thermal Sensors 347
10.10.1 Thermocouple and Thermopile 347
10.10.2 Bolometer 348
10.10.3 Pyroelectric Sensors 348
10.11 Displays 350
10.11.1 Display Characteristics 350
10.11.2 Types of Displays 350
10.12 Light-emitting Diodes 351
10.12.1 Characteristic Curves 352
10.12.2 Parameters 354
10.12.3 Drive Circuits 354
10.13 Liquid-crystal Displays 356
10.13.1 Construction 356
10.13.2 Driving LCD 357
10.13.3 Response Time 358
10.13.4 Types of LCD Displays 358
10.13.5 Advantages and Disadvantages 361
10.14 Cathode Ray Tube Displays 361
10.15 Emerging Display Technologies 362
10.15.1 Organic Light-emitting Diodes (OLEDs) 362
10.15.2 Digital Light-processing (DLP) Technology 363
10.15.3 Plasma Display Panels (PDPs) 363
10.15.4 Field Emission Displays (FEDs) 363
10.15.5 Electronic Ink Displays 363
10.16 Optocouplers 363
10.16.1 Characteristic Parameters 364
10.16.2 Application Circuits 366
10.17 Summary 370
Review Questions 372
Problems 373
Self-evaluation Exercise 374
Bibliography 377

Part IV LASER APPLICATIONS **379**

11 Lasers in Industry **381**
11.1 Introduction 381
11.2 Material-processing Applications 381
11.2.1 Classification 381
11.2.2 Important Considerations 382
11.2.2.1 Material Properties: Absorption Length and Diffusion Length 382
11.2.2.2 Laser Parameters 382
11.2.3 Common Material-processing Applications 383
11.2.4 Advantages 385
11.3 Laser Cutting 385
11.3.1 Basic Principle 385
11.3.2 Laser Cutting vs Plasma Cutting 387
11.3.3 Laser Cutting Processes 387
11.3.4 Machine Configurations 388

11.4 Laser Welding 390
11.4.1 Laser Welding Processes 390
11.4.2 Welding Lasers 390
11.4.3 Advantages 392
11.5 Laser Drilling 393
11.5.1 Basic Principle 393
11.5.2 Laser Drilling Processes 394
11.5.3 Lasers for Drilling 395
11.5.4 Advantages of Laser Drilling 396
11.6 Laser Marking and Engraving 396
11.6.1 Principle of Operation 396
11.6.2 Laser Marking Processes 397
11.6.3 Lasers for Marking and Engraving 398
11.6.4 Advantages 399
11.7 Laser Micromachining 401
11.7.1 Laser Micromachining Operations 402
11.7.2 Lasers for Micromachining 403
11.7.3 Laser Micromachining Techniques 404
11.8 Photolithography 407
11.8.1 Basic Process 408
11.8.2 Lasers for Photolithography 411
11.9 Rapid Manufacturing 411
11.9.1 Additive Versus Subtractive Manufacturing 412
11.9.2 Rapid Manufacturing Technologies 412
11.9.3 Lasers for Rapid Manufacturing 413
11.9.4 Advantages 414
11.10 Lasers in Printing 414
11.10.1 Laser Printing Process 415
11.10.2 Anatomy of Laser Printer 415
11.10.3 Choice Criteria 416
11.10.4 Laser Printers vs Inkjet Printers 417
11.11 Summary 418
Review Questions 419
Self-evaluation Exercise 420
Bibliography 421

12 Lasers in Medicine 422
12.1 Introduction 422
12.2 Light–tissue Interaction 422
12.2.1 Light–tissue Interaction for Diagnostic Applications 423
12.2.1.1 Fundamental Interaction Mechanisms 423
12.2.1.2 Optical Properties of Tissues 425
12.2.1.3 Fluence Rate Distribution 426
12.2.2 Light–tissue Interaction for Therapeutic Applications 427
12.2.2.1 Photochemical Effects 427
12.2.2.2 Photothermal Effects 428
12.2.2.3 Photomechanical Effects 429
12.3 Laser Diagnostics 430
12.3.1 Basic Principle 431
12.3.2 Comparison with Other Techniques 431
12.3.3 In Vivo Optical Diagnostic Techniques 432
12.3.3.1 White Light Imaging 432

12.3.3.2 Diffuse Optical Spectroscopy 434
12.3.3.3 Elastic Scattering Spectroscopy 434
12.3.3.4 Optical Coherence Tomography 435
12.3.3.5 Confocal Imaging 438
12.3.3.6 Fluorescence Spectroscopy and Imaging 438
12.3.3.7 Raman Spectroscopy 441
12.4 Therapeutic Techniques: Application Areas 442
12.5 Ophthalmology 443
12.5.1 Refractive Surgery 444
12.5.2 Treatment of Glaucoma 447
12.5.3 Cataract Surgery 448
12.5.4 Treatment of Retinal Detachment (Retinopexy) 449
12.5.5 Treatment of Proliferative Diabetic Retinopathy 449
12.6 Dermatology 449
12.6.1 Portwine Stains 450
12.6.2 Pigmented Lesions and Tattoos 451
12.6.3 Hair Removal 452
12.6.4 Lasers for Dermatology 453
12.7 Laser Dentistry 453
12.7.1 Considerations in Laser Dentistry 453
12.7.2 Lasers for Dentistry 454
12.8 Vascular Surgery 455
12.8.1 Conventional Treatment of Angioplasty: PTA 455
12.8.2 Laser Angioplasty 456
12.9 Photodynamic Therapy 456
12.9.1 Important Considerations 458
12.9.2 Applications of PTD 458
12.10 Thermal Therapy 459
12.10.1 Treatment of Haemorrhages of Peptic Ulcers 459
12.10.2 Treatment of Cancer 460
12.11 Summary 460
Review Questions 462
Self-evaluation Exercise 463
Bibliography 465

13 Lasers in Science and Technology **466**
13.1 Introduction 466
13.2 Optical Metrology 466
13.2.1 Interferometers 466
13.2.1.1 Michelson Interferometer 467
13.2.1.2 Twyman–Green Interferometer 468
13.2.1.3 Mach–Zehnder Interferometer 468
13.2.1.4 Fabry–Pérot Interferometer 469
13.2.1.5 Sagnac Interferometer 471
13.2.2 Length Metrology 471
13.2.3 Time and Frequency Metrology 472
13.2.3.1 Optical Clock 474
13.2.4 Measurement of Line Width 474
13.2.5 Infrared Thermometer 476
13.3 Laser Velocimetry 478
13.3.1 Laser Doppler Velocimetry 478
13.3.1.1 Operational Principle 478
13.3.1.2 Advantages 480

13.3.1.3 Applications 480
13.3.2 Particle Image Velocimetry 480
13.4 Laser Vibrometry 482
13.4.1 Operational Principle 482
13.4.2 Types of Laser Doppler Vibrometers 484
13.4.3 Applications 484
13.5 Electronic Speckle Pattern Interferometry 484
13.5.1 Operational Principle 485
13.5.2 Measurement Configurations 486
13.5.2.1 Out-of-plane Displacement Measurement 486
13.5.2.2 In-plane Displacement Measurement 487
13.5.2.3 In-plane Displacement Gradient Measurement 487
13.5.2.4 Three-dimensional (3D) Measurement 488
13.5.2.5 Vibration Measurement 490
13.6 Satellite Laser Ranging 490
13.6.1 Operational Principle 491
13.6.2 Lasers for SLR 492
13.6.3 SLR Telescopes and Stations 492
13.6.4 SLR Applications 493
13.7 Lasers in Astronomy 494
13.7.1 Adaptive Optics 494
13.7.2 Laser Guide Star 494
13.7.3 Laser Guide Star Mechanisms 496
13.7.3.1 Rayleigh Guide Star 496
13.7.3.2 Sodium Beacon Guide Star 496
13.8 Holography 496
13.8.1 Basic Principle 498
13.8.2 Types of Hologram 499
13.8.2.1 Amplitude- and Phase-modulated Holograms 499
13.8.2.2 Transmission and Reflection Holograms 499
13.8.2.3 Thin and Thick Holograms 500
13.8.2.4 Other Commonly Encountered Holograms 501
13.8.3 Applications 502
13.9 Summary 503
Review Questions 504
Self-evaluation Exercise 505
Bibliography 507

14 Military Applications: Laser Instrumentation 508
14.1 Introduction 508
14.2 Military Applications of Lasers 508
14.3 Laser-based Instrumentation 512
14.3.1 Laser Aiming Modules 512
14.3.2 Laser Rangefinders 513
14.3.2.1 Time-of-Flight Technique 514
14.3.2.2 Triangulation Technique 515
14.3.2.3 Phase Shift Technique 516
14.3.2.4 FM-CW Range-finding Technique 516
14.3.2.5 Lasers for Laser Rangefinders 518
14.3.2.6 Applications and Related Devices 519
14.3.3 Laser Target Designators 520
14.3.4 Laser Proximity Sensors 520
14.3.5 Laser Bathymetry Sensors 524

14.3.6 Laser Radar (Ladar) Sensors 526
14.3.7 Forward-looking Infrared (FLIR) Sensors 528
14.4 Guided Munitions 532
14.4.1 Guidance Techniques 532
14.4.1.1 Beam Rider Guidance 532
14.4.1.2 Command Guidance 533
14.4.1.3 Homing Guidance 536
14.4.1.4 Navigation Guidance 538
14.4.1.5 Ring Laser Gyroscope 540
14.4.1.6 Fibre-optic Gyroscope 545
14.4.2 Electro-optically Guided Precision Strike Munitions 546
14.4.2.1 Laser-guided Munitions: Operational Basics 549
14.4.2.2 IR-guided Missiles: Operational Basics 554
14.5 Laser Communication 556
14.5.1 Advantages and Limitations 556
14.5.2 Free-space Communication 557
14.5.3 Fibre-optic Communication 559
14.6 Summary 561
Review Questions 562
Problems 563
Self-evaluation Exercise 564
Bibliography 565

15 Military Applications: Directed-energy Laser Systems **566**
15.1 Introduction 566
15.2 Laser Technology for Low-intensity Conflict (LIC) Applications 566
15.2.1 Importance of Laser Technology in LIC Applications 566
15.2.2 Detection of Chemical and Biological Warfare Agents 567
15.2.2.1 Detection of Chemical Warfare Agents 567
15.2.2.2 Detection of Biological Agents 568
15.2.3 Detection of Explosive Agents 570
15.2.4 Detection of Optical and Electro-optic Devices 573
15.2.5 Disposal of Unexploded Ordnances 574
15.2.6 Non-lethal Laser Dazzlers 576
15.3 Electro-optic Countermeasures 580
15.3.1 Need and Relevance 580
15.3.2 Passive and Active Countermeasures 580
15.3.3 Types of EOCM Equipment 581
15.3.4 Infrared Countermeasures 584
15.4 Directed-energy Laser Weapons 585
15.4.1 Operational Advantages and Limitations 587
15.4.2 Operational Scenario 588
15.4.3 Components of Directed-energy Weapon Systems 589
15.4.4 International Status 590
15.5 Summary 592
Review Questions 595
Self-evaluation Exercise 596
Bibliography 598

Appendix A: Laser Safety **599**

Index **605**

Preface

Laser, an acronym for light amplification by stimulated emission of radiation (as coined by Gould in his notebooks) is a household name today. In their early stages of development and evolution, lasers were originally confined to the premises of prominent research centres such as the Bell laboratories and Hughes research laboratories and to major academic institutes such as Columbia University. More than five decades after Theodore Maiman demonstrated the first laser in May 1960 at Hughes, this is no longer the case. Lasers are undoubtedly one of the greatest inventions of the 20th century along with satellites, computers and integrated circuits. Their use in commercial, industrial, bio-medical, scientific and military applications continues to expand today.

Lasers and Optoelectronics: Fundamentals, Devices and Applications is a comprehensive treatise on the physical and engineering principles of laser operation, laser system design, optoelectronics and laser applications. It provides a first complete account of the technological and application-related aspects of the subject of lasers and optoelectronics. The book is divided into four parts: laser fundamentals; types of lasers; laser electronics and optoelectronics; and laser applications.

The first chapter of Part I (Laser Fundamentals) aims to introduce the readers to the operational fundamentals of lasers with the necessary dose of quantum mechanics. The topics discussed in Chapter 1 include the principles of laser operation; concepts of population inversion, absorption, spontaneous emission and stimulated emission; three-level and four-level lasers; basic laser resonator; longitudinal and transverse modes of operation; and pumping mechanisms. Chapter 2 discusses the special characteristics that distinguish laser radiation from ordinary light. This is followed by a discussion of the various laser parameters that interest the designers and users of laser devices and systems.

Chapters 3–5 (Part II) describe the three main types of lasers. Based on the nature of the lasing medium, lasers are classified as either solid-state lasers (Chapter 3), gas lasers (Chapter 4) and semiconductor lasers (Chapter 5). Chapter 3 is focused on the operational fundamentals of solid-state lasers, their salient features and typical applications. Gas lasers covered in Chapter 4 include helium-neon lasers, carbon dioxide lasers, metal vapour lasers, rare gas ion lasers, excimer lasers, chemical lasers and gas dynamic carbon dioxide lasers. Again, the emphasis is on operational fundamentals, salient features and typical applications of these lasers. (Dye lasers, free electron lasers and x-ray lasers are also covered in Chapter 4, although these do not belong to any of the three main categories.) Chapter 5 discusses semiconductor diode lasers, which typically emit in the visible to near-infrared bands of the electromagnetic spectrum. Optically pumped semiconductor lasers, quantum cascade lasers, lead salt lasers and antimonide lasers are also covered briefly. Topics covered in this chapter include operational fundamentals, semiconductor materials used in the fabrication of semiconductor lasers, different types of semiconductor diode lasers, characteristic parameters, handling precautions and application areas.

Part III (Chapters 7–10) provides information on the electronics that accompany most laser systems. Chapter 6 describes the basic building blocks of electronics generally used in the design of electronics packages of prominent laser sources and systems configured around them. The intention is to familiarize the readers with the operational basics of these building blocks, allowing an understanding of the specific laser electronics packages discussed in the following chapters. This chapter will particularly benefit laser and optoelectronics students and professionals who do not have a comprehensive knowledge of electronics.

Chapters 7–9 describe the electronics that feature in the three major categories of lasers. Chapter 7 deals with the design and the operational aspects of different types of power supplies, pulse repetition rate and high-voltage trigger generation circuits used in pulsed and continuous wave solid-state lasers. Chapter 8 describes the fundamentals of gas laser power supplies in terms of requirement specifications, circuit configurations and design guidelines, with particular reference to the two most commonly used gas lasers (helium-neon and carbon dioxide lasers). Power supply configurations for metal vapour lasers and excimer lasers (and to a large extent noble gas ion lasers) are similar to those used for helium-neon and carbon dioxide lasers except for minor deviations, which are discussed during the course of the text. Frequency stabilization techniques used in the case of helium-neon and carbon dioxide lasers are also discussed. Chapter 9 discusses the different topics related to semiconductor diode laser electronics. The chapter describes the common mechanisms of laser damage and the precautions which must be observed to protect them. The different topologies commonly used in the design of laser diode drive circuits and temperature controllers to meet the requirements of different applications are also described. Chapter 10 provides detailed information on the fundamentals and application circuits of different types of optoelectronic devices.

Part IV (Chapters 11–15) comprehensively covers applications of lasers and optoelectronic devices. Chapters 11, 12 and 13 cover industrial, medical and scientific applications respectively, and Chapters 14 and 15 cover military applications. The major industrial applications of lasers discussed in Chapter 11 includes cutting, welding, drilling, marking, rapid prototyping, photolithography and laser printing. The use of lasers in medical disciplines such as angioplasty, cancer diagnosis and treatment, dermatology, ophthalmology, cosmetic applications such as hair and tattoo removal and dentistry are discussed in Chapter 12. Chapter 13 describes some of the important applications of lasers in the pursuit of science and technology. Topics discussed include the use of lasers for optical metrology, laser velocimetry, laser vibrometry, electron speckle pattern interferometry, Earth and environmental studies, astronomy and holography. The use of lasers and optoelectronic devices in defence are exhaustively covered in Chapters 14 and 15.

The book is concluded with an appendix including a brief discussion on laser safety, which is of paramount importance to a wide cross-section of designers and users of laser systems.

The book covers each of the topics in its entirety from basic fundamentals to advanced concepts, thereby leading the reader logically from the basics of laser action to advanced topics in laser system design. Simple explanations of the concepts, a number of solved examples and unsolved problems and references for further reading are significant features of each chapter.

The motivation to write this book was provided by the absence of any one volume combining the technology and application-related aspects of laser and optoelectronics. The book is aimed at a wide range of readers, including: undergraduate students of physics and electronics, graduate students specializing in lasers and optoelectronics, scientists and engineers engaged in research and development of lasers and optoelectronics; and practising professionals engaged in the operation and maintenance of lasers and optoelectronics systems.

Part I
Laser Fundamentals

1

Laser Basics

1.1 Introduction

Although lasers were confined to the premises of prominent research centres such as the Bell laboratories, Hughes research laboratories and major academic institutes such as Columbia University in their early stages of development and evolution, this is no longer the case. Theodore Maiman demonstrated the first laser five decades ago in May 1960 at Hughes research laboratories. The acronym 'laser', Light Amplification by Stimulated Emission of Radiation, first used by Gould in his notebooks is a household name today. It was undoubtedly one of the greatest inventions of the second half of the 20th century along with satellites, computers and integrated circuits; its unlimited application potential ensures that it continues to be so even today. Although lasers and laser technology are generally applied in commercial, industrial, bio-medical, scientific and military applications, the areas of its usage are multiplying as are the range of applications in each of these categories.

This chapter, the first in Laser basics, is aimed at introducing the readers to operational fundamentals of lasers with the necessary dose of quantum mechanics. The topics discussed in this chapter include: the principles of laser operation; concepts of population inversion, absorption, spontaneous emission and stimulated emission; three-level and four-level lasers; basic laser resonator; longitudinal and transverse modes of operation; and pumping mechanisms.

1.2 Laser Operation

The basic principle of operation of a laser device is evident from the definition of the acronym 'laser', which describes the production of light by the stimulated emission of radiation. In the case of ordinary light, such as that from the sun or an electric bulb, different photons are emitted spontaneously due to various atoms or molecules releasing their excess energy unprompted. In the case of stimulated emission, an atom or a molecule holding excess energy is stimulated by a previously emitted photon to release that energy in the form of a photon. As we shall see in the following sections, *population inversion* is an essential condition for the stimulated emission process to take place. To understand how the process of population inversion subsequently leads to stimulated emission and laser action, a brief summary of quantum mechanics and optically allowed transitions is useful as background information.

1.3 Rules of Quantum Mechanics

According to the basic rules of quantum mechanics all particles, big or small, have discrete energy levels or states. Various discrete energy levels correspond to different periodic motions of its constituent nuclei and electrons. While the lowest allowed energy level is also referred to as the *ground state*, all other relatively higher-energy levels are called *excited states*. As a simple illustration, consider a hydrogen

Lasers and Optoelectronics: Fundamentals, Devices and Applications, First Edition. Anil K. Maini.

atom. Its nucleus has a single proton and there is one electron orbiting the nucleus; this single electron can occupy only certain specific orbits. These orbits are assigned a quantum number N with the innermost orbit assigned the number $N=1$ and the subsequent higher orbits assigned the numbers $N=2$, 3, 4 . . . outwards. The energy associated with the innermost orbit is the lowest and therefore $N=1$ also corresponds to the ground state. Figure 1.1 illustrates the case of a hydrogen atom and the corresponding possible energy levels.

The discrete energy levels that exist in any form of matter are not necessarily only those corresponding to the periodic motion of electrons. There are many types of energy levels other than the simple-to-describe electronic levels. The nuclei of different atoms constituting the matter themselves have their own energy levels. Molecules have energy levels depending upon vibrations of different atoms within the molecule, and molecules also have energy levels corresponding to the rotation of the molecules. When we study different types of lasers, we shall see that all kinds of energy levels – electronic, vibrational and rotational – are instrumental in producing laser action in some of the very common types of lasers.

Transitions between electronic energy levels of relevance to laser action correspond to the wavelength range from ultraviolet to near-infrared. Lasing action in neodymium lasers (1064 nm) and argon-ion lasers (488 nm) are some examples. Transitions between vibrational energy levels of atoms correspond to infrared wavelengths. The carbon dioxide laser (10 600 nm) and hydrogen fluoride laser (2700 nm) are some examples. Transitions between rotational energy levels correspond to a wavelength range from 100 microns (μm) to 10 mm.

In a dense medium such as a solid, liquid or high-pressure gas, atoms and molecules are constantly colliding with each other thus causing atoms and molecules to jump from one energy level to another. What is of interest to a laser scientist however is an *optically allowed transition.* An optically allowed transition between two energy levels is one that involves either absorption or emission of a photon which satisfies the resonance condition of $\Delta E = h\nu$, where ΔE is the difference in energy between the two involved energy levels, h is Planck's constant (= $6.626\,075\,5 \times 10^{-34}$ J s or $4.135\,669\,2 \times 10^{-15}$ eV s) and ν is frequency of the photon emitted or absorbed.

1.4 Absorption, Spontaneous Emission and Stimulated Emission

Absorption and emission processes in an optically allowed transition are briefly mentioned in the previous section. An electron or an atom or a molecule makes a transition from a lower energy level to a higher energy level only if suitable conditions exist. These conditions include:

1. the particle that has to make the transition should be in the lower energy level; and
2. the incident photon should have energy (= $h\nu$) equal to the transition energy, which is the difference in energies between the two involved energy levels, that is, $\Delta E = h\nu$.

If the above conditions are satisfied, the particle may make an absorption transition from the lower level to the higher level (Figure 1.2a). The probability of occurrence of such a transition is proportional to both the population of the lower level and also the related Einstein coefficient.

There are two types of emission processes, namely: *spontaneous emission* and *stimulated emission.* The emission process, as outlined above, involves transition from a higher excited energy level to a lower energy level. Spontaneous emission is the phenomenon in which an atom or molecule undergoes a transition from an excited higher-energy level to a lower level without any outside intervention or stimulation, emitting a resonance photon in the process (Figure 1.2b). The rate of the spontaneous emission process is proportional to the related Einstein coefficient. In the case of stimulated emission (Figure 1.2c), there first exists a photon referred to as the stimulating photon which has energy equal to the resonance energy ($h\nu$). This photon perturbs another excited species (atom or molecule) and causes it to drop to the lower energy level, emitting a photon of the same frequency, phase and polarization as that of the stimulating photon in the process. The rate of the stimulated emission process is proportional to the population of the higher excited energy level and the related Einstein coefficient. Note that, in the case of spontaneous emission, the rate of the emission

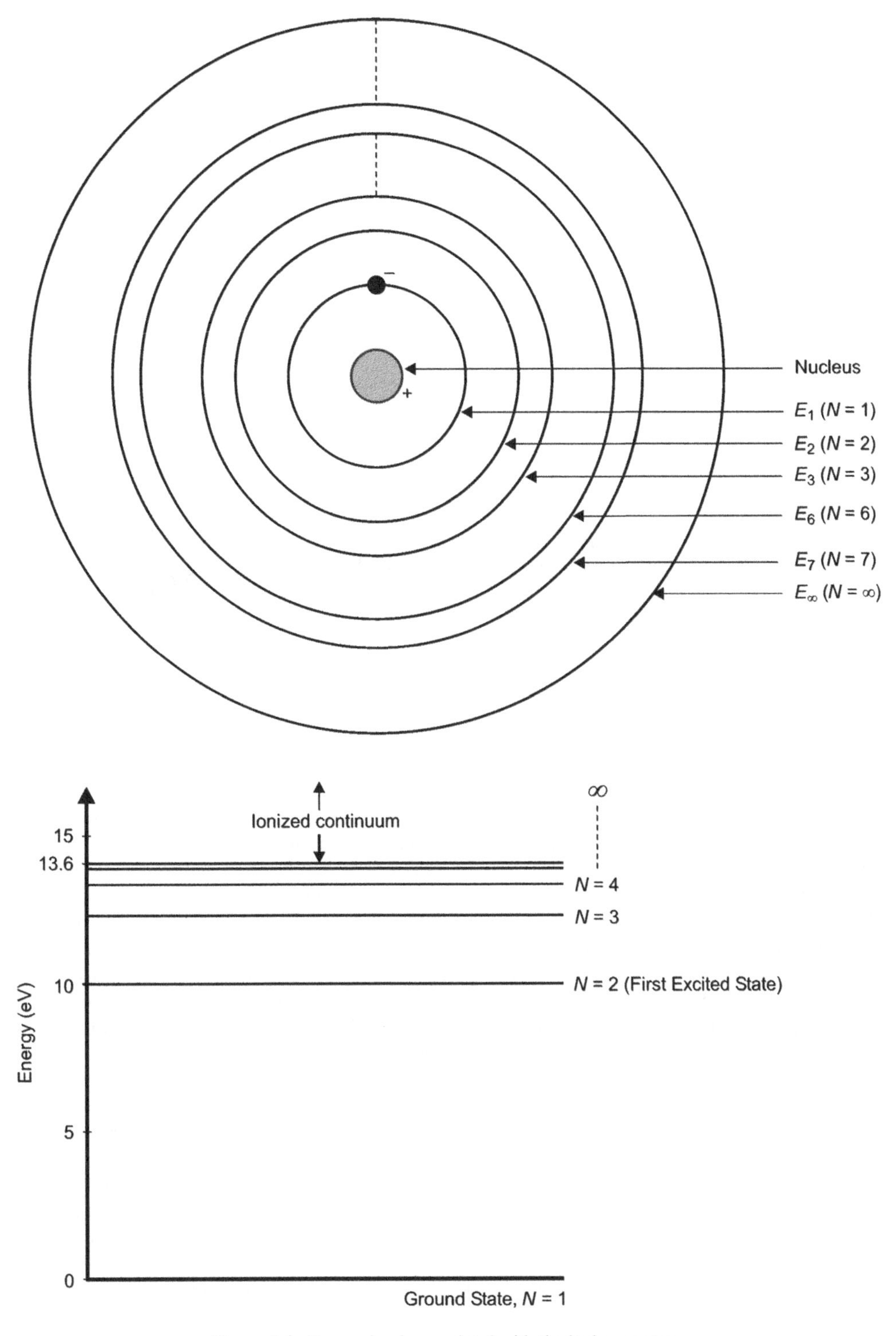

Figure 1.1 Energy levels associated with the hydrogen atom.

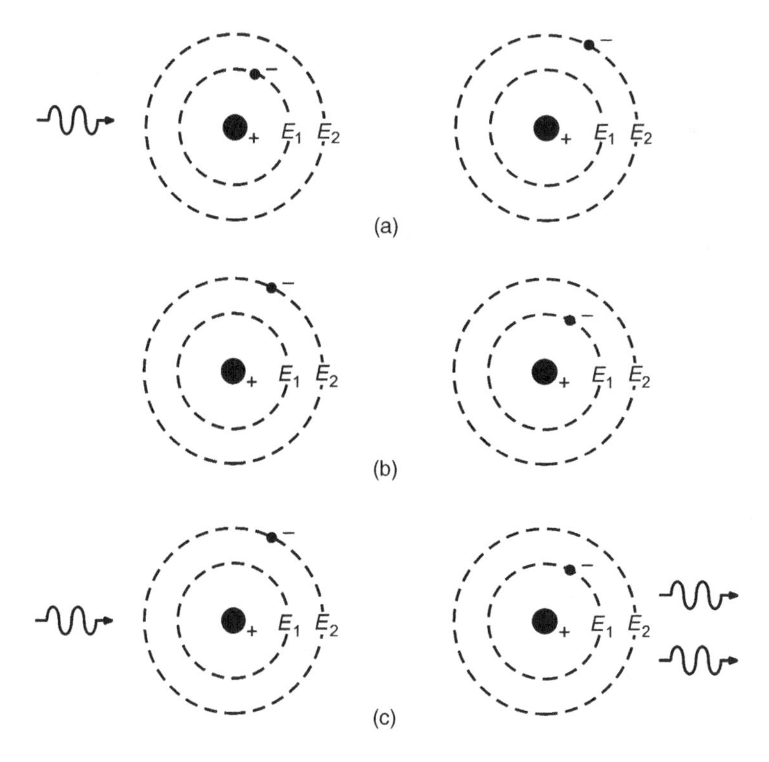

Figure 1.2 Absorption and emission processes: (a) absorption; (b) spontaneous emission; and (c) stimulated emission.

process does not depend upon the population of the energy state from where the transition has to take place, as is the case in absorption and stimulated emission processes. According to the rules of quantum mechanics, absorption and stimulated emission are analogous processes and can be treated similarly.

We have seen that absorption, spontaneous emission and stimulated emission are all optically allowed transitions. Stimulated emission is the basis for photon multiplication and the fundamental mechanism underlying all laser action. In order to arrive at the necessary and favourable conditions for stimulated emission and set the criteria for laser action, it is therefore important to analyze the rates at which these processes are likely to occur. The credit for defining the relative rates of these processes goes to Einstein, who determined the well-known 'A' and 'B' constants known as Einstein's coefficients. The 'A' coefficient relates to the spontaneous emission probability and the 'B' coefficient relates to the probability of stimulated emission and absorption. Remember that absorption and stimulated emission processes are analogous phenomenon. The rates of absorption and stimulated emission processes also depend upon the populations of the lower and upper energy levels, respectively.

For the purposes of illustration, consider a two-level system with a lower energy level 1 and an upper excited energy level 2 having populations of N_1 and N_2, respectively, as shown in Figure 1.3a. Einstein's coefficients for the three processes are B_{12} (absorption), A_{21} (spontaneous emission) and B_{21} (stimulated emission). The subscripts of the Einstein coefficients here represent the direction of transition. For instance, B_{12} is the Einstein coefficient for transition from level 1 to level 2. Also, since absorption and stimulated emission processes are analogous according to laws of quantum mechanics, $B_{12} = B_{21}$. According to Boltzmann statistical thermodynamics, under normal conditions of thermal equilibrium atoms and molecules tend to be at their lowest possible energy level, with the result that population decreases as the energy level increases. If E_1 and E_2 are the energy levels

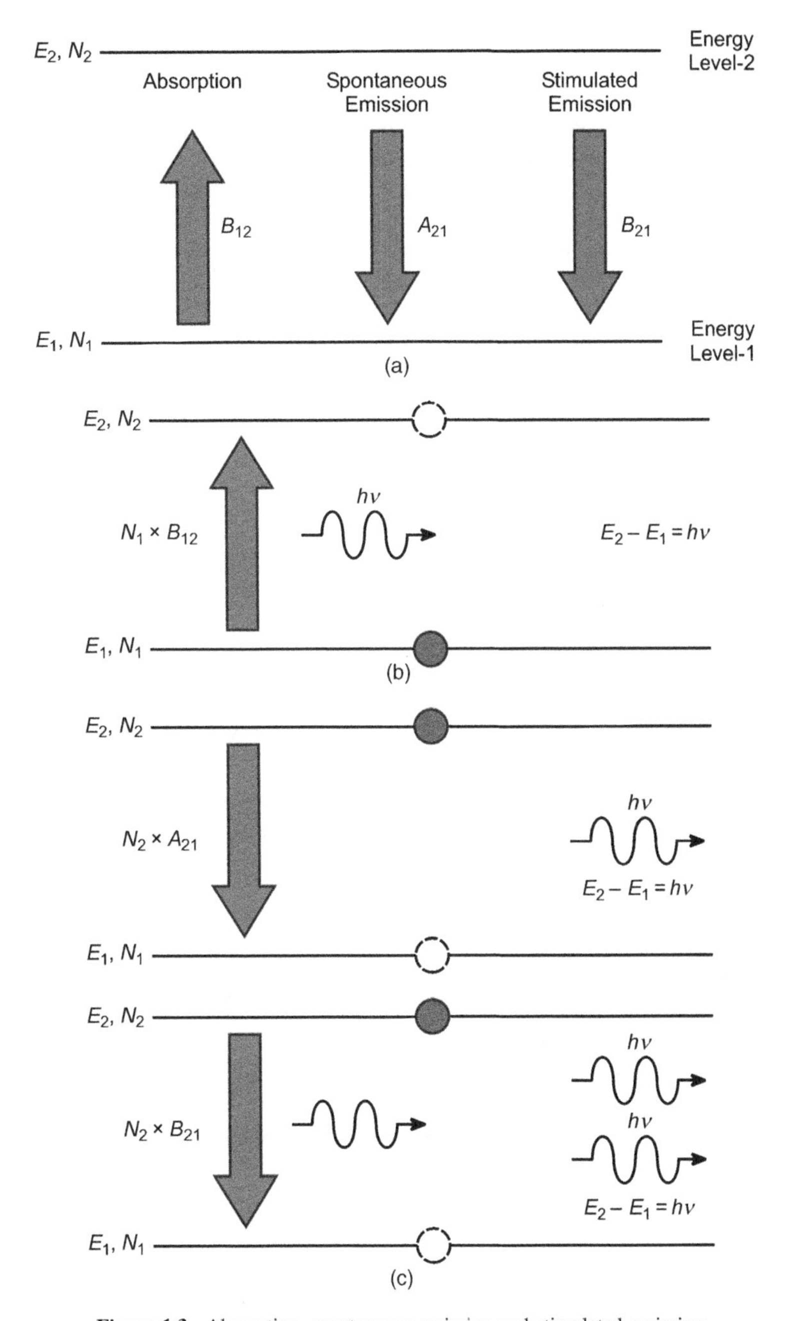

Figure 1.3 Absorption, spontaneous emission and stimulated emission.

associated with level 1 and level 2, respectively, then the populations of these two levels can be expressed by Equation 1.1:

$$\frac{N_2}{N_1} = \exp[-(E_2 - E_1)/kT] \tag{1.1}$$

where

k = Boltzmann constant = $1.38 \times 10^{-23}\,\mathrm{J\,K^{-1}}$ or $8.6 \times 10^{-23}\,\mathrm{eV\,K^{-1}}$
T = absolute temperature in degrees Kelvin

Under normal conditions, N_1 is greater than N_2. When a resonance photon ($\Delta E = h\nu$) passes through the species of this two-level system, it may interact with a particle in level 1 and become absorbed, in the process raising it to level 2. The probability of occurrence of this is given by $B_{12} \times N_1$ (Figure 1.3b). Alternatively, it may interact with a particle already in level 2, leading to emission of a photon with the same frequency, phase and polarization. The probability of occurrence of this process, known as stimulated emission, is given by $B_{21} \times N_2$ (Figure 1.3d). Yet another possibility is that a particle in the excited level 2 may drop to level 1 without any outside intervention, emitting a photon in the process. The probability of this spontaneous emission is A_{21} (Figure 1.3c). The spontaneously emitted photons have the same frequency but have random phase, propagation direction and polarization.

If we analyze the competition between the three processes, it is clear that if $N_2 > N_1$ (which is not the case under the normal conditions of thermal equilibrium), there is the possibility of an overall photon amplification due to enhanced stimulated emission. This condition of $N_2 > N_1$ is known as *population inversion* since $N_1 > N_2$ under normal conditions. We shall explain in the following sections why population inversion is essential for a sustained stimulated emission and hence laser action.

Example 1.1

Refer to Figure 1.4. It shows the energy level diagram of a typical neodymium laser. If this laser is to be pumped by flash lamp with emission spectral bands of 475–525 nm, 575–625 nm, 750–800 nm and 820–850 nm, determine the range of emission wavelengths that would be absorbed by the active medium of this laser and also the wavelength of the laser emission.

Solution

1. Referring to the energy level diagram of Figure 1.4, two edges of the absorption band correspond to energy levels of 12 500 cm^{-1} and 13 330 cm^{-1}. Corresponding wavelengths (of photons) that would have these energy levels are computed as:

   Wavelength corresponding to 12 500 cm^{-1} = (1/12 500) cm = (10^7/12 500) nm = 800 nm
   Wavelength corresponding to 13 330 cm^{-1} = (1/13 330) cm = (10^7/13 330) nm = 750.19 nm ≅ 750 nm

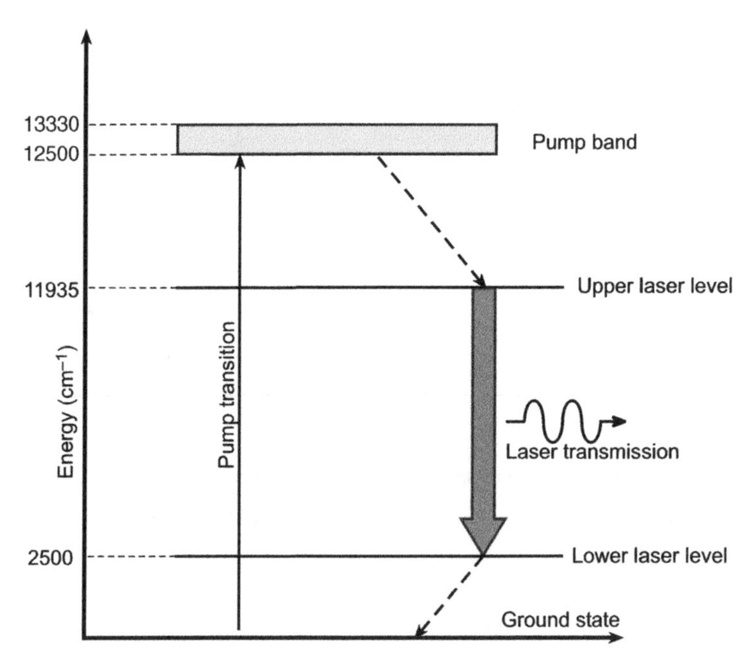

Figure 1.4 Example 1.1: Energy level diagram.

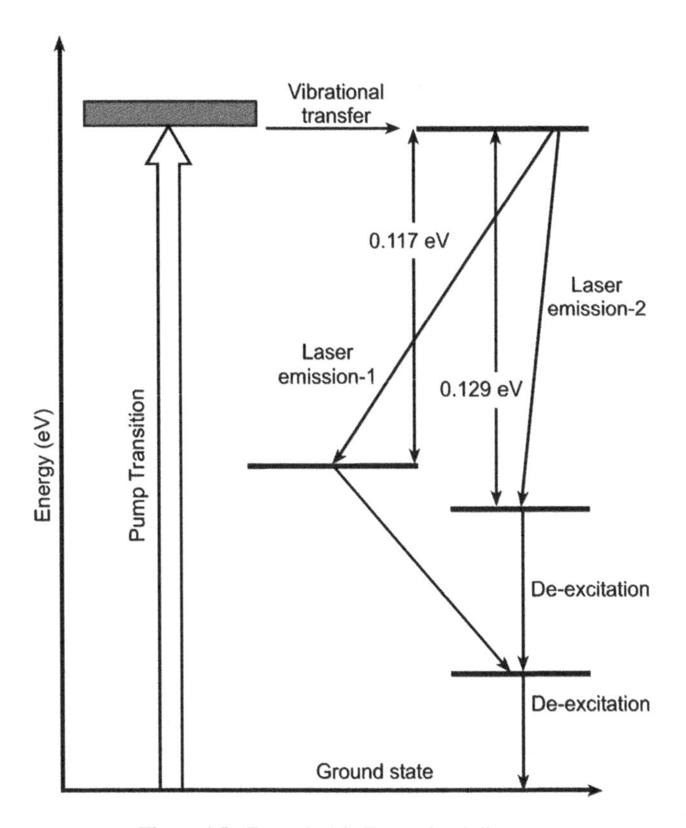

Figure 1.5 Example 1.2: Energy level diagram.

2. The absorption band of the active medium is therefore 750–800 nm. This is the band of wavelengths that would be absorbed by the active medium.
3. Lasing action takes place between metastable energy level 11 935 cm^{-1} and the lower energy level 2500 cm^{-1}. The difference between two energy levels is 11 935–2500 cm^{-1} = 9435 cm^{-1}.
4. This energy corresponds to a wavelength of (1/9435) cm = (10^7/9435) nm = 1059.88 nm ≅ 1060 nm.
5. The emitted laser wavelength is therefore = 1060 nm.

Example 1.2

Figure 1.5 shows the energy level diagram of a popular type of a gas laser. Determine the possible emission wavelengths.

Solution

1. The emission wavelength is such that the corresponding energy value equals the energy difference between the involved lasing levels.
2. **For emission 1**, the energy difference (from Figure 1.5) = 0.117 eV.

   If λ_1 is the emission wavelength, then $hc/\lambda_1 = 0.117$ where
   h = Planck's constant = $6.626\ 075\ 5 \times 10^{-34}$ J s = $4.135\ 669\ 2 \times 10^{-15}$ eV s
   $c = 3 \times 10^{10}$ cm s^{-1}
   Substituting these values, $\lambda_1 = (4.135\,669\,2 \times 10^{-15} \times 3 \times 10^{10})/0.117$ cm = 106.04×10^{-5} cm = 10 604 nm.

3. **For emission 2**, the energy difference (from Figure 1.5) = 0.129 eV

   If λ_1 is the emission wavelength, then $hc/\lambda_1 = 0.129$. Substituting these values, $\lambda_1 = (4.1\ 356\ 692 \times 10^{-15} \times 3 \times 10^{10})/0.129\ \text{cm} = 96.178 \times 10^{-5}\ \text{cm} = 9617.8\ \text{nm}$.

4. The energy level diagram shown in Figure 1.5 is that of carbon dioxide laser, which is also evident from the results obtained for the two emission wavelengths.

Example 1.3

We know that absorption and emission between two involved energy levels takes place when the photon energy corresponding to the absorbed or emitted wavelength equals the energy difference between the two energy levels. If ΔE is energy difference in eV, prove that the absorbed or emitted wavelength (in nm) approximately equals $(1240/\Delta E)$.

Solution

1. Emitted or absorbed wavelength $\lambda = hc/\Delta E$
2. In the above expression, if we substitute the value of h in eV s, c in nm s^{-1} and ΔE in eV, we obtain λ in nm.
3. Now, $h = 4.135\,669\,2 \times 10^{-15}$ eV s and $c = 3 \times 10^8\ \text{m s}^{-1} = 3 \times 10^{17}\ \text{nm s}^{-1}$

   Therefore, λ (in nm) $= 4.135\ 669\ 2 \times 10^{-15} \times 3 \times 10^{17}/\Delta E \cong 1240/\Delta E$.

1.5 Population Inversion

We shall illustrate the concept of population inversion with the help of the same two-level system considered above. If we compute the desired transition energy for an optically allowed transition, let us say at a wavelength of 1064 nm corresponding to the output wavelength of a neodymium-doped yttrium aluminium garnet (Nd:YAG) laser, it turns out to be about 1 eV (transition energy $\Delta E = h\nu$). For a transition energy of 1 eV, we can now determine the population N_2 of level 2, which is the upper excited level here, for a known population N_1 of the lower level at room temperature of 300 K from Equation 1.1. The final relationship is $N_2 = 1.5 \times 10^{-17} N_1$.

This implies that practically all atoms or molecules are in the lower level under thermodynamic equilibrium conditions. Let us not go that far and instead consider a situation where the population of the lower level is only ten times that of the excited upper level. We shall now examine what happens when there is a spontaneously emitted photon. Now there are two possibilities: either this photon stimulates another excited species in the upper level to cause emission of another photon of identical character, or it would hit an atom or molecule in the lower level and be absorbed. Since there are 10 atoms or molecules in the lower level for every excited species in the upper level, we can say that 10 out of every 11 spontaneously emitted photons hit the atoms or molecules in the lower level and become absorbed. Only 9% (1 out of every 11) of the photons can cause stimulated emission. The photons emitted by the stimulated process will also become absorbed successively due to the scarcity of excited species in the upper level. Another way of expressing this is that when the population of the lower level is much larger than the population of the excited upper level, the probability of each spontaneously emitted photon hitting an atom or molecule in the lower level and becoming absorbed is also much higher than the same stimulating another excited atom or molecule in the upper level. The same concept underlies the expressions for the probability of absorption, spontaneous emission and stimulated emission previously outlined in Section 1.4:

Probability of absorption $= B_{12} \times N_1$

Probability of spontaneous emission $= A_{12}$

Probability of stimulated emission $= B_{21} \times N_2$

If we want the stimulated emission to dominate over absorption and spontaneous emission, we must have a greater number of excited species in the upper level than the population of the lower level. Such a situation is known as *population inversion* since under normal circumstances the population of the lower level is much greater than the population of the upper level. Population inversion is therefore an essential condition for laser action. The next obvious question is that of the desired extent of population inversion. Spontaneous emission depletes the excited upper level population (N_2 in the present case) at a rate proportional to A_{21} producing undesired photons with random phase, direction of propagation and polarization. Due to this loss and other losses associated with laser cavity (discussed in Section 1.7), each laser has a certain minimum value of $N_2 - N_1$ for the production of laser output. This condition of population inversion is known as the *inversion threshold* of the laser. *Lasing threshold* is an analogous term.

Next, we shall discuss how we can produce population inversion.

1.5.1 Producing Population Inversion

That population inversion is an essential condition for laser action is demonstrated above. Population inversion ensures that there are more emitters than absorbers with the result that stimulated emission dominates over spontaneous emission and absorption processes. There are two possible ways to produce population inversion. One is to populate the upper level by exciting extra atoms or molecules to the upper level. The other is to depopulate the lower laser level involved in the laser action. In fact, for a sustained laser action, it is important to both populate the upper level and depopulate the lower level.

Two commonly used pumping or excitation mechanisms include optical pumping and electrical pumping. Both electrons and photons have been successfully used to create population inversion in different laser media. While optical pumping is ideally suited to solid-state lasers such as ruby, Nd:YAG and neodymium-doped glass (Nd:Glass) lasers, electrical discharge is the common mode of excitation in gas lasers such as helium-neon and carbon dioxide lasers.

The excitation input, optical or electrical, usually raises the atoms or molecules to a level higher than the upper laser level from where it rapidly drops to the upper laser level. In some cases, the excitation input excites atoms other than the active species. The excited atoms then transfer their energy to the active species to cause population inversion. A helium-neon laser is a typical example of this kind where the excitation input gives its energy to helium atoms, which subsequently transfer the energy to neon atoms to raise them to the upper laser level.

The other important concept essential for laser action is the existence of a *metastable state* as the upper laser level. For stimulated emission, the excited state needs to have a relatively longer lifetime of the order of a few microseconds to a millisecond or so. The excited species need to stay in the excited upper laser level for a longer time in order to allow interaction between photons and excited species, which is necessary for efficient stimulated emission. If the upper laser level had a lifetime of a few nanoseconds, most of the excited species would drop to the lower level as spontaneous emission. The crux is that, for efficient laser action, the population build-up of the upper laser level should be faster than its decay. A longer upper laser level lifetime helps to achieve this situation.

1.6 Two-, Three- and Four-Level Laser Systems

Another important feature that has a bearing on the laser action is the energy level structure of the laser medium. As we shall see in the following sections, energy level structure, particularly the energy levels involved in the population inversion process and the laser action, significantly affect the performance of the laser.

1.6.1 Two-Level Laser System

In a *two-level laser system*, there are only two levels involved in the total process. The atoms or molecules in the lower level, which is also the lower level of the laser transition, are excited to the upper level by the pumping or excitation mechanism. The upper level is also the upper laser level. Once the population inversion is achieved and its extent is above the inversion threshold, the laser action can take place. Figure 1.6 shows the arrangement of energy levels in a two-level system. A two-level system is, however, a theoretical concept only as far as lasers are concerned. No laser has ever been made to work as a two-level system.

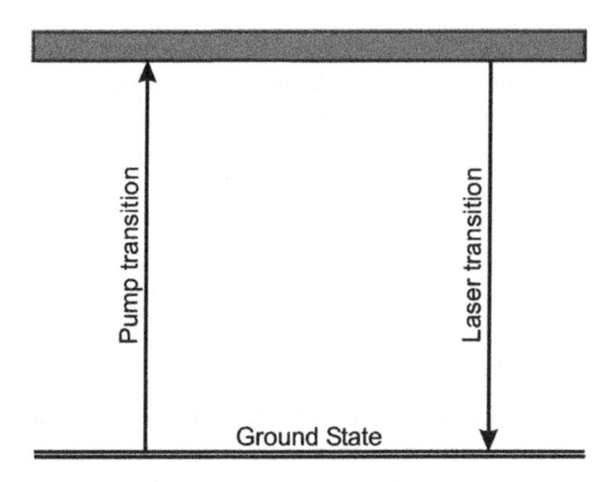

Figure 1.6 Two-level laser system.

1.6.2 Three-Level Laser System

In a *three-level laser system*, the lower level of laser transition is the ground state (the lowermost energy level). The atoms or molecules are excited to an upper level higher than the upper level of the laser transition (Figure 1.7). The upper level to which atoms or molecules are excited from the ground state has a relatively much shorter lifetime than that of the upper laser level, which is a metastable level. As a result, the excited species rapidly drop to the metastable level. A relatively much longer lifetime for the metastable level ensures a population inversion between the metastable level and the ground state provided that more than half of the atoms or molecules in the ground state have been excited to the uppermost short-lived energy level. The laser action occurs between the metastable level and the ground state.

A ruby laser is a classical example of a three-level laser. Figure 1.8 shows the energy level structure for this laser. One of the major shortcomings of this laser and other three-level lasers is due to the lower laser level being the ground state. Under thermodynamic equilibrium conditions, almost all atoms or molecules are in the ground state and so it requires more than half of this number to be excited out of the ground state to achieve laser action. This implies that a much larger pumping input would be required to exceed population inversion threshold. This makes it very difficult to sustain population inversion on a continuous basis in three-level lasers. That is why a ruby laser cannot be operated in continuous-wave (CW) mode.

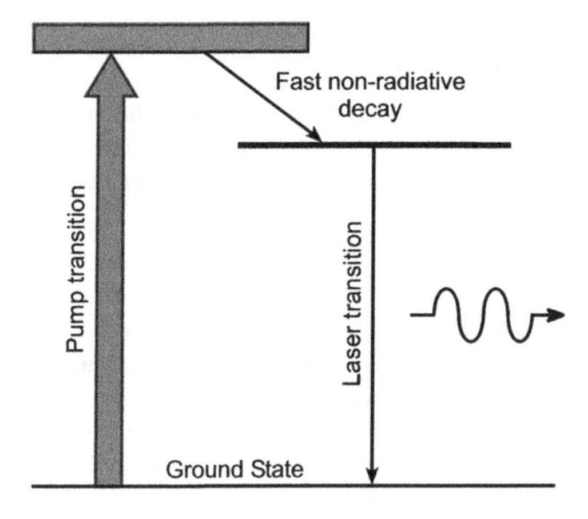

Figure 1.7 Three-level laser system.

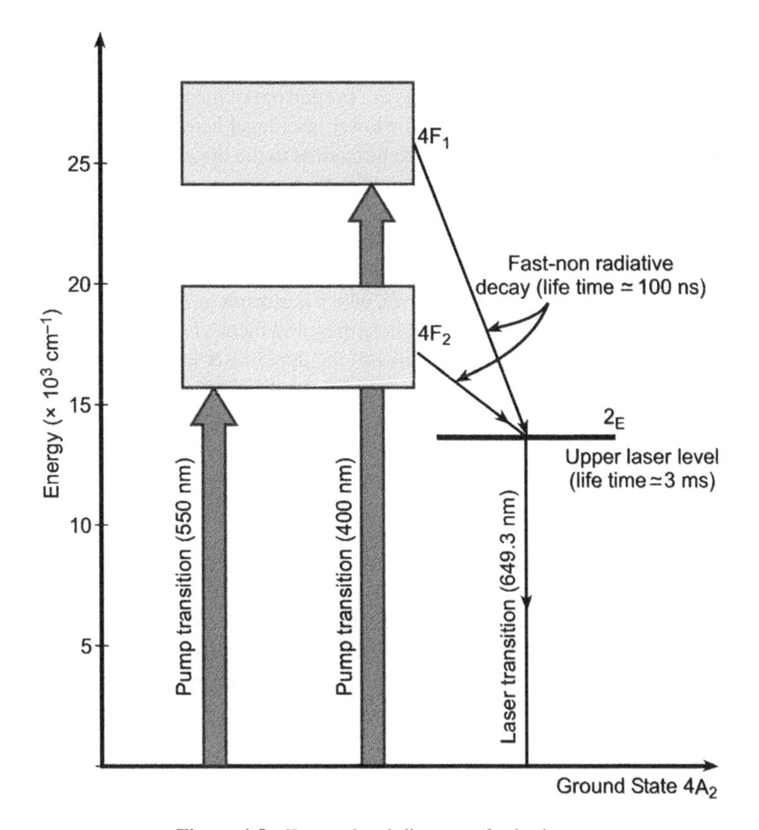

Figure 1.8 Energy level diagram of ruby laser.

An ideal situation would be if the lower laser level were not the ground state so that it had much fewer atoms or molecules in the thermodynamic equilibrium condition, solving the problem encountered in three-level laser systems. Such a desirable situation is possible in four-level laser systems in which the lower laser level is above the ground state, as shown in Figure 1.9.

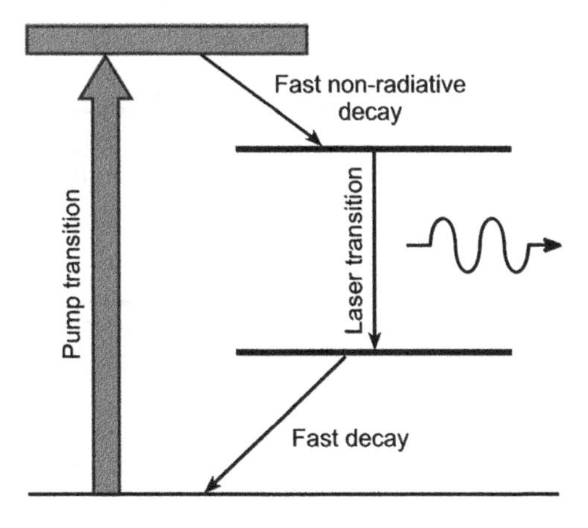

Figure 1.9 Four-level laser system.

1.6.3 Four-Level Laser System

In a *four-level laser system*, the atoms or molecules are excited out of the ground state to an upper highly excited short-lived energy level. Remember that the lower laser level here is not the ground state. In this case, the number of atoms or molecules required to be excited to the upper level would depend upon the population of the lower laser level, which is much smaller than the population of the ground state. Also if the upper level to which the atoms or molecules are initially excited and the lower laser level have a shorter lifetime and the upper laser level (metastable level) a longer lifetime, it would be much easier to achieve and sustain population inversion. This is achievable due to two major features of a four-level laser. One is rapid population of the upper laser level, which is a result of an extremely rapid dropping of the excited species from the upper excited level where they find themselves with excitation input to the upper laser level accompanied by the longer lifetime of the upper laser level. The second occurrence is the depopulation of the lower laser level due to its shorter lifetime. Once it is simpler to sustain population inversion, it becomes easier to operate the laser in the continuous-wave (CW) mode. This is one of the major reasons that a four-level laser such as an Nd:YAG laser or a helium-neon laser can be operated in the continuous mode while a three-level laser such as a ruby laser can only be operated as a pulsed laser.

Nd:YAG, helium-neon and carbon dioxide lasers are some of the very popular lasers with a four-level energy structure. Figure 1.10 shows the energy level structure of a Nd:YAG laser. The pumping or excitation input raises the atoms or molecules to the uppermost energy level, which in fact is not a single level but instead a band of energy levels. This is a highly desirable feature, the reason for which is discussed more fully in Section 1.11 on pumping mechanisms. The excited species rapidly fall to the upper laser level (metastable level). This decay time is about 100 ns. The metastable level has a

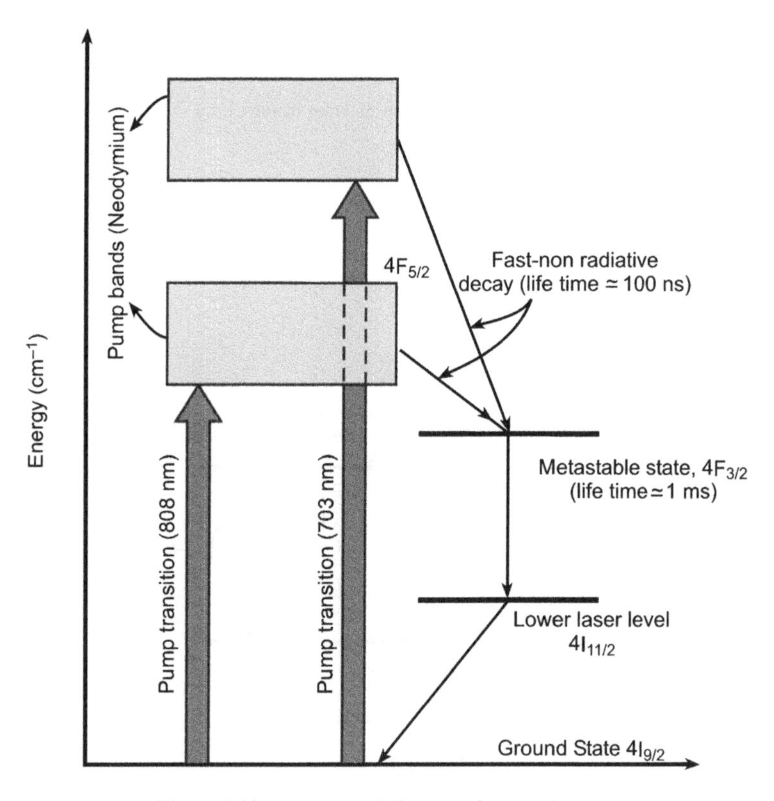

Figure 1.10 Energy level diagram of Nd:YAG laser.

metastable lifetime of about 1 ms and the lower laser level has a decay time of 30 ns. If we compare the four-level energy level structure of a Nd:YAG laser with that of a neodymium-doped yttrium lithium fluoride (Nd:YLF) laser, another solid-state laser with a four-level structure, we find that there is a striking difference in the lifetime of the metastable level. Nd:YLF has a higher metastable lifetime (typically a few milliseconds) as compared to 1 ms of Nd:YAG. This gives the former a higher storage capacity for the excited species in the metastable level. In other words, this means that a Nd:YLF rod could be pumped harder to extract more laser energy than a Nd:YAG rod of the same size.

1.6.4 Energy Level Structures of Practical Lasers

In the case of real lasers, the active media do not have the simple three- or four-level energy level structures as described above, but are far more complex. For instance, the short-lived uppermost energy level, to which the atoms or molecules are excited out of the ground state and from where they drop rapidly to the metastable level, is not a single energy level. It is in fact a band of energy levels, a desirable feature as it makes the pumping more efficient and a larger part of the pumping input is converted into a useful output to produce population inversion. The energy levels involved in producing laser output are not necessarily single levels in all lasers. There could be multiple levels in the metastable state, in the lower energy state of the laser transition or in both states. This means that the laser has the ability to produce stimulated emission at more than one wavelength. Helium-neon and carbon dioxide lasers are typical examples of this phenomenon. Figure 1.11 shows the energy level structure of a helium-neon laser.

Another important point worth mentioning here is that it is not always the active species alone that constitute the laser medium or laser material. Atoms or molecules of other elements are sometimes added with specific objectives. In some cases, such as in a helium-neon laser, the active species producing laser transition is the neon atoms. Free electrons in the discharge plasma produced as a result of electrical pumping input excite the helium atoms first as that can be done very efficiently. When the excited helium atoms collide with neon atoms, they transfer their energy to them. As another example, in a carbon dioxide laser the laser gas mixture mainly consists of carbon dioxide, nitrogen and helium. While

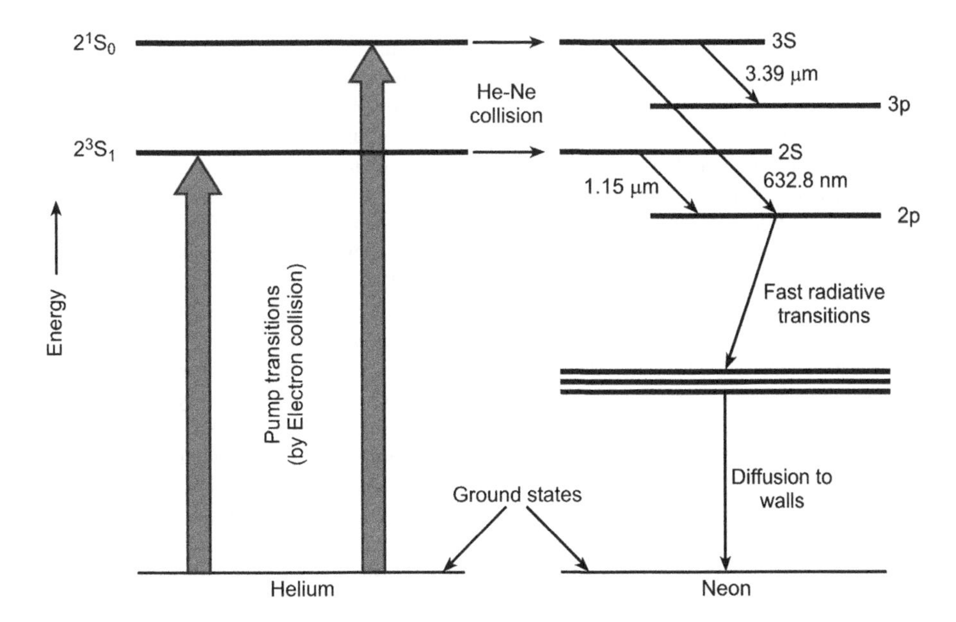

Figure 1.11 Energy level diagram of He-Ne laser.

nitrogen participates in the excitation process and plays the same role as that played by helium in a helium-neon laser, the helium in a carbon dioxide laser helps in depopulating the lower laser level.

1.7 Gain of Laser Medium

When we talk about the *gain* of the laser medium we are basically referring to the extent to which this medium can produce stimulated emission. The gain of the medium is defined more appropriately as a *gain coefficient*, which is the gain expressed as a percentage per unit length of the active medium. When we say that the gain of a certain laser medium is 10% per centimetre, it implies that 100 photons with the same transition energy as that of an excited laser medium become 110 photons after travelling 1 cm of the medium length. The amplification or the photon multiplication offered by the medium is expressed as a function of the gain of the medium and the length of the medium, as described in Equation 1.2:

$$G_{\mathrm{A}} = e^{\alpha x} \tag{1.2}$$

where

G_{A} = amplifier gain or amplification factor
α = gain coefficient
x = gain length

The above expression for gain can be re-written in the form:

$$G_{\mathrm{A}} = (e^{\alpha})^{x} = (1+\alpha)^{x} \qquad \text{for}\ \alpha \ll 1 \tag{1.3}$$

Therefore, to a reasonably good approximation, we can write

$$\text{Amplification factor} = (1 + \text{gain coefficient})^{\text{length of medium}}$$

This implies that when the medium with a gain coefficient of 100% is excited and population inversion created, a single spontaneously emitted photon will become two photons after this spontaneously emitted photon travels 1 cm of the length of the medium. The two photons cause further stimulated emission as they travel through the medium. This amplification continues and the number of photons emitted by the stimulation process keeps building up just as the principal amount builds up with compound interest. The above relationship can be used to compute the amplification. It would be interesting to note how photons multiply themselves as a function of length. For instance, although 10 photons become 11 photons after travelling 1 cm for a gain coefficient of 10% per centimetre, the number reaches about 26 for 10 cm and 1173 after travelling gain length of 50 cm, as long as there are enough excited species in the metastable state to ensure that stimulated emission dominates over absorption and spontaneous emission. On the other hand it is also true that, for a given pump input, there is a certain quantum of excited species in the upper laser level. As the stimulated emission initially triggered by one spontaneously emitted photon picks up, the upper laser level is successively depleted of the desired excited species and the population inversion is adversely affected. This leads to a reduction in the growth of stimulated emission and eventually saturation sets in; this is referred to as *gain saturation*.

Another aspect that we need to look into is whether the typical gain coefficient values that the majority of the active media used in lasers have are really good enough for building practical systems. Let us do a small calculation. If a 5 mW CW helium-neon laser were to operate for just 1 s, it would mean an equivalent energy of 5 mJ. Each photon of He-Ne laser output at 632.8 nm would have energy of approximately 3×10^{-19} J, which further implies that the above laser output would necessitate generation of about 1.7×10^{16} photons. With the kind of gain coefficient which the helium-neon laser plasma has, the required gain length can be calculated for the purpose. For any useful laser output, the solution therefore lies in having a very large effective gain length, if not a physically large gain length.

If we enclose the laser medium within a closed path bounded by two mirrors, as shown in Figure 1.12, we can effectively increase the interaction length of the active medium by making the

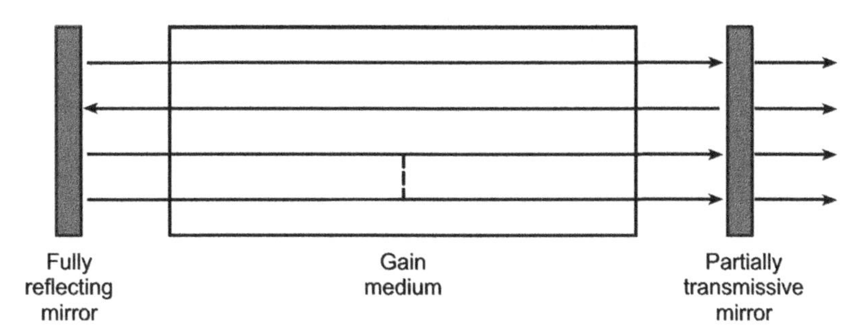

Figure 1.12 Lasing medium bounded by mirrors.

photons emitted by stimulated emission process back and forth. One of the mirrors in the arrangement is fully reflecting and the other has a small amount of transmission. This small transmission, which also constitutes the useful laser output, adds to the loss component. This is true because the fraction of the stimulated emission of photons taken as the useful laser output is no longer available for interaction with the excited species in the upper laser level. The maximum power that can be coupled out of the system obviously must not exceed the total amount of losses within the closed path. For instance, if the gain of the full length of the active medium is 5% and the other losses such as those due to absorption in the active medium, spontaneous emission, losses in the fully reflecting mirror (which will not have an ideal reflectance of 100%) and so on are 3%, the other mirror can have at the most a transmission of 2%.

In a closed system like this, the power inside the system is going to be much larger than the power available as useful output. For instance, for 1% transmission and assuming other losses to be negligible, if the output power is 1 mW the power inside the system would be 100 mW.

Example 1.4

Determine the gain coefficient in case of a helium-neon laser if a 50 cm gain length produces amplification by a factor of 1.1.

Solution

1. We have that $x = 50\,\text{cm}$ and the amplification factor $G_A = 1.1$
2. The gain coefficient α can be computed from $G_A = e^{\alpha x}$

   or $\alpha = (1/x)\ln G_A = (1/50)\ln(1.1) = 0.0019\,\text{cm}^{-1}$

1.8 Laser Resonator

The active laser medium within the closed path bounded by two mirrors as shown in Figure 1.12 constitutes the basic laser resonator provided it meets certain conditions. Resonator structures of most practical laser sources would normally be more complex than the simplistic arrangement of Figure 1.12. As stated in the previous section, with the help of mirrors we can effectively increase the interaction length of the active medium by making the photons emitted by the stimulated emission process travel back and forth within the length of the cavity. One of the mirrors in the arrangement is fully reflecting and the other has a small amount of transmission. It is clear that if we want the photons emitted as a result of the stimulated emission process to continue to add to the strength of those responsible for their emission, it is necessary for the stimulating and stimulated photons to be in phase. The addition of mirrors should not disturb this condition. For example, if the wave associated with a given photon was at its positive peak at the time of reflection from the fully reflecting mirror, it should again be at its positive peak only after it makes a round trip of the cavity and returns to the fully reflecting mirror again. If this happens, then all those photons stimulated by this photon would also satisfy this condition. This is possible if we satisfy the condition given in Equation 1.4:

$$\text{Round trip length} = 2L = n\lambda \tag{1.4}$$

where

$L =$ length of the resonator
$\lambda =$ wavelength
$n =$ an integer

The above expression can be rewritten as

$$f = \frac{nc}{2L} \tag{1.5}$$

where

$c =$ velocity of electromagnetic wave
$f =$ frequency

1.9 Longitudinal and Transverse Modes

The above expression for frequency indicates that there could be a large number of frequencies for different values of the integer n satisfying this resonance condition. Most laser transitions have gain for a wide range of wavelengths. Remember that we are not referring to lasers that can possibly emit at more than one wavelength (such as a helium-neon laser). Here, we are referring to the gain-bandwidth of one particular transition. We shall discuss in detail in Chapter 4 how gas lasers such as He-Ne and CO_2 lasers have Doppler-broadened gain curves. A He-Ne laser has a bandwidth of about 1400 MHz for 632.8 nm transition (Figure 1.13a) and a CO_2 laser has a bandwidth of about 60 MHz at 10 600 nm (Figure 1.13b).

It is therefore possible to have more than one resonant frequency, each of them called a *longitudinal mode*, simultaneously present unless special measures are taken to prevent this from happening. As is clear from Equation 1.5, the inter-mode spacing is given by $c/2L$. For a He-Ne laser with a cavity length of 30 cm for example, inter-mode spacing would be 500 MHz which may allow three longitudinal modes to be simultaneously present as shown in Figure 1.14a. Interestingly, the cavity length could be reduced

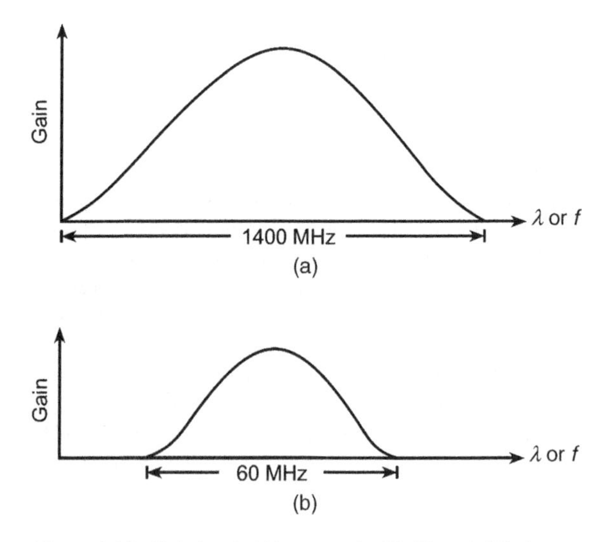

Figure 1.13 Gain-bandwidth curves for He-Ne and CO_2 lasers.

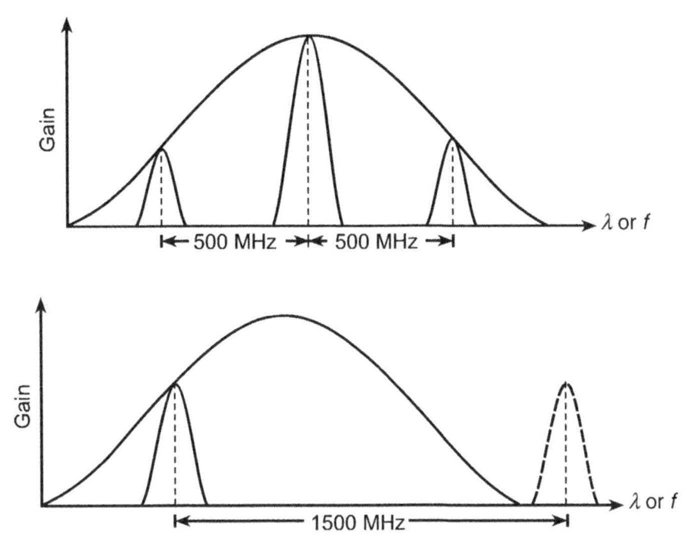

Figure 1.14 Longitudinal modes.

to a point where the inter-mode spacing exceeds the gain-bandwidth of the laser transition to allow only a single longitudinal mode to prevail in the cavity. For instance, a 10 cm cavity length leading to an inter-mode spacing of 1500 MHz would allow only a single longitudinal mode (Figure 1.14b). However, there are other important criteria that also decide the cavity length.

Another laser parameter that we are interested in and that is also largely influenced by the design of the laser resonator is the *transverse mode* structure of the laser output. We have already seen in the previous sections how the resonator length and the laser wavelength together decide the possible resonant frequencies called longitudinal modes, which can simultaneously exist. The transverse modes basically tell us about the irradiance distribution of the laser output in the plane perpendicular to the direction of propagation or, in other words, along the orthogonal axes perpendicular to the laser axis. To illustrate this further, if the z axis is the laser axis, then intensity distribution along the x and y axes would describe the transverse mode structure.

TEM_{mn} describes the transverse mode structure, where m and n are integers indicating the order of the mode. In fact, integers m and n are the number of intensity minima or nodes in the spatial intensity pattern along the two orthogonal axes. Conventionally, m represents the electric field component and n indicates the magnetic field component. Those who are familiar with electromagnetic theory should not find this difficult at all to grasp. Remember that transverse modes must satisfy the boundary conditions such as having zero amplitude on the boundaries. The simplest mode, also known as the fundamental or the lowest order mode, is referred to as TEM_{00} mode. The two subscripts here indicate that there are no minima along the two orthogonal axes between the boundaries. The intensity pattern in both the orthogonal directions has a single maximum with the intensity falling on both sides according to the well-known mathematical distribution referred to as the Gaussian distribution. The Gaussian distribution (Figure 1.15) is given by Equation 1.6:

$$I(r) = I_0 \exp\left(-2r^2/w^2\right) \tag{1.6}$$

where

$I(r)$ = intensity at a distance r from the centre of the beam
w = beam radius at $(1/e^2)$ of peak intensity point, which is about 13.5% of the peak intensity

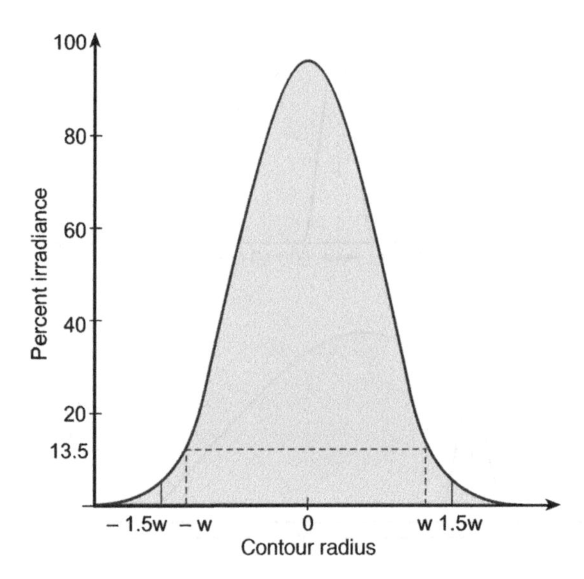

Figure 1.15 Gaussian distribution.

We also have

$$I_0 = \frac{2P}{\pi w^2} \tag{1.7}$$

where

P = total power in the beam

Before we discuss the definite advantages that the operation at lowest order or fundamental modes TEM_{00} offers, we shall have quick look at higher-order modes and also how different transverse mode appear in relation to their intensity distributions. Figure 1.16 shows the spatial intensity distribution of the laser spot for various transverse mode structures of the laser resonator.

Going back to the fundamental mode, we can appreciate that this mode has the least power spreading. To add to this, this mode has the least divergence; it has the minimum diffraction loss and therefore can be focused onto the smallest possible spot. The transverse mode structure is also critically dependent upon parameters such as laser medium gain, type of laser resonator and so on. There are established resonator design techniques to ensure operation at the fundamental mode. Often, lasers optimized to produce maximum power output operate at one or more higher-order modes. Also, lasers with low gain and stable resonator configuration can conveniently be made to operate at fundamental mode. Details are beyond the scope of this book, however.

Example 1.5

Given that the Doppler-broadened gain curve of a helium-neon laser with a 50-cm-long resonator emitting at 1.15 μm is 770 MHz, determine (a) inter-longitudinal mode spacing and (b) the number of maximum possible sustainable longitudinal modes.

Solution: Resonator length $L = 50$ cm. Therefore, inter-longitudinal mode spacing $= c/2L = 3 \times 10^{10}/100 = 300$ MHz.

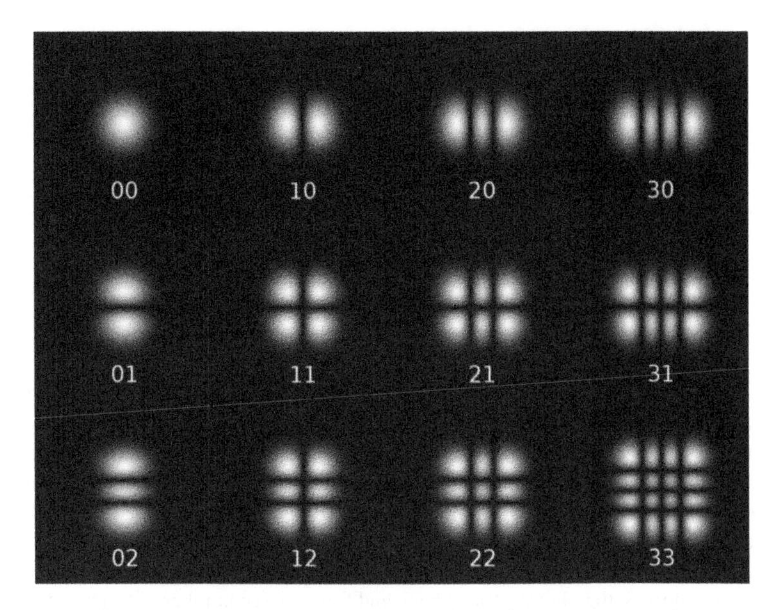

Figure 1.16 Spatial intensity distribution for various transverse modes.

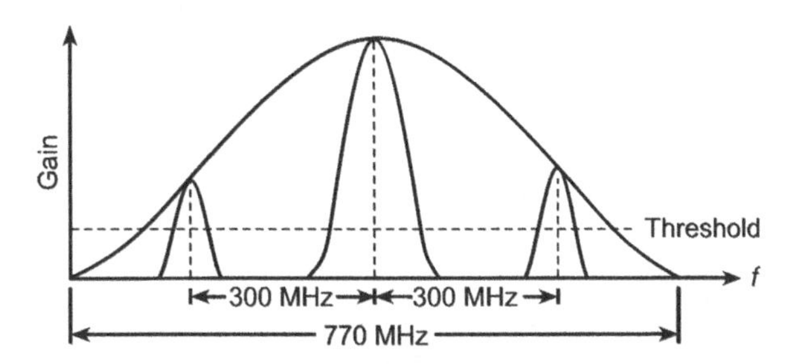

Figure 1.17 Diagram for Example 1.5.

Width of Doppler-broadened gain curve = 770 MHz. The number of longitudinal modes possible within this width = 3 (Figure 1.17).

1.10 Types of Laser Resonators

According to the type of end mirrors used and the inter-element separation, which largely dictates the extent of interaction between the emitted photons and the laser medium and also the immunity of the laser resonator to misalignment of end components, the resonators can be broadly classified as *stable* and *unstable* resonators. A *stable resonator* is one in which the photons can bounce back and forth between the end components indefinitely without being lost out the sides of the components. Due to the focusing nature of one or both components, the light flux remains within the cavity in such a resonator. A *plane-parallel resonator* (Figure 1.18) in which both end components are plane mirrors and are placed precisely at right angles to the laser axis is a stable resonator. In practice, however, this is not true. A slight misalignment of even one of the mirrors would ultimately lead to light flux escaping the laser cavity after several reflections from the two mirrors. Nevertheless, such a resonator encompasses a large volume of the active medium. It is not used in practice, as it is highly prone to misalignment.

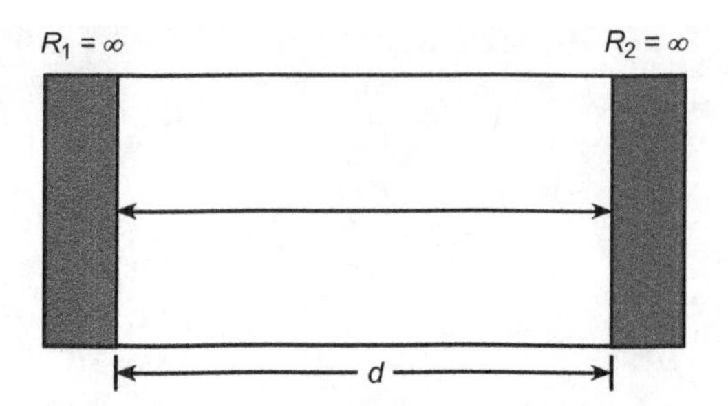

Figure 1.18 Plane-parallel resonator.

This problem can be overcome by using one plane and one curved mirror, as is the case for *hemispherical* and *hemifocal* resonators shown in Figure 1.19a and b, respectively, or two curved mirrors, as is the case for *concentric* and *confocal* resonators shown in Figure 1.20a and b, respectively.

Although the problem of sensitivity of the plane-parallel resonator to misalignment of cavity mirrors is largely overcome by the use of different stable resonator configurations discussed above (Figures 1.19 and 1.20), not all of them have emitted photons interacting with a large volume of the excited species, which is also equally desirable. It is also true that in the case of low-gain media with consequent very low transmission output mirrors, the photons travel back and forth a large number of times within the cavity before their energy appears at the output. This makes the resonator alignment more critical. That is why a plane-parallel resonator will never be the choice for a low-gain laser medium.

On the other hand, in a high-gain medium a certain amount of light flux leakage can be tolerated. This fact is made use of in an *unstable resonator* configuration, which otherwise achieves interaction of the emitted photons with a very large volume of the excited species. Figure 1.21 shows one possible type of unstable resonator. Note that photons escape from the sides of the mirror after one or two passes within the cavity. This light leakage, which also constitutes the useful laser output, is more than compensated for by a high-gain medium and large interaction volume. Further, since the photons have to make

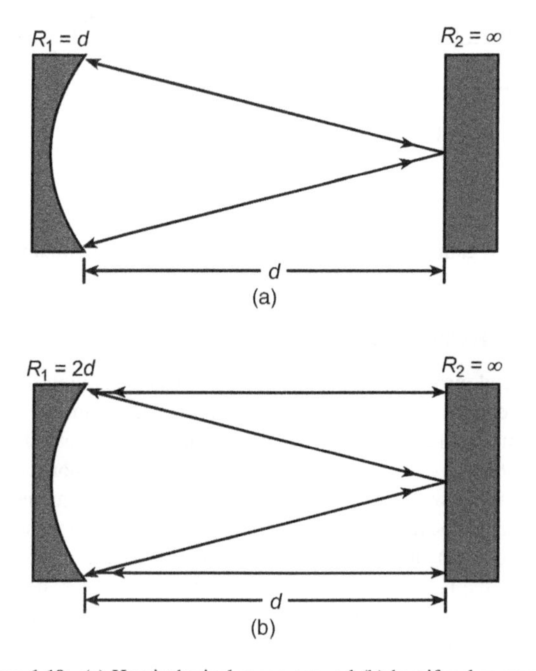

Figure 1.19 (a) Hemispherical resonator and (b) hemifocal resonator.

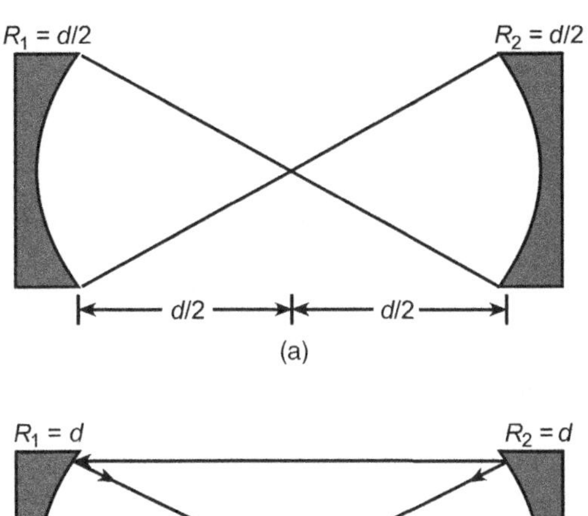

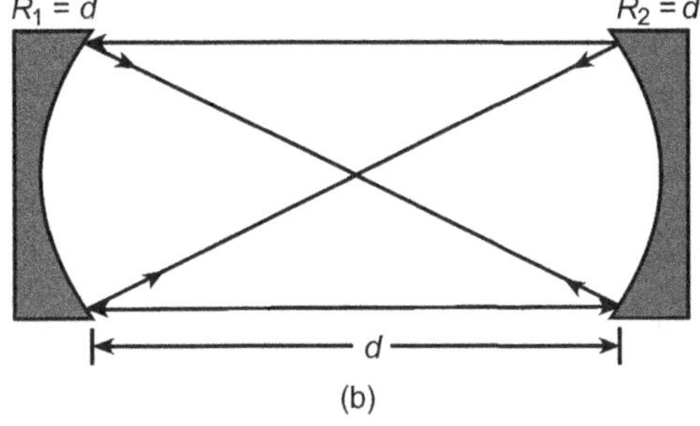

Figure 1.20 (a) Concentric resonator and (b) confocal resonator.

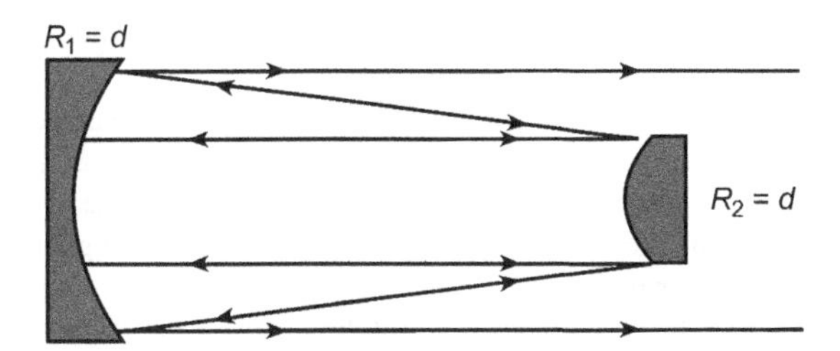

Figure 1.21 Unstable resonator.

relatively fewer passes within the cavity as compared to a low-gain stable resonator configuration before drifting out, the alignment becomes much less critical.

1.11 Pumping Mechanisms

By pumping mechanism, we mean the mechanism employed to create population inversion of the lasing species. Commonly employed pumping mechanisms include:

1. optical pumping;
2. electrical pumping; and
3. other mechanisms such as pumping by chemical reactions, electron beams and so on.

One aspect that is common to all pumping mechanisms is that the pumping energy/power must be greater than the laser output energy/power. When applied to optical pumping, it is obvious that the optical pump wavelength must be smaller than the laser output wavelength. This has to be true as the

Figure 1.22 Linear flash lamps.

lasing species are first excited to the topmost level from where they drop to the upper laser level. Since the energy difference between the ground state and the topmost pump level is always greater than the energy difference between the two laser levels, the wavelength of the pump photon must be less than the wavelength of the laser output. Another aspect that is common to all schemes is that pumping efficiency largely affects the overall laser efficiency. For instance, if the energy difference for the pump transition is much greater than that of the laser transition, the laser efficiency is bound to be relatively poorer. An argon-ion laser is a typical example. Yet another aspect that is common to all pumping mechanisms is that the topmost pump level is not a single energy level but rather a band of closely spaced energy levels with allowed transitions to a single and, in some cases, more than one metastable level. When applied to optical pumping, this allows the use of optical sources such as flash lamps with broadband outputs.

1.11.1 Optical Pumping

Optical pumping is employed for those lasers that have a transparent active medium. Solid-state and liquid-dye lasers are typical examples. The most commonly used pump sources are the flash lamp in the case of pulsed and the arc lamp in the case of continuous-wave solid-state lasers.

Flash lamps are pulsed sources of light and are widely used for the pumping of pulsed solid-state lasers. These are available in a wide range of arc lengths (from a few centimetres to as large as more than a metre, although arc length of 5–10 cm is common), bore diameter (typically in the range of 3–20 mm), wall thickness (typically 1–2 mm) and shape (linear, helical). Figures 1.22 and 1.23 depict the constructional features of typical linear (Figure 1.22) and helical (Figure 1.23) flash lamps.

Flash lamps for pumping solid-state lasers are usually filled with a noble gas such as xenon or krypton at a pressure of 300–400 torr. Two electrodes are sealed in the envelope that is usually made of quartz. An electrical discharge created between the electrodes leads to a very high value of pulsed current, which further produces an intense flash. The electrical energy to be discharged through the lamp is stored in an energy storage capacitor/capacitor bank.

Xenon-filled lamps produce higher radiative output for a given electrical input as compared to krypton-filled lamps. Krypton however offers a better spectral match, more so with Nd:YAG. That is, the emission spectrum of a krypton flash lamp is better matched to the absorption spectrum of Nd:YAG. Emission spectra in the case of xenon- and krypton-filled lamps are depicted by Figures 1.24 and 1.25, respectively. The absorption spectrum of a Nd:YAG laser is given in Figure 1.26.

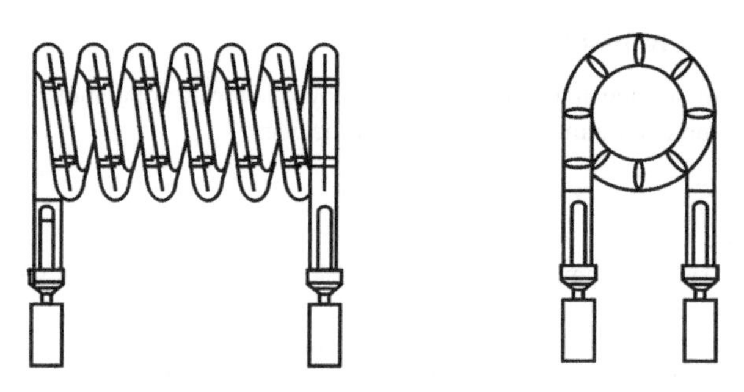

Figure 1.23 Helical flash lamp.

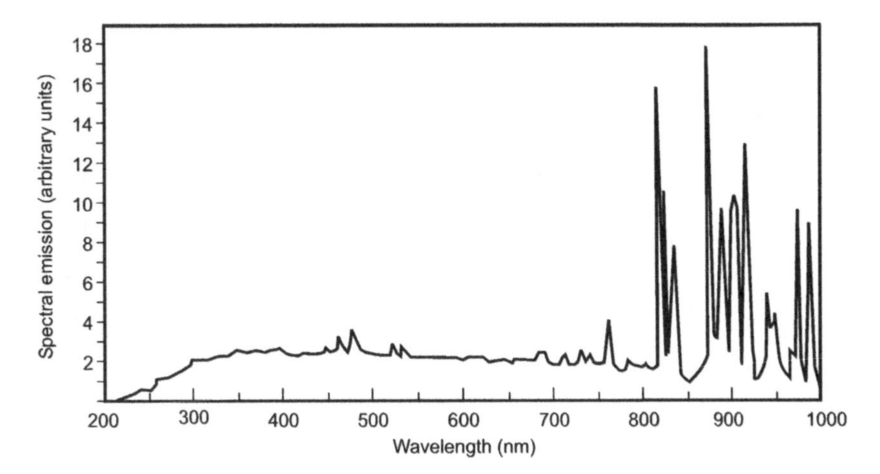

Figure 1.24 Emission spectrum of xenon-filled flash lamp.

Major electrical parameters include the flash lamp impedance parameter, maximum average power, maximum peak current, minimum trigger voltage and explosion energy. Impedance characteristics of a flash lamp are extremely important as they determine the energy transfer efficiency from energy storage capacitor, where it is stored, to the flash lamp.

Table 1.1 gives typical values of various characteristic parameters of xenon-filled and krypton-filled pulsed flash lamps from Heraeus Noblelight Ltd. The type numbers chosen for the purpose include both air-cooled as well as liquid-cooled flash lamps of different bore diameter and arc length. This assortment of flash lamps highlights the variation of the electrical parameters with bore diameter and arc length for a

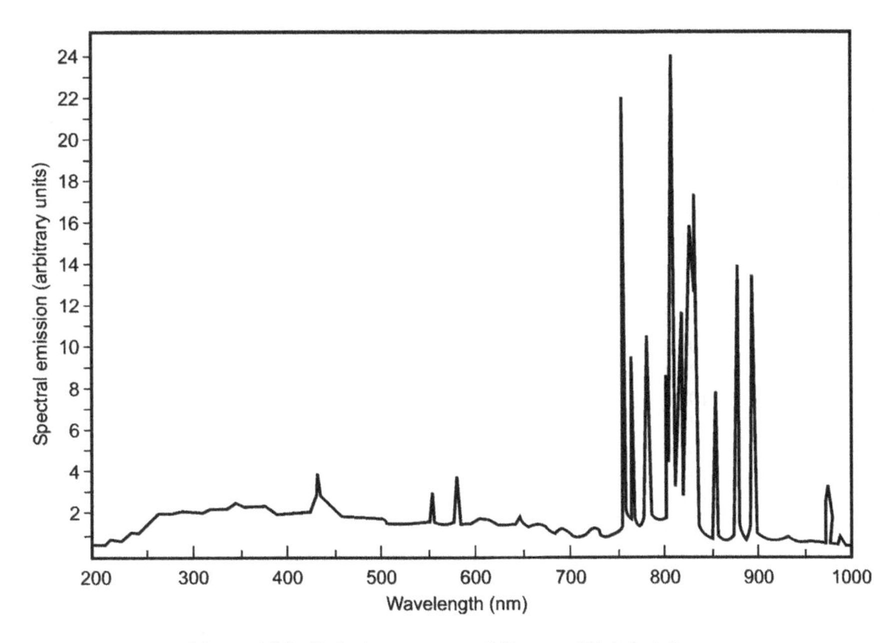

Figure 1.25 Emission spectrum of Krypton-filled flash lamp.

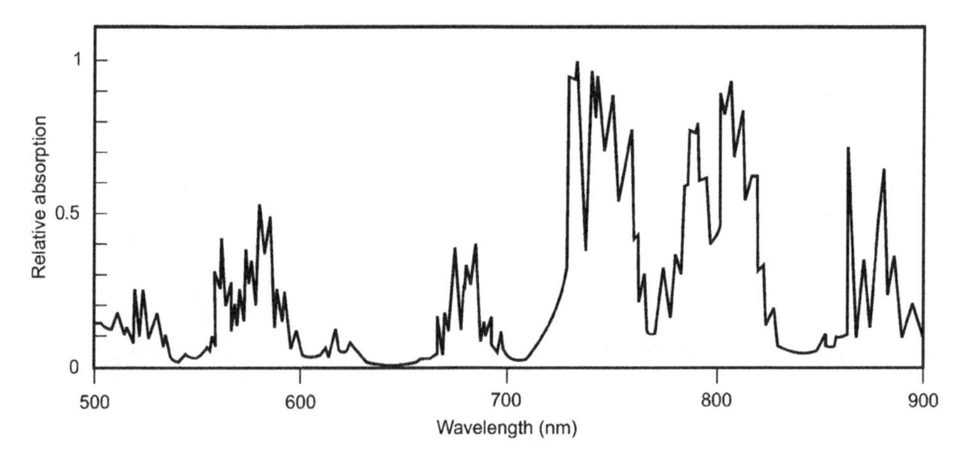

Figure 1.26 Absorption spectrum of Nd:YAG.

Table 1.1 Characteristic parameters of linear flash lamps. (In the case of maximum average power specification of air-cooled lamps, the listed value is for forced-air cooling. In the case of convection air-cooled, it is half of the value given for forced-air cooling.)

Flash lamp number	Bore diameter (mm)	Arc length (mm)	Impedance constant (K_0) ($\Omega A^{1/2}$)	Explosion energy constant ($Ws^{1/2}$)	Maximum average power (W)	Minimum trigger voltage (kV)	Minimum trigger pulse width (μs)
3 × 25XAP[1]	3	25.4	10.8	1.87×10^4	72	16	0.2
3 × 76XAP	3	76.2	32.3	5.62×10^4	214	16	0.6
3 × 25XFP[2]	3	25.4	11.2	1.87×10^4	479	16	0.2
3 × 76XFP	3	76.2	33.7	5.62×10^4	1436	16	0.6
3 × 25KAP[3]	3	25.4	8.9	1.87×10^4	72	18	0.2
3 × 76KAP	3	76.2	26.7	1.87×10^4	214	18	0.6
3 × 25KFP[4]	3	25.4	9.0	1.87×10^4	479	18	0.2
3 × 25KFP	3	76.2	29.2	1.87×10^4	1436	18	0.6
5 × 51XAP	5	50.8	12.9	6.25×10^4	238	16	0.4
5 × 102XAP	5	101.6	25.8	1.25×10^5	478	16	0.8
5 × 51XFP	5	50.8	13.5	6.25×10^4	1595	16	0.4
5 × 102XFP	5	101.6	27.0	1.25×10^5	3190	16	0.8
5 × 51KAP	5	50.8	10.6	6.25×10^4	238	18	0.4
5 × 102KAP	5	101.6	24.4	1.25×10^5	478	18	0.8
5 × 51KFP	5	50.8	10.9	6.25×10^4	1595	18	0.4
5 × 102KFP	5	101.6	21.7	1.25×10^5	3190	18	0.8
8 × 76XAP	8	76.2	12.1	1.50×10^5	574	18	0.6
8 × 102XAP	8	101.6	16.1	2.00×10^5	764	18	0.8
8 × 76XFP	3	76.2	12.7	1.50×10^5	3830	18	0.6
8 × 102XFP	3	101.6	16.9	2.00×10^5	5106	18	0.8
8 × 76KFP	8	76.2	10.2	1.50×10^5	3830	20	0.6
8 × 102KFP	8	101.6	9.9	2.00×10^5	5106	20	0.8
13 × 102XAP	13	101.6	14.9	3.25×10^5	1244	20	0.8
13 × 152XAP	13	152.4	16.1	4.87×10^5	1866	22	1.2
13 × 102XFP	13	101.6	10.4	3.25×10^5	8299	20	0.8
13 × 152XFP	13	152.4	15.6	4.87×10^5	12 448	25	1.2
13 × 102KFP	13	101.6	8.3	3.25×10^5	8299	25	0.8
13 × 152KFP	13	152.4	12.5	4.87×10^5	12 448	25	1.2

[1]Xenon-filled air-cooled; [2]Xenon-filled liquid cooled; [3]Krypton-filled air-cooled; [4]Krypton-filled liquid-cooled.

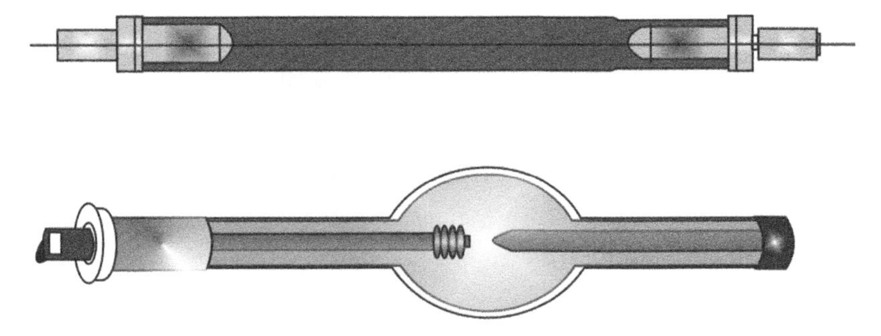

Figure 1.27 Construction of linear arc lamp.

given category of flash lamps, and also the range of values for bore diameter and arc length with the different categories of flash lamps.

Arc lamps are used for CW pumping of solid-state lasers. Like flash lamps, arc lamps are also gas-discharge devices. Arc lamps suitable for solid-state laser pumping are linear lamps (Figure 1.27), which are very much like linear flash lamps except for electrode design. As evident from Figure 1.27, arc lamps use pointed cathodes rather than the rounded cathodes used in flash lamps. Arc lamps are filled with xenon or krypton at a pressure of 1–3 atmospheres. Krypton-filled linear arc lamps are more common because of their relatively better spectral match to the Nd:YAG absorption band. Bore diameters of 4–7 mm and arc lengths in the range of 50–150 mm are common.

Table 1.2 provides typical values for various characteristic parameters in the case of linear krypton-filled arc lamps for different values of bore diameter and arc length. The information given in the table is based on the technical data of linear krypton-filled arc lamps from EG&G Electro-optics.

However, the efficiency with which pump output is usefully transferred to excite the lasing species is definitely lower in the case of the broadband optical pumping provided by flash lamps and arc lamps. Optical pumping at a single wavelength in a laser with an absorption level corresponding to that wavelength in the pump band achieves a relatively higher pumping efficiency, which leads to higher overall laser efficiency. Optical pumping of solid-state lasers by semiconductor lasers in what are better known as diode-pumped solid-state lasers achieves an efficiency that is 25–30 times that currently achievable in the case of flash lamp pumped solid-state lasers.

Laser diode arrays for solid-state laser pumping are available in various package configurations. The basic element in these arrays, also called stacks, is the laser diode bar (Figure 1.28). Each bar has multiple emitters. Laser bars are available in both conduction-cooled as well as liquid-cooled varieties. State-of-the-art bars offer up to 100 W of CW power. Stacks of these bars are also available for higher

Table 1.2 Characteristic parameters of linear krypton-filled arc lamps

Bore diameter (mm)	Arc length (mm)	Maximum average input power (W)	Maximum steady-state voltage (V)	Maximum steady-state current (A)	Maximum starting anode voltage (V)	Maximum trigger voltage (kV)
4	48	2200	104	20	1800	25
4	51	2100	95	20	2000	25
4	76	3200	160	20	2500	25
6	76	3000	120	40	2500	30
6	102	5500	140	38	3000	30

Figure 1.28 Laser diode bar.

pump power requirement (Figure 1.29). However, the maximum pump power available from diode laser arrays is still much lower than that possible from flash lamps.

1.11.2 Electrical Pumping

Pumping by electrical discharge is common in gas lasers. The excited electrons in the gas-discharge plasma transfer their energy to the lasing species either directly or indirectly through the atoms

Figure 1.29 Laser diode stack.

or molecules of another element. A helium-neon laser is a typical example of an indirect transfer of pump energy. The electrons first transfer the energy to helium atoms and then the excited helium atoms transfer the energy to neon atoms. A high voltage initially ionizes the gas and, once the discharge is struck, it can be sustained by a relatively much lower voltage and current. In a typical He-Ne laser, initiating voltage is of the order 8–10 kV while the sustaining voltage is around 1.5–2 kV.

Diode lasers are also electrically pumped, but not in the same way as gas lasers. In the case of diode lasers, the electrical current in the forward-biased diode frees electrons to create electron-hole pairs. The electrons and holes recombine to emit photons. In doing so, electrons drop back to the lower state.

1.11.3 Other Methods of Pumping

Some of the other methods of pumping or creating population inversion, which are specific to certain types of lasers, include excitation by *combustion reaction* as in gas dynamic CO_2 lasers, *chemical reaction* as in chemical lasers such as hydrogen fluoride (HF) laser, deuterium fluoride (DF) laser and chemical oxygen iodine laser (COIL) and electron acceleration as in free electron lasers. In the case of a gas dynamic laser for example, a combustion reaction produces a high-temperature high-pressure mixture of CO_2 and other gases required in a CO_2 laser. This gas mixture is then rapidly expanded through a set of nozzles to a very low-pressure low-temperature condition. Although the temperature and pressure drop rapidly a large number of molecules still remain in the excited state, thus creating population inversion.

1.12 Summary

- Lasers were undoubtedly one of the greatest inventions of the second half of 20th century – along with satellites, computers and integrated circuits – and continue to be so today due to their unlimited application potential.
- The basic principle of operation of a laser device is based on stimulated emission of radiation. In the case of ordinary light, such as that from the sun or an electric bulb, different photons are emitted spontaneously due to various atoms or molecules releasing their excess energy. In the case of stimulated emission, an atom or a molecule holding excess energy is stimulated by another previously emitted photon to release that energy in the form of a photon.
- A laser scientist is interested in an *optically allowed transition* between two energy levels, which involves either absorption or emission of a photon satisfying the resonance condition of $\Delta E = h\nu$ where ΔE is the difference in energy between the two involved energy levels, h is Planck's constant ($= 6.626\,075\,5 \times 10^{-34}$ J s or $4.135\,669\,2 \times 10^{-15}$ eV s) and ν is the frequency of the photon emitted or absorbed.
- There are two types of emission processes: spontaneous emission and stimulated emission. The emission process involves transition from a higher excited energy level to a lower energy level. Spontaneous emission is the phenomenon in which an atom or molecule undergoes a transition from an excited higher energy level to a lower level without any outside intervention or stimulation, emitting a resonance photon in the process.
- Under thermodynamic equilibrium conditions, practically all atoms or molecules are in the lower level. A condition of population inversion is said to be achieved when the population N_2 of a higher energy level is greater than the population N_1 of a lower energy level. Population inversion is an essential condition for the laser action.
- There are two possible ways to produce population inversion. One is to populate the upper level by exciting extra atoms or molecules to the upper level. The other is to depopulate the lower laser level involved in the laser action. In fact, for a sustained laser action, it is important to both populate the upper level and depopulate the lower level.

- Energy level structure of the laser medium has an important bearing on the laser action and associated characteristics. All lasers operate as either three-level (e.g. ruby laser) or four-level lasers (e.g. Nd: YAG, He-Ne, CO_2).
- The gain of the medium is defined as gain coefficient, which is the gain expressed as a percentage per unit length of the active medium. The gain of the laser medium refers to the extent to which this medium can produce stimulated emission. The amplification or the photon multiplication offered by the medium is expressed as a function of the gain of the medium and the length of the medium by: Amplification $= (1 + \text{gain})^{\text{Length of medium}}$.
- A resonator is the active laser medium within the closed path bounded by two mirrors, providing it meets certain conditions. One of the mirrors in the arrangement is fully reflecting and the other has a small amount of transmission. A laser resonator satisfies: Round trip length $= 2L = n\lambda$ where L, n and λ denote length of the resonator, an integer and wavelength, respectively.
- The resonators can be broadly classified as stable and unstable resonators. A stable resonator is one in which the photons can bounce back and forth between the end components indefinitely without being lost out the sides of the components. In an unstable resonator photons escape from the sides of the mirror after one or two passes within the cavity. An unstable resonator is usually chosen with laser media that have a very high gain, as alignment in this resonator type is much less critical.
- It is possible to have more than one resonant frequency (each referred to as a longitudinal mode) to be simultaneously present unless special measures are taken to prevent this from happening. The inter-mode spacing is given by $c/2L$.
- The transverse modes basically tell us about the irradiance distribution of the laser output in the plane perpendicular to the direction of propagation or, in other words, along the orthogonal axes perpendicular to the laser axis.
- Commonly employed pumping mechanisms include optical pumping and electrical pumping. Other pumping mechanisms include by chemical reactions and electron beams.

Review Questions

1.1. Differentiate between the processes of absorption, spontaneous emission and stimulated emission with particular reference to lasers.

1.2. In light of the fact that laser emission is nothing but stimulated emission of radiation, briefly explain why population inversion is an essential condition for the laser action to take place.

1.3. Compare and contrast three-level and four-level laser systems with reference to lasing threshold, conversion efficiency and ability to operate as a CW laser.

1.4. Differentiate between a stable resonator and an unstable resonator and explain why an unstable resonator is better when the laser medium has high gain.

1.5. What are longitudinal and transverse modes? What is their bearing on the laser characteristics such as beam divergence, coherence and directionality?

1.6. Briefly describe gain of laser medium. What do you understand by gain coefficient and what is its significance in laser resonator design?

Problems

1.1. Figure 1.30 depicts the energy level diagram of a popular four-level laser. Determine the emission wavelength.
[1064 nm]

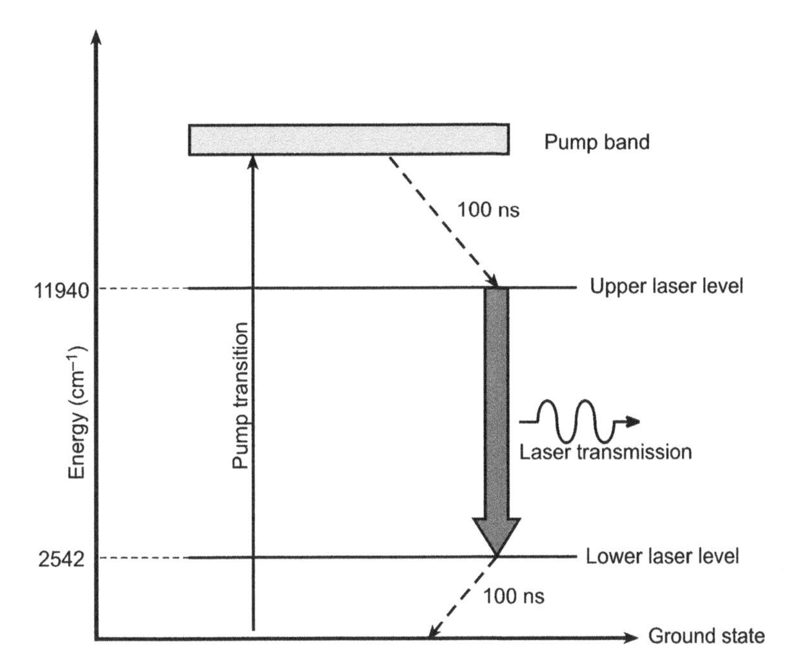

Figure 1.30 Problem 1.1.

1.2. After travelling 10 cm through the laser medium, 1000 photons become 2718 photons. Determine the gain coefficient of the medium.
[$0.1\,\mathrm{cm}^{-1}$ or $10\%\,\mathrm{cm}^{-1}$]

1.3. Given that the Doppler-broadened gain curve of a helium-neon laser with a 30-cm-long resonator emitting at 632.8 nm is 1400 MHz, determine the number of maximum possible sustainable longitudinal modes.
[3]

Self-evaluation Exercise

Multiple-choice Questions

1.1. Laser power input to a 10 cm long gain medium is 2 W. If the gain coefficient is 10% per centimetre, output power will be
a. 5.436 W
b. 2.718 W
c. 3 W
d. none of these

1.2. Which of the following is a stable resonator configuration?
a. plane-parallel resonator
b. confocal resonator
c. hemispherical resonator
d. concentric resonator
e. all of the above

1.3. When we say that gain coefficient of a laser medium is 10% per centimetre, it implies that 100 photons having the same transition energy as that of excited laser medium after travelling 1 cm will increase to
a. 101 photons
b. 110 photons
c. 1000 photons
d. none of these

1.4. What attributes does the fundamental transverse mode have?
a. least power spreading
b. minimum diffraction loss
c. can be focused to smallest possible spot
d. all of the above

1.5. A helium-neon laser having a resonator length of 10 cm and a Doppler-broadened gain curve of 1400 MHz and emitting at 633 nm can have
a. only one longitudinal mode
b. two longitudinal modes
c. any number of longitudinal modes
d. only one transverse mode

1.6. The transverse mode that is associated with the least beam divergence is
a. TEM_{00} mode
b. TEM_{01} mode
c. TEM_{10} mode
d. TEM_{03} mode

1.7. Flash lamps suitable for solid-state laser pumping are usually filled with
a. xenon
b. krypton
c. xenon or krypton
d. a mixture of xenon and krypton

1.8. Lasers used for optical pumping of laser media by another laser
a. are laser diode arrays
b. are pulsed solid-state lasers
c. include diode lasers, pulsed and CW solid-state lasers, excimer lasers, metal vapour lasers and so on.
d. none of these

1.9. A dye laser emitting in the visible wavelength band could possibly be pumped by
a. a diode laser emitting in near-infrared
b. an excimer laser emitting in ultraviolet
c. a frequency doubled Nd:YAG laser
d. any of the above lasers

1.10. An unstable resonator is associated with
a. high-gain laser medium
b. large interaction volume
c. less critical alignment
d. all of the above

Answers

1. (a) 2. (e) 3. (b) 4. (d) 5. (a) 6. (a) 7. (c) 8. (c) 9. (b) 10. (d)

Bibliography

1. *Laser Fundamentals*, 2008 by William Thomas Silfvast, Cambridge University Press.
2. *Fundamentals of Light Sources and Lasers*, 2004 by Mark Csele, Wiley-Interscience.
3. *Lasers: Fundamentals and Applications*, 2010 by K. Thyagarajan and Ajoy Ghatak, Springer.
4. *The Laser Guidebook*, 1999 by Jeff Hecht, McGraw Hill.
5. *Principles of Lasers*, 2009 by Orazio Svelto, Plenum Press.
6. *Understanding Lasers: An Entry Level Guide*, 2008 by Jeff Hecht, IEEE Press.
7. *Introduction to Laser Physics*, 1986 by Koichi Shimoda, Springer-Verlag.
8. *Introduction to Lasers and their Applications*, 1977 by Donald C. O'Shea, W. Russell Callen and William T. Rhodes, Addison-Wesley Publishing Co.
9. *Laser Handbook*, 1972 by F.T. Arechhi and E. O. Schulz-Dubois, Amsterdam North-Holland Publishing Co.
10. *Lasers and Light: Readings from Scientific American*, 1969 by A.L. Schawlaw, San Francisco, W. H. Freeman.
11. *Lasers and Optical Engineering*, 1990 by P. Das, Springer-Verlag.
12. *Lasers Theory and Applications*, 1981 by K. Thyagarajan and A.K. Ghatak, Plenum Press.
13. *Lasers*, 1986 by A.E. Siegman, University Science Books.
14. *Handbook of Laser Technology and Applications*, 2003 Volume I by Collin E. Webb and Julian D. C. Jones, Institute of Physics Publishing.

2

Laser Characteristics

2.1 Introduction

In the previous chapter were discussed operational fundamentals of lasers with emphasis on the mechanisms key to laser action. In the present chapter, we shall discuss the special characteristics that distinguish laser radiation from ordinary light and that have been responsible for the explosive growth in applications of laser devices since its invention in 1960. This is followed by a discussion of the different laser parameters that the designers and users of laser devices and systems are interested in. Some possible methods for measurement of these parameters are presented towards the end of the chapter.

2.2 Laser Characteristics

Laser radiation can be distinguished from the light from conventional sources on the basis of its special characteristics and the effects it is able to produce because of these characteristics. It is these characteristics that have led to explosive growth in the usage of laser devices since the invention of this magic source of light in 1960. These include:

1. monochromaticity
2. coherence, temporal and spatial
3. directionality.

We shall discuss each of these in some detail in the following sections.

2.2.1 Monochromaticity

Monochromaticity refers to the single frequency or wavelength property of the radiation. Laser radiation is monochromatic and this property has its origin in the stimulated emission process by which laser emits light. In describing the process of stimulated emission, we stated that the stimulated photon has the same frequency, phase and polarization as that of the stimulating photon. As we shall see in the following sections, monochromaticity is one of the essential requirements for the laser radiation to be coherent. Although not all monochromatic radiation is necessarily coherent, we shall see that coherent radiation is necessarily monochromatic.

2.2.1.1 Line-broadening Mechanisms

Laser radiation is not perfectly monochromatic. There could be various factors responsible for the spread in the frequency or wavelength, called its line width. The uncertainty principle causes slight variations in the wavelength of different photons emitted during the stimulated emission process. This implies that the

Lasers and Optoelectronics: Fundamentals, Devices and Applications, First Edition. Anil K. Maini.

frequency spread in the case of pulsed lasers depends upon the pulse width of radiation. Shorter pulses would have a larger spread. This also means that CW laser outputs are likely to be more monochromatic than the pulsed laser outputs.

The interaction of lasing species with other atoms and molecules is another cause of line broadening. In the case of gas lasers, the dependence of line width on gas pressure is an example of this phenomenon. Line width is observed to increase with gas pressure. Increased pressure decreases the time interval between successive collisions, which in turn affects energy transfer. Line broadening due to increased pressure is known as pressure broadening or collisional broadening.

Yet another factor contributing to the increase in line width is the inherent random motion of atoms and molecules. The emitted wavelength corresponding to a randomly moving atom or molecule is such that its Doppler-shifted component matches the nominal transition wavelength. The random motion of atoms and molecules causes frequency spread. As an example, gas lasers have a Doppler-broadened gain curve. The frequency spread in the case of a multimode He-Ne laser at 632.8 nm could be as much as 1400 MHz. The same figure for a CO_2 laser at 10 600 nm is 60 MHz.

Another reason for frequency spread could be the simultaneous oscillation of more than one longitudinal mode within the laser resonator. The spread in the case of a semiconductor laser is much greater, of the order 6×10^6 MHz.

Fortunately, we have techniques (some of which are discussed in Section 8.8) which can be used to stabilize the frequency at one point on the broad gain-bandwidth curve. We have techniques by which we can ensure that the laser oscillates at a single longitudinal mode.

Table 2.1 lists the line width of some well-known lasers; note that the magnitudes of line width listed in the table represent natural line width. There are techniques for narrowing the line width. As an example, an intra-cavity etalon in an Nd:YAG laser could be employed to reduce its line width to less than 0.2 cm^{-1}. As another example, frequency-stabilized helium-neon and carbon dioxide lasers have line widths that are several orders of magnitude narrower than that of unstabilized counterparts.

There are standard techniques and equipment available for precise measurement of wavelength and line width of laser radiation. Some of the common methods include: optical spectrum analysis based on diffraction gratings; conversion of frequency variation into intensity variations with the help of an unbalanced interferometer or a high-finesse reference cavity; self-heterodyne technique; heterodyne technique using two independent lasers; and frequency combs. These are briefly described in Section 2.4.5.

Table 2.1 Line width associated with lasers

Type of laser	Wavelength (nm)	Line width (cm^{-1})	Line width (nm)	Line width (GHz)
Ruby laser	694.3	11	0.53	330
Ruby laser	692.9	11	0.53	330
Nd:YAG laser	1064	1–5	0.1–0.5	25–150
Nd:Glass laser (Phosphate)	1054	180	20	5400
Nd:Glass laser (Silicate)	1062	245	27.7	7370
Helium neon laser	632.8	0.05	1.9×10^{-3}	1.4
Helium cadmium laser	441.6	0.1	0.002	3
Carbon dioxide laser	9000–11 000 (main 10 600 nm)	0.002	0.022	0.06
Alexandrite laser	720–800 (tunable)	50 at 755 nm	3 at 755 nm	1580 at 755 nm
Titanium Sapphire	680–1130 (tunable)			
GaAlAs laser	750–900			
InGaAsP laser	1200–1600			
Excimer laser (XeF)	351	3328	41	99 836
Excimer laser (XeCl)	308	3331	31.6	99 932
Excimer laser (ArF)	193	0.335	0.001 25	10
Excimer laser (KrF)	248	0.3	0.001 85	9
Copper vapour laser	510.5	0.077	0.002	2.3

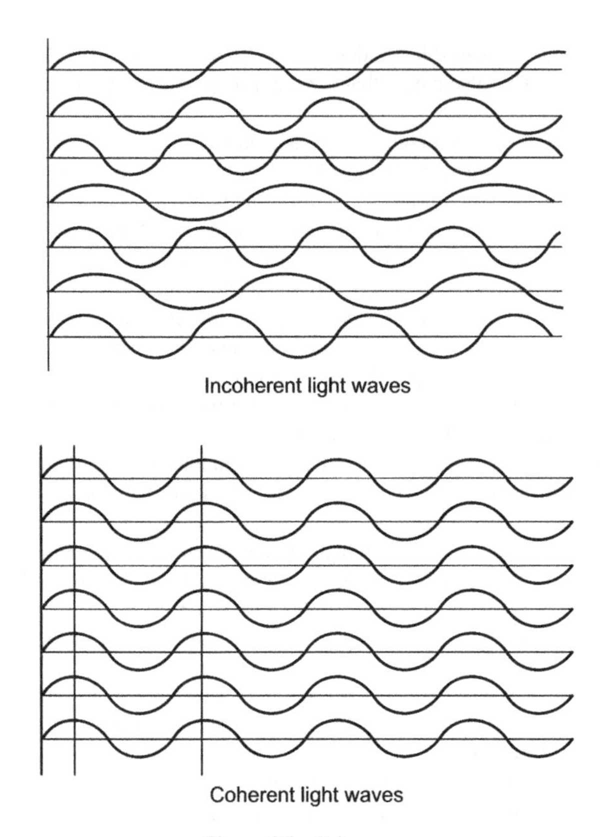

Figure 2.1 Coherence.

It may also be mentioned here that standard beam diagnostic equipment that makes use of the above-mentioned techniques is available for the purpose of laser beam analysis in terms of wavelength, line width and modal profile. These include wave meters, optical spectrum analyzers and beam profilers. Some of this commercially available diagnostic equipment are briefly mentioned in Section 2.5.

2.2.2 Coherence

The property that best distinguishes laser radiation from ordinary light is *coherence*. Light is said to be coherent when different photons (or the waves associated with these photons) have the same phase and this phase relationship is preserved as a function of time (Figure 2.1). That is, this phase relationship is preserved as the radiation wave front travels with time. There are two types of coherence: *temporal coherence* and *spatial coherence*.

2.2.2.1 Temporal Coherence

Temporal coherence is the preservation of the phase relationship with time, which is generally what is meant or assumed by the term *coherence*. *Spatial coherence*, which is the preservation of phase across the width of the beam, is discussed in Section 2.2.2.2.

The necessary conditions for temporal coherence are that all photons should be emitted with same phase and should have the same wavelength. If the starting phase is the same and all have the same wavelength, the phase will be preserved irrespective of time. Both these conditions are ideally ensured by the stimulated emission process, but in reality things are never perfect. The reasons for wavelength spread were explained in Section 2.2.1 (monochromaticity). If all photons originate with the same phase, temporal coherence would then depend upon the wavelength or frequency spread. Temporal coherence is measured as *coherence length* or *coherence time*. The two are interrelated as given in Equation 2.1:

$$\text{Coherence length} = c \times \tau_c \tag{2.1}$$

where

τ_c = coherence time
c = speed of light

Coherence length can be computed from the known value of wavelength spread from Equation 2.2:

$$\text{Coherence length} = \frac{\lambda^2}{2\Delta\lambda} \tag{2.2}$$

The same relationship can be written in terms of frequency spread by substituting $\lambda = c/f$ and $\Delta\lambda = \left(c/f^2\right) \times \Delta f$ to obtain

$$\text{Coherence length} = \frac{c}{2\Delta f} \tag{2.3}$$

A simple calculation tells you that the coherence length for ordinary light may be of the order of a fraction of a micron. On the other hand, it could be tens of kilometres for an actively frequency-stabilized CO_2 laser. For instance, a multimode He-Ne laser emitting at 632.8 nm with a frequency spread of 1400 MHz would have a coherence length of about 10 cm. The same laser, if frequency stabilized to say 1 MHz, would have a coherence length of 150 m. A CO_2 laser actively stabilized to within 10 kHz, which is possible with some of the techniques, will have a coherence length of 15 km. Ordinary light, that emits all wavelengths from 400–900 nm, has an expected coherence length of 0.36 μm assuming an average wavelength of 600 nm.

Interference fringes formed in a Michelson Interferometer are the result of temporal coherence. Figure 2.2 shows the schematic arrangement in a Michelson interferometer. As can be seen from the figure, the interferometer comprises two mirrors, one of these fixed. The laser beam is split into two paths with the help of a beam splitter. After splitting, one of the beams strikes a fixed mirror while the other strikes a movable mirror. After being reflected from their respective mirrors, the two beams recombine at the beam splitter to form an interference pattern. A compensator plate of the same thickness and material as that of the beam splitter is inserted in the path of the fixed mirror to ensure that the light beams in the two paths encounter the same thickness of glass. It may be mentioned here that, even for the same physical path length travelled by the two beams, in the absence of a compensator plate the beam moving towards the movable mirror encounters a travel path through glass that is three times the thickness of the beam splitter before it recombines with the other beam. On the other hand, the second beam travelling towards the fixed mirror encounters only one glass thickness before recombination. The addition of the compensation plate restores parity.

If the two path lengths differ by a whole number (including 0) of wavelengths, there is constructive interference; in this case, the two beams arrive back at the beam splitter in phase, assuming that the two beams at the time of splitting were in phase. On the other hand, if the two path lengths differ by a whole-number-and-a-half wavelengths, there is destructive interference; in this case the beams arrive at the beam splitter out of phase with each other. In other words, for constructive interference the path length difference is $m\lambda$ and for destructive interference, path length difference is $[(2m+1)/2]\lambda$ where m is a positive or negative integer (including 0) and λ is the wavelength. An interference pattern is formed as long as the path length difference remains less than or equal to the coherence length of the laser radiation.

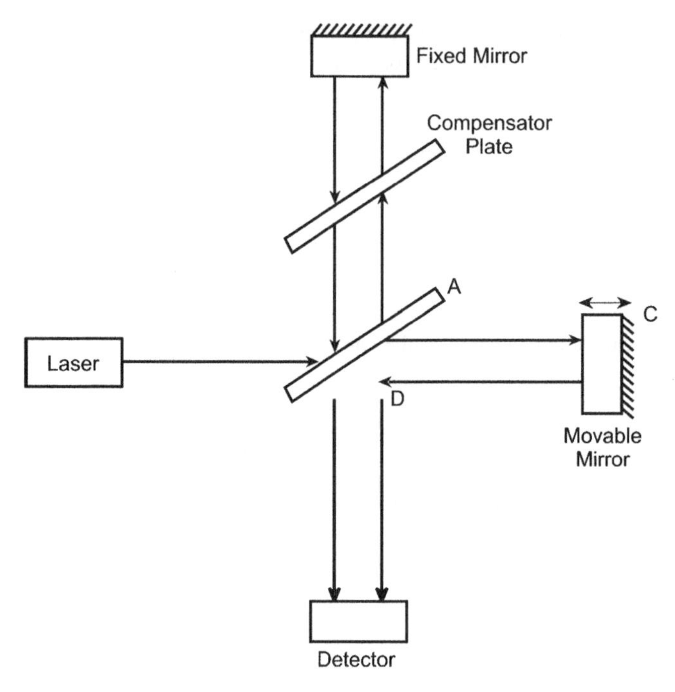

Figure 2.2 Michelson interferometer.

2.2.2.2 Spatial Coherence

The other type of coherence is *spatial coherence*, which describes the correlation in phase of different photons transverse to the direction of travel. It is the area in the plane perpendicular to the direction of travel over which the radiation preserves the coherence. The spatial coherence depends upon the transverse mode discrimination property of the laser resonator. Laser radiation operating in the lowest-order mode (TEM_{00}) will certainly be more spatially coherent than a multimode laser radiation. When a laser is operating in a single transverse mode, the radiation will be spatially coherent across the diameter of the beam over reasonable propagation distances. Young's double slit experiment and formation of fringes is the best illustration of the phenomenon of spatial coherence.

Figure 2.3 shows the schematic arrangement of Young's double slit experiment. As shown in Figure 2.3b, the incident light beam is split into beams. The beam in this case splits up after passing through two slits that are a distance d apart. The two beams strike a screen at a distance L from the plane carrying the slits. In the case d is much smaller than length L, the difference in the path lengths r_1–r_2 travelled by the two beams is approximately equal to $d\sin\theta$.

Again, if the two beams were in phase when they passed through the slits, which they would be if the incident beam were spatially coherent, then for constructive interference to occur on the screen $d\sin\theta = m\lambda$. For destructive interference, $d\sin\theta = [(2m+1)/2]\lambda$ where m is a positive or negative integer including zero. For $m = 0$ and $\theta = 0°$, we get a bright fringe which is known as central maximum. On both sides of central maximum, we get bright fringes at angles θ equal to $\pm\sin^{-1}(\lambda/d)$, $\pm\sin^{-1}(2\lambda/d)$, $\pm\sin^{-1}(3\lambda/d)$ and so on. Similarly, dark fringes occur at angles θ equal to $\pm\sin^{-1}(3\lambda/2d)$, $\pm\sin^{-1}(5\lambda/2d)$, $\pm\sin^{-1}(7\lambda/2d)$ and so on.

Temporal and spatial coherence are independent of each other. While monochromaticity leads to temporal coherence, it is the transverse mode discrimination characteristics of the resonator that determines the spatial coherence. A given laser may be temporally coherent and spatially incoherent, or *vice versa*.

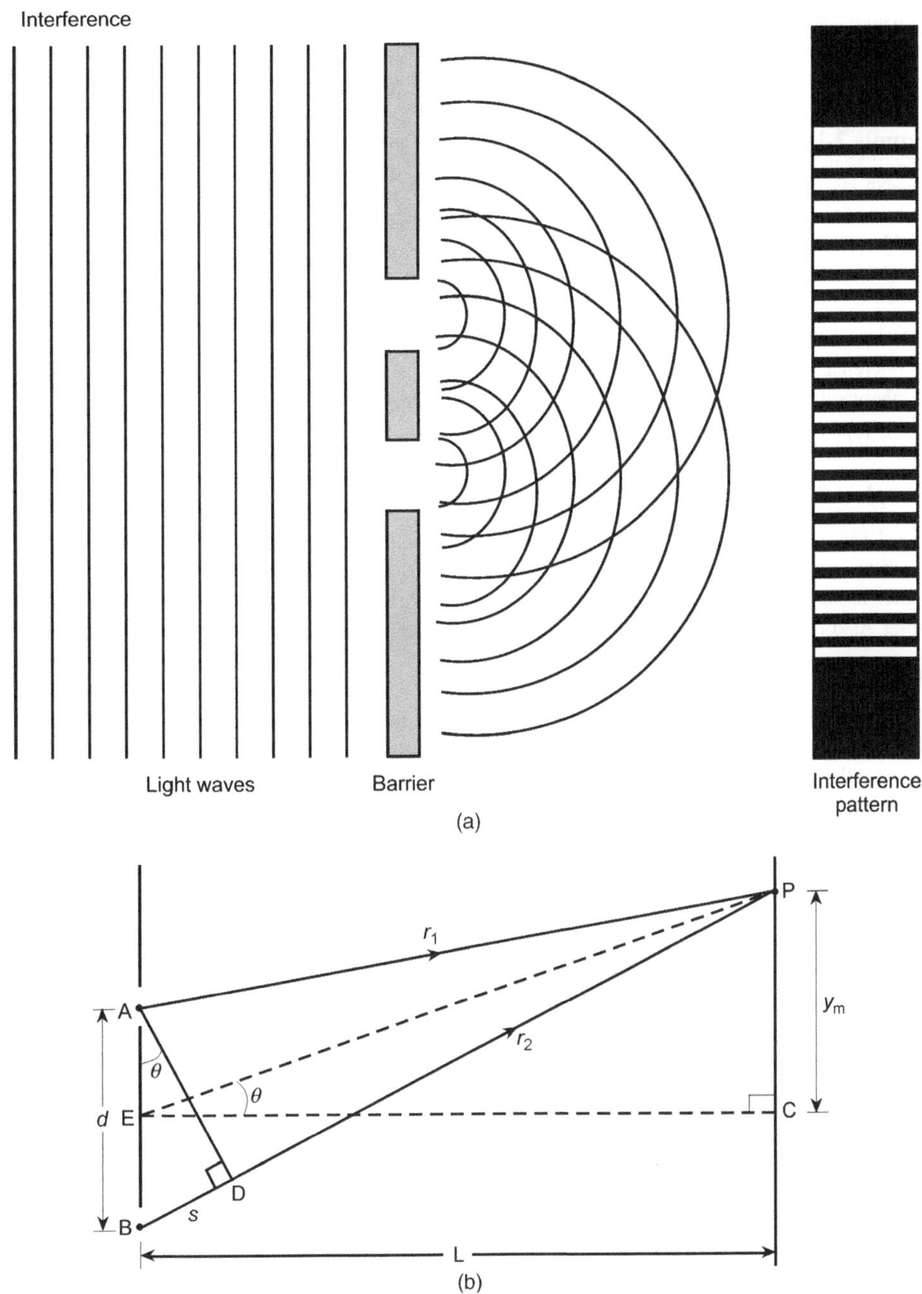

Figure 2.3 Young's double-slit experiment.

2.2.3 Directionality

The *directionality* of laser radiation has its origin in the coherence of the stimulated emission process. All photons emitted as a result of stimulated emission processes have the same frequency, phase, direction and polarization. When emitted, these photons carry no information regarding the location of the excited atom or molecule responsible for its emission. It appears as if all photons were emitted from a tiny volume with dimensions that are of the order of a wavelength. If a photon is emitted off-axis, spatial coherence makes it appear as if it were emitted from the axis. Similarly, for a photon that is emitted away from the beam waist on the same axis, temporal coherence makes it appear as if it were emitted from the beam waist.

Example 2.1

A He-Ne laser emitting at 633 nm has a line width of 0.002 nm. Determine its coherence length. Also determine the coherence length if the same laser was frequency stabilized to a frequency uncertainty of 100 kHz.

Solution

1. Coherence length $= \frac{\lambda^2}{2\Delta\lambda}$ where $\lambda = 633$ nm and $\Delta\lambda = 0.002$ nm.
2. Coherence length therefore $= (633)^2/(2 \times 0.002)$ nm $= 10^8$ nm $= 10$ cm.
3. Coherence length can also be computed from $\frac{c}{2\Delta f}$ where $\Delta f = 100$ kHz $= 100\,000$ Hz.
4. Coherence length therefore $= (3 \times 10^8)/(2 \times 100\,000) = 1500$ m $= 1.5$ km.

Example 2.2

A CO_2 laser emitting at 10 600 nm has a Doppler-broadened gain curve that is 60 MHz wide. What is the coherence length of such a laser? If the frequency of this laser is stabilized around the centre of its Doppler-broadened gain curve to an accuracy of ±1 MHz, what would be the new value of coherence length? Also determine the line width (in nm) of the laser in the two cases.

Solution

1. Coherence length $= \frac{c}{2\Delta f} = \frac{\lambda^2}{2\Delta\lambda}$ where the terms have their usual meaning.
2. Initially $\Delta f = 60$ MHz; so coherence length $= (3 \times 10^8)/(2 \times 60 \times 10^6) = 2.5$ m.
3. When the laser is frequency stabilized, $\Delta f = 2$ MHz. The new coherence length $= (3 \times 10^8)/(2 \times 2 \times 10^6) = 75$ m.
4. Line width can be computed from $\Delta\lambda = \lambda^2/(2 \times \text{coherence length})$. In the case of an unstabilized laser, $\Delta\lambda = (10\ 600 \times 10^{-9})^2/(2 \times 2.5)$ m $= 0.022\,47$ nm. In the case of a stabilized laser, $\Delta\lambda = [(10\,600 \times 10\,600 \times 10^{-18})/(2 \times 75)]$ m $= 0.00075$ nm.

Example 2.3

In Young's double slit experiment, the separation between two successive bright fringes was precisely measured to be equal to 3.164 mm. If the slit separation and slit-screen distance are 0.1 mm and 50 cm respectively, determine the laser wavelength.

Solution

1. Refer to Figure 2.3b. For small values of angle θ,

   $$\sin\theta \cong \theta = (m\lambda/d) \text{ and also } \tan\theta \cong \theta = (y_m/L)$$

   where y_m is the distance of the mth bright fringe from the central maximum.
   This gives $y_m = (m\lambda L/d)$
2. This also gives a distance of the $(m+1)$th bright fringe from the central maximum as $[(m+1)\lambda L/d]$. Therefore, the separation between the two successive fringes equals $(\lambda L/d)$.
3. Substituting the given values of L, d and the distance between two successive fringes, we get

   $$3.164 \times 10^{-3} = 50 \times 10^{-2} \times \lambda/0.1 \times 10^{-3} = 5000\lambda$$

   which gives $\lambda = 3.164 \times 10^{-3}/5000 = 632.8$ nm.

Example 2.4

In a Michelson interferometer experimental set-up, the interference pattern was observed to move by 200 fringes with respect to a reference point when a 0.1-mm-thick glass plate having a refractive index of 1.546 was inserted in the path of one of the beams. Determine the wavelength of the laser source.

Solution

1. Let us assume that the thickness of glass plate inserted in the optical path is T.
2. The phase difference in terms of number of wavelengths introduced by this insertion is therefore $(2T/\lambda_n)$, where λ_n is the wavelength of the laser in the glass, equal to λ/n where λ is the free-space wavelength or the wavelength in air and n is the refractive index of the material.
3. An equivalent air space of length T would have caused a phase difference of $2T/\lambda$ in terms of the number of wavelengths.
4. Therefore, additional phase difference in terms of the number of wavelengths introduced due to thickness T of glass equals: $(2nT/\lambda) - (2T/\lambda) = 2T(n-1)/\lambda$.
5. If N is the number of fringes observed to move with respect to a reference point because of the insertion of the glass plate, then $N = 2T(n-1)/\lambda$.
6. This gives $\lambda = 2T(n-1)/N$.
7. Substituting the given values, $\lambda = 2 \times 0.1 \times 10^{-3} \times (1.546 - 1)/200 = 546\,\text{nm}$.

2.3 Important Laser Parameters

In many applications including industrial, medical, military and scientific, the measurement of laser power or energy is often not adequate and it becomes necessary to measure many other beam parameters related to its shape and intensity profile. In addition, we are also interested in knowing how these parameters change as the laser beam propagates through atmosphere. For example, in the case of a laser target designator, in addition to laser energy per pulse and pulse width, beam divergence is another important parameter that determines the operational range of the system. On the other hand, in the case of a laser intended for spectroscopy applications, the line width and wavelength stability are the parameters that demand special attention. In the following sections we discuss the important parameters which both designers and users of laser systems are interested in:

1. wavelength
2. CW power (CW lasers)
3. peak power (pulsed lasers)
4. average power (pulsed lasers)
5. pulse energy (pulsed lasers)
6. repetition rate (pulsed lasers)
7. pulse width (pulsed lasers)
8. duty cycle (pulsed lasers)
9. rise and fall times
10. irradiance
11. radiance
12. beam divergence
13. spot size
14. M^2 value
15. wall-plug efficiency.

2.3.1 Wavelength

Wavelength is of course the first and the foremost parameter with which the laser is identified. It is in a way laser-specific. There are lasers which can possibly emit at more than one wavelength, however. While Nd:YAG is always associated with 1064 nm, a He-Ne laser can emit at 632.8 nm, 543 nm, 1150 nm

Table 2.2 Wavelengths of common lasers

Laser type	Wavelength (nm)	Laser type	Wavelength (nm)
Nd:YAG	1064	Helium-cadmium	441.6
Nd:YLF (Polarized)	1053	Helium-cadmium	325
Nd:YLF (Unpolarized)	1047	Carbon dioxide	9600
Nd:YVO4	1064	Carbon dioxide	10 600
Nd:Phosphate Glass	1054	Copper vapour	511
Nd:Silicate Glass	1062	Copper vapour	578
Nd:Fused Silica Glass	1080	Gold vapour	628
Ruby (R1-line)	694.3	Argon-ion	488
Ruby (R2-line)	692.9	Argon-ion	514.5
Alexandrite	701–826	Krypton-ion	647
Titanium-Sapphire	660–986	Nitrogen	337.1
Cr-GSGG	742–842	Argon fluoride	193
Erbium-YAG	2940	Krypton fluoride	249
Erbium-YLF	1730	Xenon chloride	308
Erbium-Glass	1540	Xenon fluoride	350
Helium-Neon	632.8	Diode laser (GaAs)	904
Helium-Neon	1150	Diode laser (GaAlAs)	720–900
Helium-Neon	1523	Diode laser (InGaAs)	1060
Helium-Neon	3390	Diode laser (InGaAsP)	1300–1550

and 3390 nm. There are some lasers that can be tuned across a band. The vibronic class of solid-state lasers, dye lasers and free electron lasers belong to this category. Table 2.2 lists the wavelengths of some well-known lasers.

2.3.2 CW Power

CW power is the power available from the CW laser. The typical laser output power level may vary from a fraction of a milliwatt in a He-Ne laser or a semiconductor diode laser used in laser pointers to hundreds of kilowatts or even several megawatts in high-power lasers. Other power parameters are the peak power and average power defined with reference to pulsed lasers.

2.3.3 Peak Power

Peak power is the highest instantaneous optical power in the laser pulse (Figure 2.4). Although peak power occurs only for a time duration which is a small fraction of the pulse width, it is considered to be present during the entire pulse width. This greatly simplifies average power and energy calculations.

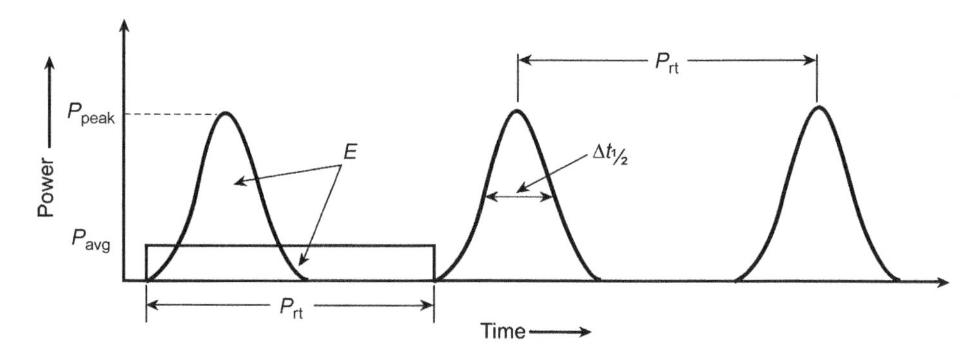

Figure 2.4 Laser pulse parameters.

Peak power is of particular relevance to laser systems such as laser range finders and laser target designators, as it is one of the important parameters that determine the maximum operational range of these systems. The peak power specification in the case of military laser range finders and target designators configured around Q-switched (where Q stands for the quality factor of the resonant cavity) solid-state lasers is in the range of 1–10 MW.

2.3.4 Average Power

Average power is the product of peak power and the duty cycle (Figure 2.4). It can also be written as the product of pulse energy and the repetition rate. A laser target designator producing 100 mJ of pulse energy in a 20 ns pulse at a repetition rate of 20 pulses per second (pps) will have peak power and average power specifications of 5 MW and 2 W, respectively.

2.3.5 Pulse Energy

Pulse energy is defined with respect to pulsed lasers. It is in fact the area under the power versus time curve representing the laser pulse. If the laser pulse is considered as rectangular with amplitude equal to the peak power, the pulse energy is the product of peak power and the pulse width. If the laser pulse (which has a Gaussian profile) is approximated as an isosceles triangle as shown in Figure 2.5 and the pulse width is measured as the full width at the points of half of the peak power, the area under the curve is the product of peak power and the pulse width.

2.3.6 Repetition Rate

Repetition rate, also referred to as *pulse repetition frequency*, of a pulsed laser is the number of laser pulses produced per second. It is equal to the reciprocal of the time interval between two successive laser pulses (Figure 2.4). Repetition rate is of particular significance in laser systems such as laser target designators used for guided-weapon delivery applications.

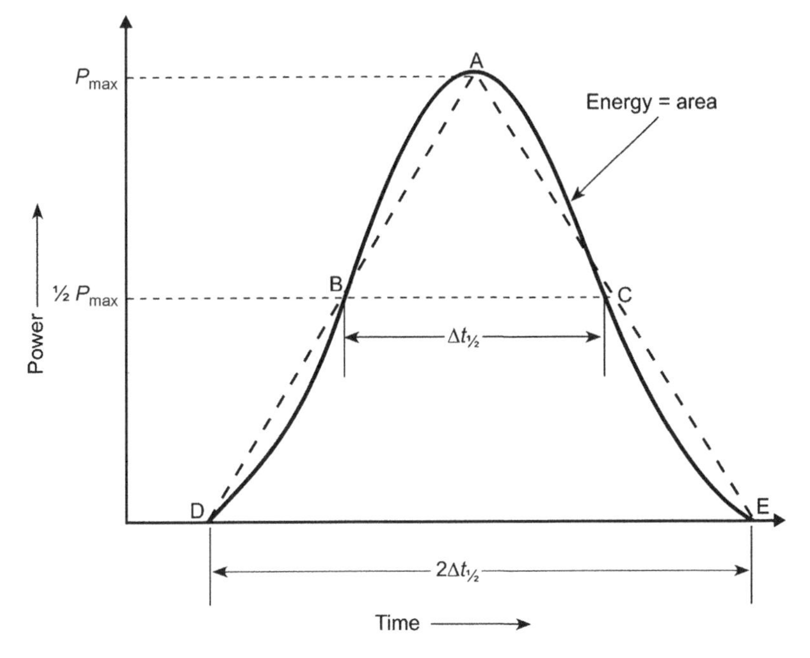

Figure 2.5 Computation of pulse energy of a Gaussian pulse.

2.3.7 Pulse Width

Pulse width (or *pulse duration* in the case of a pulsed laser) is usually measured as full width at half maximum (FWHM) as shown in Figure 2.5. Pulse width is intimately related to bandwidth. The narrower the pulse width, the higher is the required bandwidth. In fact, Heisenberg's uncertainty principle sets a limit on the minimum possible laser pulse width for a given value of available bandwidth. That is:

$$\text{Minimum pulse width} = 0.441/\text{bandwidth} \tag{2.4}$$

We also have:

$$\text{Minimum bandwidth} = 0.441/\text{pulse width} \tag{2.5}$$

Pulse width could be few femtoseconds (10^{-15} of a second) to hundreds of femtoseconds in the case of some mode-locked lasers, a few nanoseconds to several tens of nanoseconds in the case of Q-switched solid-state lasers, and a few tens of microseconds to hundreds of microseconds in case of free-running lasers.

2.3.8 Duty Cycle

Duty cycle is the ratio of pulse width to the time interval between two successive pulses (Figure 2.4). Duty cycle in the case of a laser target designator producing 20-ns-wide laser pulses at a repetition rate of 20 pps is 0.000 000 4.

Peak power, average power, pulse width, pulse energy, repetition rate and duty cycle are interrelated. The following mathematical expressions summarize their inter-relationship:

$$\text{Pulse energy} = \text{peak power} \times \text{pulse width} \tag{2.6}$$

$$\text{Average power} = \text{peak power} \times \text{duty cycle} = \text{pulse energy} \times \text{repetition rate} \tag{2.7}$$

$$\text{Peak power} = \text{pulse energy}/\text{pulse width} \tag{2.8}$$

$$\text{Duty cycle} = \text{pulse width} \times \text{repetition rate} \tag{2.9}$$

Example 2.5

Determine the following:

1. peak power of a Q-switched Nd:YAG laser producing 100 mJ, 20 ns pulses,
2. average power of a 120 mJ, 50 ns Nd:Glass laser operating at 10 pps and
3. pulse energy of a Q-switched Nd:YAG laser producing 10 MW peak power, 10 ns laser pulses.

Solution

1.
   - Pulse energy = 100 mJ
   - Pulse width = 20 ns
   - Peak power = $100 \times 10^{-3}/20 \times 10^{-9}$ W = 5 MW
2.
   - Pulse energy = 120 mJ
   - Pulse width = 50 ns
   - Pulse repetition frequency = 10 pps
   - Average power = $120 \times 10^{-3} \times 10$ W = 1.2 W

3.
- Peak power = 10 MW
- Pulse width = 10 ns
- Pulse energy = $10 \times 10^6 \times 10 \times 10^{-9}$ J = 100 mJ.

Example 2.6

An Nd:YAG laser emits 1064 nm with a bandwidth of 0.1 nm. Determine the minimum possible laser pulse duration achievable from this laser without using any special techniques.

Solution: Wavelength, $\lambda = 1064$ nm

Bandwidth, $\Delta\lambda = 0.1$ nm
Frequency, $f = c/\lambda = 3 \times 10^8/1064 \times 10^{-9}$ Hz $= 2.8 \times 10^{14}$ Hz
This gives $\Delta f = (0.1 \times 10^{-9}/1064 \times 10^{-9}) \times 2.8 \times 10^{14}$ Hz $= 2.63 \times 10^{10}$ Hz = 26.3 GHz
Minimum possible pulse duration $= 0.441/\Delta f = (0.441/26.3 \times 10^9)$ s = 0.0168 ns = 16.8 ps.

2.3.9 *Rise and Fall Times*

Rise and *fall times* refer to the time duration between 10% and 90% of the peak amplitude of the pulse during rising and falling portions of the laser pulse, respectively. Pulse rise time becomes particularly important while designing opto-electronic front-end circuits for converting a laser radiation pulse into an equivalent electrical signal. The bandwidth of the current-to-voltage converter needs to be commensurate with the rise time specification. For example, a 10 ns laser pulse with a rise time of 2 ns not only needs an optical sensor with a rise time specification of 2 ns or better, but the trans-impedance amplifier used to transform photocurrent into an equivalent voltage pulse also needs to have a bandwidth of 175 MHz (= 350/2) or better to faithfully reproduce the pulse.

2.3.10 *Irradiance*

Irradiance, also referred to as *power density*, is defined as the power per unit area of the laser radiation falling on the target. It is expressed in units of W m^{-2}. This parameter is particularly important when the laser radiation is used to illuminate a receiving system. The laser power actually entering the receiver system depends upon the receiving aperture and the power density available at that plane. A typical example where the irradiance parameter assumes importance is the use of a high-energy laser system in a countermeasure mode to saturate or damage distant electro-optic sensor systems such as laser range finders and laser target designators. In such a situation, the laser source needs to produce a certain minimum irradiance at the target plane to be able to produce the desired effect.

2.3.11 *Radiance*

Radiance, also referred to as *brightness*, is usually defined with respect to the laser source. It is the power emitted per unit area per unit solid angle, expressed as W m^{-2} sr^{-1}. For small values of angle, planar angle θ is related to solid angle Ω by $\Omega = (\pi/4)\theta^2$. Quite obviously, a small exit beam diameter and lower divergence mean higher radiance or brightness.

Example 2.7

Determine the radiance of a 5 mW helium-neon laser having an output diameter of 2 mm and a divergence of 1 mrad.

Solution

1. Divergence = 1 mrad
2. Corresponding solid angle $= (\pi/4)\ (10^{-3})^2 = 0.785 \times 10^{-6}$ sr

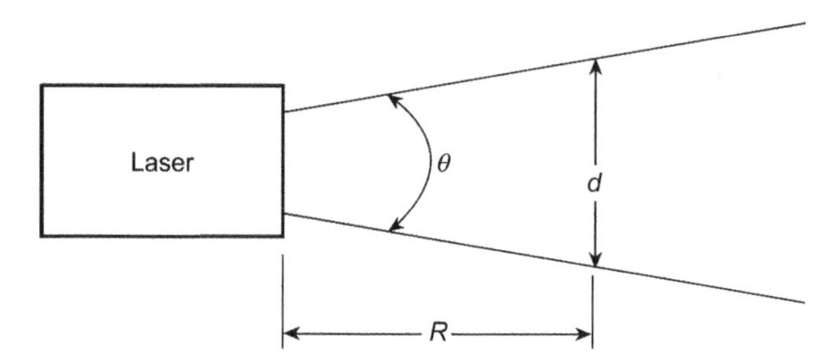

Figure 2.6 Definition of beam divergence.

3. Laser spot area = $(\pi/4)\ (2 \times 10^{-3})^2 = 3.14 \times 10^{-6}\,\text{m}^2$
4. Power = 5×10^{-3} W
5. Therefore, radiance = $(5 \times 10^{-3})/(3.14 \times 10^{-6}) \times (0.785 \times 10^{-6}) = 2 \times 10^{9}\,\text{W m}^{-2}\,\text{sr}^{-1}$.

2.3.12 Beam Divergence

Beam divergence is an indicator of the spread in the laser beam spot as it travels away from the source. It is a function of the wavelength λ and size of output optics. If D is diameter of output optics, then

$$\theta = (1.27\lambda)/d \tag{2.10}$$

where θ is the divergence in radians.

For a given λ, larger D leads to smaller divergence. This of course is the minimum value of divergence the laser can have, assuming that it is transmitting the fundamental transverse mode TEM_{00}. In the presence of higher-order transverse modes, the beam divergence increases more rapidly. The divergence parameter determines the spot size in the far field. If θ is full-angle divergence in radians and d the spot diameter at a distance R (m) (Figure 2.6), then $\theta = (d/R)$ assuming that the value of θ is very small. If θ is measured in milli-radians and R is in kilometres, then d is in metres. In fact, the exact expression for the spot diameter d is given by:

$$d = D + R\theta \tag{2.11}$$

which approximates to $d = R\theta$ for large values of R.

Beam divergence is one of the important parameters that determine the maximum operational range of laser systems such as laser range finders and laser target designators. Divergence determines the laser spot diameter at the target and therefore the power density for a given peak power. It is the power density and not the absolute value of power that determines the power entering the receiver channel of these systems. When the divergent laser beam is made to fall on a positive lens, the diameter of the focused spot is the product of the focal length of the lens and the divergence of the incident laser radiation. For a given focal length, the focused spot size is therefore a function of the divergence of the incident laser beam.

Example 2.8

A laser range finder produces a laser spot of diameter 4 m at a target located at a distance of 8 km from the source. Determine the full angle divergence of the laser beam.

Solution

Full angle divergence in milliradians is given by spot diameter (in m)/target range (in km) = 4/8 = 0.5 mrad.

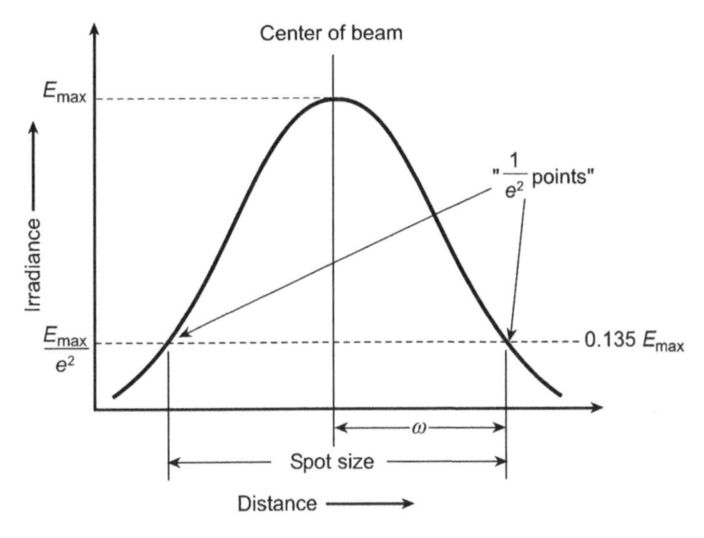

Figure 2.7 Spot size.

Example 2.9

Determine the beam divergence of a helium-neon laser beam emitting at 632.8 nm with an exit aperture of 2 mm. If this beam is made to fall on an appositive lens of 5 cm focal length, determine the minimum expected value of the focused spot size.

Solution

1. Wavelength $\lambda = 632.8$ nm and exit aperture = 2 mm. Therefore, divergence $\theta = 1.27 \times 632.8 \times 10^{-9}/2 \times 10^{-3} = 0.4$ mrad.
2. Focal length of positive lens = 5 cm. Therefore, diameter of focused spot = $5 \times 10^{-2} \times 0.4 \times 10^{-3} = 20\,\mu$m.

2.3.13 Spot Size

Spot size or the beam diameter is defined as the distance across the centre of the beam for which the irradiance equals 0.135 (= $1/e^2$) times the maximum value at the centre (Figure 2.7). This implies that if the laser beam were made to fall on a circular aperture of diameter equal to the laser beam diameter as defined above with the centre of the beam coinciding with the centre of the aperture, not all laser power is transmitted through the aperture. In fact, fractional transmission through the aperture can be computed from:

$$T = 1 - \exp\left[-2(r/w)^2\right] \tag{2.12}$$

where r is the radius of aperture and w is the spot radius.

Example 2.10

A laser beam is centred upon a circular aperture of diameter that is twice the laser beam diameter. Prove that after transmission through the aperture the laser beam will be attenuated by 0.033%.

Solution

1. $T = 1 - \exp\left[-2(r/w)^2\right] = 1 - \exp\left[-2(2)^2\right] = 1 - \exp(-8) = 0.99967$
2. Therefore, fractional attenuation = 0.000 33 = 0.033%.

Example 2.11

The power of a 5 mW helium-neon laser reduces to 2 mW after passing through a circular aperture of 2 mm diameter. Determine the diameter of the incident laser beam.

Solution

1. Fractional transmission through the aperture = 2/5 = 0.4. Also, (Aperture diameter/laser beam diameter)^2 = – [ln(1 – T)]/2 = – [ln(1 – 0.4)]/2 = 0.51/2 = 0.255.
2. Therefore, Aperture diameter/Laser beam diameter = $\sqrt{(0.255)}$ = 0.505.
3. This gives laser beam diameter = 2/0.505 = 3.96 mm.

2.3.14 *M^2 Value*

M^2 value is a measure of beam quality. When the laser beam propagates through space, the divergence in the case of an unfocused pure Gaussian beam is given by $4\lambda/\pi D$, where D is the diameter of the beam waist. In the case of real beams the divergence is higher due to various factors, such as presence of additional modes, and the equation for divergence is usually written as $M^2 \times 4\lambda/\pi D$ with $M^2 > 1$. M^2 is therefore defined as the ratio of the divergence of the real beam to that of a theoretical diffraction-limited beam of the same waist size with a Gaussian beam profile (TEM_{00} mode). It is also referred to as the beam propagation ratio as per ISO-11146 standard. Also, the closer a real beam is to a diffraction-limited beam, the more tightly it can be focused and the greater the depth of field and smaller the size of beam handling optics. The focal spot size in the case of real beam is M^2 times the focal spot size of the pure Gaussian beam. Also, the angular size of the real beam in the far field is M^2 times that of a perfect Gaussian beam (Figure 2.8).

2.3.15 *Wall-plug Efficiency*

Wall-plug efficiency is the overall efficiency of the laser system. It is the ratio of laser power produced (CW power or average power as applicable) to the power drawn from the source of input. As an example, a military laser designator producing 20 ns, 100 mJ pulses at 20 pps and drawing 12.5 A at 24 V DC would have a wall-plug efficiency of $(100 \times 10^{-3} \times 20)/(24 \times 12.5) = 2/300 = 0.67\%$.

Example 2.12

In a solid-state laser system, power conversion efficiency of the power supply is 80%. 70% of this electrical energy is converted into optical output energy of the flash lamp. Only 30% of this optical energy is absorbed by the active medium. Further, 30% of the absorbed energy causes the excited atoms

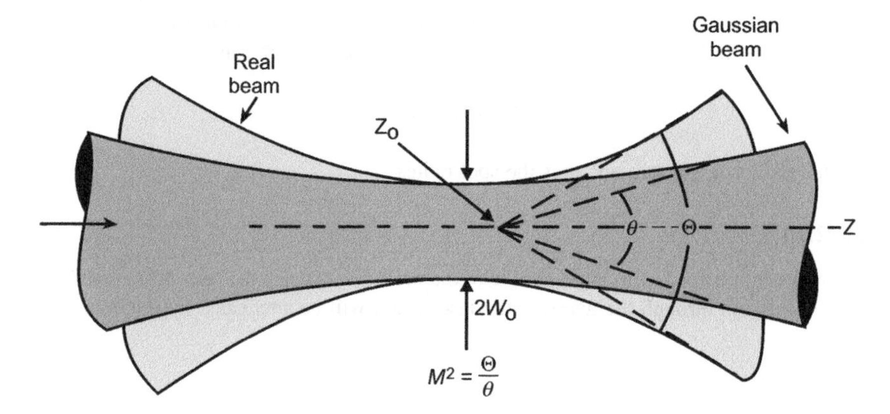

Figure 2.8 Propagation of real and perfect Gaussian beams.

in the metastable state to be available for lasing action. Finally, 40% of this produces laser output. Determine the wall-plug efficiency of this laser.

Solution:

Wall-plug efficiency is given by $0.8 \times 0.7 \times 0.3 \times 0.3 \times 0.4 = 0.02 = 2\%$.

2.4 Measurement of Laser Parameters

Section 2.3 describes the important parameters of laser radiation. In this section we discuss the experimental techniques and diagnostic instrumentation available for the measurement of these parameters. Measurement solutions are discussed in the following sections for the following parameters:

1. power (CW lasers)
2. average power and pulse energy (pulsed lasers)
3. repetition rate (repetitively pulsed lasers)
4. spot size
5. divergence
6. M^2 value
7. line width.

2.4.1 Measurement of Power, Energy and Repetition Rate

A wide range of laser power/energy meters is commercially available from a host of manufacturers covering the wavelength range from short-ultraviolet to far-infrared, CW power levels from nanowatts to kilowatts, pulse energies from nanojoules to hundreds of joules and repetition rate up to several kilohertz. Meters are also available for making these measurements on both collimated and widely diverging beams such as those from diode lasers.

These measurement systems are invariably modular in nature, comprising a sensor head connected to a meter/display unit. Almost all manufacturers offer meters that are compatible with a range of sensor heads, enabling the user to have a wide range of measurement capability from a single meter and multiple sensor heads.

Measurement of CW power, average power, pulse energy and repetition rate are interrelated functions; most of the state-of-the-art laser power/energy meters are capable of measuring these parameters. Measurement of these parameters is therefore covered under a common heading.

2.4.1.1 Choosing the Appropriate Sensor

Choosing the appropriate power/energy measurement solution is usually a three-step process involving selection of the appropriate sensor head, understanding the required meter capabilities and ensuring compatibility of the chosen sensor head with the selected meter.

Selection criteria for the sensor head are based on the nature and range of expected values of the laser parameters to be measured. There are three broad categories of sensor heads commonly used for the purpose: *thermal* or, more precisely, *thermopile* sensors, *photodiode* sensors and *pyroelectric* sensors. Each of the three types has its own characteristic features suiting a particular measurement requirement.

In the case of thermopile sensors, the incident laser radiation is absorbed and converted to heat. The heat ultimately flows to a heat sink maintained at a near-constant temperature by either water or air cooling. The differential temperature between the absorber and that of the heat sink represents the amplitude of absorbed laser energy. A thermocouple junction converts this temperature difference into an equivalent electrical signal.

These sensors are characterized by a broad spectral response from ultraviolet to far-infrared, wide dynamic range from tens of microwatts to several kilowatts and a uniform spatial response. Their response is not affected by beam size, position or uniformity. The disadvantage of these sensors is their sluggish response. Response time ranges from 1 to 50 seconds depending on sensor size, increasing with larger sensor size. These sensors are therefore best suited for measurement of CW power, average power in repetitively pulsed lasers and energy of long laser pulses.

In the case of photodiode sensors, photons in the incident laser radiation generate charge carriers which can be sensed either as a current or a voltage. These are characterized by a limited spectral response (typically from 200 nm to 1800 nm), high sensitivity (typically a few nanowatts), low noise, fast response time (typically a few nanoseconds) and relatively lower spatial uniformity. Lower spatial uniformity particularly affects measurements of non-uniform beams and beams that wander over the detector active area between successive measurements. These sensors particularly suit low-power measurements in CW lasers. These sensors saturate above a power density of about 1 mW cm^{-2}, necessitating the use of optical attenuators for the measurement of higher power levels.

Pyroelectric sensors are also a type of thermal sensor like the thermopile sensor; the difference is that pyroelectric sensors respond to a rate of change of temperature rather than an absolute value of temperature difference. These sensors essentially act like capacitors and therefore integrate pulses to produce a signal with a peak proportional to laser energy. They are ideally suited to measurement of parameters of pulsed lasers. As for thermopile sensors, these are also characterized by a broad spectral response.

2.4.1.2 Choosing the Appropriate Meter

As outlined earlier, state-of-the-art meters offer both power and energy measurement options and are compatible with all three categories of sensor heads. When coupled to different sensors, these meters offer measurement of CW power, average power, pulse energy and repetition rate. One such meter is FieldMaxII-Top by Coherent (Figure 2.9). It is capable of power measurement in the range of 10 μW to 30 kW when used a thermopile sensor head and 1 nW to 300 mW when used with a photodiode sensor head. Energy measurement in the range of 1 nJ to 300 J is possible with a pyroelectric sensor head. Maximum repetition rate in the case of pulse energy measurement is 300 Hz.

2.4.2 Measurement of Spot Size

A simple technique for measuring spot size is by measuring fractional transmission through a known aperture placed in the path of the beam (Figure 2.10). The aperture is placed in the path of the laser beam and adjusted in such a way that the power meter records maximum power. Aperture size should be such that the recorded power is in the range of 50–90% of the power recorded without aperture. Beam radius can then be computed from Equation 2.13:

$$(r/w) = \sqrt{|\ln(1-T)|/2} \tag{2.13}$$

where

r = aperture radius
T = fractional transmission through aperture = power with aperture/power without aperture
w = beam radius

2.4.3 Measurement of Divergence

Beam divergence can be computed by measuring the beam diameter at two points at known distances as shown in Figure 2.11. Full angle beam divergence in this case is given by Equation 2.14:

$$\theta \approx (d_2 - d_1)/(R_2 - R_1) \tag{2.14}$$

Figure 2.9 FieldMaxII-Top meter by Coherent (Reproduced by permission of Coherent Inc.).

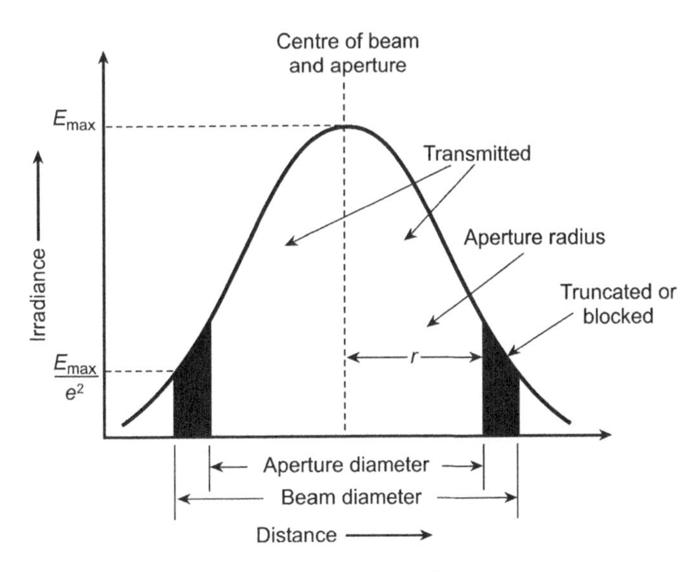

Figure 2.10 Measurement of spot size.

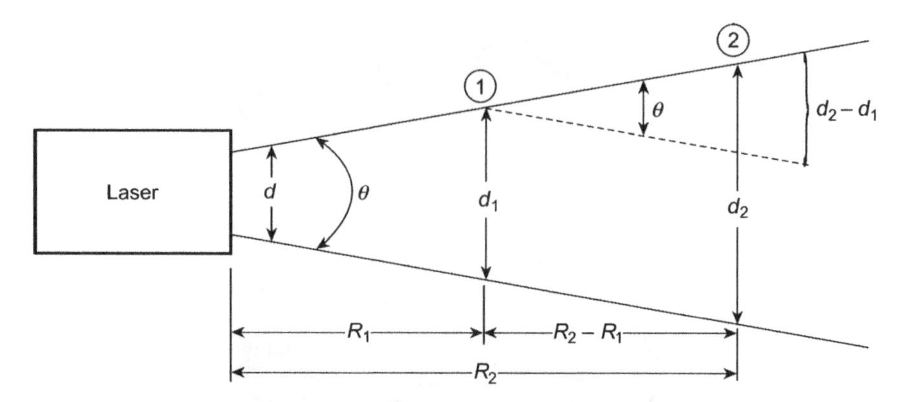

Figure 2.11 Measurement of beam divergence.

The expressions for computing beam divergence given above are valid for the region in the *far field* defined by Equation 2.15:

$$\text{Far field} \geq (100 \times D^2)/\lambda \tag{2.15}$$

where D is the diameter of output optics and λ is the wavelength of the laser beam.

There is another region referred to as *near field* defined by Equation 2.16:

$$\text{Near field} \leq D^2/\lambda \tag{2.16}$$

where D is the diameter of the output optics and λ is the wavelength of laser beam.

It may be mentioned here that near-field divergence may be different from far-field divergence; an explanation of this is beyond the scope of the book.

2.4.4 Measurement of M^2 Value

In the case of real beams, the M^2 value is measured by focusing the beam with the help of a lens of known focal length. A beam profiler arrangement configured around either moving knife edges or special cameras is used to measure the beam waist. The M^2 value cannot be determined from a single beam profile measurement; it is computed from the data generated from a series of measurements. Figure 2.12

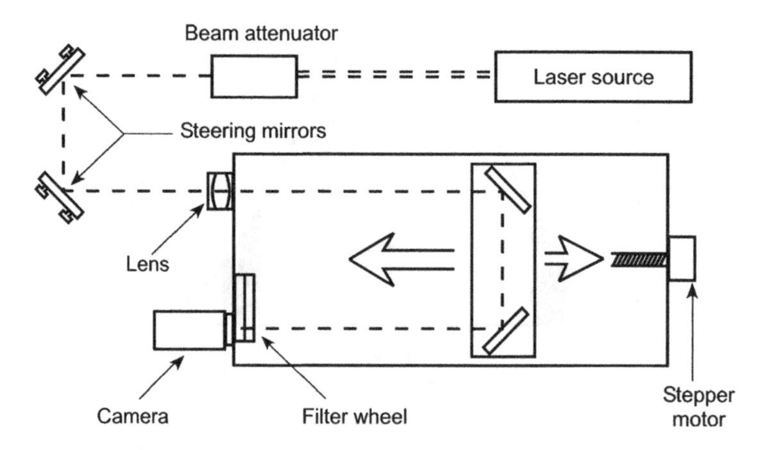

Figure 2.12 M^2 value measurement set-up using fixed position lens and moving detector.

shows one such set-up that uses a camera-based beam-profile measurement. The set-up uses a fixed position lens and a moving detector (camera in this case) to make multiple measurements in the beam waist region. This design is used in the Spiricon M^2-200 Beam Propagation Analyzer.

An alternative set-up uses a fixed position detector and a scanning lens assembly (Figure 2.13). The Mode Master PC M^2 Beam Propagation Analyzer from Coherent is built around this design philosophy. It uses a dual knife-edge beam profiler integrated with a diffraction-limited precision scanning lens. The focusing lens creates an internal beam waist and the two orthogonal knife edges mounted on a rotating drum measure the beam diameter and the corresponding beam axis location at multiple planes along the beam waist. The M^2 value is then computed from the multiple beam width measurements along with their location on the beam axis and the known characteristics of the focusing lens.

2.4.5 Measurement of Line Width

Line width of laser radiation can be measured by a variety of techniques. These include traditional techniques of optical spectrum analysis, for instance use of: diffraction gratings; a frequency discriminator configured around an unbalanced interferometer or a high-finesse reference cavity to transform frequency variations to intensity variations; self-heterodyne technique; extended self-heterodyne technique using a recirculating fibre loop with an internal fibre amplifier; heterodyne technique using a reference laser with significantly better frequency stability than that of the laser under test; and frequency combs for line width measurements over a wide spectral range.

The diffraction-grating-based optical spectrum analysis technique is suitable for measurement of large line widths exceeding 10 GHz, as obtained in presence of multiple modes. For smaller line widths the self-heterodyne technique is often used, which involves recording a beat note between the output of the laser under test and a frequency-shifted and delayed version of it. For sub-kHz line widths, use of a recirculating fibre loop with an internal fibre amplifier may extend the capability of self-heterodyne technique. For even smaller line widths, the heterodyne technique that records a beat note between the output of the laser under test and that from a reference laser having a relatively much higher frequency

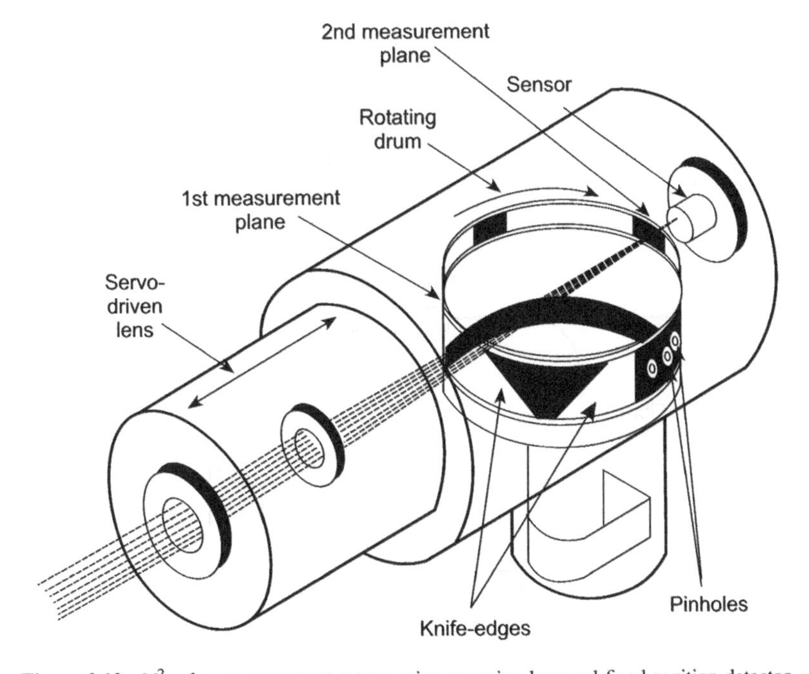

Figure 2.13 M^2 value measurement set-up using scanning lens and fixed position detector.

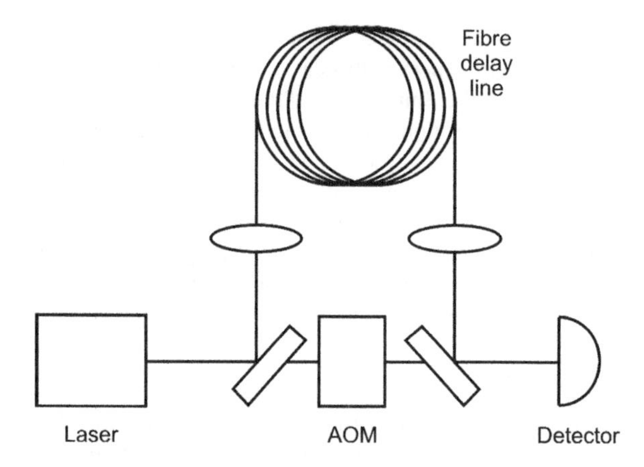

Figure 2.14 Self-heterodyne method of line width measurement.

stability is the preferred solution. In the case where the line width is to be measured over a wide spectral range, frequency comb laser sources may be used as a source of a reference laser.

Figure 2.14 shows the schematic arrangement of a self-heterodyne method for measuring line width. It is basically an interferometric technique which is particularly suitable for the measurement of relatively smaller line widths such as those from single frequency lasers. A beam splitter splits the output of the laser under test into two beams. One part of the laser beam travels through a long optical fibre, which provides some time delay. The other part passes through an acousto-optic modulator, which shifts all frequency components present in the beam under test by some tens of MHz. The time delay introduced by the fibre should be larger than the coherence length of the laser under test. The two beams are then superimposed on a beam splitter and the resulting beat note centred on the frequency of the acousto-optic modulator is recorded to compute the line width. Time delay longer than the coherence length ensures that the superimposed beams are essentially uncorrelated and the beat spectrum is a simple self-convolution of the laser output spectrum from which the laser line width can be easily retrieved. However, it may become impractical to fulfil the condition of time delay being longer than coherence length in the case of lasers with very narrow line widths of the order of sub-kHz. A recirculating fibre loop with an internal fibre amplifier overcomes this limitation.

Figure 2.15 shows the test set-up. A recirculating fibre loop is a fibre-optical arrangement which allows the laser light to do many rounds in the fibre, thus simulating a much longer effective fibre length

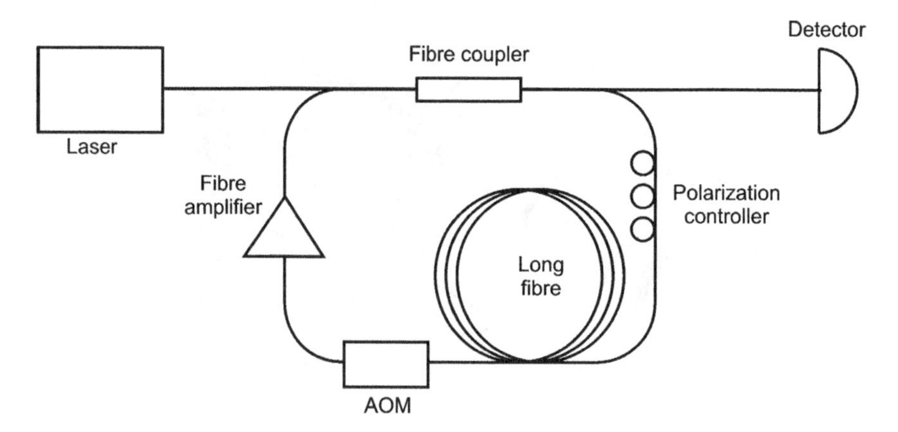

Figure 2.15 Extended self-heterodyne technique by using recirculating fibre loop.

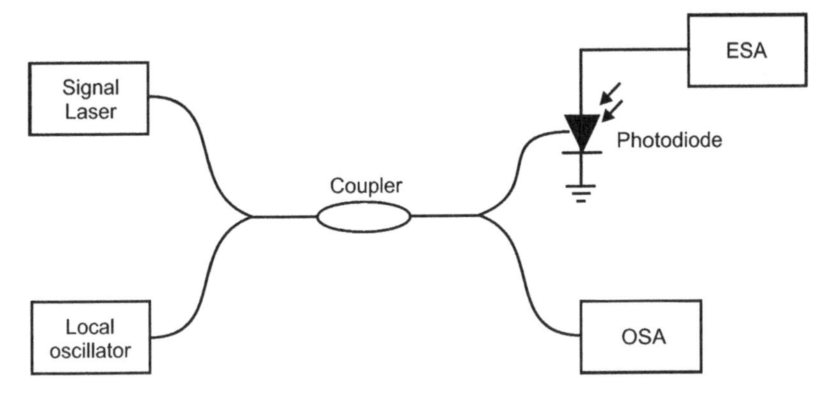

Figure 2.16 Heterodyne set-up with a reference laser.

than would be available in the case of a single traverse. The acousto-optic modulator keeps the light from round trips well separated in frequency. The modulator shifts the optical frequency in each trip by about 100 MHz. The beat spectrum contains beat notes of the original laser light with different frequency-shifted components. Loss of signal in the fibre and also the acousto-optic modulator due to multiple round trips strongly limits the number of round trips. The insertion of fibre amplifier in the loop overcomes this limitation.

Figure 2.16 shows the heterodyne set-up that uses a separate reference laser instead of extracting the second laser beam required for heterodyne action from the laser beam under test.

A frequency comb laser source can be used to extend the spectral range of the measurement of line width. A frequency comb (Figure 2.17) is an optical spectrum with equidistant lines, and can be used to measure unknown frequencies by measuring the beat notes between the unknown frequency and the comb frequencies. Frequency comb laser sources are now commercially available from multiple sources and are being increasingly used for laser metrology.

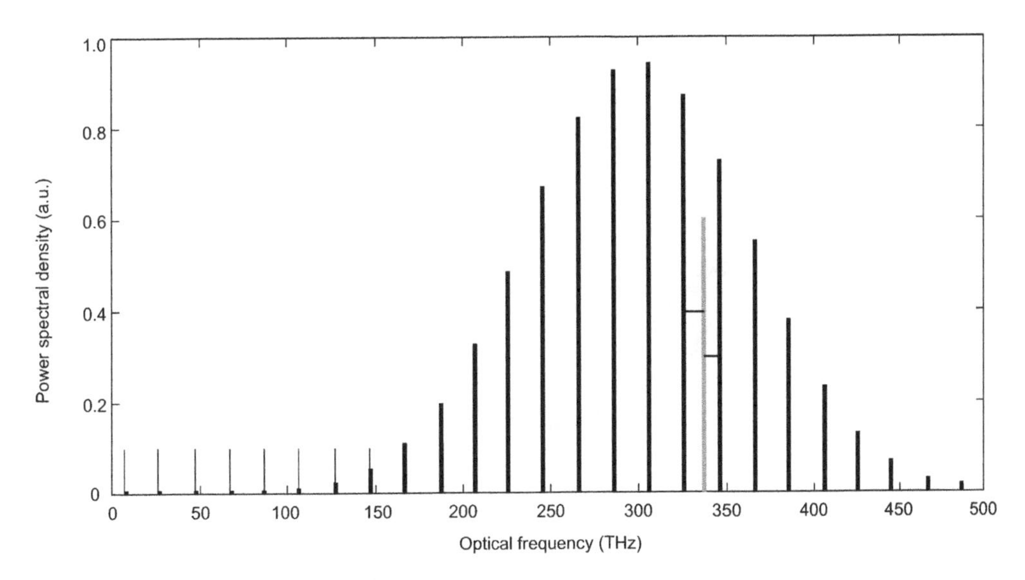

Figure 2.17 Frequency comb output spectrum.

2.5 Laser Beam Diagnostic Equipment

Section 2.4 briefly described the basic techniques available to us for the measurement of important laser parameters. A wide range of laser power and energy meters are commercially available to meet different measurement requirements of laser power, energy and other related parameters. Alternatively, laser beam diagnostic equipment allows precise measurement of parameters related to its shape and profile. These test and measuring instruments include wavelength meters, laser spectrum analyzers, beam profilers and beam propagation analyzers. The functions performed by each of these along with other salient features available in the modern diagnostic equipment are briefly described in the following sections.

2.5.1 Wavelength Meter

A wavelength meter is used for the measurement of wavelength of the laser radiation. One such equipment is the WaveMaster^{TM} Laser Wavelength meter from Coherent (Figure 2.18). It measures the wavelength of both CW and pulsed lasers of any pulse repetition frequency (PRF) in the wavelength range 380–1095 nm. Wavelength reading can be displayed in nanometres, wave numbers or GHz. Measurement of wavelength in both air and vacuum are possible. Wavelength is measured with 0.001 nm resolution and 0.005 nm accuracy. The equipment comes with a built-in RS-232 (recommended standard) and an optional general-purpose input-output (GPIB) interface.

2.5.2 Laser Spectrum Analyzer

A laser spectrum analyzer is used to carry out modal analysis of lasers operating in more than one longitudinal or transverse mode of the laser cavity. The spectrum analyzer typically uses a scanning Fabry-Perot interferometer cavity. The resonant frequency of the interferometer is scanned by varying the spacing between the cavity mirrors with the help of a piezoelectric spacer segment placed between the two mirrors. Laser light transmitted through the cavity to a detector has only those line spectra that match the resonant spectral frequency of the cavity. An oscilloscope synchronized to the cavity scan rate then displays the detected spectra of different laser lines.

One such piece of equipment is SAPlus Laser Spectrum Analyzer from Artisan Scientific (Figure 2.19). The equipment provides modal analysis in the wavelength range of 450–1800 nm. It offers free spectral range choice of 2 GHz or 8 GHz. It is capable of measuring the line width, longitudinal mode structure and frequency stability of narrow band lasers.

Figure 2.20 depicts the spectrum analyzer. As shown in the figure, it utilizes a scanning confocal Fabry-Perot interferometer cavity comprising two concave mirrors, with the spacer separating the two mirrors having a piezoelectric section. When driven electrically from the controller, the piezoelectric

Figure 2.18 WaveMaster wavelength meter by Coherent (Reproduced by permission of Coherent Inc.).

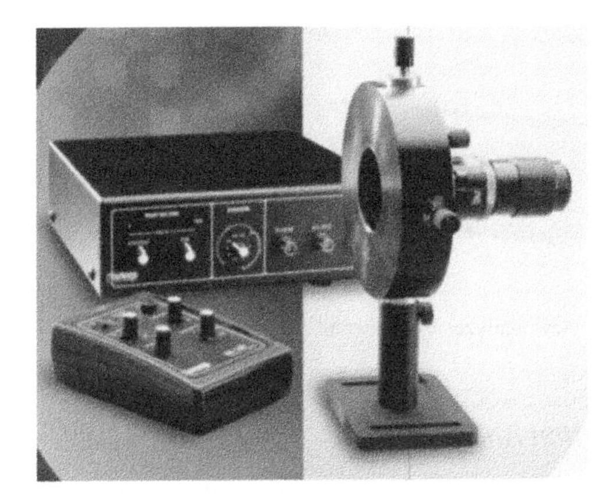

Figure 2.19 Laser spectrum analyzer by Artisan Scientific (Reproduced from Artisan Scientific).

section can be used to vary the mirror spacing and thus the resonance frequency of the cavity. Free spectral range in this case is given by $c/4d$ where d is the mean mirror separation. The purpose of the lens assembly is to match the laser beam to the input aperture of the confocal cavity. The laser beam can be directed straight into the spectrum analyzer or through a beam splitter that allows only a fraction of the laser beam (about 10% in this case) to the spectrum analyzer.

2.5.3 *Laser Beam Profiler*

A laser beam profiler is used for the measurement of spatial intensity profile and other related parameters of the laser beam. As the laser beam propagates, the width and spatial intensity profile of the beam changes. The spatial intensity distribution of the laser beam is a very important parameter when it comes to studying its behaviour in any application.

There are two types of profilers in use. One of the types uses special cameras as detectors and offers an excellent solution to fast and detailed analyses of CW and pulsed lasers. The other type uses moving knife edges, which is particularly attractive for carrying out measurements on small and focused beams and if a large dynamic range is required. The Beam View Analyzer (Figure 2.21) and Beam Master Profiler (Figure 2.22) from Coherent are examples of camera-based and moving knife-edge-based laser beam profilers.

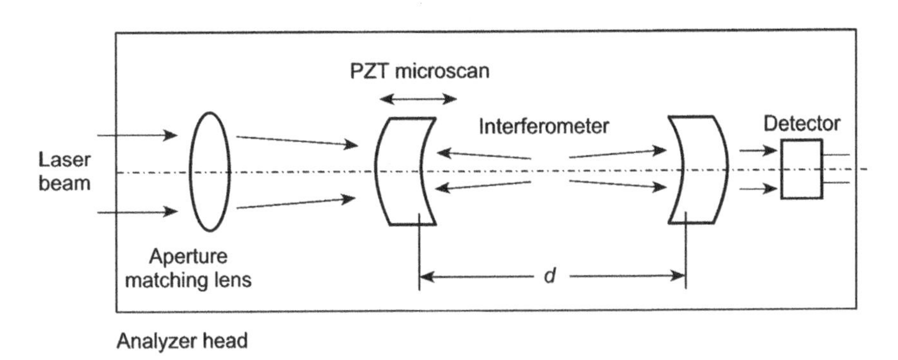

Figure 2.20 Schematic arrangement of spectrum analyzer.

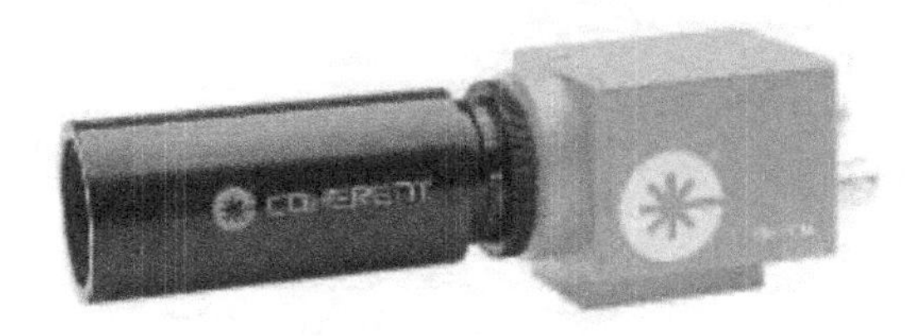

Figure 2.21 Beam view analyzer by Coherent (Reproduced by permission of Coherent Inc.).

2.5.4 Beam Propagation Analyzer

A laser beam propagation analyzer is used for measurement of beam quality (M^2 value) and other beam propagation analysis functions for CW lasers. State-of-the-art beam propagation analyzers offer measurement functions which include beam quality (M^2 value), beam diameter, second moment diameter, divergence, divergence asymmetry, astigmatism, astigmatism asymmetry, pointing stability, waist diameter and location, waist asymmetry, beam profiles, power density and Rayleigh range. ModeMaster PC (Figure 2.23) is one such beam propagation analyzer from Coherent. It comprises a dual knife-edge beam profiler integrated with servo-controlled diffraction-limited scanning lens. The lens focuses the beam to create an internal beam waist and the two orthogonal knife edges mounted on a rotating drum measure beam diameter and corresponding beam axis location at 256 planes along the beam waist. The software provided with the system determines different parameters listed above from the data generated during the course of a large number of measurements.

Figure 2.22 Beam master profiler by Coherent (Reproduced by permission of Coherent Inc.).

Figure 2.23 ModeMaster PC beam propagation analyzer by Coherent (Reproduced by permission of Coherent Inc.).

2.6 Summary

- Three important characteristics of lasers include monochromaticity, directionality and coherence. Monochromaticity refers to the single frequency or wavelength property of the radiation. Laser radiation is monochromatic and this property has its origin in the stimulated emission process by which lasers emit light. Light is said to be coherent when different photons (or the waves associated with those photons) have the same phase and this phase relationship is preserved as a function of time. There are two types of coherence: temporal coherence and spatial coherence. Temporal coherence is preservation of phase relationship with time and spatial coherence is preservation of phase across the width of the beam. The directionality of laser radiation has its origin in the coherence of the stimulated emission process.
- The important laser parameters include wavelength, output power (in case of CW lasers), output energy/pulse (in the case of pulsed lasers), irradiance, radiance, pulse width, duty cycle, repetition rate (in the case of pulsed lasers), beam divergence, spot size, M^2 value and wall-plug efficiency.
- CW power is the continuous power available from the laser. Peak power and average power are the other related parameters. Peak power is the highest instantaneous power in the laser pulse. Average power is the product of peak power and the duty cycle, where duty cycle is the ratio of the pulse width to the time interval between two successive laser pulses.
- Pulse energy is defined with respect to pulsed lasers and is equal to the area under the power versus time curve representing the laser pulse. If the laser pulse is considered as rectangular with amplitude equal to the peak power, the pulse energy is the product of peak power and the pulse width.

- Repetition rate is the number of laser pulses produced per second. The product of pulse energy and repetition rate gives average power.
- Pulse width or pulse duration is usually measured as FWHM. It is intimately related to bandwidth. The narrower the pulse width, the higher is the required bandwidth.
- Beam divergence is an indicator of the spread in the laser beam spot as it travels away from the source. It is a function of the wavelength λ and size of output optics.
- Irradiance, also referred to as power density, is defined as the power per unit area of the laser radiation falling on the target.
- Radiance, also referred to as brightness, is usually defined with respect to the laser source. It is the power emitted per unit area per unit solid angle.
- Spot size or the beam diameter is defined as the distance across the centre of the beam for which the irradiance equals 0.135 ($= 1/e^2$) times the maximum value at the centre.
- M^2 value is a measure of beam quality. It is defined as the ratio of the divergence of the real beam to that of a theoretical diffraction-limited beam of the same waist size with a Gaussian beam profile (TEM_{00} mode).
- Wall-plug efficiency is the overall efficiency of the laser system. It is the ratio of laser power produced (CW power or average power as applicable) to the power drawn from source of input.
- For measurement of power, energy and other related parameters, a wide range of laser power/energy meters is commercially available from a host of reputable manufacturers covering wavelength range from short-ultraviolet to far-infrared, CW power levels from nanowatts to kilowatts, pulse energies from nanojoules to hundreds of joules and a repetition rate to several kilohertz. Meters are also available for making these measurements on both collimated as well as widely diverging beams such as those from diode lasers.
- Although sophisticated beam diagnostic instruments are available for the measurement of laser beam profile and related parameters, a simple technique for measuring spot size is to measure fractional transmission through a known aperture placed in the path of the beam.
- A simple experimental set-up to measure beam divergence is to measure the beam diameter at two points at known distances.
- Line width of a laser radiation can be measured by a variety of techniques. These include the traditional techniques of optical spectrum analysis, for instance using diffraction gratings, a frequency discriminator configured around an unbalanced interferometer or a high-finesse reference cavity to transform frequency variations to intensity variations, self-heterodyne technique, extended self-heterodyne technique using a re-circulating fibre loop with an internal fibre amplifier, heterodyne technique using a reference laser with significantly better frequency stability than that of the laser under test and frequency combs for line width measurements over a wide spectral range.
- Laser beam diagnostic equipment allow precise measurement of parameters related to the shape and profile of the beam. These instruments include wavelength meters, laser spectrum analyzers, beam profilers and beam propagation analyzers.
- A wavelength meter is used for measurement of wavelength. A laser spectrum analyzer is used to carry out modal analysis of lasers operating in more than one longitudinal or transverse mode of the laser cavity. A laser beam profiler is used for measurement of spatial intensity profile and other related parameters of the laser beam. Finally, a laser beam propagation analyzer is used for measurement of beam quality (M^2 value) and other beam propagation analysis functions for CW lasers.

Review Questions

2.1. What are the three most important inherent characteristics of laser radiation? Briefly describe each of them.

2.2. Differentiate between temporal and spatial coherence. What is the effect of these characteristics on the monochromatic nature of the radiation and distant spot diameter?

2.3. Define the following:
a. beam divergence
b. peak power
c. wall-plug efficiency
d. spot diameter
e. M^2 value

2.4. With respect to laser radiation, differentiate between:
a. peak power and average power
b. pulse width and line width
c. irradiance and radiance

2.5. Write down mathematical expressions relating
a. peak power, duty cycle and average power
b. average power, pulse energy and repetition rate
c. fractional transmission, beam diameter and aperture diameter
d. M^2 value, divergence of pure Gaussian beam and divergence of real beam
e. minimum achievable pulse width and line width

2.6. List any three types of laser beam diagnostic equipment and briefly describe functions performed by each of them.

Problems

2.1. A Q-switched solid-state laser produces a pulsed output with 80 mJ/pulse at a repetition rate of 1200 pulses/minute. Determine peak and average power of this laser if the pulse duration is 40 ns.
[1.6 W, 0.21 MW/m^2]

2.2. A laser target designator produces 100 mJ/pulse, 20 ns output having a full angle divergence of 0.5 mrad. If the atmospheric attenuation factor is 0.2 km^{-1}, determine the laser spot size and the peak power density at a target distance of 6 km.
[3 m, 1.5 MW]

2.3. A portable laser target designator produces 120 mJ/pulse at a repetition rate of 20 pps. The system operates from a 24 V battery and draws 14 A of current during the operation of the system. Determine the wall-plug efficiency of the system.
[0.714%]

2.4. A helium-neon laser with an effective output aperture diameter of 1.5 mm emits 5 mW power at 633 nm. If this laser radiation enters a human eye with a lens focal length of 1.7 cm, determine the irradiance of the focused spot on the retina.
[7.674 kW cm^{-2}]

2.5. The power of a certain helium-neon laser reduces to half of its value after passing through an aperture of 5 mm diameter. Determine the spot size of the laser beam.
[4.25 mm]

2.6. A laser beam has a diameter of 2 cm at a distance of 10 m from the laser and a diameter of 4 cm at a distance of 30 m from the laser. Determine the full angle beam divergence.
[1 mrad]

2.7. Determine radiance of a 1 mW helium-neon laser having a full angle divergence of 1 mrad and exit spot diameter of 1 mm.
[1.6×10^9 W m^{-2} sr^{-1}]

2.8. The line width of a Nd:YAG laser emitting at 1064 nm is 0.5 nm. Determine its coherence length.
[1.132 mm]

Self-evaluation Exercise

Multiple-choice Questions

2.1. A semiconductor diode laser emitting at 900 nm has a bandwidth of 2 nm. Its coherence length is
a. 0.2 cm
b. 0.2 m
c. 0.2 mm
d. 902 mm

2.2. Coherence length of a frequency-stabilized carbon dioxide laser emitting at 10 600 nm and having a bandwidth of 100 kHz is
a. 10.6 μm
b. 1500 m
c. 750 m
d. 1500 mm

2.3. The spot size produced by a helium-neon laser having a full angle divergence of 100 mrad at a distance of 100 km is
a. 1 m
b. 100 m
c. 10 m
d. 1000 m

2.4. The focal length of the lens required to produce a diffraction-limited focused spot of 50 μm for a laser beam of full angle divergence of 1 mrad is
a. 50 mm
b. 50 cm
c. 5 mm
d. None of these

2.5. The coherence length of an Nd:YAG laser having a fluorescent line width of 30 GHz lasing over this entire frequency range is
a. 5 cm
b. 10 cm
c. 10 mm
d. 5 mm

2.6. A Q-switched Nd:Glass laser produces 100 mJ/pulse, 50 ns pulses. If the average power of the laser radiation is 2 W, its pulse repetition frequency is
a. 10 pps
b. 20 pps
c. 50 pps
d. indeterminate from given data

2.7. The M^2 value in the case of a real laser beam is always
a. < 1
b. $= 1$
c. > 1
d. $= 0$

2.8. Irradiance produced by a laser beam on a certain target plane is
a. the same as its power density on the target plane
b. the same as its power density per unit solid angle
c. the total power received at the target site
d. directly proportional to beam divergence

2.9. Which of the following interferometer cavities is commonly used in laser beam spectral analyzer equipment?
a. Michelson interferometer
b. Fabry-Perot interferometer
c. Mach-Zehnder interferometer
d. None of these

2.10. While passing through an aperture of diameter equivalent to the laser beam spot diameter, a laser beam
a. will be attenuated by more than 10%
b. will go through almost unattenuated
c. will become collimated
d. will lose its coherence properties

Answers

1. (c) 2. (b) 3. (c) 4. (a) 5. (d) 6. (b) 7. (c) 8. (a) 9. (b) 10. (a)

Bibliography

1. *Laser Fundamentals*, 2008 by William Thomas Silfvast, Cambridge University Press.
2. *Fundamentals of Light Sources and Lasers*, 2004 by Mark Csele, Wiley-Interscience.
3. *Principles of Lasers*, 2009 by Orazio Svelto, Plenum Press.
4. *Understanding Lasers: An Entry Level Guide*, 2008 by Jeff Hecht, IEEE Press.
5. *Introduction to Laser Physics*, 1986 by Koichi Shimoda, Springer-Verlag.
6. *Introduction to Lasers and their Applications*, 1977 by Donald C. O'Shea, W. Russell Callen and William T. Rhodes, Addison-Wesley Publishing Co.
7. *Laser Handbook*, 1972 by F.T. Arechhi and E. O. Schulz-Dubois, Amsterdam North-Holland Publishing Co.
8. *Lasers and Light: Readings from Scientific American*, 1969 by A.L. Schawlaw, W. H. Freeman.
9. *Lasers*, 1986 by A.E. Siegman, University Science Books.
10. *Spatiotemporal Characteristics of Laser Emission*, 1994 by M.V. Pyatakhin and A.F. Suchkov, Nova Science Publishers.
11. *Handbook of Laser Technology and Applications*, 2003 Volume I by Collin E. Webb and Julian D. C. Jones, Institute of Physics Publishing.

Part II
Types of Lasers

3

Solid-state Lasers

3.1 Introduction: Types of Lasers

Lasers are classified on the basis of various parameters, such as the nature of the active medium:

1. solid-state lasers
2. semiconductor lasers
3. gas lasers.

In addition, there are a large number of other varieties of lasers that do not fit any of the above-mentioned broad categories (dye lasers, excimer lasers, metal vapour lasers, free electron lasers, x-ray lasers, chemical lasers and gas dynamic lasers).

Further, lasers can be classified according to the pumping mechanism (optically pumped lasers, gas dynamic lasers, electrically pumped lasers) or the nature of the laser output in terms of power/energy level or wavelength (visible lasers, infrared or IR lasers, continuous wave or CW lasers, pulsed lasers, Q-switched lasers). Solid-state lasers are almost invariably pumped optically, gas lasers are excited by an electrical discharge while semiconductor lasers are pumped by an electrical current flowing through a forward-biased diode junction.

The most important method of classifying lasers is on the basis of the nature of the lasing medium. Solid-state lasers are discussed in this chapter, including the operational fundamentals of solid-state lasers, their salient features and typical applications, and semiconductor and gas lasers are discussed in Chapters 4 and 5, respectively.

3.2 Importance of Host Material

The active medium in the case of solid-state lasers is the lasing species embedded into a crystalline or a glass host material. The characteristics of the host material are no less important than those of the lasing species. The host material should have the required optical, mechanical and thermal properties in order to favour homogenous propagation of light through the crystal: good beam quality (optical), high average power operation and the ability to withstand severe operating conditions of practical laser systems (mechanical and thermal properties). For instance, the host material should be reasonably transparent to the pump radiation. It should not absorb radiation at either the pumping wavelength or laser wavelength. It should have high thermal conductivity in order to allow CW operation or operation at high repetition rates in the case of pulsed operation, leading to high average power operation.

Relatively poorer thermal conductivity in case of a glass host disallows the operation of an Nd:Glass laser in CW mode, even at high repetition rates. On the other hand, an Nd:YAG laser, where the host material is yttrium aluminium garnet (YAG, $Y_3Al_5O_{12}$), can be made to operate in CW due to the far

Lasers and Optoelectronics: Fundamentals, Devices and Applications, First Edition. Anil K. Maini.

superior thermal conductivity of the host. Another point worth mentioning in connection with hosts of solid-state lasers is the interaction of light-emitting species with the host material. This influences the energy level structure of the active species, thus modifying the energy levels involved in the laser action. As an illustration, the neodymium in the YAG host of a Nd:YAG laser emits at 1064 nm while the same neodymium when doped in phosphate-based glass (Nd:Glass) acting as the host emits at 1054 nm. Going a step further, neodymium in silicate glass emits at 1062 nm, and in fused silica glass at 1080 nm.

Yttrium lithium fluoride (YLF) is another important host material popular with neodymium-based solid-state lasers. It has fewer heat-related problems as compared to YAG. Also, Nd:YLF can store more energy than its Nd:YAG counterpart and is therefore capable of generating higher-energy Q-switched laser pulses. Due to its birefringent nature, it can produce laser output at 1047 nm and 1053 nm, each having its own polarization orientation. Birefringence is a property of polarization-dependent refractive index.

Synthetic garnets other than YAG include popular host materials such as gadolinium gallium garnet (GGG, $Gd_3Ga_3O_{12}$) and gadolinium scandium gallium garnet (GSGG, $Gd_3Sc_2Ga_3O_{12}$). These host materials have good thermal properties, are hard and optically isotropic. Good thermal properties permit operation at high average power levels. Yttrium doped vanadate (YVO_4) is yet another important host material. Neodymium-doped YVO_4 has a very large stimulated emission cross-section which makes it a high-gain laser material, and a strong broadband absorption around 808 nm which makes it a highly suitable material for laser diode pumping.

This interaction of active species with the host material sometimes assumes interesting proportions in case of a special class of solid-state lasers called Vibronic (Vibrational-Electronic) lasers where the electronic energy levels of the light emitting species interact with vibrational levels of the host and get broadened. The lasing levels instead of being single levels get transformed to upper and lower lasing bands. This feature makes this class of lasers tunable over a range of wavelengths. Important solid-state lasers in this category include Titanium doped *Sapphire* (Sapphire is the host) and Alexandrite, which is Chromium doped *Chrysoberyl* (Chrysoberyl is the host).

3.2.1 Lasing Species

Neodymium and chromium are the most widely exploited lasing species. Chromium is used in ruby (chromium-doped aluminium oxide), alexandrite (chromium-doped chrysoberyl, $BeAl_2O_4$) and chromium-doped GSGG. Titanium is used in titanium sapphire (titanium-doped Al_2O_3) lasers. Neodymium is used in Nd:YAG and Nd:Glass lasers. Erbium is another lasing species, which in YAG and glass hosts give the active medium for a rapidly emerging class of solid-state lasers referred to as 'eye-safe' lasers. These lasers produce output at 1540 nm, which presents a much lower risk to the eye compared to neodymium lasers. Neodymium lasers produce output at around 1064 nm, which is very dangerous to the eye.

Figure 3.1 shows the arrangement of different components of a typical solid-state laser comprising active medium (usually in the shape of rod or slab), pumping source (flash lamp, arc lamp, laser diode) and a resonant structure. A specially contoured enclosure called a cavity houses the laser medium and the pumping source, and a set of mirrors placed at either end of the medium transform this cavity into a resonant cavity. In addition, the electronics drive the pumping source. Electronics accompanying a solid-state laser are separately discussed at length in Chapter 8.

3.3 Operational Modes

Operational mode or methodology refers to the different resonator designs leading to different laser output formats. The range of available output formats includes the following:

1. continuous wave (CW)
2. free-running
3. Q-switched
4. cavity-dumped
5. mode-locked.

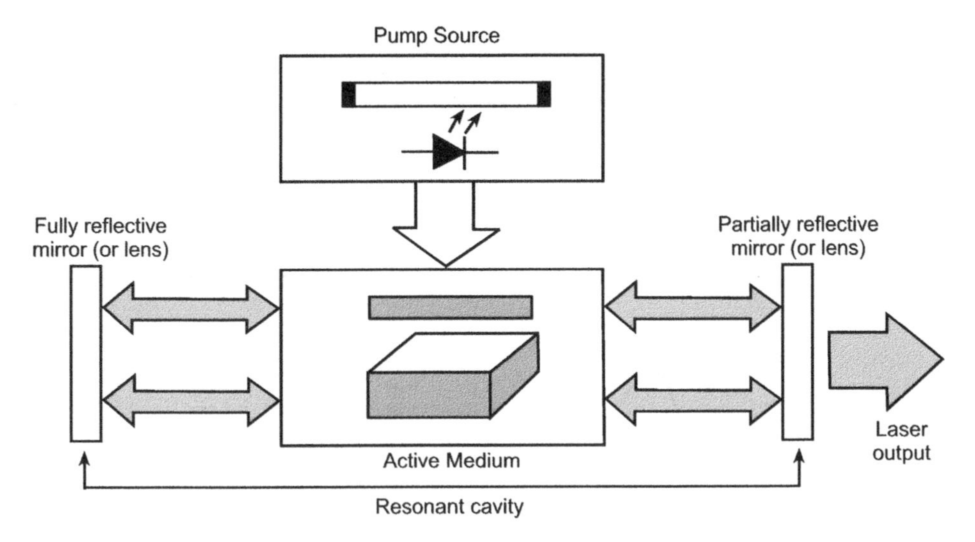

Figure 3.1 Components of a typical solid-state laser.

3.3.1 CW Output

In this operational mode, which is the simplest mode of operating a laser, the laser produces a continuous output as long as the laser is pumped. Pumping needs to be continuous in this case. The laser resonator simply comprises the active medium and the resonator mirrors, and does not contain any switches. In a CW laser, the steady-state gain equals the loss and the excess pump power is converted into laser output. Output of such a laser will fluctuate due to the many ways in which different longitudinal and transverse modes couple to each other via the active medium, unless means are used to control longitudinal and transverse modes. A laser producing a single longitudinal mode and a single transverse mode may have output fluctuation of less than 1%.

3.3.2 Free-running Output

Free-running output a quasi-CW mode of operation in which the laser operates in the CW mode for a time period of the order of a few hundreds of microseconds to a few milliseconds. This time period is generally equal to or is slightly longer than the storage time of the active medium, which in turn is of the same order as the pump input pulse. Refer to Figure 3.2 which shows the timing sequence of production of a free-running pulsed output, that is, application of pump input and gain and loss response of the laser resonator.

As is clear from the timing diagram, in the case of free-running output the steady-state gain equals the loss and the excess pump power appears as the laser output. As the pump input exceeds a certain threshold (where gain equals loss), laser radiation begins to appear at the output. Both gain and laser output respond with overshoots, which ultimately damp out after a few cycles, and the output settles down to the CW-like output as shown. Overshoots depend upon the rise time of the pump input. Faster rise time leads to a larger overshoot.

3.3.3 Q-switched Output

Q-switching is a mechanism of producing short laser pulses with a pulse width of the order of a few nanoseconds. The term Q-switching refers to rapidly switching the quality factor of the resonator cavity – the ratio of energy stored per cycle to the energy lost per cycle – from a low to a high value. Refer to the

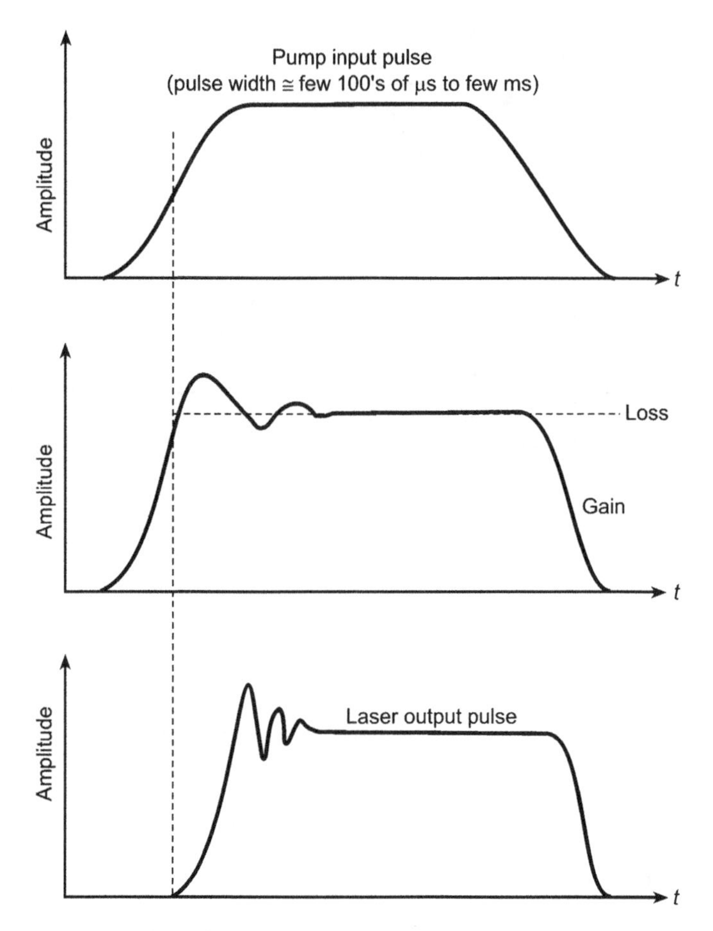

Figure 3.2 Timing diagram of free-running output.

timing diagram of Figure 3.3. Initially, the Q-factor of the resonator cavity is kept at a very low value by introducing an optical switch such as a Pockels cell and a polarizer. When the laser is pumped, it is prohibited from lasing as it does in the case of a CW output. The obvious consequence of this is that the active medium builds up a much higher population inversion density or gain than would have been possible in the case of a CW laser. When the inversion density almost reaches its peak value the Q-factor of the resonator cavity is rapidly switched to the high value, leading to a steep fall in the loss value. This manifests itself in the production of a short laser pulse at the output as shown. The output laser pulse begins to appear at the time instant where the loss becomes less than the gain and reaches the peak value when the diminishing gain equals the loss. The gain then becomes less than the loss, which manifests itself in the form of a falling laser output. These time instants can be observed more clearly if we expand the timescale around the point where the Q-factor is switched from a low value to a high value (Figure 3.4).

The pulse duration of a Q-switched pulse output depends upon resonator round-trip time, reflectivity of the output coupler and the extent to which the active medium is pumped above the threshold. More specifically, the pulse width depends primarily on the resonator cavity's round-trip time and the resonator gain to loss ratio. In an optimized Q-switched laser oscillator, the output laser pulse width is of the order of a few times the cavity round-trip time, and may be typically in the range 10–50 ns. To a good

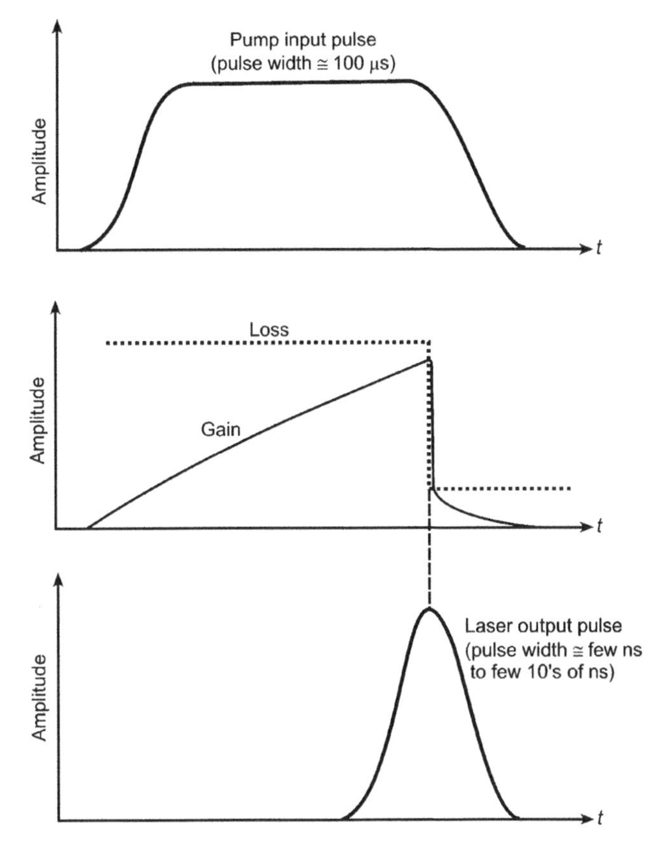

Figure 3.3 Timing diagram of Q-switched output.

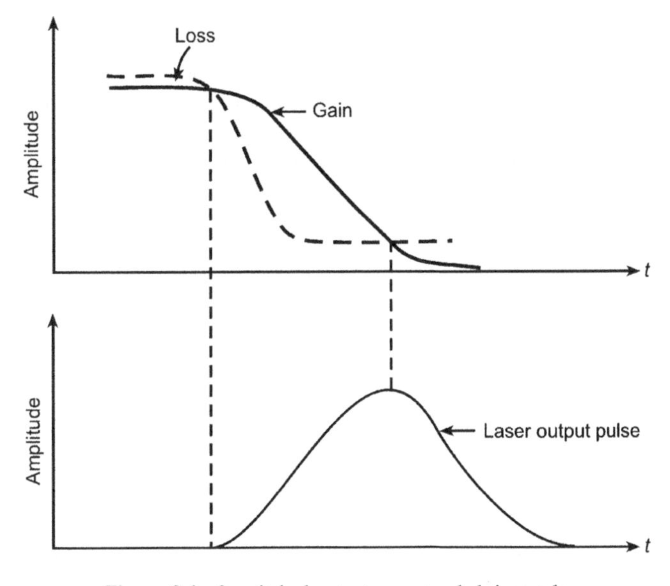

Figure 3.4 Q-switched output on expanded timescale.

approximation, Q-switched pulse width can be computed from Equation 3.1:

$$t_{pq} = \frac{t}{(1 - R)} \tag{3.1}$$

where

t_{pq} = Q-switched pulse width
t = laser cavity round-trip time
R = reflectivity of output mirror

Laser cavity round-trip time can in turn be computed from Equation 3.2:

$$t = \frac{2Ln}{c} \tag{3.2}$$

where

L = length of laser cavity
n = refractive index of laser medium

3.3.4 Cavity-dumped Output

The phenomenon of cavity dumping is a slight variation of the Q-switching process with the difference that, in case of the former, both of the resonator cavity mirrors are 100% reflective. When the active medium is pumped, energy is initially stored in the population inversion as in the case of Q-switching. During this time, the loss is at a very low value. Since both of the resonator mirrors are 100% reflective, the amplified light remains trapped within the cavity. As peak irradiance is reached, the loss is again switched to a high value thus ejecting the intra-cavity circulating energy in the form of a pulse. One of the advantages of the cavity-dumping technique is that the output pulse width depends upon the cavity length only and is independent of the gain characteristics of the active medium. More specifically, the pulse width equals the round-trip cavity transit time provided that the Q-switch employed for switching the quality factor of the cavity is also switched within the same time period. This technique allows generation of output pulses of a few nanoseconds duration; the lower limit of the pulse width achievable with this technique is 2–3 times smaller than that achievable with Q-switching. Figure 3.5 shows the timing diagram of generation of a cavity-dumped output pulse.

In the case of cavity-dumped output, pulse width can be computed from Equation 3.3:

$$t_{pc} = \frac{2Ln}{c} \tag{3.3}$$

where

t_{pc} = Q-switched pulse width
L = Length of laser cavity
n = Refractive index of laser medium.

3.3.5 Mode-locked Output

Minimum pulse width achievable from the laser is intimately related to its frequency bandwidth. Heisenberg's uncertainty principle (Section 2.3.7) sets a limit on the minimum possible laser pulse width

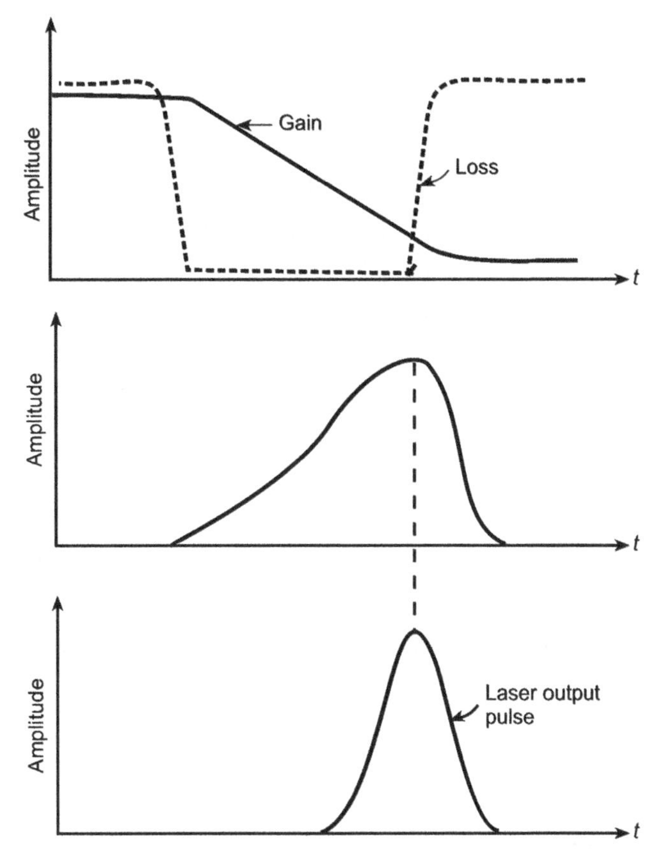

Figure 3.5 Timing diagram of cavity-dumped output.

for a given value of available bandwidth. That is,

$$t_{\mathrm{p}} = \frac{0.441}{B} \tag{3.4}$$

where

t_{p} = Minimum pulse width
B = Bandwidth

It is not practicable to reach this limit with every pulse-forming technique, however. For instance, in the case of Q-switching, minimum achievable pulse width is of the order of 10 ns or so because of the required pulse build-up time. Cavity dumping overcomes this shortcoming to some extent and pulse widths of the order of 1–2 ns are achievable; this is still nowhere close to what is theoretically achievable for a given solid-state laser. For instance, the frequency bandwidth of an Nd:YAG is 150 GHz for a homogeneously broadened line. For a Gaussian pulse, the minimum achievable pulse width would be (0.441/150) ns $\approx$ 3 ps. Mode locking helps achieve pulse widths approaching the theoretical limit.

In the absence of special measures, a laser usually oscillates at several transverse modes and typically hundreds of longitudinal modes. The number of oscillating longitudinal modes depends upon the inter-mode spacing and gain bandwidth of the transition line. Each mode oscillates independent of the others

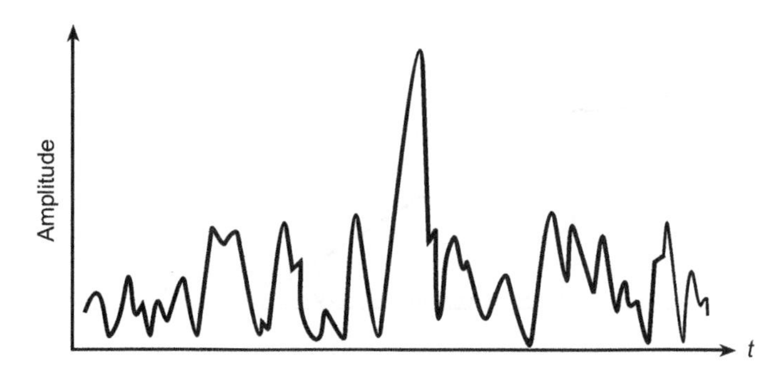

Figure 3.6 CW laser output in the absence of mode locking.

and their phases are randomly distributed within the range of $-\pi$ to $+\pi$ radians. As a result of oscillations at a large number of longitudinal modes and various modes having random amplitude and phase relationship, the laser output is a time-averaged statistical mean value and is observed to fluctuate in amplitude with respect to time as shown in Figure 3.6. The process of mode locking forces different longitudinal modes to oscillate with a fixed phase relationship with respect to each other, which produces an ultra-short pulse with well-defined amplitude as a function of time. Figure 3.7 shows a mode-locked pulse generated from phase locking of different longitudinal modes. In the case of an ideal mode-locked laser pulse, the intensities of different longitudinal modes follow a Gaussian distribution and the spectral phases are identically zero as shown in Figure 3.7a and b. Figure 3.7c shows the ideal mode-locked pulse in the time domain, which is a single Gaussian pulse.

Phase locking is achieved by introducing into the resonator cavity a suitable non-linear element such as a saturable absorber or an externally driven optical modulator. While the former is referred to as the

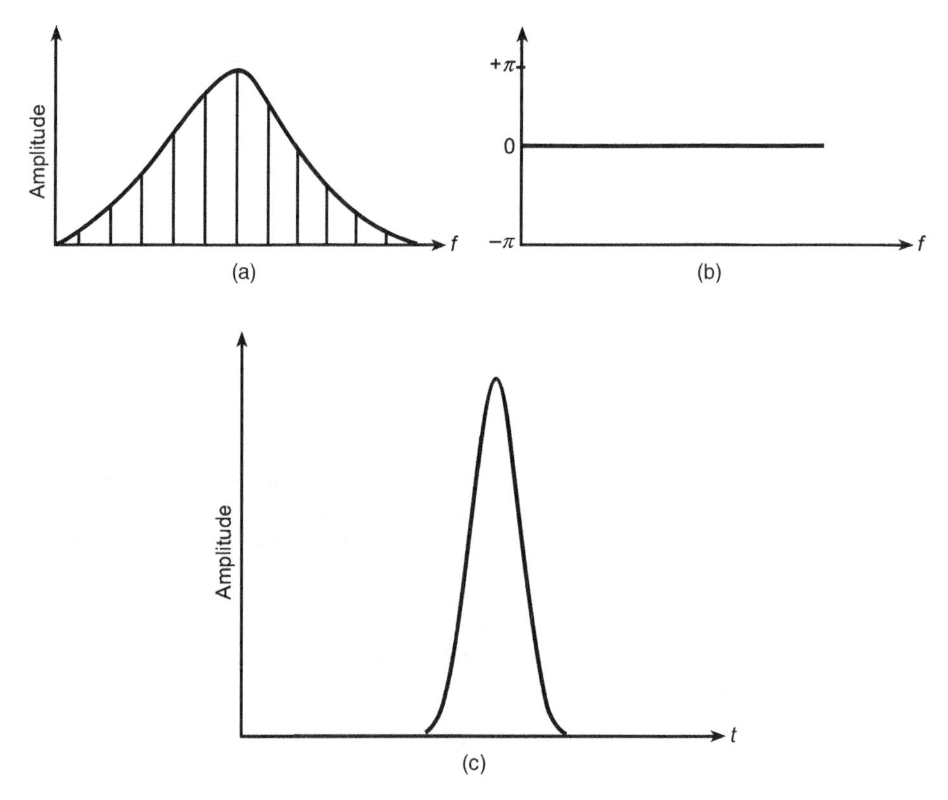

Figure 3.7 Mode-locked pulse.

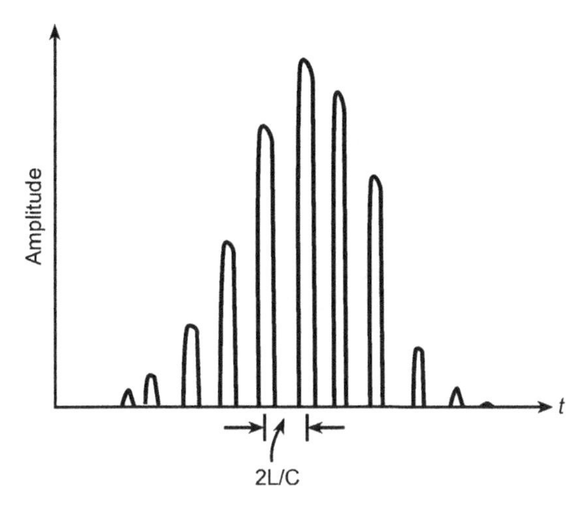

Figure 3.8 Mode-locked pulses with Q-switch envelope.

process of passive mode locking, the latter is called active mode locking. The phenomenon of mode locking is a process that locks together a cluster of photons in phase. The mode-locking element transmits this cluster every time it passes through it while bouncing back and forth between the cavity mirrors. The repetition rate of the mode-locked pulses is equal to the round-trip transit time of the resonator cavity. That is,

$$\text{PRF} = \frac{1}{\Delta T} = \frac{c}{2L} \tag{3.5}$$

It is also possible to combine the processes of Q-switching and mode locking or the processes of cavity dumping and mode locking. Q-switching and mode locking can be achieved simultaneously by introducing Q-switching elements into the cavity in addition to the mode-locking element. The output in this case has the pulse envelope of a Q-switched pulse, comprising individual short pulses obtained from mode-locking process as shown in Figure 3.8.

Example 3.1

A certain laser has a bandwidth of 22 GHz. Determine the theoretically possible shortest mode-locked pulse width it can generate.

Solution

1. Bandwidth = 22 GHz = 22×10^9 Hz.
2. Shortest possible pulse width according to the transform limit = 0.441/bandwidth = $0.441/(22 \times 10^9) = (441 \times 10^{-12})/22$ s = 20 ps.

Example 3.2

What would be the repetition rate of the mode-locked pulses in Example 3.1 if the laser cavity is 0.6 m long?

Solution

1. Repetition rate of mode-locked pulses = $c/2L$.
2. $L = 0.6$ m and $c = 3 \times 10^8$ m s^{-1}.
3. Therefore, repetition rate = $(3 \times 10^8/1.2)$ Hz = 250 MHz.

Example 3.3

The round-trip time of the laser cavity in a certain solid-state laser is 2 ns. What should be the approximate value of Q-switched pulse width at the output if the output coupling mirror had 10% reflectivity? How would the pulse width change if the reflectivity of the coupling mirror is changed to 20%?

Solution

1. Q-switched pulse width = round-trip time/$(1 - R) = 2/(1 - 0.1)$ ns = 2/0.9 ns = 2.22 ns.
2. For $R = 20\%$, pulse width = $2/(1 - 0.2)$ ns = 2/0.8 ns = 2.5 ns.

3.4 Ruby Lasers

The active medium of a ruby laser is sapphire, that is, aluminium oxide (Al_2O_3) doped with 0.01–0.5% chromium. The doping process replaces the small percentage of Al^{+3} by Cr^{+3} by adding small amounts of Cr_2O_3 to the melt of highly purified Al_2O_3. Figure 3.9 shows the energy level diagram of a ruby laser. As can be seen from the figure, it has two pump absorption bands centred around 400 nm (blue band) and 550 nm (green band), both having a lifetime of 100 ns. There are two closely spaced metastable levels with a lifetime of 3 ms from where they drop to the ground state. The ground state here acts as the lower laser level. The laser emits at 694.3 nm (R_1 line). Another possible emission line is at 692.9 nm (R_2 line) but the former is the predominant emission line. The three-level energy structure of ruby is the reason for its poor efficiency (0.1–1%). Laser-related properties of ruby are outlined in Table 3.1.

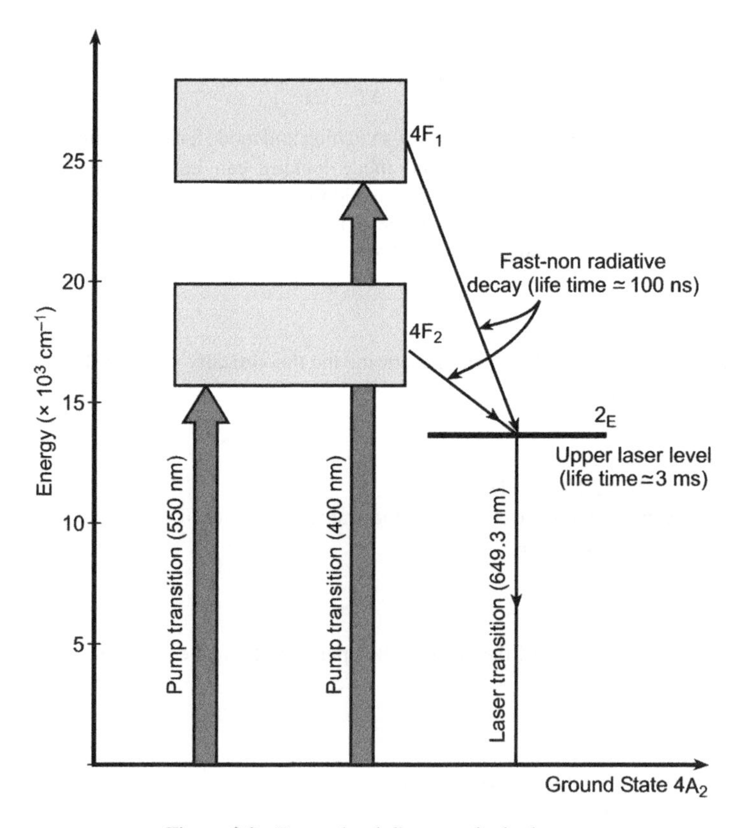

Figure 3.9 Energy level diagram of ruby laser.

Table 3.1 Properties of ruby laser

Cr_2O_3 doping	0.05 wt %
Cr^{+3} ion concentration	1.6×10^{19} ions cm^{-3}
Fluorescent lifetime	3 ms at 300 K
Spectral line width	0.53 nm (11 cm^{-1})
Output wavelength	694.3 nm (R_1-line) and 692.9 nm (R_2-line)
Major pump bands	Blue (404 nm) and green (554 nm)
Stimulated emission cross-section	2×10^{-20} cm^2
Thermal conductivity	46.02 W m^{-1} K^{-1} at 0 °C
Quantum efficiency	70%
Refractive index (at 694.3 nm)	1.763 (ordinary ray); 1.755 (Extraordinary ray)

Its properties rapidly degrade with increasing temperature, with the result that ruby lasers are operated as pulsed lasers at relatively lower repetition rates. The fact that ruby lasers are three-level lasers means that it cannot be operated as a CW laser. Figure 3.10 depicts the Sinon Q-switched ruby laser system from Quantel-Derma Gmbh. Major performance specifications include wavelength 694 nm, pulse width in Q-switched mode 20 ns, pulse width in long pulse mode 4 ms, pulse repetition rate 0.5–2.0 Hz and an adjustable beam diameter of 3–6 mm. The system is primarily applied in dermatology and intended for the removal of pigmentation and tattoos.

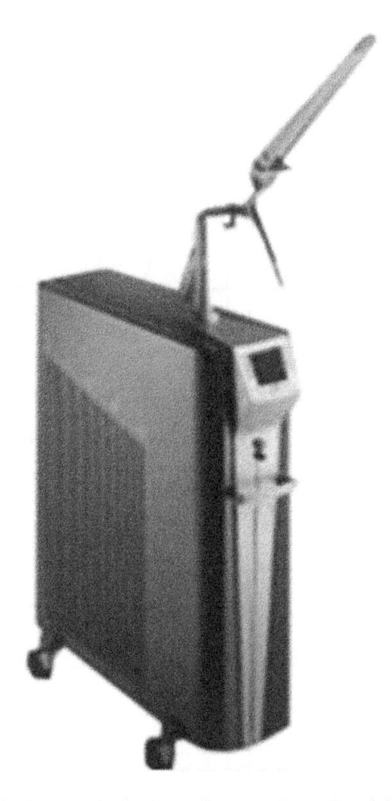

Figure 3.10 Q-switched output Ruby Laser System (Reproduced from Alma Lasers Inc.).

3.5 Neodymium-doped Lasers

Neodymium-doped solid-state lasers are the most widely used type of laser in general (not just solid-state laser). Neodymium-doped lasers differ in terms of the host structure that is doped with neodymium. YAG, YLF ($YLiF_4$) and yttrium-doped vanadate (YVO_4) are the most commonly used crystalline hosts while silicate, phosphate and fused silica are the popular glass hosts. Figure 3.11 shows the energy level diagram of a Nd:YAG laser. All neodymium-doped lasers, whether crystalline-host based or glass-host based have an energy level diagram similar in structure to that shown in Figure 3.11. The interaction of neodymium with the host may lead to a slight change in the output wavelength by around 1% from one neodymium-doped laser to another. For instance, Nd:YAG, Nd:YLF, Nd:YVO_4, Nd:Glass (silicate), Nd: Glass (phosphate) and Nd:Glass (fused silica) have wavelengths of 1064 nm, 1047–1053 nm, 1064 nm, 1062 nm, 1054 nm and 1080 nm respectively.

While Nd:YAG is mainly known for its high gain and good thermal and mechanical properties, Nd:Glass is capable of being grown in large sizes with diffraction limited optical quality. As a result, Nd:YAG is the laser of choice in a wide range of applications requiring either CW or high repetition rate pulsed operation. Due to the poorer thermal conductivity of Nd:Glass lasers they can only be operated as pulsed lasers, and at very low repetition rates. The minimum achievable mode-locked pulse width in a given laser is given by the empirical formula:

$$\tau = \frac{0.44}{\Delta \nu} \tag{3.6}$$

where

$\Delta \nu =$ Line width

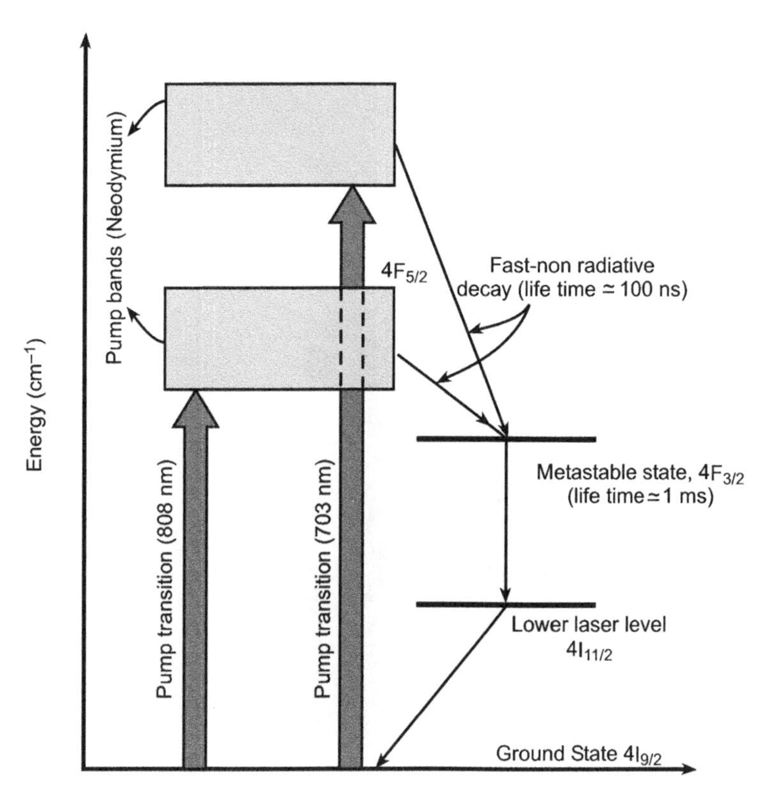

Figure 3.11 Energy level diagram of Nd:YAG laser.

Table 3.2 Properties of Nd:YAG laser

Neodymium doping	0.725 wt %
Neodymium atoms concentration	1.38×10^{20} atoms cm^{-3}
Fluorescent lifetime	0.23 ms
Spectral line width	0.45 nm
Output wavelength	1064 nm
Stimulated emission cross-section	2.8×10^{-19} cm^2
Thermal conductivity (at 300 K)	0.14 W cm^{-1} K^{-1}
Quantum efficiency	70%
Refractive index (at 1.0 μm)	1.82

3.5.1 Nd:YAG Lasers

Among the neodymium-doped lasers, Nd:YAG is the most important and most widely used solid-state laser because of its high gain and good thermal and mechanical properties. In Nd:YAG, trivalent neodymium replaces trivalent yttrium. Some of its noteworthy properties include hardness, stability of structure from very low temperatures to up to the melting point, good optical quality and high thermal conductivity of the YAG host. In addition, the cubic structure of the YAG host favours a narrow line width leading to high gain and low lasing threshold. Nd:YAG has absorption peaks around 540 nm, 590 nm, 750 nm and 810 nm. Pump bands around 540 nm and 590 nm are relevant for flash lamp pumping as they emit a large amount of radiation in the visible region; 750 nm and 810 nm for pumping by CW arc lamps; and 810 nm diode-pumped Nd:YAG lasers. The output laser wavelength is 1064 nm. Major laser-related properties of Nd:YAG are listed in Table 3.2.

Due to its good thermal and optical properties, Nd:YAG lasers can be used both in CW as well as high repetition rate Q-switched pulsed mode. Average and peak powers in excess of 1 kW and 100 MW, respectively, can be achieved in these lasers. Figure 3.12 depicts an arc-lamp-pumped CW Nd:YAG laser of the LCY-series from Litron Lasers UK, intended for a range of industrial applications such as laser marking, engraving, hole drilling, micromachining and materials processing. The series offers pure CW and Q-switched pulsed CW outputs. Multimode power output up to 90 W and TEM_{00} output up to 18 W are available in different models. Figure 3.13 is a photograph of the LDY-series laser from Litron Lasers UK. This is a diode-pumped Nd:YAG laser producing Q-switched CW multimode output power in the range of 10–90 W in multimode and 3–20 W in TEM_{00} mode in different models. Repetition rate is in the range of 1–100 kHz.

High repetition rate (up to 20 Hz) Q-switched (5–20 ns pulse width) Nd:YAG lasers find extensive use in a variety of battlefield applications such as range finding and target designation. Figure 3.14 depicts a portable lightweight laser target designator model AN/PEQ-17 from Elbit Systems, typically used for LGB (laser-guided bomb) delivery applications. The laser produces 1064 nm wavelength, 50–70 mJ/pulse Q-switched laser pulses of output beam divergence 0.3 mrad. The system operates on pulse repetition frequencies of NATO Stanag band I/II and offers a designation range of 5 km.

3.5.2 Nd:YLF Lasers

The advantages of Nd:YLF lasers include fewer heat-related problems due to a smaller change in its refractive index with temperature and higher energy storage capability leading to its ability to generate higher-energy Q-switched pulses. The fluorescence lifetime is in fact twice as long as in the case of Nd:YAG. In addition, Nd:YLF crystal is birefringent, which allows generation of two output wavelengths at 1047 nm and 1053 nm, each with its own polarization orientation. Figure 3.15 shows the simplified energy level diagram of a Nd:YLF laser. Depending upon the polarization, two lines are possible around both 1.05 and 1.3 μm. Both emissions (1.05 and 1.3 μm) originate from the same metastable upper laser level as shown in Figure 3.15. Nd:YLF is predominantly used to produce either of

Figure 3.12 LCY-series CW Nd:YAG laser (Courtesy of Litron Lasers).

Figure 3.13 LDY-series diode-pumped Q-switched CW Nd:YAG laser (Courtesy of Litron Lasers).

Figure 3.14 Laser target designator Model No. AN/PEQ-17 (Courtesy of Elbit Systems of America).

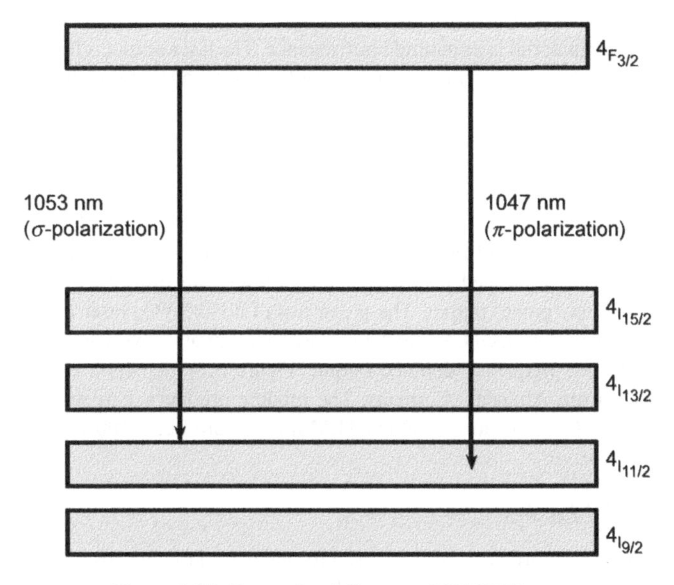

Figure 3.15 Energy level diagram of Nd:YLF laser.

the two lines around 1.05 μm. An intra-cavity polarizer can be used to select either of the two emission lines at 1053 nm (ordinary) or 1047 nm (extraordinary). Laser-related properties of an Nd:YLF laser are listed in Table 3.3.

Emission at 1053 nm matches the peak gain of neodymium-doped phosphate and fluorophosphate glasses. This interesting coincidence makes it highly suitable as an oscillator for pumping Nd:Glass-based laser amplifier chains used to generate the very high peak power Q-switched pulses needed for fusion research.

Figure 3.16 depicts the LDY300-series diode-pumped Q-switched frequency-doubled Nd:YLF laser that generates laser pulse energy in the range of 10–30 mJ at a repetition rate of 0.2–20 kHz. Output

Table 3.3 Properties of Nd: YLF Laser

Neodymium doping	1% (atomic neodymium)
Fluorescent lifetime:	0.48 ms
Stimulated emission cross-section	$1.2 \times 10^{-19}\,cm^2$ (ordinary), $1.8 \times 10^{-19}\,cm^2$ (extraordinary)
Thermal conductivity	$0.06\,W\,cm^{-1}\,K^{-1}$
Output wavelength	1053 nm (ordinary), 1047 nm (extraordinary)
Major pump bands	792 nm, 797 nm and 806 nm
Refractive index (at 1.06 μm)	1.4481 (ordinary), 1.4704 (extraordinary)

wavelength is 527 nm and typical applications include particle imaging velocimetry (PIV), particle sizing and pumping of Ti-Sapphire lasers.

3.5.3 Nd:YVO$_4$ Lasers

Neodymium-doped yttrium vanadate is important as a laser material because of several properties that make it particularly attractive for laser diode pumping. These include a large stimulated emission cross-section and a strong broadband absorption around 808 nm. The high gain of Nd:YVO_4 coupled with strong absorption of laser diode pump radiation around 808 nm obviate the need for a large crystal size, which was a serious problem in the early stages of development of this material. Yet another significant characteristic of this laser material is its natural birefringence. The laser output is linearly polarized in the extraordinary direction and the absorption coefficient for the laser diode pump radiation polarized in the same direction is much larger (about four times) in Nd:YVO_4 than in the case of Nd:YAG.

Some of the disadvantages of Nd:YVO_4 as a laser material include a shorter fluorescence lifetime and a slightly poorer thermal conductivity when compared to Nd:YAG. In fact, a higher stimulated emission cross-section, which contributes in making it a low lasing threshold material, is partially offset by a shorter fluorescence lifetime. It may be mentioned here that the lasing threshold depends upon the product of fluorescence lifetime and the stimulated emission cross-section. Nd:YVO_4 is particularly attractive for laser diode end-pumped CW lasers. These lasers are quite often internally frequency doubled to produce 532 nm (green) output. The properties of a Nd:YVO_4 laser are listed in Table 3.4.

Figure 3.17 depicts an end-pumped CW Nd:YVO_4 laser configuration with an intra-cavity non-linear optical element to produce frequency-doubled output. Figure 3.18 depicts a compact Nd:YVO_4 Q-switched laser module from Alphalas, Germany. The module produces 1-ns-wide 1.5 mJ pulses at a repetition rate 0–100 Hz. Other features include TEM_{00} beam profile, >100:1 polarization ratio and beam diameter of 0.3 mm.

3.5.4 Nd:Cr:GSGG Lasers

Nd:Cr:GSGG is primarily important as laser material for the reason that its absorption band has a better spectral match to the emission from flash lamps than Nd:YAG; Nd:Cr:GSGG lasers therefore offer significantly higher electrical input to laser output efficiency as compared to Nd:YAG. In the case of

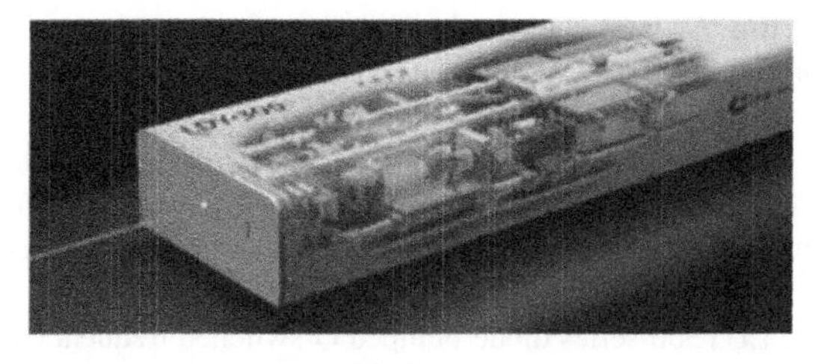

Figure 3.16 LDY 300-series diode-pumped Q-switched Nd:YLF laser (Courtesy of Litron Lasers).

Table 3.4 Properties of $Nd{:}YVO_4$ laser

Neodymium doping	1% (atomic neodymium)
Fluorescent lifetime	0.1 ms
Stimulated emission cross-section	$15.6 \times 10^{-19}\,cm^2$
Thermal conductivity	$0.05\,W\,cm^{-1}\,K^{-1}$
Spectral line width	0.8 nm
Output wavelength	1064.3 nm
Peak absorption wavelength	808.5 nm
Peak absorption coefficient at 808 nm	$37\,cm^{-1}$ (extraordinary polarization)

Nd:Cr:GSGG, the pump energy is absorbed by the broad absorption bands of chromium, which is further transferred non-radiatively to neodymium atoms. This concept was initially tested in a YAG host but was unsuccessful due to the inefficient energy transfer process. Inefficiency was primarily due to a prohibitively large transfer time of about 6 ms, which should be shorter than the fluorescence decay time of 0.23 ms. The problem was overcome by using GSGG as the host material, yielding a transfer time of 17 μs.

The higher efficiency of Nd:Cr:GSGG lasers does not necessarily mean that they are better laser systems than Nd:YAG and solid-state lasers. Nd:Cr:GSGG not only has a significantly poorer thermal conductivity than Nd:YAG, but the additional absorption bands of chromium in blue and red regions that lead to higher pump efficiency produce a significant quantum defect heating due to the large difference in pump photon and laser photon energies. This phenomenon leads to a more severe thermal lensing and thermally induced birefringence.

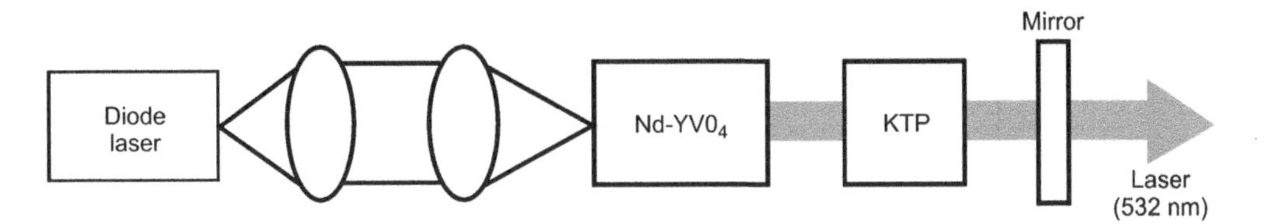

Figure 3.17 Schematic arrangement of end-pumped CW $Nd{:}YVO_4$ laser.

Figure 3.18 $Nd{:}YVO_4$ laser module Type No. PULSELAS-P-1064-150-HE (Courtesy of ALPHALAS GmbH).

3.5.5 *Nd:Glass Lasers*

Nd:Glass is the other important neodymium-doped solid-state laser. While Nd:YAG is mainly known for its high gain and good thermal and mechanical properties, Nd:Glass is capable of being grown in large sizes (rod sizes of 5–6 cm diameter and 100 cm length are not uncommon) with diffraction-limited optical quality. Nd:Glass laser is the preferred laser for high-energy and high-peak-power applications requiring either single pulse or low repetition rate operation, such as in laser fusion research. Glass lasers also have a broad line width, which makes them more adaptable to mode-locking phenomenon in order to generate much shorter pulse widths of the order of picoseconds. Minimum pulse width that can be achieved with mode locking in this case is of the order of a few tens of femtoseconds.

The host materials for Nd:Glass lasers are silicate, phosphate and fused silica glasses, with silicate and phosphate being more common. Silicate and phosphate glasses are based on SiO_2 (Silicon dioxide) and P_2O_5 (Phosphorus pentoxide) materials, respectively, with the latter being the material of choice due to its large emission cross-section and lower non-linear coefficient.

As for other neodymium-doped crystalline lasers, Nd:Glass laser is a four-level laser. Figure 3.19 shows the simplified energy level diagram of a Nd:Glass laser. As shown in the diagram, the laser transition is from the lower-lying level of $^4F_{3/2}$ to a lower-lying level of $^4I_{11/2}$. The emitted wavelength is typically 1062 nm in the case of silicate glasses and 1054 nm for phosphate glasses.

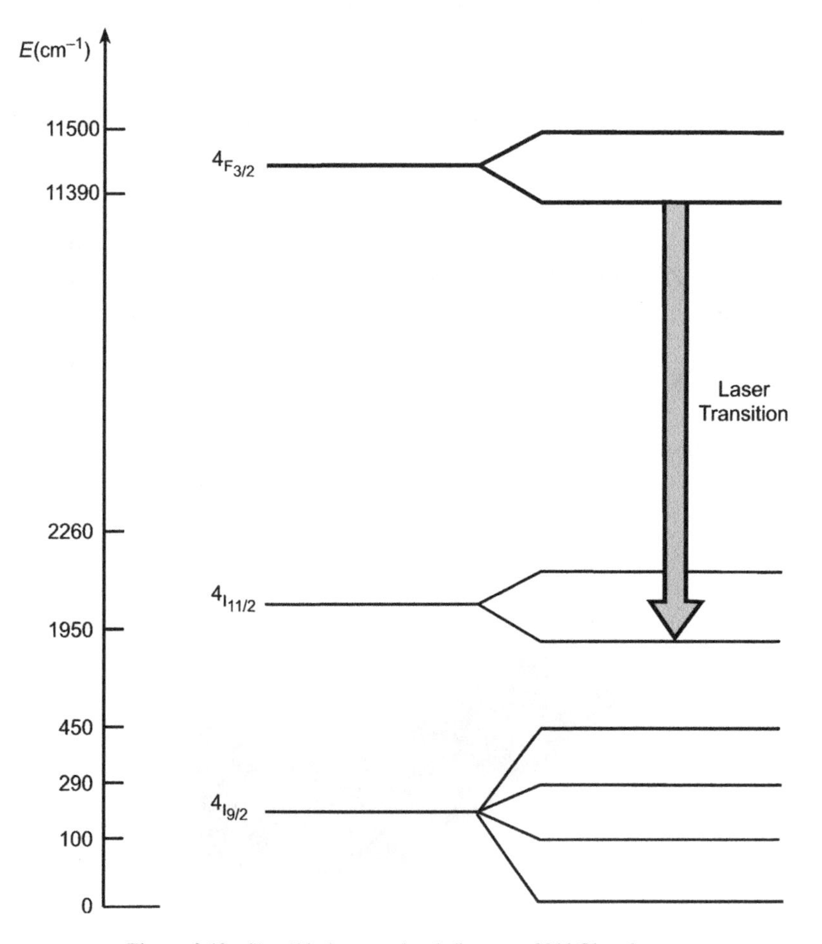

Figure 3.19 Simplified energy level diagram of Nd:Glass laser.

Table 3.5 Properties of Nd:Glass (Silicate) laser

Nd_2O_3 doping	2 wt %
Fluorescence lifetime	0.3 ms
Stimulated emission cross-section	$2.9 \times 10^{-20}\,cm^2$
Thermal conductivity	$0.013\,W\,cm^{-1}\,K^{-1}$
Spectral line width	27 nm
Output wavelength	1062 nm
Peak absorption wavelength	808.5 nm

Table 3.6 Properties of Nd:Glass (Phosphate) laser

Nd_2O_3 doping	3 wt %
Fluorescent lifeime	0.3 ms
Stimulated emission cross-section	$4 \times 10^{-20}\,cm^2$
Thermal conductivity	$0.01\,W\,cm^{-1}\,K^{-1}$
Spectral line width	20 nm
Output wavelength	1054 nm
Peak absorption wavelength	808.5 nm

The fluorescent line width in the case of Nd:Glass lasers is insensitive to temperature variations, which makes it practically feasible to operate these lasers over a wide temperature range (e.g. from −100 °C to +100 °C) with very little change in performance characteristics. Tables 3.5 and 3.6 list the characteristics of neodymium-doped silicate and phosphate glass lasers, respectively.

3.6 Erbium-doped Lasers

Erbium-doped lasers possess medical and military applications because of the two wavelengths they are capable of generating when doped in YAG and glass hosts. These wavelengths are 2940 nm (Er:YAG) and 1540 nm (Er:Glass) and their importance arises from the water absorbent characteristics of these wavelengths. While 2940 nm holds promise for medical applications in the field of plastic surgery due to its extremely large absorption by water in tissue, 1540 nm is attractive as an eye-safe alternative to neodymium-doped YAG (or Glass) based military laser rangefinders and laser target designators. Neodymium lasers with emission around 1064 nm present a serious eye hazard, so eye-safe lasers are definitely a much better option for training exercises and war games.

3.6.1 Er:YAG Laser

Figure 3.20 depicts the energy level diagram for Er:YAG, which produces an output wavelength of 2940 nm. The lasing action occurs between an upper laser level with a fluorescence lifetime of 0.1 ms and a lower laser level with a relatively much larger lifetime of 2 ms. The larger lifetime of the lower level is in fact a big disadvantage as accumulation of population at the lower level inhibits laser action and also disallows Q-switching operation. The properties of a Er:YAG laser are listed in Table 3.7. Some of the other erbium-doped crystalline hosts that have been exploited for laser action include Er:YLF, Er:$YAlO_3$ and Er:Cr:YSGG. These materials produce wavelengths in the range 2710–2920 nm.

3.6.2 Er:Glass Laser

Figure 3.21 shows the energy level diagram of erbium-doped glass, which produces output at 1540 nm. The three-level behaviour of erbium leads to a low laser efficiency. The problem is worsened by the weak absorption of pump radiation by erbium ions. In order to overcome these shortcomings, ytterbium (Yb^{+3}) and chromium (Cr^{+3}) ions are added. Ytterbium acts as a sensitizing agent and helps in absorbing pump radiation in the wavelength region (0.9–1 μm) where erbium is more or less transparent.

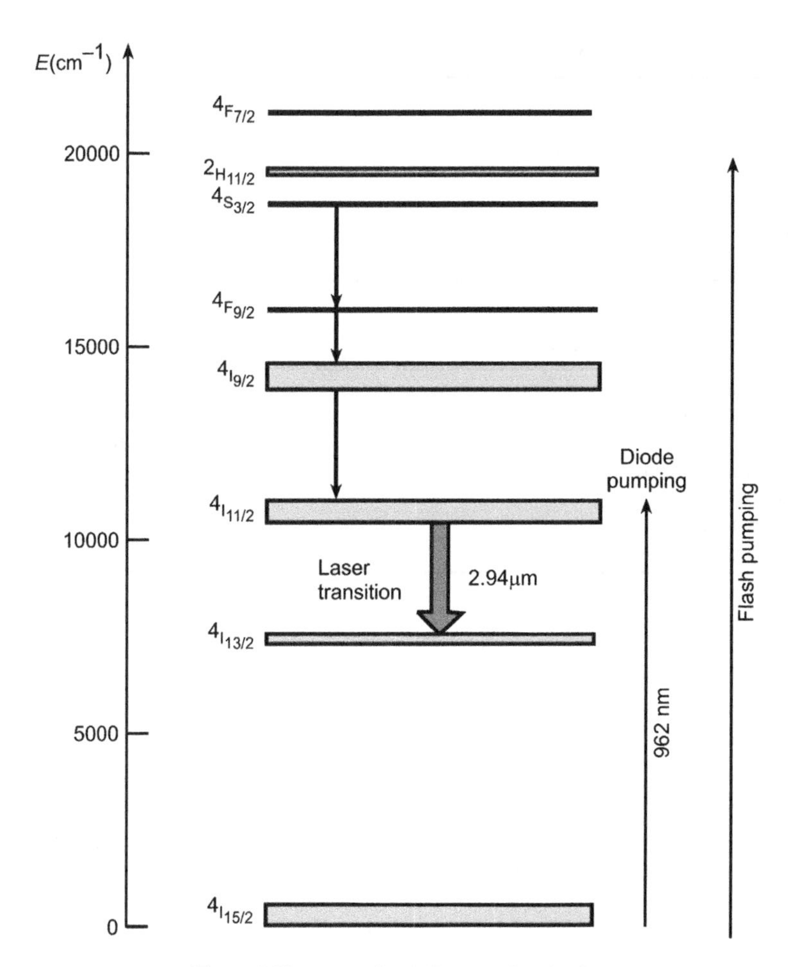

Figure 3.20 Energy level diagram of Er:YAG.

Chromium ions also do a similar job. They help in matching the emission spectrum of flash lamps with the absorption spectrum of ytterbium: erbium: glass. The properties of erbium-doped glass (phosphate) are listed in Table 3.8.

Q-switched Er:Glass lasers find their main application in eye-safe handheld laser rangefinders. These are lasers with low repetition rates, typically in the range 5–20 pulses per minute although in some cases it may be as high as 2 Hz. A large number of manufacturers offer handheld Er:Glass laser rangefinders. The LH-40 Eye-safe Laser Rangefinder from Eloptro South Africa and LRB-21K and LRB-25000

Table 3.7 Properties of Er:YAG laser

Erbium doping	1% (atomic erbium)
Density	$5.35\,\text{g}\,\text{cm}^{-3}$
Fluorescent lifetime	0.1 ms
Stimulated emission cross-section	$15.6 \times 10^{-19}\,\text{cm}^2$
Thermal conductivity	$0.12\,\text{W}\,\text{cm}^{-1}\,\text{K}^{-1}$
Spectral line width	0.8 nm
Output wavelength	2940 nm

Table 3.8 Properties of Er:Glass (Phosphate) laser

Density	$3.15\,g\,cm^{-3}$
Fluorescent lifetime	8 ms
Stimulated emission cross-section	$0.8 \times 10^{-20}\,cm^2$
Thermal conductivity	$0.007\,W\,cm^{-1}\,K^{-1}$
Spectral line width	0.8 nm
Output wavelength	1540 nm
Refractive index (1540 nm)	1.531

(Figure 3.22) Eye-safe Laser Rangefinders from Newcon Optik are some examples. Most of the devices in this category have similar performance specifications in terms of operational range, range accuracy and pulse repetition rate. The LRB-25000 has a maximum operational range of 25 km, range accuracy of ± 5 m and pulse repetition frequency of 0.15 Hz. High-repetition-rate military lasers of the eye-safe variety currently employ Nd:YAG lasers whose output is wavelength shifted using an optical parametric

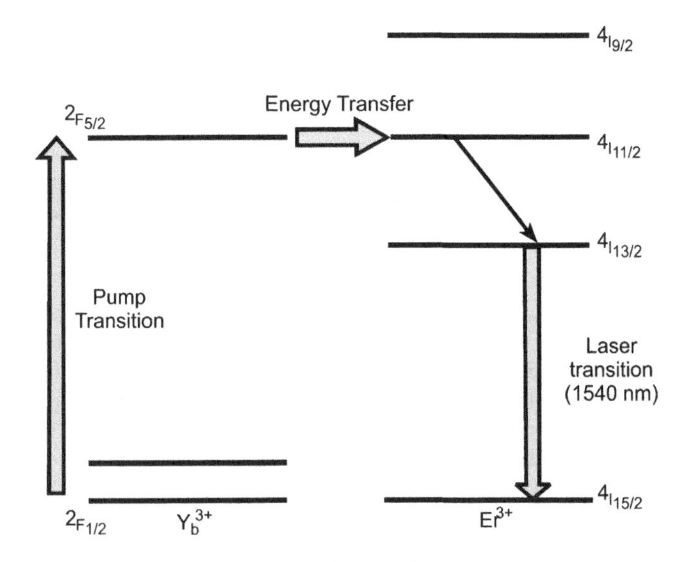

Figure 3.21 Energy level diagram of Er:Glass (phosphate).

Figure 3.22 Er:Glass Laser Rangefinder, Type LRB-25000 (Courtesy of Newcon Optik).

Figure 3.23 OPO shifted Nd:YAG Laser Rangefinder, Type G-TOR (Courtesy of SAAB Sweden).

oscillator (OPO). Laser rangefinder LDM-38 from Carl Zeiss Optronics and laser rangefinder G-TOR from SAAB Sweden (Figure 3.23) are examples. Both these laser rangefinders are configured around OPO-shifted Nd:YAG lasers emitting at 1570 nm. G-TOR offers a pulse repetition rate as high as 25 Hz.

3.7 Vibronic Lasers

What distinguishes a vibronic laser from the more commonly used and better-known neodymium-doped lasers and ruby lasers is that their energy level structure has a lower lasing level in the form of a band rather than a single discrete energy level. This energy band results from the interaction of the electronic energy levels of the active species with the vibrational levels of the crystalline lattice, giving rise to vibrational-electronic sub-levels. The laser transition occurs between the lowest level of the upper band to anywhere in the lower band. This characteristic feature of vibronic lasers makes this class of lasers tunable.

These lasers are important not only for the tunability aspect associated with them, but also due to their ability to generate wavelengths not available from other solid-state laser media. Alexandrite (chromium-doped $BeAl_2O_4$), Cr-GSGG (chromium-doped $Gd_3Sc_2Go_3O_{12}$) and titanium-sapphire (titanium-doped Al_2O_3), with tunability ranges of 700–850 nm, 740–850 nm and 660–1180 nm, respectively, are popular varieties of vibronic lasers.

3.7.1 Alexandrite Laser

With its intense red output coupled with tenability, the alexandrite laser is a good candidate for developing a laser dazzler, a device employed for causing temporary dazzling of human eyes. Alexandrite is the common name for chromium-doped chrysoberyl, with chromium concentration in the range 0.1–0.4 atomic percent. Alexandrite has optical and mechanical properties similar to that of ruby. It has many of the physical and chemical properties of a good laser host material, including high thermal conductivity, hardness, chemical stability and a high thermal fracture limit. The thermal conductivity and thermal fracture limit of Alexandrite is about twice and five times that of Nd:YAG, respectively, which enables Alexandrite to be pumped at high average powers without thermal fracture.

Figure 3.24 depicts the energy level diagram of Alexandrite. Vibronic lasing action occurs between the upper level $^4T_2$ and any of the excited vibronic states near the ground level $^4A_2$. This characteristic makes it a tunable laser, and the output wavelength is tunable in the range of 700–850 nm. As shown in the energy level diagram, there is another energy level at about 800 cm^{-1} below the upper level. This energy level ^2E is in thermal equilibrium with the upper level. $^4T_2$ and ^2E levels have lifetimes of 6.6 μs

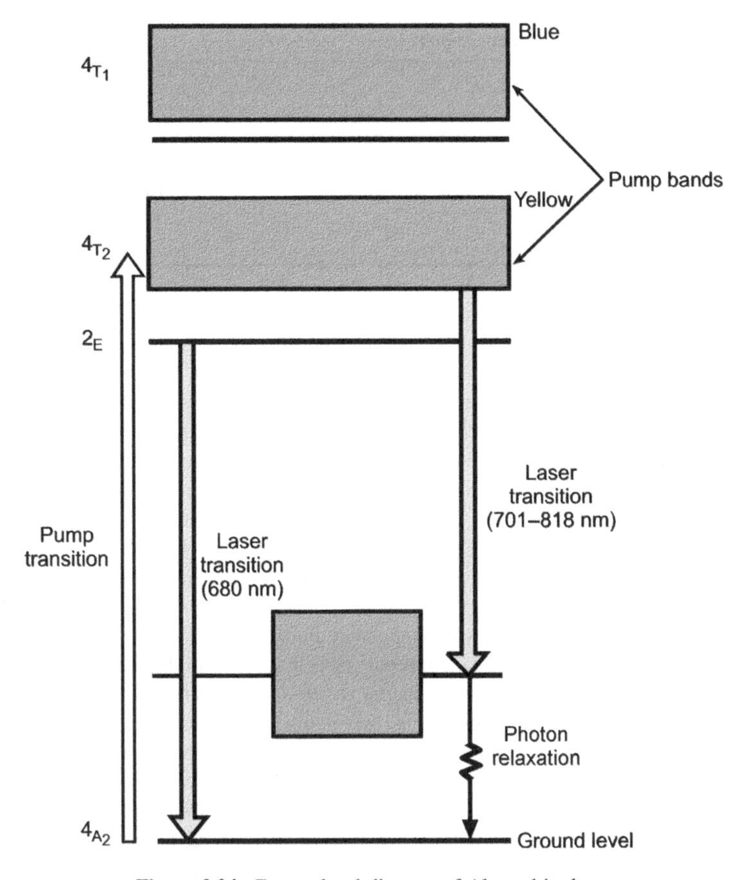

Figure 3.24 Energy level diagram of Alexandrite laser.

and 1.54 ms, respectively. The effective fluorescent lifetime of the upper level here is dependent upon the combined influence of $^4T_2$ and ^2E states, and is approximately 260 μs.

Another interesting property of Alexandrite is the way its performance is influenced by temperature. The stimulated emission cross-section, and hence the gain of Alexandrite, increases with increase in temperature. The gain peak shifts to a longer wavelength due to the population of terminals levels, particularly those closer to ground level, which decreases the population inversion for lower emission wavelengths. The net result is that laser performance is positively affected with increasing temperature for larger wavelengths in the tuning range. The other effect of increase in temperature is reduced fluorescent lifetime, which decreases from about 260 μs at room temperature to half that at 100 °C. The properties of Alexandrite are listed in Table 3.9. The ability of Alexandrite lasers to sustain high gain at elevated temperatures, coupled with tenability, make them different from conventional solid-state lasers.

Table 3.9 Properties of Alexandrite laser

Chromium doping	0.1–0.4% (atomic erbium)
Fluorescent lifetime	0.26 ms
Stimulated emission cross-section	1.0×10^{-20} cm^2
Thermal conductivity	0.23 W cm^{-1} K^{-1}
Spectral line width	100 nm
Output wavelength	700–850 nm
Refractive index (750 nm)	1.7421(for E-vector parallel to b-axis)

Table 3.10 Properties of titanium-sapphire laser

Titanium doping	0.1%
Fluorescent lifetime	3.2 μs
Stimulated emission cross-section (at 795 nm) parallel to c-axis	$2.8 \times 10^{-19}\,cm^2$
Thermal conductivity	0.3–0.5 $W\,cm^{-1}\,K^{-1}$
Spectral line width	180 nm
Output wavelength	660–1180 nm
Refractive index	1.76

3.7.2 Titanium-sapphire Laser

Titanium-sapphire is the most widely used tunable solid-state laser because of its wide tunability and good material characteristics. Output wavelength is tunable from 660 nm to 1180 nm with the peak of the gain curve located around 800 nm. The host material (sapphire) has very high thermal conductivity, mechanical rigidity and exceptionally high chemical inertness. The concentration of titanium ions in the laser crystal is about 0.1%. Properties of Ti-sapphire lasers are listed in Table 3.10.

Titanium-sapphire lasers cannot be efficiently pumped by flash lamps due to their too-short fluorescence lifetime. It may be mentioned here that the population inversion required in a laser to exceed the lasing threshold is inversely proportional to the product of fluorescence lifetime and stimulated emission cross-section. An extremely small fluorescence lifetime of 3.2 μs in titanium-sapphire makes this product very small. The consequence is that a very high pump flux would be needed to achieve the threshold of lasing. The absorption band has a peak at around 500 nm, which makes it again very difficult to use laser diode pumping as this is too short a wavelength for diode lasers.

In view of the above reasons, titanium-sapphire lasers are usually pumped by frequency-doubled neodymium lasers (Nd:YAG, Nd:YLF for pulsed output), argon ion lasers (for CW output) and copper vapour lasers (pulsed output). The neodymium lasers in turn may be diode pumped. The strong emission of argon ion lasers at 488 nm and 514.5 nm, frequency-doubled YAG lasers at 532 nm, frequency-doubled YLF lasers at 527 nm and copper vapour lasers at 510 nm match with the absorption band of titanium-sapphire.

Figure 3.25 depicts the simplified energy level diagram of a titanium-sapphire laser. Argon ion laser-pumped Ti-sapphire lasers producing a few watts of CW output, as well as frequency-doubled Nd:YAG or Nd:YLF laser-pumped Ti-sapphire lasers producing several watts of CW power output or pulsed output of several millijoules at 1 kHz or in excess of 100 mJ at 20 PPS, are commercially available. Figure 3.26 depicts a tunable CW Ti-Sapphire laser of the MBR-110 of the MBR ring series from Coherent Inc. The laser produces >3.5 W of output power tunable from 700 nm to 1030 nm with an ultra-low line width of <75 kHz (root mean square or RMS). The laser is ideally suited to atomic and molecular spectroscopy and fluorescence spectroscopy.

3.8 Colour Centre Lasers

Colour centre lasers are solid-state lasers in which the active medium is a thin crystal doped with some selected impurities that create microscopic defects in the crystal lattice structure. These microscopic defects absorb light, thus colouring the otherwise colourless crystal, hence the name colour centres. The phenomenon of absorption of light by the colour centres is used for pumping the crystal. Although laser action has been demonstrated in colour centre lasers near 400 nm and in the range 800–4000 nm, commercially established types operate near 1500 nm and beyond 2000 nm. Colour centre lasers are pumped by another laser such as an ion laser, a dye laser or Nd:YAG laser. These lasers normally operate in continuous wave mode but can be synchronously pumped by mode-locked lasers to produce a pulsed output.

Some of the common active media used in colour centre lasers include lithium-doped potassium chloride, lithium-doped rubidium chloride, sodium-doped potassium chloride and sodium-doped rubidium chloride.

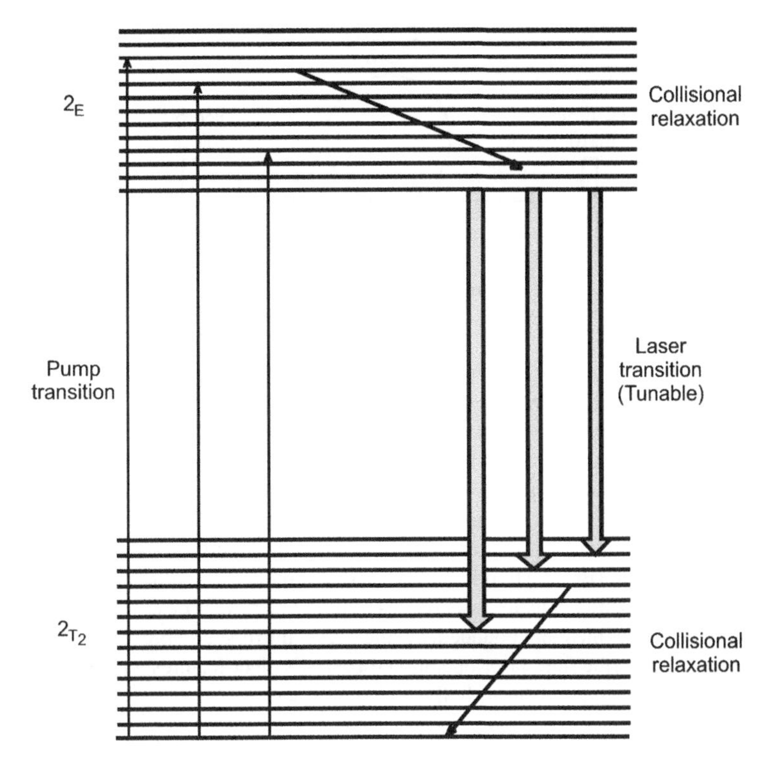

Figure 3.25 Simplified energy level diagram of Ti-Sapphire laser.

Figure 3.26 Tunable CW Ti-Sapphire laser, Type MBR-110 (Reproduced by permission of Coherent Inc.).

A large number of possible vibrational states in the colour centre crystals lead to homogeneous broadening of the laser transition, which permits tuning of the emission wavelength over a broad wavelength range. Wavelength tuning by as much as 15–20% of the centre wavelength is possible.

Colour centre lasers operate at liquid nitrogen temperature. A Dewar that maintains operation at 77 K cools the laser crystal and the pumping chamber. Colour centre lasers find application in high-resolution spectroscopy, where their high spectral purity is a significant advantage. They operate in single longitudinal mode with a line width of the order of 1 MHz; active stabilization could improve this to 10–20 kHz.

3.9 Fibre Lasers

Fibre lasers are a type of solid-state laser where the gain medium is a fibre doped with rare earth ions, rather than a rod or slab. They are an alternative to bulk solid-state lasers and have a wide range of industrial and military applications requiring high power levels with high beam quality in a compact and

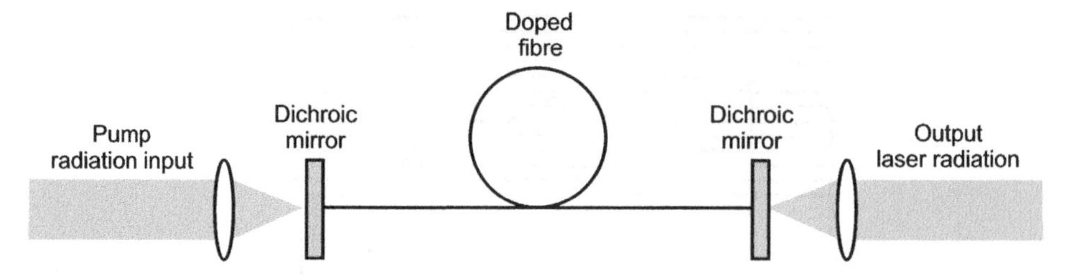

Figure 3.27 Basic fibre laser.

rugged package configuration. Inherent in the fibre laser design and operational regimes are excellent performance characteristics, which include a high level of immunity to misalignment, high beam quality, compactness and long-term stability. In addition to these, fibre lasers exhibit outstanding thermo-optical properties due to their large surface-area-to-volume ratio. Advantages and limitations of fibre lasers are discussed in detail in the following sections.

The gain medium in the case of fibre lasers is a glass fibre doped with rare earth element ions such as neodymium (Nd^{3+}), erbium (Er^{3+}), ytterbium (Yb^{3+}), thulium (Tm^{3+}), holium (Ho^{3+}) or praseodymium (Pr^{3+}). Fibre coupled semiconductor laser diodes or fibre lasers are used as the pump source and the resonator comprises active medium bounded by dielectric mirrors or fibre Bragg gratings. In the following sections we describe fibre laser operational basics, salient features, operational regimes and typical applications and compare them to bulk solid-state lasers.

3.9.1 Basic Fibre Laser

In its simplest form, a fibre laser comprises fibre doped with rare earth element ions as the active medium, a fibre coupled semiconductor diode laser or another fibre laser as the pump source, and dielectric mirrors or fibre Bragg gratings to form the resonant cavity. Figure 3.27 depicts the simplified arrangement of different components of the basic fibre laser. Both the pump and laser radiation are guided through the waveguide structure constituted by the core and the cladding of the single-clad fibre. The dielectric mirrors and the single-clad fibre constitute the resonant cavity in the arrangement of Figure 3.27. In practical fibre lasers, in most cases, fibre Bragg gratings are used instead.

A fibre Bragg grating is a type of distributed Bragg reflector constructed in a short segment of fibre as shown in Figure 3.28a . The grating is created by a periodic variation of the refractive index of the core and is designed in order to reflect a particular wavelength and transmit all other wavelengths as illustrated in Figure 3.28b. The fibre core doped with rare earth ion guides the light, so the pump radiation must be spatially coherent.

As the power available from single mode semiconductor diode lasers is usually limited to a few watts, such a configuration cannot be used to build relatively higher output power lasers. This limitation is overcome by using a double-clad fibre design. In this case, the active doped core is surrounded by a second waveguide structure in the form of inner cladding, also known as the pump core. Figure 3.29 depicts the arrangement.

Double-clad fibre laser design allows the use of multimode semiconductor diode lasers as the pump source. Pump radiation in this case is launched in the inner cladding. The pump radiation is gradually absorbed over the entire length of the fibre which is converted into single-mode high-brightness laser radiation by laser action. Unlike bulk solid-state lasers where the intensity is limited to Rayleigh length by diffraction, in the case of a double-clad fibre laser the intensity is maintained over the entire fibre length due to confinement of both pump and laser radiation. Consequently, the gain of the active medium (the product of light intensity in the gain medium and the interaction length) is significantly higher than that in the case of a bulk solid-state laser. This property gives fibre lasers a high single-pass gain and a low pump threshold value.

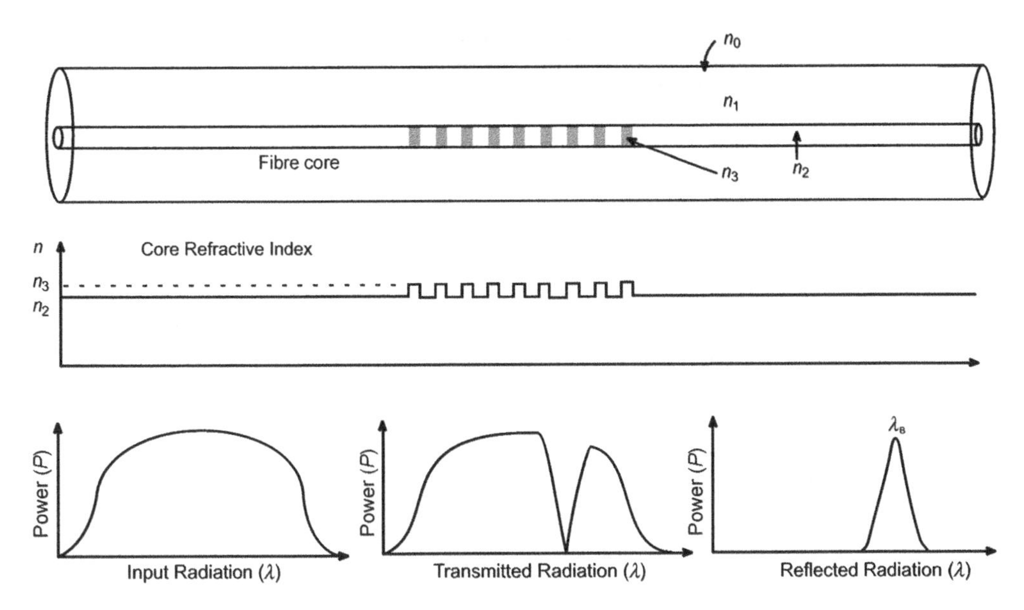

Figure 3.28 Principle of fibre Bragg grating.

Wavelengths emitted by common fibre lasers are in the region of 1.0–1.1 μm from ytterbium-doped lasers, 1.52–1.57 μm from erbium-doped lasers and 1.9–2.1 μm from thulium- and holmium-doped lasers. Figure 3.30a–d show simplified energy level diagrams of ytterbium-, erbium-, thulium- and holmium-doped active media. Because of a low quantum defect ($<10\%$) ytterbium is the dopant material of choice, particularly in high-average-power fibre lasers.

One drawback of the double-clad fibre laser concept is reduced absorption of pump radiation by the active-ion-doped core due to the existence of intensity distributions in the inner cladding that has no overlap with the active medium. As a result of this, optical-to-optical efficiency reduces. This problem is overcome by breaking the cylindrical symmetry of inner cladding, usually by D-shaped or rectangular pump core geometries. Periodic bending when used in symmetrical fibres also enhances absorption of pump radiation.

3.9.2 Fibre Lasers versus Bulk Solid-state Lasers

In the following, we compare fibre lasers with bulk solid-state lasers in terms of their performance characteristics, applications, strengths and weaknesses.

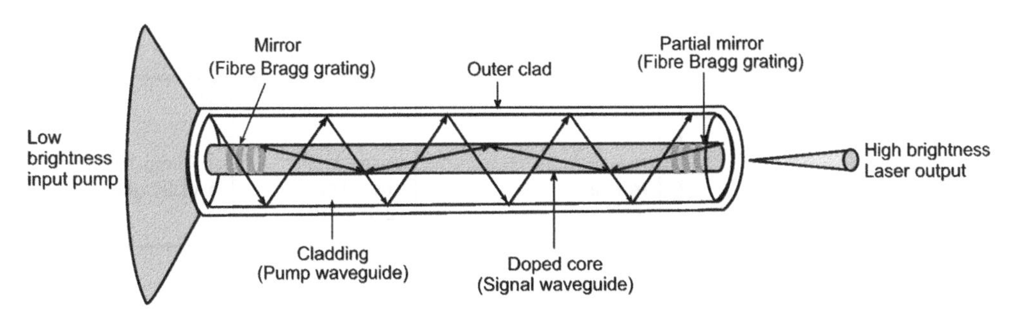

Figure 3.29 Basic double-clad fibre laser.

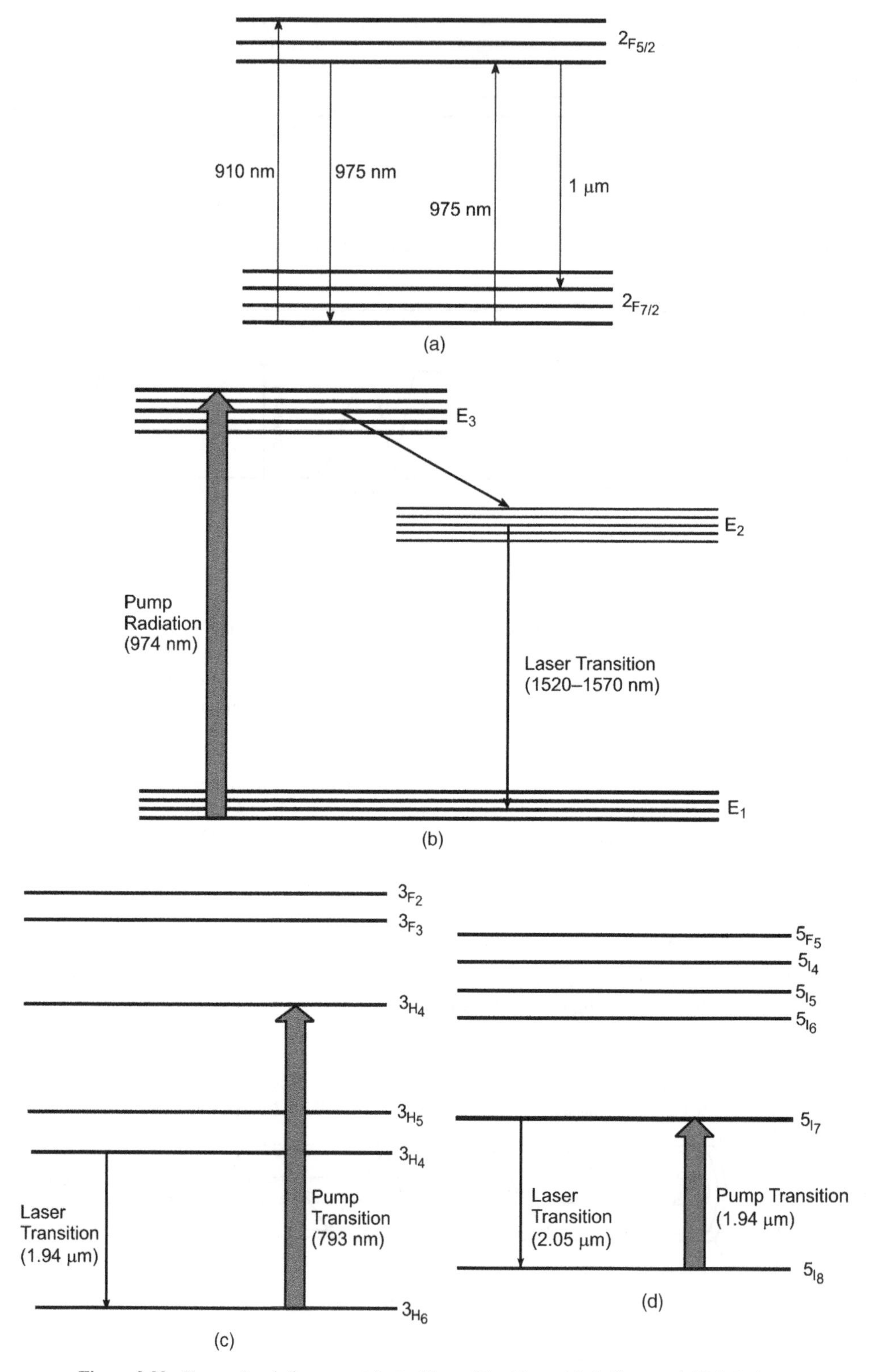

Figure 3.30 Energy level diagrams: (a) ytterbium; (b) erbium; (c) thulium; and (d) holmium.

1. Fibre lasers exhibit relatively much higher wall-plug efficiency compared to bulk solid-state lasers. Wall-plug efficiency approaching 50% is achievable in the case of fibre lasers, as against a figure of 20–30% for solid-state lasers. However, fibre lasers need to use pump diodes producing a higher beam quality than the beam quality required in the case of bulk solid-state lasers. Bulk lasers can also employ pump sources with poor beam quality. As an illustration, an extremely high pulse energy/ peak power pulse of laser with a moderate repetition rate such as that used in laser fusion research could be built by using a rod laser side-pumped by high-energy flash lamps. This would not be possible for a fibre laser.
2. Fibre lasers produce a laser with higher beam quality at higher average output power levels as compared to bulk solid-state lasers. The fibre structure and the associated waveguide effect coupled with large surface-area-to-volume ratio give fibre lasers excellent thermo-optical properties. As a result, fibre lasers are able to produce high output power levels with high beam quality. Although both fibre and bulk solid-state lasers can generate multikilowatt power levels, it is far more difficult in the case of the latter to achieve the beam quality of the former for a given output power level.
3. Fibre lasers inherently offer relatively higher gain bandwidth. The glass host broadens the optical transition in rare earth ion dopant. This makes fibre lasers continuously tunable in the near-infrared spectral region from 1–2 μm by using a suitable dopant from ytterbium, erbium and thulium. Higher gain bandwidth also allows fibre lasers to generate ultra-short pulses by using passive mode locking, although sometimes it is not feasible to fully exploit their large gain bandwidth to achieve the shortest possible pulse width due to excessive non-linearity.
4. Fibre lasers have broad absorption bands with good absorption, which makes the pumping process less critical in terms of pump wavelength. This further implies that the pump diode lasers could possibly be used without temperature stabilization.
5. Fibre lasers are significantly cheaper than the bulk lasers with comparable performance specifications. This is more so in the case of all fibre lasers which do not employ any free space optics. One of the reasons for the low cost of fibre lasers is that processes such as fusion splicing used for integrating various fibre-based components of the laser can be largely automated. Also, in all fibre lasers the absence of any free space optics makes them immune to any misalignment and renders robustness and long-term stability.
6. While fibre lasers offer a far superior performance when it comes to high average power lasers with high beam quality, bulk solid-state lasers have the edge in high-pulse-energy, high-peak-power applications.
7. Fibre lasers suffer from excessive Kerr non-linearity arising from the long length and small mode area of the fibre. Non-linearity often limits the shortest achievable pulse width and single-frequency operation of the laser. Non-linear effects are much less severe in the case of bulk solid-state lasers.
8. Fibre lasers can be operated on difficult laser transitions. This is made possible by the waveguide structure of the fibre laser, which allows high pump radiation intensities to be applied over long fibre lengths. Up-conversion lasers is an example of this.

3.9.3 Operational Regimes

Common operational regimes of a fibre laser include CW, Q-switched and mode-locked modes of operation. The *continuous wave* fibre laser mode of operation, with output power levels extending from milliwatts to kilowatts, is the most sought-after variety with a wide range of industrial and military applications. The Q-switched mode of operation can be used to generate pulse energies in the range of several millijoules to several tens of millijoules with pulse widths in the range of several tens of nanoseconds to hundreds of nanoseconds. The mode-locked regime is used to generate ultra-short pulses with pulse widths of the order of tens to hundreds of femtoseconds. Both active as well as passive mode locking is possible. Another operational regime that is attractive in the case of fibre lasers is the *up-conversion mode* of operation, in which lasing action is achieved in difficult-to-operate lasing transitions.

Single-frequency low-gain operation is also possible in fibre lasers by using a short-length resonator with narrow-bandwidth fibre Bragg gratings as reflectors on the two ends. These single-frequency fibre

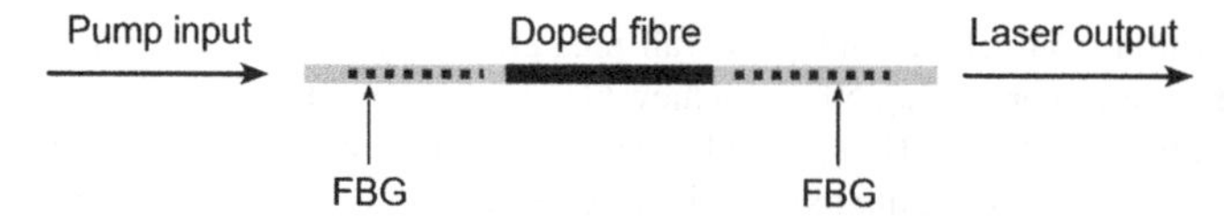

Figure 3.31 DBR-based single-frequency fibre laser design.

Figure 3.32 DFB-based single-frequency fibre laser design.

lasers can be used to generate CW power levels of a few milliwatts to a few tens of milliwatts; a power level approaching 1 W has also been achieved. Single-frequency operation is also possible using the distributed feedback (DFB) laser concept. In a DFB laser, a single-fibre Bragg grating with a phase shift in the middle is used.

In the distributed Bragg reflector (DBR) based single-frequency laser design, depicted in Figure 3.31, a short cavity length of the order of a few centimetres is used between two narrow bandwidth fibre Bragg gratings with a spectral line width of a fraction of a nanometre. Short cavity length ensures a longer separation between adjacent longitudinal modes, forcing only a single longitudinal mode to propagate. The shorter spectral line width of fibre Bragg grating reflectors ensures that only a single longitudinal mode experiences gain above lasing threshold.

In the case of DFB laser design, depicted in Figure 3.32, a longer fibre Bragg grating (FBG) with a $\pi/2$ phase jump is used to construct the resonator. Due to its precise phase control over the long FBG, DFB operational mode combined with $\pi/2$ phase jump gives a much lower spectral line width and single-frequency operation.

3.9.4 Photonic Crystal Fibre Lasers

In comparison to the conventional step-index fibre that derives its wave-guiding properties from a spatially varying glass composition, the basic photonic crystal fibre (PCF) is an optical fibre with a structured pattern of air holes (also called voids) that run parallel to the axis all along its length (Figure 3.33). Unlike conventional fibres, core and cladding are made from the same material in the case of PCF. Most PCFs are made from pure fused silica, which is highly compatible with PCF fabrication

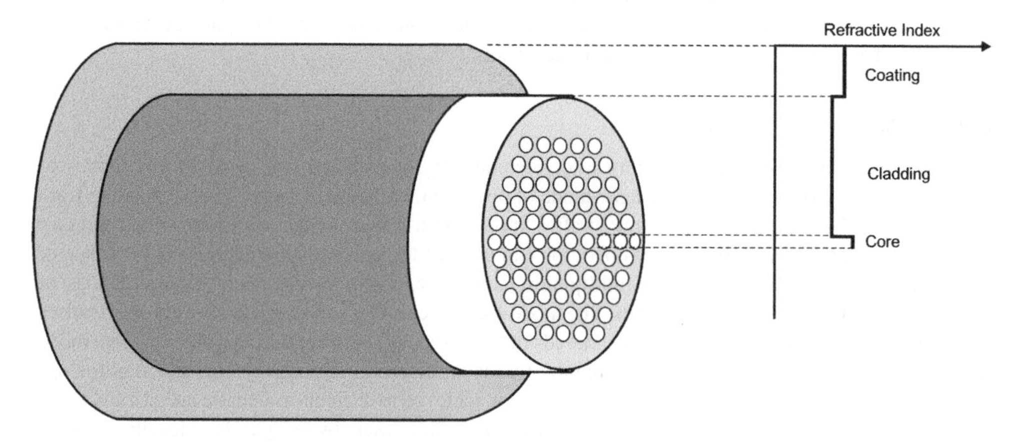

Figure 3.33 Construction of photonic crystal fibre.

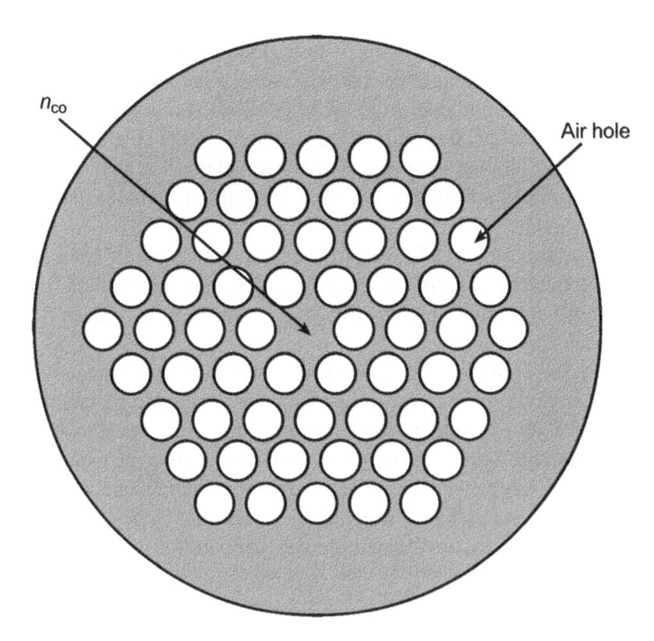

Figure 3.34 Basic photonic crystal fibre.

techniques, although PCFs have also been made of other materials such as heavy metal soft glasses and polymers. One of the methods of fabrication of PCFs, known as stacked tube technique, involves using a preform with larger holes made by stacking capillary and/or solid tubes and inserting them into a larger tube. The preform is first drawn to a diameter of approximately 1 mm, which is subsequently further drawn to a fibre of final diameter of the order of 100 μm.

The most common type of PCF has a hexagonal array of air holes with one or more missing holes in the centre forming the core (Figure 3.34). The region with the missing hole has a higher refractive index, analogous to the core of a conventional step-index fibre.

3.9.4.1 Guiding Mechanisms in PCF

There are two distinct mechanisms of guiding light in PCFs: *photonic band-gap guiding* and *index guiding*. In the case of the photonic band-gap guiding mechanism, the wavelengths falling within the band gap of the photonic crystal cannot propagate out and are therefore confined to the core. The index guiding mechanism is similar to conventional step-index fibre operation.

A typical index-guided PCF has a solid core and a structured pattern of air holes surrounding the core. The presence of air holes reduces the refractive index. In comparison to a conventional step-index fibre, an index-guided PCF structure offers much finer and more accurate control over the magnitude of the refractive index. The guiding mechanism is due to total internal reflection.

3.9.4.2 Subclasses of PCFs

The properties of fibre in both index- and band-gap-guided PCFs can be varied by controlling parameters such as the size, shape and arrangement of air holes. These parameters can be suitably manipulated to obtain an optimum performance for a specific application. Many subclasses of PCFs are available, supporting a wide range of applications. Some of these types include large numerical aperture (NA), large mode area, highly non-linear, low index core and air-guiding hollow core PCFs.

One of the subclasses of PCFs known as *endlessly single mode* is characterized by a triangular matrix of circular holes with one or more (3, 5 or 7) missing holes. Analysis of this structure shows that the fibre supports a single mode for ratio of hole diameter to spacing greater than a certain value, regardless of operating wavelength for core diameter, approaching tens of microns. In other words, this PCF offers a single-mode operation for large mode areas over a wide wavelength range in sharp contrast to conventional step-index fibre, which offers a single-mode operation for relatively much smaller core diameter over a narrow wavelength range. Single-mode operation over a large mode area reduces power density, thus minimizing non-linear effects and allowing operation at high power levels.

Large NA fibre is another subcategory of PCF. An increased volume of air holes lowers the refractive index of the cladding with the resulting fibre having a large NA; NA approaching 0.7 is practical. Large NA coupled with large mode area allows highly efficient coupling of light from an extended source such as high-power laser diodes.

Highly non-linear fibres are constructed by having a small core of approximately 1.0 μm surrounded by a honeycomb array of large air spaces to achieve a large core to cladding refractive index ratio. Such fibres are particularly useful, for example in Raman amplification, four-wave mixing, super-continuum generation and optical parametric amplification. A common type of photonic band-gap fibre consists of a hollow core surrounded by closely packed triangular matrix of circular holes. Depending upon structure, these fibres offer a limited spectral bandwidth of the order of 10–20% of nominal transmission wavelength. Fibres with transmission wavelength centred around 800 nm, 1064 nm and 1550 nm are commercially available.

3.9.4.3 PCF Lasers

Conventional high-power fibre lasers use double-clad step-index fibres with a polymer outer cladding and a core doped with rare earth ions such as ytterbium or erbium. This configuration has an upper limit to achievable pump and output power levels while maintaining a single-mode output. This is due to the fact that operation at single transverse mode necessitates use of small core size of the order of a few microns, resulting in higher power densities at high pump and output power levels. High power density leads to further detrimental non-linear effects, although these effects can be minimized to an extent by increasing the core size while still maintaining a single-mode output. The techniques used include controlling index and initial excitation profiles, and introducing microbending losses to preferentially inhibit propagation of higher-order modes.

Dual-clad fibre laser architecture based on PCF technology provides a good alternative to scale the output power to higher levels. In the double-clad PCFs (Figure 3.35), the pump cladding is surrounded by an air cladding region. Due to a large refractive index contrast the pump cladding has a very high NA, typically in the range 0.6–0.7. Large NA significantly lowers the pump source requirements with respect to beam quality and brightness, and permits efficient pumping with relatively lower-cost large-emitting-area pump diodes. Within the inner cladding is another microstructured rare-earth-doped core. The single-mode core can be expanded to a large-mode area to facilitate operation at higher power levels with single-mode output. With such PCF designs, problems due to non-linearities are substantially reduced and there is a good overlap between the pump mode and laser mode.

3.9.5 Applications

High-power fibre lasers with average power levels in the range of tens of watts to multikilowatt levels are important in a wide range of industrial and military applications. Figure 3.36 depicts a CW fibre laser from SPI lasers, UK. The R4 series lasers produce output at 1070 ± 10 nm and are capable of generating both CW and modulated output power up to 500 W. The lasers are intended for a wide range of industrial applications which includes cutting, welding, bending, bonding, sintering, annealing and surface texturing. Until the 1990s, the telecommunications industry was the largest consumer of fibre lasers. Fibre laser growth, in particular the high-power fibre laser, has benefited immensely from technological developments during the days of telecommunications industry expansion. Many of the components including the single-emitter fibre coupled pump diodes used in high-power fibre lasers today were developed during the 1980s and 1990s.

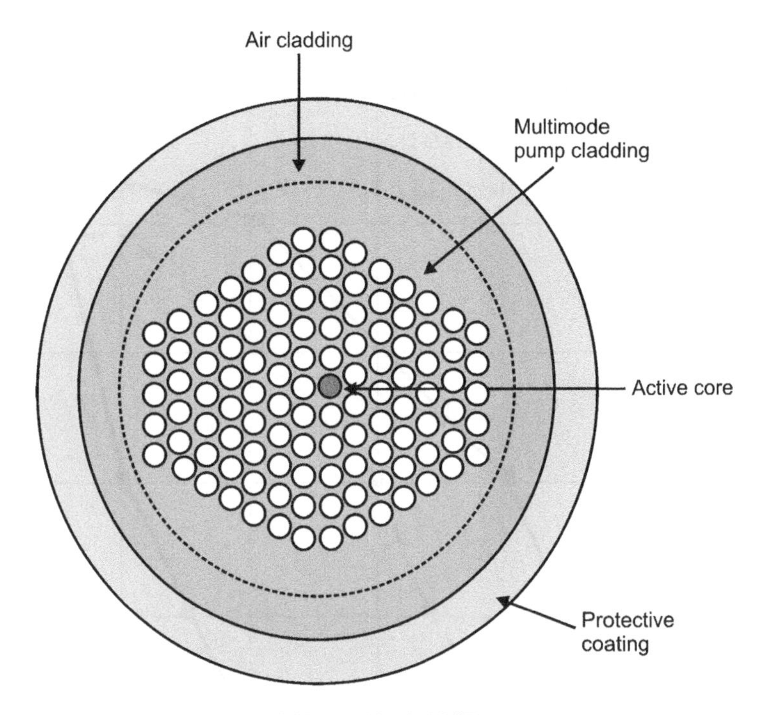

Figure 3.35 Double-clad PCF structure.

Salient features of fibre lasers and the advantages they offer compared to bulk solid-state lasers were discussed in Section 3.9. In the case of fibre lasers, power scalability while retaining the beam quality has been the key factor in their increasing popularity and acceptance over carbon dioxide and bulk solid-state lasers used earlier in the majority of industrial and military applications. While the highest achieved power level in the case of carbon dioxide and Nd:YAG lasers saturated after the year 2000, fibre lasers continue to evolve with a single-mode fibre laser achieving 10 kW of power level and broadband multimode fibre lasers scaled up to a power level of 50 kW. Figure 3.37 shows the power levels achieved

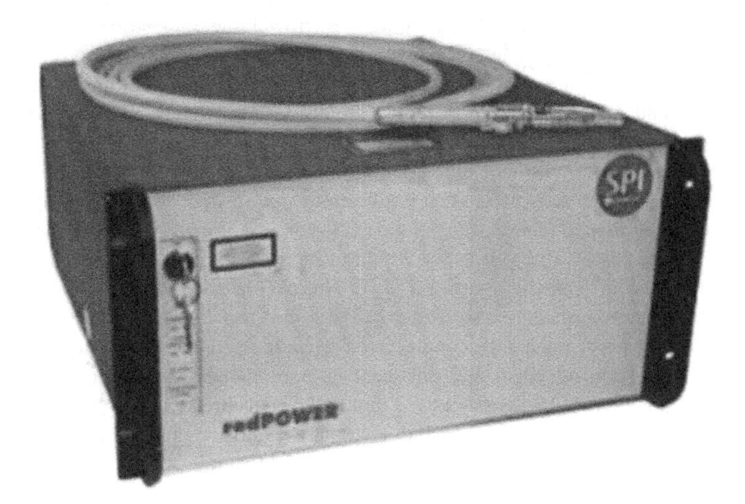

Figure 3.36 CW fibre laser, type R4 (Courtesy of SPI Lasers).

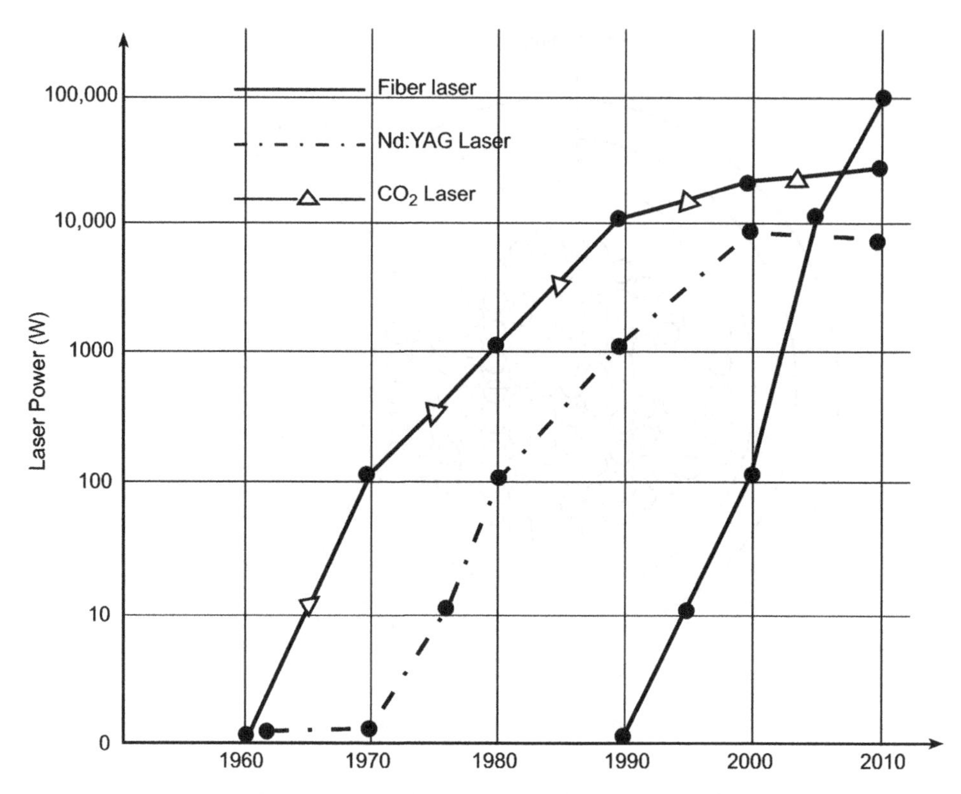

Figure 3.37 Power levels achieved in Nd:YAG, CO_2 and fibre laser.

in Nd:YAG, CO_2 and fibre lasers over the last six decades. The power level is continuing to grow for fibre lasers.

High-power pulsed and CW fibre lasers are fast replacing CO_2 and Nd:YAG lasers for a wide range of industrial applications in automotive, aerospace and medical industries, where they are mostly used for cutting, welding and cladding operations. Single-mode fibre lasers with moderate power levels in the range of tens to hundreds of watts are ideally suited to microwelding applications and for cutting thin materials. Fibre lasers with average power levels of a few hundred watts and peak power levels of a few kilowatts are very attractive for a host of industrial applications including cutting, welding, marking, engraving, sintering and soldering. These operations are discussed in detail in Chapter 11 on industrial applications of lasers.

Single-mode and low-order multimode fibre lasers with power levels of 1 kW or more have found wide acceptance in a range of materials processing applications due to their more reliable and consistent performance, low operating and maintenance costs and faster rate of materials processing. Fibre lasers are preferred by companies as a retrofit as the resulting increase in production is found to offset the additional costs involved in replacing existing CO_2 or Nd:YAG lasers.

Another major application of high-power fibre lasers is in directed-energy weapon applications in defence. High power at high beam quality, compactness, ruggedness, reliability and fibre delivery are the key features that give them an edge over chemical lasers and bulk solid-state lasers for military applications. Fibre lasers have already been used in a number of directed-energy weapon technology demonstrators for a range of applications including counter explosive devices, counter RAM (rocket artillery mortar) and counter UAV (unmanned aerial vehicle) applications. Directed-energy laser weapons for ballistic missile defence and anti-satellite applications have not been ruled out for the future. Military applications of lasers are described at length in Chapters 14 and 15.

3.10 Summary

- Based on the type of lasing medium used, the three major categories of lasers are solid-state lasers (this chapter), gas lasers (Chapter 4) and semiconductor lasers (Chapter 5). In addition, there are a large number of other varieties of laser such as dye lasers, excimer lasers, metal vapour lasers, free electron lasers, x-ray lasers and chemical lasers.
- Solid-state lasers are almost invariably optically pumped; gas lasers are excited by an electrical discharge while semiconductor lasers are pumped by an electrical current flowing through a forward-biased diode junction.
- Neodymium and chromium are the most widely used lasing species in solid-state lasers. While chromium is used in ruby, alexandrite and chromium-doped GSGG lasers, neodymium is used in Nd:YAG, Nd:Glass, Nd:YLF and Nd:YVO_4 lasers. Erbium is the lasing species for the eye-safe class of solid-state lasers. Titanium is used in titanium-sapphire lasers.
- In the case of solid-state lasers (although this is also true for any other laser), the characteristics of the host material are no less important than those of the lasing species. A good host material in the case of solid-state lasers should have optical, mechanical and thermal properties to favour the homogeneous propagation of light through it and thus a good beam quality, high average power operation and the ability to withstand severe operating conditions of practical laser systems. Hosts can either be of crystalline format (e.g. YAG, YLF, GGG, GSGG, YVO_4, sapphire and chrysoberyl) or glass (e.g. silicate and phosphate).
- Lasers can also be defined in terms of operation mode: CW output; free-running output, Q-switched output, cavity-dumped output and mode-locked output.
- In the case of CW output, the laser produces a continuous output as long as it is pumped. Free-running output is a quasi-CW mode of operation in which the laser operates in CW mode for a time period equal to or slightly longer than the storage time of the active medium, which in turn is of the same order as the pump input pulse length.
- Q-switching is a mechanism of producing short laser pulses of the order of a few nanoseconds, achieved by switching the quality factor of the resonator cavity from a low to a high value at the time instant when the population inversion density almost reaches its peak value.
- Cavity-dumping is a slight variation of the Q-switching process, where output pulse width is independent of the gain characteristics of the active medium and equals round-trip cavity transit time.
- Mode locking helps to achieve short pulse widths approaching the theoretical limit set by the bandwidth. The process forces different longitudinal modes to oscillate with a fixed phase relationship with respect to each other.
- In the case of common solid-state lasers, output wavelength is: 694.3 nm (R_1 line) and 692.9 nm (R_2 line) for ruby laser; 1064 nm for Nd:YAG and YVO_4 lasers; 1053 nm (ordinary) and 1047 nm (extraordinary) for Nd:YLF laser; 2940 nm for Er:YAG laser; 1540 nm for Er:Glass laser; 700–850 nm for Alexandrite laser; and 680–1180 nm for titanium-sapphire laser.
- A fibre laser is a type of solid-state laser where the gain medium is fibre doped with rare earth ions, rather than a rod or slab. Inherent in the fibre laser design and operational regimes are excellent performance characteristics which include a high level of immunity to misalignment, high beam quality, compactness and long-term stability. In addition to these, fibre lasers exhibit outstanding thermo-optical properties as a result of large surface-area-to-volume ratio.
- In its simplest form, a fibre laser comprises of fibre doped with rare earth element ions as the active medium, a fibre coupled semiconductor diode laser or another fibre laser as the pump source and dielectric mirrors or fibre Bragg gratings to form the resonant cavity.
- Relative to bulk solid-state lasers, fibre lasers exhibit higher wall-plug efficiency, higher beam quality at higher average output power levels and offer relatively higher gain bandwidth. While fibre lasers offer far superior performance when it comes to high average power lasers with high beam quality, bulk solid-state lasers have the edge in high pulse energy, high peak power applications.

- In comparison to the conventional step-index fibre that derives its wave-guiding properties from a spatially varying glass composition, the basic PCF is an optical fibre having a structured pattern of air holes or voids that run parallel to the axis all along its length.

Review Questions

3.1. Name some common lasing species and host materials in the case of solid-state lasers. What laser parameters or characteristics can the host material influence in solid-state lasers? Give examples in support of your answer.

3.2. With reference to solid-state lasers, distinguish between
a. continuous wave and free-running outputs
b. Q-switched and cavity-dumped outputs

3.3. What are the various pumping mechanisms used in the case of solid-state lasers? Why does diode-pumped solid-state laser have relatively much higher wall-plug efficiency than the flash-lamp-pumped counterpart?

3.4. Describe the vibronic class of solid-state lasers. What phenomenon makes this class of lasers tunable?

3.5. Briefly describe the reasons for the following:
a. Nd:Glass lasers cannot be operated at high repetition rates.
b. Nd:YLF lasers can generate output either at 1047 nm or at 1053 nm.
c. Nd:YVO_4 lasers are particularly attractive for laser diode pumping.
d. OPO-shifted Nd:YAG laser configuration is preferred over Erbium:Glass laser configuration for building high–repetition-rate laser rangefinders at the eye-safe wavelength of 1540 nm.

Problems

3.1. A laser has a bandwidth of 441 GHz. Determine the shortest theoretically possible mode-locked pulse width it can generate. What would be the repetition rate of the mode-locked pulses if the laser cavity were 30 cm long?
[1 ps, 500 MHz]

3.2. A certain Q-switched Nd:YAG laser has a cavity length of 15 cm and an output coupling mirror with reflectivity of 10%. If the refractive index of the active medium is taken as 1.8 at the operating wavelength, what is the expected Q-switched pulse width from this laser?
[2 ns]

Self-evaluation Exercise

Multiple-choice Questions

3.1. In order that a pulsed solid-state laser operates satisfactorily at a high pulse repetition rate, the host material used in the active medium should
a. have high thermal conductivity
b. have low thermal conductivity
c. be highly transparent to pump radiation
d. be birefringent

3.2. Which of the following solid-state laser host materials is particularly suitable for diode pumping:
a. yttrium aluminium garnet (YAG)
b. yttrium lithium fluoride (YLF)

c. yttrium vanadate (YVO_4)
d. phosphate glass.

3.3. Of the following, which is the lasing species used in the case of solid-state lasers?
a. neodymium
b. chromium
c. titanium
d. erbium
e. all the above.

3.4. Which of the following operational modes is likely to produce the shortest pulse width?
a. Q-switched
b. cavity-dumped
c. quasi-CW
d. mode locked.

3.5. One of the following produces output at 1053 nm:
a. Nd:YAG
b. Nd:YLF
c. Nd:YVO4
d. Er:YAG.

3.6. Which of the following lasing species is used in the laser material of a tunable solid-state laser?
a. titanium
b. erbium
c. neodymium
d. none of these.

3.7. Identify the three-level solid-state laser from the following:
a. Er:Glass
b. Nd:YLF
c. ruby
d. both (a) and (c).

3.8. Identify the false statement from the following:
a. Nd:YLF generates two laser transitions at 1047 nm and 1053 nm due to its birefringent nature.
b. Er:Glass laser cannot operate at higher repetition rates because of its poor thermal conductivity and the fact that it is a three-level laser.
c. Nd:YVO_4 is better suited to diode pumping because of its higher fluorescence lifetime.
d. The tunability of the vibronic class of solid-state lasers comes from their lower laser level being a band rather than discrete energy level.

3.9. Identify the correct match:
a. neodymium-doped silicate glass: 1054 nm
b. neodymium-doped phosphate glass: 1062 nm
c. erbium-doped YAG: 1540 nm
d. neodymium-doped YVO_4: 1064 nm.

3.10. One of the following is not the laser material for a tunable solid-state laser.
a. chromium-doped $BeAl_2O_4$
b. neodymium-doped $Gd_3Sc_2Ga_3O_{12}$ (GSGG)
c. titanium-doped Al_2O_3
d. lithium-doped KCl.

Answers

1. (a) 2. (c) 3. (e) 4. (d) 5. (b) 6. (a) 7. (d) 8. (c) 9. (d) 10. (b)

Bibliography

1. *Solid State Laser Engineering*, 2006 by W. Koechner, Springer.
2. *Solid State Lasers and Applications*, 2010 by Alphan Sennaroglu, CRC Press.
3. *Solid State Lasers: A Graduate Text*, 2003 by Walter Koechner and Michael Bass, Springer-Verlag.
4. *Advanced Solid State Lasers*, 1976 by Walter R. Bosenberg and Martin Michael Fejer, Optical Society of America.
5. *Solid State Lasers: Properties and Applications*, 2008 by Thomas O. Hardwell, Nova Publishers.
6. *Solid State Lasers and Applications*, 1991 by Brian J. Thompson, Taylor and Francis Group, LLC.
7. *Handbook of Solid State Lasers*, 1989 by Peter K. Cheo, M. Dekker.
8. *Introduction to Laser Diode-Pumped Solid State Lasers*, 2002 by Richard Scheps, SPIE Press.
9. *The Laser Guidebook*, 1999 by Jeff Hecht, McGraw Hill.
10. *Principles of Lasers*, 2009 by Orazio Svelto, Plenum Press.
11. *Handbook of Laser Technology and Applications*, 2003 Volume II by Collin E. Webb and Julian D. C. Jones, Institute of Physics Publishing.

4

Gas Lasers

4.1 Introduction to Gas Lasers

In a continuation of our discussion on types of lasers, this chapter is on gas lasers. Gas lasers covered in this chapter include helium-neon lasers, carbon dioxide lasers, metal vapour lasers, rare gas ion lasers, excimer lasers, chemical lasers and gas dynamic carbon dioxide lasers. Dye lasers, free-electron lasers and x-ray lasers are also covered towards the end of the chapter as these do not fit either of the three main categories (solid-state, gas and semiconductor diode lasers). However, a separate chapter on these three types of laser would have not been justified. Again, the emphasis is on operational fundamentals, salient features and typical applications of these lasers.

The family of gas lasers, solid-state and semiconductor diode lasers, has widely varying characteristics including the wavelength range, power levels and to some extent even the pumping mechanism. The available output powers from gas lasers vary from a fraction of a milliwatt in a small helium-neon laser used for optical alignment to a megawatt level in a gigantic high-power chemical laser used as a weapon. The wavelength range also spans almost the entire optical spectrum from ultraviolet to far-infrared with thousands of laser wavelengths discovered throughout the region. Another significant feature that is all gas lasers have Doppler-broadened gain versus frequency curve. The practical implication of this is that it allows more than one longitudinal mode to oscillate simultaneously in the laser cavity, unless special measures are taken for mode selection. As an illustration, a helium-neon laser emitting at 632.8 nm has a Doppler-broadened gain curve with FWHM (full width at half maximum) of 1400 MHz. One longitudinal mode in this case typically has 1 MHz bandwidth with the result that the gain curve may accommodate three longitudinal modes assuming a 500 MHz inter-mode spacing, which is the case for a 30-cm-long laser cavity. Doppler broadening occurs in gas lasers due to the random motion of atoms. Such a broadening mechanism is however absent from solid-state and semiconductor diode lasers, as the atoms are fixed in these lasers.

4.1.1 The Active Media

The active medium in the case of a gas laser is almost invariably a mixture of more than one gas, with gases other than the actual lasing species performing certain subtle functions such as: acting as an intermediate step during transfer of energy from pump source to lasing species (helium in helium-neon lasers and nitrogen in carbon dioxide lasers); assisting in heat transfer (helium in carbon dioxide lasers) and depopulating the lower lasing level (helium in helium-neon laser).

The gas mixture is contained in a tube and held at a pressure that again depends upon a number of parameters. Low pressure of the order of a small fraction of atmospheric pressure, which is suitable for stable discharge for longer periods, is mostly utilized in continuous wave (CW) lasers. In the case of pulsed lasers, where the discharge stability is required for a shorter period, the laser gas mixture could be

Lasers and Optoelectronics: Fundamentals, Devices and Applications, First Edition. Anil K. Maini.

Table 4.1 Popular gas laser types

Laser type	Wavelength (nm)	Transition	Power Range (W)[a]	Mode
Argon fluoride excimer	193	Electronic	1–5	Pulsed
Krypton fluoride excimer	249	Electronic	1–100	Pulsed
Argon-ion	275–305	Electronic	0.001–2	CW
Argon-ion	333–364	Electronic	0.001–7	CW
Krypton-ion	335–360	Electronic	0.001–2	CW
Krypton-ion	406–416	Electronic	0.001–3	CW
Krypton-ion	647	Electronic	0.001–7	CW
Helium-cadmium	325	Electronic	0.002–0.01	CW
Helium-cadmium	354	Electronic	0.001–0.02	CW
Helium-cadmium	442	Electronic	0.001–0.1	CW
Copper vapour	510. 578	Electronic	1–120	Pulsed
Gold vapour	628	Electronic	1–10	Pulsed
Helium-neon	543	Electronic	0.0001–0.001	CW
Helium-neon	632.8	Electronic	0.0001–0.05	CW
Helium-neon	1153	Electronic	0.001–0.015	CW
Hydrogen fluoride	2600–3000[b]	Vibrational	MW	Pulsed or CW
Deuterium fluoride	3600–4000[b]	Vibrational	MW	Pulsed or CW
Carbon dioxide	9000–11 000[b]	Vibrational	Watts to hundreds of kilowatts	Pulsed or CW
Carbon monoxide	5000–6500[b]	Vibrational	—	Pulsed or CW

[a]In case of pulsed lasers, it is the average power.
[b]There are many discrete wavelengths in the given range.

filled at a pressure close to atmospheric pressure and sometimes at >1 atm. The optimum gas pressure for lasers of a given type also depends upon the laser design.

The active media of different gas lasers may not be in the same form. For example, in the case of argon-ion and krypton-ion lasers, the active medium is the ionized atoms of the rare gases argon and krypton, respectively. In the case of metal vapour lasers such as copper vapour and gold vapour lasers, the active medium is the hot metal vapour. In the helium-cadmium laser, the metal vapour is also ionized.

4.1.2 Inter-level Transitions

In the case of most gas lasers, inter-level transitions are electronic except for carbon dioxide (CO_2) lasers, hydrogen fluoride (HF) lasers, deuterium fluoride (DF) lasers and carbon monoxide (CO) lasers that involve vibrational transitions. Some far-infrared lasers producing wavelengths greater than 30 μm have vibrational or rotational transitions. Table 4.1 lists some of the popular gas lasers along with the type of transition, corresponding wavelength, mode of operation (pulsed or CW) and typical power level range.

4.1.3 Pumping Mechanism

Most of the gas lasers are excited by electrical discharge as shown in the generalized arrangement of a gas laser in Figure 4.1. The active medium is usually excited either by passing an electric discharge current along the length of the tube known as longitudinal excitation (Figure 4.1), or by an electric discharge perpendicular to the length of the laser tube known as transverse excitation (Figure 4.2). The former is used in relatively low-power CW lasers while the latter is employed in high-power pulsed or CW lasers.

In the case of CW lasers, a high DC voltage is initially required to ionize the gas. Once the ionization has taken place, the DC voltage is brought to a much lower value needed to sustain the plasma. In the case of a pulsed laser, a large capacitor is charged to the required DC voltage and then made to discharge through the laser medium.

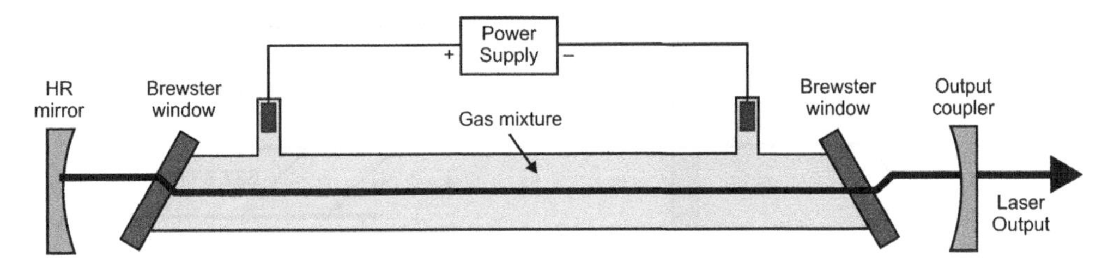

Figure 4.1 Generic gas laser: longitudinal excitation.

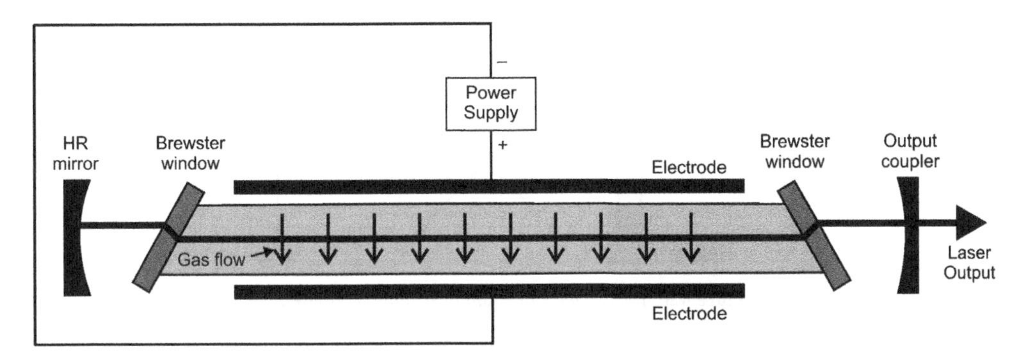

Figure 4.2 Generic gas laser: transverse excitation.

Some gas lasers such as those generating far-infrared wavelengths are optically pumped. Another gas laser of shorter wavelength usually pumps these lasers. A CO_2 laser may be used to pump such a laser.

We briefly describe the two most commonly used gas lasers in Sections 4.2 and 4.3: helium-neon (He-Ne) and CO_2 lasers. Sections 4.4–4.8 are dedicated to the discussion of other varieties of gas lasers which are important although less common, and Section 4.9 provides information on dye, free-electron and x-ray lasers, which do not belong to any of the three main types of laser (solid-state, gas or semiconductor).

4.2 Helium-neon Lasers

The helium-neon laser is one of the most commonly used types of gas lasers. Although we are more familiar with the red beam from a He-Ne laser at 632.8 nm, lasing action is also possible at the infrared wavelengths of 1.153 μm and 3.391 μm and the visible wavelength of 543.5 nm. In fact, first successful operation of a CW laser was achieved in a He-Ne laser at 3.391 μm by Javan, Benett and Herriott at Bell Labs following the first lasing action demonstrated by Maiman in a ruby laser. Incidentally, He-Ne lasers are usually identified by red output; the lasing transition in He-Ne that that has the highest gain is at 3.391 μm.

Figure 4.3 depicts the energy level structure of a He-Ne laser. The active medium is the laser gas mixture of helium and neon, which is predominantly helium with only 10–20% neon. The gas mixture is excited by an electrical discharge. Initially, a high voltage of the order of 10 kV ionizes the gas mixture. The voltage is then brought down to about 2 kV or so, which is sufficient to sustain the discharge by supplying a few milli-amperes of current. The free electrons in the plasma excite helium atoms to higher energy levels. The excited helium atoms then transfer their energy to neon atoms, which have energy levels (4s and 5s levels of neon) coincident with excited helium levels (2s levels of helium) as shown in the figure. The transfer of energy is through a collisional process. The 4s and 5s levels of neon are metastable levels having a relatively much larger lifetime, which facilitates laser action. The metastable lifetime here is of the order 1 ms and the laser action takes place between metastable levels (4s or 5s of neon) and the lower lasing levels (3p or 4p of neon). These combinations of numerals and letters (s, p, d,

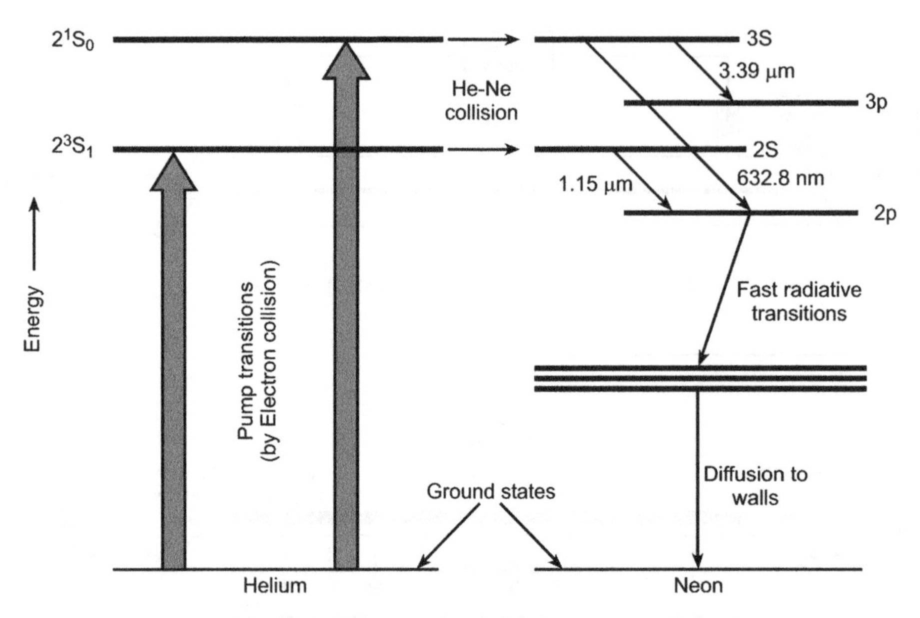

Figure 4.3 Energy level diagram of helium-neon laser.

etc.) are spectroscopic notations indicating electronic energy levels. In ascending order of energy quantum, these energy levels are designated 1s, 2s, 2p, 3s, 3p, 4s, 3d, 4p, 5s, etc.

A helium-neon laser has a low gain; cavity losses therefore need to be very low to make the laser action possible. Efficiency is also very low in the range of 0.01–0.1% due to the involved lasing levels being far above the ground state, as is clear from the energy level diagram of Figure 4.3.

A helium-neon laser has a Doppler-broadened gain curve which is about 1400 MHz at 632.8 nm (Figure 4.4); recall that Doppler broadening is proportional to frequency. Superimposed on the Doppler-broadened gain curve is the cavity resonance function with inter-longitudinal mode spacing given by $\Delta D = c/2L$, L being the length of the cavity. A short cavity, for example 15 cm long, would support only a single longitudinal mode as shown in Figure 4.4; in that case, the inter-mode spacing would be 1000 MHz. A longer cavity opens up the possibility of more than one longitudinal mode being sustained simultaneously in the cavity. The uncertainty in frequency of single longitudinal mode would be dictated by the Doppler-broadened line width due to temperature changes causing cavity length variation. The same phenomenon causes power fluctuation in the output beam. However, the single longitudinal mode could be forced to occur at the centre of the gain curve and stay there, irrespective of temperature and other factors causing cavity length variation, by employing what is known as active means of frequency stabilization. These are discussed at length in Chapter 8 on gas laser electronics.

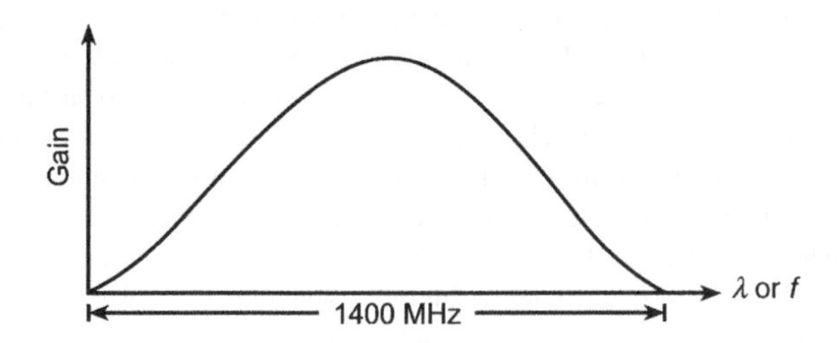

Figure 4.4 Doppler-broadened gain curve of helium-neon laser at 632.8 nm.

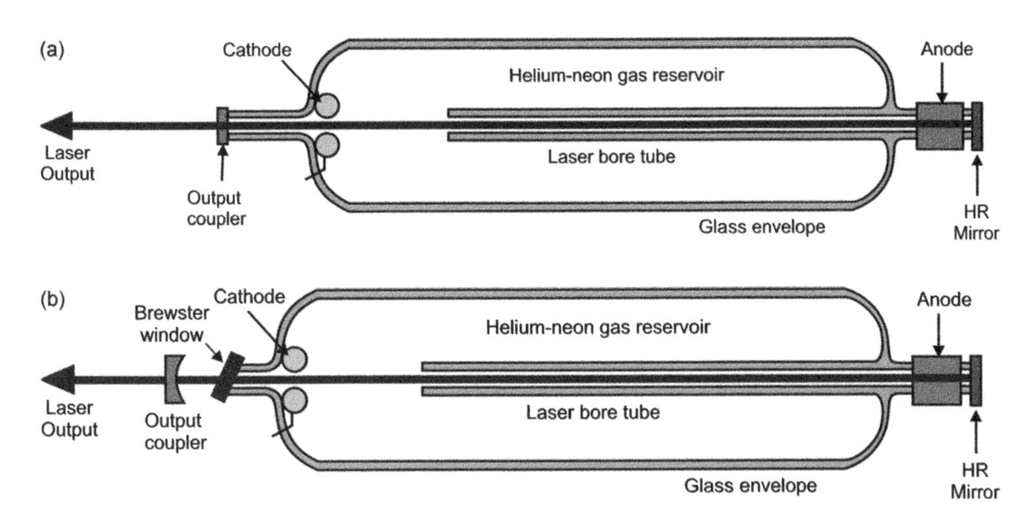

Figure 4.5 Helium-neon laser construction: (a) both mirrors sealed; and (b) Brewster window sealed.

Figure 4.5a and b show the constructional features of a typical helium-neon laser. In one of the geometries (Figure 4.5a), mirrors are directly bonded to the two ends of the tube by a high-temperature process called hard sealing. This minimizes helium leak, which is vital for the long life of the laser. One of the disadvantages of direct bonding of laser mirrors is that it exposes the mirror coatings to the discharge, which could adversely affect the life of the laser. This could however be overcome by using mirrors with high damage threshold and an improved laser geometry. In another type as shown in Figure 4.5b, one of the ends of the tube is sealed with a Brewster window which practically eliminates any reflection losses for the plane polarized light. In plane polarized light, the plane of polarization is parallel to the plane of incidence. The other mirror component, output coupler or the high reflectivity mirror is placed outside the tube in this case, and may or may not be part of the overall housing. This proposition is more expensive but is definitely a better option for relatively high-power helium-neon lasers producing output power in the range of 10–60 mW. It also allows the selection of different wavelengths. Lower-power lasers invariably employ the former type of construction. Figure 4.6 depicts a 2.5 mW He-Ne laser tube from Melles Griot.

Regarding helium-neon laser tube construction, the mirrors need to have low losses to compensate for the low gain of the laser. One of the mirrors is 100% reflective. The output mirror transmits a small percentage of irradiance inside the cavity and reflects most of it back into the laser cavity. The output mirror has a transmission of typically 2%. Another notable feature of laser tube construction is that the electric discharge applied to the ends of the tube is concentrated in a narrow bore of a few millimetres in diameter. This raises laser excitation efficiency and improves beam quality. The rest of the tube volume acts as a gas reservoir and contains extra helium and neon. Gas pressure within the tube is about 2–3 torr, which is 0.3–0.4% of atmospheric pressure.

The power output of a helium-neon laser depends upon several parameters, predominantly tube length, gas pressure and the discharge bore diameter. The figure-of-merit is defined as the product of gas pressure (in torr) and bore diameter (in mm), which has an optimum value of 3.5–4. That is, the output power for a given tube length is highest when the figure-of-merit equals approximately 3.5 to 4. As an illustration, helium-neon laser tubes with bore diameter in the range of 1.0–1.5 mm have gas fill pressure of the order of 3.0–4.0 torr.

He-Ne lasers are available both as sealed-off tubes as well as packaged types. The package types are also available as lasers with a separate power supply as well as self-contained lasers for original equipment manufacturer (OEM) applications. Figure 4.7 depicts different types of He-Ne laser including self-contained lasers (Figure 4.7a) and those available with separate power supplies (Figure 4.7b). He-Ne

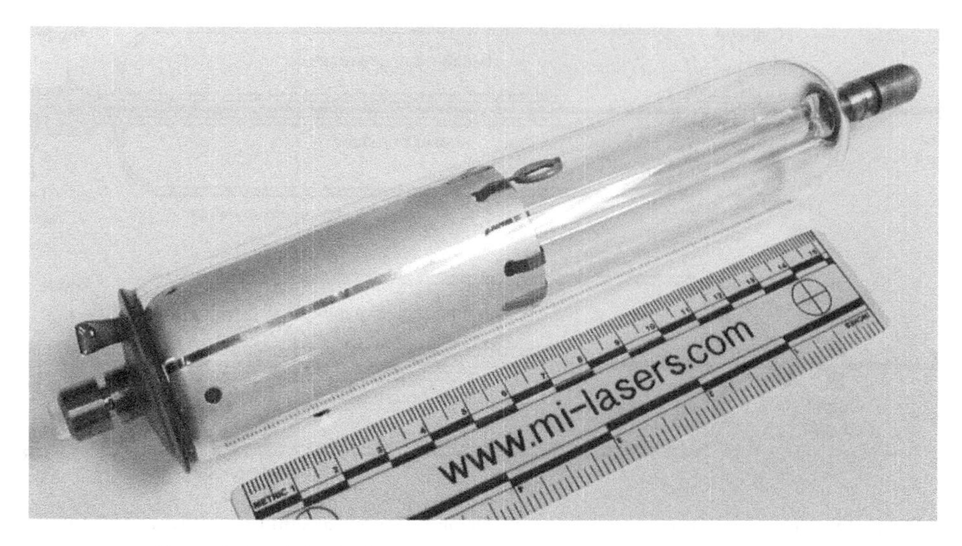

Figure 4.6 2.5 mW He-Ne laser tube, Model P-122 (Courtesy Melles Griot).

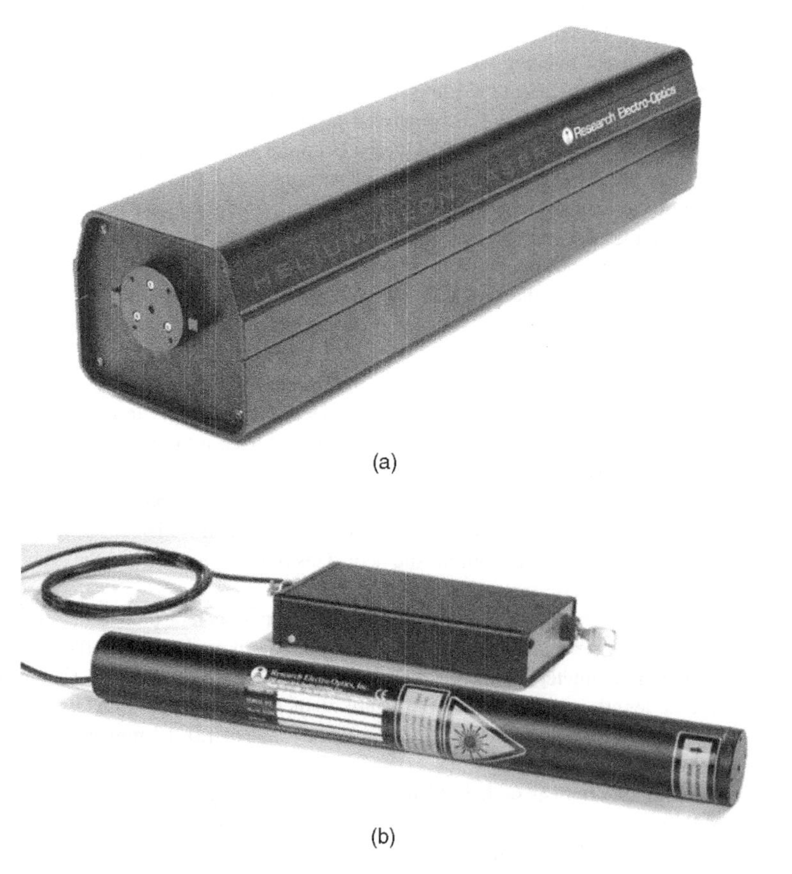

(a)

(b)

Figure 4.7 Helium-neon lasers (Courtesy Newport Corporation).

lasers are increasingly being replaced by semiconductor diode lasers, particularly in applications where the high-beam quality of a helium-neon laser is not a necessity.

4.3 Carbon Dioxide Lasers

Carbon dioxide (CO_2) lasers are the most widely used and diversely exploited type of gas laser. The CO_2 laser has a myriad of applications in industry, medical diagnosis and treatment, science and technology and warfare.

The laser medium in this case is a gas mixture of carbon dioxide (CO_2), helium (He) and nitrogen (N_2). CO_2 is the lasing species and laser transitions correspond to the energy levels associated with molecular vibration and rotation modes of the CO_2 molecule. Nitrogen participates in the process of creation of population inversion by acting as an intermediate step in the same way as helium does in the case of a He-Ne laser. In the case of a CO_2 laser, helium helps in depopulating the lower laser level.

Unlike gas lasers such as He-Ne lasers, the energy levels responsible for laser action in a CO_2 laser do not correspond to excitation and de-excitation of electrons; instead, they correspond to vibrational/rotational levels of CO_2 molecule. The CO_2 molecule has three types of vibrations which are (in descending order of energy level): asymmetric stretching, symmetric stretching and bending (Figure 4.8). The free electrons in the gas-discharge plasma transfer their energy very efficiently to N_2 molecules, which raise themselves to an appropriate level coincident with CO_2 energy level corresponding to asymmetric stretching. Figure 4.9 shows the energy level diagram in the case of a CO_2 laser. The N_2 molecules transfer the energy to CO_2 molecules, which is the upper lasing level. Laser transitions correspond to the CO_2 molecules dropping from higher-energy asymmetric stretching mode to the lower-energy symmetric stretching or bending modes. Transition to a lower level corresponding to symmetric stretching produces 10.6 μm output while its transition to another lower level corresponding to symmetric bending produces 9.6 μm output.

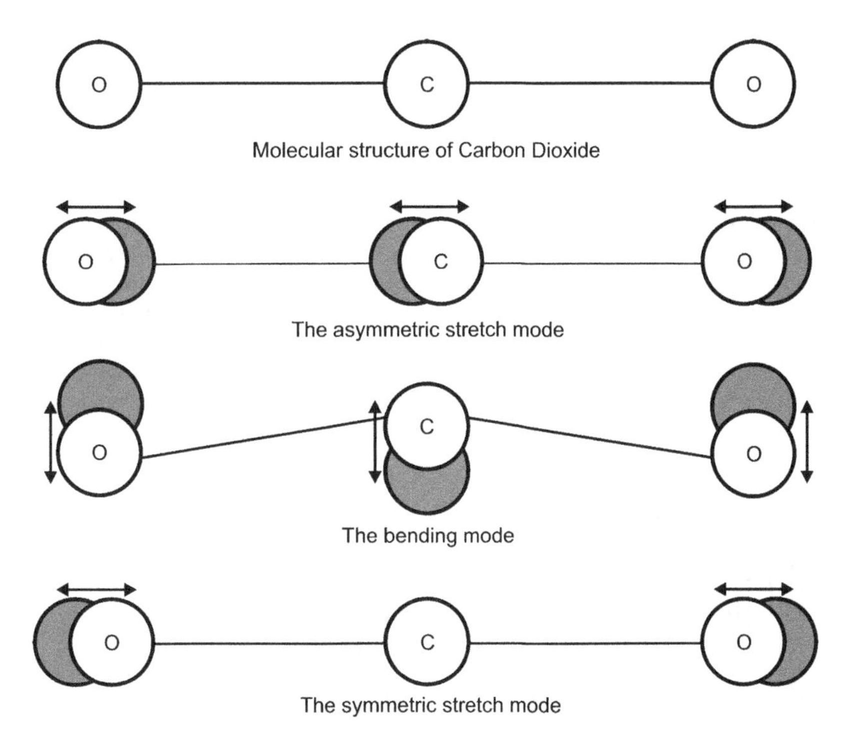

Figure 4.8 Vibrational modes in CO_2 molecule.

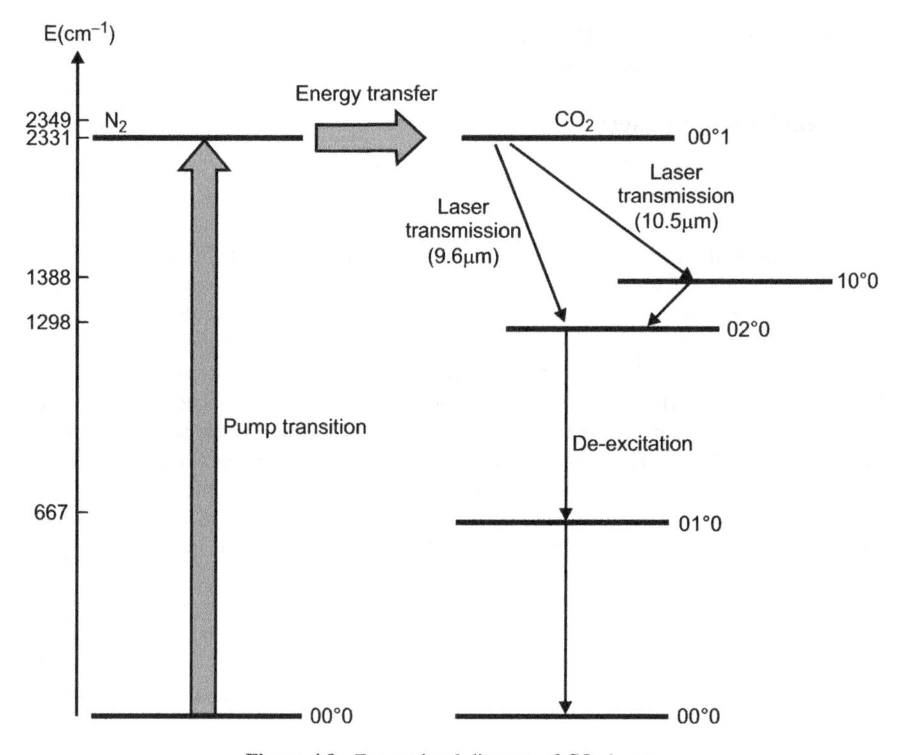

Figure 4.9 Energy level diagram of CO_2 laser.

In fact, around both 9.6 μm and 10.6 μm outputs, there are a large number of closely spaced lines as shown in Figure 4.10. This is the outcome of rotational motion of the molecule. The rotational transition energy is however much smaller than either the thermal or the vibrational transition energy. The molecule can either speed up or slow down a little in its rotational motion while moving between the vibrational levels. When it speeds up it takes energy from the vibrational transition, and when it slows down it gives energy to the vibrational transition. In the two cases, the effective transition energy either increases (in the case of slowing down of rotational motion) or decreases (in the case of speeding up of rotational motion), thus changing the emitted wavelength around the nominal value.

A carbon dioxide laser also has a Doppler-broadened gain curve that is 60 MHz wide at 10.6 μm output (Figure 4.11). The frequency uncertainty in the case of an unstabilized laser could be as much as 60 MHz. The laser could however be frequency stabilized either on the centre of gain curve, or anywhere on the gain curve using some of the established frequency stabilization techniques to an accuracy of

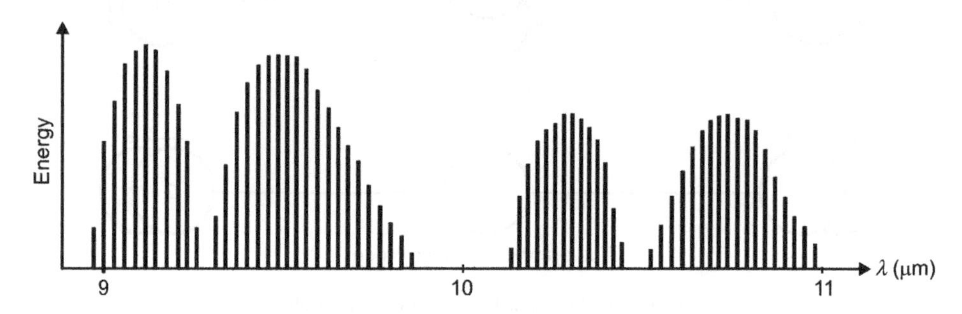

Figure 4.10 Different wavelengths emitted by CO_2 laser.

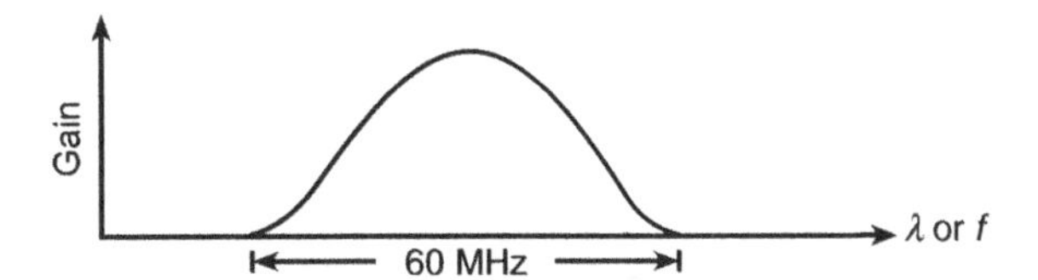

Figure 4.11 Doppler-broadened gain curve of CO_2 laser.

better than ±100 kHz. Some techniques allow the laser to be stabilized down to ±10 kHz, making it an ideal tool for carrying out Doppler-free spectroscopic studies. The frequency stabilization techniques are discussed at length in Chapter 8 on gas laser electronics.

Carbon dioxide lasers employ either a sealed tube or flowing gas construction. The gas mixture is excited by an electric discharge. Both direct current (DC) as well as radio-frequency (RF) excitation has been successfully and widely used in the case of CO_2 lasers. As the electric discharge breaks down CO_2 molecules to form carbon monoxide and oxygen, a catalyst is added to regenerate CO_2. Figure 4.12 depicts a sealed-off CO_2 laser of 48-series from Synrad Inc. The laser employs robust sealed-off construction. Three different models in the series produce CW output power levels of 10, 25 and 50 W with output beam divergence of 4 mrad and mode quality factor $M^2 < 1.2$.

The flowing-gas type of CO_2 lasers may be of longitudinal flow or transversal flow. In the former case the gas flows along the axis of the cavity, while in the latter case it flows perpendicular to the cavity axis. In longitudinal flow lasers (Figure 4.13) the gas pressure is low, the output power is also relatively lower and lasers operate as CW lasers. Transverse flow lasers (Figure 4.14) are normally used to obtain higher power outputs. Gas pressures in such lasers can be higher. In the case of longitudinal flow lasers, the gas is usually recycled with some fresh gas added. In the case of transverse flow lasers, the gas flow through the laser cavity is much faster, which removes waste heat and also the contaminants.

Pressure broadening of the line width also allows for mode-locked operation, thus enabling the generation of pulses as short as 1 ns. For pressures approaching 10 atm, the line broadening is large enough to allow near-continuous tuning across most of the wavelength range. Transverse flow is usually used in relatively higher output power lasers. While power in excess of 50 W per metre of gain length is achievable in the case of sealed-off and longitudinal flow carbon dioxide lasers, as much as 10 kW per metre of gain length is achievable in the case of transverse flow lasers.

Another type of CO_2 laser is one that is transversely excited and where the gas pressure is about 1 atm. The CO_2 lasers described above have low gas pressures and usually produce CW output. CW lasers

Figure 4.12 Sealed-off CO_2 laser (Courtesy of Synrad Inc.).

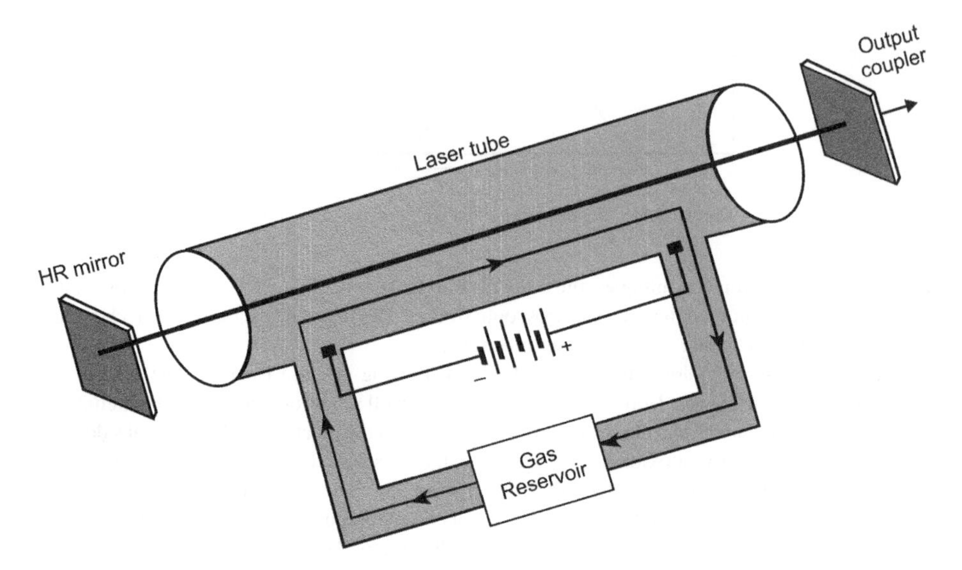

Figure 4.13 Longitudinal gas flow CO_2 laser.

cannot have high-pressure gas mixture as it is not practical to have a stable continuous discharge at pressures above about 0.1 atm. Pulsed electric discharge is however possible at higher pressures and works very well if the discharge is transverse to the laser axis; this transversely excited atmospheric pressure (TEA) CO_2 laser (Figure 4.15) is a high-power pulsed CO_2 laser. The gas pressure is around 1 atm and the discharge is transverse to the laser axis. TEA CO_2 lasers invariably use a form of pre-ionization to uniformly ionize the space between the electrodes. The primary attraction of a TEA CO_2

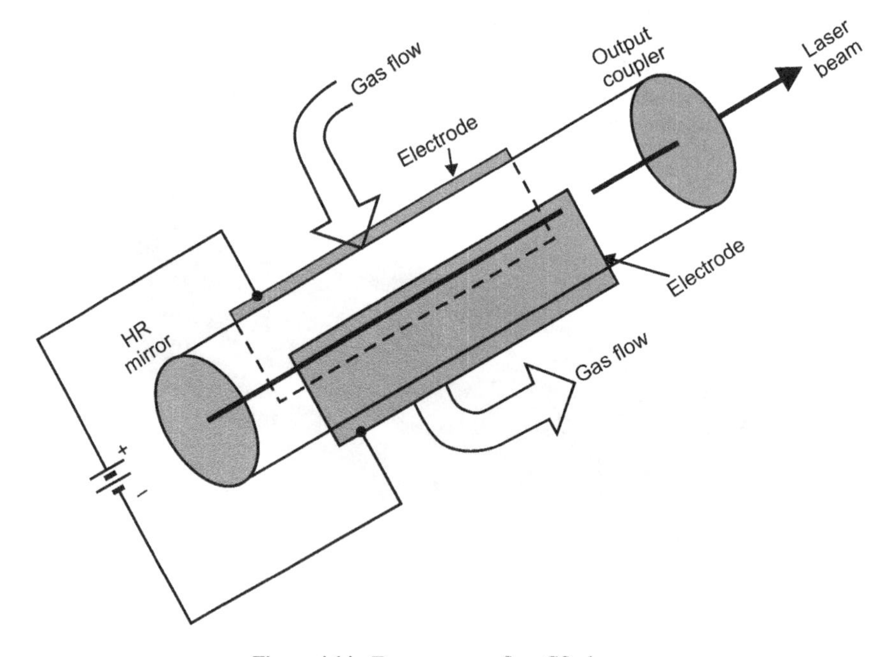

Figure 4.14 Transverse gas flow CO_2 lasers.

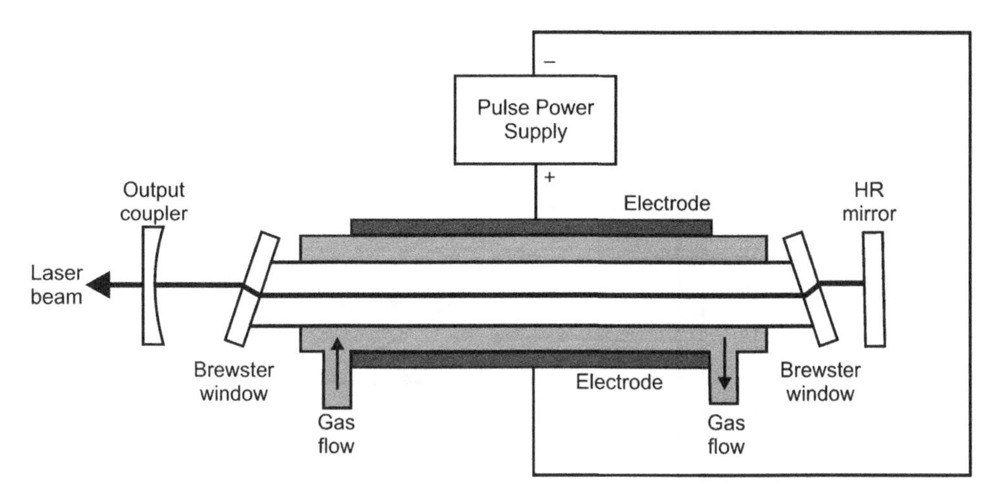

Figure 4.15 TEA CO_2 laser.

Figure 4.16 RF-excited waveguide carbon dioxide laser (Courtesy of Synrad Inc.).

laser is its ability to generate short intense pulses and extraction of high power per unit volume of laser gas mixture. Pulse durations of a few tens of nanoseconds to few microseconds and pulse energies from a few millijoules to hundreds of joules at repetition rates of up to a few hundred hertz are achievable.

Waveguide CO_2 laser is another type of CO_2 laser that renders compactness. RF-excited sealed-off waveguide CO_2 lasers produce several watts to several tens of watts of CW output, demonstrating great promise. Figure 4.16 depicts one such sealed-off RF-excited waveguide CO_2 laser, Model No. v30 from Synrad Inc. The laser produces 30 W of CW output with a beam divergence and beam quality factor M^2 better than 7 mrad and 1.2, respectively.

4.4 Metal Vapour Lasers

The active medium in the case of metal vapour lasers is either ionized metal atoms in the helium-cadmium family of lasers or neutral atoms in copper vapour and gold vapour lasers. The helium-cadmium lasers produce emissions at 325 nm and 353.6 nm (ultraviolet) and 441.6 nm (blue). The neutral metal vapour lasers emit in the visible spectrum. Copper vapour lasers emit at 511 nm (green) and 578 nm (yellow). Gold vapour laser emits at 628 nm (red). Pumping takes place through electric discharge.

4.4.1 Helium-cadmium Laser

In the case of helium-cadmium lasers, electrons in the discharge plasma excite helium atoms to higher energy levels. Excited helium atoms transfer their energy to cadmium atoms, ionizing them and raising them to upper laser levels in the process. It is much easier to ionize cadmium than helium, due to the former having only two electrons in its incompletely filled outer shell and the latter having a filled outer shell.

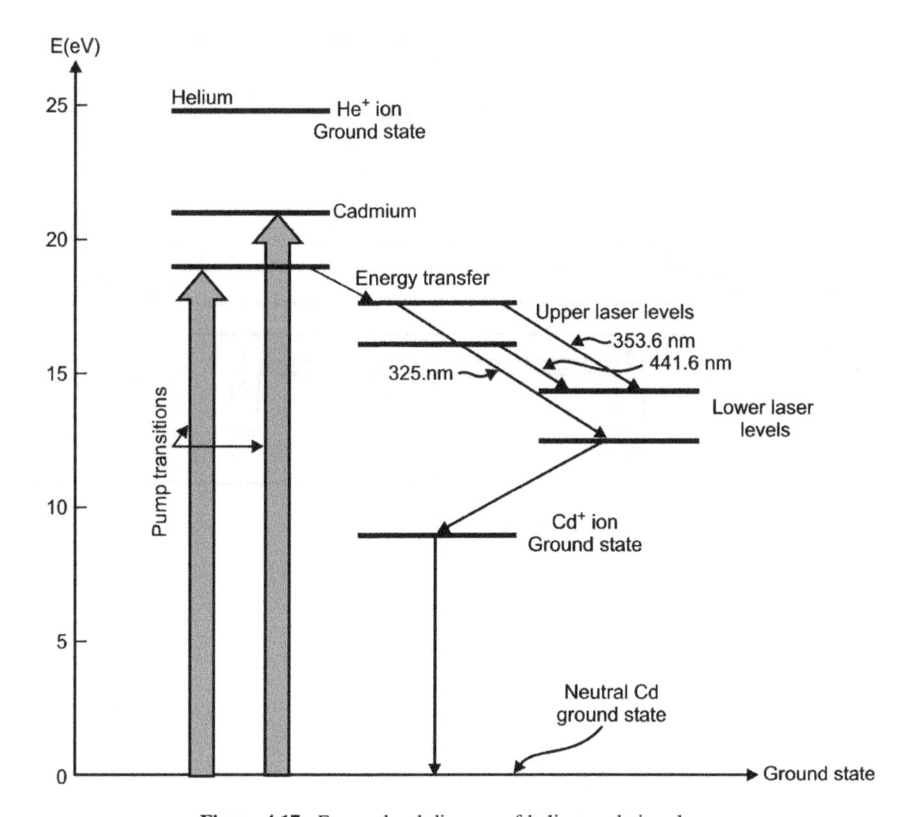

Figure 4.17 Energy level diagram of helium-cadmium laser.

Figure 4.17 depicts the energy level diagram of a helium-cadmium laser. As shown in the diagram, cadmium atoms are trapped in two metastable states. One of them produces the two ultraviolet lines with the two lower laser levels and the other metastable state produces the blue line with one of the lower laser levels.

Cadmium is in solid form at room temperature and is heated to about 250 °C to vaporize it to a pressure of the order of a few milli-torr required for laser operation. Helium pressure is of the order of a few torr, about a thousand times higher than that of cadmium. Helium-cadmium tubes usually have a helium gas reservoir to replace helium gas that may leak out. Cadmium vapours may condense on relatively cooler parts of the tube during laser operation. The tube is designed in order to have certain portions that can collect surplus cadmium so that it does not deposit on the optical surfaces.

As for helium-neon lasers, helium-cadmium lasers have one of two optical designs: either resonator mirrors sealed at the ends of the discharge tube or the tube sealed with Brewster windows and the resonator mirrors placed outside the tube. The emitted wavelength depends upon the chosen optics. Figure 4.18 depicts a helium-cadmium laser of the 74-series from Melles Griot. The 74-series lasers are capable of producing up to 55 mW of CW output at 355 nm and up to 130 mW of CW output at 442 nm. The single transverse mode lasers in this series offer beam divergence of 1.5 mrad and beam quality factor $M^2 < 1.2$.

4.4.2 Copper Vapour and Gold Vapour Lasers

Important members of the family of metal vapour lasers where the lasing species is the neutral atom vapour are the copper vapour and gold vapour lasers. As mentioned above, copper vapour lasers have two emission lines at 511 nm and 578 nm and the gold vapour laser emits at 628 nm. Copper vapour and gold vapour lasers are pulsed lasers with typical pulse repetition rates in the range of several kilohertz and achievable average power of the order of tens of watts.

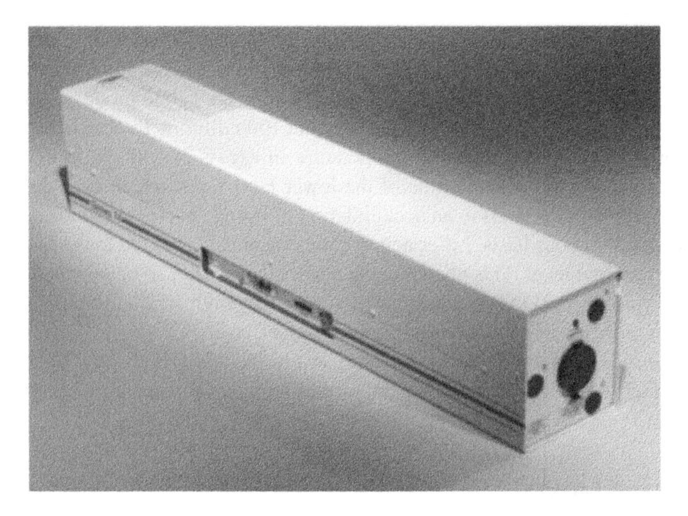

Figure 4.18 Helium-cadmium laser (Courtesy of Melles Griot).

Figure 4.19 depicts the energy level diagram of a copper vapour laser; a gold vapour laser has a similar energy level structure. The energy level structure of a copper vapour laser can be used to explain why such a laser cannot be used to produce a CW output; the description is equally valid for a gold vapour laser. When an electric discharge passes through the metal vapour, electrons in the discharge plasma collide with the metal vapour atoms (copper vapour atoms with reference to the energy level diagram of

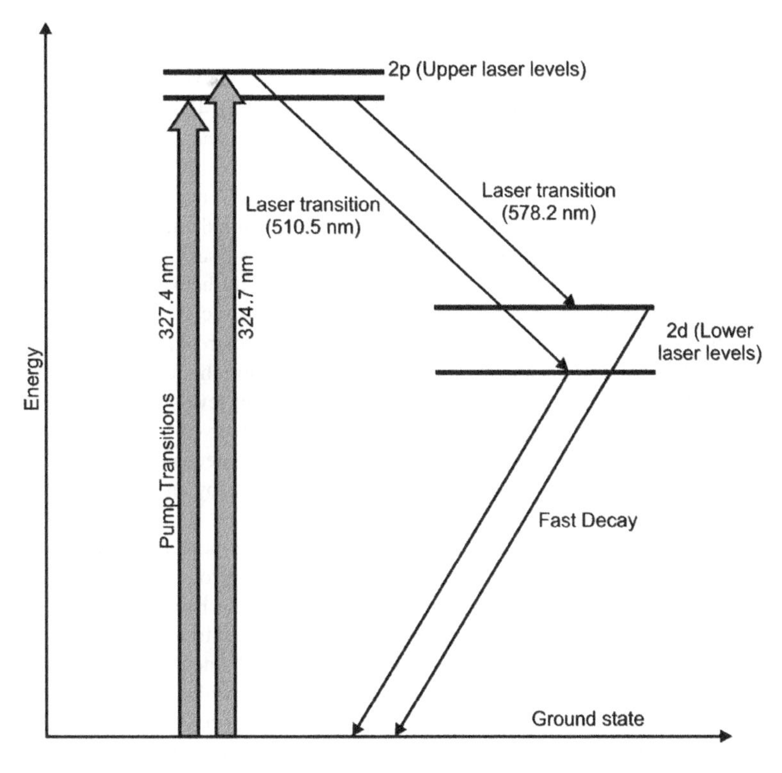

Figure 4.19 Energy level diagram of copper vapour laser.

Figure 4.19) and raise them to an excited state. The excited state has a lifetime of the order of 10 ns or so if the metal vapour pressure is low. This small lifetime is not long enough to produce laser action. The effective lifetime can be increased to about 10 ms (which is long enough to produce stimulated emission and laser action) by increasing the vapour pressure to about 100 millitorr. To obtain this vapour pressure, metallic copper or gold must be heated to a temperature in excess of 1500 °C in the laser tube.

The problem begins with the long lifetime of the lower laser level, which is of the order of tens to hundreds of microseconds. This leads to an accumulation of lasing species at the lower level, inhibiting the laser action on a continuous basis. Laser action can be resumed after the lower laser level becomes depopulated to the required level. This phenomenon allows these lasers to be operated only on a pulsed basis, although the pulse repetition frequency can be sufficiently high. The addition of a rare gas such as helium, neon or argon helps in depopulating the lower laser level.

Some of the significant features of metal vapour lasers include a high gain of the laser medium (10–30% per cm), high laser efficiency (typically several tenths of a percent) and high average power (approaching 100 W).

4.5 Rare Gas Ion Lasers

Argon-ion and krypton-ion lasers are popular members of the family of rare gas ion lasers. The laser emission in case of argon-ion and krypton-ion lasers occurs from the transitions made by ionized argon and krypton atoms. These lasers emit a range of wavelengths from ultraviolet through the entire visible spectrum. Shorter wavelengths (< 400 nm) are emitted from doubly ionized argon (Ar^{+2}) and krypton (Kr^{+2}) atoms. Longer wavelengths are emitted from singly ionized argon (Ar^{+}) and krypton (Kr^{+}) atoms. These lasers can generate output powers from a few milliwatts to a few tens of watts, much higher than their CW gas laser counterparts such as helium-neon and helium-cadmium lasers.

Although both argon-ion and krypton-ion lasers emit a range of wavelengths from ultraviolet through visible to near-infrared, argon-ion laser has strong emission at 488 nm (blue) and 514.5 nm (green). A krypton-ion laser emits strongly at 647.1 nm (red). Possible emissions from argon-ion lasers include 275.4 nm, 300.3 nm, 302.4 nm, 305.5 nm, 334 nm, 351.1 nm, 363.8 nm, 454.6 nm, 457.9 nm, 465.8 nm, 472.7 nm, 476.5 nm, 488 nm, 496.5 nm, 501.7 nm, 514.5 nm, 528.7 nm and 1090 nm. In case of krypton-ion lasers, possible emission lines are 337.4 nm, 350.7 nm, 356.4 nm, 406.7 nm, 413.1 nm, 415.4 nm, 468 nm, 476.2 nm, 482.5 nm, 520.8 nm, 530.9 nm, 568.2 nm, 647.1 nm, 676.4 nm, 752.5 nm and 799.3 nm. The two gases can be put into a single tube and excited simultaneously to obtain laser lines throughout the visible spectrum. Such mixed-gas lasers are used in laser light shows.

In the case of rare gas ion lasers, laser action can be explained with the help of the energy level diagram of an argon-ion laser, as shown in Figure 4.20. A similar energy level structure also exists for krypton-ion lasers. A high initial voltage ionizes the gas. Electrons in the electric current passing through the ionized gas transfer energy directly to argon atoms, ionizing them and raising the resulting ions to a cluster of high energy levels. Ions drop from populated metastable upper laser levels to a cluster of lower laser levels. The emission wavelength depends upon the pair of levels involved in the transition. In fact, if allowed by optics, several different wavelengths can be made to oscillate simultaneously. The ions drop from the lower laser level with very short lifetime to ion ground state by emitting light at the extreme ultraviolet wavelength of 74 nm.

Construction of these lasers is not very different from that of helium-neon lasers. These lasers usually have Brewster window sealed plasma tubes with external cavity mirrors. The gas pressure is about 0.7–0.8 torr. Wavelength selection optics can be inserted between the Brewster window and the rear mirror, if so desired. As for helium-neon lasers, the discharge is confined to a narrow bore diameter in the centre of the tube. A series of metal disks with central holes may also define the bore in these lasers. A gas reservoir is also a part of the structure to replace the depleted gas. Wall-plug efficiency of these lasers is in the range of 0.01–0.001%. Because of lower efficiency, high-output-power rare ion lasers need to be cooled, either by forced air cooling (in case of medium-power lasers) or by liquid cooling (in case of higher-output-power lasers.

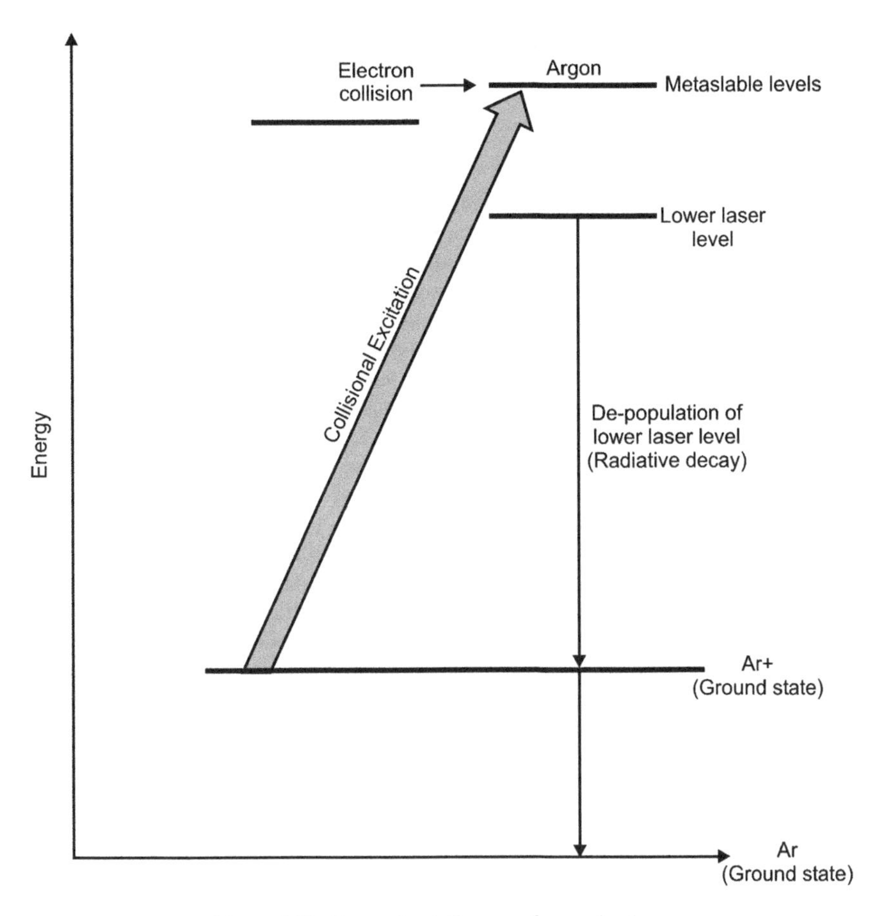

Figure 4.20 Energy level diagram of argon-ion laser.

Figure 4.21 depicts an argon-ion laser of Innova-90C series from Coherent Inc. The laser is available in the output power range of 3–6 W predominantly at 488 nm and 514.5 nm wavelengths. One of the most common applications of argon-ion lasers is in laser light shows. They have been largely replaced by diode-pumped solid-state lasers, as the latter are inherently smaller, more efficient and reliable and increasingly more versatile and less expensive than rare gas ion lasers.

Figure 4.21 Argon-ion laser (Reproduced by permission of Coherent Inc.).

Due to their broad line width of about 5 GHz, argon-ion lasers adapt themselves well to mode locking. Mode-locked laser pulses from these lasers are suitable for pumping titanium-sapphire and tunable dye lasers.

4.6 Excimer Lasers

Excimer lasers are pulsed lasers capable of providing pulse energies of the order of a few joules. These are the most powerful practical lasers emitting in the ultraviolet region. With the exception of the fluorine excimer laser, all excimer lasers have an active medium consisting of molecules of one rare gas atom such as argon, xenon or krypton and one halogen such as fluorine, chlorine or bromine. The active medium is a rare gas halide. Some of the more popular and practical excimer lasers include F_2 (157 nm), ArF (193 nm), KrCl (222 nm), KrF (249 nm), XeCl (308 nm) and XeF (350 nm).

The process of population inversion creation and subsequent laser emission in the case of excimer lasers is unique. The two atoms constituting the molecule are bound to each other only when the molecule is in the excited state, which is the upper laser level. The molecule falls apart when it drops to the ground state, which is the lower laser level. Figure 4.22 shows the energy level diagram of a rare gas halide molecule, depicting its behaviour both in the excited state where it is a molecule as well as in the ground state where the molecule breaks apart as individual atoms. Figure 4.22 shows the energy level as a function of the spacing between the rare gas atoms and the halogen atoms. The absence of any dip in the energy level indicates that the atoms are not bound to each other and that their energy decreases as the inter-atomic separation increases. The region in the excited state with a dip in the energy level is the one indicating that the rare gas and halogen atoms are bound to each other to form molecules. These excited molecules can occupy any of the several vibrational energy levels as shown by the horizontal lines. These molecules fall apart as they fall to the ground state, as there is no bonding energy available in the relatively lower-energy ground state to hold them together.

The gas mixture comprises the derived rare gas, the halogen and the buffer gas such as helium or neon. In fact, more than 90% of the gas mixture is the buffer gas. The mixture has a small percentage of rare gas. Present in an even smaller fraction is the source of halogen atoms. As pure halogens can pose serious

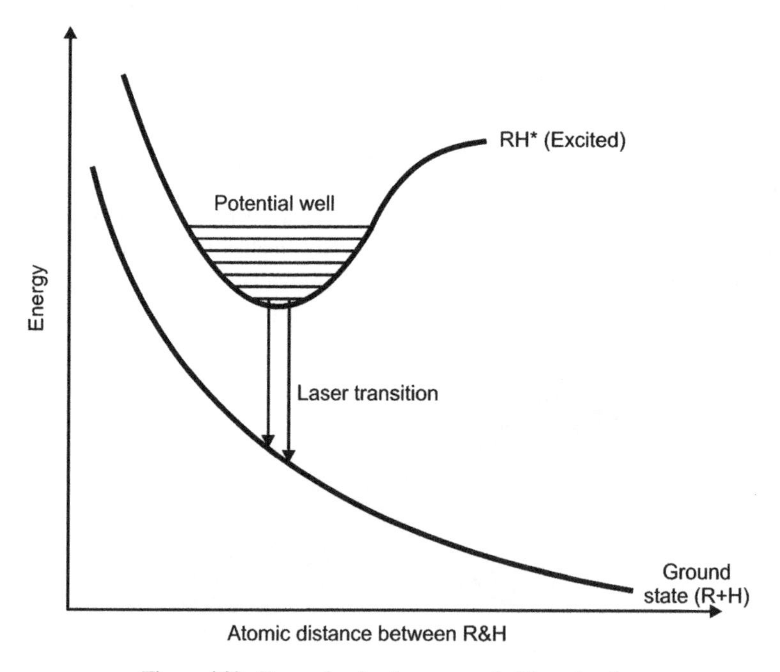

Figure 4.22 Energy levels of a rare gas halide molecule.

handling problems due to their high reactivity, it may be supplied in the form of molecules containing desired halogen atoms. It is more so for fluorine, which is considered to be the most dangerous to handle of all the halogens and therefore may be supplied as nitrogen trifluoride (NF_3). The gas mixture is excited by passing a short intense electrical pulse through it. Electrons in the plasma transfer energy to the gas mixture. The halogen molecules break up to form electronically excited molecules and create population inversion. The upper state lifetime is about 10 ns. The molecules drop to the ground state and break up, emitting ultraviolet radiation in the process.

Excimer lasers are the best-available pulsed lasers and have many applications in the semiconductor manufacturing industry, research and medicine. Pulse energies from several microjoules to a few joules are available at repetition rates of tens to a few hundreds of hertz. Repetition rates of a few kilohertz are achievable, but the pulse energies available at those repetition rates are relatively lower. Krypton-fluoride (KrF) and xenon-chloride (XeCl) are the more energetic of all excimer lasers, and the average power in these lasers can be in excess of 200 W.

4.7 Chemical Lasers

A chemical laser derives the energy required to produce population inversion and consequent laser emission from a chemical reaction. The importance of chemical lasers lies in their ability to generate CW output power level reaching several megawatts. This output power, combined with the fact that the range of wavelengths generated by these lasers is well absorbed by metals, makes them suitable for cutting and drilling operations in industry and more so as high-power laser sources for directed-energy weapon applications.

There are two categories of chemical lasers. One category has hydrogen fluoride (HF) and deuterium fluoride (DF) and the other has two iodine lasers including chemical oxygen iodine laser (COIL) and all gas-phase iodine laser (AGIL). Both HF/DF and COIL systems have the capability to be operated at megawatt-class CW output power. Lasing action has been demonstrated in AGIL and efforts are underway to scale the output to higher power levels.

In a chemical laser, a suitable chemical reaction produces a stream of gas rich in excited atoms or molecules. Another gas is then injected into the stream, which reacts with the excited particles in the gas stream to produce the excited species for laser action. In case of an HF/DF chemical laser, this stream of gas is abundant in fluorine atoms. Excited fluorine atoms then react with hydrogen or its heavier isotope deuterium to produce vibrationaly excited hydrogen fluoride (HF^*) or deuterium fluoride (DF^*) molecules (the asterisk denotes the excited state of the molecules). The laser emission is then due to the transition of excited HF or DF molecules. In the case of COIL, the initial chemical reaction produces singlet-delta oxygen. It transfers its energy to iodine to generate excited atomic iodine, which subsequently produces laser emission at the wavelength of transition of atomic iodine. Both types of chemical lasers are briefly described in the following sections.

4.7.1 Hydrogen Fluoride/Deuterium Fluoride (HF/DF) Lasers

In the HF laser, fluorine atoms are produced by pre-mixing fluorine gas with helium used as a buffer gas and some other gases. Helium is added to stabilize the reaction and control the temperature. Some other gases may be added to control production of fluorine atoms. Since fluorine gas could be very nasty to handle, it is usually supplied in the form of some other molecule such as sulphur hexafluoride (SF_6) or nitrogen trifluoride (NF_3). The free excited fluorine atoms produced in the combustor are accelerated through supersonic nozzles into a laser cavity. In the cavity, under low-temperature low-pressure conditions, fluorine atoms mix and react with hydrogen to form excited hydrogen fluoride (HF^*) molecules. A suitable resonator configuration produces stimulated emission of photons from these excited HF molecules. The output wavelength is in the region of 2.6–3.0 μm. Figure 4.23 shows a typical block schematic arrangement of an HF laser.

Similar to the HF laser is the deuterium fluoride (DF) laser. In this laser, deuterium replaces hydrogen and reacts with fluorine atoms to produce excited DF molecules (DF^*). Since deuterium is heavier than hydrogen, a DF laser produces output at the longer wavelength range of 3.6–4.0 μm. While a HF laser

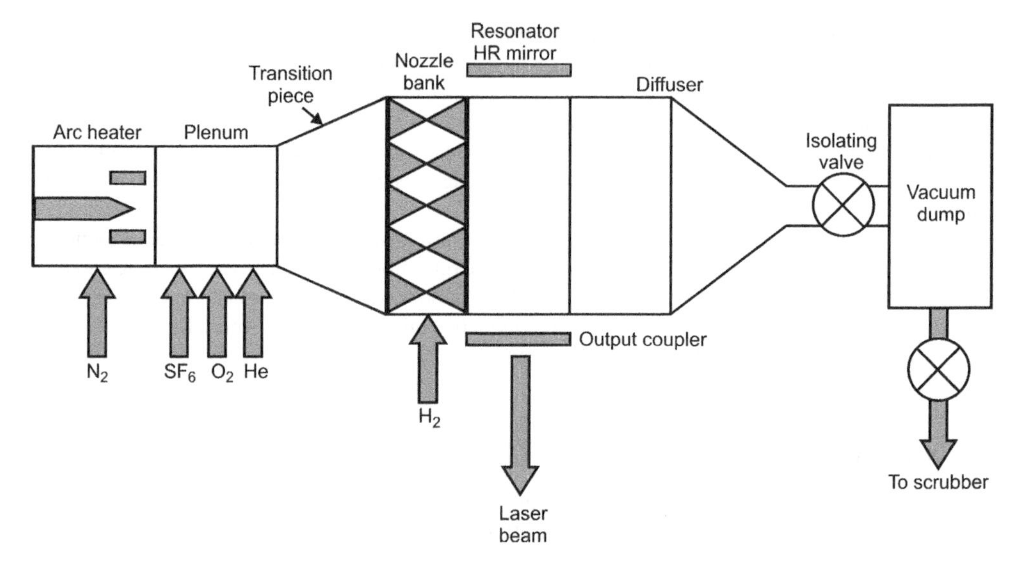

Figure 4.23 Block schematic of HF laser.

has the advantage of lower wavelength, a DF laser propagates relatively much better through atmosphere. As a result, HF is better suited to space-borne platforms. Some of the better known HF/DF laser systems include Alpha laser, MIRACL (mid-infrared advanced chemical laser) and THEL (tactical high-energy laser; Figure 4.24). While ALPHA is a HF laser, MIRACL and THEL systems are configured around DF lasers.

Figure 4.24 THEL system (Reproduced from public domain (http://en.wikipedia.org/wiki/File:THEL-ACTD.jpg)).

MIRACL was the first megawatt-class CW chemical laser. In the MIRACL system, free excited fluorine atoms are generated by the combustion reaction of a fuel (ethylene, C_2H_4) and an oxidizer (nitrogen trifluoride, NF_3). Deuterium and helium are injected into the exhaust downstream of the nozzle to produce excited DF (DF^*) molecules. Excited DF molecules produce stimulated emission and laser action in the resonator cavity, and the output is spread over several lasing lines in the wavelength range of 3.6–4.2 μm.

4.7.2 Chemical Oxygen Iodine Laser (COIL)

The three main parts of a COIL system are: the singlet oxygen generator (SOG); the supersonic nozzle; and the laser cavity. Singlet-delta oxygen is produced through a reaction of gaseous chlorine and basic hydrogen peroxide (BHP), which is a mixture of hydrogen peroxide and potassium hydroxide. This is a highly exothermic reaction, releasing most of the energy as heat into the BHP solution. The rest of the energy is used to generate singlet-delta oxygen, which is an electronically excited state of oxygen. Potassium chloride is produced as a by-product of this reaction. Molecular iodine is injected into the gas flow in the plenum just upstream of the supersonic nozzle. The molecular iodine dissociates into atoms through a series of energy transfer reactions with singlet-delta oxygen. Singlet-delta oxygen transfers its energy to the iodine molecules injected into the gas stream in a collisional process. The transfer of energy from singlet-delta oxygen to iodine is nearly resonant and thus very rapid. The excited iodine undergoes stimulated emission in the resonator cavity, producing a laser output at 1.315 μm. Figure 4.25 shows the block schematic arrangement of a COIL system.

COIL works at a relatively lower gas pressure. The pressure downstream of the nozzle is of the order of a few torr, which makes its discharge to the atmosphere a difficult proposition due to the large pressure differential involved. This is not the case if the laser is to work on an airborne platform, however. COIL is therefore the laser of choice for a laser-based directed-energy weapon (DEW) on an airborne platform.

The airborne laser (ABL) program of the US Air Force uses a high-power COIL system. ABL is one part of a layered ballistic missile defence system that addresses the ever-increasing ballistic missile threat. ABL is designed to destroy the hostile missile while it is still in the vulnerable boost phase of its flight (Figure 4.26). ABL uses an onboard surveillance system to detect and track the missile after launch. The beam control system locks on to the target and then fires the high-energy laser. The entire system (i.e. the high-energy laser, the surveillance system and the beam control/fire control system), is

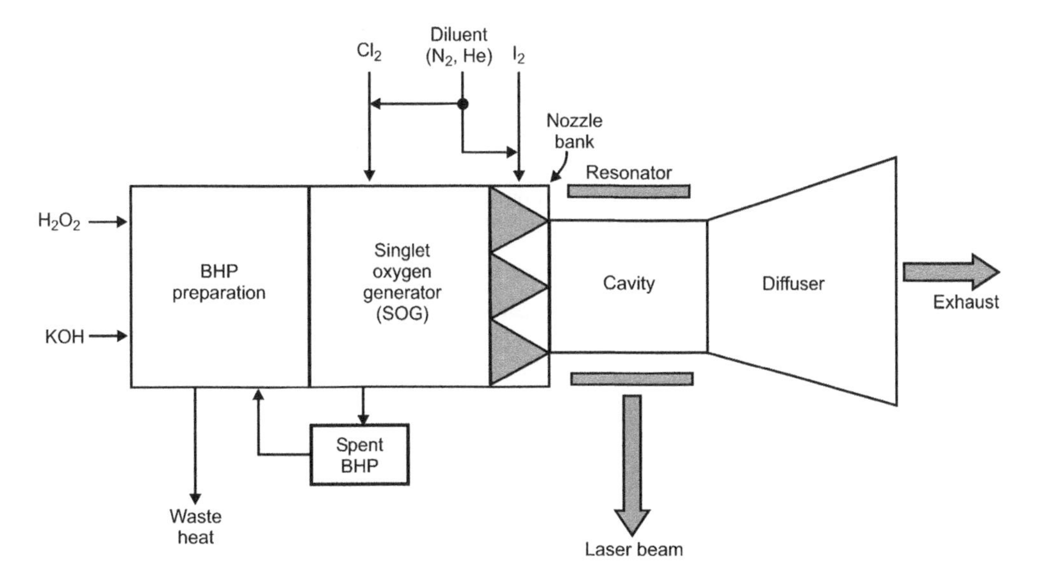

Figure 4.25 Block schematic of COIL system.

Figure 4.26 ABL concept (Reproduced from public domain (http://en.wikipedia.org/wiki/File:USAF_Airborne_laser.jpg)).

configured on a Boeing 747–400 fighter aircraft (Figure 4.27). ABL made its maiden flight in July 2002. In January 2010, the high-energy laser was used in-flight to intercept (but not destroy) a test missile in the boost phase of flight. In February 2010, the system successfully destroyed a liquid-fuel boosting ballistic missile. In December 2011, the ABL program was reportedly shelved as it was considered to be operationally unviable.

4.7.3 All Gas-Phase Iodine Laser (AGIL)

In AGIL, chlorine gas and gaseous hydrogen azide (HN_3) are mixed to produce excited nitrogen chloride (NCl) molecules. Nitrogen chloride then transfers energy to atomic iodine in the same manner as for

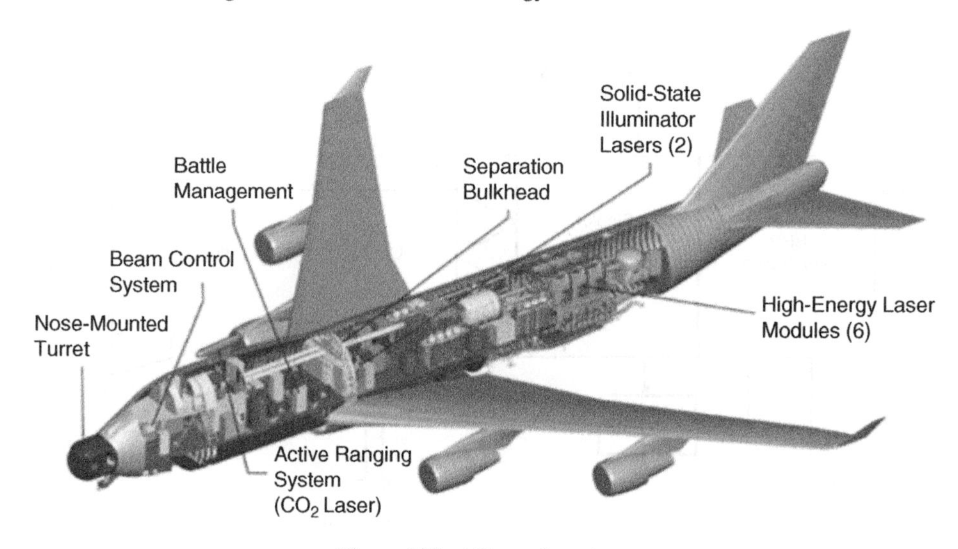

Figure 4.27 ABL configuration.

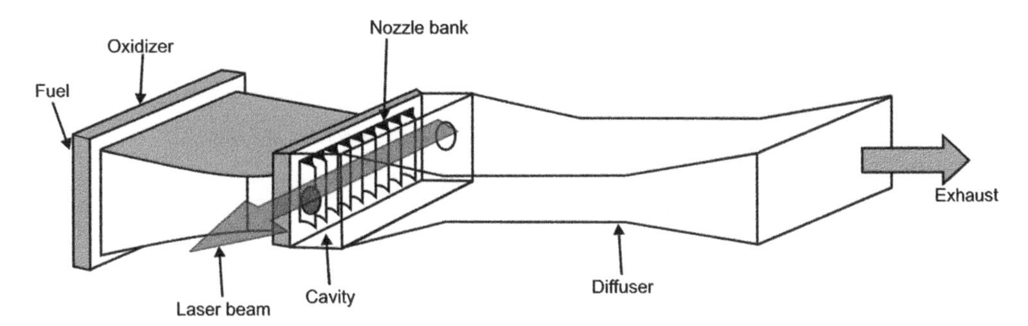

Figure 4.28 Functional schematic of a combustion-driven gas dynamic laser.

singlet-delta oxygen in COIL. This all gas-phase laser overcomes the disadvantage of the heavy aqueous-based chemistry of a COIL device and the undesirable wavelength range of HF laser that is strongly absorbed by the atmosphere. If scaled to higher power levels, it could be a potent laser technology for the space-borne laser programme.

4.8 Carbon Dioxide Gas Dynamic Lasers

The gas dynamic laser (GDL) derives its energy from the combustion of a suitable fuel–oxidizer mixture, which means that it does not require any electrical energy for its operation. A gas dynamic laser achieves population inversion by the rapid expansion of a high-temperature high-pressure laser gas mixture produced during combustion to a near vacuum in an adiabatic process through an integrated supersonic nozzle bank. Although the expansion reduces the gas temperature, a large number of excited molecules are still in the upper laser level. Population inversion is created if the reduction in pressure and temperature downstream of the nozzle bank takes place over a time that is much shorter than the vibrational relaxation time of the upper laser level, corresponding to an asymmetric stretching mode of carbon dioxide coupled with nitrogen. The cavity axis is transverse to the direction of gas flow. Since the high-temperature high-pressure gas mixture is created in a combustion reaction, the laser is referred to as combustion-driven gas dynamic laser (CD-GDL). Figure 4.28 shows the functional schematic arrangement of a CD-GDL showing its various constituent subsystems. This type of laser, capable of producing hundreds of kilowatts of CW power at 10.6 μm, was initially important as a potential high-power laser (HPL) weapon. It has however been superseded by HF/DF and COIL systems due to the heavy absorption of the 10.6 μm wavelength by water vapour in the atmosphere.

4.9 Dye Laser

Dye lasers are popular devices because of their inherent ability to produce a coherent output that is tunable over a wide range of wavelengths from near-ultraviolet through visible to the near-infrared region. A single dye cannot account for wavelength tunability from ultraviolet to infrared however; dyes need to be switched. The vibronic class of solid-state lasers is the only close competitor when looking for a tunable laser.

Dye lasers are particularly important in scientific research and some medical applications where a large wavelength tuning range and their inherent ability to produce ultra-short pulses are an advantage. While tunable output from dye lasers was initially considered for use in spectroscopic applications, applications of dye lasers have expanded over the years to include medical applications for both diagnostic and therapeutic purposes.

The tunability property of dye lasers is a result of the energy level diagram of the active medium. The active medium is an organic dye characterized by upper and lower laser levels that are bands of a large number of sublevels rather than single energy levels. Their ability to produce ultra-short pulses results from the wide gain-bandwidth profile characteristic of dye lasers. Titanium-sapphire lasers, members of the vibronic class of solid-state lasers, also have a wide gain-bandwidth profile and are therefore capable of

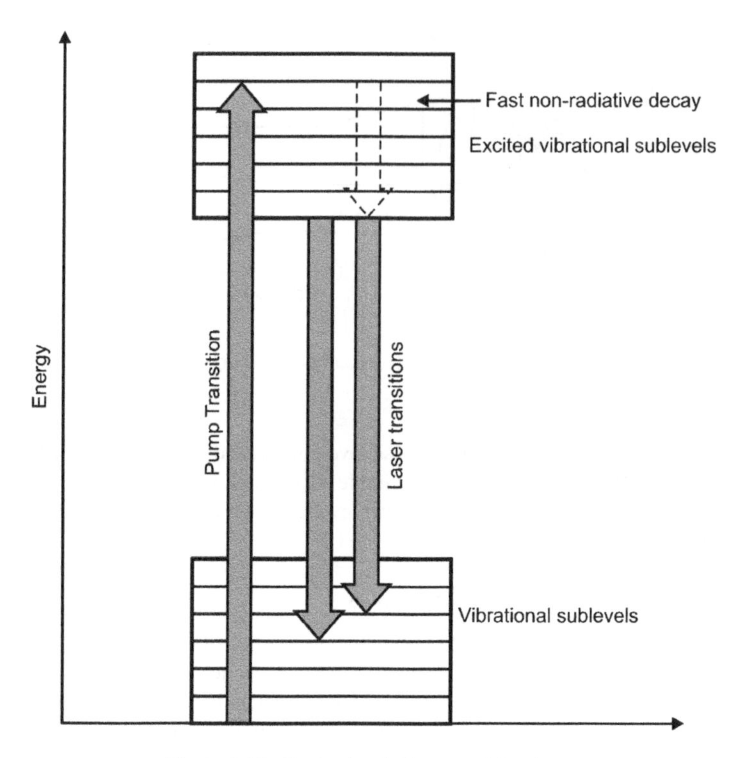

Figure 4.29 Energy level diagram of dye laser.

producing short pulses. Dye lasers however have an advantage that they cover the wavelength range from ultraviolet to near-infrared; titanium-sapphire lasers can only produce output in the near-infrared.

The minimum pulse length producible from the laser is a function of gain-bandwidth profile, a phenomenon described in Chapter 1. Pulse length in nanoseconds is given by 0.441/bandwidth, where bandwidth is in MHz.

4.9.1 Active Medium

The active medium in a dye laser is an organic dye. Organic dyes come from a family of complex molecules characterized by electronic-vibrational levels. The complex sets of electronic and vibrational levels associated with these molecules gives them the bright colour and hence the name dyes. The vibrational levels in fact create sublevels of the excited electronic levels, thus leading to a band of upper and lower laser levels in the same way as can be found in the vibronic class of solid-state lasers. The similarity between the two is clearly visible in the energy level diagram of an organic dye as shown in Figure 4.29. The dye is usually dissolved in a liquid solvent, which acts as the host material in the same way as crystals and glasses in the case of solid-state lasers.

In the case of low-power CW lasers and pulsed lasers with modest average power, the laser dye solution is contained in a sealed glass container called cuvette. In the case of higher power levels, the dye solution is usually made to flow rapidly through the cell to prevent it from becoming heated and subsequently degraded. The dye solution may also be made to flow in the form of an unconfined jet with the help of a specially designed nozzle.

Some examples of laser dyes include rhodamine-6G, fluorescein and coumarin. Rhodamin-6G is a commonly used laser dye pumped by a frequency-doubled Nd:YAG laser beam with a very high photostability, high quantum yield, low cost and, more importantly, close proximity to the absorption maximum (approximately 530 nm) of the dye. Figure 4.30 shows the absorption spectrum of a rhodamine-6G laser dye. The tunable range is 555–585 nm with a maximum at 566 nm.

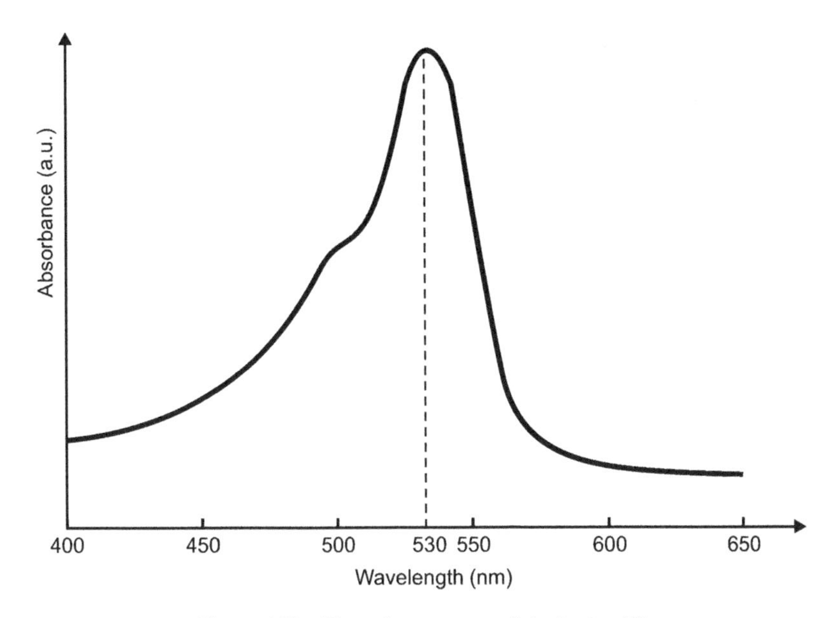

Figure 4.30 Absorption spectrum of rhodamine-6G.

4.9.2 Pump Mechanisms

Dye lasers are optically pumped. They are either flash-lamp pumped or pumped by pulsed lasers. In case of flash-lamp pumped dye lasers, there are two types. In one case, a linear flash lamp pumps a parallel linear dye cell in the same way as it pumps the laser rod in a solid-state laser. In the other case, a coaxial flash lamp is used and the dye flows through the centre of the flash lamp. Flash-lamp-pumped dye lasers produce a pulsed laser output with a pulse repetition rate in the range of a few pulses per minute (in the case of high-energy pulses) to a few hundreds of pulses per second (in the case of low-energy pulses).

Pulsed lasers are very usefully employed for the purpose of pumping dye lasers. Most lasers suitable for dye laser pumping emit at the short wavelength end of the visible spectrum and in the ultraviolet, as dye lasers mostly emit in visible wavelengths. Some of these lasers include frequency-doubled neodymium lasers (532 nm), frequency-tripled neodymium lasers (355 nm), frequency-quadrupled neodymium lasers (266 nm), copper vapour laser (510 nm and 578 nm), krypton-fluoride excimer laser (249 nm), xenon-chloride excimer laser (308 nm) and xenon-fluoride excimer laser (351 nm). CW lasers are also used for pumping dye lasers to achieve CW outputs. Argon, krypton and neodymium lasers are used for this purpose.

4.9.3 Wavelength Selection

Wavelength is selected with the help of a tuning element such as a diffraction grating, a prism or some other wavelength-selective element such as a tunable etalon in the laser resonator. Diffraction gratings and prisms disperse light at different angles depending upon the wavelength, which allows either these elements or some other optical components to be used for the purpose of wavelength tuning. Etalon on the other hand, with its two reflective surfaces, functions like a miniature optical cavity and is used to limit oscillations to a narrow range of wavelengths. Wavelength is selected by altering the distance between the two reflective surfaces.

4.10 Free-electron Lasers

A free-electron laser is unique in terms of its process of light amplification. Unlike conventional lasers that rely on bound atomic or molecular states, free-electron lasers use a relativistic electron beam as the active medium. In the case of a free-electron laser, the lasing medium is a beam of free electrons completely unattached to any atoms.

A relativistic particle is a particle moving at a speed close to the speed of light, such that the effects of special relativity are important for its behaviour. Mass-less particles such as photons always move with the speed of light and are therefore always relativistic. Particles with some mass are considered relativistic when their kinetic energy is comparable to or greater than the energy mc^2 corresponding to their rest mass. This implies that their speed is close to that of light. Such particles are generated in particle accelerators. In the context of free-electron laser, a beam of electrons is accelerated to relativistic speeds and then made to pass through a periodic (or, more precisely, alternating) transverse magnetic field. The transverse magnetic field is produced with the help of an array of magnets with alternating poles placed along the beam path. The array of magnets is sometimes called a wiggler (and the magnetic field produced by it as the wiggler field) as it forces the electron beam to assume a sinusoidal path. As the beam travels through the magnetic field, it releases some of its energy as light before it exits from the other end of the field. The emitted wavelength is given by Equation 4.1:

$$\lambda = \frac{p}{2(1 - v^2/c^2)} \tag{4.1}$$

where

p = period of Wiggler field
v = velocity of electrons

The same relationship may also be written in the form:

$$\lambda = \frac{0.131p}{(0.511 + E)^2} \tag{4.2}$$

where

E = electron energy in MeV

The free-electron laser is tunable over a wide range of wavelengths in the region from x-rays to microwaves by varying E. The accelerated beam of electrons with the desired energy can be obtained from well-established technology of charged particle accelerators. Figure 4.31 depicts the simplistic arrangement of various components in a free-electron laser.

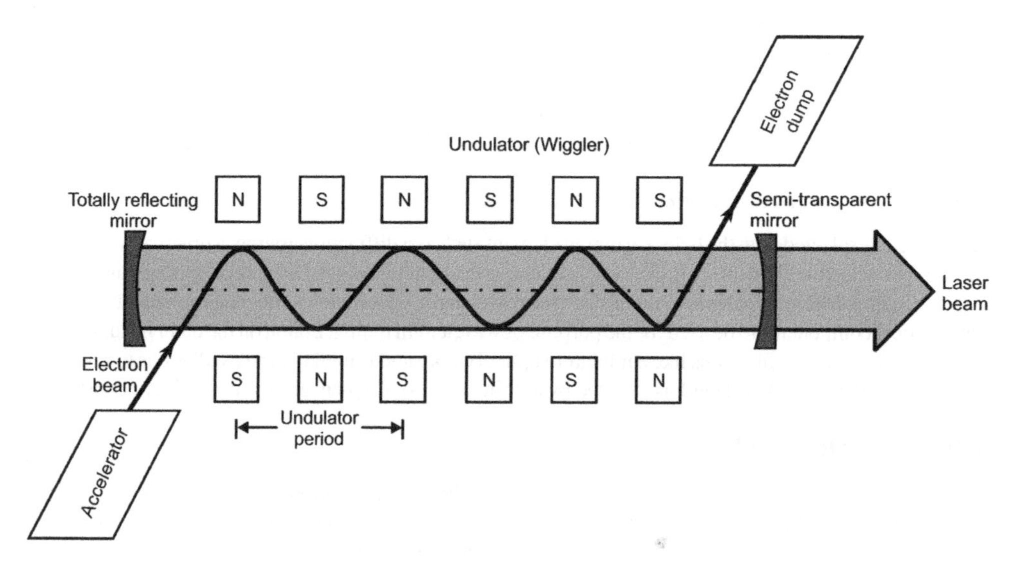

Figure 4.31 Free-electron laser.

Although thought of as potential sources for DEW application, free-electron lasers have their problems when it comes to generating higher powers at shorter wavelengths. Smaller free-electron lasers with power output up to 100 W are particularly important for research and medical applications, however.

4.11 X-Ray Lasers

The level of complexity involved in generating a coherent beam of x-rays is evident from a simple calculation: an energy level difference of the order of 100 eV or so is needed between the two laser levels in order to generate a 10 nm laser output. Creating a population inversion in an x-ray laser is a gigantic task. That is why these lasers are either excited by nuclear explosions or by massive Q-switched Nd: Glass lasers meant for fusion research. The development of x-ray lasers was initiated during the 'Star Wars' program but, due to technological complexities and infrastructure requirements, x-ray lasers are not considered an attractive proposition.

4.12 Summary

- Gas lasers have widely varying characteristics including wavelength range, power levels and, to some extent, pump mechanisms. Power level varies from a fraction of a milliwatt (in low-power He-Ne lasers) to the megawatt level (in weapon-class high-power lasers). Thousands of gas laser wavelengths have been discovered from ultraviolet to far-infrared. Gas lasers have Doppler-broadened gain versus frequency curve and most gas lasers are excited by electrical discharge.
- The active medium in a gas laser is almost invariably a mixture of more than one gas with gases other than the lasing species performing certain subtle functions (e.g. assisting in heat transfer as helium in carbon dioxide laser or depopulating the lower laser level as helium in helium-neon laser). The active media in different gas lasers may not be in the same form; they could be in the form of ionized atoms (argon-ion and krypton-ion lasers) or hot metal vapour (copper vapour and gold vapour lasers).
- Excimer lasers are pulsed lasers capable of providing pulse energies of the order of joules and are the most powerful lasers emitting in ultraviolet.
- Chemical lasers derive their pump energy from a chemical reaction. There are two categories of chemical laser: (1) HF/DF lasers and (2) COIL and AGIL.
- Gas dynamic lasers derive their pump energy from the combustion of a suitable fuel–oxidizer mixture, which means that they do not require any electrical energy for operation.
- Dye lasers (which are *not* gas lasers) have an inherent ability to produce coherent output tunable over a wide range of wavelengths from near-ultraviolet through visible to near-infrared region. Their wide gain-bandwidth profile means they are capable of producing ultra-short pulses. They are optically pumped either by flash lamp or by another suitable laser. Free-electron lasers use a relativistic electron beam as active medium, tunable over a wide range of wavelengths in the region from x-rays to microwaves.

Review Questions

4.1. Briefly describe the reasons for the following:
   a. Dye lasers can be used to produce ultra-short pulses.
   b. TEA carbon dioxide lasers can produce near-continuous tunable output across most of the wavelength band.
   c. Argon-ion and krypton-ion lasers are particularly attractive as sources for laser light shows.

4.2. Briefly describe the operational principle of the following lasers:
   a. chemical lasers
   b. gas dynamic lasers
   c. free-electron lasers.

4.3. Briefly explain why
   a. helium is added to helium-neon laser gas mixture
   b. helium is added to CO_2 laser gas mixture
   c. nitrogen is added to CO_2 laser gas mixture

4.4. Briefly describe the merits and demerits of the two gas laser plasma tube constructions, namely (a) mirrors sealed off to the plasma tube and (b) plasma tube sealed off with Brewster windows and mirrors placed outside the tube.

4.5. Briefly describe
   a. metal vapour lasers
   b. rare gas ion lasers
   c. excimer lasers.

Self-evaluation Exercise

Multiple-choice Questions

4.1. Which of the following gas lasers emits at the shortest wavelength?
   a. carbon dioxide laser
   b. helium-neon laser
   c. helium-cadmium laser
   d. argon-fluoride laser

4.2. Which of the following gas lasers is particularly suitable for use in laser light shows?
   a. rare gas ion lasers
   b. excimer lasers
   c. metal vapour lasers
   d. helium-neon lasers.

4.3. The shortest pulse width that can be generated from a dye laser with a bandwidth of 20 GHz is
   a. 22 ps
   b. 44 ps
   c. indeterminate from given data
   d. none of these.

4.4. Which of the following lasers is not pumped by any of the conventional techniques such as optical pumping, electric discharge pumping and so on:
   a. excimer lasers
   b. rare gas ion lasers
   c. metal vapour lasers
   d. combustion-driven carbon dioxide gas dynamic laser

4.5. Which of the following cannot be categorized as a chemical laser:
   a. HF laser
   b. combustion-driven carbon dioxide gas dynamic laser
   c. chemical oxygen iodine laser
   d. DF laser.

4.6. Identify the gas lasers that produce only pulsed output:
   a. argon-ion laser
   b. copper vapour laser
   c. krypton-fluoride laser
   d. both (b) and (c).

4.7. Identify gas lasers where inter-level transitions are electronic.
a. argon ion laser
b. helium-neon laser
c. copper vapour laser
d. all of the above.

4.8. Increased pressure in TEA laser leads to:
a. broadening of line width
b. narrowing of line width
c. shortening of its lifetime
d. none of these.

4.9. Identify the correct laser/lasing species combination
a. helium-neon laser/neon
b. argon-fluoride laser/argon
c. helium-cadmium laser/cadmium
d. both (a) and (c).

4.10. Identify the laser that produces output in the red region of wavelength:
a. helium-neon laser
b. gold vapour laser
c. krypton-ion laser
d. all of the above.

Answers

1. (d) 2. (a) 3. (a) 4. (d) 5. (b) 6. (d) 7. (d) 8. (a) 9. (d) 10. (d)

Bibliography

1. *Gas Lasers,* 2006 by Masamori Endo and Robert F. Walter, CRC Press.
2. *CRC Handbook of Laser Science and Technology*, 1994 Volume 2, by Marvin J. Weber, CRC Press.
3. *Principles of Gas Lasers*, 1967 by Leslie Allen, David Geoffrey and Conway Jones, Butterworths.
4. *Gas Lasers*, 1968 by Arnold L. Bloom, John Wiley and Sons.
5. *Introduction to Gas Lasers: Population Inversion Mechanisms*, 1974 by Collin S. Willett, Pergamon Press.
6. *The Laser Guidebook*, 1999 by Jeff Hecht, McGraw Hill.
7. *Principles of Lasers*, 2009 by Orazio Svelto, Plenum Press.
8. *Handbook of Laser Technology and Applications*, 2003 Volume II by Collin E. Webb and Julian D. C. Jones, Institute of Physics Publishing.
9. *Fundamentals of Light Sources and Lasers*, 2004 by Mark Csele, John Wiley & Sons.
10. *CO_2 Lasers Effects and Applications*, 1976 by W.W. Duley, Academic Press.

5

Semiconductor Lasers

5.1 Introduction

In continuation of our discussion on types of lasers, the focus in this chapter is on semiconductor lasers. Most semiconductor lasers are diode lasers that are pumped by an electrical current; these are also known as injection diode lasers. However, there are also optically pumped semiconductor lasers, either from a direct-coupled single emitter or fibre-coupled laser diode array, and quantum cascade lasers that utilize intra-band rather than inter-band transitions of the conventional semiconductor lasers for laser action. Most of the discussion in this chapter is centred on semiconductor diode lasers, which typically emit in visible to near-infrared bands of electromagnetic spectrum. Optically pumped semiconductor lasers, quantum cascade lasers, lead salt lasers and antimonide lasers are also briefly discussed. Topics covered in this chapter include operational fundamentals, semiconductor materials used for making semiconductor lasers, different types of semiconductor diode lasers, characteristic parameters, handling precautions and application areas.

5.2 Operational Basics

The active medium in a semiconductor laser is a semiconductor material. The optical gain in this case is usually achieved by a process of stimulated emission at an inter-band transition triggered by prevailing conditions of high carrier density in the conduction band. In the case of diode lasers, high carrier density in the conduction band is caused by an injection current. The emission of radiation is due to the recombination of holes and electrons in a forward-biased P-N junction diode. An *injection diode laser* powered by an injected electrical current is the most practical form of diode laser, distinguishable from an optically pumped diode laser.

Figure 5.1 illustrates the operational basics of a semiconductor diode laser action. Figure 5.1a and b depict no forward bias and applied forward bias greater than the band gap energy conditions, respectively. As shown in the diagrams, energy levels up to the Fermi level are occupied by electrons. The Fermi level is continuous across the junction under zero-applied forward-bias conditions. When a forward bias with corresponding energy greater than the band gap energy E_g is applied across the junction, the barrier potential reduces to zero. Fermi level on the N-side (E_{FN}) and P-side (E_{FP}) are separated by eV_F where e is electronic charge and V_F is forward bias voltage. There are more electrons in the conduction band near E_C than in the valence band near E_V. This leads to population inversion near E_C and E_V energy levels. This region of population inversion along the junction is called an inversion layer or active region. Due to the presence of electrons and holes in the active region, there is a likelihood of their recombination leading to spontaneous emission of a photon. The energy of an emitted photon equals the difference in energy levels of electron and hole states involved in the recombination. In the process, the electron may re-occupy the energy state of the hole. The injected electrons and holes

Lasers and Optoelectronics: Fundamentals, Devices and Applications, First Edition. Anil K. Maini.

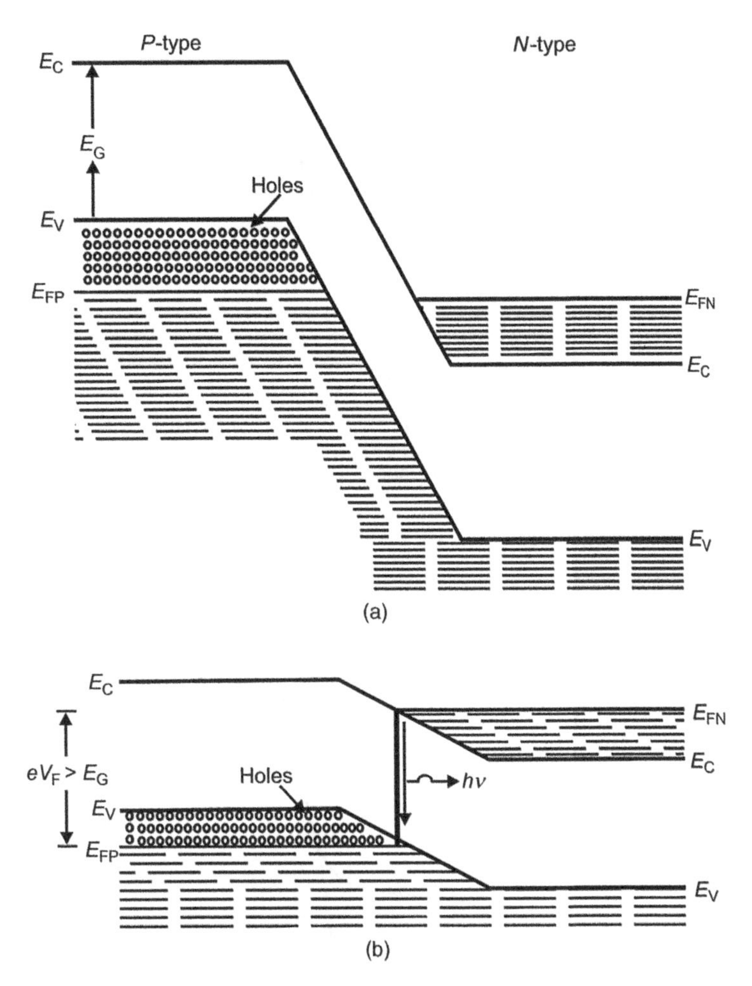

Figure 5.1 Semiconductor diode laser basics: (a) zero-applied forward bias; and (b) applied forward bias greater than E_g.

comprise the injection current and those involved in the recombination process constitute the spontaneously emitted photon output.

If the injection current exceeds a certain minimum value called the *lasing threshold*, the number of electrons and holes available for recombination becomes sufficiently large such that a spontaneously emitted photon with energy equal to the involved recombination energy stimulates an electron-hole pair. These recombine to emit a photon of the same frequency, phase and polarization as that of the stimulating photon, which is referred to as stimulated emission. Surrounding the recombination region (also called gain region) by a suitable optical cavity facilitates the process of stimulated emission. In the case of a diode laser, the cavity is made by cleaving the two ends of the crystal to form perfectly smooth parallel edges forming a Fabry–Pérot resonator. Since the semiconductors have a high refractive index, the smooth surfaces offered by cleaved ends reflect about 30% of the light back into the material to achieve sustained laser action in a high-gain semiconductor laser material. The stimulated emission produces light amplification as the photons travel back and forth between the two end-faces of the cavity. When the gain due to stimulated emission exceeds the losses due to absorption or imperfect reflections, sustained lasing action is produced.

Light-emitting diodes (LEDs) operate in the same way, with a major difference in the forward-biased current. While the current in the case of an LED is of the order of a few milliamperes, the current in the

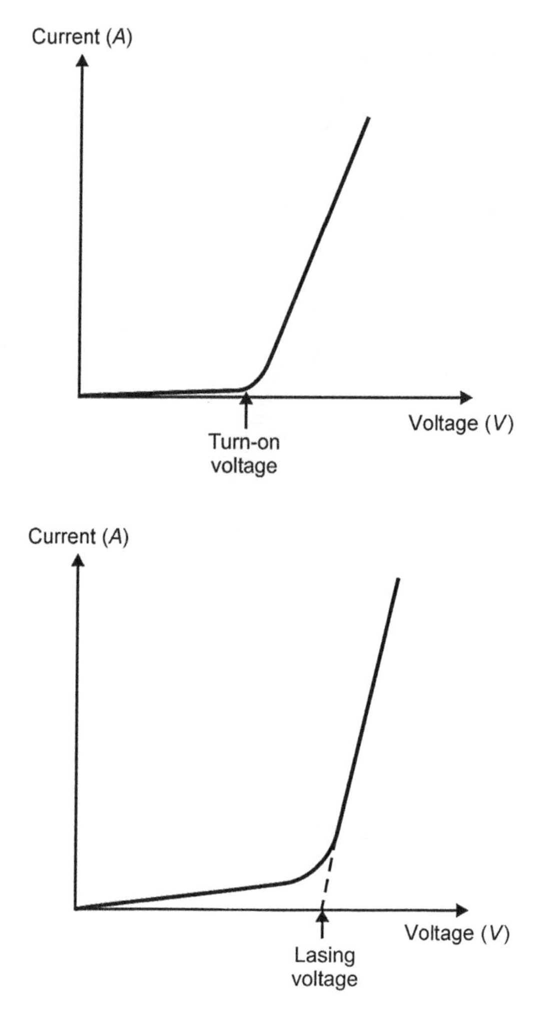

Figure 5.2 *I–V* characteristics of (a) LED; and (b) semiconductor diode laser.

case of laser diodes emitting few milliwatts of laser power is of the order 80–100 mA. At low levels of drive current, spontaneous emission predominates. When the drive current is more than the lasing threshold, the light output is predominantly due to stimulated emission. Figure 5.2a and b show the current–voltage (*I–V*) characteristics of a typical LED and a semiconductor diode laser. As shown in the figure, the LED turns on for a forward-bias voltage greater than or equal to the turn-on voltage and the current flowing through the device is nearly zero. In the case of a laser diode, although the device is turned on and a small magnitude of current does start flowing, the lasing action starts only after the forward voltage exceeds the lasing threshold voltage.

This phenomenon can also be explained with the help of output optical power versus drive current characteristics of the laser diode shown in Figure 5.3. As shown in the figure, stimulated emission becomes predominant, thereby leading to sustained lasing action only after the drive current exceeds a certain lasing threshold current value. For drive currents less than the threshold value, spontaneous emission predominates and an LED-like emission is obtained.

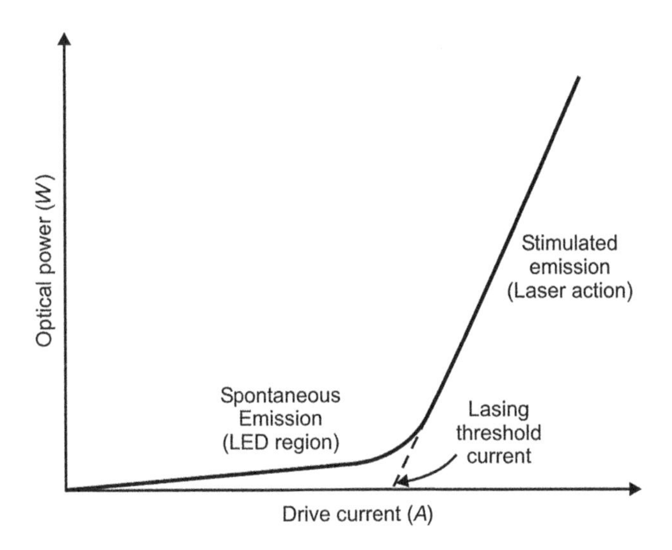

Figure 5.3 Optical power versus drive current characteristics of semiconductor diode laser.

5.3 Semiconductor Laser Materials

Only direct band gap semiconductor materials are suitable for building diode lasers and LEDs, although some compound semiconductors with indirect band gap can be used to make LEDs. In the case of direct band gap materials such as gallium arsenide (GaAs), gallium nitride (GaN) and gallium aluminium arsenide (GaAlAs), a transition from conduction band to valence band releases all its energy in the form of light. In the case of indirect band gap materials such as silicon, some of it goes into vibrations of the crystal lattice which makes it difficult for this class of semiconductor materials to emit light by recombination of charge carriers. Compound semiconductors are used for making laser diodes. For laser applications, the most important of these are those comprising equal amounts of elements from III-A (gallium, aluminium and indium) and V-A (arsenic, antimony, phosphorus and nitrogen) groups of the periodic table. Compounds with three elements are referred to as ternary compounds, and compounds with four elements are called quaternary compounds. The generalized formula for the most commonly used ternary compounds is $(Ga_{1-x}\,Al_x\,As)$, where $0 < x < 1$.

Other important ternary compounds for semiconductor diode laser fabrication include indium gallium nitride (InGaN), indium gallium arsenide (InGaAs) and gallium antimonide arsenide (GaSbAs).

There are two types of commonly used quaternary compounds with generalized formulae $(In_{1-x}\,Ga_x\,As_{1-y}\,P_y)$ and $(In_{1-x-y}\,Ga_x\,Al_yP)$ where $0 < x,\ y < 1$.

From the generalized expressions of the two types of compound semiconductors, it is evident that elements from groups III-A and V-A of the periodic table are present in equal quantities. In the first case, the quantity of gallium and aluminium (Group III-A elements) together is the same as that of arsenic (Group V-A). Similarly, in the case of one type of quaternary compounds, the quantity of indium and gallium (Group III-A) put together is the same as that of the total quantity of arsenic and phosphorus (Group V-A). In the other type of quaternary compound, the total quantity of indium, gallium and aluminium (Group III-A) is the same as that of the total quantity of phosphorus (Group V-A).

To summarize, commonly used semiconductor material compositions for semiconductor diode laser fabrication, with examples of common laser wavelengths, include: gallium arsenide (905 nm at room temperature), gallium nitride (405 nm), gallium aluminium arsenide (785 nm and 808 nm), aluminium gallium arsenide (1064 nm), indium gallium arsenide (980 nm), indium gallium nitride (405 nm and 445 nm), gallium antimonide arsenide (1877 nm, 2004 nm, 2330 nm, 2680 nm, 3030 nm and 3330 nm), indium gallium arsenide phosphide (1310 nm, 1480 nm, 1512 nm, 1550 nm and 1625) and aluminium gallium indium phosphide (635 nm, 657 nm, 670 nm and 760 nm).

5.4 Types of Semiconductor Lasers

Depending upon the structure of various semiconductor materials used to fabricate semiconductor lasers, these can be categorized as either:

1. homojunction and heterojunction lasers
2. quantum well diode lasers
3. distributed-feedback (DFB) lasers
4. vertical-cavity surface-emitting lasers (VCSELs)
5. vertical external-cavity surface-emitting lasers (VECSELs)
6. external-cavity semiconductor diode lasers
7. optically pumped semiconductor lasers
8. quantum cascade lasers
9. lead salt lasers.

These are all described briefly in the following sections.

5.4.1 Homojunction and Heterojunction Lasers

The boundaries between the active layer and the adjacent layers play an important role in determining some of the vital parameters such as efficiency of the lasers. In the case of *homojunction lasers*, all layers are of the same semiconductor material. One such example is GaAs/GaAs laser. However, this simple laser diode structure is highly inefficient and can only be used to demonstrate pulsed operation. In the case of *heterojunction lasers*, the active layer and either one or both of the adjacent layers are of different material. If only one of the adjacent layers is of a different material, it is called a simple heterojunction; if both are different, it is called a double heterojunction.

Some of the popular semiconductor laser types and the corresponding emission wavelength band are listed in Table 5.1.

In heterostructure diode lasers, a layer of low band gap material is sandwiched between layers of a high band gap material as shown in Figure 5.4. One such commonly used pair of materials is gallium arsenide (GaAs) and aluminium gallium arsenide ($Al_xGa_{1-x}As$). In these devices, the active region (the region where free electrons and holes exist simultaneously) is confined to the thin middle layer. As a result of this, a relatively much larger number of electron–hole pairs can contribute to the amplification process. Light is also reflected from the heterojunction, which helps to confine the emitted photons to the active region. These factors contribute to the high efficiency of heterostructure diode lasers.

5.4.2 Quantum Well Diode Lasers

For semiconductor layers that are only a few nanometers thick, the assumption that the material is a continuum is no longer valid and the quantum mechanical properties of atoms and electrons become important. When a thin layer of a semiconductor material with a relatively smaller band gap is sandwiched between two thick layers with larger band gaps, a structure known as a *quantum well* is formed and the electrons passing through the semiconductor are captured in the thin layer. Although

Table 5.1 Emission bands of common semiconductor lasers

Semiconductor material	Type of laser	Wavelength range (nm)
AlGaInP/GaAs	Heterojunction	620–680
$Ga_{0.5}In_{0.5}P$/GaAs	Heterojunction	670–680
GaAlAs/GaAs	Heterojunction	750–870
GaAs/GaAs	Homojunction	904
InGaAsP/InP	Heterojunction	1100–1650

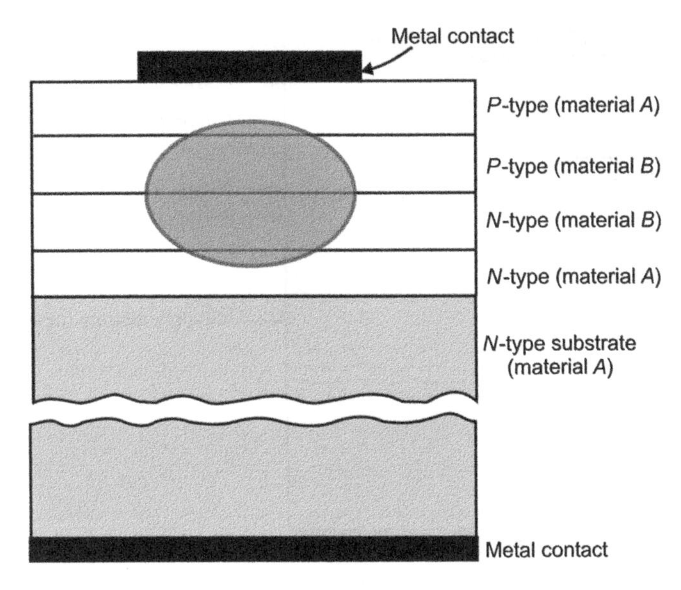

Figure 5.4 Heterojunction diode laser structure.

these electrons have sufficient energy to be free in the small band gap quantum well layer, it is not enough to allow them to enter the large band gap thicker layers. The thicker layers therefore help in confining the captured electrons to the quantum well. If the quantum well is placed in the semiconductor junction, the concentration of electrons as they recombine with holes to emit photons leads to an increase in efficiency and a decrease in lasing threshold. The fact that the quantum well layer and the outer thicker layers are made from semiconductor materials of different refractive indices further helps the cause by confining the emitted photons to that narrow region. Figure 5.5a shows the basic structure of a simple quantum well semiconductor diode laser.

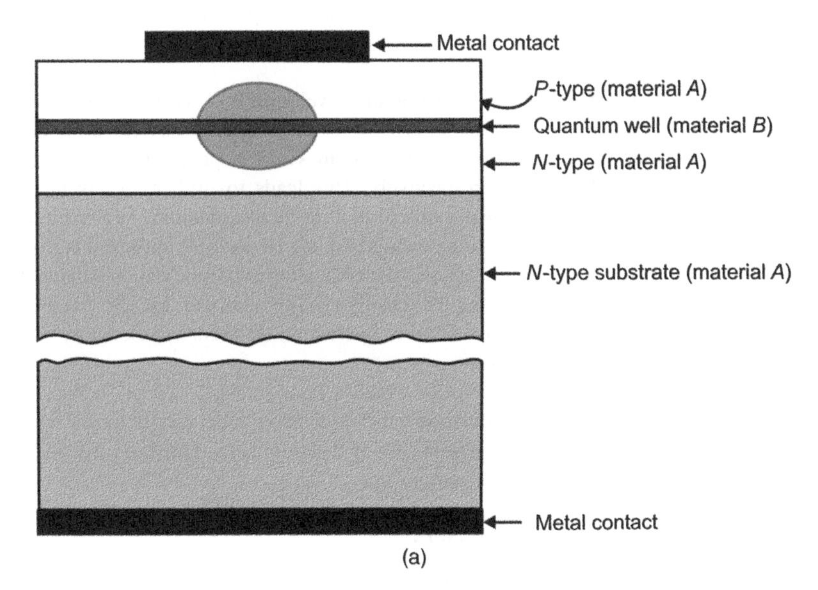

Figure 5.5 Quantum well semiconductor diode laser: (a) simple quantum well; and (b) SCH structure.

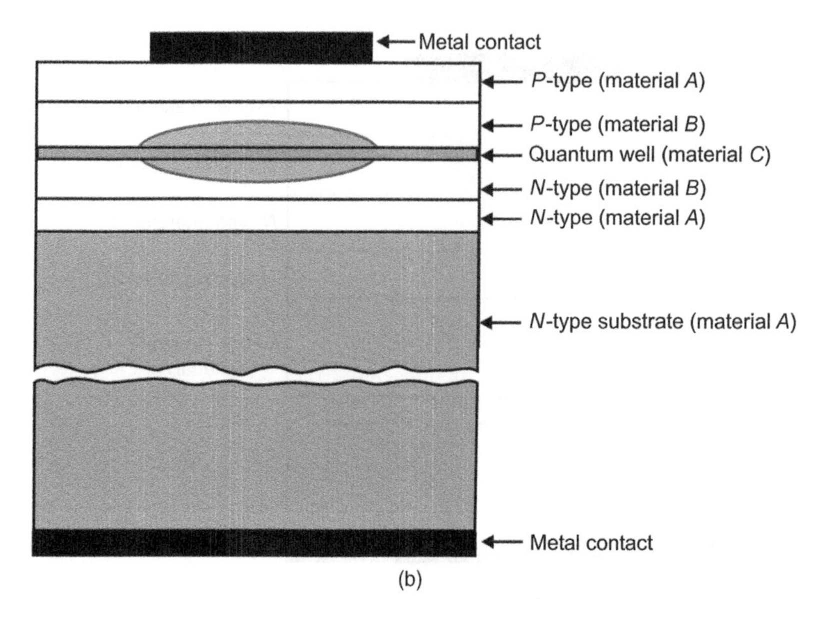

Figure 5.5 *(Continued)*

One of the drawbacks of the simple quantum well structure of Figure 5.5a is that the thin layer is too small to effectively confine the light. A modification of this basic quantum well structure is shown in Figure 5.5b. In this, two additional layers are added outside three existing layers. The additional layers made of material A in Figure 5.5b have a refractive index that is smaller than the refractive indices of the centre layers made of materials B and C. Such an arrangement succeeds in confining the light with greater efficacy than is feasible with a simple quantum well structure. Semiconductor diode lasers having this modified structure are referred to as separate confinement heterostructure (SCH) lasers. Most practical diode lasers have an SCH quantum well structure.

Lasers with more than one quantum well layer are known as multiple quantum well lasers. These are nothing but a stacked arrangement of quantum wells and confinement layers. Multiple quantum well structures can generate relatively higher power outputs.

An extension of the quantum well concept has led to the development of what are known as *quantum wire* and *quantum dot* structures. While a quantum well structure can confine electrons in only one dimension, that is, in the plane of quantum well layer, quantum wire and quantum dot structures provide confinement in two and three dimensions, respectively. This leads to further improvement in laser efficiency. A quantum dot laser uses quantum dot structures as the active medium. As a result of the tight confinement of charge carriers, quantum dot semiconductor diode lasers offer improved performance as compared to lasers built on bulk or quantum well active media in terms of modulation bandwidth, lasing threshold, relative intensity noise and temperature sensitivity. The quantum dot active region may be engineered by varying the dot size and composition to operate on different wavelengths, which is not possible with semiconductor laser technology.

Yet another type is a *quantum cascade laser* in which a laser transition occurs between different energy levels of the quantum well instead of the band gap, which allows operation at longer wavelengths. The wavelength can be tuned by changing the thickness of quantum layer. Quantum cascade lasers are described in Section 5.4.8.

5.4.3 Distributed-feedback (DFB) Lasers

Diode lasers have a gain-bandwidth curve that is broad enough to accommodate several longitudinal modes. In a conventional diode laser, due to changes in temperature and other operating conditions,

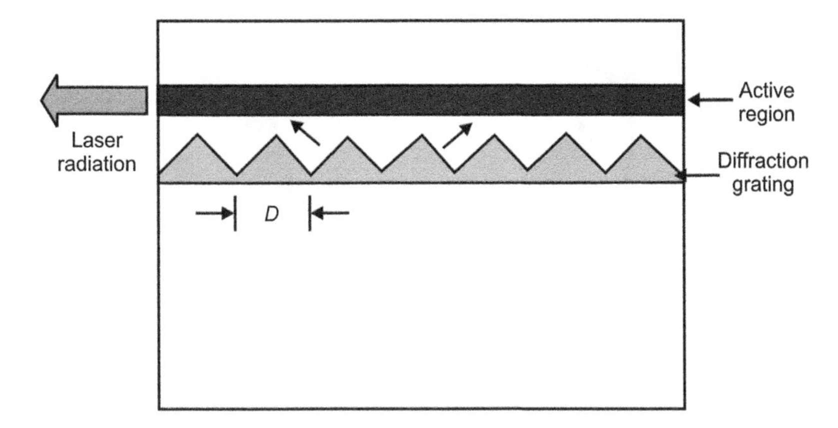

Figure 5.6 DFB laser.

the dominant longitudinal mode may therefore hop from one value to the next adjacent value, leading to instability in output power and wavelength. There are applications where this instability is highly undesirable. One such example is fibre-optic communication where instability in output wavelength causes chromatic dispersion in the fibre, thus limiting its utility.

DFB lasers offers a mechanism of ensuring a very narrow output wavelength band. In this, a diffraction grating is etched very close to the active layer (Figure 5.6). The grating provides wavelength-selective feedback to the gain region. The feedback causes interference and only the narrow wavelength region for which the interference is constructive is allowed to sustain. DFB lasers offer single longitudinal and transverse mode operation. For a diffraction grating with pitch D, wavelength λ is given by Equation 5.1:

$$\lambda = \frac{2nD}{m} \tag{5.1}$$

where n is the refractive index of the medium and m is a positive integer. Typical values of m are 1 or 2, indicating the order of distributed feedback coupling. A slight shift in output wavelength is possible due to temperature dependence of n.

Since the grating provides the feedback required for lasing in a DFB laser, end faces are not required to do that. That is why one of the end faces is anti-reflection coated. Because of their narrow line width and wavelength tunability, DBF lasers find extensive use in optical communications, spectroscopy, gas sensing and metrology. In fact, DFB lasers today are the workhorse of fibre-optic communication systems. Figure 5.7 depicts some DFB laser types from Eagleyard Photonics, Germany. These lasers are available in the wavelength range 760–1083 nm and output power levels 10–150 mW. The FWHM spectral width, temperature coefficient of wavelength ($d\lambda/dT$) and current coefficient of wavelength ($d\lambda/dI$) of this series of DFB laser diodes are 2.0 MHz, 0.06 nm K^{-1} and 0.003 nm A^{-1}, respectively. These are ideally suited to applications such as spectroscopy, metrology and gas sensing.

Example 5.1

Determine the pitch of the first-order diffraction grating in a distributed feedback laser, if the laser were to emit at 1000 nm and the laser medium had a refractive index of 2.5.

Solution

1. Pitch = $m\lambda/2n$
2. For a first-order grating, $m = 1$
3. Therefore, pitch = $\lambda/2n$ = 1000/5 nm = 200 nm.

Figure 5.7 DFB lasers.

5.4.4 Vertical-cavity Surface-emitting Laser (VCSEL)

In the diode laser structures discussed so far, the optical cavity is perpendicular to the direction of current flow. In the case of VCSELs, the optical cavity is along the direction of flow of injection current as shown in Figure 5.8. The laser beam in this case emerges from the surface of the wafer rather from its edges. Having mirrors on both sides of the active medium forms the resonant structure. Mirrors are usually of the distributed Bragg reflector type formed by a multilayer structure of alternating low and high refractive index semiconductor materials. Mirrors in this case need to be highly reflective as the overall gain is low due to the short length of the gain medium. Compared to edge emitting lasers, these lasers produce relatively lower output power levels. However, VCSEL require a very small chip area, typically a few tens of squared microns, with the result that a large number of such lasers can be tightly packaged in an array structure on a single chip. These lasers are characterized by a very small threshold current, which could be as low as 1 μA.

5.4.5 Vertical External-cavity Surface-emitting Lasers (VECSEL)

In the case of vertical-cavity surface-emitting lasers (VCSELs), the two mirrors are either grown epitaxially as a part of the diode structure or grown separately and then bonded to the semiconductor chip having the active region. Vertical external-cavity surface-emitting lasers (VECSELs) are a variant of VCSEL where the resonator is completed with a mirror placed external to the diode structure, thus introducing a free space region in the resonant cavity as shown in Figure 5.9. Compared with other types of semiconductor diode lasers, VECSELs are capable of generating much higher optical powers with high beam quality. When mode area is large, an external resonator is necessary for achieving high beam quality. Although VCSELs have a large mode area, they are not suitable for generating high-power levels with high beam quality.

In the case of VECSELs, the resonant cavity may contain some additional optical elements such as an optical filter to facilitate wavelength tuning or a saturable absorber for passive mode locking to generate picosecond- and femtosecond-pulses at a repetition rate of few gigahertz. The insertion of a non-linear

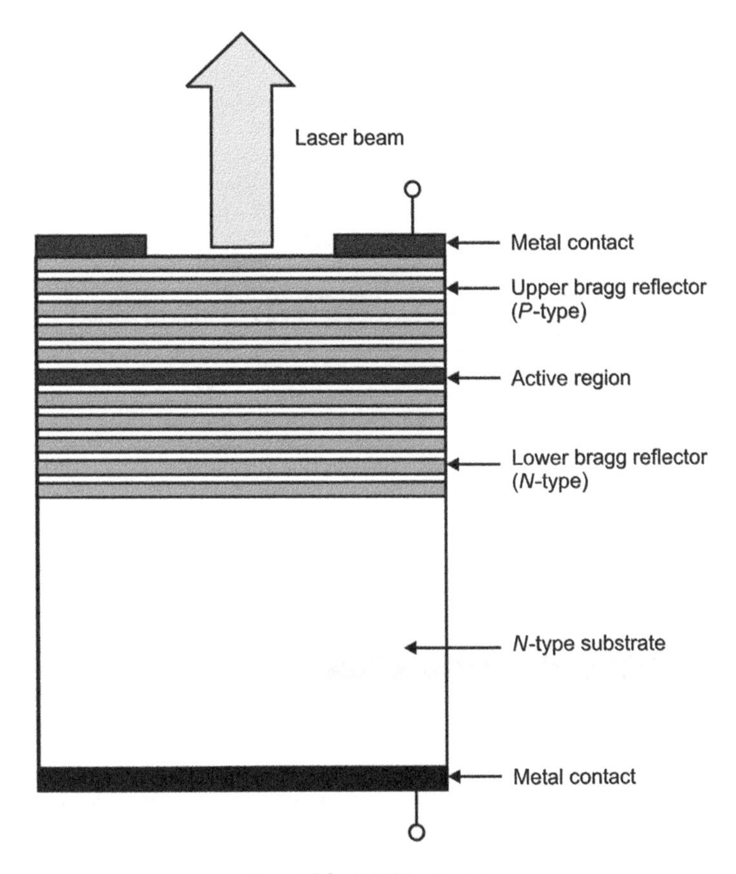

Figure 5.8 VCSEL structure.

crystal in the cavity allows frequency doubling which, for example, could lead to fabrication of devices with red, green and blue outputs. A VECSEL is a low-gain device and the intra-cavity power is much higher than the output power. This facilitates intra-cavity frequency doubling.

A typical VECSEL has a gain structure with a Bragg reflector and a gain medium with multiple quantum wells. Typical material combinations used to generate different wavelength regions include a GaAs wafer, InGaAs quantum wells in the gain medium and a Bragg reflector grown from AlAs and GaAs. This structure is suitable for generating wavelengths in the region 960–1030 nm. Emission in the 1.5 μm region is possible with indium phosphide (InP) as the wafer material, indium gallium arsenide phosphide (InGaAsP) quantum wells and indium phosphide/indium gallium aluminium arsenide (InP/InGaAlAs) combination for Bragg mirror. For longer wavelengths such as 2.0 μm, gallium indium antimonide (GaAlSb) quantum wells could be used on gallium antimonide (GaSb) wafers.

5.4.6 External-cavity Semiconductor Diode Lasers

The basic operational concept of an external-cavity diode laser (ECDL) is similar to that of VECSEL, which is an external-cavity semiconductor laser.

An ECDL is a semiconductor diode laser whose resonator is completed with one or more optical components outside the diode laser chip. In the simplest form, the diverging output from one of the anti-reflection-coated end faces is collimated by an external lens. The collimated beam is made to fall on a partially reflecting mirror that provides optical feedback and acts as an output coupler. An ECDL allows

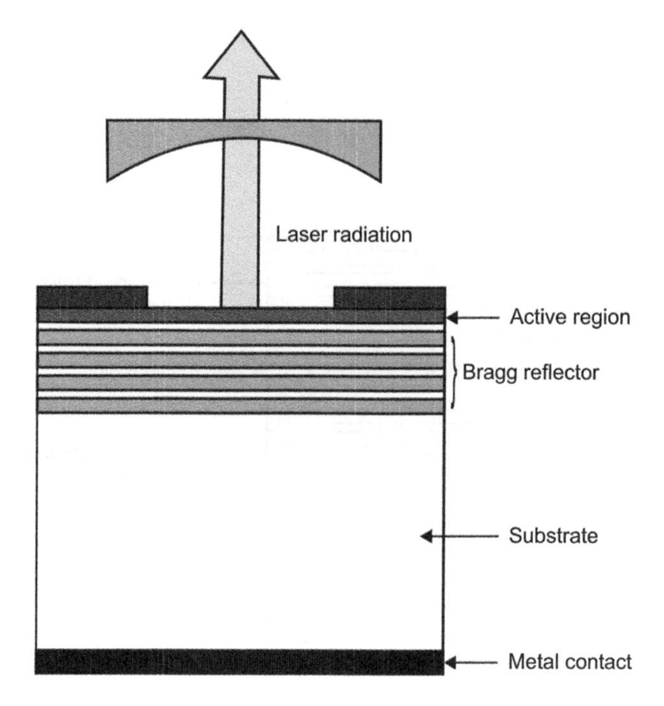

Figure 5.9 VECSEL structure.

for a longer resonator, which in turn leads to lower phase noise and narrow emission line width resulting from the increased damping time of intra-cavity light. The external resonator also opens up the possibility of introducing suitable intra-cavity optical elements for wavelength selectivity and tuning and mode locking.

A diffraction grating is usually employed for wavelength selectivity and tuning. Two common configurations are *Littrow configuration* and *Littman–Metcalf configuration*. Figure 5.10 shows the basic schematic arrangement of the Littrow configuration: a diffraction grating is used as the end mirror with a first-order diffracted beam providing optical feedback. Wavelength tuning is achieved by changing grating orientation. This configuration has the drawback that the direction of output beam

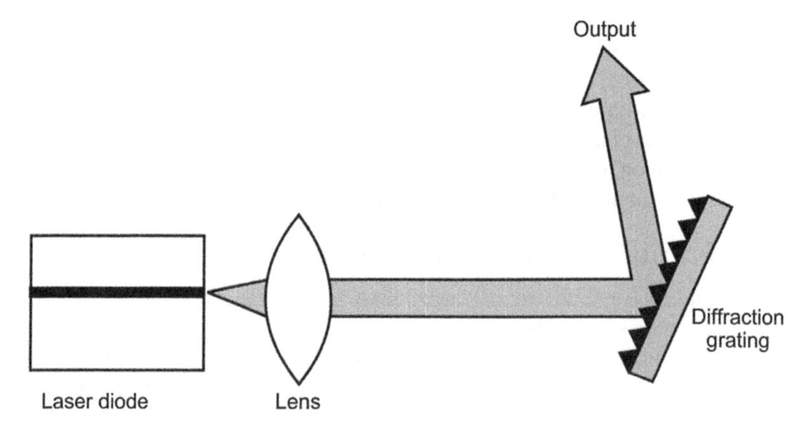

Figure 5.10 External cavity diode laser: Littrow configuration.

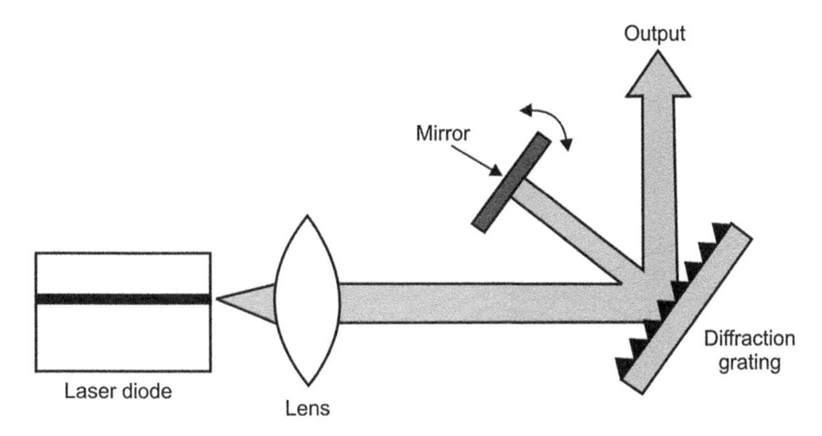

Figure 5.11 External cavity diode laser: Littman–Metcalf configuration.

changes with wavelength tuning. This problem is overcome in a Littman–Metcalf configuration shown in Figure 5.11, where grating orientation is fixed and an additional mirror is used to provide first-order feedback. The mirror is rotated for wavelength tuning. The direction of output beam in this case is fixed. The beam undergoes wavelength-dependent diffraction twice per resonator round-trip, resulting in higher wavelength selectivity and a narrower line width.

External-cavity diode lasers find extensive use in applications such as spectroscopy, where a tunable narrow line width source is important.

5.4.7 Optically Pumped Semiconductor Lasers

Some of the problems of the edge-emitting conventional semiconductor lasers such as high divergence, asymmetry and astigmatism are largely overcome by VCSELs. The features of VCSELs are further improved by introducing an external cavity to widen the horizon of its capabilities in the case of VECSELs. However, the conventional electrically pumped VCSELs and VECSELs cannot match the output power capability of edge-emitting devices as it is not practical to flood a large area with charge carriers through electrical pumping without using extended electrodes. Extended electrodes introduce too much of loss, and are therefore not desirable. The solution lies in employing optical pumping by another diode laser as shown in the simplified schematic arrangement of Figure 5.12.

An optically pumped semiconductor laser (OPSL) is simply a vertical external-cavity semiconductor laser that is pumped optically. The active region in this case comprises alternate layers of a binary semiconductor material and tertiary semiconductor material quantum wells. The emission wavelength depends upon the stoichiometry and physical dimensions of quantum wells.

As is evident from Figure 5.12, the pump radiation enters the chip at an angle. The off-axis pump geometry facilitates the use of multiple pumps arranged azimuthally for the generation of higher output power. The off-axis pump geometry also ensures that the intra-cavity beam is not obstructed by pump beam relay optics. The non-collinear pumping fills the mode volume and introduces very little ellipticity.

OPSLs offer several advantages over their diode-pumped solid-state laser counterparts, in terms of far less stringent requirement of pump source wavelength band and absence of thermal lensing. Their biggest advantage however is in their ability to be readily customized over a wide range of emission wavelengths and their superior longitudinal mode characteristics. For example, intra-cavity frequency-doubled and -tripled outputs in an optically pumped InGaAs gain chip cover almost the full visible band (355–577 nm).

The above-mentioned features mean that OPSLs find extensive use in medical therapeutics, life sciences, forensics, light shows and scientific research. OPSL technology allows the designer to think of many new applications or improve many existing applications previously limited by the fixed

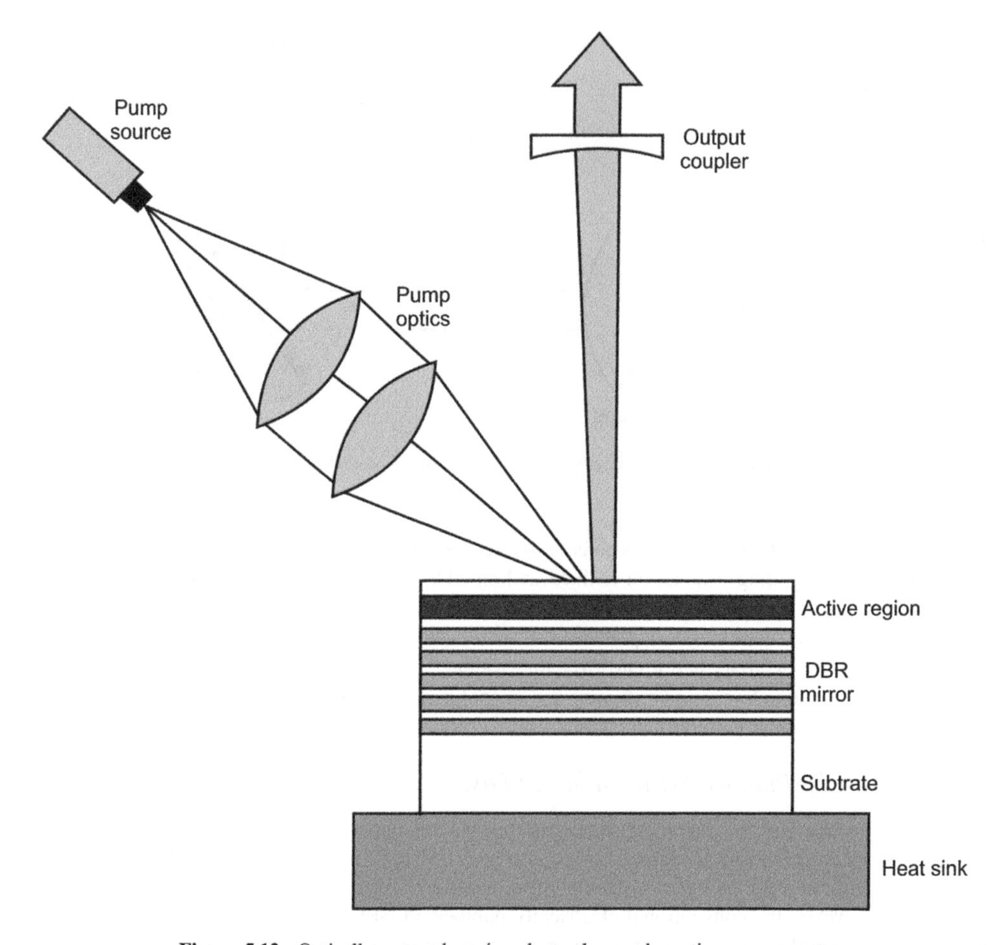

Figure 5.12 Optically pumped semiconductor laser schematic arrangement.

wavelength feature of conventional electrically pumped semiconductor lasers. They use the best wavelength for the intended application, rather than having to work around the closest-fit wavelength.

Figure 5.13 depicts OPSLs of Genesis MX-series from Coherent Inc. The Genesis MX-series is a range of high-power CW optically pumped semiconductor lasers. Variants of this series offer output powers in the range of 500 mW to 20 W and emission wavelength band of 460–1154 nm. Genesis MX-series lasers are intended for use in medicine, life sciences, defence and scientific research applications.

Figure 5.13 Genesis MX-series OPSLs (Reproduced by permission of Coherent Inc.).

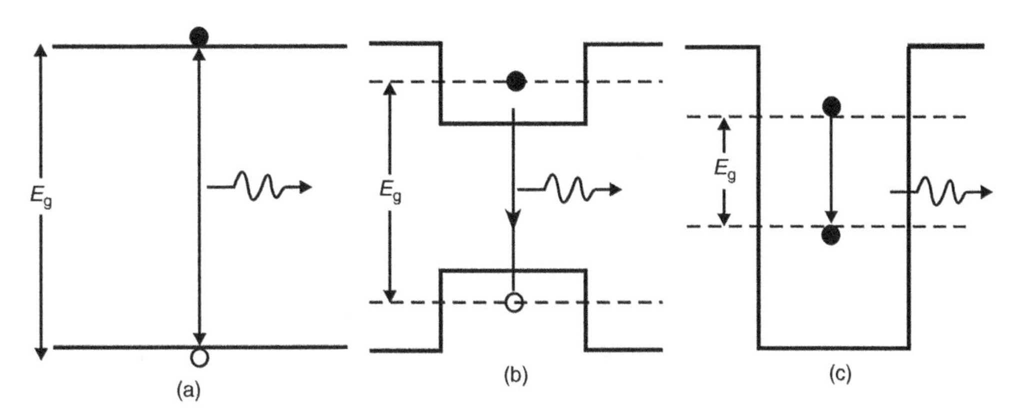

Figure 5.14 Inter-band and intra-band laser transitions in semiconductor lasers.

5.4.8 Quantum Cascade Lasers

Quantum cascade lasers (QCLs) are compact high-power wavelength-agile semiconductor lasers that emit in the mid-infrared to far-infrared wavelength band. The upper limit of emission wavelength can even extend to the terahertz region. The semiconductor lasers described in earlier sections are all inter-band devices where laser radiation is emitted due to the recombination of electrons in the conduction band and holes in the valence band across the band gap of the semiconductor material. The earliest double heterostructure devices had an operating wavelength that was exclusively dependent on the band gap energy as shown in Figure 5.14a. In the case of quantum well structures, as shown in Figure 5.14b, carriers are confined to energy levels within these wells, introducing the possibility of lower transition energies thereby extending the operating wavelength. Semiconductor diode lasers are bipolar devices; quantum cascade lasers on the other hand are unipolar devices. Laser emission in the latter case occurs across inter-subbands, which are also called intra-band transitions, of electrons in the conduction band as shown in Figure 5.14c.

The gain region comprises a periodic structure of an active region and an injector region as shown in Figure 5.15. The overall downward trend in energy levels is due to an applied electric field. In the case of a conventional bipolar semiconductor laser, electrons and holes are annihilated after they undergo recombination across the band gap and can therefore play no further role in photon generation. In the case of unipolar QCL, after the electron has undergone an intra-band transition and generated a photon in a given period of superlattice it can tunnel into the next period of the structure, causing emission of another photon. As the electron traverses through the QCL structure, it causes emission of multiple photons. This increases optical gain, leading to generation of higher output power. The intra-band transition energy in the case of QCL depends upon the design parameters, such as layer thickness of quantum wells, rather than the material system properties. As a result, emission wavelength can be tailored to anywhere within a wide spectral band from mid-infrared to far-infrared for the same semiconductor material.

In order to make a light emitting device the gain region is confined in an optical waveguide, enabling the photon emission to be directed into a collimated beam and a laser resonator to be built. Two types of optical waveguides – ridge waveguide and buried heterostructure waveguide – are in common use. A detailed description of the two types of waveguides is beyond the scope of this text, however.

Several material systems are in use for the fabrication of QCLs. One of the earlier QCLs was fabricated by using a InGaAs/InAlAs material system in an InP substrate. The quantum well depth of 0.52 eV of the InP devices emits in the mid-infrared spectral band. InP-based QCLs have achieved high levels of performance with the development of high-power CW devices operating above room temperature. Another common material system used for fabrication of QCLs is GaAs/AlGaAs. This material system offers variable quantum well depth depending on the aluminium fraction in the barriers. These devices have been successfully used to generate output in the terahertz region. Yet another

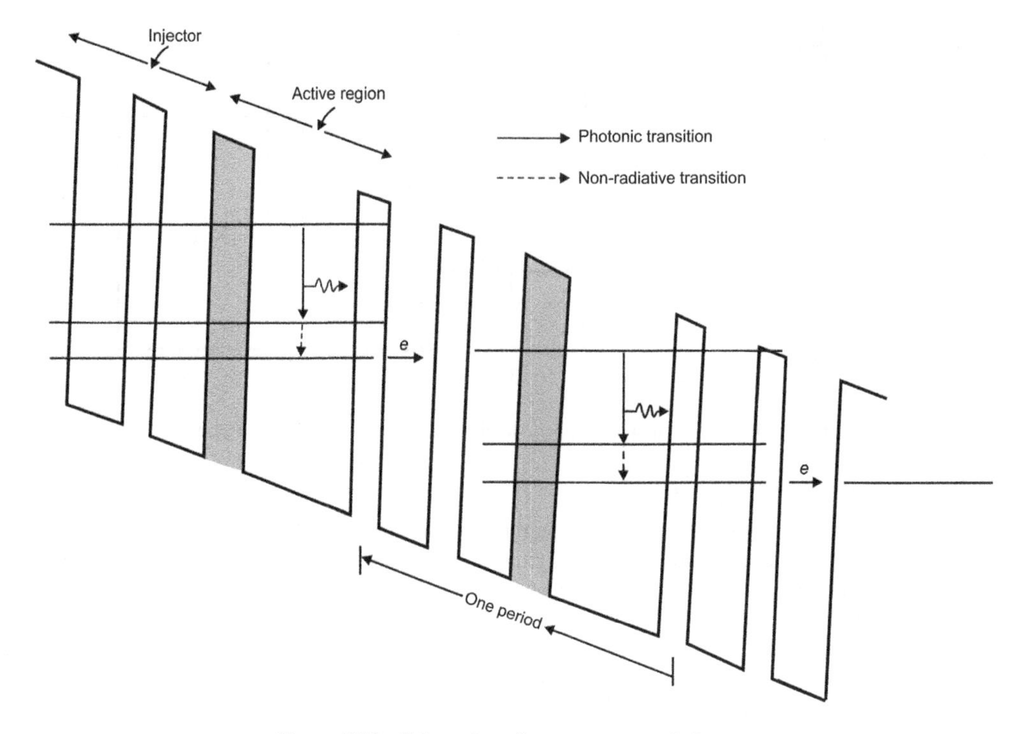

Figure 5.15 Gain region of a quantum cascade laser.

material system in use for QCLs is InGaAs/AlAsSb. With a quantum well depth of 1.6 eV, QCL devices emitting at 3.0 μm have been fabricated. The emission wavelength is further reduced to about 2.5 μm in InAs/AlSb QCLs, which have a quantum well depth of 2.1 eV.

Intra-band optical transitions are independent of the relative momentum of conduction band and valence band minima, with the result that QCLs based on indirect band gap materials are feasible. Si/SiGe is one such material system that has shown promise for fabricating QCL devices. The alternating layers of two different semiconductor materials forming the quantum heterostructure are grown onto a substrate using various methods such as molecular beam epitaxy (MBE), metal organic vapour phase epitaxy (MOVPE) and metal organic chemical vapour deposition (MOCVD).

Different types of resonators are used to build QCL devices. These include Fabry–Pérot lasers, distributed-feedback lasers and external cavity lasers. Fabry–Pérot cavity QCL is the simplest. The ends of the crystalline semiconductor device are cleaved on either end of the waveguide to form two parallel mirrors. The residual reflectivity offered by the cleaved facets creates the resonant cavity. Fabry–Pérot QCLs are capable of generating high powers, although the output tends to be multimode at higher operating currents. The emission wavelength can be tuned by changing the operating temperature of the device.

The DFB QCL has a distributed Bragg reflector built on the top of the waveguide to prevent it from emitting at anything other than the desired wavelength; otherwise it is similar to Fabry–Pérot configuration. This allows the DFB QCL to operate at a single wavelength even at higher operating currents. The laser wavelength can be tuned by changing device temperature. Another mode of operation of DFB QCL is the chirped mode in which the laser is pulsed and the wavelength of laser is rapidly chirped during the course of the pulse. This facilitates rapid scanning of a given wavelength band.

In the case of external-cavity QCL, one or both of the cleaved facets are given anti-reflection coating. As a result, cleaved facets no longer provide cavity action. The optical cavity is created by mirrors placed external to the device. A frequency selective optical element such as a diffraction grating may be used in the external cavity to achieve both single-wavelength operation and wavelength tuning (Figure 5.16).

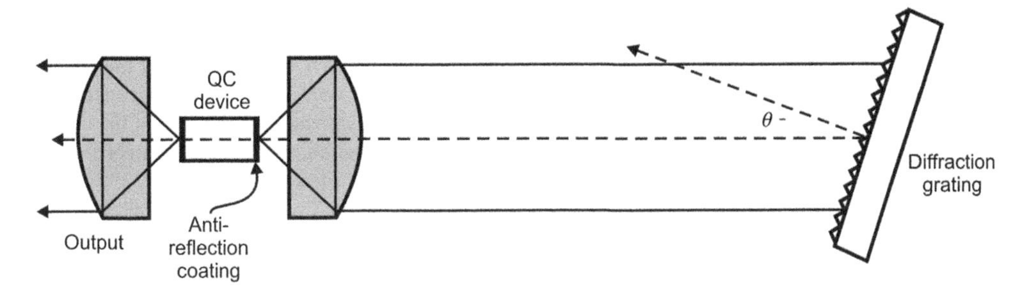

Figure 5.16 External-cavity quantum cascade laser.

High output power in the mid-infrared to far-infrared spectral region extending to terahertz, tunability and room-temperature operation make quantum cascade lasers highly suitable for a wide range of applications in remote sensing of environmental gases and atmospheric pollutants, vehicular cruise control and collision avoidance in poor visibility conditions, medical diagnostics, industrial process control and homeland security.

5.4.9 Lead Salt Lasers

Lead salt semiconductor lasers are semiconductor diode lasers generating wavelength tunable pulsed and continuous wave (CW) output in the mid-infrared spectral band, particularly useful in high-resolution absorption spectroscopy used for detection and identification of trace gases with high sensitivity and selectivity. In general, tunable CW mid-infrared laser sources are grouped as class A and class B sources. Class A sources include those that generate tunable mid-infrared laser radiation directly from the gain media, which could be a gas discharge, a semiconductor material, rare-earth- and transition-metal-doped bulk solid-state material or optical fibre. Class B includes those laser sources where mid-infrared output is generated by the optical parametric frequency conversion of a near-infrared radiation. Most of the mid-infrared laser sources using solid-state, gas and semiconductor lasing media have been covered earlier in this chapter and in Chapters 3 and 4 on solid-state lasers and gas lasers, respectively.

A lead salt semiconductor laser is a P-N junction diode laser that consists of a single crystal of lead telluride (PbTe), lead selenide (PbSe), lead sulphide (PbS) or their alloys with themselves or with strontium selenide (SnSe), strontium telluride (SnTe), cadmium sulphide (CdS) and other materials. The structure of laser cavity is similar to the conventional semiconductor diode laser employing a Fabry–Pérot resonator cavity comprising two parallel end faces. The injection current populates the near-empty conduction band and laser radiation is emitted by the process of stimulated emission across the band gap. In the case of lead salt semiconductor lasers the band gap energy is very small, in the range 0.25–0.30 eV. These lasers require cryogenic cooling for population inversion and laser action. The band gap energy depends on semiconductor composition and temperature. The band gap energy and hence the operating wavelength may be tailored by either varying the stoichiometry between lead and other elements or by using different alloys of lead.

Wavelength tuning over 100 cm^{-1} by changing temperature or over 10 cm^{-1} by changing injection current is possible. Temperature tuning is basically due to a change in band gap energy; injection current tuning is due to a change in refractive index of the active region. Temperature tuning rates of 2.0–5.0 $(cm\,K)^{-1}$ and injection current tuning rates of 0.02–0.07 cm^{-1} are achievable. The temperature tuning mechanism is relatively sluggish as it involves changing the temperature of the entire laser package. On the other hand, current tuning process is very rapid.

Lead salt lasers emit in the 3–30 μm band and are characterized by large beam divergence and astigmatism, which puts particularly stringent requirements on the first optical element of collection optics. Typical CW power level from lead salt lasers is in the range 0.1–0.5 mW. CW operation has been reported for temperatures above 200 K and pulsed operation to above 80 °C.

5.5 Characteristic Parameters

Important characteristic parameters of diode lasers relate either to their *I–V* characteristics or the output beam characteristics. These include threshold current, slope efficiency and linearity of laser operation in the case of the former, and beam divergence, line width and beam polarization in the case of the latter. Most of the characteristic parameters are sensitive to changes in temperature. Different parameters and their sensitivity to temperature changes are briefly described in the following sections.

5.5.1 Threshold Current

Threshold current is the minimum forward-biased injection current needed to achieve sustained laser action. When the injection current is below the threshold value, most of the input electrical energy is dissipated as heat and conversion to light output is highly inefficient. Higher threshold current therefore means that more electrical energy must be dissipated as heat energy. Current density is defined as ratio of threshold current to the active area and measured in units of A cm^{-2}. It is also referred to as the figure-of-merit of the diode laser, as it determines the lifetime of the laser. Higher-threshold current density shortens the lifetime of the laser diode. *I–V* characteristics of the laser diode indicating threshold current are highlighted in Figure 5.3, and the light power versus drive current characteristics of a semiconductor diode laser indicating lasing threshold current are depicted in Figure 5.3.

Threshold drive current is a function of temperature and increases rapidly with increasing temperature. The change in threshold current I_s (ΔT) due to incremental change in temperature ΔT is computed from Equation 5.2:

$$I_s(\Delta T) = I_s(T) \times \left[e^{(\Delta T/T_0)} - 1\right] \tag{5.2}$$

where $I_s(T)$ is the threshold current at absolute temperature T and T_0 is a substrate-specific characteristic temperature (e.g. T_0 lies within 120–230 K for GaAlAs diode laser and 60–80 K for InGaAsP lasers). The lower the value of T_0, the more sensitive is the laser to changes in temperature.

A shift in threshold current occurs primarily due to the temperature dependency of the carrier concentration in the active layer and the probability of non-radiative recombination processes. Figures 5.17 and 5.18 show the effect of change in temperature on *I–V* characteristics and light power versus drive current characteristics of semiconductor lasers, respectively. An increase in temperature also adversely affects the lifetime of the laser as it leads to an increase in threshold current density. Lifetime approximately doubles for every decrease in chip temperature by 10 °C. This is taken into account by mounting the chip on a heat sink.

Example 5.2

The temperature of a certain InGaAsP laser chip increases by 30 K. If the characteristic temperature figure for this laser is taken to be 60 K, prove that the threshold current would undergo a 65% increase from its value before the occurrence of temperature change.

Solution:

1. Percentage increase in threshold current can be computed from

$$[I_s(\Delta T)/I_s(T)] \times 100 = \left[e^{(\Delta T/T_0)} - 1\right] \times 100$$

2. Therefore, $I_s(\Delta T)/I_s(T) = [e^{(30/60)} - 1] \times 100 = (e^{0.5} - 1) \times 100 = (1.65 - 1) \times 100 = 65\%$.

5.5.2 Slope Efficiency

Slope efficiency is determined by the slope of the characteristic *I–V* curve above the threshold current and is measured in mW mA^{-1} (or W A^{-1}). In the case of a pigtailed device, slope is reduced by a factor depending upon the coupling efficiency of laser power into the fibre.

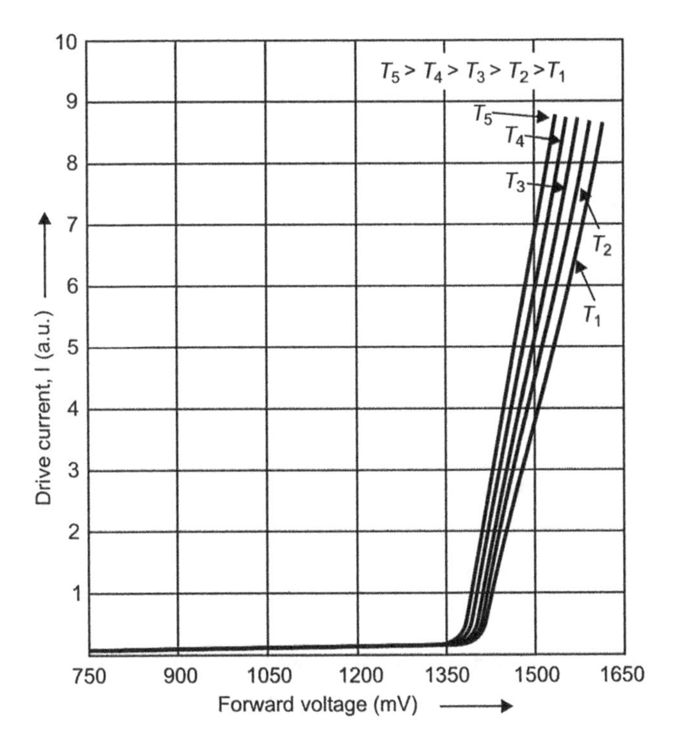

Figure 5.17 Temperature dependence of *I–V* characterristics.

Slope efficiency is strongly dependent upon temperature and decreases with increasing temperature, as shown in Figure 5.18.

Example 5.3

Refer to the family of curves shown in Figure 5.19. Determine the percentage change in slope efficiency of the laser diode as the operating temperature changes from 25 °C to 50 °C.

Solution:

1. For the output power versus drive current at 25 °C, above lasing threshold,

   Output power for a drive current of 50 mA = 2.5 mW
   Output power for a drive current of 60 mA = 5 mW.

2. Therefore, slope efficiency = (5 – 2.5)/(60 – 50) mW mA^{-1} = 0.25 mW mA^{-1}
3. For the output power versus drive current at 50 °C, above lasing threshold,

   Output power for a drive current of 60 mA = 2 mW
   Output power for a drive current of 70 mA = 4 mW.

4. Therefore, slope efficiency = (4 – 2)/(70 – 60) mW mA^{-1} = 0.2 mW mA^{-1}
5. Percentage decrease in slope efficiency = (0.25 – 0.20/0.25) × 100% = 20%.

5.5.3 Beam Divergence

Laser diodes characteristically produce a highly diverging laser beam with the exception of surface-emitting diode lasers. Higher divergence is primarily influenced by the diffraction of light waves as they are coupled out of the chip. Also, due to the rectangular-shaped active light-emitting area with

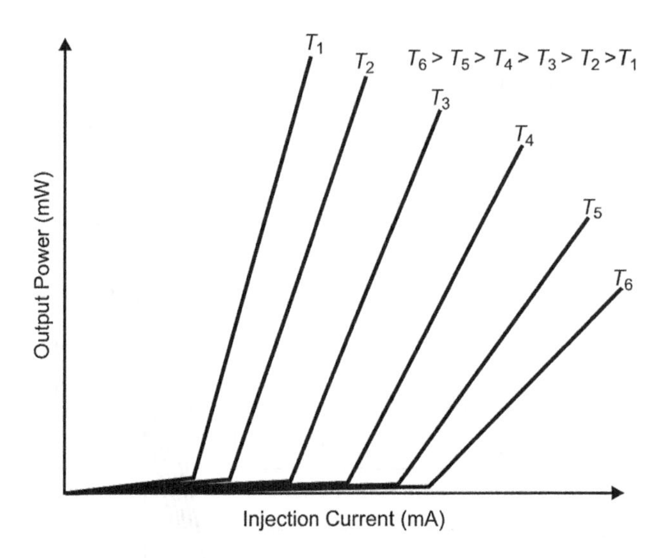

Figure 5.18 Temperature dependence of light power versus drive current characteristics.

strongly differing edge lengths, the divergence in the two orthogonal planes is different as shown in Figure 5.20. Divergence in the plane parallel to the plane of the active layer is relatively much smaller than that in the plane perpendicular to the active layer. As a result, the laser beam appears as an elliptical spot at some distance from the laser. If required, it is possible to circularize the elliptical beam with the help of a cylindrical lens, which refracts the light in parallel direction. VCSELs produce a relatively more symmetrical beam and the beam divergence is also relatively lower due to their large emitting area.

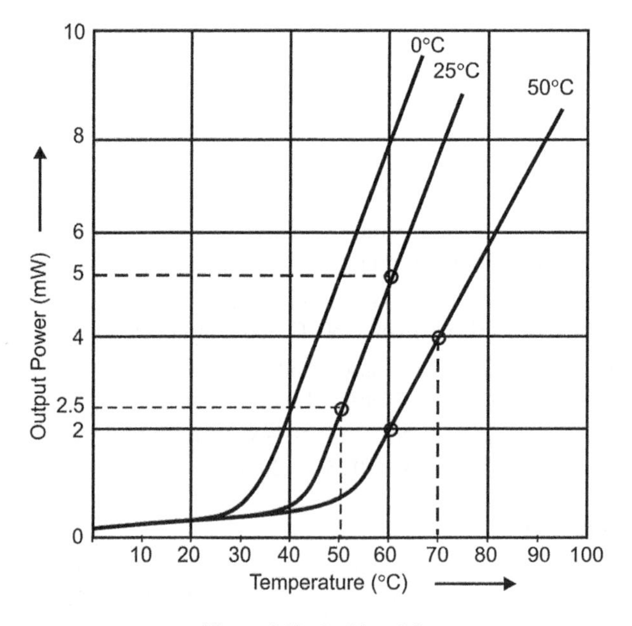

Figure 5.19 Problem 5.3.

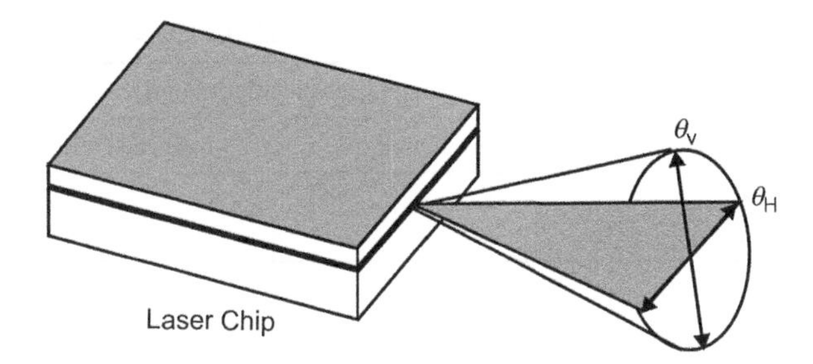

Figure 5.20 Beam divergence in two orthogonal planes in edge-emitting semiconductor lasers.

5.5.4 Line Width

Line width is another important specification. In the case of gain-guided lasers (described in Section 5.6), the envelope of gain profile typically has a 3 dB width of 2–3 nm, which corresponds to a frequency range of about 1000 GHz at 800 nm. Such a gain-bandwidth profile can support a large number of longitudinal modes. In the case of an index-guided laser (also described in Section 5.6), one spectral line is dominant with the result that the line width is much narrower, typically 10^{-2} nm. This corresponds to a frequency spread of a few GHz at 800 nm. In the case of DFB lasers, which is also index guided, the line width is still smaller; it is typically of the order 10^{-4} nm, which corresponds to a frequency spread of a few MHz at 800 nm.

The spectral profile of semiconductor lasers is also strongly affected by temperature changes. Both the gain profile as well as individual lines shift to longer wavelengths with an increase in temperature. Figure 5.21 shows the effect of change in temperature on gain profile. Effect on individual lines is due to

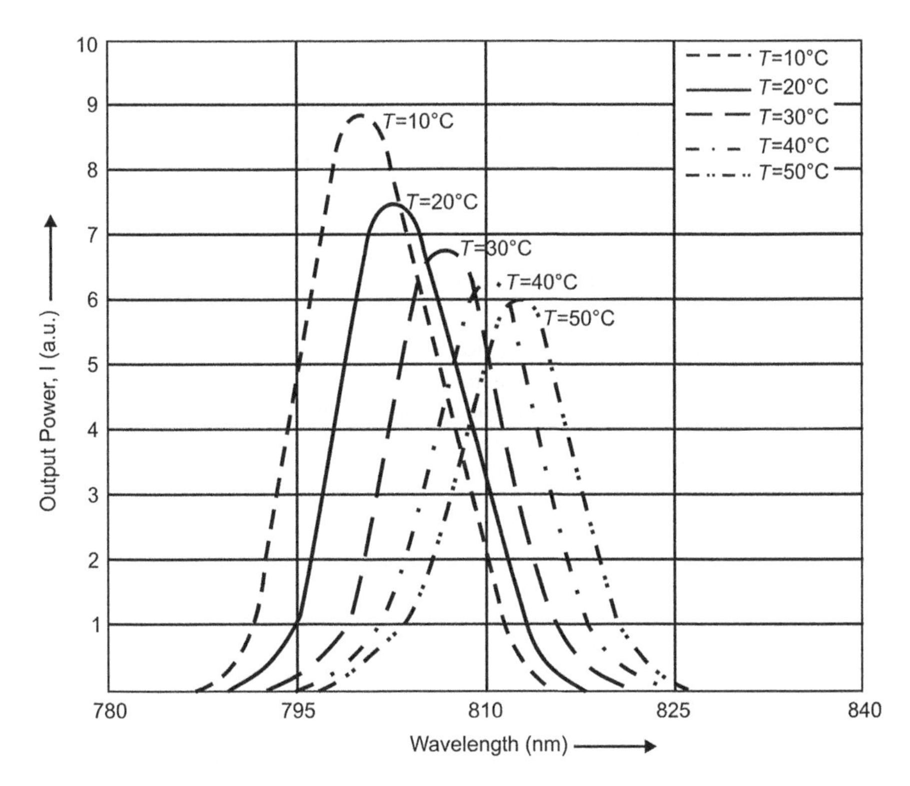

Figure 5.21 Effect of change in temperature on gain profile.

Table 5.2 Temperature coefficients of common semiconductor lasers

Type of laser	Temperature coefficient (nm K^{-1})	
	Envelope variation	Individual line variation
GaAlAs	0.24	0.12
InGaAsP	0.30	0.08

an increase in resonator length accompanied by an increase in refractive index. Typical values of temperature coefficient are listed in Table 5.2.

5.5.5 Beam Polarization

Beam polarization is another important parameter. Diode lasers emit almost linearly polarized light if driven above threshold, which is influenced by polarization dependency of the reflection factor of the emission area of the crystal.

5.6 Gain- and Index-guided Diode Lasers

The narrow stripe of the active layer that is responsible for generating laser output defines whether the laser is gain guided or index guided. In the case of *gain-guided lasers*, an insulating layer on the top of the laser chip confines the flow of injection current to a narrow stripe in the active layer, with the result that only this region has sufficient gain to produce sustained laser action. There is no gain at the sides so those regions do not emit light, even though no physical boundary separates the stripe from the rest of the active layer.

In the case of *index-guided lasers*, the stripe from where laser emission takes place is defined by semiconductor composition and associated change in refractive index. The current flows only through the central 'mesa' and the stripe of the active layer buried below it. The layers that border the sides of the active layer have a different composition and different refractive index, trapping the light in the active stripe as the adjacent layers confine light in the active layer in a double-heterostructure laser. Index-guided lasers produce better beam quality and are used in most laser applications.

5.7 Handling Semiconductor Diode Lasers

Diode lasers exhibit high reliability and a long lifetime exceeding 100 000 hours providing certain precautions are taken while handling them and also in the design of the driver circuits used to power them. Diode lasers are particularly sensitive to electrostatic discharge, short duration electric transients such as current spikes, injection current exceeding the prescribed limit and reverse voltage exceeding the breakdown limit. Damage to the laser diode often manifests itself in the form of reduced output power, a shift in the value of threshold current and its inability to be focused to a sharp spot. The electrostatic discharge caused by human touch is the most common cause of premature failure of a diode laser.

In order to protect the diode lasers from the above failure modes, the driver circuit should be carefully designed and should have all the features recommended by the manufacturer of the laser. The driver should be a constant current source with inbuilt features such as soft start, protection against transients, interlock control for the connection cable to the laser and safe adjustable limit for injection current. If the laser is to be operated in the pulsed mode, the injection current should be pulsed between two values above the lasing threshold, rather than between cut-off and lasing mode. Leading manufacturers of diode lasers offer a wide variety of suitable current sources for low-, medium- and high-power laser diodes. They also offer temperature controllers to stabilize the output wavelength. Drivers and temperature controllers for diode lasers are discussed in Chapter 9.

5.8 Semiconductor Diode Lasers: Application Areas

Diode lasers find extensive use in commercial, industrial, medical, military and scientific applications. Depending on application, one or more of the inherent characteristics of the laser beam is exploited, including directed energy due to low divergence, coherence and monochromaticity (or narrow bandwidth of emission spectrum).

5.8.1 Directed Energy

The applications that utilize the directed energy property of the laser light include laser printing, barcode reading, image scanning, target illumination and designation, laser surgery, laser-based ignition of combustion reaction, laser-based ignition of explosives and optical data recording.

In the case of laser printing, the laser beam modulated by the information to be printed scans across a rotating drum that is coated with some kind of light-sensitive material. In doing so, it writes a pattern of charged and uncharged areas corresponding to the printed page. The drum picks up the toner and transfers the image to the plane paper. In the case of barcode reading with lasers, the laser beam illuminates the object. A receiver circuit then reads the reflected beam containing the information. Low-power visible and infrared diode lasers are used on small arms to illuminate targets to improve accuracy of aim in low light and dark conditions. The laser beam is bore-sighted with the barrel of the weapon. Infrared laser diodes are also used as target illuminators in night vision devices to enhance their performance. Laser-based ignition is achieved by focusing the laser beam to the region of interest. Optical data recording is also performed by a modulated laser beam containing the information to be stored by scanning across a light-sensitive surface.

5.8.2 Coherence

Applications that exploit the coherence property of laser light include distance measurement with an interferometer, holography and coherent communication. An interferometer makes use of the phenomenon of interference of waves for accurate measurement of small distances. Accuracy of the order of less than the wavelength of the laser used is possible with this method. Holography allows us to record three-dimensional images of objects. Both amplitude as well as phase information of the light reflected from the object is recorded on the photographic plate to create a hologram. In order to record the hologram of a three-dimensional object, the coherence length of the laser used should be at least equal to the dimensions of the object. Initially, laser-created holograms could only be viewed in laser light; advances have made it possible to view them in white light. The phenomenon of holography combined with interferometry can be used to study the deformation behaviour of objects under stress.

5.8.3 Monochromaticity

Applications that make use of monochromaticity mainly include telecommunications, spectroscopy and a large number of biomedical diagnostic and therapeutic applications. Spectroscopy helps in the study of materials by measuring wavelengths absorbed and emitted by them. The monochromaticity of lasers allows precise quantification of absorption and emission phenomena.

Laser diodes are extensively used in telecommunications because of the ease with which they can be modulated and coupled to optical fibres. Biomedical diagnostics is simply the application of laser-based spectroscopy. Laser-based fluorescence and absorption spectroscopy is used in diagnostic instruments to identify abnormalities, count cells and make many more measurements. Lasers are widely used in a variety of therapeutic applications which exploit the wavelength-dependent interaction of laser light and body tissues. Laser ophthalmology, laser dermatology, laser surgery including fibre-optic surgery, laser dentistry and many more applications such as laser-based photodynamic therapy for cancer treatment are very common. Another potential application of laser-based spectroscopy is in the detection and identification of chemical and biological contaminants in the atmosphere.

The applications described here are only a summary of the major application areas; these and many others are covered in detail in Part IV of the book covering applications of lasers and optoelectronics.

5.9 Summary

- In a semiconductor diode laser, emission of radiation is due to the recombination of electrons and holes in a forward-biased P-N junction. Only direct band gap semiconductor materials are suitable for making diode lasers. Compound semiconductors are used for making diode lasers; the most important of these comprise equal amount of elements from the III-A and V-A periodic table groups. Both ternary and quaternary compounds are used.
- Depending upon the structure of semiconductor materials used, diode lasers are categorized as homojunction and heterojunction lasers, quantum well lasers, DFB lasers, VCSELs, VECSELs, external-cavity semiconductor diode lasers, optically pumped semiconductor lasers, quantum cascade lasers and lead salt lasers.
- All layers in a homojunction laser are of the same semiconductor material. One such example is GaAs/GaAs lasers. However, this simple laser diode structure is highly inefficient and can only be used to demonstrate pulsed operation.
- The active layer and either one or both of the adjacent layers in a heterojunction laser are of different materials. If only one of the adjacent layers is of a different material it is a simple heterojunction; if both are different, it is called a double heterojunction.
- A quantum well layer is placed in the semiconductor junction of a quantum well diode laser. The concentration of electrons as they recombine with holes to emit photons leads to an increase in efficiency and a decrease in lasing threshold. The fact that the quantum well layer and the outer thicker layers are made from semiconductor materials of different refractive indices contributes to performance by confining the emitted photons to that narrow region.
- Lasers with more than one quantum well layer are known as multiple quantum well lasers, a stacked arrangement of quantum wells and confinement layers. Multiple quantum well structures can generate relatively higher power outputs.
- The extension of the quantum well concept has led to the development of quantum wire and quantum dot structures. While a quantum well structure can confine electrons in only one dimension (i.e. in the plane of quantum well layer), quantum wire and quantum dot structures provide confinement in two and three dimensions, respectively. This leads to a further improvement in laser efficiency.
- In a quantum cascade laser, the laser transition occurs between different energy levels of quantum well and not the band gap. This allows operation at longer wavelengths, which can be tuned by changing the thickness of the quantum layer. QCLs are compact high-power wavelength-agile semiconductor lasers that emit in the mid-infrared to far-infrared wavelength band.
- Different types of resonators are used to build QCL devices: Fabry–Pérot lasers, distributed-feedback lasers and external-cavity lasers. High output power, tunability and room temperature operation make QCLs highly suitable for a wide range of applications. DFB lasers ensure a very narrow output wavelength band by use of a diffraction grating which provides a wavelength-selective feedback to the gain region. The feedback causes interference and only the narrow wavelength region for which the interference is constructive is sustained.
- In the case of VCSEL, the optical cavity is along the direction of flow of injection current. The laser beam in this case emerges from the surface of the wafer rather from its edges. The two mirrors are either grown epitaxially as a part of the diode structure or grown separately and then bonded to the semiconductor chip having the active region. A VECSEL is a variant of VCSEL, where one of the mirrors is placed externally to the diode structure thus introducing a free-space region in the resonant cavity.
- Lead salt semiconductor lasers are semiconductor diode lasers generating wavelength tunable pulsed and CW output in mid-infrared spectral band, particularly useful in high-resolution absorption spectroscopy used for detection and identification of trace gases with high sensitivity and selectivity. A lead salt semiconductor laser is a P-N junction diode laser that consists of a single crystal of PbTe, PbSe, PbS or their alloys with themselves or with SnSe, SnTe, CdS and other materials.
- Lead salt lasers emit in the 3–30 μm band and are characterized by large beam divergence and astigmatism, which puts particularly stringent requirements on first optical element of collection

optics. Typical CW power level from lead salt lasers is in the range 0.1–0.5 mW. CW operation has been reported for temperatures above 200 K and pulsed operation to above 80 °C.

- Important characteristic parameters of diode lasers relate either to their *I–V* characteristics (threshold current, slope efficiency and linearity of laser operation) or the output beam characteristics (beam divergence, line width and beam polarization). Most of the characteristic parameters are sensitive to changes in temperature.
- Both threshold current and slope efficiency are adversely affected by an increase in temperature. While threshold current increases with increasing temperature, slope efficiency decreases with increasing temperature. The spectral profile of diode lasers is also strongly affected by temperature changes. Both the gain profile as well as individual lines shift to longer wavelengths with increasing temperature.
- In the case of gain-guided lasers, an insulating layer on the top of the laser chip confines the flow of injection current to a narrow stripe in the active layer, with the result that only this region has sufficient gain to produce sustained laser action. In the case of index-guided lasers, the stripe from where laser emission takes place is defined by semiconductor composition and an associated change in refractive index. Index-guided lasers produce a better beam quality and are used in most laser applications.
- Diode lasers are particularly sensitive to electrostatic discharge, short duration electrical transients such as current spikes, injection current exceeding prescribed limit and reverse voltage exceeding breakdown limit.

Review Questions

5.1. Briefly describe the principle of operation of a diode laser. Define lasing threshold, slope efficiency and beam divergence with respect to diode lasers.

5.2. What precautions should you observe while handling and using diode lasers, if you were concerned about a long life and reliable performance? In what way is deterioration in the performance of diode laser usually manifested?

5.3. How does temperature affect the following parameters of diode lasers?
a. slope efficiency
b. threshold current
c. gain-bandwidth envelope
d. different longitudinal modes within the gain envelope.

5.4. Write down two characteristic features of each of the following:
a. distributed-feedback (DFB) lasers
b. vertical-cavity surface-emitting lasers (VCSEL)
c. heterojunction lasers.

5.5. Which laser property plays a key role in each of the following applications? Give one or two reasons for your answer.
a. biomedical diagnostics
b. spectroscopy
c. laser metrology using interferometry
d. holography
e. laser designation
f. laser pointing
g. fibre-optic communication.

Problems

5.1. It is desired that a distributed feedback laser produce output at 1200 nm. Determine the pitch of the first-order diffraction grating if the semiconductor has a refractive index of 3 at the desired wavelength.
[200 nm]

5.2. The temperature of a certain InGaAsP laser chip increases by 25 K. If the characteristic temperature figure for this laser is taken to be 75 K, determine percentage change in the threshold current with respect to its value before the occurrence of temperature change.
[39.6%]

5.3. In a laser diode, drive current of 50 mA produces an output power of 5 mW. When the drive current is increased by 20%, the output power increases by 100%. Determine slope efficiency.
[0.5 mW mA^{-1}]

Self-evaluation Exercise

Multiple-choice Questions

5.1. The parameter that can possibly be used to stabilize output wavelength in the case of a semiconductor laser is:
a. drive current
b. diode temperature
c. both (a) and (b)
d. none of the above.

5.2. The generalized formula for the most commonly used ternary compounds in semiconductor diode lasers is:
a. $Ga_{1-x}Al_xAs$
b. $Ga_xAl_{1-x}As$
c. $Ga_{1-x}AlAs_x$
d. Ga_xAlAs_{1-x}.

5.3. In the case of diode lasers,
a. slope efficiency increases with an increase in temperature
b. gain profile shifts towards shorter wavelengths with an increase in temperature
c. slope efficiency decreases with an increase in temperature
d. threshold current increases with an increase in temperature
e. both (c) and (d).

5.4. In the case of ternary and quaternary compound semiconductors used for making laser diodes,
a. the total quantity of group III-A elements is more than the total quantity of group V-A elements
b. the total quantity of group III-A elements is less than the total quantity of group V-A elements
c. the total quantity of group III-A elements is equal to the total quantity of group V-A elements
d. none of these.

5.5. Slope efficiency of a laser diode is measured in
a. AV^{-1}
b. VA^{-1}
c. WA^{-1}
d. WA^{-1} or mW mA^{-1}.

5.6. The diode laser type that employs a diffraction grating etched very close to its active layer is the
a. quantum well laser
b. vertical-cavity surface-emitting laser
c. distributed-feedback laser
d. tunable diode laser.

5.7. The reason for a double heterostructure semiconductor laser having a higher efficiency than a homostructure semiconductor laser is:
a. lower threshold current in the case of the former
b. higher levels of spontaneous emission in the case of the latter

c. better confinement of stimulated emission to the active layer in the case of the former
d. restriction of current flow to the active layer in the case of former.

5.8. Identify the type of semiconductor laser that emits a single longitudinal mode from the following:
a. distributed-feedback laser
b. homojunction diode laser
c. heterojunction diode laser
d. vertical-cavity surface-emitting laser.

5.9. Which of the following types of laser diode produces output in red?
a. GaAlAs
b. AlGaInP
c. InGaAsP
d. none of these.

5.10. Identify the false statement from the following.
a. Only direct band gap semiconductor materials are suitable for making laser diodes.
b. Distributed-feedback lasers produce output in a narrow bandwidth.
c. InGaAsP is a quaternary III-V semiconductor
d. InGaAsP lasers emit in the 750–870 nm wavelength band.

Answers

1. (c) 2. (a) 3. (e) 4. (c) 5. (d) 6. (c) 7. (c) 8. (a) 9. (b) 10. (d)

Bibliography

1. *Handbook of Laser Technology and Applications*, 2003 Volume II by Collin E. Webb and Julian D. C. Jones, Institute of Physics Publishing.
2. *Semiconductor Lasers*, 2013 by Ohtosubo Junji, Springer.
3. *Long Wavelength Infrared Semiconductor Lasers*, 2004 by Hong K. Choi, John Wiley & Sons.
4. *Physics of Semiconductor Laser Devices*, 1980 by G. H. B. Thompson, John Wiley & Sons.
5. *Semiconductor Lasers-II: Materials and Structures*, 1999 by Eli Kapon, Academic Press.
6. *Semiconductor Lasers-I: Fundamentals*, 1999 by Eli Kapon, Academic Press.
7. *Semiconductor Laser Fundamentals: Physics of the Gain Materials*, 1999 by Weng W. Chow and Stephen W. Koch, Springer.
8. *The Blue Laser Diode: The Complete Story*, 2000 by Shuji Nakamura, Gerhard Fasol and Stephen J. Pearton, Springer.
9. *High Power Laser Handbook*, 2011 by Hagop Injeyan and Gregory D. Goodno, McGraw-Hill.
10. *Overview of Laser Diode Characteristics: Application Note # 5*, 2005 by Tyll Herstens, ILX Lightwave Photonic Test and Measurement.

Part III

Laser Electronics and Optoelectronics

6

Building Blocks of Laser Electronics

6.1 Introduction

Chapter 1–5 have discussed the fundamentals such as operational basics (Chapter 1) and characteristics of lasers (Chapter 2) followed by the three main types of lasers in Chapters 3–5. In the present chapter we discuss the basic building blocks of electronics commonly used to design most of the laser sources and systems configured around them. The intention is to familiarize the readers with the operational basics of these building blocks and make it easier to understand the specific laser electronics packages discussed in the following chapters. This chapter will particularly benefit laser and optoelectronics students and professionals who do not have comprehensive knowledge of electronics. Chapters 7–9 are then devoted to extensive coverage of electronics packages of each of the three major types of lasers (solid-state, gas and semiconductor).

6.2 Linear Power Supplies

Every electronics system, whether an entertainment gadget or a test-and-measurement device, requires one or more DC voltages for operation. It is essential in most cases and almost always desirable that these DC voltages are filtered and well regulated. The power supply provides required DC voltages from available AC mains in the case of mains-operated systems and from DC input in the case of portable systems. Power supplies are often classified as linear power supplies or switched-mode power supplies, depending upon the nature of regulation circuit. In the following sections we discuss the basic constituents of a linear power supply, including transformers, rectifiers, filters and voltage regulators. Switched-mode power supplies are discussed in Section 6.3.

6.2.1 Constituents of a Linear Power Supply

A linear power supply essentially comprises a *mains transformer*, a *rectifier circuit*, a *filter circuit* and a *regulation circuit* (Figure 6.1). The transformer provides voltage transformation and produces the AC voltage(s) across its secondary winding(s) required for producing desired DC voltages. It also provides electrical isolation between the input power supply, that is, AC mains and the DC output. Step-down transformers required for generating common DC voltages and load current ratings are commercially available. Step-up transformers for generating higher output voltages could be custom designed.

The rectifier circuit changes the AC voltage appearing across the transformer secondary to DC or, more precisely, a unidirectional output. Commonly used rectifier circuits include the half-wave rectifier, conventional full-wave rectifier (requiring a tapped secondary) and the full-wave bridge rectifier.

Lasers and Optoelectronics: Fundamentals, Devices and Applications, First Edition. Anil K. Maini.

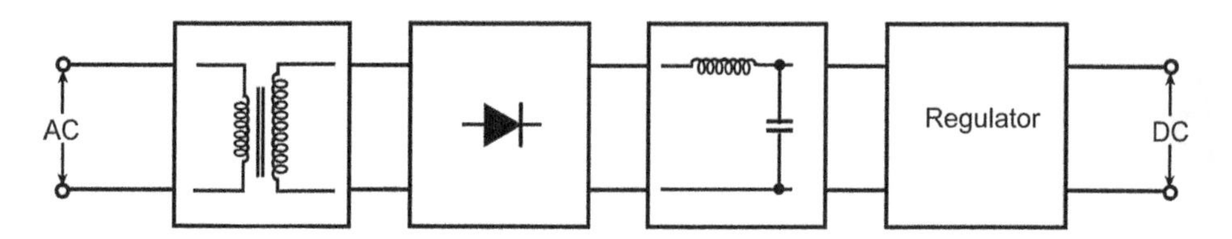

Figure 6.1 Constituents of a linear power supply.

The rectifier voltage will always have some AC content, known as power supply ripple. The filter circuit smoothes the ripple of the rectified voltage. The regulator circuit is a type of feedback circuit that ensures that the output DC voltage does not change from its nominal value due to a change in line voltage or load current.

In a linearly regulated power supply the active device, usually a bipolar junction transistor, is operated anywhere between cut-off and saturation. Commonly used regulator circuit configurations include an emitter-follower regulator, a series-pass transistor regulator and a shunt regulator. An emitter-follower regulator is in fact now available in an integrated circuit (IC) package in both fixed-output voltage as well as variable-output voltage varieties. These three terminal regulators are discussed in Section 6.2.4.

All power supplies have in-built protection circuits. Common protection features include current limit, short-circuit protection, thermal shut down and crowbarring.

6.2.2 Rectifier Circuits

Figures 6.2 and 6.3 show the three rectifier circuits for positive and negative output voltages, respectively, along with input and output waveforms. The rectifier circuits shown in Figures 6.2a and 6.3a are half-wave rectifier circuits. Circuits shown in Figures 6.2b and 6.3b are conventional full-wave rectifier circuits. Figures 6.2c and 6.3c show full-wave bridge rectifier configurations. The circuits are self explanatory. Some common terms used with reference to rectifier circuits are defined in the following.

Ratio of rectification is the ratio of DC power delivered to the load to the AC power rating of the transformer secondary. It is the lowest (=0.287) in the case of half-wave rectifier, 0.693 for a conventional full-wave rectifier and highest (=0.812) for a bridge rectifier.

Ripple frequency in the case of a half-wave rectifier is f and $2f$ for full-wave rectifiers (conventional and bridge), where f is the frequency of AC mains.

Ripple factor is the ratio of the RMS amplitude of the AC component to the the DC component. It is 1.21 for half-wave and 0.482 for full-wave rectifiers.

Peak inverse voltage that appears across the rectifier diodes in the three cases is V_m for a half-wave rectifier, $2V_m$ for a conventional full-wave rectifier and V_m for a bridge rectifier, where V_m is the peak value of the AC voltage appearing across the secondary.

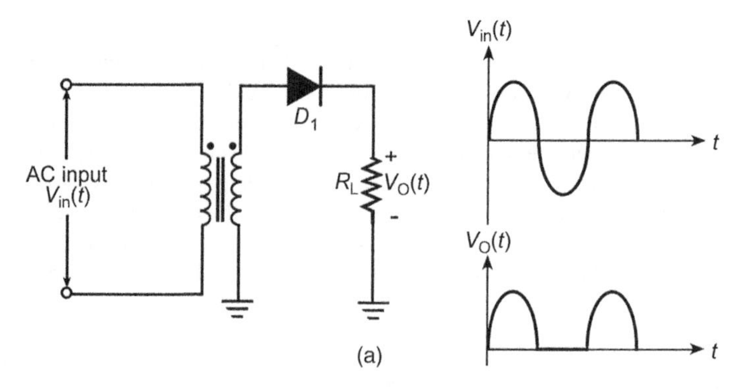

Figure 6.2 Rectifier circuits for positive output voltages: (a) half-wave rectifier; (b) conventional full-wave rectifier; and (c) full-wave bridge rectifier.

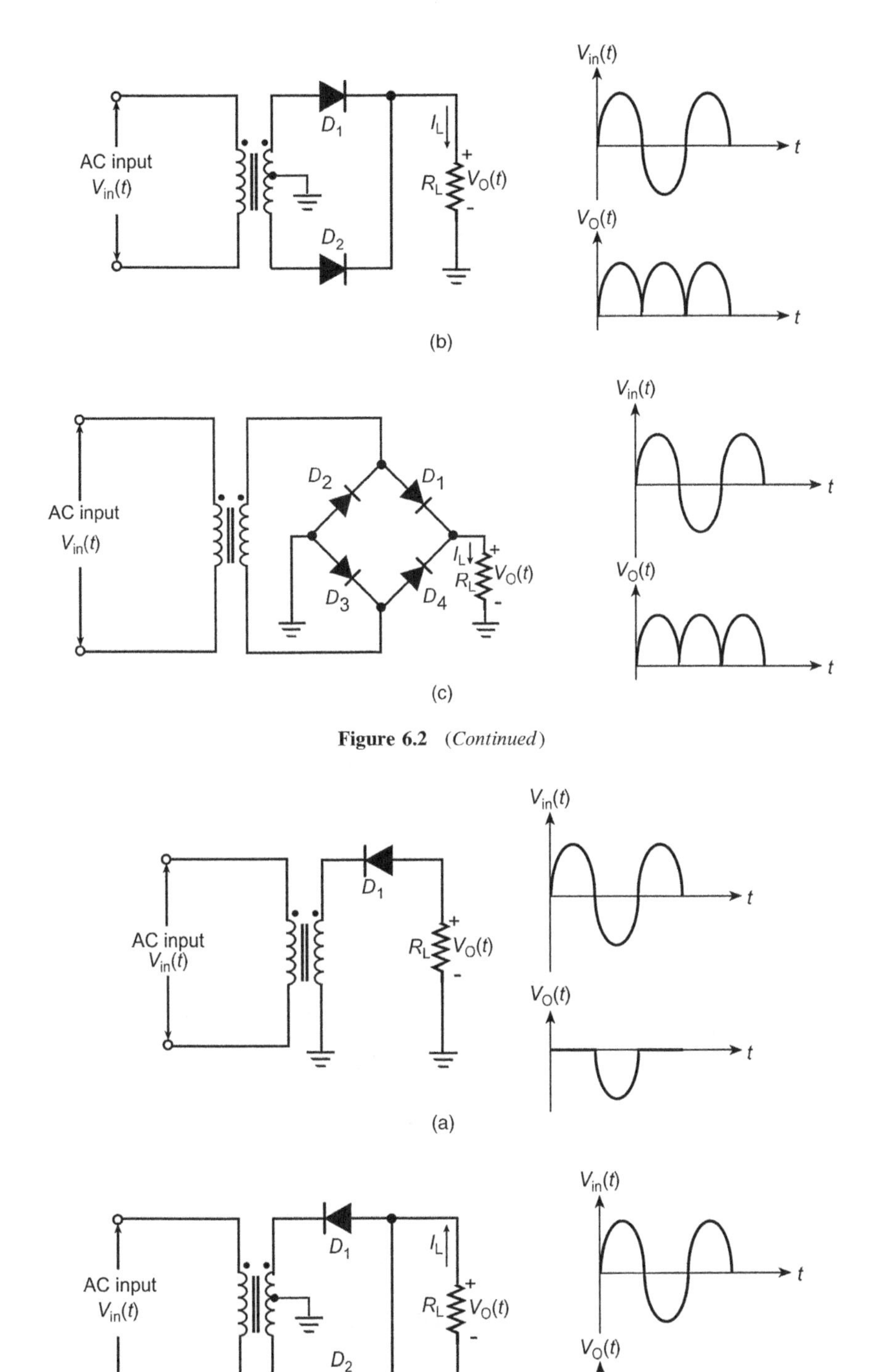

Figure 6.2 *(Continued)*

Figure 6.3 Rectifier circuits for negative output voltages: (a) half-wave rectifier; (b) conventional full-wave rectifier; and (c) full-wave bridge rectifier.

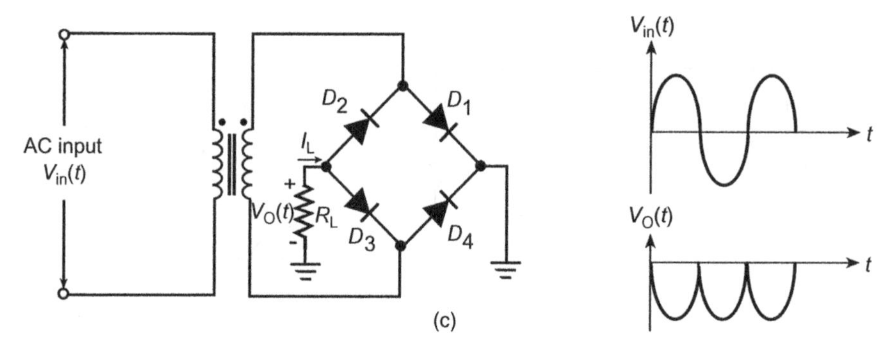

Figure 6.3 *(Continued)*

6.2.3 Filters

The filter in a power supply helps to reduce the ripple content which, in the rectified waveform, is so large that the waveform can hardly be called a DC. Inductors, capacitors and inductor–capacitor combinations are used for the purpose of filtering. The fact that an inductor offers high reactance to AC components is the basis of filter action provided by inductors. In the case of an inductor filter, also referred to as a choke filter, the ripple factor r is given by Equation 6.1:

$$r = \frac{R_L}{3\sqrt{2}(\omega L)} \tag{6.1}$$

where R_L is load resistance in ohms (Ω), L is inductance in henries (H) and ω ($=2\pi f$) is angular frequency in radians. The above equation simplifies to $R_L/1330L$ for a power line frequency of 50 Hz. Figure 6.4 shows an inductor filter used at the output of a full-wave bridge rectifier, along with input and output waveforms.

As is clear from the above expression, the ripple factor r is directly proportional to load resistance R_L. That is, ripple content increases with an increase in load resistance. In other words, a choke filter is not

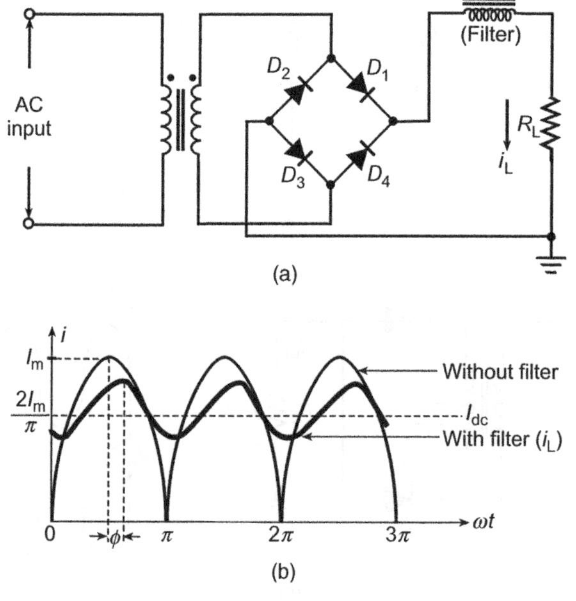

Figure 6.4 Inductor (choke) filter.

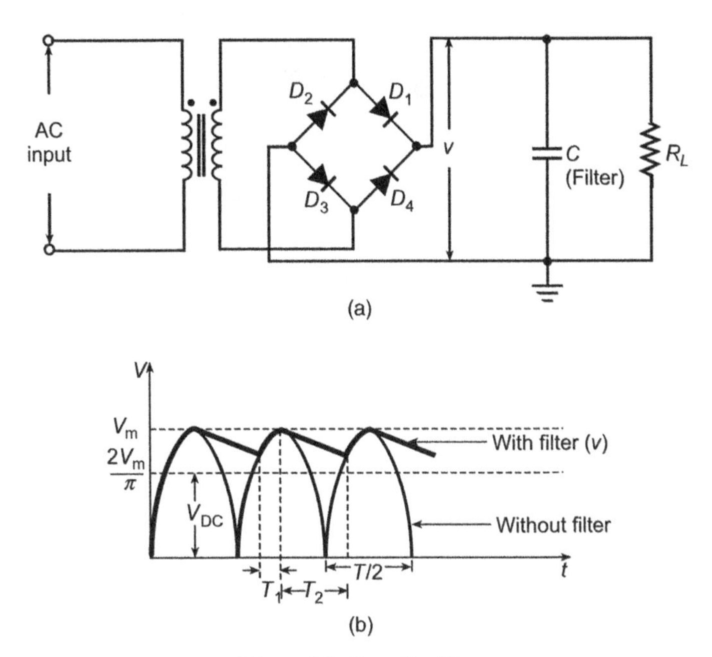

Figure 6.5 Capacitor filter.

effective for light loads (or high values of load resistance) and is preferably used for relatively higher load currents.

The filtering action of a capacitor connected across the output of the rectifier comes from the fact that it offers a low reactance to AC components. Figure 6.5a shows the capacitor filter connected across the output of a bridge rectifier. The AC components are bypassed to ground through the capacitor and only the DC is allowed to go through to the load. The capacitor charges to the peak value of the voltage waveform during the first cycle and, as the voltage in the rectified waveform is decreasing, the capacitor voltage is not able to follow the change as it can only discharge at a rate determined by the time constant CR_L where C is capacitance. In the case of light loads (or high values of load resistance), the capacitor would discharge only a little before the voltage in the rectified waveform exceeds the capacitor voltage, thus charging it again to the peak value (Figure 6.5b). The ripple content is inversely proportional to C and R_L.

Ripple can be reduced by increasing C for a given of R_L. For heavy loads when R_L is small, even a large capacitance value may not be able to provide ripple within acceptable limits. Ripple factor r is also given by Equation 6.2:

$$r = \frac{1}{4\sqrt{3}(fCR_L)} \tag{6.2}$$

where f is the power line frequency. This simplifies to $2886/CR_L$ for a power line frequency of 50 Hz. Here capacitance C is in microfarads (μF) and R_L is in ohms (Ω).

We have seen that while a choke filter is effective only for heavy load currents, a capacitor filter provides adequate filtering only for light loads. The performance of inductor and capacitor filters deteriorates fast as load resistance is increased in the case of the former or decreased in the case of the latter. An appropriate combination of L and C could give us a filter that would provide adequate filtering over a wide range of R_L values. For all practical purposes, the ripple factor in a choke input L–C filter is independent of R_L and is given by $1.2/LC$ for a power line frequency of 50 Hz. In this expression, L is in henries and C is in microfarads. Figure 6.6 shows an LC filter connected across the output of a full-wave rectifier.

The inductance value in the choke input LC filter should be such that the current I through the inductance never falls to zero, which amounts to saying that the negative peak of the ripple current should never exceed the DC value of the current. The minimum inductance that achieves this is known as critical

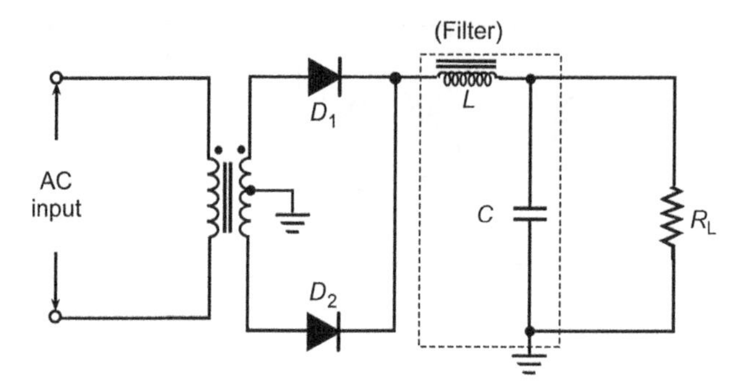

Figure 6.6 LC (choke input) filter.

inductance. The chosen value of inductance should be greater than the critical inductance. Critical inductance is given by $L_C = R_L/755$ for a power line frequency of 50 Hz, where R_L is in ohms and L_C is in henries.

6.2.4 Linear Regulators

As outlined earlier, the regulator circuit in the power supply ensures that the load voltage (in the case of voltage-regulated power supplies) or the load current (in the case of current-regulated power supplies) is constant, irrespective of variations in line voltage or load resistance. In this section we discuss the different types of voltage regulator circuits. Constant-current sources are discussed in Section 6.4. The three basic types of linear voltage regulator configurations are: emitter-follower regulator, series-pass regulator and shunt regulator. Each of these is briefly described in the following sections.

6.2.4.1 Emitter-follower Regulator

Figure 6.7 shows the basic positive-output emitter-follower regulator. The emitter voltage, which is also the output voltage, remains constant as long as the base voltage is held constant. A Zener diode connected between the base and ground terminals ensures this. The regulated output voltage in this case is given by Equation 6.3:

$$V_O = V_Z - 0.6 \tag{6.3}$$

where V_Z is the voltage of the Zener diode in volts.

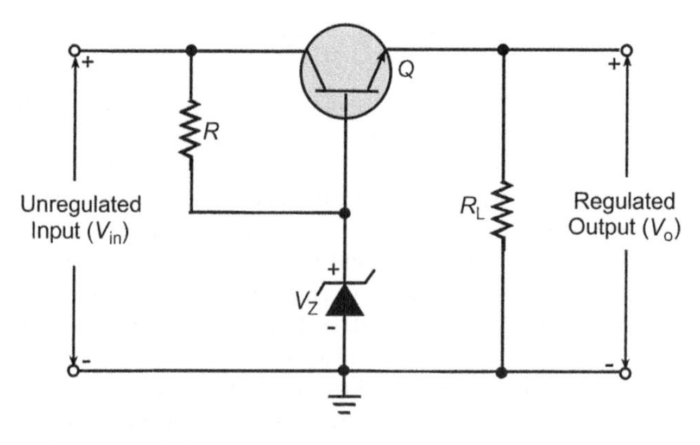

Figure 6.7 Emitter-follower regulator for positive output voltages.

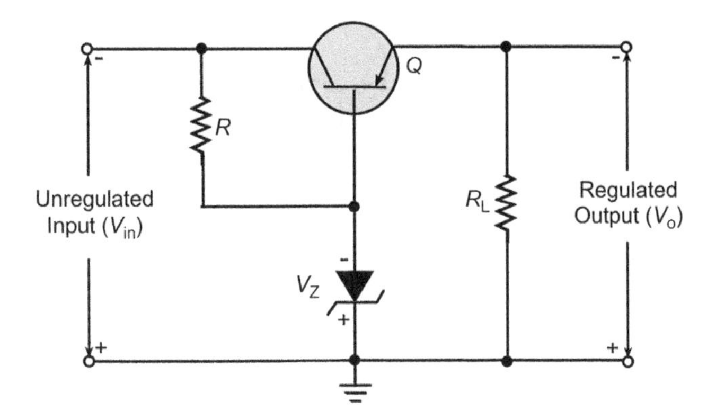

Figure 6.8 Emitter-follower regulator for negative output voltages.

Due to high inherent current gain of the series-pass transistor, a low-power Zener diode can be used to regulate the high value of load current. The base current in this case only needs to be ($1/h_{FE}$) times the load current, where h_{FE} is the current gain of the transistor in the common emitter configuration. Figure 6.8 shows the emitter-follower regulator circuit for negative output voltages. If the load current is so large that it is beyond the capability of a Zener diode to provide the requisite base current, a Darlington combination can be used instead of a single-transistor series-pass element (Figures 6.9 and 6.10).

6.2.4.2 Series-pass Regulator

Figure 6.11 shows the basic components of a series-pass linear regulator. The series-pass element, a bipolar transistor in the circuit shown, works like a variable resistance with the conduction of the transistor dependent upon the base current. The regulator circuit functions as follows.

A small fraction of the output voltage is compared with a known reference DC voltage and their difference is amplified in a high-gain DC amplifier. The amplified error signal is then fed back to the base of the series-pass transistor to alter its conduction in order to maintain a constant output voltage. The regulated output voltage in this case is given by Equation 6.4:

$$V_O = V_{Ref} \times (R_1 + R_2)/R_2 \tag{6.4}$$

where V_{Ref} is reference voltage.

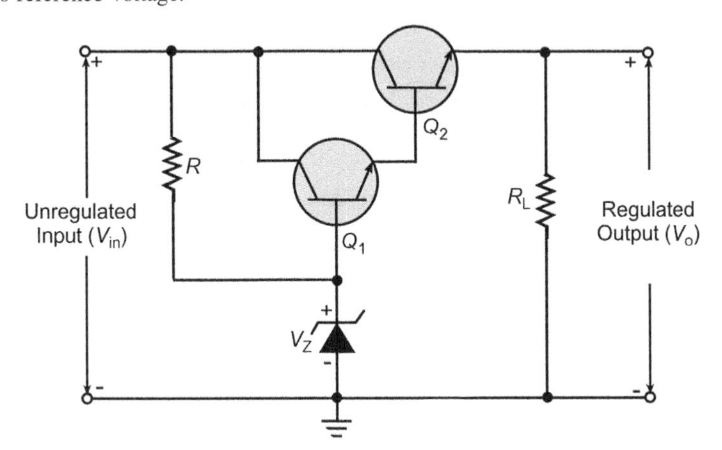

Figure 6.9 Emitter-follower regulator with Darlington element (positive output voltage).

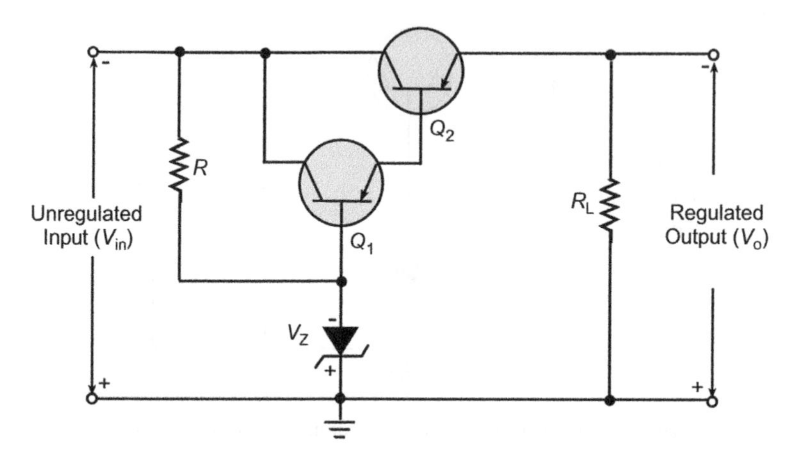

Figure 6.10 Emitter-follower regulator with Darlington element (negative output voltage).

As the output voltage tends to decrease due to a decrease in input voltage or an increase in load current, the error voltage produced as a result of this causes the base current to increase. The increased base current increases transistor conduction, thus reducing its collector emitter voltage drop, which compensates for the reduction in the output voltage.

Similarly, when the output voltage tends to increase due to an increase in input voltage or a decrease in load current, the error voltage produced as a consequence is of the opposite sense. It tends to decrease transistor conduction, thus increasing its collector-emitter voltage drop and maintaining a constant output voltage. The regulation provided by this circuit depends upon the stability of the reference voltage V_{ref} and the gain of the DC amplifier. A typical series-pass regulator circuit using a bipolar junction transistor as the error amplifier is shown in Figure 6.12.

The power dissipated in the series-pass transistor is the product of its collector-emitter voltage and the load current. As the load current increases within a certain range, the collector-emitter voltage decreases due to the feedback action, keeping the output voltage as constant. The series-pass transistor is chosen in order to safely dissipate the power under normal load conditions. If there is an overload condition for some reason, the transistor is likely to be damaged if such a condition is allowed to persist for long. In the worst case, if there was a short circuit at the output the whole of the unregulated input would appear across the series-pass transistor, increasing the power dissipation to a prohibitively large magnitude and

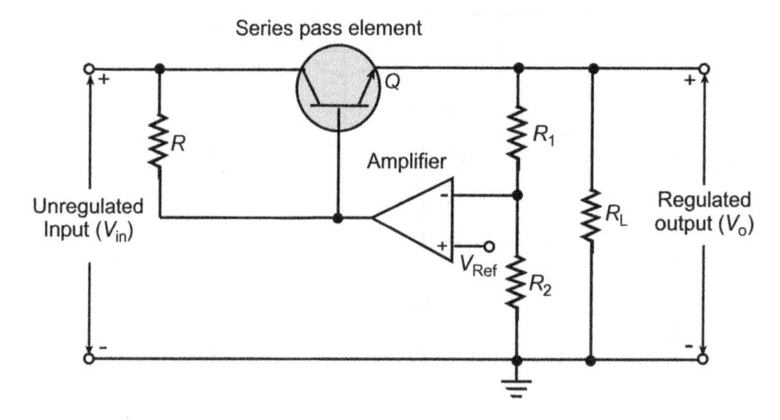

Figure 6.11 Series-pass linear regulator.

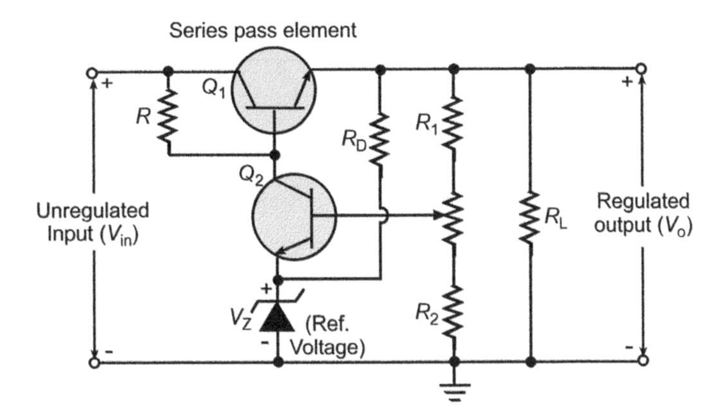

Figure 6.12 Series-pass linear regulator using bipolar junction transistor as the error amplifier.

eventually destroying the transistor. Even a series-connected fuse does not help in such a case, as the thermal time constant of the transistor is much smaller than that of the fuse. It is therefore always desirable to build overload protection or short-circuit protection in the linearly regulated power supply design. One such configuration is shown in Figure 6.13.

Under normal operating conditions, transistor Q_3 is in saturation. It therefore offers very little resistance to the load current path. In the event of an overload or a short circuit, diode D_1 conducts thus reducing the base drive to transistor Q_3. Transistor Q_3 offers an increased resistance to the flow of load current. In the event of a short circuit, the whole of input voltage would appear across Q_3. Transistor Q_3 should be chosen in order to safely dissipate the power given by the product of worst-case unregulated input voltage and the limiting value of current. Diode D_1 and transistor Q_3 should preferably be mounted on the same heat sink so that the base-emitter junction of Q_3 and the diode's P-N junction are equally affected by temperature rise and the short-circuit-limiting current is the preset value.

Other possible circuit configurations can also provide the desired protection, but a comprehensive discussion of all is beyond the scope of this book. Other types of protection features that are usually built into power supplies include crowbarring and thermal shutdown. Crowbarring is a type of over-voltage protection and thermal shutdown disconnects the input to the regulator circuit in the event of the active device(s) temperature exceeding a certain upper limit. Control and protection functions are usually provided by an integrated circuit (IC) in a modern power supply. A wide range of control ICs is available for both linear and switched-mode power supplies.

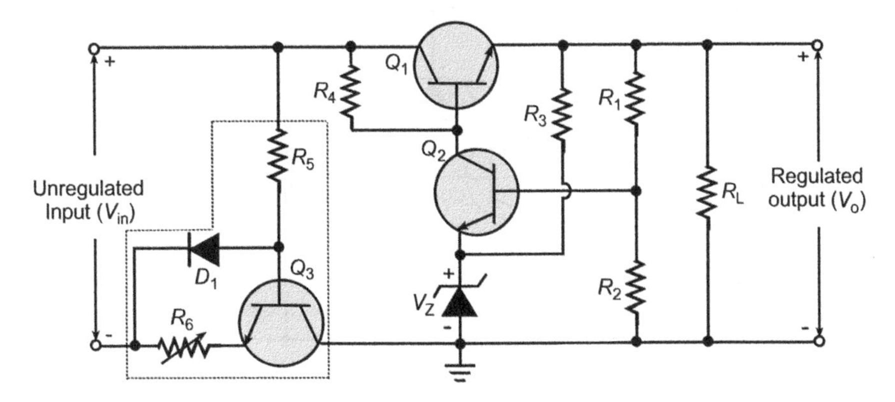

Figure 6.13 Series-pass linear regulator with overload protection.

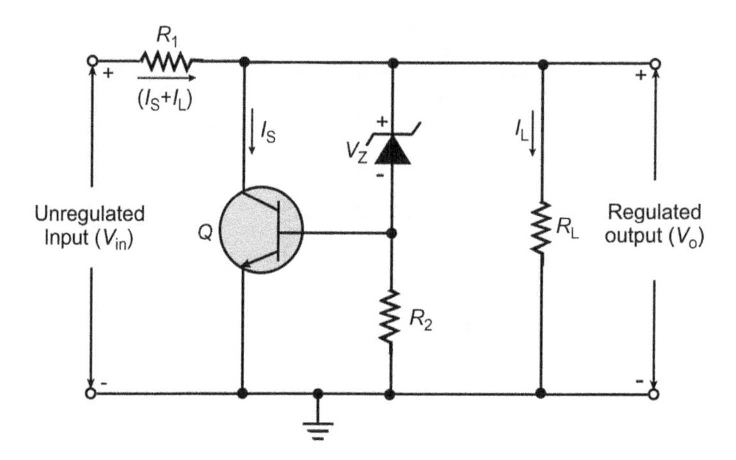

Figure 6.14 Shunt regulator.

6.2.4.3 Shunt Regulator

In a series-type linear regulator, the pass element is connected in series with the load and any decrease or increase in the output voltage is accompanied by a decrease or increase in the collector-emitter voltage of the series-pass element. In the case of a shunt-type linear regulator (Figure 6.14), regulation is provided by a change in the current through the shunt transistor to maintain a constant output voltage. The regulated output voltage in a shunt-regulated linear power supply is the unregulated input voltage minus drop across a resistance R_1. The current through R_1 is the sum of load current I_L and current through shunt transistor I_s. As the output voltage decreases, the base current through the transistor reduces with the result that its collector current I_s also reduces. This reduces drop across R_1, and the output voltage is restored to its nominal value. Similarly, any tendency of the output voltage to increase is accompanied by an increase in current through the shunt transistor, consequently increasing voltage drop across R_1 which in turn maintains a constant output voltage. A Darlington combination in place of a shunt transistor enhances the current capability (Figure 6.15).

A shunt regulator is not as efficient as a series regulator for the simple reason that the current through the series resistor in the case of the former is the sum of load current and shunt transistor current, and it

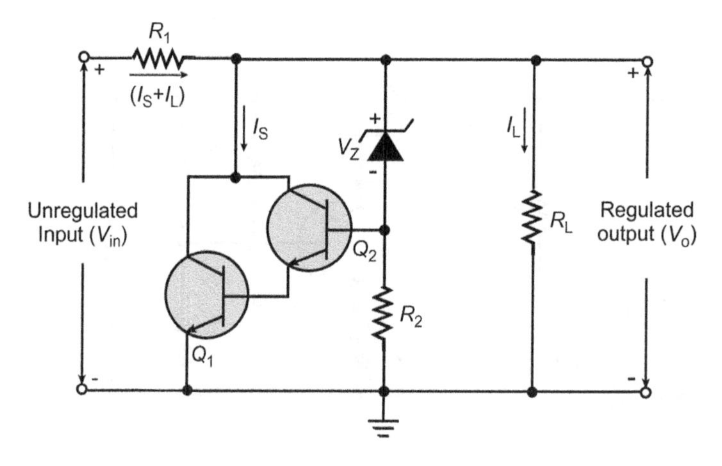

Figure 6.15 Shunt regulator with Darlington arrangement.

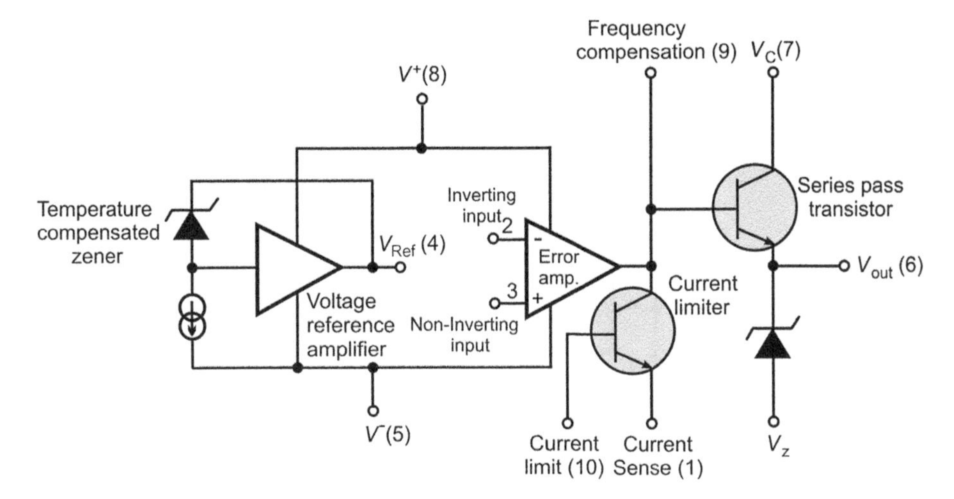

Figure 6.16 Internal schematic arrangement of regulator IC 723.

dissipates more power than the series-pass regulator with same unregulated input and regulated output specifications. In a shunt regulator, the shunt transistor also dissipates power in addition to the power dissipated in the series resistor. The only advantage of a shunt regulator is its simplicity and protection against overload condition.

6.2.4.4 Linear IC Voltage Regulators

In the previous section we discussed series and shunt regulator circuits designed with discrete components. Present-day regulating circuits are almost exclusively configured around one or more integrated circuits known as IC voltage regulators. IC voltage regulators are available to meet a wide range of requirements. Both fixed-output voltage (positive and negative) and adjustable-output voltage (positive and negative) IC regulators are commercially available in a wide range of voltage, current and regulation specifications. These have built-in protection features such as current limit and thermal shutdown.

The IC 723 is an example of a general-purpose adjustable-output voltage regulator that is capable of being operated in positive or negative power supplies as a series, shunt and switching regulator. The internal schematic arrangement of IC 723 resembles the typical circuit for a series-pass linear regulator and comprises a temperature-compensated reference, an error amplifier, a series-pass transistor and a current limiter with access to remote shutdown. Figure 6.16 shows the internal schematic arrangement of regulator IC 723.

Figures 6.17 and 6.18 show the basic circuits for building low positive-output voltage (2–7 V) and high positive-output voltage (7–37 V) regulator circuits. In the case of Figure 6.17, the regulated output voltage is given by Equation 6.5:

$$V_O = V_{Ref} \times R_2/(R_1 + R_2) \tag{6.5}$$

In the case of the circuit arrangement of Figure 6.18, the output voltage is given by Equation 6.6:

$$V_O = V_{Ref} \times (R_1 + R_2)/R_2 \tag{6.6}$$

In both cases, the recommended value of R_3 is $R_1R_2/(R_1 + R_2)$ and $R_{SC} = 0.6/I_{SC}$, where I_{SC} is the short-circuit limiting value.

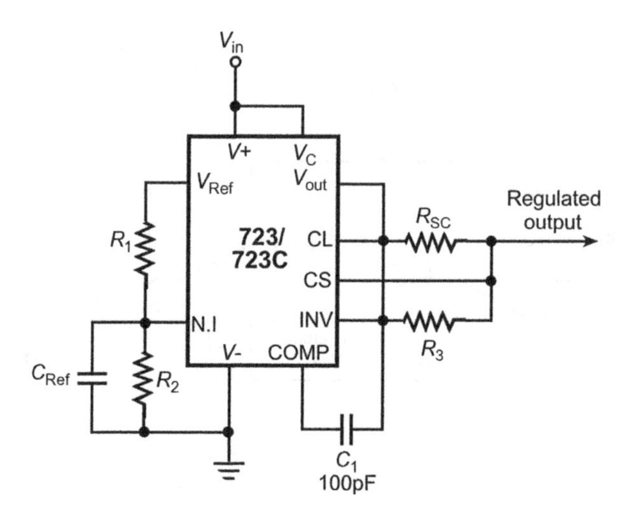

Figure 6.17 Low positive output voltage regulator using IC 723.

In their basic operational mode, three-terminal regulators require virtually no external components. These are available in both fixed-output voltage (positive and negative) as well as adjustable-output voltage (positive and negative) types in current ratings of 100 mA, 500 mA, 1.5 A and 3.0 A. Popular fixed positive-output voltage three-terminal regulators include LM/MC 78XX-series and LM 140-XX/340-XX series. LM117/217/317 is a common adjustable positive-output voltage regulator. Popular fixed negative-output voltage three-terminal regulators include LM/MC 79XX-series and LM 120-XX/320-XX series. LM137/237/337 is a common adjustable negative output voltage regulator. The two-digit number in place of 'XX' indicates the regulated output voltage.

Figure 6.19 shows the basic application circuits using LM/MC 78XX-series and LM/MC 79XX-series three-terminal regulators. C_1 and C_2 are decoupling capacitors. C_1 is generally used when the regulator is located far from the power supply filter. Typically, a 0.22 μF ceramic disc capacitor is used for C_1. C_2 is typically a 0.1 μF ceramic disc capacitor. LM 140XX/340XX series and LM 120XX/320XX series regulators are also used in the same manner. In the case of fixed-output voltage three-terminal regulators, if instead of being grounded the common terminal was applied a DC voltage, the regulated output voltage

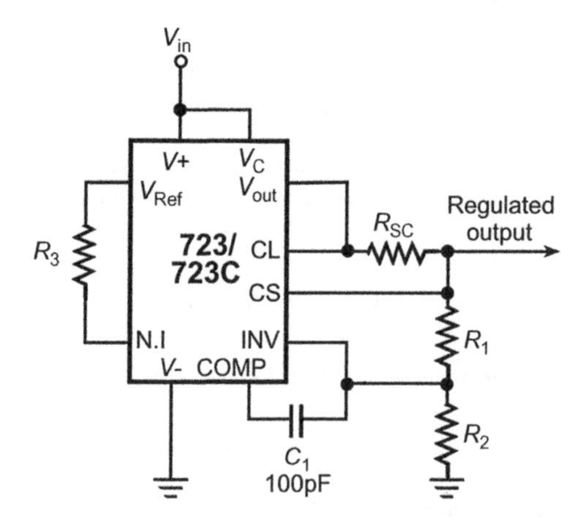

Figure 6.18 High positive output voltage regulator using IC 723.

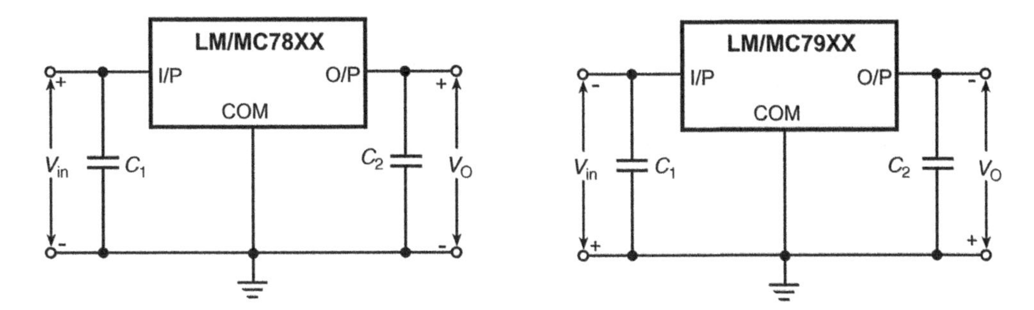

Figure 6.19 Basic application circuits using three-terminal regulators.

would be greater than the expected value by a quantum equal to the voltage applied to the common terminal. For details on the salient features and performance specifications of three-terminal regulators and their application circuits, refer to the data sheets and application notes provided by manufacturers of these devices.

6.3 Switched-mode Power Supplies

Based on the regulation concept, power supplies are classified as either linear or switched-mode. Conventional AC/DC power supplies comprising a transformer, rectifier, filter and regulator (series or shunt type) constitute the linear power supply. The active device in the regulator circuit of a linear power supply is always operated in the active or linear region of its output characteristics. Any change in the output voltage due to a change in the input voltage or load current results in a change in the voltage drop across the regulator transistor (in the case of a series regulator) or a change in the current through the regulator transistor (in the case of a shunt regulator), in order to maintain a constant output voltage across the load.

DC/DC converters and DC/AC inverters are members of the switched-mode power supply (SMPS) category. There are also switching supplies which operate from AC mains, referred to as off-line switching supplies. An off-line switching supply can be distinguished from a conventional AC/DC supply as the AC mains is rectified and filtered without using an input transformer in the former. The DC voltage so obtained is then used as input to a switching-type DC/DC converter.

In a switching power supply, the active device that provides regulation is always operated in the switched mode, that is, it is operated either in the cut-off or saturation region of the output characteristics. The input DC is chopped at a high frequency (typically 10–100 kHz) using an active device (bipolar junction transistor, power metal-oxide semiconductor field-effect transistor or MOSFET, insulated gate bipolar transistor or IGBT or silicon-controlled rectifier or SCR) and the converter transformer. The transformed chopped waveform is rectified and filtered. A sample of the output voltage is used as a feedback signal for the drive circuit of the switching device to achieve regulation.

6.3.1 Linear versus Switched-mode Power Supplies

Linear power supplies are well known for their extremely good line and load regulation, low output voltage ripple and almost negligible radio frequency interference (RFI) or electromagnetic interference (EMI). Switching power supplies on the other hand have a much higher efficiency (typically 80–90% compared to 50% in the case of linear supplies) and reduced size/weight for a given power-delivering capability. Quite often, compactness and efficiency are two major selection criteria. An improved efficiency and reduced size/weight are significant when designing a power supply for a portable system, particularly when there is a requirement for a number of different regulated output voltages. Also, unlike linear supplies, efficiency in switching supplies is not reduced as the unregulated input to regulated output differential becomes large.

In portable systems operating from battery packs and requiring higher DC voltages for their operation, the switching supply is the only option. We cannot use a linear regulator to change a given unregulated input voltage to a higher regulated output voltage.

6.3.2 *Different Types of Switched-mode Power Supplies*

Switched-mode power supplies are designed in a variety of circuit configurations depending upon the intended application. Almost all switching supplies belong to one of the following three broad categories:

1. flyback converters
2. forward converters
3. push-pull converters.

There are variations in the circuit configuration within each of these categories of switched-mode power supplies. For instance, in the category of flyback converters, we have the self-oscillating and the externally driven flyback converters. Also within the category of externally driven flyback converters, there are isolation and non-isolation configurations as well as DC/DC and off-line flyback converters.

Similarly, there are different circuit configurations within the other two categories of switching supplies. Although these configurations differ to an extent, the basic operational principle and the design criteria for different types belonging to one category remain more or less the same.

6.3.2.1 Flyback Converters

The self-oscillating flyback DC/DC converter is the most basic converter based on the flyback principle (Figure 6.20). A switching transistor, a converter transformer, a fast recovery rectifier and an output filter capacitor make up a complete DC/DC converter. It is a constant output power converter.

During the conduction time of the switching transistor, the current through the transformer primary starts ramping up linearly with a slope equal to V_{in}/L_p where V_{in} is the unregulated input voltage and L_p is the inductance of the primary winding of the transformer. The voltages induced in the secondary and the feedback windings make the fast recovery rectifier reverse-biased and hold the conducting transistor 'on'.

When the primary current reaches its peak value I_P where the core begins to saturate, the current tends to rise very sharply. This sharp rise in current cannot be supported by the fixed base drive provided by the feedback winding. As a result, the switching transistor begins to come out of saturation.

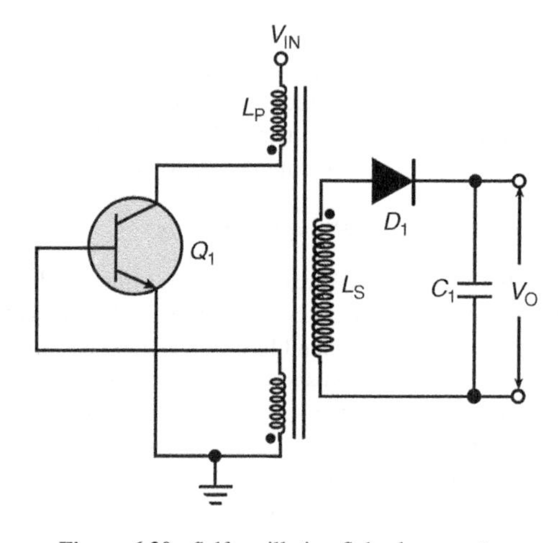

Figure 6.20 Self-oscillating flyback converter.

This is a regenerative process with the transistor being switched off. The magnetic field produced by the current flowing in the primary winding collapses, thus reversing the polarities of the induced voltages. The fast recovery rectifier is forward biased and the stored energy is transferred to the capacitor and the load through the secondary winding. Energy is therefore stored during the ON-time and transferred during the OFF-time of the switching transistor.

The output capacitor supplies the load current during the ON-time of the transistor, when no energy is being transferred from the primary side. It is a constant output power converter. The converter can deliver an output power given by Equation 6.7:

$$P_O = \frac{1}{2}\eta L_P I_P^2 f \tag{6.7}$$

where L_P is the primary inductance; I_P is the peak value of primary current; f is the switching frequency; and η is the conversion efficiency.

The output voltage reduces as the load increases and vice versa. Utmost care should be taken to ensure that the load is not accidentally taken off the converter. In this case, the output voltage would rise without limit until the converter components become damaged. It is suitable for low-output power applications due to its inherent nature of operation, and may be used with advantage up to an output power of 150 W. It is characterized by a high-output voltage ripple.

A variation of this circuit is the externally driven flyback converter. Figure 6.21 shows the circuit diagram of the basic externally driven flyback type DC/DC converter. The basic principle remains the same. Energy is stored during ON-time of the active device and transferred during OFF-time. The feedback loop, consisting of a comparator and the resistance divider, provides the voltage sense as well as some degree of regulation.

Extension of the converter circuit of Figure 6.21 is the externally driven flyback converter with pulse width modulation (PWM) control to achieve regulation, depicted in Figure 6.22. Pulse width modulation is the most widely used control technique in conjunction with flyback converters. As the load current increases, the output voltage tends to fall. The PWM control senses the change and increases the ON-time of the active device in order to increase the power-delivering capability (increased ON-time means increased stored energy) and restores the output voltage. Similarly, an increase in the output voltage causes a reduction in the ON-time of the active device.

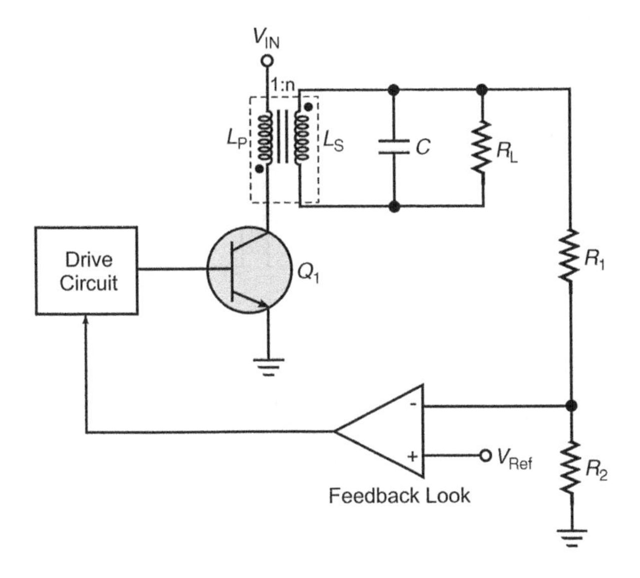

Figure 6.21 Basic externally driven flyback converter.

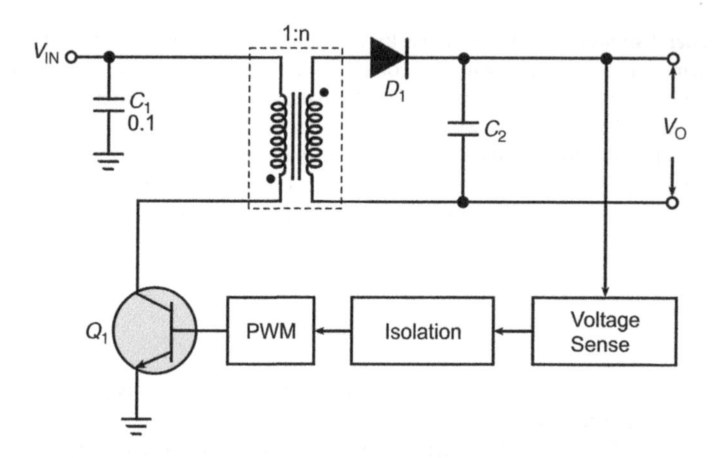

Figure 6.22 Externally driven flyback converter with PWM control.

There are other control circuitries which provide regulation by changing the OFF-time rather than the ON-time. A reduced OFF-time increases the drive frequency and hence the power-delivering capability and vice versa.

A number of integrated circuits have been developed to provide drive and control functions for DC/DC converters. Some of these ICs provide PWM control while others offer constant ON-time and variable frequency operations. These ICs have built-in features such as over-voltage protection and current limit. Such ICs (e.g. TL497, TL494, TL594 and SG3524 to name a few) have considerably simplified the drive and control circuit design.

Most switching supplies used in consumer and industrial systems are off-line. Figure 6.23 shows an off-line externally driven flyback DC/DC converter. It is called off-line because the input voltage to the transistor switch is developed right from the AC line without first going through the 50/60 Hz transformer. The bridge rectifier and the filter capacitor accomplish this in the circuit of Figure 6.23. The feedback loop in an off-line supply must have isolation so that the DC output is isolated from the AC line. A small transformer or an opto-isolator usually accomplishes this. Most switching supplies are

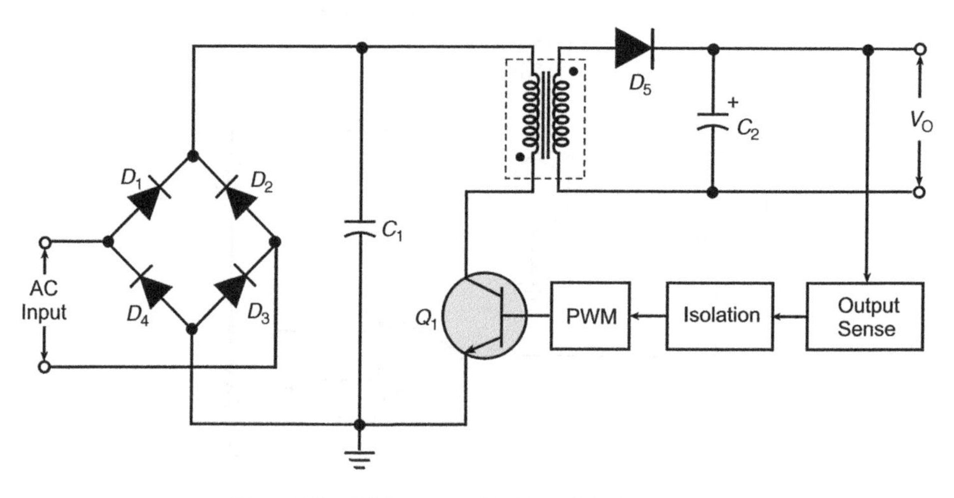

Figure 6.23 Off-line externally driven flyback converter.

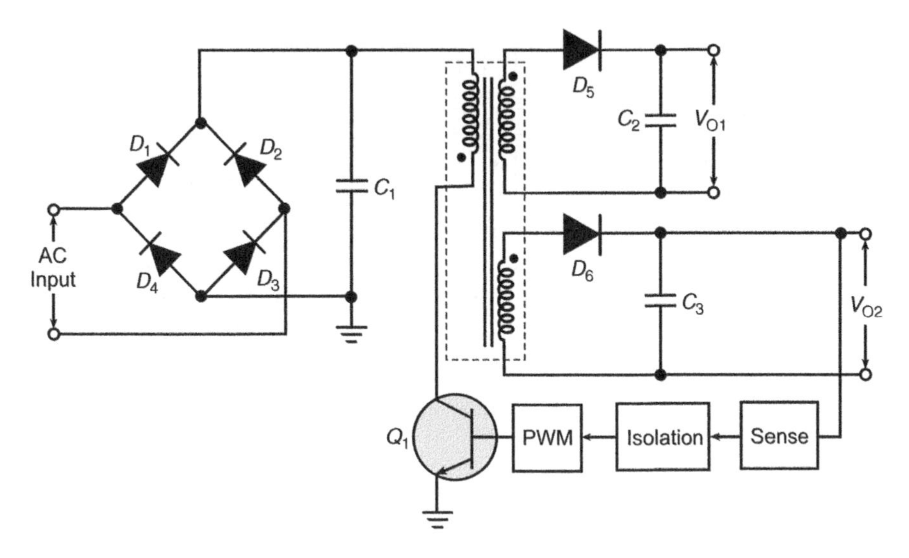

Figure 6.24 Off-line flyback DC/DC converter with multiple outputs.

required to produce more than one regulated DC voltage. Figure 6.24 shows an off-line multiple output flyback DC/DC converter.

If a more stringent regulation is required in respect of one or more outputs, a linear post-regulator can be used as shown in Figure 6.25. Three-terminal IC regulators have been used here for this purpose.

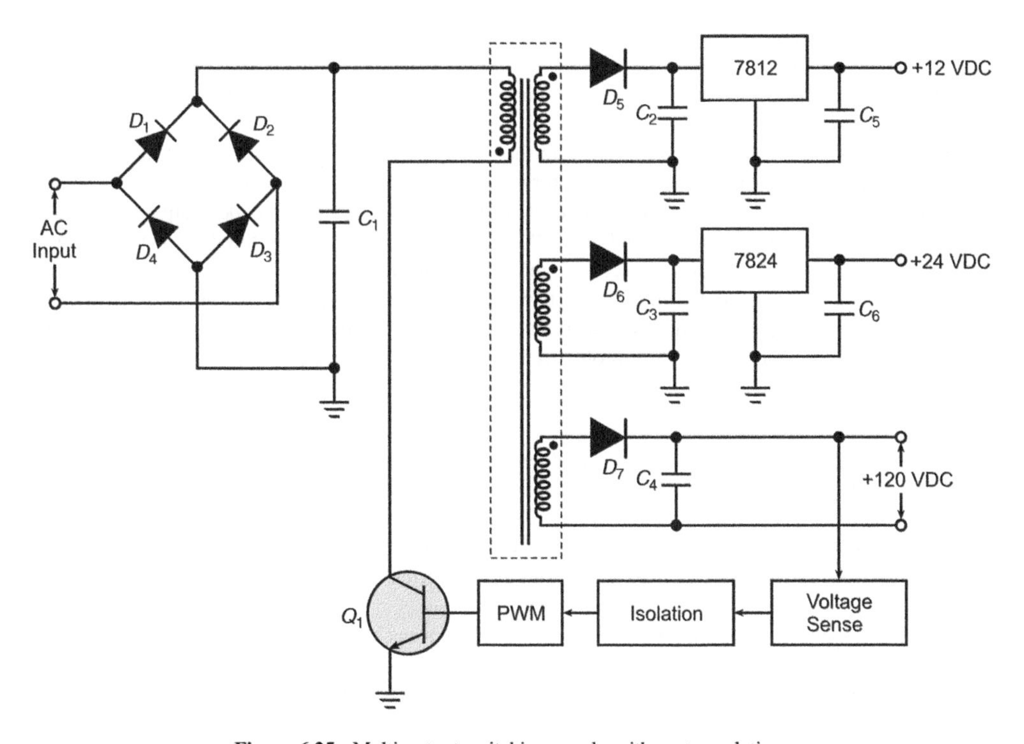

Figure 6.25 Multi-output switching supply with post regulation.

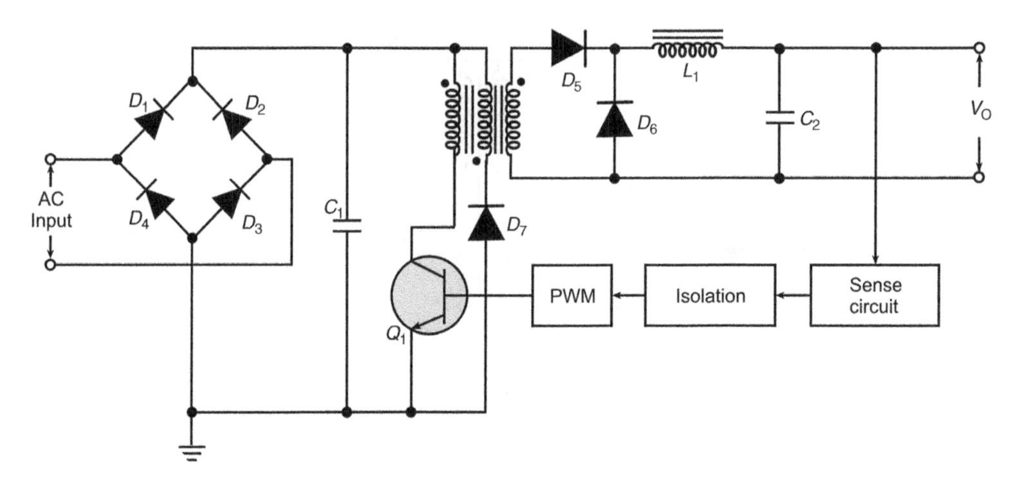

Figure 6.26 Forward converter.

6.3.2.2 Forward Converter

Forward converter is another popular switching supply configuration. Figure 6.26 shows the basic circuit diagram of an off-line forward converter. There are some fundamental differences between a flyback converter and a forward converter. In the case of the circuit diagram shown in Figure 6.26, when the transistor Q_1 is switched on the polarities of the transformer windings (as indicated by the position of dots) are such that diode D_5 is forward biased and diodes D_6 and D_7 are reverse biased. Most of the energy in a forward converter is stored in the output inductor rather than the transformer primary used to store energy in a flyback converter. When the transistor switch is turned off, the magnetic field collapses. Diode D_5 is reverse biased and diodes D_6 and D_7 are forward biased. As the current through an inductor cannot change instantaneously, the output current continues to flow through the output and the forward-biased diode D_6 provides the current path.

Unlike a flyback converter, current flows from the energy storage element during both halves of the switching cycle. For the same output power, a forward converter therefore has much less output ripple than a flyback converter. Controlling the duty cycle of the transistor switch provides output regulation.

In the absence of the third winding and diode D_7, a good fraction of energy stored in the transformer primary is lost. This effect is more severe at higher switching frequencies. The third winding and the forward-biased diode D_7 return the energy, which would otherwise be lost, and reset the transformer core after each operating cycle. This not only increases converter efficiency but also makes the converter transformer core immune to saturation problems.

6.3.2.3 Push-pull Converter

A push-pull converter is the most widely used switching supply belonging to the family of forward converters. There are several different circuit configurations within the push-pull converter subfamily. These circuits differ only in the mode in which the transformer primary is driven. These include the conventional two-transistor, one-transformer push-pull converter (both self-oscillating and extremely driven type) two-transistor, two-transformer push-pull converter, half-bridge converter and full-bridge converter.

Figure 6.27 shows the conventional self-oscillating type of push-pull converter. Its operation can be explained by considering it equivalent to two alternately operating self-oscillating flyback converters. When transistor Q_1 is in saturation, energy is stored in the upper half of the primary winding. When the linearly rising current reaches a value where the transformer core begins to saturate the current tends to rise sharply, which is not supported by a more or less fixed base bias. The transistor starts to come out of

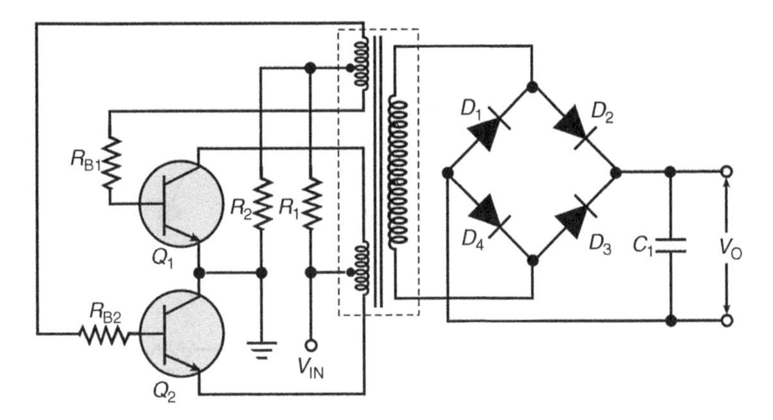

Figure 6.27 Self-oscillating push-pull converter.

saturation. This is a regenerative process which leads to transistor Q_1 being switched off and transistor Q_2 being switched on. Transistors Q_1 and Q_2 therefore switch on and off alternately. When Q_1 is on, energy is being stored in the upper half of the primary winding. The energy stored in the immediately preceding half cycle in the lower half of the primary winding (when transistor Q_2 was on) is transferred. Energy is therefore stored and transferred at the same time. The voltage across the secondary is a symmetrical square waveform, which is then rectified and filtered to obtain the DC output.

As the primary is centre-tapped, and only half of the primary winding is active at one time, the main transformer is not utilized as well as in case of other forms of push-pull converter (e.g. half-bridge and full-bridge converters). In a push-pull converter, switching transistors operate at collector stress voltages of at least twice the DC input voltage. As a result, a push-pull converter is not a highly recommended choice for off-line operation. A push-pull converter that has wider applications than its self-oscillating counterpart is the externally driven push-pull converter. Figure 6.28 shows the block schematic arrangement of externally driven push-pull converter. This has been possible due to the availability of a variety of SMPS drive and control ICs.

Self-oscillating push-pull converters are frequently used along with a voltage multiplier chain to design a high-voltage low-current power supply (Figure 6.29). This configuration is particularly useful

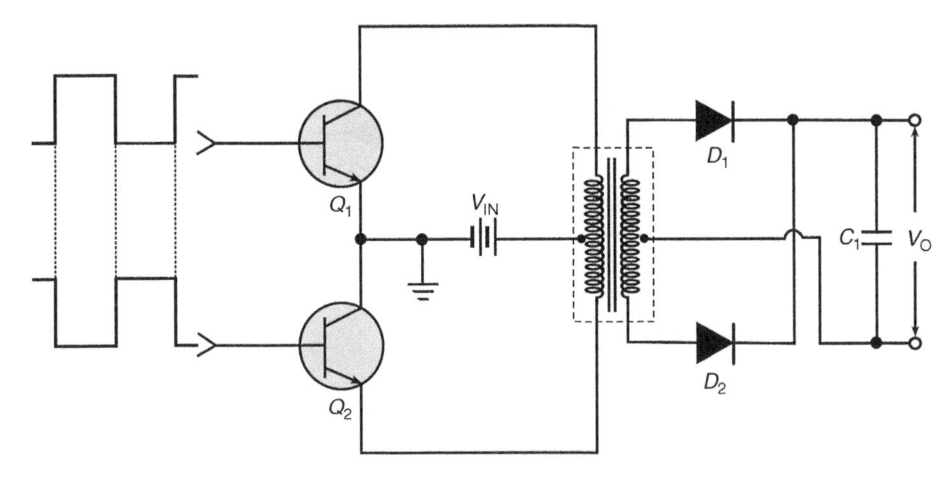

Figure 6.28 Externally driven push-pull converter.

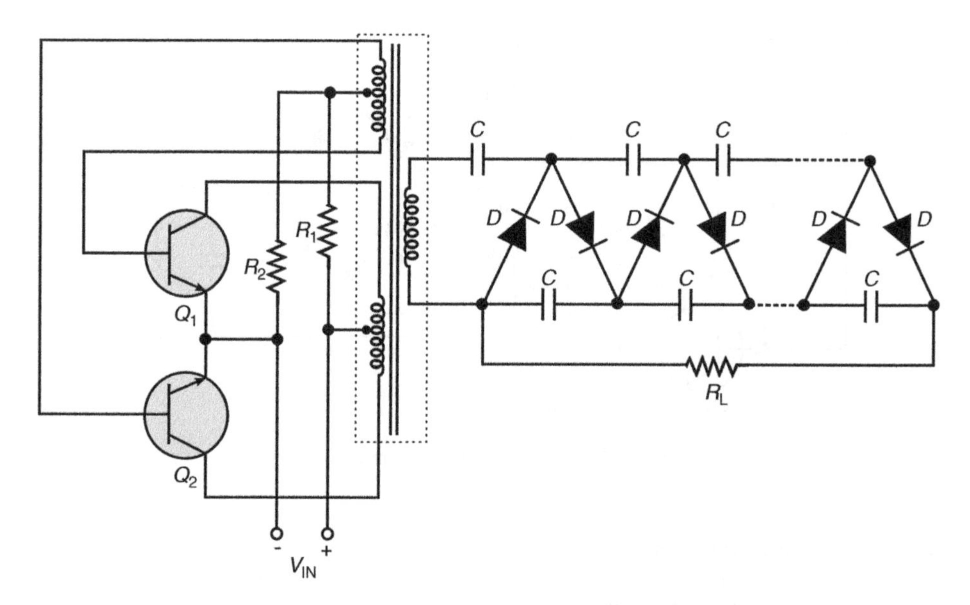

Figure 6.29 Push-pull converter with voltage multiplier chain.

for designing helium-neon laser power supplies. The basic push-pull converter converts the low DC input voltage to a stepped-up square waveform, which is then multiplied using a chain of diodes and capacitors. Voltage multiplier circuits are discussed in detail in Chapter 8 on gas laser electronics.

In the self-oscillating two-transistor one-transformer push-pull converters, the transformer provides both power transformation as well as power switching. This circuit has some disadvantages. Firstly, as the power switching is carried out at output power levels, the converter efficiency lowers quite a bit in the case of a high power converter. Secondly, the peak collector current depends upon the available base voltage, transistor gain and input characteristics and is dependent on load. As there is a wide variation in the input characteristics from device to device, the circuit performance depends upon the chosen device. The transformer core must also be constructed of expensive square-loop material with a large maximum flux density rating.

These problems are overcome in the two-transformer two-transistor push-pull converter (Figure 6.30). Power switching is performed at base power level and the output transformer performs power

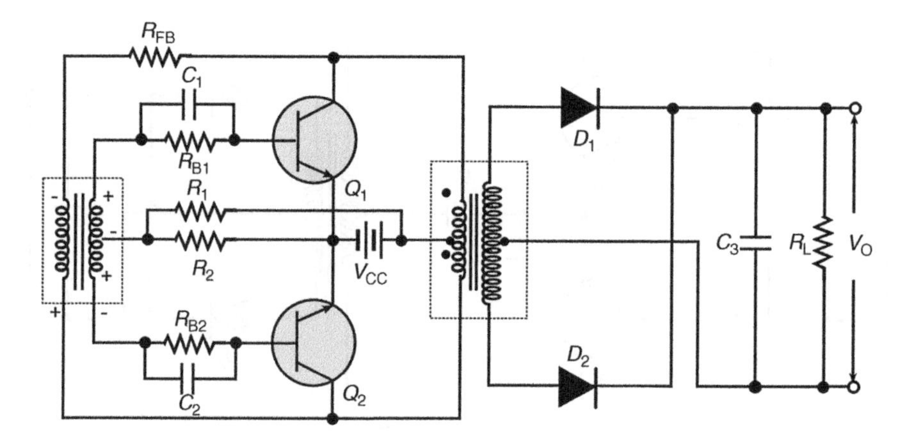

Figure 6.30 Two-transformer two-transistor push-pull converter.

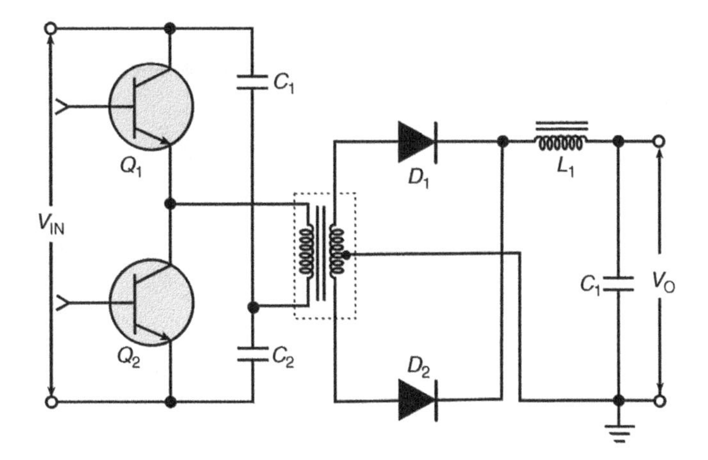

Figure 6.31 Half-bridge converter.

transformation only. Capacitors C_1 and C_2 are the speed-up capacitors (also known as commutating capacitors) used to achieve a faster turn-off of the respective transistors.

The half-bridge converter shown in Figure 6.31 is recommended for high-power applications. Transistors Q_1 and Q_2 operate alternately. The half-bridge converter has the advantage that it allows the use of transistors with lower breakdown voltages.

The full-bridge converter as shown in Figure 6.32 has the advantage that the highest voltage any transistor is subjected to is only V_{in} as opposed to $2V_{in}$ in the case of a push-pull converter. Due to reduce voltage and stress on the transistors, a full-bridge converter offers a great reliability.

6.3.2.4 Switching Regulators

Commonly used switching regulator configurations include a step-down or buck regulator, step-up or boost regulator and an inverting regulator. Figure 6.33 shows the basic buck regulator. It resembles the conventional forward converter discussed in Section 6.3.2.2 except for the fact that it does not use a transformer and there is no input–output isolation. Output voltage is always less than the input voltage

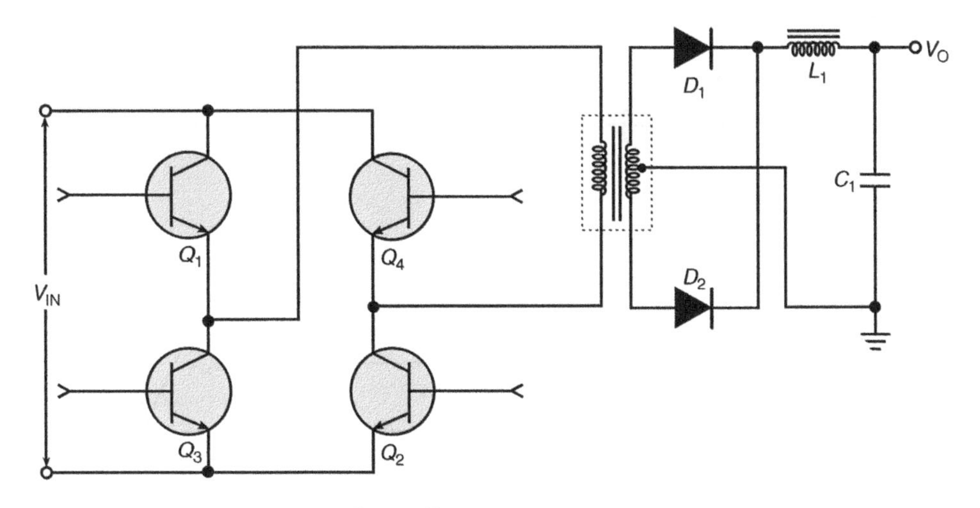

Figure 6.32 Full-bridge converter.

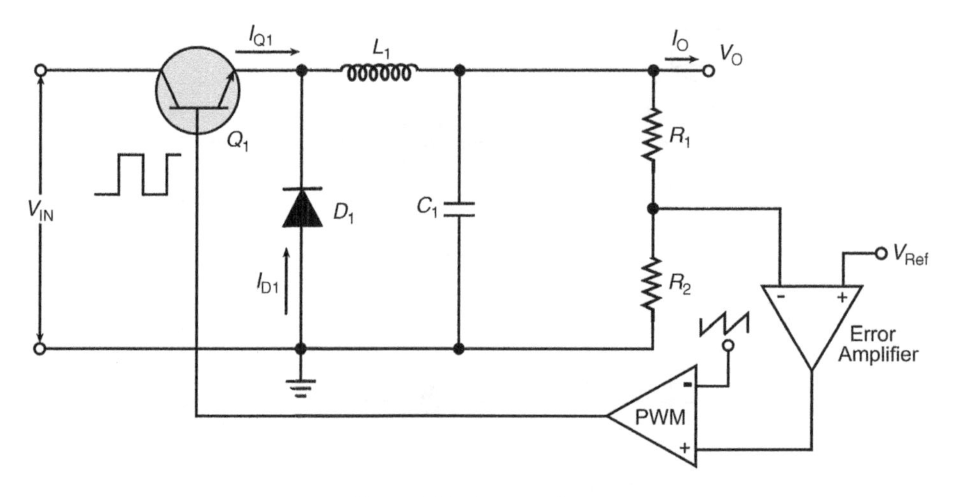

Figure 6.33 Buck regulator.

and is given by Equation 6.8:

$$V_O = DV_{in} \tag{6.8}$$

where D is the duty cycle of the drive waveform to the transistor switch, defined by T_{on}/T where T_{on} is the on-time of the drive waveform and T is the period of the drive waveform.

Regulation is achieved by pulse width modulation of the transistor switch. It is a very popular circuit configuration for fabrication of high-efficiency three-terminal switching regulators.

The step-up switching regulator, also called the boost regulator (Figure 6.34), is based on the flyback principle. It resembles the basic flyback converter except that it is of non-isolating type. The energy storage and transfer element in this case is an inductor rather than a transformer. The energy is stored in the magnetic field of the inductor during the conduction time of the switching transistor. Energy stored is equal to $0.5 \times (L_P I_P^2)$. As the diode is reverse biased during conduction, the energy cannot be transferred

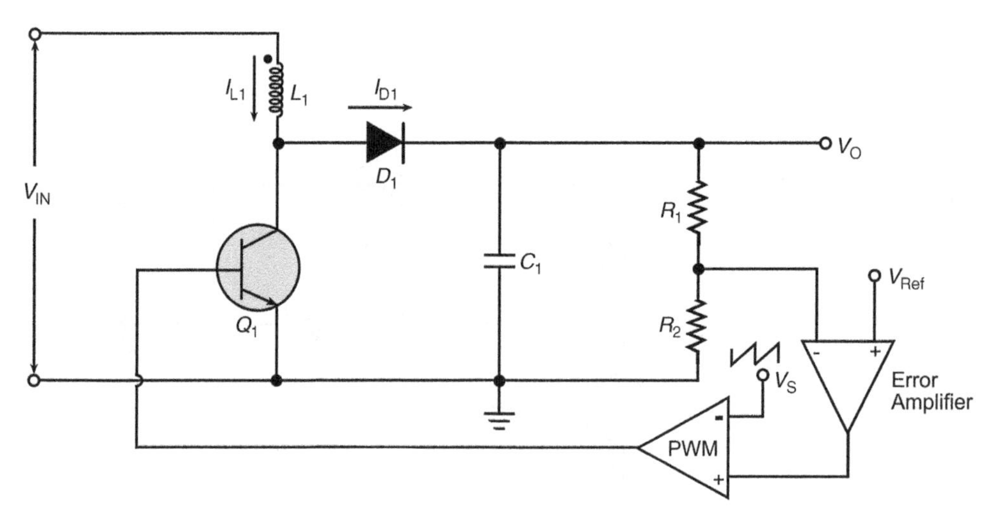

Figure 6.34 Boost regulator.

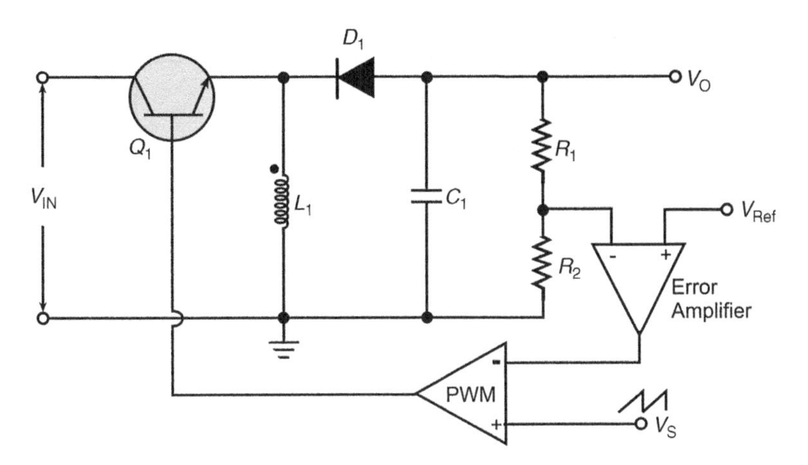

Figure 6.35 Inverting regulator.

while it is being stored. When the switching transistor is driven to cut-off, the diode becomes forward biased and the stored energy is delivered to the load along with the energy from the DC input voltage. The voltage across the load equals the DC input voltage plus the voltage due to the energy stored in the inductor. The output voltage in this case is given by Equation 6.9:

$$V_O = \frac{V_{in}}{(1-D)} = V_{in}\frac{T}{T_{off}} \tag{6.9}$$

where D is the duty cycle and T is the total time period equal to $T_{on} + T_{off}$.

The power output capability of this circuit is given by Equation 6.10:

$$P_O = \frac{1}{2}L_P I_P^2 f \tag{6.10}$$

where f is the switching frequency.

As the DC input voltage is not electrically isolated from the output, the output voltage in this case cannot be exclusively determined from the power rating of the circuit.

An inverting regulator (Figure 6.35) is another circuit configuration based on the flyback converter principle. For a positive input, it produces a negative output. Energy is stored in inductor L during the conduction time of the transistor. The diode is reverse biased during this time period. The stored energy is transferred during off-time. The circuit delivers a constant output power to the load. The output voltage is given by Equation 6.11:

$$V_O = -\sqrt{P_O R_L} = -V_{In}\frac{T_{ON}}{T_{OFF}} \tag{6.11}$$

Regulation is achieved by controlling the duty cycle of the drive waveform. In the inverting regulator configuration, it is possible to have an output voltage that is either less than or greater than the input. It is also sometimes referred to as a buck-boost regulator.

6.3.2.5 Three-terminal Switching Regulators

The basic step-down (buck) regulator of Figure 6.26 has been widely exploited in the form of a three-terminal switching regulator. Figure 6.36 shows the typical circuit configuration found inside such a

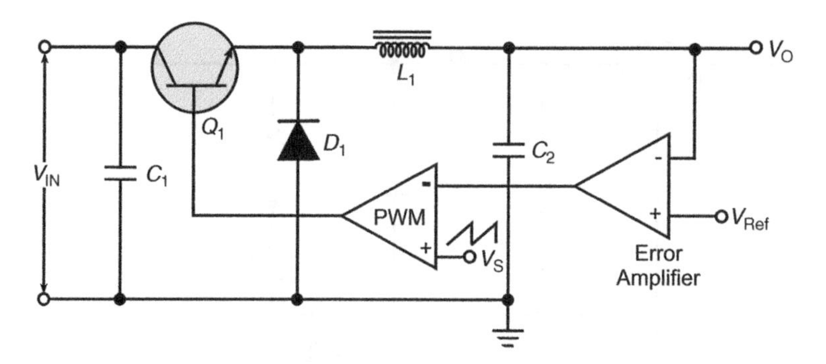

Figure 6.36 Three-terminal switching regulator.

regulator. Except for the switching transistor and the output inductor, all other component blocks are integrated on the chip.

The output voltage is compared with a reference voltage and the difference is amplified to drive a pulse width modulator, which in turn operates the switch. The three-terminal regulator can be used to construct a step-down switching supply that works very well for a wide input voltage range (typical ratio of maximum to minimum voltage is 4 : 1). Output power levels of 300 W are conveniently achievable.

6.3.3 Connecting Power Converters in Series

Power converters can generally be connected in series, but it is advisable to check the specifications of the converters to be connected in series. It is possible that one output affects the feedback loop of another. Another limitation on the series connection of the two converters is that the total output voltage of the series-connected converters should not exceed the working breakdown voltage of any one of the power converters.

In order to protect each output from the reverse voltage of the other output in the event of a shorted-load condition, reverse-biased diodes may be connected across the output of each series-connected converter as shown in Figure 6.37. Series connection can be used to obtain a higher output voltage. In a typical application, a dual output supply can be series connected to realize a single-ended supply with double output voltage (Figure 6.38).

6.3.4 Connecting Power Converters in Parallel

Power converters should be connected in parallel only when they have been specifically designed for that purpose or when the manufacturer recommends a parallel operation. The biggest problem in parallel

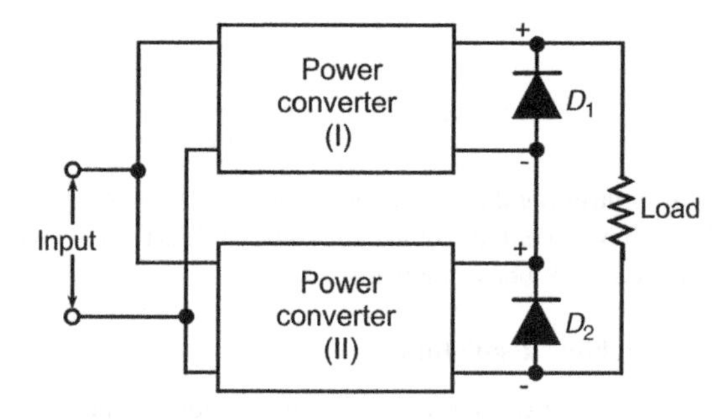

Figure 6.37 Series connection of power converters.

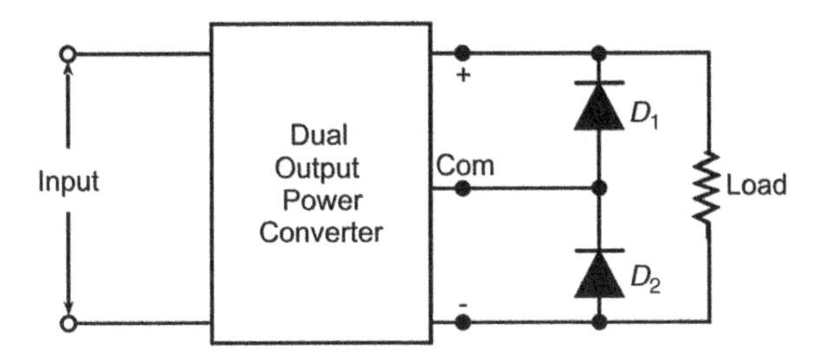

Figure 6.38 Series connection of dual-output switching supply.

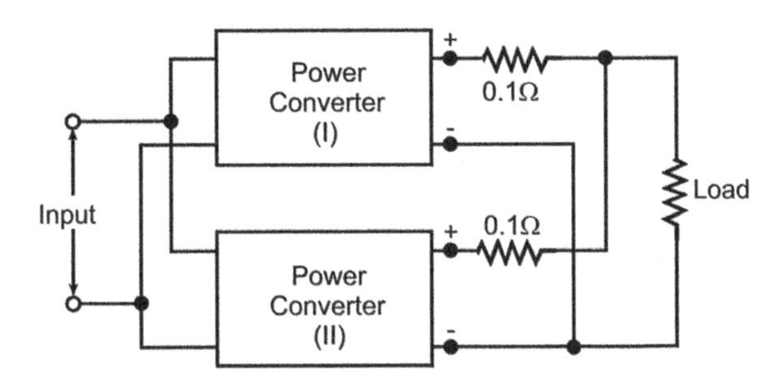

Figure 6.39 Parallel connection of converters.

connection of two converters arises from unequal load sharing, which occurs as a result of output voltages not being precisely equal. The converter with the greater output voltage will tend to provide the entire load current. Even if the output voltages are adjusted to be precisely equal, a difference in the output impedance and also its drift with time and temperature will cause the loads to become imbalanced.

One method used to overcome this unequal load sharing is to use small individual series resistors as shown in Figure 6.39. While this type of parallel connection could be useful in some applications, it must be borne in mind that the series resistors degrade the output regulation seriously. Also, the circuit will always have a current imbalance. If the two converters in Figure 6.39 had a nominal output voltage of 5 V, a 50 mV difference in the output voltages may cause a current imbalance of 25%. In such a case, each supply should not be capable of providing just 50% of the load current, but 75% of it.

A good reason for parallel connection of power converters is to provide redundancy. The output may be connected in parallel through two diodes (Figure 6.40). For 100% redundancy, each power converter must be capable of supplying the total load. In this case it does not matter whether the load is shared equally or not, although it is desirable that each output provides at least a part of the load current.

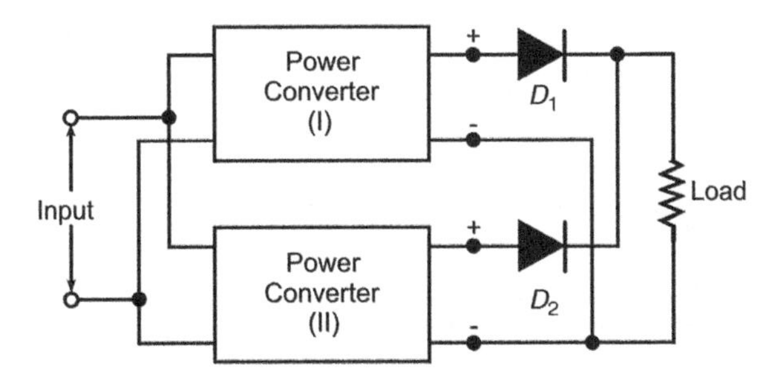

Figure 6.40 Parallel connection to provide redundancy.

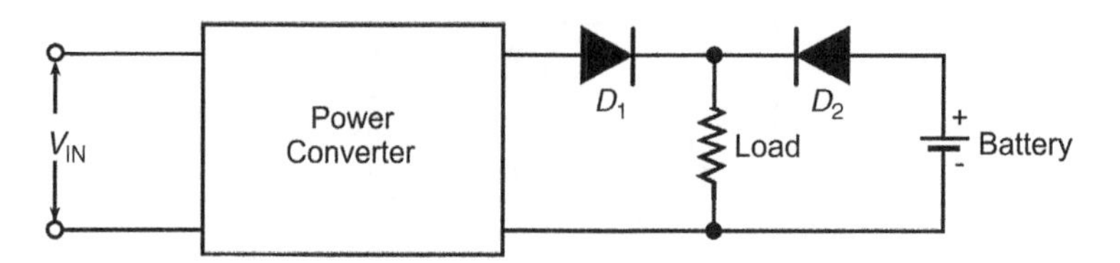

Figure 6.41 Uninterrupted DC power supply.

The diodes permit one output to fail without affecting the other, which continues to supply the power to the load. Such a system is useful in applications where power supply failure is not tolerable, or where a high degree of reliability is required. If one of the power converters is replaced by a DC battery source of the same voltage, it becomes an uninterrupted DC power supply (Figure 6.41).

6.4 Constant Current Sources

A constant current source delivers a predefined constant current to a load despite variations in the supply voltage and the load resistance. In this section, we shall discuss some common constant current sources configured around active devices; a constant current source is an important building block of diode laser electronics.

6.4.1 *Junction Field-effect-transistor-based Constant Current Source*

A junction field-effect transistor (JFET) with a constant gate-source voltage acts like a constant source if the drain-source voltage is greater than the pinch-off voltage. This is evident from the drain current versus drain voltage characteristics of the JFET. Figure 6.42 shows the drain current versus drain voltage characteristics of a typical N-channel JFET. In the saturation region, the drain current is almost independent of the drain-source voltage. For a gate-source voltage of zero, the drain current is a maximum I_{DSS}. In the case of a JFET used as a constant current source, if gate terminal were shorted to the source terminal the constant current obtained would be I_{DSS}.

Current regulating diodes are simply JFETs with gate terminal internally shorted to source terminal. Figure 6.43 shows the internal circuit of a current regulating diode. Inclusion of source resistor R_S as shown in Figure 6.43 allows the magnitude of constant current to be tuned to the desired value.

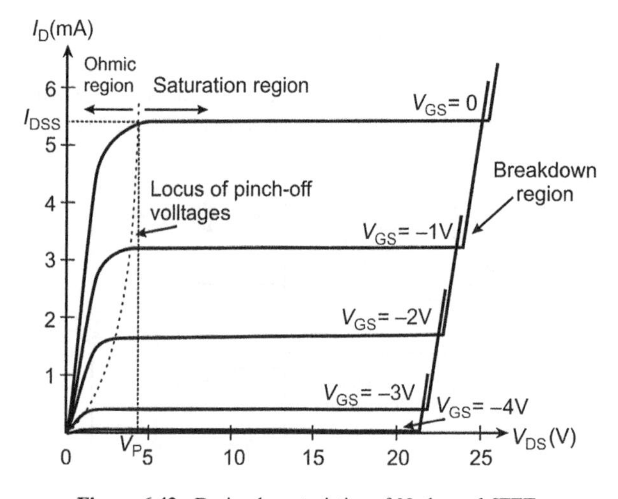

Figure 6.42 Drain characteristics of N-channel JFET.

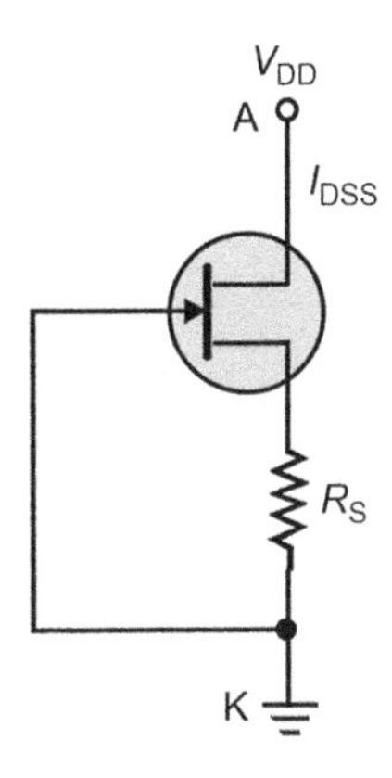

Figure 6.43 Current regulating diode.

6.4.2 *Transistor-based Constant Current Source*

Figure 6.44 shows a simple constant current source configured around a bipolar transistor. The base voltage of the bipolar transistor (equal to the breakdown voltage of the Zener diode, V_Z) and hence its emitter voltage (equal to $V_Z - V_{BE}$, where V_{BE} is base to emitter voltage) remain constant irrespective of the supply voltage variations as long as current through the Zener diode is more than a certain minimum value. Resistor R_1 supplies both the Zener diode current as well as the base current for the transistor. Current through resistor R_2 is given by $(V_Z - V_{BE})/R_2$. Since V_{BE} (typically 0.6–0.7 V for a silicon transistor) is also constant for a given temperature, emitter current is constant. If the h_{FE} of the transistor is sufficiently large, emitter current is nearly equal to the collector current or the load current. The load current in the circuit of Figure 6.44 is therefore constant irrespective of the variations in the load resistance value and the supply voltage. The circuit behaves as constant current source. Varying R_2 changes the value of the constant current.

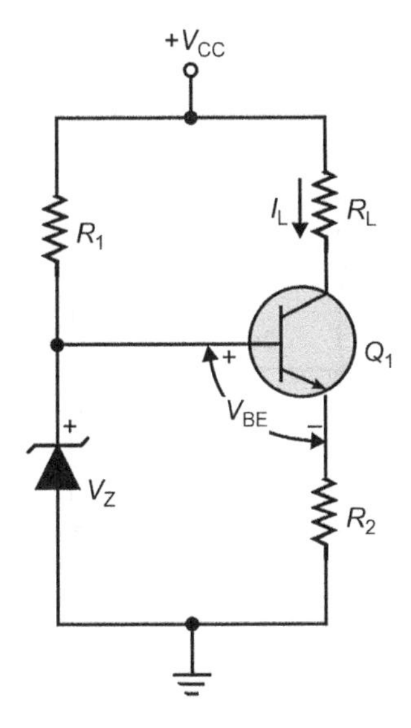

Figure 6.44 Constant current source configured around bipolar transistor.

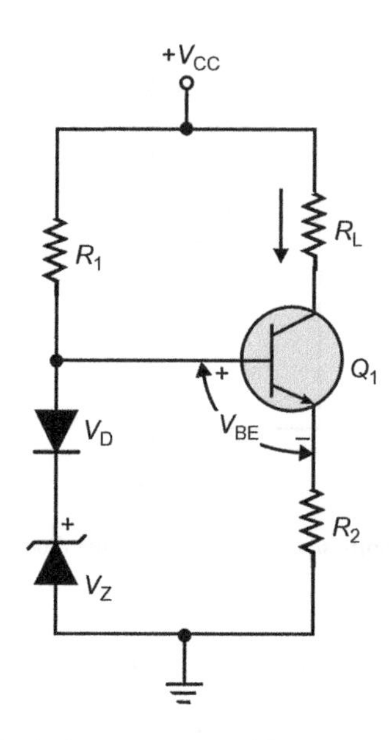

Figure 6.45 Constant current source with temperature compensation.

In the circuit of Figure 6.44, the load current is likely to vary with changes in the operating temperature due to temperature dependence of V_{BE}. This can be taken care of by connecting a diode (made of the same semiconductor material as the transistor) in series with the Zener diode as shown in Figure 6.45. In this case, the emitter current equals $(V_Z + V_D - V_{BE})/R_2$, which equals V_Z/R_2 for $V_D = V_{BE}$. Another possible circuit configuration makes use of a light-emitting diode (LED) in place of a Zener diode (Figure 6.46). This configuration has the advantage that the voltage across the LED tracks the variation in V_{BE} due to change in temperature, thus eliminating the need for the series-connected diode. The emitter current and hence the load current in this case can be computed from $(V_{LED} - V_{BE})/R_2$.

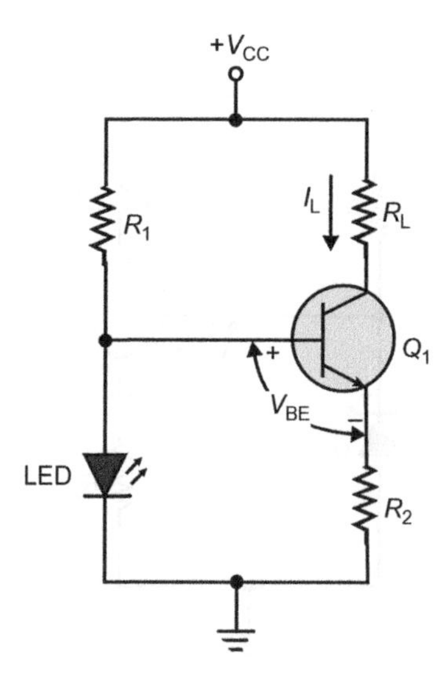

Figure 6.46 Constant current source with LED as reference voltage source.

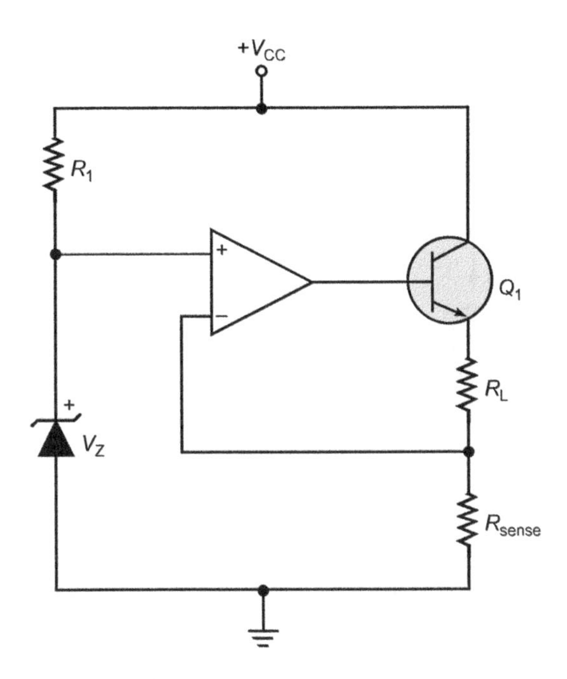

Figure 6.47 Opamp-based constant current source.

6.4.3 Opamp-controlled Constant Current Source

Another constant current source that holds promise is configured around an operational amplifier (opamp); Figure 6.47 depicts such a circuit. The non-inverting terminal of the opamp is applied a constant reference voltage. The opamp output drives the transistor to conduction and, due to virtual earth phenomenon, the voltage at the inverting input of the opamp and hence across the sense resistor R_{sense} equals the reference voltage. The constant current through the load resistance then equals V_Z/R_{sense} and is independent of V_{BE} of the transistor.

6.4.4 Constant Current Source Using Three-terminal Regulators

Fixed- and adjustable-output three-terminal IC voltage regulators can be conveniently used to construct constant current sources. Figure 6.48 shows the basic circuit. The load current in this case is given by

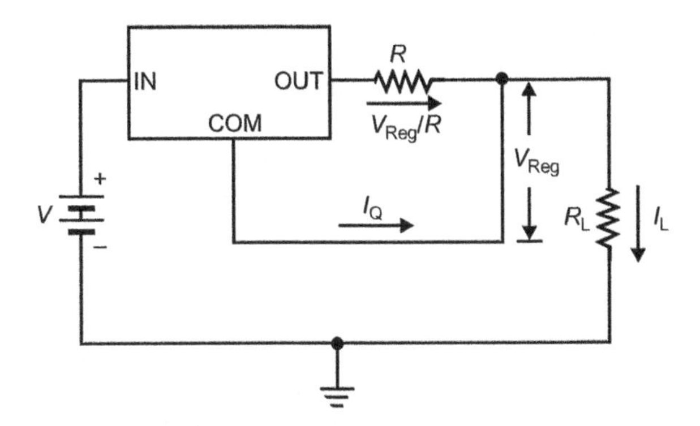

Figure 6.48 Constant current source with three-terminal regulator.

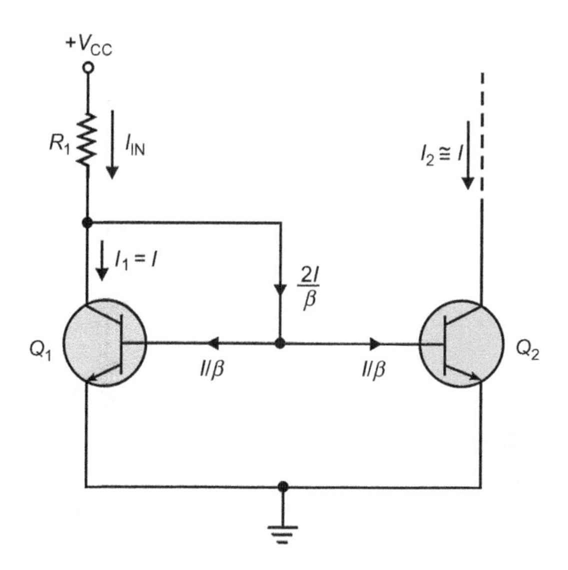

Figure 6.49 Current mirror implemented with bipolar transistor.

$(V_{Reg}/R) + I_Q$ where V_{Reg} is the regulated output voltage and I_Q is the quiescent current of the three-terminal regulator.

6.4.5 *Current Mirror Configurations*

A current mirror circuit mirrors or copies the current flowing through an active device by controlling the current flowing through another active device. Conceptually, an ideal current mirror is simply an ideal current amplifier with a unity gain. There are various current mirror circuits exploited for designing constant current sources. The basic current mirror circuits and the variations to the basic current mirror, including Widlar and Wilson current sources, are described in the following sections.

6.4.5.1 Basic Current Mirror

Figure 6.49 shows the basic current mirror circuit. If the two transistors were perfectly matched, they could be assumed to have the same values of base-emitter voltage drops and DC current gains. This is possible to achieve when the two transistors are fabricated on the same IC chip at the same time. Transistor Q_1 with its collector terminal shorted to its base-terminal is wired like a diode. A semiconductor diode has not been used as that would have made the matching of characteristics of the diode and the base-emitter junction diode of the transistor Q_2 difficult to achieve. With the help of simple mathematics, it can be proved that current I flowing through Q_2 equals the current I_N flowing through R_1.

Since the two transistors have identical values of DC current gain β and base current I_B, they would have identical collector and emitter currents as shown in Figure 6.49. Since

$$I_{IN} = I + (2I/\beta) = [(\beta + 2)/\beta]I \cong I,$$

the current I_{IN} set by V_{CC} and R_1 is mirrored as the collector current of Q_2. It may be mentioned here that more than one matched transistor can be connected to the base of Q_1. The current through the collector of each of the transistors equals the current flowing through the collector of the left-half transistor. The higher the value of β, the closer the mirrored current approximates the set value. Figure 6.50 shows another current mirror circuit in which a JFET current-regulating diode sets the constant current. The set

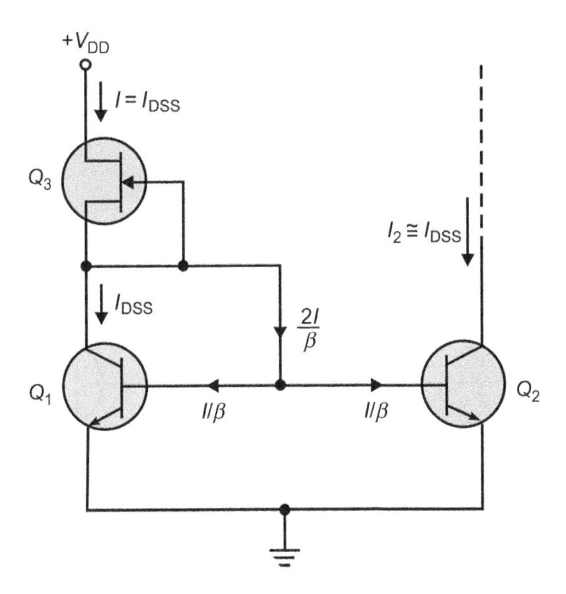

Figure 6.50 Current mirror with JFET providing constant current.

value of current in this case is I_{DSS}, which means that $I = I_{DSS}$. It may be mentioned here that in the circuit arrangement of Figure 6.49, current through R_1 would vary with variation in the supply voltage with the result that the current I though a mirror image of current I_{IN} will not be constant. The current mirror configuration of Figure 6.50 where transistor Q_1 is driven by a constant current source overcomes this problem.

6.4.5.2 Widlar Current Source

A constant current source should ideally have an infinite incremental output impedance. A Widlar current source as shown in Figure 6.51 is an improvement over the basic current mirror circuits of Figures 6.49 and 6.50, and offers much higher incremental output impedance. This is achieved by connecting a resistor (R_2 in this case) in series with the emitter terminal of Q_2. This resistor introduces a current feedback around Q_2, thus increasing its output impedance by a factor $(1 + G)$ where G is the loop gain defined by $1 + g_{m2}R_2$ where g_{m2} is the transconductance of Q_2. In order to have a constant load current, the circuit must be driven by a constant current rather than R_1.

6.4.5.3 Wilson Current Source

As shown in Figure 6.52, a Wilson current source has the advantage of virtually eliminating the base current mismatch of the conventional current mirror configuration. The other advantage is its higher output impedance.

6.5 Integrated-circuit Timer Circuits

In this section, we shall discuss IC-based timer circuits configured around digital integrated circuits such as the 74 121 from the TTL family and linear IC timers such as the IC timer 555.

6.5.1 Digital IC-based Timer Circuits

Some of the commonly used digital ICs that can be used as monostable multivibrators include the 74121 (single monostable multivibrator), 74221 (dual monostable multivibrator), 74122 (single retriggerable monostable multivibrator) and the 74123 (dual retriggerable monostable multivibrator) of the TTL

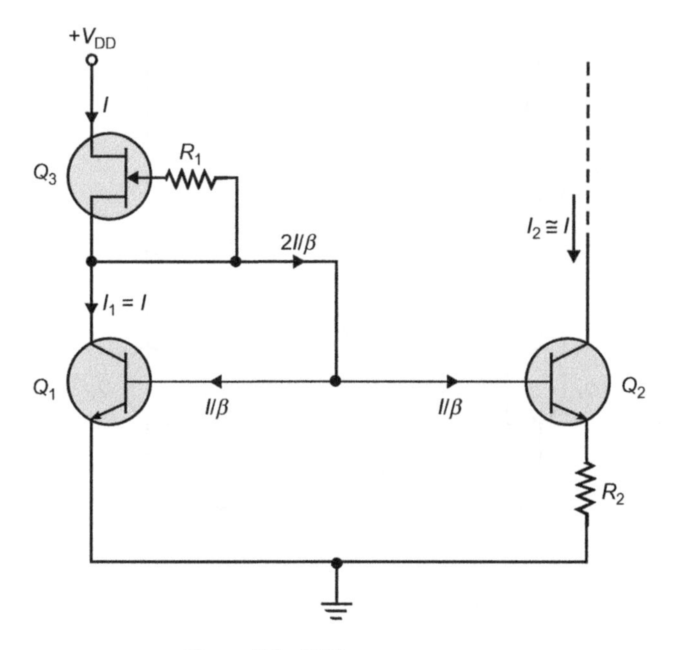

Figure 6.51 Widlar current source.

family and the 4098B (dual retriggerable monostable multivibrator) of the CMOS family. Figure 6.53 shows the use of IC 74121 as a monostable multivibrator along with trigger input. The IC provides features for triggering on either LOW-to-HIGH or HIGH-to-LOW edges of the trigger pulses. Figures 6.53a and 6.53b show a possible application circuits for HIGH-to-LOW and LOW-to-HIGH edge triggering, respectively. Output pulse width depends on external R and C. The output pulse width T

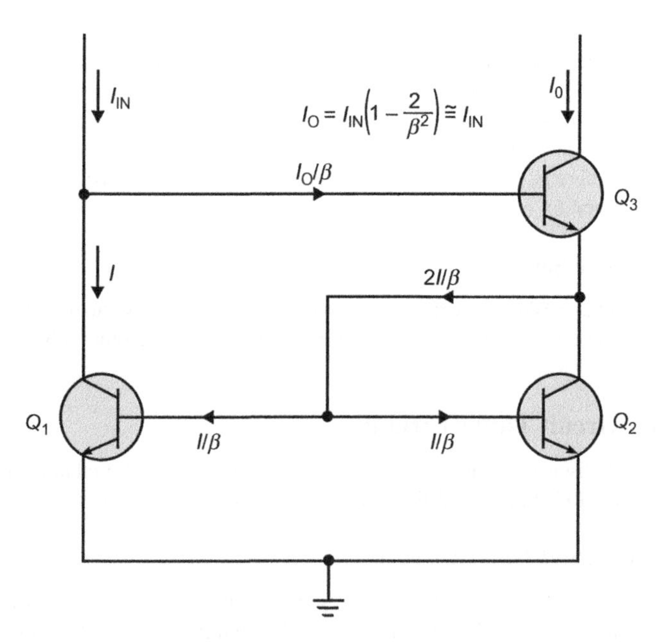

Figure 6.52 Wilson current source.

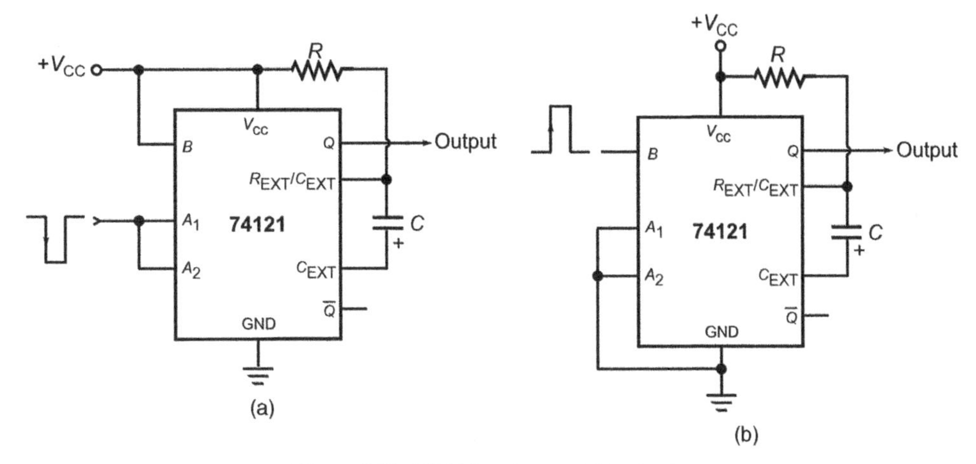

Figure 6.53 74121-based monostable circuits.

can be computed from $T = 0.7RC$. Recommended ranges of values for R and C are 4–40 kΩ and 10 pF to 1000 μF, respectively. IC provides complementary outputs, that is, we have stable LOW or HIGH state and corresponding HIGH or LOW state available on Q and $\bar{Q}$ outputs.

Figure 6.54 shows the use of 74123, a retriggerable monostable multivibrator. Like 74121, this IC also provides features for triggering on either LOW-to-HIGH or HIGH-to-LOW edges of the trigger pulses. Output pulse width depends on external R and C, calculated as $T = 0.28RC[1 + (0.7/RC)]$, where R and C are in kiloohms and picofarads, respectively, and T is in nanoseconds. This formula is valid for $C > 1000$ pF. The recommended range of values for R is 5–50 kΩ. Figure 6.54a and b depict application circuits for HIGH-to-LOW and LOW-to-HIGH triggering, respectively. It should be mentioned that there are other triggering circuit options for both LOW-to-HIGH and HIGH-to-LOW edge-triggering of monoshot.

6.5.2 *IC Timer-based Multivibrators*

IC timer 555 is one of the most commonly used general-purpose linear integrated circuits. The simplicity with which monostable and astable multivibrator circuits can be configured around this IC is one of the main reasons for its wide use. Figure 6.55 shows the internal schematic of timer IC 555. It comprises two opamp comparators, a flip-flop, a discharge transistor, three identical resistors and an output stage. The resistors set the reference voltage levels at the non-inverting input of the lower comparator and inverting input of the upper comparator at + $V_{CC}/3$ and +2 $V_{CC}/3$, respectively. Outputs of two comparators feed

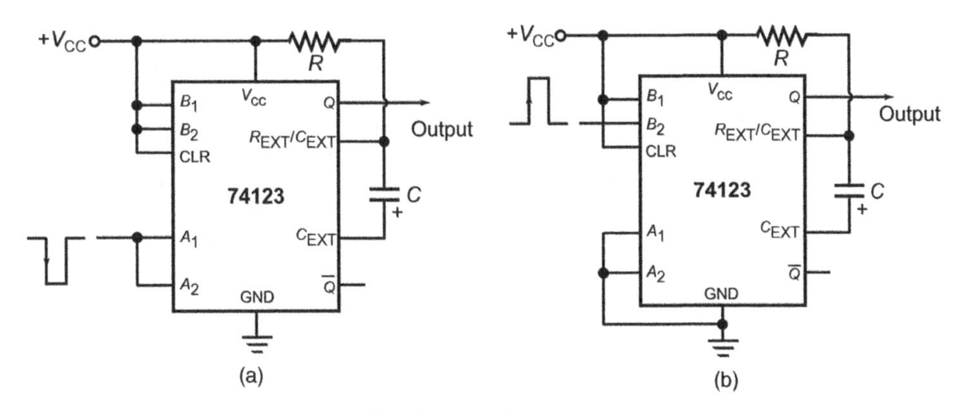

Figure 6.54 74123-based monostable circuits.

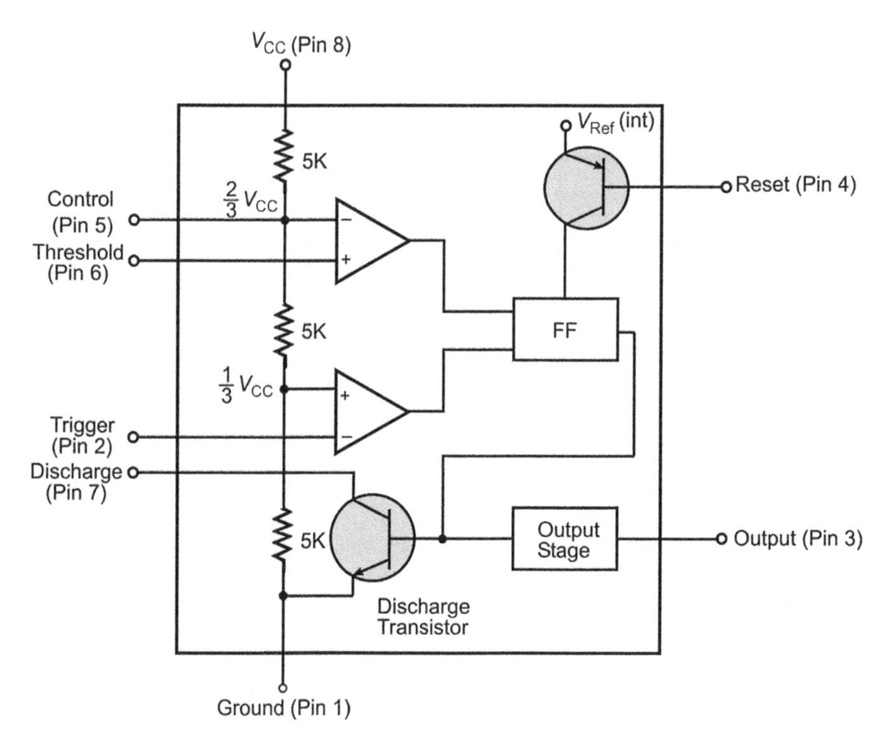

Figure 6.55 Internal schematic arrangement of IC timer 555.

the SET and RESET input of the flip-flop and determine the logic status of its output and subsequently the final output. The flip-flop's complementary outputs feed the output stage and the base of the discharge transistor. This ensures that the output is HIGH when the discharge transistor is OFF, and LOW when the discharge transistor is ON. Different terminals of the timer 555 are designated as Ground (Pin 1), Trigger (Pin 2), Output (Pin 3), RESET (Pin 4), Control (Pin 5), Threshold (Pin 6), Discharge (Pin 7) and $+V_{CC}$ (Pin 8). With this background, we shall now describe the astable and monostable circuits configured around timer 555.

6.5.2.1 Astable Multivibrator Using Timer IC 555

Figure 6.56 depicts the basic 555 timer-based astable multivibrator circuit along with relevant waveforms. Initially, capacitor C is fully discharged, which forces output to go to the HIGH state. An open discharge transistor allows the capacitor C to charge from $+V_{CC}$ through R_1 and R_2. When the voltage across C exceeds $+2V_{CC}/3$, the output goes to the LOW state and the discharge transistor is switched ON at the same time. Capacitor C begins to discharge through R_2 and the discharge transistor inside the IC. When the voltage across C falls below $+V_{CC}/3$, output goes back to HIGH state. The charge and discharge cycles repeat and the circuit behaves like a free-running multivibrator. Terminal 4 of the IC is the RESET terminal, which is usually connected to $+V_{CC}$. If the voltage at this terminal is driven below 0.4 V, output is forced to the LOW state overriding command pulses at terminal 2 of the IC. The HIGH-state and LOW-state time periods are governed by the charge $+V_{CC}/3$ to $+2V_{CC}/3$ and discharge $+2V_{CC}/3$ to $+V_{CC}/3$ timings.

$$\text{HIGH-state time period } T_{\text{HIGH}} = 0.69C(R_1 + R_2)$$

$$\text{LOW-state time period } T_{\text{LOW}} = 0.69R_2C$$

Remember that when the astable multivibrator is powered, the first cycle HIGH-state time duration is about 30% longer as the capacitor is initially discharged and it charges from 0 (rather than $+V_{CC}/3$) to $+2V_{CC}/3$.

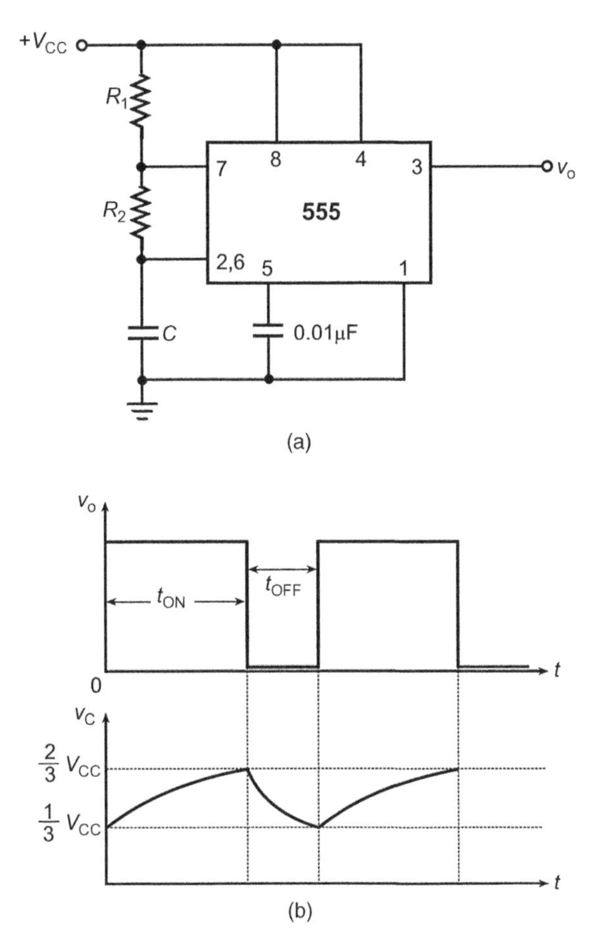

Figure 6.56 Basic 555 timer-based astable multivibrator.

In the case of the astable multivibrator circuit of Figure 6.56a, the HIGH-state time period is always greater than the LOW-state time period. Figure 6.57 depicts the modified circuit where HIGH-state and LOW-state time periods can be chosen independently. For the astable multivibrator circuit of Figure 6.57, the two time periods are given by:

$$\text{HIGH-state time period } T_{\text{HIGH}} = 0.69R_1C$$

$$\text{LOW-state time period } T_{\text{LOW}} = 0.69R_2C$$

$$\text{for } R_1 = R_2 = R, T = 1.38RC \text{ and } f = (1/1.38RC).$$

6.5.2.2 Monostable Multivibrator Using Timer IC 555

Figure 6.58a shows the basic monostable multivibrator circuit configured around timer 555 along with relevant waveforms as shown in Figure 6.58b. Trigger pulse is applied to pin 2 of the IC, which should initially be kept at $+V_{CC}$. A HIGH at pin 2 forces the output to LOW state. A HIGH-to-LOW trigger pulse at pin 2 sets the output in the HIGH state and simultaneously allows the capacitor C to charge from $+V_{CC}$ through R. Remember that the LOW-level of the trigger pulse needs to go at least below $+V_{CC}/3$. When the capacitor voltage exceeds $+2V_{CC}/3$, the output goes back to the LOW state. We need to apply

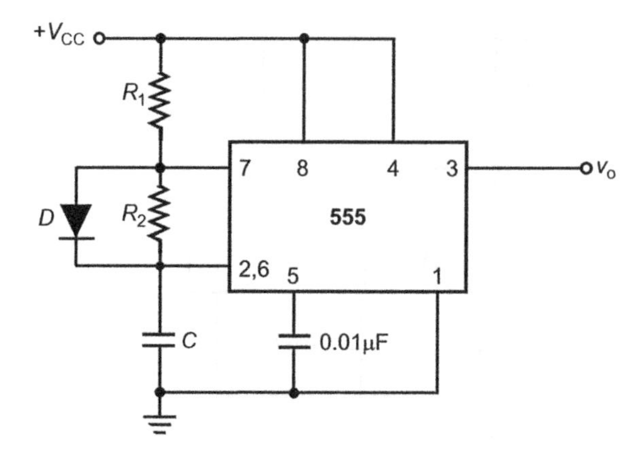

Figure 6.57 Modified 555 timer-based astable multivibrator.

another trigger pulse to pin 2 to make the output return to the HIGH state. Every time the timer is appropriately triggered, the output goes to the HIGH state and stays there for the time taken by the capacitor to charge from 0 to $+2V_{CC}/3$. This time period, which equals the monoshot output pulse width, is given by $T = 1.1RC$.

It is often desirable to trigger a monostable multivibrator either on the trailing (HIGH-to-LOW) or leading edges (LOW-to-HIGH) of the trigger waveform. In order to achieve that, we need an external circuit between the trigger waveform input and pin 2 of timer 555. The external circuit ensures that pin 2 of the IC gets the required trigger pulse corresponding to the desired edge of the trigger waveform. Figure 6.59a shows the monoshot configuration that can be triggered on the trailing edges of the trigger waveform. Figure 6.59b shows relevant waveforms. R_1–C_1 constitutes a differentiator circuit. One of the terminals of resistor R_1 is tied to $+V_{CC}$ with the result that the amplitudes of differentiated pulses are $+V_{CC}$ to $+2V_{CC}$ and $+V_{CC}$ to ground, corresponding to the leading and trailing edges of the trigger waveform, respectively. Diode D clamps the positive-going differentiated pulses to about +0.7 V.

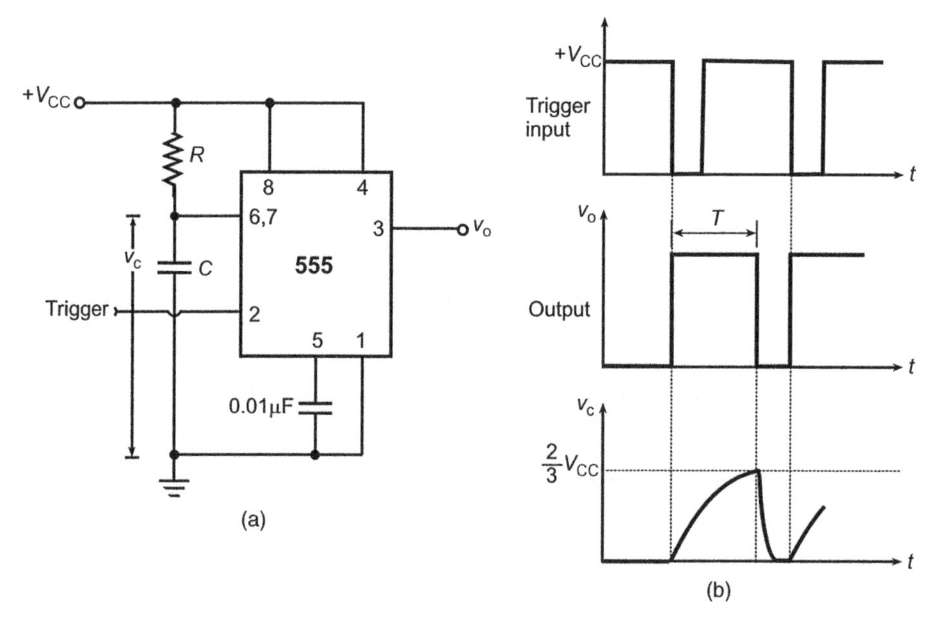

Figure 6.58 Basic 555 timer-based monostable multivibrator circuit.

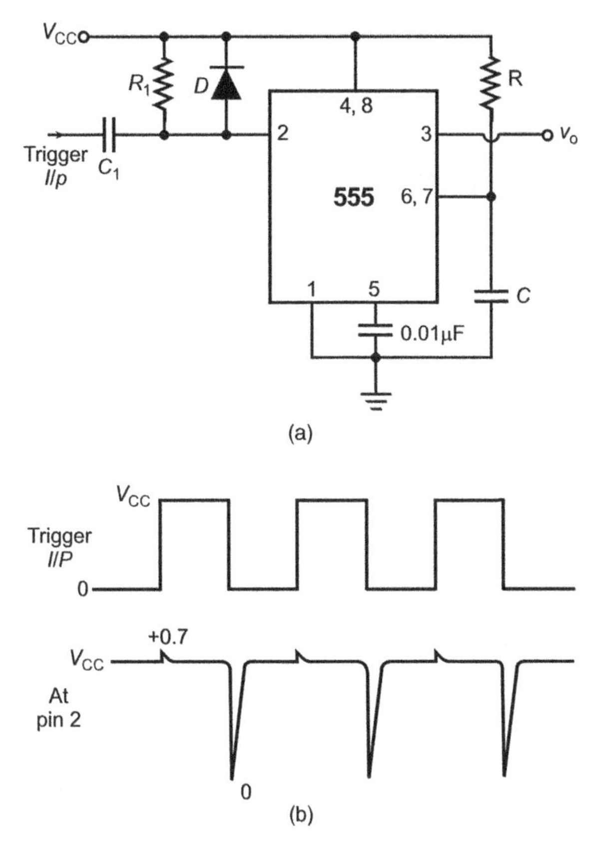

Figure 6.59 555-timer monostable configuration triggering on high-to-low edges.

The net result is that the trigger terminal of timer 555 gets the required trigger pulses corresponding to the HIGH-to-LOW edges of the trigger waveform.

Figure 6.60a shows the monoshot configuration that can be triggered on the leading edges of the trigger waveform and Figure 6.60b shows the relevant waveforms. The R_1–C_1 combination constitutes the differentiator producing positive and negative pulses corresponding to LOW-to-HIGH and HIGH-to-LOW transitions of the trigger waveform. Negative pulses are clamped by the diode and the positive pulses are applied to the base of a transistor switch. The collector-terminal of the transistor feeds the required trigger pulses to pin 2 of the IC.

For the circuits shown in Figures 6.59 and 6.60 to function properly, values of R and C for the differentiator should be chosen carefully. Firstly, the differentiator time constant should be much smaller than the HIGH time of the trigger waveform for proper differentiation. Secondly, differentiated pulse width should be less than the expected HIGH time of the monoshot output.

6.6 Current-to-voltage Converter

There are many situations where the signal of interest is current rather than voltage. As an example, the electrical signal produced by a photosensor as a result of light falling on it is a current. In order to process the signal to extract the intended information, more often than not the current needs to be converted into an equivalent voltage. This function is performed with a current-to-voltage converter circuit, which is also known as transimpedance amplifier for obvious reasons. Figure 6.61 shows the basic opamp-based current-to-voltage converter fed at its input by a current source I. The analysis of the circuit is very simple. Since the non-inverting input of the opamp is tied to ground and the circuit has negative feedback

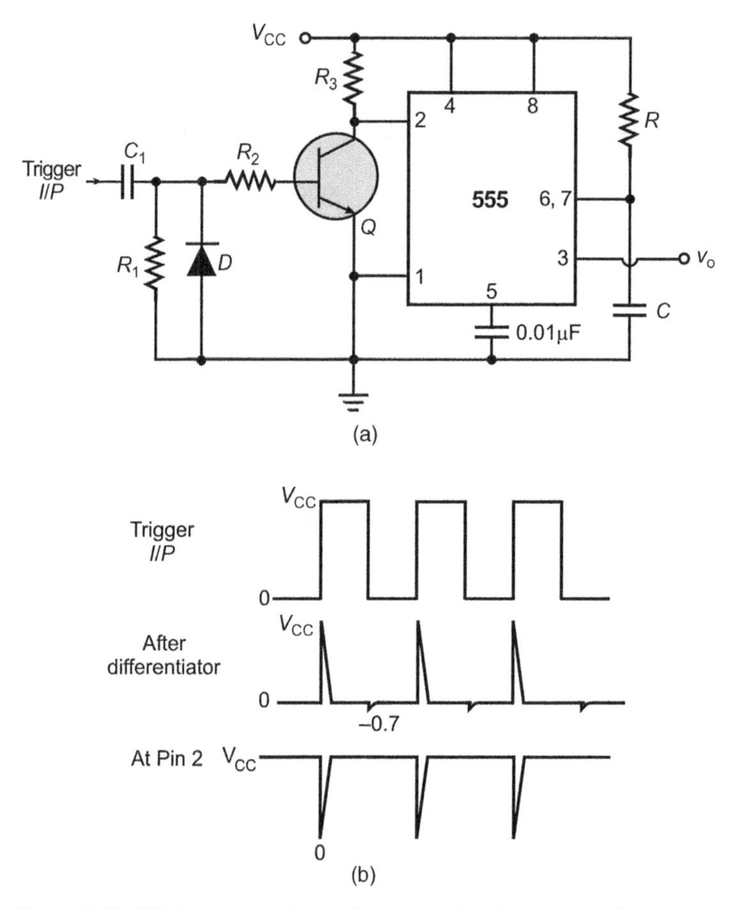

Figure 6.60 555-timer monostable configuration triggering on low-to-high edges.

due to virtual earth phenomenon, the inverting input also behaves as if it were grounded. In fact, the high open loop gain of the opamp coupled with negative feedback forces the two inputs to remain at the same potential for a finite output. With this background, the magnitude of current flowing through the feedback resistor R would be V_O/R. Since no current flows into the opamp due to its infinite input impedance (under ideal conditions), applying Kirchoff's current law at the inverting terminal node yields $I + (V_O/R) = 0$ which gives $V_O = -IR$.

A current-to-voltage converter constitutes an important building block of laser electronics. All laser receiver circuits, light meters, laser rangefinders, laser warning sensors and laser communication

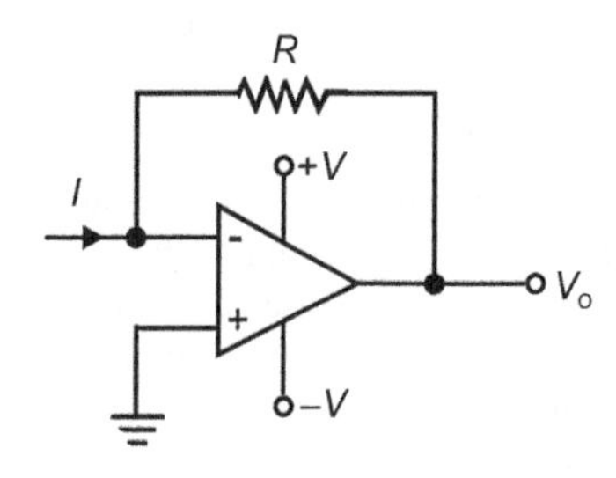

Figure 6.61 Basic current-to-voltage converter.

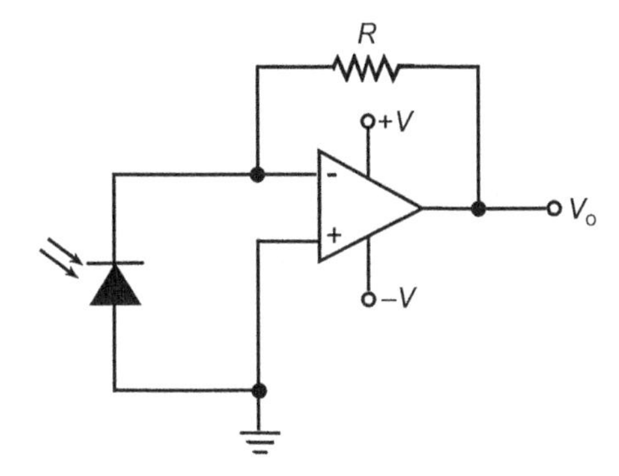

Figure 6.62 Photodiode current-to-voltage converter.

receivers use some kind of photosensor. The desired information is in the form of a continuous or pulsed electric current, which needs to be converted into an equivalent voltage signal for further processing. A current-to-voltage converter circuit is therefore invariably used as a front end in all laser detection systems. Figure 6.62 shows the basic current-to-voltage converter fed at its input from a photodiode. The photodiode here is used in photovoltaic mode, that is, the bias voltage is zero. More current-to-voltage converter circuits that convert photodiode current into an equivalent voltage are discussed in Chapter 10 on photosensors.

6.7 Peak Detector

A peak detector produces at its output a voltage equal to the peak value of the signal applied at its input. The circuit could be designed to detect either the positive or the negative peak. Figure 6.63 shows a typical positive peak detector circuit. In the case of negative peak detector circuit, polarity of the diode is reversed. The circuit functions as described below.

In the case of a sinusoidal input or any other bi-directional input, the diode is forward biased during the positive half-cycles. The capacitor rapidly charges to the positive peak from the output of the opamp through relatively much smaller ON-resistance of the forward-biased diode. In the case of a positive pulse input, the same happens during the duration of the pulse. As the input begins to decrease beyond the peak value the diode becomes reverse biased, thus isolating the capacitor from the output of the opamp. The capacitor can discharge only through the parallel combination of the two resistors. One of these is the resistor connected across the capacitor and the other is the series combination of the feedback resistor and the input resistor of the opamp at the inverting input. This effective discharge path resistance is relatively much larger than the ON-resistance of the forward-biased diode. It may be mentioned here that

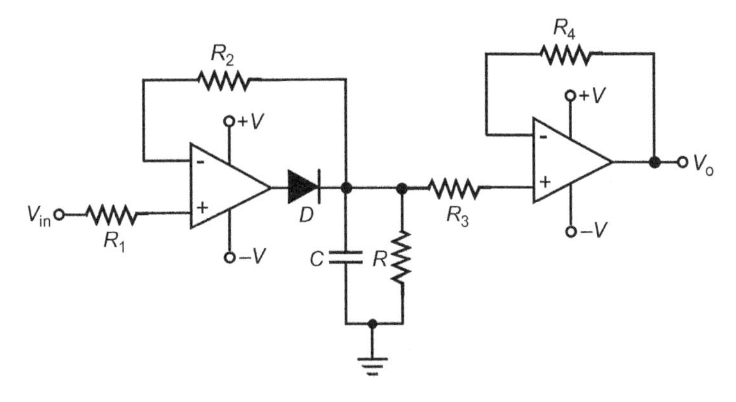

Figure 6.63 Positive-peak detector circuit.

the resistor across the capacitor is there to allow a discharge path for the circuit to respond to the decreasing amplitude of the signal peaks. The unity gain buffer at the output is to prevent capacitor discharge due to the loading effects of the following circuit.

The parallel *R–C* circuit time constant is typically 100 times the time period corresponding to minimum frequency of operation. The *R–C* time constant also controls the response time to changing amplitude; higher *R–C* time constant gives lower ripple but a sluggish response while a lower *R–C* time constant gives a faster response at the cost of increased ripple. The chosen value is a trade-off between two conflicting requirements.

Slew rate is the main opamp parameter to be considered while choosing an opamp for this application. The desired slew rate should be such that the slew-rate-limited frequency, which is a function of the peak-to-peak output swing and the slew rate, is at least equal to the highest frequency of operation. Peak-to-peak voltage swing at the output of the opamp is equal to $V_P - (-V_{SAT}) = V_P + V_{SAT}$. In the case of pulse input, the slew rate should at least equal V_O/T volts per microsecond. V_P is the peak amplitude of the pulse and T is its width.

As for a current-to-voltage converter, peak detector is also an important building block of laser receiver electronics in systems such as laser warning sensors, laser power and energy meters. It is particularly important while designing front-end detection circuits for Q-switched laser pulses. The peak detector circuit in that case acts as a pulse stretcher. The peak of the stretched pulse equals the peak of the input pulse. The stretched pulse allows the designer to use relatively lower speed processing circuitry subsequent to the front-end. This is discussed further in Chapter 10 on photosensors and Chapter 14 on military applications of lasers and optoelectronics.

6.8 High-voltage Trigger Circuit

A high-voltage trigger circuit is another important building block of laser electronics. It is an essential component of the electronics package with any flash-lamp-pumped solid-state laser, where it performs the function of initiating an electrical discharge in the gas inside the tube to create a low resistance path between the two electrodes. The amplitude of the high-voltage pulses is typically in the range 5–20 kV and the pulse shape is not critical. High-voltage pulses are also required to operate the Q-switch in the case of electro-optically Q-switched solid-state lasers. The amplitude of the high-voltage pulse required for the Q-switch is typically of the order 2–3 kV and the pulse has a well-defined shape in terms of its rise time and pulse width.

Figure 6.64a shows a typical high-voltage trigger pulse generation circuit used for flash lamp triggering. A bipolar junction transistor, an enhancement MOSFET or an insulated gate bipolar transistor

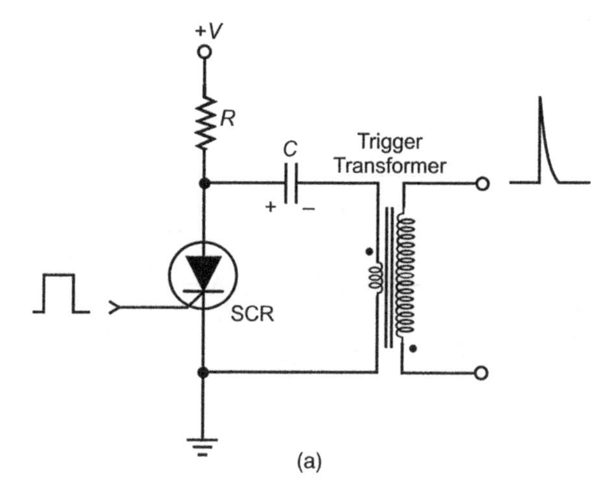

Figure 6.64 High-voltage trigger pulse generator.

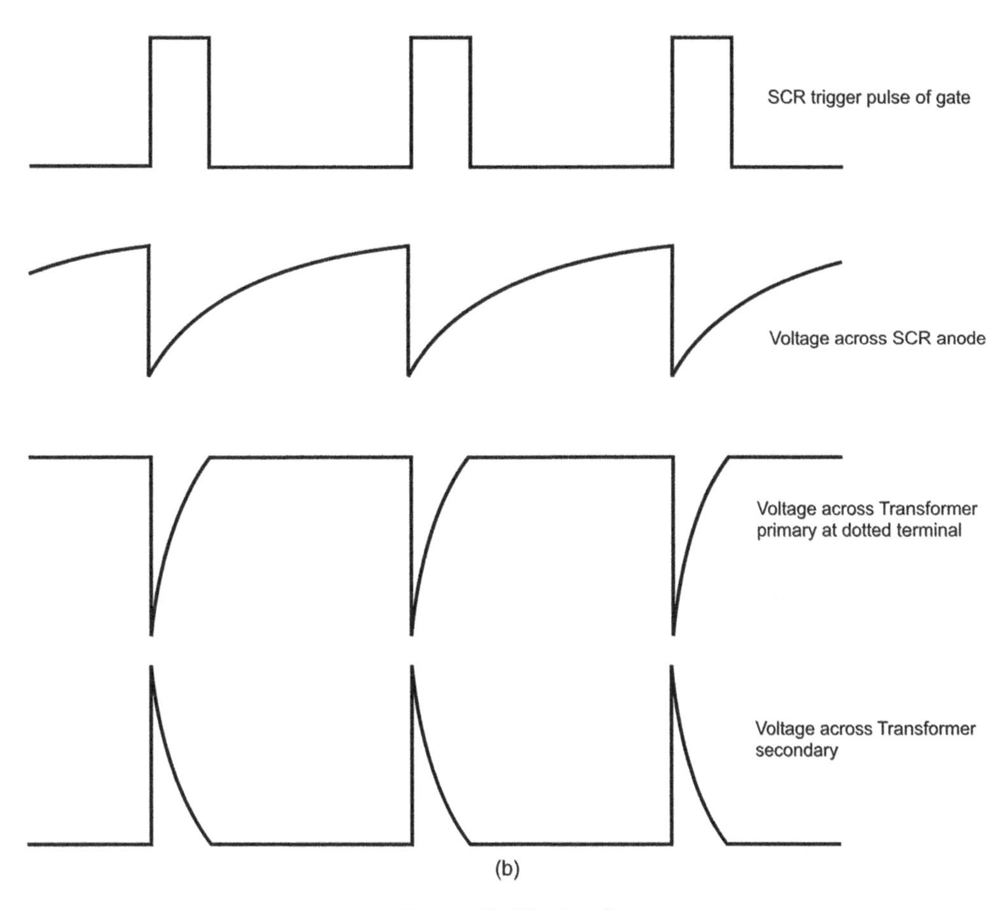

Figure 6.64 (*Continued*)

(IGBT) can also be used as the electronic switch in place of SCR. When the SCR is in the OFF-state, capacitor *C* charges through resistor *R* and the transformer primary. When the capacitor is fully charged to the applied DC voltage *V*, the SCR is triggered to the ON-state by application of a suitable trigger pulse at its gate-terminal. The capacitor rapidly discharges through the primary winding of the transformer, generating a pulse across the primary winding. This pulse is stepped up to produce the desired amplitude of high voltage pulse across the secondary winding. The polarity of the pulse depends upon the winding polarity. In the case of the circuit shown in Figure 6.64a, the polarity of the pulse at the dotted end of the primary winding is negative which means that the polarity of the pulse on the dotted end of the secondary winding will also be negative. As a result, we get a positive polarity output pulse. Figure 6.64b shows the relevant waveforms.

Value of resistance *R* should be such that the current V/R is less than the holding current value of the SCR. Otherwise, once turned ON, the SCR will continue to conduct even after the capacitor has fully discharged through it and the primary winding. Figure 6.65a and b show the same circuit implemented with an N-channel enhancement MOSFET and relevant waveforms. Note that in the case of circuit Given that the Doppler-broadened gain curve of a helium-neon laser with a 50-cm-long resonator emitting at 1.15 mm is 770 MHz, determine (a) inter-longitudinal mode spacing and (b) the number of maximum possible sustainable longitudinal modes of Figure 6.65a, the value of resistance *R* has no lower limit as in the case of an SCR circuit. In the case of a MOSFET-based trigger circuit, MOSFET remains ON during the duration of the input pulse and the capacitor starts charging only after the MOSFET goes back to the OFF-state. The IGBT-based circuit functions similarly to the MOSFET-based circuit.

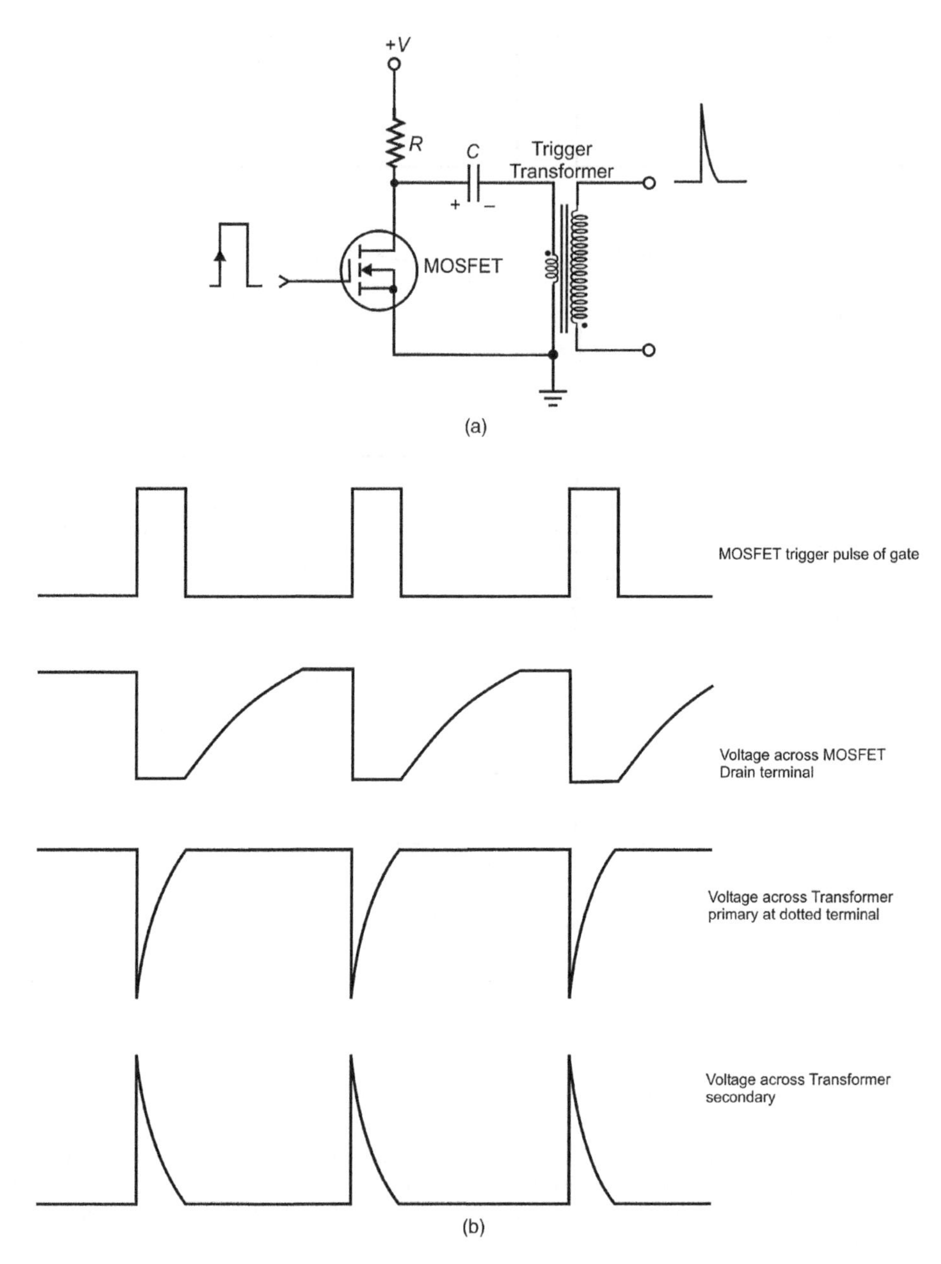

Figure 6.65 MOSFET-based high-voltage trigger circuit.

6.9 Summary

- Power supply is an essential building block of all laser electronics packages. It does the job of providing required DC voltages from available AC mains (for mains-operated systems) and DC input (for portable systems). Power supplies are often classified as linear power supplies or switched-mode power supplies, depending upon the nature of regulation circuit.
- A linear power supply comprises: a mains transformer, a rectifier circuit, a filter circuit and a regulation circuit.

- The transformer transforms the mains AC input to the desired value of AC. The rectifier circuit converts the AC into a unidirectional waveform. The filter smoothes the ripple and the regulator circuit ensures that the load voltage (in the case of voltage-regulated power supplies) or the load current (in the case of current-regulated power supplies) is constant irrespective of variations in line voltage or load resistance.
- There are three basic types of linear voltage regulator configurations: emitter-follower regulator, series regulator and shunt regulator. For an emitter-follower regulator, the emitter voltage (which is also the output voltage) remains constant as long as the base voltage is held constant. A Zener diode connected at the base ensures this. For a series linear regulator, the pass element is connected in series with the load and any decrease or increase in the output voltage is accompanied by a decrease or increase in the collector-emitter voltage of the series-pass transistor. A shunt regulator provides a change in the current through the shunt transistor, thereby changing the voltage drop across a series-connected resistor to maintain a constant output voltage.
- Contemporary regulating circuits are almost exclusively configured around one or more ICs known as IC voltage regulators. IC voltage regulators are available to meet a wide range of output voltage and load current requirements. Both fixed-output voltage (positive and negative) and adjustable-output voltage (positive and negative) IC regulators are commercially available in a wide range of voltage, current and regulation specifications. These have built-in protection features such as current limit and thermal shutdown.
- Linear power supplies are well known for their extremely good line and load regulation, low output voltage ripple and almost negligible RFI/EMI. Switching power supplies on the other hand have much higher efficiency (typically 80–90% compared to 50% in the case of linear supplies) and reduced size/weight for a given power-delivering capability.
- Switched-mode power supplies are designed in a variety of circuit configurations depending upon the intended application. Almost all switching supplies belong to one of the following three broad categories: flyback converters, forward converters and push-pull converters.
- A constant current source delivers a predefined constant current to a load despite variations in supply voltage and load resistance. There are various constant current source configurations using JFETs, bipolar junction transistors and operational amplifiers. In addition, there are mirror-based constant current source configurations.
- Timer circuits can be designed with the help of general-purpose digital integrated circuits and linear integrated circuits.
- A current-to-voltage converter is an important component of laser electronics where the desired information is in the form of a continuous or pulsed electric current which needs to be converted into an equivalent voltage signal for further processing. All laser receiver circuits, light meters, laser rangefinders, laser warning sensors and laser communication receivers use some kind of photosensor.
- A peak detector is an important component of laser receiver electronics in systems such as laser warning sensors, laser power and energy meters. It is particularly important while designing detection circuits for Q-switched laser pulses. Peak detector circuits to detect either positive or negative peak can also be configured around simple operational amplifier circuits.
- A high-voltage trigger circuit is an essential component of the electronics package that accompanies any flash-lamp-pumped solid-state laser, where it performs the function of initiating an electrical discharge in the gas inside the tube to create a low-resistance path between the two electrodes.

Review Questions

6.1. Name the basic building blocks of electronics that are commonly used to construct electronics packages of solid-state, gas and semiconductor diode lasers. Briefly describe the functions performed by these building blocks with reference to the type of laser these are used for.

6.2. Briefly describe salient features of linearly regulated power supplies and switched-mode power supplies. Why does a switched-mode power supply seem to be the only choice in portable systems?

6.3. In what type of laser is the constant current source the main power supply? With the help of relevant circuit diagram, briefly describe the functional principle of an opamp-based constant current source.

6.4. What do you understand by a current mirror? What are its applications?

6.5. Briefly describe the functional principle of (a) a flyback converter and (b) a push-pull converter. Which topology would you prefer for a high-output-voltage low-current requirement?

6.6. What are switching regulators? Briefly describe buck, boost and inverting type of switching regulators with the help of relevant circuit diagrams.

6.7. What are three-terminal switching regulators? Which basic switching regulator configuration is commonly used to build these regulators?

6.8. With the help of relevant circuit diagrams, briefly explain how (a) a JFET can be used as a constant current source without any external component and (b) a three-terminal positive output linear regulator can be used to generate a regulated output voltage higher than it is intended for?

6.9. With the help of relevant circuit diagram, briefly explain the operational principle of a positive edge-triggered monostable multivibrator configured around 555 timer IC.

6.10. Why is a current-to-voltage converter block important for laser electronics? What function does it perform? Briefly describe the operational principle of a current-to-voltage converter configured around an opamp.

6.11. Draw the basic circuit schematics of opamp-based positive and negative peak detector circuits. If the input to the peak detector circuit is a Q-switched pulse, how would you go about choosing the right opamp for this application?

6.12. Where do we usually need a high-voltage trigger pulse generation circuit in lasers? With the help of a basic circuit drawing, briefly describe the operation of an SCR-based high-voltage trigger circuit, utilizing capacitive charge–discharge phenomenon.

Problems

6.1. Find the regulated output voltage for the emitter-follower regulator circuit of Figure 6.66. Also determine the current I_Z if transistors Q_1 and Q_2 in the Darlington pair had β of 10 and 100, respectively. Assume forward voltage drop of D_1 and D_2 and V_{BE} (Q_1) and V_{BE} (Q_2) to be 0.6 V each. [$V_O = -24$ V, $I_Z = 10$ mA]

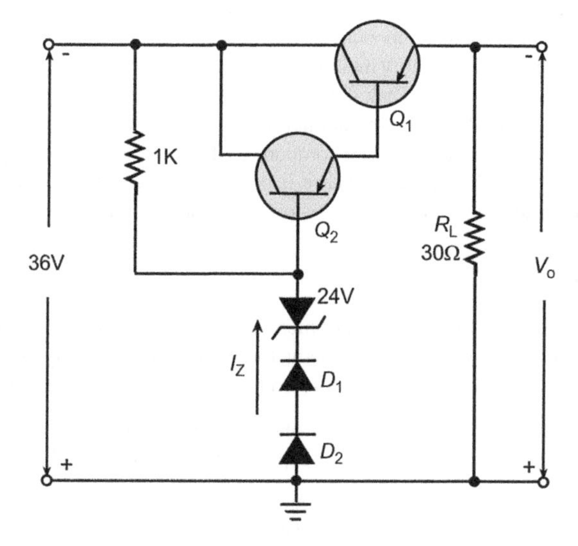

Figure 6.66 Problem 6.1.

6.2. Refer to the three-terminal regulator circuit of Figure 6.67. Determine the regulated output voltage and power dissipated in the regulator.
[−7.5 V, 2.25 W]

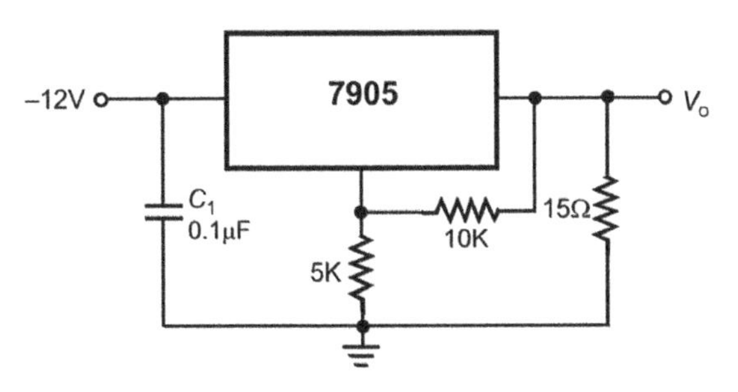

Figure 6.67 Problem 6.2.

6.3. Refer to the step down switching regulator circuit of Figure 6.68. If the circuit produces a regulated output voltage of 12 V from an unregulated input voltage range of 18–24 V, and the switching transistor was switched at 20 kHz, determine the on time of the switching transistor for unregulated input voltages of 18 and 24 V.
[33.3 μs, 25 μs]

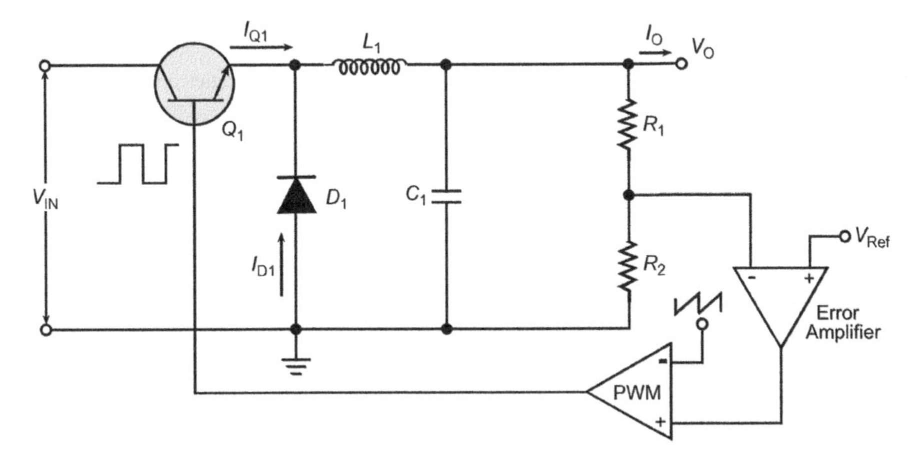

Figure 6.68 Problem 6.3.

6.4. Design an opamp-based current-to-voltage converter circuit with a trans-resistance gain of 100 000.
[Figure 6.69]

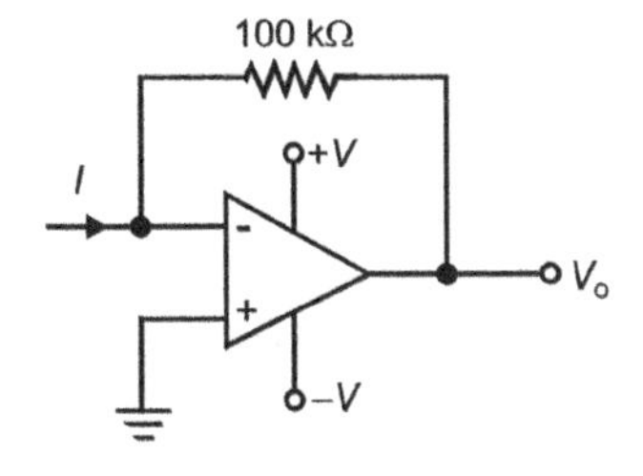

Figure 6.69 Answer to Problem 6.4.

Self-evaluation Exercise

Multiple-choice Questions

6.1. Identify the false statement.

a. Power supply with an emitter-follower regulator can be categorized as a linearly regulated power supply.
b. A shunt-regulated power supply is inherently protected against current overloads.
c. A series-pass element regulated power supply is likely to be less efficient than a shunt-regulated power supply.
d. A fixed-output-voltage linear three-terminal regulator can be used to produce a variable regulated output voltage.

6.2. In the case of switched-mode power supplies using a bipolar junction transistor as the active device, the transistor is

a. always operated in the saturation region of its output characteristics.
b. always operated in the active region of its output characteristics.
c. switched between the active and saturation regions of its output characteristics.
d. switched between the cut-off and saturation regions of its output characteristics.

6.3. A regulated power supply operates from 12 ± 3 VDC and produces a regulated output voltage of 24 VDC. The power supply is possibly using a

a. series-pass linear regulator
b. switching regulator
c. shunt regulator
d. emitter-follower regulator

6.4. In which of the following types of switched-mode power supply topologies is energy transferred to the load only during the turn-off time of the switching device?

a. flyback converter
b. push-pull converter
c. buck regulator
d. forward converter

6.5. In the case of a JFET-based constant current source,

a. gate-terminal is shorted to source-terminal.
b. gate-terminal is shorted to drain-terminal.
c. source-terminal is shorted to drain-terminal.
d. none of these.

6.6. The type of laser that is driven by a constant current source is

a. solid-state laser
b. semiconductor diode laser
c. gas laser
d. none of these.

6.7. The current-to-voltage converter is also known by the name of

a. transconductance amplifier
b. operational amplifier
c. impedance transformer
d. transimpedance amplifier.

6.8. A current-to-voltage converter configured around an operational amplifier produces an output voltage of 1 V for an input current of 1 μA. The value of resistance connected between the output and the inverting input is

a. 1 MΩ
b. 2 MΩ

c. indeterminate from given data
d. none of these.

6.9. Which of the following specifications of the operational amplifier needs to be paid particular attention while designing a peak detector circuit?
a. open-loop gain
b. input and output offsets
c. slew rate
d. CMRR.

6.10. Identify the group of electronics building blocks likely to be found in the electronics package of a flash-lamp-pumped Q-switched solid-state laser:
a. capacitor charging switched-mode power supply, high-voltage trigger circuit, three-terminal regulators, low-voltage pulse generator
b. constant current source, low-voltage regulated power supply, peak detector circuit, three-terminal regulators
c. current-to-voltage converter, peak detector, constant current source, high-voltage trigger circuit
d. push-pull converter, voltage multiplier chain, high-voltage trigger circuit.

Answers

1. (c) 2. (d) 3. (b) 4. (a) 5. (a) 6. (b) 7. (d) 8. (a) 9. (c) 10. (a)

Bibliography

1. *OpAmp Applications Handbook*, 2006 by Walt Jung (ed.), Analog Devices Inc.
2. *Operational Amplifiers*, 2003 by George Clayton and Steve Winder, Newness.
3. *OpAmps for Everyone*, 2003 by Ron Mancini, Texas Instruments.
4. *Optoelectronics*, 1998 by J. Wilson and J.B. Hawkes, Prentice Hall.
5. *Optoelectronic Devices and Circuits*, 2007 by A.K. Ganguly, Alpha Science International Ltd.
6. *Electronic Devices and Circuits*, 2009 by Anil K. Maini and Varsha Agrawal, Wiley India Pvt Ltd.
7. *Switched Mode Power Supply Handbook*, 2010 by Keith H. Billings, McGraw-Hill.
8. *An Introduction to Power Electronics*, 1983 by B.M. Bird and K.G. King, John Wiley & Sons.
9. *Power Supply Cookbook*, 1990 by Marty Brown, Academic Press.
10. *DC Power Supplies: A Technician's Guide*, 1996 by Joseph J. Carr, TAB Books.
11. *Switched Mode Power Supply Design Handbook*, 1986 by P.R.K. Chetty, TAB Professional and Reference Books.
12. *Understanding DC Power Supplies*, 1983 by B. Davis, Prentice-Hall.
13. *Power Supplies, Part A and Part B*, 1987 by Odon Ferenczi, Elsevier.
14. *Power Electronics Handbook*, 1997 by F.F. Mazda, Newnes.
15. *Switching Power Supply Design*, 1998 by A.I. Pressman, McGraw-Hill.

7

Solid-state Laser Electronics

7.1 Introduction

There has been a dramatic increase in commercial, industrial, medical, defence and research applications of lasers since the early 2000s. A variety of lasers covering a large portion of the electromagnetic spectrum and producing a range of output power levels are available at affordable prices. Significant developments have also been made in the field of laser- or laser-based-system electronics. The role of an electronics engineer working in the field of lasers is far more challenging today than it was in the early stages of development of lasers. In this chapter and in Chapters 8 and 9 the *electronics component* of different types of laser sources and the systems built around them are discussed. After a brief introduction to the wide spectrum of electronics circuits that form part of various laser and optoelectronics systems, the design and operational aspects of electronics that accompany solid-state lasers are described. Electronics packages for gas lasers and semiconductor lasers are discussed in Chapters 8 and 9, respectively.

7.2 Spectrum of Laser Electronics

In this section we briefly outline the role of electronics in lasers, laser-based systems and optoelectronic systems. With reference to laser sources and laser-based systems, we usually talk about only those subsystems that are essential from the operational viewpoint of the laser in question. However, if we confined the discussion only to power supplies for different types of lasers, then we would probably only be covering 30% of the electronics that concerns contemporary lasers and laser-based systems. In the following sections, we shall briefly look at the range of electronics circuitry that accompany solid-state, semiconductor diode and gas lasers. In addition to the significant role that the electronics plays in the case of laser sources and systems configured around them, electronics also plays a major role in the design and development of optoelectronic systems related to the testing and evaluation of laser systems and optoelectronic sensor systems. Systems used for testing and evaluation include those used for measurement of different laser source parameters such as pulse energy, CW power and beam divergence as well as the operational parameters of the laser system such as power/energy density or the beam profile at a certain distance from the laser source. Optoelectronic sensor systems are generally used for the detection of laser radiation along with other parameters, including laser warning, laser position sensing and laser guidance, depending on the application in question.

7.2.1 Solid-state Lasers

Let us take the case of a flashlamp-pumped pulsed solid-state laser such as Nd:YAG or Nd:Glass. When discussing an electronics package for such a laser source, it is the power supply unit needed to charge the energy storage capacitor to store the requisite quantum of energy that comes to mind first. It is not only

Lasers and Optoelectronics: Fundamentals, Devices and Applications, First Edition. Anil K. Maini.

the most complex of the electronics circuit modules used in the case of flashlamp-pumped solid-state lasers; its electrical conversion efficiency plays a key role in determining the wall-plug efficiency and the size/weight of the laser. Capacitor-charging power supplies intended for flashlamp-pumped solid-state laser applications are commercially available both as bench-top models as well as modular units for original equipment manufacturers (OEM) applications. These capacitor-charging power supplies are available for a range of input voltage (AC or DC), DC output voltage and charging rate specifications. DC output voltage from a few hundreds of volts to several kilovolts and charging rate in the range of tens–thousands of joules per second are common. In addition, these power supplies offer many control and protection features relevant to flashlamp-pumped laser power supplies. Some of these features include end-of-charge status indication, peak output voltage hold, output voltage monitor and over voltage and over temperature protection. Figure 7.1 depicts one such capacitor-charging power supply suitable for flashlamp-pumped solid-state lasers from Kaiser Systems, Inc.

Less important are other circuit modules such as the simmer power supply, used in high-repetition-rate flashlamp-pumped solid-state lasers, the Q-switch driver, used in Q-switched lasers and the pulse-forming network (PFN) that ensures a critically damped current pulse through the flashlamp when the energy storage capacitor is made to discharge through it. When the same laser is used as a source to build a laser rangefinder, a lot more is added to the electronics component of the overall system. In addition to the circuit blocks outlined above, which constitute laser source electronics, the rangefinder system also needs to have a receiving channel capable of processing the laser pulses received after being scattered from the intended target. Most military rangefinders work on the time-of-flight measurement principle and compute the target range by measuring the time interval between the transmitted and the received laser pulses. The receiving channel typically comprises circuit blocks such as low-noise optoelectronic front-end, having the requisite bandwidth to process the received laser pulses, and high-speed counting circuitry to determine the time interval and control logic.

7.2.2 *Semiconductor Diode Lasers*

As outlined in Chapter 5 during the discussion on the handling of diode lasers, it was emphasized that certain guidelines needed to be strictly adhered to in both the handling of these lasers and in the design of the driver and control circuits used to power them in order to achieve the prescribed life and reliability performance. Diode lasers are particularly sensitive to electrostatic discharge, short-duration electric transients such as current spikes, injection current exceeding the prescribed limit and reverse voltage exceeding the breakdown limit.

In order to protect the diode lasers from the above-mentioned failure modes, the driver circuit should be carefully designed and should have all the features recommended by the diode laser manufacturer. The driver should be a constant current source with inbuilt features such as soft start, protection against transients, interlock control for the connection cable to the laser and safe adjustable limit for injection current. In case the laser is to be operated in the pulsed mode, the injection current should be pulsed between two values above the lasing threshold rather than between cut-off and lasing mode.

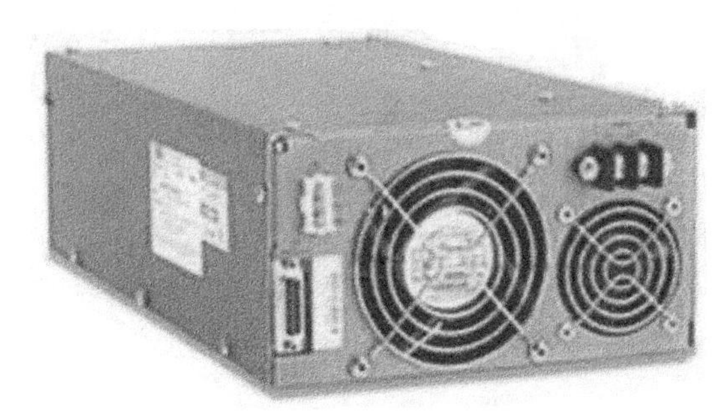

Figure 7.1 Capacitor-charging power supply (Courtesy of Excelitas).

When used in a laser printer, a laser pointer or even a compact disk player, a laser diode may need a conventional constant current source without too stringent a requirement on the current stabilization to do the job. Drive current and diode temperature stabilization to a high degree for stabilizing the output wavelength become very important in applications such as telecommunication, spectroscopy and biomedical diagnostics. The concepts of drive current and diode temperature stabilization have been put to use very effectively in tuning the diode laser output wavelength also. In fact, it is the ability to stabilize the parameter that allows it to be varied.

Leading manufacturers of diode lasers offer a wide range of current sources for low-, medium- and high-power laser diodes to suit different requirements. Both general-purpose bench-top models and modular units for OEM applications are commercially available from a fairly large number of manufacturers. They also offer temperature controllers to stabilize the output wavelength. Figure 7.2 depicts a bench-top precision laser diode driver of LDX-3200 series from ILX Lightwave Optics. The LDX-3200 series consists of laser diode drivers with current ranges of 50–100 mA (LDX-3210), 200–500 mA (LDX-3220) and 2–4 A (LDX-3232) covering a wide range of laser diode drive and control applications. The series offers operation in both constant current and constant power modes and has inbuilt protection features including adjustable current and voltage limits and intermittent contact protection. Also, output shorting relays are designed into each model. It may be mentioned here that these are typical features in most diode laser drivers available from leading manufacturers.

Temperature controllers for diode laser temperature control are also available in a wide range of performance parameters to suit different requirement specifications. Both bench-top units and OEM modules are commercially available. Figure 7.3 depicts a typical bench-top thermoelectric laser diode

Figure 7.2 LDX-3200 series laser diode driver (Courtesy of Newport Corporation).

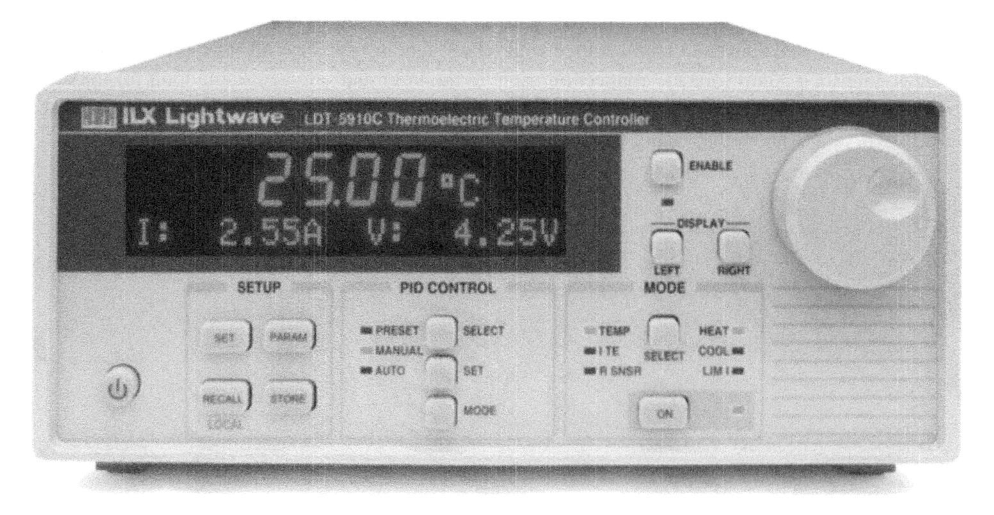

Figure 7.3 LDT-5900 series thermoelectric diode laser temperature controller (Courtesy of Newport Corporation).

temperature controller of LDT-5900 series from ILX Light Wave Optics. These temperature controllers are capable of operating in constant temperature, constant power or constant current modes with temperature stability of better than 0.003 °C. Integrated laser diode controllers offering both current drive as well as thermoelectric temperature control in a single instrument are also commercially available from a number of manufacturers.

7.2.3 Gas Lasers

A high-voltage DC power supply used to initiate and then sustain plasma is the electronics that accompany a helium-neon (He-Ne) or a carbon dioxide (CO_2) laser, although RF-excited versions of these lasers are also available. Figure 7.4 shows the use of a high-voltage DC power supply along with a He-Ne laser plasma tube. Figure 7.5 depicts He-Ne power supplies of 1200-series (Figure 7.5a) and 101T-series (Figure 7.5b) from JDS Uniphase. These power supplies are available in both 115–230 V AC input and 12 V DC input versions.

The current stabilization that the He-Ne laser power supply needs to provide depends on the intended application. There are applications where a high degree of current stabilization is not an essential requirement. One such application is its use as an alignment laser. However, when it comes to using the same laser in an inertial-grade rotation rate sensor such as a ring laser gyroscope (RLG),

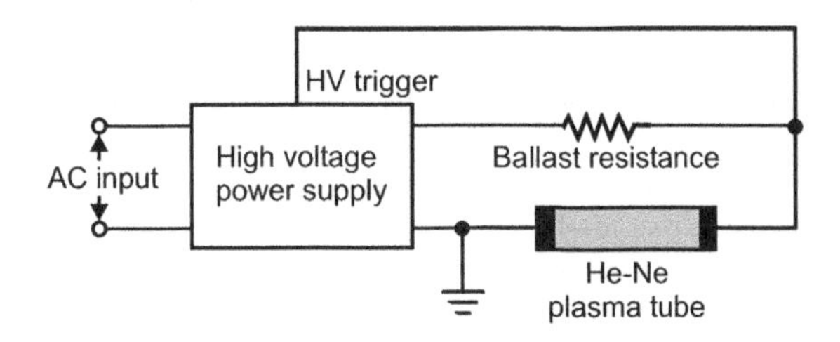

Figure 7.4 Power supply feeding He-Ne laser plasma tube.

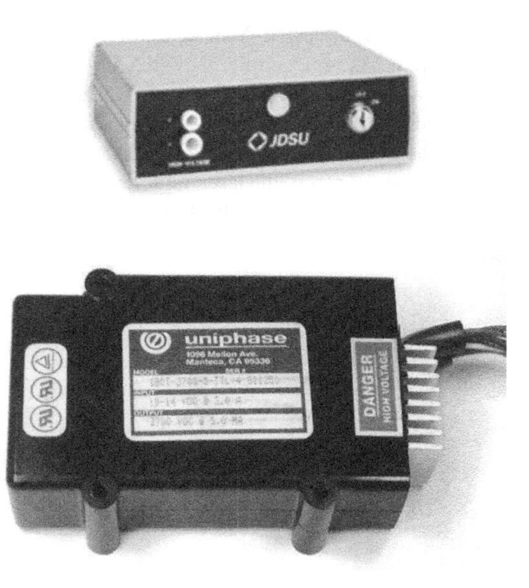

Figure 7.5 He-Ne laser power supplies (Courtesy of JDS Uniphase Corporation).

the current would need to be stabilized to a level better than 100 parts per million (PPM). The drive electronics in this case becomes fairly complex as is evident in Section 8.4.4. In addition, for such applications frequency also needs to be stabilized better than ± 1 MHz on its Doppler-broadened gain curve, that is, about 1400 MHz wide for 632.8 nm output wavelength. In fact, frequency stabilization of gas lasers is a complete field in itself. Scientists have worked in this area for decades to discover new methods of stabilizing the laser frequency using active means or improve upon those already existing. Different frequency stabilization schemes commonly used with gas lasers are also described in Section 8.9.

7.2.4 Testing and Evaluation of Lasers

Measurement of laser parameters such as power, energy, pulse width and wavelength is yet another area where electronics plays an important role. We have all kinds of lasers producing CW, pulsed or Q-switched pulsed laser outputs. While we are mainly interested in the output power in CW lasers such as gas lasers, it is the energy and the pulse width that is of interest in the case of pulsed lasers. In case of Q-switched pulsed solid-state lasers, we need to measure energy per pulse, average power and also the peak power. Testing and measuring equipment capable of measuring one or more of these parameters are commercially available. The equipment usually comes in the form of a display unit and a family of sensor heads to cater for different power level, pulse energy and wavelength range requirements.

Figure 7.6 depicts a power/energy meter, model FieldMaxII-TOP from Coherent Inc. The meter is used along with a thermopile and optical sensor heads for measurement of power and a pyroelectric sensor head for measurement of pulse energy. The meter is capable of measuring CW power in the range of 1 nW–300 mW with an optical sensor and 10 μW–30 kW with thermopile sensor. It can measure pulse energy in the range of 1 nJ–300 J up to a maximum repetition rate of 300 Hz with a pyroelectric sensor. All parameters are measured with a resolution of 0.1% of full scale.

Although the above diagnostic gadgets are commercially available, this does not mean that the designers of laser electronics packages do not need to know anything about the design of such gadgets. There are a large variety of laser systems where measurement of laser power or energy forms an integral part of the total system. Commercially available testing and measurement instruments are certainly of immense use during the development phase of a laser system and also during the performance evaluation of the finished system. It is highly likely that the designers of laser electronics packages will encounter situations where they would be required to design dedicated power- or energy-measuring subsystems that are a part of the overall system. As an example, the Nd:YAG laser system meant for eye care invariably has an energy meter as an integral part of the system. The energy-measuring subsystem here forms a critical part of the system due to the importance of knowing the selected energy level for various surgical exercises. As another example, in the case of a laser diode or a He-Ne laser where the laser output power is to be maintained constant, part of the laser diode driver or power supply circuit monitors the output power and changes the drive current in a feedback loop in order to keep the laser power constant.

Figure 7.6 Laser power and energy meter, Model FieldMaxII-TOP (Reproduced by permission of Coherent Inc.).

There are other areas of diagnostics where no commercially available gadgets can be used. One such area is measurement of power density or energy density rather than power or energy. Such a situation arises during the field evaluation of the electro-optic countermeasure (EOCM) class of laser systems, where the effect of the laser beam on the target sensor depends upon the energy/power density available at the target location. It is important to measure the power/energy density parameters as it is affected by atmospheric attenuation and also by increase in laser spot diameter due to beam divergence.

Photosensors play a very important role in the design and development of diagnostic gadgets meant for measurement of laser-related parameters. There is a large variety of such sensors catering to different wavelength bands and different power and energy levels. Photosensors are covered in detail in Chapter 10.

7.2.5 Laser Sensor Systems

In a large number of optoelectronic systems, in particular battlefield optoelectronic systems, the primary function is detection of laser radiation with or without its important parameters depending upon the intended application. Such systems include laser warning sensors, laser position sensors and laser seekers. A laser warning sensor system is an essential constituent of both passive and active EOCM systems. Different battlefield applications demand laser warning sensor systems with different functional features, thus having different design complexities. In the simplest form, a laser warning sensor may be required to give an indication of the existence of a laser threat without giving any information on the direction of arrival of the threat. In another case, it may also have the direction-sensing feature with the resolution of direction sensing also varying with the intended application. In addition, the operational wavelength band may also vary from one application to another. More complex laser warning systems indicate the type of laser threat, the wavelength and the direction of arrival of the threat with an angular resolution approaching less than a degree. Figure 7.7 depicts such a sensor from Goodrich ISR. Different variants of laser threat detection systems capable of characterizing threats from laser rangefinders, laser target designators and laser beam riders are available for mounting on helicopters, main battle tanks and light armoured vehicles.

Laser position sensors have extensive application in laser-based research. A laser seeker, which is also a kind of position sensor, is the heart of the guidance system of a laser-guided weapon such as a laser-guided bomb or missile. Figure 7.8 a depicts a Raytheon's GBU 24 Paveway-III class of laser-guided bomb integrated with a laser seeker head. Figure 7.8b provides a close-up view of a gimballed laser seeker head of the laser-guided bomb. Battlefield laser sensor systems are discussed in detail in Chapter 14.

7.3 Electronics for Solid-state Lasers

The most commonly used and widely exploited solid-state lasers include Nd:YAG and ND:Glass lasers. These lasers are either flashlamp pumped or pumped by semiconductor laser diode arrays. The electronics for these two types of lasers are similar. The electronics required for diode-pumped solid-state lasers is the same as used in the case of semiconductor diode lasers. In the following sections the electronics package that accompanies flashlamp-pumped pulsed and CW solid-state lasers is

Figure 7.7 Laser warning sensor system from Goodrich (Reproduced from Goodrich ISR).

Figure 7.8 (a) Raytheon's GBU 24 Paveway-III class of laser-guided bomb (Reproduced from public domain (http://en.wikipedia.org/wiki/File:GBU-24_xxl.jpg)); and (b) laser seeker Fig 7.8 (a) and (b) Public domain images (Reproduced from public domain (http://en.wikipedia.org/wiki/File:Paveway_III_laser_guided_bomb_seeker_head.jpg)).

discussed. The electronics components of diode-pumped solid-state lasers are not discussed under the heading of solid-state lasers, but are covered under the heading of Laser Diode Electronics in Chapter 9.

7.4 Electronics for Pulsed Solid-state Lasers

Pulsed solid-state lasers are operated in either of the two basic operational modes: free-running and Q-switched modes. When compared to a free-running laser, an electronics package for a Q-switched laser has an additional Q-switch driver module. For cavity-dumped and mode-locked lasers, the electronics package is similar to that of a free-running laser. In the following sections the electronics package of a Q-switched flashlamp-pumped solid-state laser is described, as this covers all operational modes. The discussion begins with an introduction in the form of a block schematic of the electronics package, and a brief outline of the functions of each of the modules with particular reference to the importance of each module in the overall package. This is followed by a detailed description of each of the important modules along with preferred schematic options and design guidelines.

7.4.1 Electronics for Q-switched Solid-state Lasers

Figure 7.9 shows the basic building blocks of an actively Q-switched, flashlamp-pumped Nd:YAG or Nd: Glass laser. A Q-switch driver is not required in the case of passive Q-switching, as is a simmer module in the case of non-simmer mode operation. Figure 7.10 shows the block schematic arrangement in the case of non-simmer mode of operation.

The heart of the system is the *main power supply*, which is invariably a switched mode used to charge an energy storage capacitor to a voltage in order to store the required quantum of energy per pulse to be delivered to the flashlamp. The main power supply is also called the *capacitor-charging power supply*. The capacitor must charge to the desired voltage in a certain time, which is at the most equal to the reciprocal of the repetition rate of the laser. In practice, it should be slightly less, allowing for some minimum time for flashlamp quenching. The average power that this supply is expected to deliver at its output is the product of the energy per pulse and the repetition rate. The power supply accounts for more than 90% of the total electrical input to the system. The efficiency of this supply is therefore the prime determinant factor for the overall electrical efficiency of the laser. The conversion efficiency also directly affects the size and weight of the overall system, a parameter particularly important in the military applications of Q-switched, flashlamp-pumped solid-state lasers.

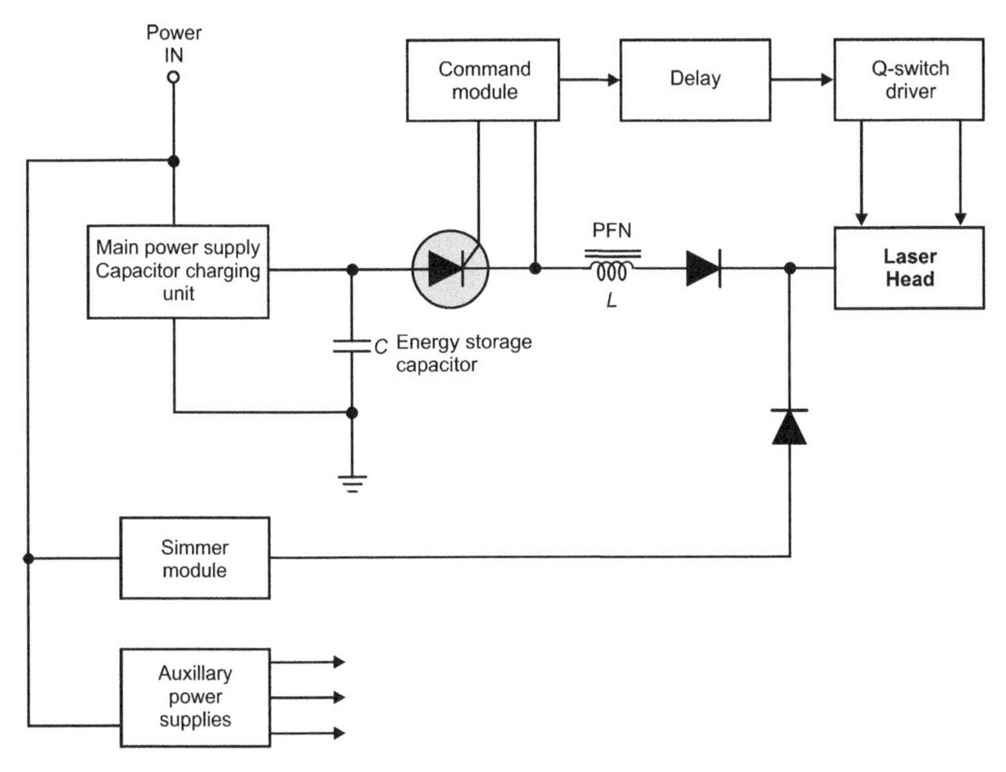

Figure 7.9 Block schematic of Q-switched flashlamp-pumped solid-state laser electronics for simmer mode of operation.

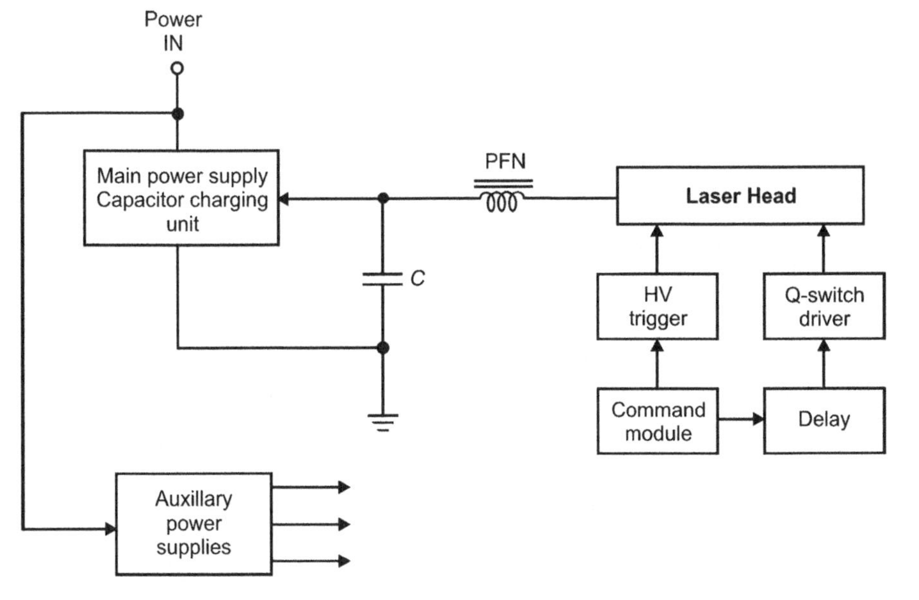

Figure 7.10 Block schematic of Q-switched flashlamp-pumped solid-state laser electronics for non-simmer mode of operation.

The *simmer module* maintains a relatively low-amplitude keep-alive current through the flashlamp at all times, irrespective of whether the lamp is flashing or not. The current varies typically from a few tens of milliamperes to several hundreds of milliamperes depending upon the characteristics of the flashlamp. This mode of operation, referred to as the *simmer mode*, has many advantages. From the operational viewpoint, it allows a low-voltage (transistor–transistor logic or TTL, complementary metal oxide semiconductor or CMOS, etc.) trigger pulse to be used to transfer the energy stored in the capacitor to the flashlamp. Secondly, it significantly enhances the flashlamp life, offers tremendous improvement on the pulse-to-pulse jitter, and overcomes most of the electromagnetic interference problems present in non-simmer mode of operation. It may be mentioned here that in the non-simmer mode of operation of the flashlamp, the flashlamp is triggered by applying high-voltage trigger pulses with amplitude of the order of 10–15 kV. These pulses appear at a rate equal to the repetition rate of the laser and are the major source of electromagnetic interference. Operation in simmer mode overcomes this shortcoming.

There is also a *pseudo simmer mode* of operation, which is a slight variation from the traditional simmer mode of operation. In this mode, the simmer current flows for a short time, starting a little ahead of the energy discharge operation. It has all the advantages of simmer mode operation but also saves on the power at the expense of added circuit complexity. Simmer and pseudo-simmer modes of operation are discussed in Sections 7.4.3 and 7.4.4, respectively.

A *Q-switch driver* is another important module for solid-state lasers and generates the drive signal for electro-optic Q-switch. The driver needs to generate a high-voltage pulse with the desired amplitude (typically 2.5–3.5 kV), pulse width that could be in the range of 200–500 ns and a rise time that should not be more than a few tens of nanoseconds. These parameters significantly affect the Q-switch performance and consequently the shape of the Q-switched laser pulse.

The *pulse-forming network (PFN)* ensures that the discharge current pulse through the flashlamp has the desired pulse width and is critically damped, thus providing the optimum energy transfer. In this case, the energy storage capacitor is exposed to the least voltage reversal while discharging through the flashlamp.

The *command module* generates flashlamp-firing command pulses and also the delayed trigger pulses for the Q-switch driver module. It may be mentioned that Q-switch drive pulse is applied after a certain time delay from the time instant of the application of the flashlamp trigger command pulse, to allow for the population inversion to build up to its peak value. The flashlamp command pulses are low-voltage pulses (TTL, CMOS etc.) in the case of simmer mode and high-voltage trigger pulses in the case of non-simmer mode. The delayed trigger pulses for the Q-switch driver are always low-voltage pulses and the Q-switch driver produces the desired high-voltage pulses for the electro-optic Q-switch. In addition, there is an *auxiliary module* that generates the regulated low-voltage DC power supplies from the input source of power for the operation of different circuit modules.

7.4.2 Capacitor-charging Power Supply

The power supply required to charge the energy storage capacitor is perhaps the most important of all the modules. In addition, designing an efficient capacitor-charging unit for a high-repetition-rate flashlamp-pumped laser is not so simple an exercise, as demonstrated in the following. The problem arises from the capacitive nature of the load. The power supply output needs to charge a high-value capacitor, typically 20–50 μF in the case of designators and rangefinders and as high as thousands of microfarads in high-power pulsed lasers producing laser pulse energies of several kilojoules meant for electro-optic countermeasures (EOCM) and laser weapon applications. In the following, we examine some circuit configurations that could possibly be used for designing the capacitor-charging power supplies.

The simplest capacitor-charging power supply may be configured around a transformer-rectifier arrangement of an AC-DC power supply as shown in Figure 7.11. Resistor R_1 is the current-limiting resistor to limit the charging current at the start of the charging process when the energy storage capacitor C_1 is fully discharged. The semiconductor switch, typically an SCR, is used to turn off the charging process when the capacitor is charged to the desired output voltage. The turning on and off of this switch is controlled by the output of the comparator. One of the inputs to the comparator is tied to a reference

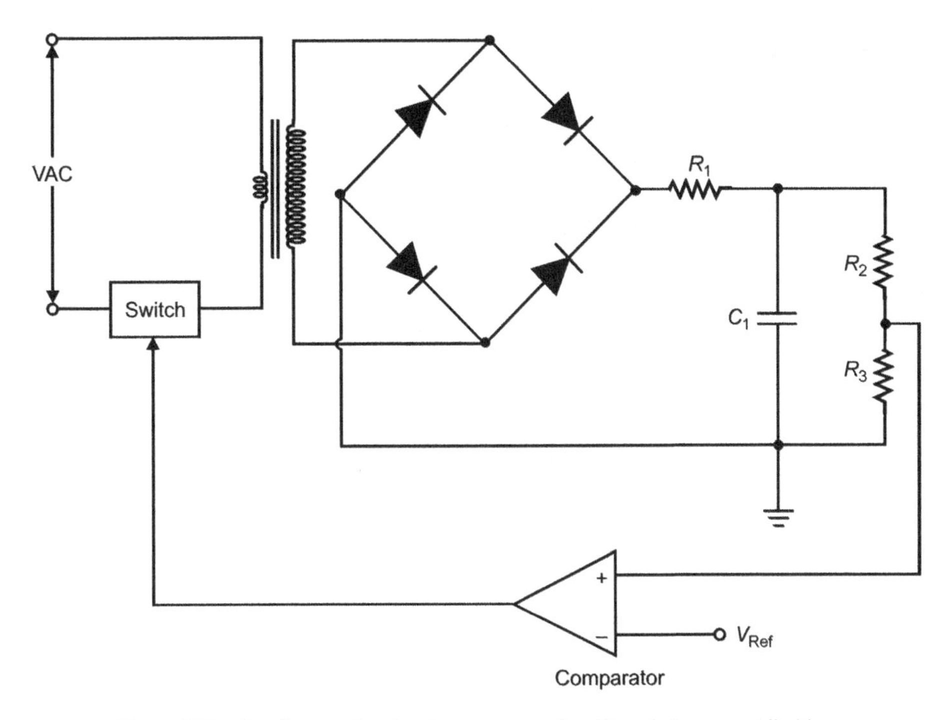

Figure 7.11 Simple capacitor-charging power supply with resistive current limiting.

voltage and the other input is applied a fraction of the output voltage through a potential divider arrangement of R_2 and R_3. The desired output voltage can be changed by changing either the reference voltage or the potential divider ratio.

Such a circuit becomes increasingly inefficient due to increased losses in the current-limiting resistor when the laser is to operate at a higher pulse repetition frequency (PRF). At higher PRF, the charging current needs to be higher to charge the energy storage capacitor faster due to a relatively smaller inter-pulse period. One way to overcome this problem is to use inductive current limiting, as shown in Figure 7.12 . Here, the inductive reactance is used to limit the charging current.

The circuit configurations of Figures 7.11 and 7.12 become ineffective when PRF becomes still higher and approaches the power line frequency. In this case, few cycles of power line frequency fall within the charging period; this means that the charging current is in the form of current surges, one current surge for each half cycle. This renders the charging process erratic and irreproducible due to fluctuations in the number of current surges appearing in the inter-pulse time period.

A resonant converter overcomes this problem; Figure 7.13 shows the basic circuit. In this case, C_1 is much larger than the energy storage capacitor C_2 and acts like a filter capacitor. The energy storage capacitor is charged to twice the source voltage in the first half cycle of the resonant frequency. Diode D_1 prevents any flow of current in the reverse direction when the capacitor is charged to a voltage higher than the source voltage. The diode also allows the designer to choose the discharge rate of the capacitor independent of the resonant frequency.

In the case of circuit configurations discussed in the previous, the AC to DC conversion occurs at power line frequency which makes them bulkier and inefficient. Modern capacitor-charging power supplies invariably use switched-mode power supply concepts to derive the benefits inherent to these supplies. These include relatively much higher conversion efficiency and reduced size and weight. In addition, the source of input power could be either AC in off-line switching supplies or DC depending upon the intended application. DC as input source of power is particularly attractive for portable and military laser systems. An overview of different switching supply configurations along with their

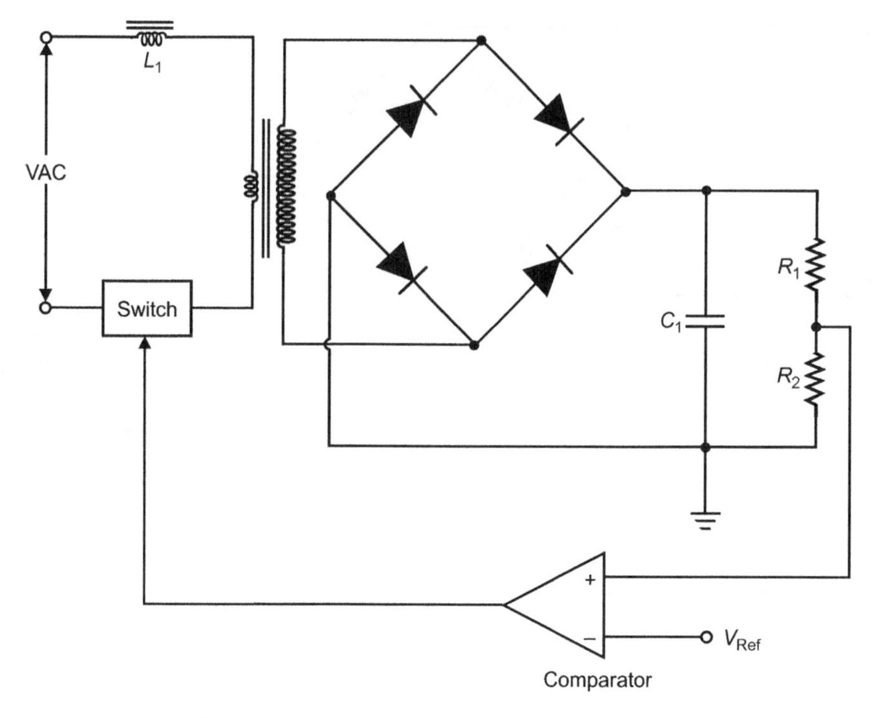

Figure 7.12 Simple capacitor-charging power supply with inductive current limiting.

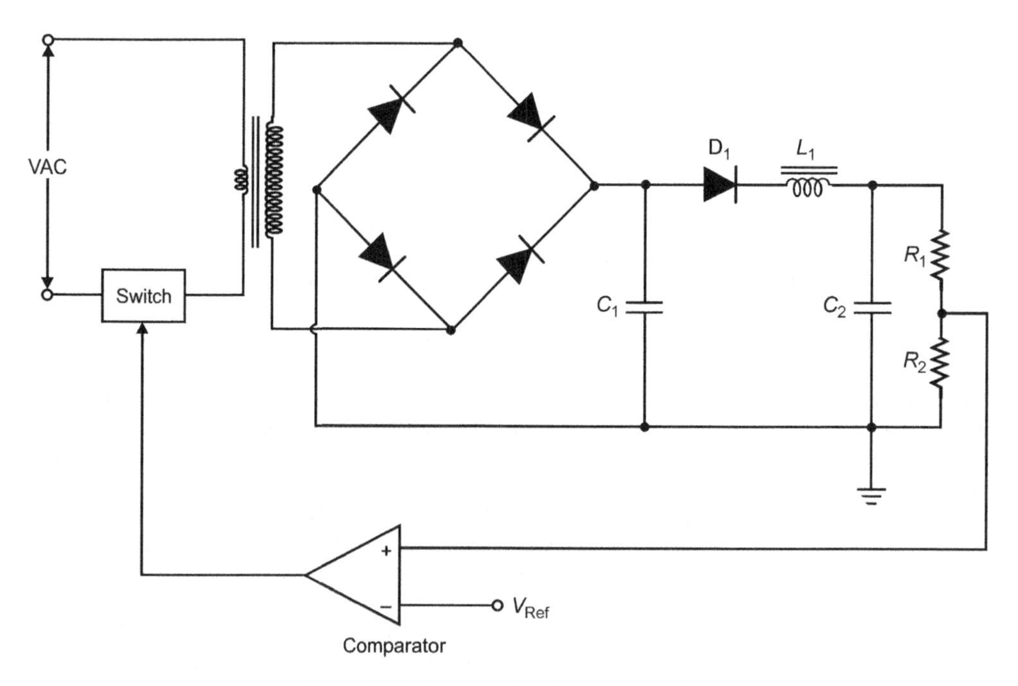

Figure 7.13 Resonant charging power supply.

merits/demerits was given in Chapter 6. An externally driven flyback converter best suits the capacitor-charging power supply requirements as it allows the designer an independent control on the on and off times of the switching device. This feature is an asset as evident from the following discussion.

Figure 7.14 shows the basic circuit of an externally driven DC/DC converter. In such a case, where a flyback converter of the type shown in Figure 7.14 is to charge an energy storage capacitor, the design is not as straightforward as it would be had there been a resistive load. The problem arises for two reasons: (1) the fundamental fact that the voltage across a capacitor cannot change instantaneously and (2) for a given energy to be transferred to a capacitor, the quantum by which the capacitor voltage would increase depends upon the initial voltage across the capacitor just before the energy transfer. It is inversely proportional to the initial voltage across the capacitor, which implies that the time required to transfer a given energy to a capacitor would decrease with the build-up of voltage across the capacitor. This prevents a capacitor-charging unit being designed around the fixed switching frequency.

To create an optimum design, the changing pattern of the voltage quantum needs to be known and a circuit built that simulates these conditions. This problem is very specific to laser power supplies and no available switched-mode power supply (SMPS) controller chip can be used to do the job. It becomes more severe in the case of portable systems where high conversion efficiency is of utmost important.

There are various circuit topologies available for designing a DC/DC converter, such as flyback converters, forward converters and push-pull converters. However, forward and push-pull converters are only suitable for loads that are predominantly resistive, such as those encountered in the power supplies designed for TVs, music systems, personal computers and testing equipment.

These types cannot be used in solid-state laser capacitor-charging power sources where the voltage across the energy storage capacitor builds up in voltage packets of continuously varying sizes with the energy transferred in each packet remaining the same. The first voltage packet is the largest and the last packet that takes the capacitor voltage to the desired value the smallest.

Such a load necessitates an independent control on the energy transfer time in accordance with the pattern of voltage build-up across the capacitor for an optimum power conversion. It is theoretically feasible to have such a control in a flyback converter where the energy is stored during the on-time of the switching device and transferred during its off-time.

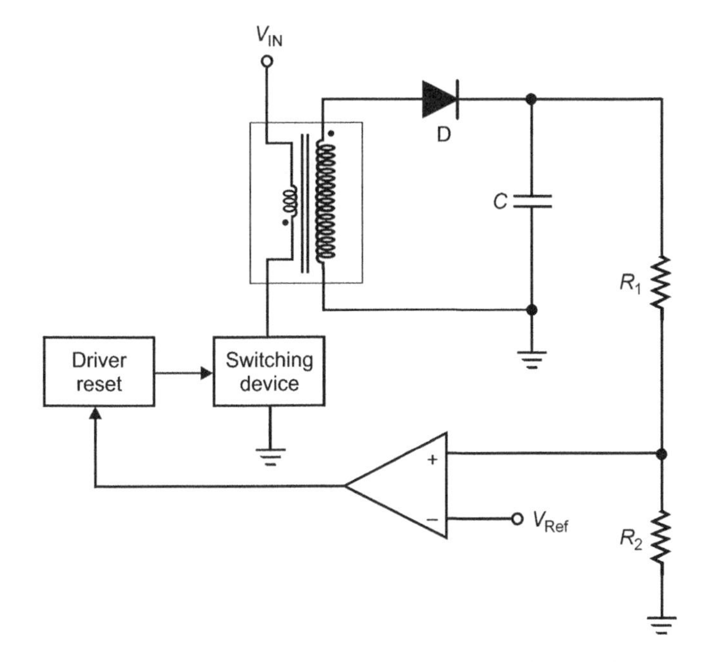

Figure 7.14 Simple externally flyback DC/DC converter as capacitor-charging power supply.

While in the case of resistive loads it is the voltage regulation that is of prime concern and is achieved by some form of duty-cycle control, usually pulse width modulation (PWM), in the case of capacitive loads the energy transfer process is a critical factor. An externally driven flyback converter seems to be the best bet for this application; however, the entire energy storage and transfer process must be known, particularly the energy transfer times required in different discrete packets. An appropriate control circuit that would simulate the desired switching waveform then needs to be designed; what the desired switching waveform looks like is explained in the following.

In a flyback converter, during the on-time of the switching device (bipolar junction transistor or MOSFET), the primary winding of the switching transformer stores energy equal to $LI_P^2/2$ where L is the primary inductance and I_P is the peak primary current. This energy is transferred to the energy storage capacitor to charge it to a certain voltage during the switch-off time of the switching device. In a lossless and complete transfer of energy, the magnetic energy stored in the primary equals the electrostatic energy stored in the capacitor. The number of switching cycles required to charge the capacitor to the desired output voltage V_O can be computed from the ratio of energy stored in the capacitor at the final voltage $(CV_O^2/2)$ to the energy stored in each switching cycle $(LI_P^2/2)$. The size of the voltage packet received by the capacitor follows a decreasing pattern beginning with the largest in the first cycle, as is evident from the following simple calculations.

The voltage packet received by the capacitor in the first cycle when it is fully discharged can be computed from Equation 7.1:

$$LI_P^2/2 = CV_1^2/2 \tag{7.1}$$

where V_1 is the size of the first voltage packet. This gives

$$V_1 = I_P \times \sqrt{L/C} \tag{7.2}$$

This is also the voltage across the energy storage capacitor at the end of the first cycle. The voltage packet received by the capacitor during the second cycle when it is initially charged to voltage V_1 is given by Equation 7.3:

$$LI_P^2/2 = CV_2^2/2 - CV_1^2/2 \tag{7.3}$$

With the help of simple mathematics, it can be shown that the voltage across the capacitor at the end of second cycle is given by Equation 7.4:

$$V_2 = V_1 \times \sqrt{2} \tag{7.4}$$

Similarly, the voltage across the capacitor at the end of third, fourth and Nth cycles is given by Equations 7.5–7.7 respectively:

$$V_3 = V_1 \times \sqrt{3} \tag{7.5}$$

$$V_4 = V_1 \times \sqrt{4} \tag{7.6}$$

$$V_O = V_1 \times \sqrt{N} \tag{7.7}$$

where N is the number of cycles to charge the capacitor to voltage (V_O). The value of N can be calculated from Equation 7.7 and is given by Equation 7.8:

$$N = \left(\frac{V_O}{V_1}\right)^2 \tag{7.8}$$

We also have:

$N = \left(\frac{CV_O^2}{LI_P^2}\right) = \left(\frac{V_O}{V_1}\right)^2$, which validates the expression.

The sizes of different voltage packets are given by Equations 7.9–7.12:

$$\text{Size of first packet} = V_1 \tag{7.9}$$

$$\text{Size of second packet} = \sqrt{2}V_1 - V_1 = \left(\sqrt{2} - \sqrt{1}\right)V_1 \tag{7.10}$$

$$\text{Size of third packet} = \left(\sqrt{3} - \sqrt{2}\right)V_1 \tag{7.11}$$

$$\text{Size of } N\text{th packet} = \left[\sqrt{N} - \sqrt{(N-1)}\right]V_1 \tag{7.12}$$

It is evident from the above equations that the voltage packet size reduces in successive cycles with the build up of voltage across the energy storage capacitor. The time period required to transfer energy in each cycle depends upon the corresponding voltage packet imparted to the capacitor; a fixed switching frequency therefore cannot be used. Ideally, the drive waveform must have a variable off-time beginning with the largest value and then decreasing in a well-defined pattern. It is because of this that the flyback configuration of Figure 7.14 is not effective. A relatively lower switching frequency to allow complete energy transfer in the few initial cycles where large off-time is required leads to low conversion efficiency, as in that case the converter is sitting idle for most of the time in the later cycles. The use of a relatively higher switching frequency with the intention of increasing the conversion efficiency leads to incomplete energy transfer in the initial cycles, which in turn reduces the conversion efficiency. In addition, incomplete energy transfer in a flyback converter could cause damage to the switching device. Figure 7.15 shows the modified schematic of a flyback-converter-based capacitor-charging power supply that takes care of the variable off-time requirement. The circuit operates as described in the following.

The basic difference between the block schematic of Figure 7.15 and the conventional externally driven flyback converter configuration of Figure 7.14 lies in the mechanism of generating the drive signal waveform for the switching device. The *drive portion* of the converter hardware comprises a cascaded arrangement of a voltage-controlled oscillator (VCO), a monoshot circuit and a drive circuit. The output of the VCO feeds the trigger input of the monoshot. The pulse width of the monoshot is chosen to be equal to the desired on-time of the switching device. The frequency of monoshot output and hence the off-time of the waveform is governed by the frequency of the VCO output, which in turn depends upon the voltage applied to its control input. The drive circuit provides the required drive current and/or voltage depending upon the type of switching device used.

The *feedback circuit* comprises two independent potential divider arrangements, a subtractor circuit and a comparator circuit. Potential divider R_1–R_2 along with the comparator constitutes the voltage sense loop. The output of the comparator circuit resets the monoshot when the energy storage capacitor has charged to the desired voltage. Due to this feedback loop, the voltage across the capacitor remains around the desired value, with the output ripple amplitude depending upon the minimum input differential required by the comparator to change state at the output and the dividing factor of the potential divider R_1–R_2.

As an example, for an output voltage of 1000 V, a dividing factor of 200 may be used to bring it down to 5 V; an input differential of 5 mV leads to peak-to-peak ripple amplitude of 1 V. Potential divider R_3–R_4 along with the subtractor constitutes the energy transfer sense loop. That is, it is instrumental in generating the desired variable off-time pattern in the drive waveform to ensure complete or almost-complete energy transfer in different cycles. The subtractor output feeds the control terminal of the VCO. The voltage present at the output of the subtractor is a function of the output voltage. Output frequency of VCO depends upon the control voltage. The components of the frequency-determinant network of the

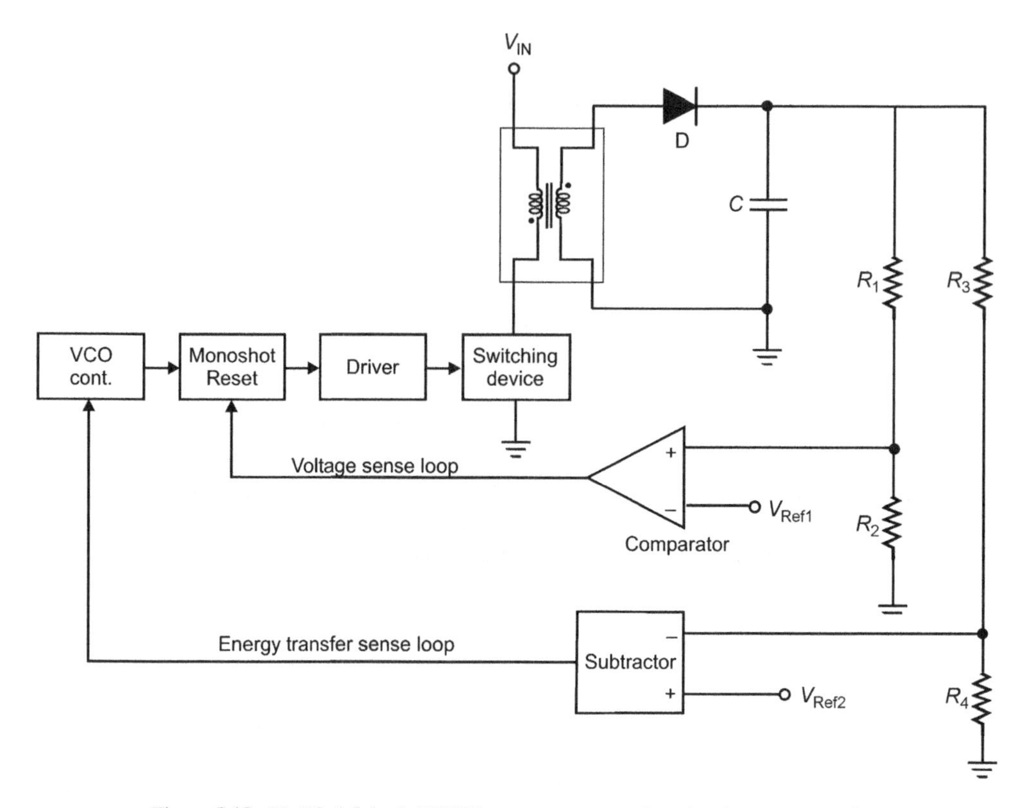

Figure 7.15 Modified flyback DC/DC converter as capacitor-charging power supply.

VCO are chosen such that the energy storage capacitor is fully discharged and the voltage present at the subtractor output produces VCO output of desired frequency that in turn produces the desired off-time for the first cycle of energy transfer. As the voltage across the capacitor builds up, the output of subtractor circuit changes in such a manner as to change the drive frequency in the correct direction. That is, the VCO output frequency reduces in a pattern identical to the pattern of voltage build-up across the capacitor. In a modification of the above design, the comparator, subtractor, VCO and the monoshot can be replaced by an embedded processor to generate the output waveform to be fed to the drive element.

The above paragraphs are the qualitative description of the concept of having a variable off-time drive waveform in accordance with the mode of energy transfer in a capacitor-charging power supply configured around an externally driven flyback converter. The author has patented this design philosophy for use with capacitor-charging power supplies for flashlamp-pumped pulsed solid-state lasers. This design methodology has been used by the author to design a range of high-efficiency capacitor-charging modules with different joules/second delivery ratings for various laser applications. The detailed design equations are however beyond the scope of this book.

7.4.3 Simmer Power Supply

The purpose of the simmer power supply, as outlined in Section 7.2.1, is to maintain a relatively low-amplitude keep-alive current through the flashlamp at all times irrespective of whether the lamp is flashing or not. The simmer power supply maintains a steady-state partial ionization of the flashlamp during the time the lamp is not flashing. A simmer module must be designed with due consideration to *I–V* characteristics of the flashlamp.

Initially, the simmer module generates a high-voltage trigger pulse, typically 10–15 kV, to create pre-ionization before the simmer power supply can take over and deliver the keep-alive current through the flashlamp. The simmer power supply is a high-voltage DC power supply producing an output voltage in the range of 800–1500 V, depending upon the characteristics of the flashlamp and the required magnitude of simmer current. The output of the simmer power supply is applied to the flashlamp through a series resistor called ballast resistor. The magnitude of the simmer current therefore depends upon the difference between the simmer supply output voltage, the voltage across the flashlamp in the simmer mode (partial ionization state) and the value of the ballast resistance. The value of the ballast resistance should be slightly higher than the negative impedance offered by the flashlamp in the simmer regime. It may be mentioned here that the flashlamp offers negative impedance while operating in such a regime. The voltage v across the lamp in this condition varies as given by Equation 7.13:

$$v \propto i^{-0.3} \tag{7.13}$$

where i is the simmer current.

The above equation indicates the existence of I–V characteristics with negative impedance. Figure 7.16 shows a simplified arrangement depicting simmer mode of operation. The power supply portion can be configured around a flyback converter topology.

Ideally, the status of the flashlamp should be monitored and the high-voltage trigger-generation part of the simmer module disabled as soon as the flashlamp starts simmering. Readers may consider delivery of a single high-voltage (HV) pulse to the flashlamp, assuming that the simmer operation would be established with the first pulse itself, and get rid of the additional circuitry needed to monitor flashlamp status; this is not advisable, however. The flashlamp may not strike with the first pulse and, even if it does so, it can extinguish during operation. In that case, a fresh HV trigger would be required to restore the simmer mode.

Figure 7.17 shows the block schematic of a simmer module that avoids any eventuality of the lamp becoming extinguished during operation. The ballast resistance connected from the output of the simmer supply in series with the lamp is split as two series-connected resistors and a part is used to monitor current through the flashlamp. The 'off' and 'on' conditions of the lamp are used to enable or disable an astable multivibrator whose output, in turn, is used to trigger the SCR-based HV trigger-generating circuit. The repetition rate of the trigger pulses could be anywhere between 20 and 50 pps.

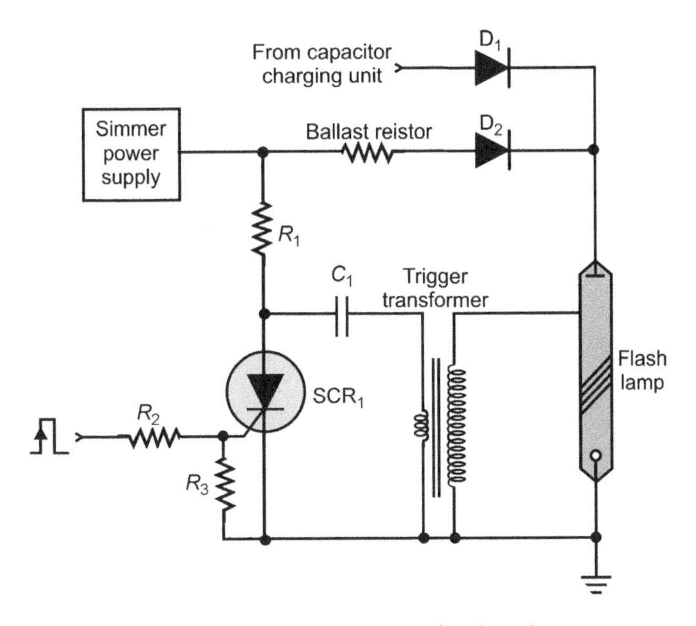

Figure 7.16 Simmer power supply schematic.

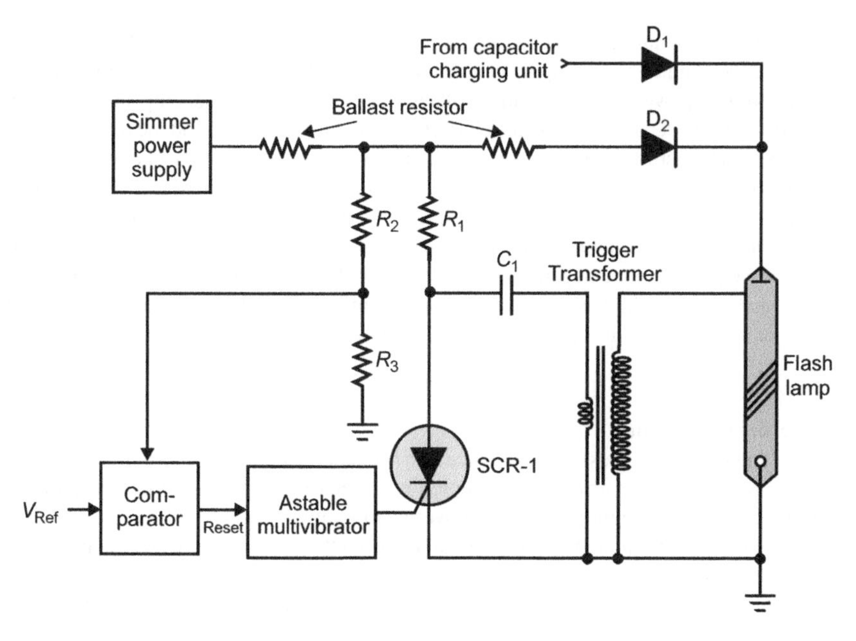

Figure 7.17 Modified simmer module.

7.4.4 Pseudo-simmer Mode

For pulse repetition frequencies that are not very high, the simmer mode of operation leads to significant power loss during the long time intervals between the two flashes. This is why the simmer mode of operation is usually preferred for pulse repetition frequencies in excess of 50 Hz. For relatively lower PRF, pseudo-simmer mode of operation is employed. It has all the advantages of the simmer mode of operation with some added circuit complexity. In this, the lamp remains in the non-conducting state for most of the time between two consecutive flashes and the partial state of ionization (or simmer mode) is activated about 100–200 μs before every flash trigger pulse. Figure 7.18 shows the typical schematic

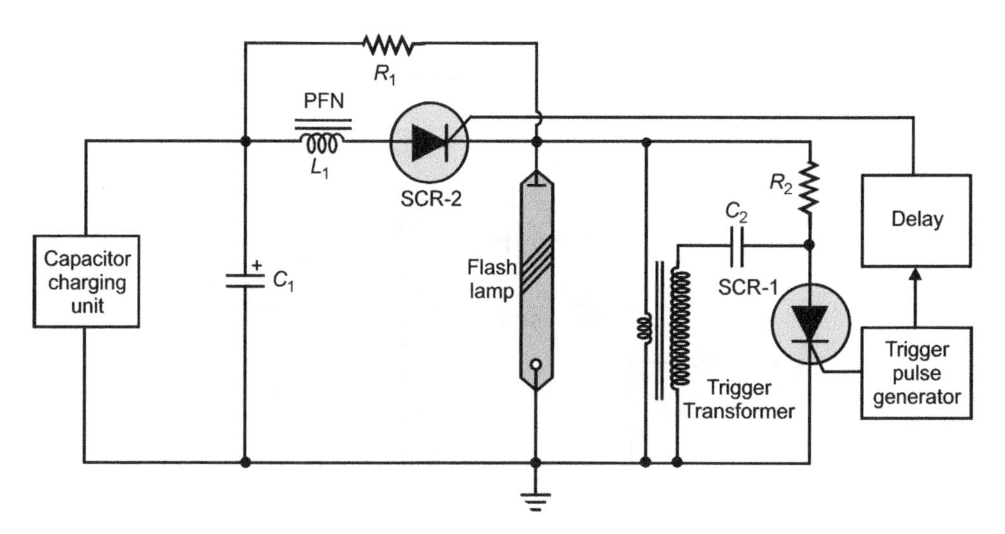

Figure 7.18 Pseudo-simmer module.

arrangement. A simmer initiation trigger pulse activates the partial discharge state. A delayed trigger pulse fires the SCR, allowing the energy stored in the capacitor to discharge through the lamp producing a flash.

Example 7.1

Refer to the experimental set-up of Figure 7.19 used to measure the flashlamp impedance in the simmer mode. Two sets of readings are taken while the flashlamp is simmering.

I. Power supply output voltage = 800 V
Ballast resistance value = 25 kΩ
Voltage across flashlamp = 300 V.

II. Power supply output voltage = 800 V
Ballast resistance value = 30 kΩ
Voltage across flashlamp = 350 V

Calculate the flashlamp resistance in the simmer mode.

Solution

1. For the first set of readings, simmer current = $(800 - 300)/(25 \times 10^3)$ mA = 20 mA
2. For the second set of readings, simmer current = $(800 - 350)/(30 \times 10^3)$ mA = 15 mA
3. Magnitude of the change in the simmer current is therefore = 5 mA
4. Corresponding change in magnitude of flashlamp voltage = 50 V
5. Magnitude of flashlamp impedance = $50/(5 \times 10^{-3}) = 10$ kΩ
6. This resistance is negative as a decrease in simmer current has resulted in an increase in flashlamp voltage.

7.4.5 Pulse-forming Network

The pulse-forming network (PFN) shapes the flashlamp current pulse and is designed to produce a critically damped current pulse. A single-section PFN can be used as shown in Figure 7.20 a or a multiple-mesh PFN as shown in Figure 7.20b. The PFN functions as follows.

As is evident from the single-section PFN, which is also the most commonly used PFN configuration, it is basically a series *R–L–C* network with the capacitor charged to a certain DC voltage, where *R* represents the resistance offered by the flashlamp. When the capacitor is made to discharge through the

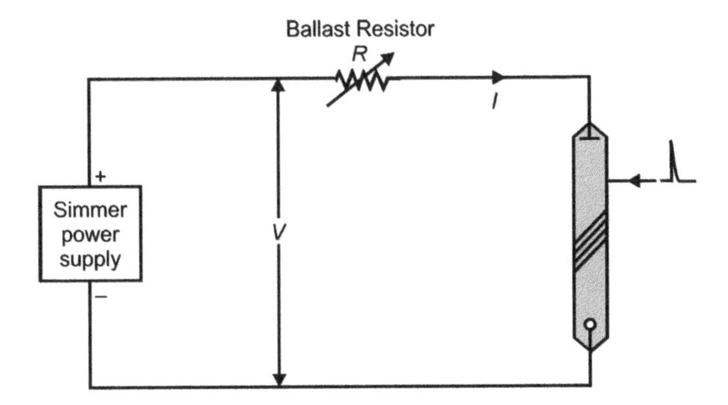

Figure 7.19 Example 7.1.

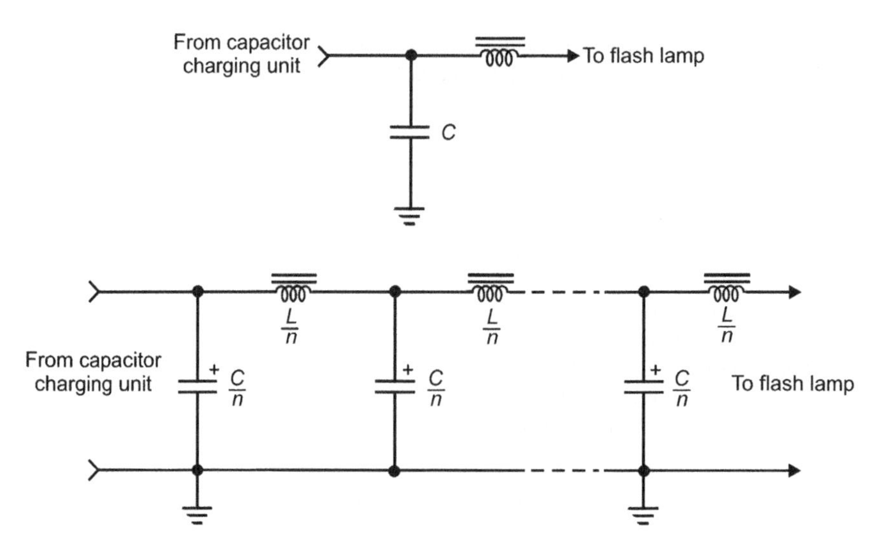

Figure 7.20 Pulse -frming network: (a) single-section PFN; and (b) multiple-mesh PFN.

flashlamp, the shape of the current pulse through the flashlamp is governed by one of the following three conditions:

1. $R < 2\sqrt{L/C}$
2. $R > 2\sqrt{L/C}$
3. $R = 2\sqrt{L/C}$

In the case of the first condition, the current pulse is said to be under-damped and the value of the damping factor α is between 0.2 and 0.8. The current pulse waveform is oscillatory with cyclic negative values of current. Current reversals lead to highly inefficient pumping. In addition, flashlamps are also prone to damage due to repetitive current reversals as they are designed for current flow in one direction only.

In the case of the second condition, the current pulse is said to be over-damped with value of damping factor $\alpha > 0.8$ (typically $0.8 < \alpha < 3$). Although there are no current reversals in this case, the current pulse is characterized by a reasonably fast rise but a very slow decrease with time up to the flash extinction. Also, the peak value of the current is never reached. This is an inefficient method of pumping the flashlamp.

In the case of the third condition, the current pulse is said to be critically damped which corresponds to damping factor $\alpha = 0.8$. There are no current reversals and the rise time is faster (slower) than in the case of the over- (under-) damped condition. Current pulse rise time is more or less equal to the current decay time, which results in better efficiency and optimal peak intensities. A slightly under-damped current pulse may be used to achieve shorter rise time and consequent higher value of peak current, providing care is taken to keep current reversal well within the prescribed limits of the flashlamp and energy storage capacitor. Figure 7.21 shows a comparison of current pulse shape in the three cases discussed above.

The above analysis however assumes that the flashlamp resistance is constant during the entire duration of the current pulse. In practice, the flashlamp resistance exhibits a time-varying behaviour. Flashlamp resistance is given by Equation 7.14:

$$R(t) = \frac{K_0}{\sqrt{I(t)}} \tag{7.14}$$

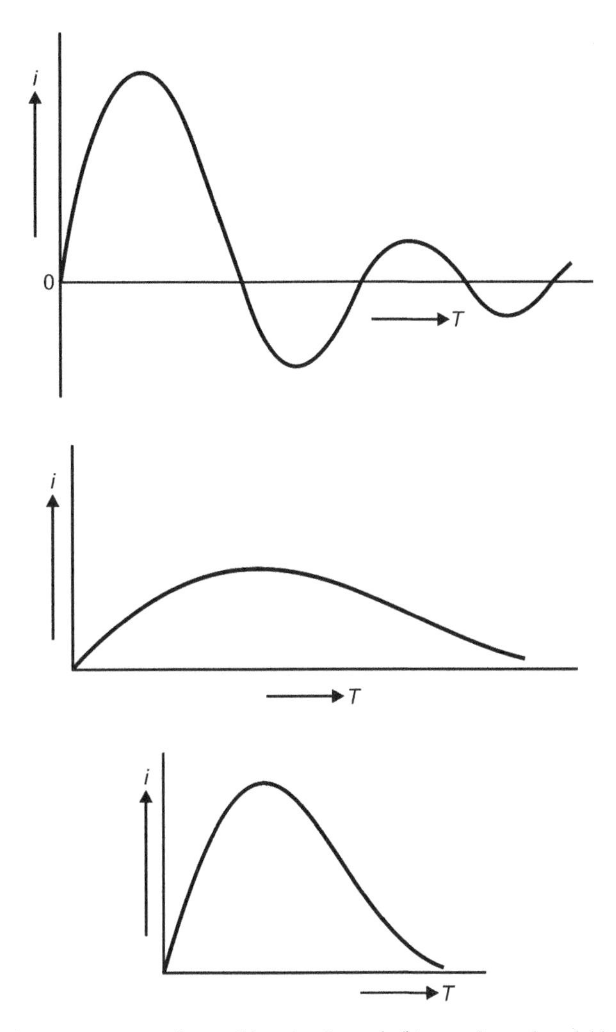

Figure 7.21 Discharge current waveforms: (a) under-damped; (b) over-damped; and (c) critically damped.

where K_0 is the lamp resistance parameter. The value of K_0 can be calculated using Equation 7.15:

$$K_0 = 1.28 \times \left(\frac{l}{d}\right) \times \left(\frac{P}{P_G}\right)^{0.2} \tag{7.15}$$

where

l = arc length
d = bore diameter
P = gas fill pressure in torr
P_G = constant = 450 (for xenon) and 805 (for krypton)

The functional relationship between voltage V and current I in a flashlamp discharge is given by Equation 7.16:

$$V = \pm K_0 \times \sqrt{I} \tag{7.16}$$

The design guidelines listed below should be followed to ensure a critically damped current pulse. These guidelines have been derived from extensive study of the behaviour of pulse-forming networks containing a flashlamp and their associated time-varying impedance.

1. The first step is to calculate the value of energy storage capacitor from the known values of pulse width t_p, energy to be discharged through the flashlamp E_O and the flashlamp impedance parameter K_0. For critically damped current pulse, the value of energy storage capacitor in farads (C) is computed from: Equation 7.17:

$$C = \left(2E_O\alpha^4T^2K_0^{-4}\right)^{1/3} \tag{7.17}$$

For critically damped condition, $\alpha = 0.8$ and $T = t_p/3$ and the above expression reduces to

$$C = \left(0.09E_O t_p^2 K_0^{-4}\right)^{1/3} \tag{7.18}$$

where $T = \sqrt{(LC)}$.

2. Having computed the value of C, the next step is to calculate the value of PFN inductance L to determine the desired pulse width. L is computed from Equation 7.19:

$$L = \frac{t_p^2}{9C} \tag{7.19}$$

where t_p is the pulse width measured at one-third of the peak value of current pulse.

3. The voltage V to which the capacitor should be charged can then be computed from Equation 7.20:

$$V = \sqrt{\frac{2E_O}{C}} \tag{7.20}$$

4. The rise time in the case of critically damped current pulse is given by Equation 7.21:

$$t_r = 1.25\sqrt{LC} \tag{7.21}$$

The amplitude of peak current I_p in the case of critically damped current pulse is given by Equation 7.22:

$$I_p = \frac{V}{2Z_O} \tag{7.22}$$

where

$$Z_O = \sqrt{L/C}$$

In the case of multiple-mesh PFN (Figure 7.20b), if C_T and L_T are the total capacitance and total inductance respectively, then

$$C_T = \frac{C}{N}$$

and

$$L_T = \frac{L}{N}$$

where N is the number of meshes.

Characteristic impedance Z_O of the PFN is given by Equation 7.23:

$$Z_O = \sqrt{\frac{L_T}{C_T}} \tag{7.23}$$

The rise time t_r and fall time t_f of the current pulse are given by Equations 7.24 and 7.25, respectively, where:

$$t_r = \frac{\sqrt{L_T C_T}}{N} \tag{7.24}$$

$$t_f = 3t_r \tag{7.25}$$

As is evident from the above equations, rise and fall times in the case of multiple-mesh PFN are relatively much shorter than their single-mesh counterparts. The current pulse also has a relatively flatter top, but with a reduced value of current peak as shown in Figure 7.22. The advantage of such a current pulse shape is that it provides more constant pumping of the laser medium throughout the duration of the pulse.

Designing a multiple-mesh PFN is an iterative process. The first step is to estimate the flashlamp resistance from an assumed value of peak current. The characteristic impedance Z of the PFN is set equal to the flashlamp resistance and the values of C_T and L_T are computed from Equations 7.26 and 7.27, respectively:

$$C_T = \frac{t_p}{2Z_O} \tag{7.26}$$

$$L_T = \frac{t_p Z_O}{2} \tag{7.27}$$

where t_p in this case is taken to be the pulse width measured at 70% of the peak value.

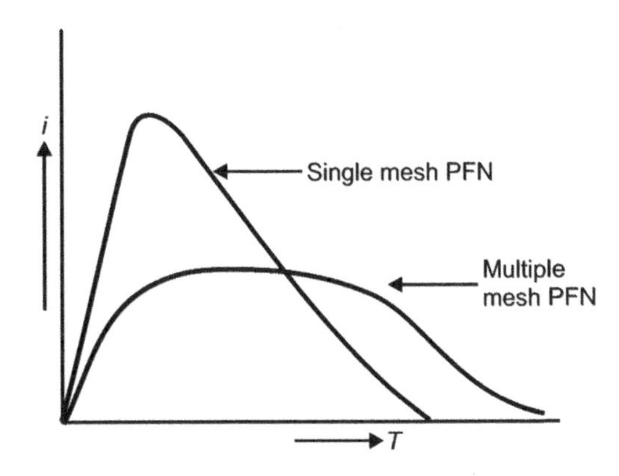

Figure 7.22 Comparison of single-mesh and multiple-mesh PFN current pulses.

Voltage V is calculated from Equation 7.28:

$$V = \sqrt{\frac{2E_O}{C_T}} \tag{7.28}$$

The new value of peak current is now calculated and compared with the assumed value on the basis of which C_T, L_T and V have been computed. If they do not match, a new value of peak current is assumed and the PFN components re-determined. The process continues until convergence takes place.

Example 7.2

It is desired to discharge energy of 10 J through a flashlamp in a critically damped pulse of 100 μs duration. If the flashlamp impedance parameter K_0 is 15 Ω $A^{1/2}$, determine the values of L and C of the pulse-forming network that would achieve the objective. Also determine the DC voltage which the capacitor should be charged to for the desired quantum of pulse energy and the resulting peak current.

Solution

1. Capacitance value can be computed from $C = \left(0.09E_O t_p^2 K_0^{-4}\right)^{1/3} = [0.09 \times 10 \times (100 \times 10^{-6})^2 \times 15^{-4}]^{1/3}$ F = 56.2 μF
2. $L = \frac{t_p^2}{9C} = (100 \times 10^{-6})^2/(9 \times 56.2 \times 10^{-6}) = 19.77$ μH
3. DC voltage, $V = \sqrt{\frac{2E_O}{C}} = \surd(2 \times 10)/(56.2 \times 10^{-6}) = 596$ V
4. Peak current $I_p = \frac{V}{2Z_O}$ where $Z_O = \sqrt{L/C}$
5. $Z_O = \surd(19.77 \times 10^{-6})/(56.2 \times 10^{-6}) = 0.593$ Ω
6. Therefore, $I_p = 596/(2 \times 0.593)$ A = 502.5 A.

Example 7.3

During flash discharge in a lamp after the peak current had been achieved, the instantaneous voltage across the lamp and the instantaneous current through it at a certain time were observed to be 400 V and 400 A, respectively. After some time had elapsed, at another time instant during the flash duration, the instantaneous current was observed to be 324 A. Determine the voltage across the lamp at that instant of time. Also determine lamp resistance in the two conditions.

Solution:

1. At the first time instant,
   Voltage across lamp $V = 400$ V and current through the lamp $I = 400$ A
2. During flash discharge, voltage and current are interrelated by

$$V = K_0\sqrt{I}$$

   where K_0 is the lamp resistance parameter.
3. Substituting the values of V and I,
   $400 = K_0 \times \surd 400 = 20\,I$, which gives $K_0 = 400/20\ \mathrm{VA}^{-0.5} = 20\ \mathrm{VA}^{-0.5}$
4. At the second time instant, current through the lamp $I = 324$ A
   Therefore $V = 20 \times \surd 324$ V = 360 V
5. Also, lamp resistance R is given by
   $R = K_0/\sqrt{I} = 20/\surd 400\ \Omega = 1\ \Omega$ (in the first case)
   R = 20/√324 Ω = 1.11 Ω (in the second case).

7.4.6 Flashlamp Trigger Circuit

Simmer and non-simmer modes of operation were discussed earlier in Section 7.4.1. In the case of non-simmer mode operation, the energy storage capacitor is connected across the lamp and there is no isolation between the capacitor and the flashlamp (Figure 7.10). The capacitor is made to discharge through the flashlamp by application of a high-voltage pulse of the order of 10–20 kV. In the case of simmer mode operation, the energy storage capacitor is kept isolated from the lamp by a series-connected high-voltage switch such as an SCR or a triggered spark gap (Figure 7.9). In this case, the trigger pulse is applied to this switch to turn it on, thus allowing the energy storage capacitor to discharge through the flashlamp. The trigger pulse is a low-voltage pulse in the case of an SCR and a high-voltage pulse in the case of triggered spark gap. In the case of simmer mode operation, a high-voltage trigger circuit is also needed for initiating a state of partial ionization for the simmer supply to take over. Different high-voltage trigger circuits used for flashlamp triggering are discussed in the following.

There are four types of high-voltage trigger circuits that can possibly be used for the purpose, each having its own merits and demerits.

1. over-voltage triggering
2. external triggering
3. series triggering
4. parallel triggering.

In the case of an *over-voltage trigger circuit* (Figure 7.23), the DC voltage across the energy storage capacitor is sufficient to break down the gas inside the flashlamp and initiate the main discharge. The DC voltage appears across the flashlamp once the series-connected high-voltage switch is turned on. The switch may be an SCR, a MOSFET or a triggered spark gap. Although the trigger circuit is simple, it is not common with solid-state lasers.

In the case of an *external-trigger circuit* (Figure 7.24), the high-voltage trigger pulse is applied to a trigger wire wrapped around the outside envelope of the flashlamp. In the case of liquid-cooled applications, the trigger pulse may be applied to the metal laser cavity. However, caution should be exercised in this case because of the high voltages involved. The main advantage of the external trigger is that it does not interfere with the main energy discharge circuit. The disadvantage is that the high-voltage trigger point is exposed and therefore needs to be properly isolated from the environment to avoid

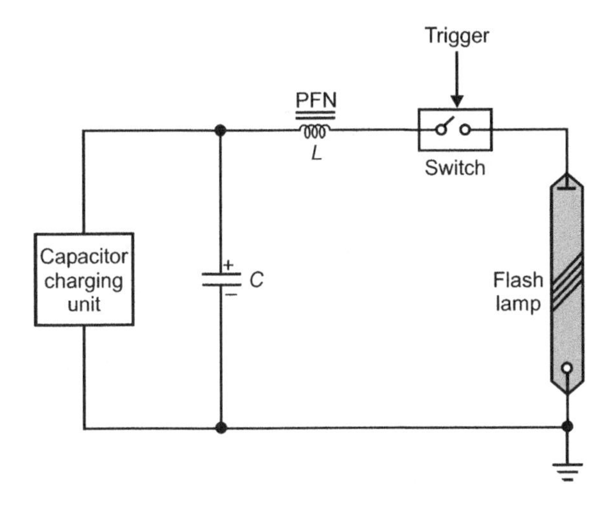

Figure 7.23 Over-voltage triggering of flashlamp.

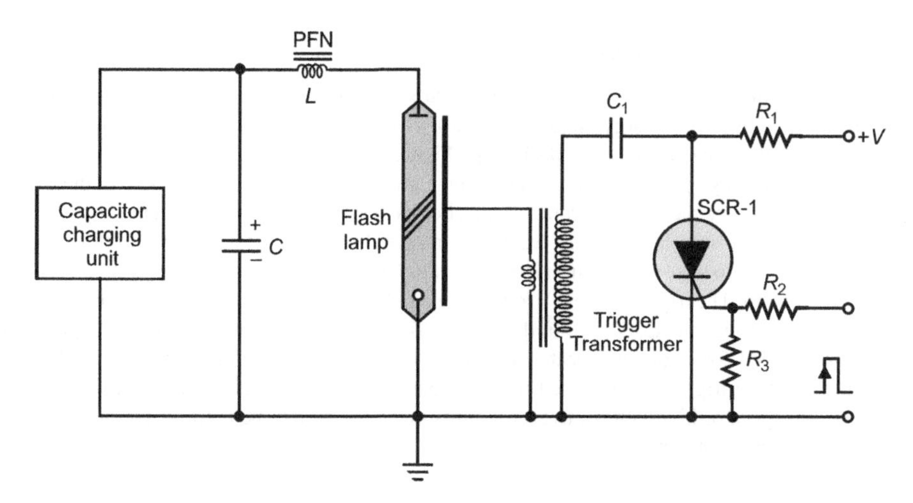

Figure 7.24 External triggering of flashlamp.

problems in conditions of high altitude or humidity. External triggering is recommended for low repetition rate low-energy systems where the flashlamp is air cooled.

In the case of a *series-trigger circuit* (Figure 7.25), the secondary winding of the trigger transformer is in series with the energy storage capacitor. The secondary winding carries the flashlamp discharge current. The series trigger transformer is designed such that the transformer core saturates and the saturated secondary winding inductance also performs the function of the PFN inductor. Series triggering offers the advantages of reliable and reproducible triggering. Triggering is reliable even for low-energy storage capacitor voltages. The reliability of the triggering is enhanced by using a ground plane near the lamp. In the case of air-cooled lamps, this may be done by wrapping a wire around the lamp and connecting one of the ends to the ground terminal. In the case of liquid-cooled lamps, a metal laser cavity may be used as the ground plane. Disadvantages include the large trigger transformer

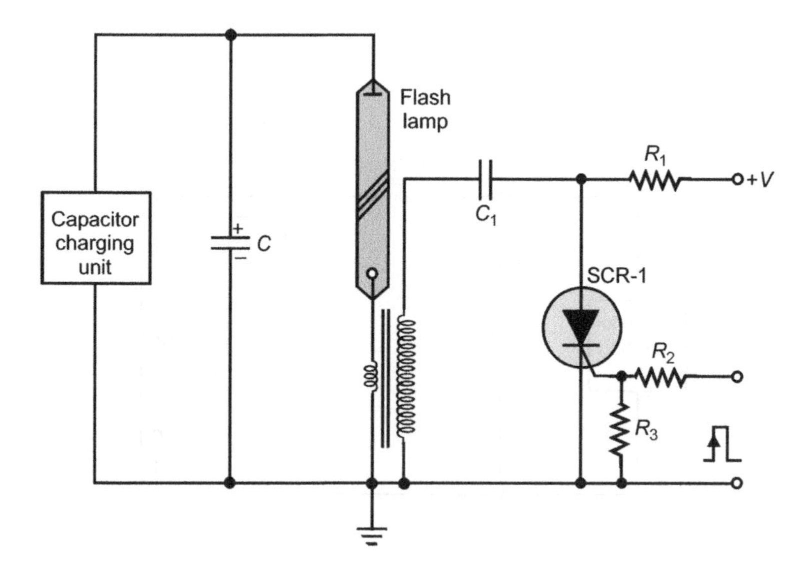

Figure 7.25 Series triggering of flashlamp.

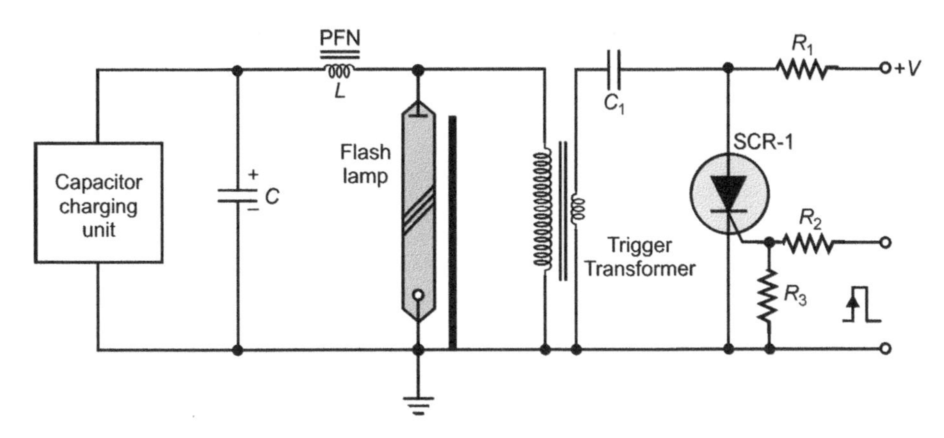

Figure 7.26 Parallel triggering of flashlamp.

required to obtain the desired saturated secondary inductance. The secondary winding also uses a thicker wire in order to be able to carry a high value of discharge current.

In the case of a *parallel-trigger circuit* (Figure 7.26), the secondary of the trigger transformer is connected in parallel. The circuit has all the advantages of series triggering and also uses a small trigger transformer. In this case, however, the secondary of the transformer needs to be isolated from the energy storage capacitor by using a capacitor or a diode not shown in Figure 7.26. This method is rarely used due to the prohibitively high cost of protection components.

In the different triggering circuits discussed above, with the exception of over-voltage triggering, the trigger pulse is generated by discharging a capacitor into the primary winding of a trigger transformer using an SCR as a switch. The trigger voltage amplitude appearing across the secondary of the trigger transformer is in the range of 10–20 kV. Trigger pulse energy of the order of 100 mJ and trigger capacitor in the range of 0.1–1.0 μF charged to a voltage in the range of 300–800 V serves the purpose. The trigger pulse duration is typically a few microseconds. A trigger pulse duration of a minimum of 60 ns/cm of arc length of flashlamp is recommended.

7.5 Electronics for CW Solid-state Lasers

Continuous wave (CW) solid-state lasers are pumped by arc lamps. Arc lamps are also used to pump solid-state lasers operating at very high repetition rates of the order of tens of kilohertz. In that case, the pumping is continuous and the pulsed operation of the laser is accomplished by using an optical switch in the laser cavity. Before discussing the power supplies for operating arc lamps, a brief description of the types and characteristic parameters of arc lamps is presented in the following sections.

7.5.1 Arc Lamps

Like flashlamps, arc lamps are also gas-discharge tubes of a similar design with the difference that they are designed to operate in CW mode. There are two types of arc lamps: linear arc lamps and short arc lamps. It is the linear arc lamp that is well suited to act as the pump source for solid-state lasers. Structurally, linear arc lamps are similar to linear flashlamps except for the electrode design. Arc lamps use pointed cathodes while flashlamps use rounded cathodes.

Although arc lamps may be filled with different gases, including krypton, xenon, mercury and mercury-xenon, it is the krypton-filled linear arc lamp that suits solid-state lasers (Nd:YAG lasers in particular) the best. Xenon arc lamps have a relatively broad emission spectrum and are not well matched to the absorption bands of Nd:YAG. Mercury arc lamps have strong emission in ultraviolet. Mercury-xenon arc lamps are characterized by an emission spectrum that is a combination of mercury lines in ultraviolet and xenon lines in infrared. The emission spectrum of a krypton arc lamp is in the wavelength

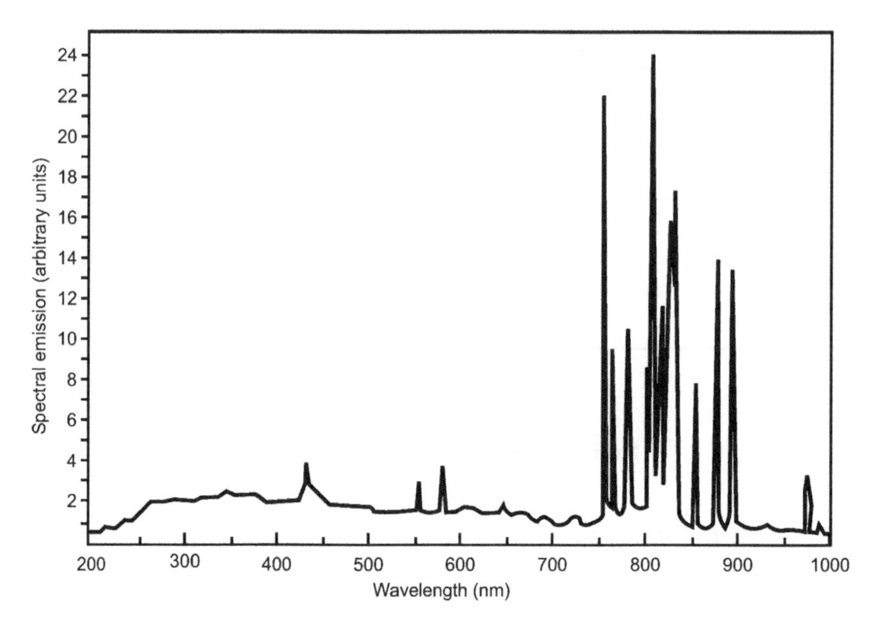

Figure 7.27 Emission spectrum of krypton arc lamp.

band of 750–900 nm (Figure 7.27), which matches the peak absorption of Nd:YAG in 800 nm region. In fact, under favourable conditions as much as 40% of the light output from the krypton arc lamp can be used for excitation of Nd:YAG. The gas is filled at a pressure in the range 1–3 atmospheres. The useful output power of krypton arc lamps used for pumping Nd:YAG lasers increases rapidly with input power, gas-fill pressure and bore diameter.

7.5.2 Electrical Characteristics

Figure 7.28 shows typical *I–V* characteristics of an arc lamp. The characteristic curve exhibits a negative resistance region characterized by a steep fall in voltage with a corresponding small increase in current. This is followed by a positive resistance region with a relatively much lower slope. The operation of an arc lamp can be divided into three regimes: initial arc formation, unconfined discharge regime and confined and wall-stabilized discharge regime.

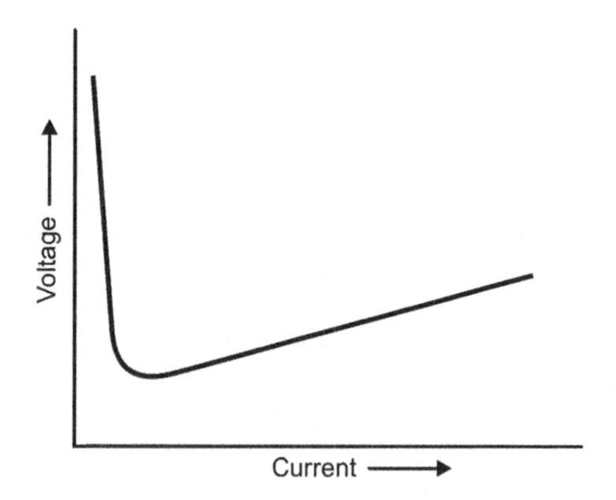

Figure 7.28 Typical *I–V* characteristics of an arc lamp.

The triggering process initiates the electrical discharge in the gas forming a thin stream of spark in the gap between the electrodes, thus creating a conductive path between them. Triggering is initiated with the application of a high-voltage trigger pulse in the range of 20–30 kV. If the power supply output voltage is greater than the voltage drop across the conducting path, current begins to flow through the lamp. The discharge region is initially very thin and said to be unconfined. The discharge region begins to expand with a resultant rapid fall in voltage. When the discharge expands to fill the tube, it is stabilized by the presence of envelope of the lamp. While the arc lamp exhibits negative resistance in the unconfined discharge region, it is positive in the wall-stabilized regime. In the positive resistance region of *I–V* characteristics, a large increase in lamp current produces a relatively much smaller increase in lamp voltage, indicating a low value of positive impedance. This is in contrast to the unconfined discharge regime where a rapid fall in voltage is associated with a small increase in current, indicating a large negative resistance.

7.5.3 Arc Lamp Power Supply

The power supply for an arc lamp has three components:

1. trigger circuit to initiate discharge
2. starting circuit, also called boost power supply, for driving the lamp during arc expansion
3. power supply for supplying current during wall-stabilized region.

Figure 7.29 shows a typical power supply arrangement for driving an arc lamp. The trigger circuit generates a high-voltage trigger pulse to initiate the discharge. Usually the energy in the trigger pulse is not sufficient to take the discharge current and lamp voltage to a point on the *I–V* characteristics where the load line of the main power supply intersects it in the negative resistance region.

Figure 7.30 shows the *I–V* characteristics along with the load line (marked load line I) of a typical main power supply. This load line intersects the *I–V* characteristics at points A and B. If the trigger pulse fails to take the discharge state to point A, the main power supply will not be able to drive it to point B in the stabilized regime. Any attempt to design the main power supply to be able to do the job is not practical as it would require a much higher output voltage than is actually required for running the arc lamp. It is because of this reason that we need another power supply, with relatively higher output voltage, capable of supplying the desired current for a short time during the discharge expansion regime. It may be

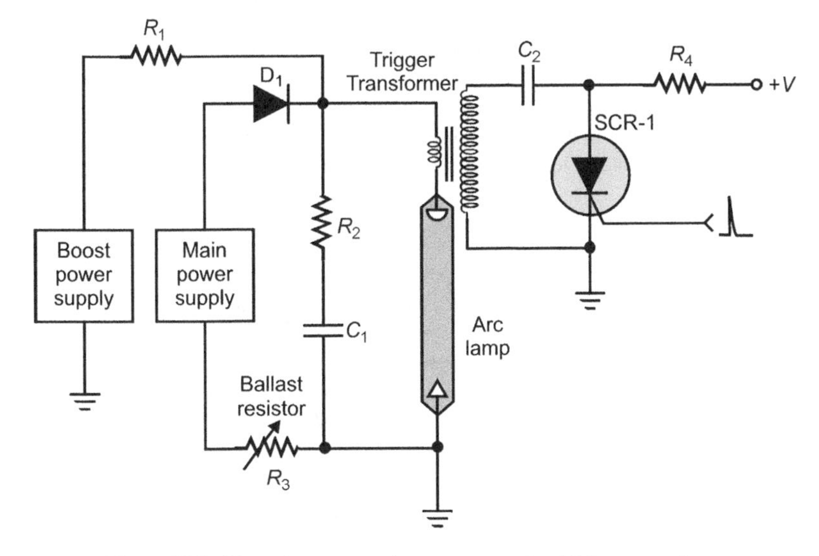

Figure 7.29 Typical power supply arrangement for driving an arc lamp.

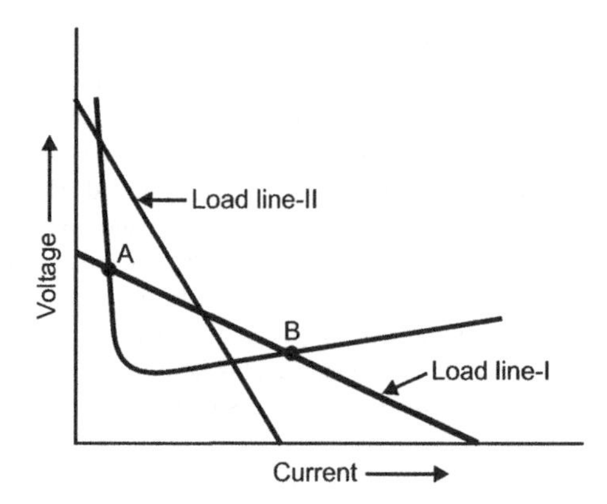

Figure 7.30 *I–V* characteristics with load line.

mentioned here that this power supply has a relatively low output current delivery capability of the order of a few tens of milliamperes at an output voltage of the order of 1.5–2 kV. The output of the power supply is used to charge a capacitor that can supply the required current for a short time. This power supply is characterized by a load line (marked load line II in Figure 7.30. This power supply, referred to as boost power supply, begins to pass current and drive the lamp through the discharge expansion regime. When the voltage across the lamp falls below the output voltage of the main power supply, the main power supply takes over and starts operating the lamp in the stabilized regime.

Referring to the circuit schematic of Figure 7.29, the boost power supply output charges capacitor C_1 through resistors R_1 and R_2 when the power supply is switched on. Diode D_1 isolates the boost power supply output from the relatively much lower output voltage of main power supply. Capacitor C_1 and R_2–C_1 time constant are selected in order to provide the desired current for the desired time period during discharge expansion. R_1 limits the current that can be drawn from the boost power supply output. With a much lower resistance value, R_2 provides the desired current for a short time period. Ballast resistor R_3 is used in series with the output of the main power supply to achieve the desired value of the lamp current in the stabilized regime.

7.5.4 Modulated CW and Quasi-CW Operation of Arc Lamp

In the case of *quasi-CW operation*, the arc lamp is driven by a current that has a sinusoidal wave shape with a DC value as shown in Figure 7.31. The difference between the peak value of the drive current and the DC value, called depth of modulation, is usually not allowed to exceed 50% of the DC value. While using this mode of operation, the maximum power rating of the arc lamp must not be exceeded. The frequency of modulation is chosen to be either 50 Hz or 100 Hz, simplifying the power supply design.

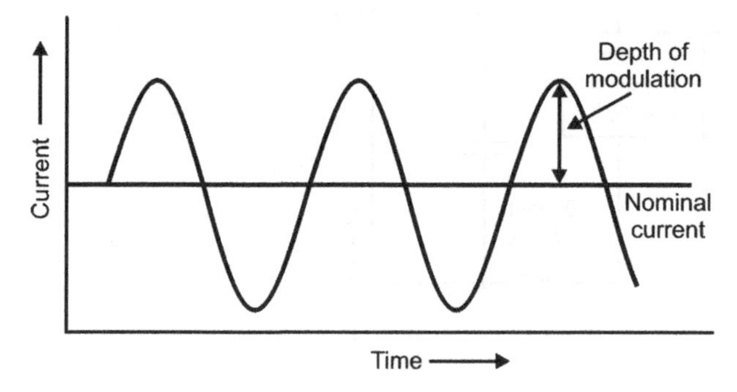

Figure 7.31 Quasi-CW operation of arc lamp.

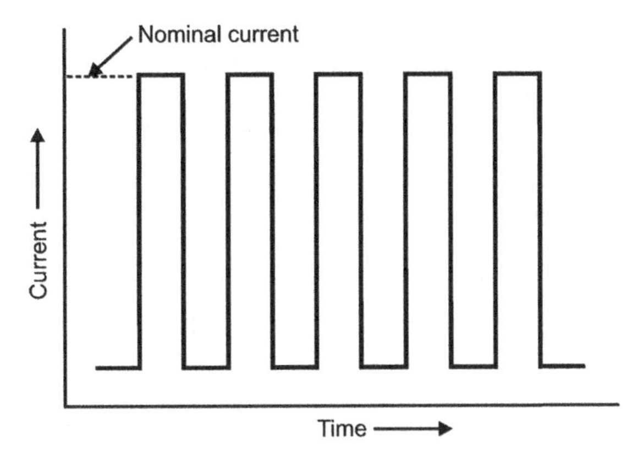

Figure 7.32 Modulated CW operation of arc lamp.

In the case of *modulated CW operation*, the arc lamp current and consequently its light output is switched between two values with upper- and lower-level time periods being in the range of a few seconds. The upper-level current is kept equal to the nominal current and the lower-level current is chosen in order to be above the minimum recommended current level. The waveform is shown in Figure 7.32. While using this mode of operation, it must be ensured that the average power rating of the arc lamp is not exceeded.

7.6 Solid-state Laser Designators and Rangefinders

Laser rangefinders are used by the armed forces to measure ranges of potential targets. These are used on ground as well as on airborne platforms to range both ground and airborne targets. In the modern battlefield scenario, a rangefinder is hooked directly onto the gunnery computer in what is known as the integrated fire control system. Rangefinders are also used by astronomers (for measuring the distances to planets) and by surveyors. The application-related aspects of laser rangefinders are described in Chapter 14.

A laser target designator on the other hand is used to mark or identify a target to smart ammunition (bomb, projectile, missile, etc.) by sending a series of coded laser pulses at a high repetition rate (typically 10–20 pps). The bomb or missile that has the corresponding decoder can then home in on the target.

The basic building blocks of electronics of a laser rangefinder or a designator are similar to those used in the case of a typical Q-switched solid-state laser source, except that the circuit that generates the flashlamp firing commands has an encoder. They invariably have a receiving channel that is used to determine the range of the designated target; both laser rangefinders and target designators therefore have additional building blocks used in the receiving channel. The capacitor-charging power supply in the case of a laser designator is rated for relatively much higher joules/second output delivery capability due to both the higher pulse energy (100–150 mJ) and the pulse repetition frequency (10–20 pps). On the other hand, laser rangefinders with relatively much lower pulse energy (5–10 mJ) and pulse repetition frequency (10–20 pulses per minute) need a capacitor-charging power supply with much lower joules/second rating.

Figure 7.33 shows the building blocks of the receiving channel of the laser rangefinder. To avoid repetition, the laser source electronics is omitted here. One of the most critical building blocks of the receiver section is the optoelectronic front-end. The function of the front-end circuit is to transform the received laser pulse, which could be anywhere in the range of 5–50 ns, to an equivalent electrical signal. The peak power of the received laser pulse could be as low as a few tens of nanowatts when ranging a far-off target and as high as a few tens of milliwatts when the target is close by. This implies that the amplifier

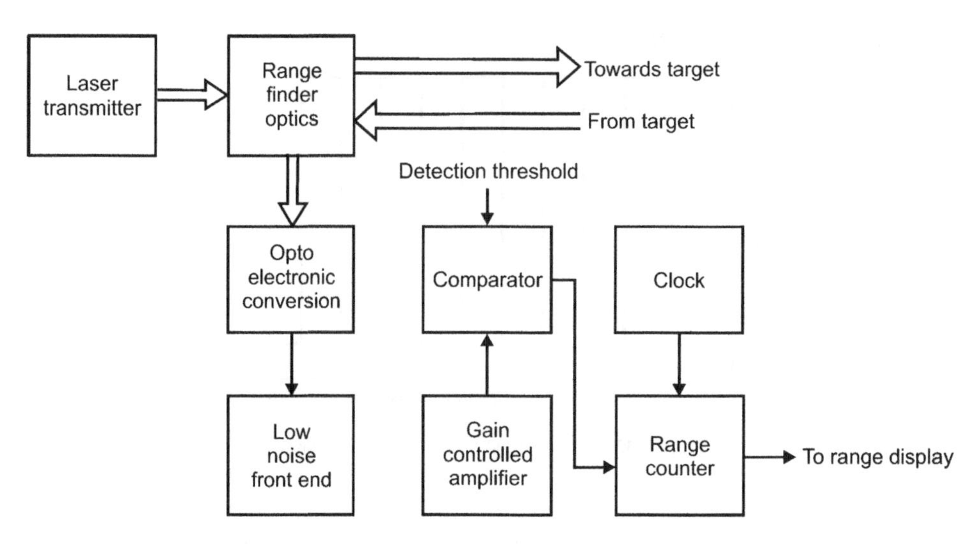

Figure 7.33 Laser rangefinder receiver channel electronics.

portion of the front-end needs to have a dynamic range as high as 100–110 dB. This is usually achieved partly in the avalanche photodiode (APD) by controlling the responsivity of the device through its reverse-bias variation and partly in the gain-controlled amplifier stage. Laser pulse width determines the bandwidth that the front-end optoelectronic conversion and the programmable amplifier circuits that follow it must have. The required bandwidth is inversely proportional to the rise time of the received laser pulses and can be computed from Equation 7.29:

$$\mathrm{BW} = \frac{350}{t_\mathrm{r}} \tag{7.29}$$

where BW is the bandwidth in MHz and t_r is the rise time in ns.

The design of range counter determines range resolution and accuracy. The conventional method of counting a fixed clock frequency between the start and stop laser pulses has its accuracy limited by the clock frequency. Range measurement accuracy equals $\pm[c/(2 \times f_\mathrm{clock})]$. This technique can be conveniently used up to a range accuracy of ±5 m. A clock frequency of 150 MHz is needed to achieve an accuracy of ±1 m. For higher accuracy and resolution, other counting techniques must be used. Advances in technology have made sub-metre accuracy rangefinders built around solid-state lasers possible.

7.7 Summary

- Only 30% of the electronics packages of contemporary lasers and laser-based systems are concerned with required power supplies.
- In the case of flashlamp-pumped pulsed solid-state lasers, in addition to the capacitor-charging module used to pump the flashlamp, important circuit modules include the simmer module, the pulse-forming network, the flashlamp trigger and the Q-switch driver modules.
- The simmer module is used to maintain a low current discharge through the lamp when the lamp is not flashing. Simmer mode of operation has advantages of better pulse-to-pulse reproducibility, absence of requirement of high-voltage trigger pulse every time the energy stored in the capacitor is to be discharged through the lamp, enhanced flashlamp life and reduced electromagnetic interference. In the case of pseudo-simmer mode of operation, the low-current simmer discharge state is not maintained throughout the inter-pulse period. Instead, it is initiated about 100–200 μs before every flashlamp

firing pulse. This method of operation leads to reduced power loss due to simmer current while retaining all the advantages of simmer mode operation.

- Different methods of flashlamp triggering include over-voltage triggering, external triggering, series triggering and parallel triggering. Series and external triggering schemes are the most commonly used.
- The PFN is used to obtain the desired flashlamp discharge current pulse shape. It is designed to ensure a critically damped current pulse, which leads to most optimum conversion efficiency.
- CW solid-state lasers are pumped by arc lamps. The power supply for an arc-lamp-pumped solid-state laser includes a trigger module for initiating the arc discharge, a boost power supply to drive the lamp through the discharge expansion regime and the main power supply that operates the lamp in the stabilized regime.
- In addition to the CW mode of operation of arc lamps, other operational modes include the modulated CW and quasi-CW modes. In the case of quasi-CW operation, the arc lamp is driven by a current that has a sinusoidal wave shape with a DC value. The depth of modulation in this case is not allowed to exceed 50% of the DC value. In the case of modulated CW operation, the lamp current and consequently its light output is switched between two values with upper- and lower-level time periods being in the range of few seconds. The upper-level current is kept equal at the nominal value and the lower-level current is kept above the minimum recommended current value.
- If a solid-state laser is to function as a rangefinder, building blocks such as high-speed optoelectronic conversion, high bandwidth amplifier with controllable gain, high-speed digital counting circuit and an appropriate interface/display console are involved.

Review Questions

7.1. Looking at the nature of electronics packages needed for different types of laser sources and related equipment, it is very clear that laser electronics involves a vast range of building blocks of electronics and is not confined to merely power supplies. Justify this statement.

7.2. Name the different building blocks of the electronics package of a flashlamp-pumped Q-switched ND:YAG laser source operating in the simmer mode. Briefly describe the primary function of each building block.

7.3. Name the different building blocks of the electronics package of a flashlamp-pumped time-of-flight Nd:YAG laser rangefinder. Briefly describe the primary function of each building block.

7.4. Explain why the power converter that charges the energy storage capacitor to the required voltage in a flashlamp-pumped pulsed solid-state laser presents many design challenges due to the capacitive nature of the load. Give arguments in support of your answer.

7.5. What are the advantages of operating the flashlamp in the simmer mode? With the help of a block schematic arrangement, briefly describe the operation of the simmer module. How does the simmer mode operation differ from the pseudo-simmer mode of operation?

7.6. Briefly describe different methods of triggering flashlamps. What are the preferred methods?

7.7. What do you understand by different regimes of operation of CW arc lamps? With the help of a suitable circuit diagram, briefly describe the operation of the power supply used for operating arc lamps.

7.8. Describe the quasi-CW and modulated-CW modes of operation of arc lamps.

7.9. With the help of relevant mathematical expressions, describe the different steps involved in designing a PFN in a pulsed flashlamp-pumped solid-state laser to ensure a critically damped current pulse for given values of energy to be discharged through the lamp, pulse width of the critically damped current pulse and lamp parameters.

7.10. Outline the criteria for selecting the value of ballast resistance
a. for simmer mode operation of flashlamp in a pulsed solid-state laser
b. for the main power supply of the arc lamp in a CW solid-state laser.

Problems

7.1. In the case of simmer mode of operation of flashlamp, when the ballast resistance is changed by 5 kΩ, voltage drop across the ballast resistor reduces by 100 V while voltage across the lamp increases by 300 V. Calculate the magnitude of flashlamp resistance in the simmer mode.
[15 kΩ]

7.2. It is desired to discharge energy of 100 J through a flashlamp in a critically damped pulse of 60 μs duration. If the flashlamp impedance parameter K_0 is 13 Ω $A^{-0.5}$, calculate the values of L and C of the pulse-forming network and the DC voltage across the capacitor for the desired quantum of pulse energy. Also determine the resulting peak current.
[$C = \sim 200\ \mu F$, $L = 140\ \mu H$, $V = \sim 1000\ V$, $i_p = 597\ A$]

7.3. During flash discharge in a lamp after the peak current had been achieved, the instantaneous current through the lamp at a certain time was observed to be 625 A. After some time had elapsed, at another time instant during the flash duration, the instantaneous current was observed to be 324 A. Determine the percentage change in voltage across the lamp along with polarity of change in voltage.
[Voltage decreases by 28%]

Self-evaluation Exercise

Multiple-Choice Questions

7.1. In the simmer mode of operation, flashlamp is operated in
a. negative resistance region of its *I–V* characteristics
b. positive resistance region of its *I–V* characteristics
c. unconfined regime of gas discharge
d. both (a) and (c).

7.2. Arc lamps are operated in
a. unconfined regime of gas discharge
b. wall-stabilized regime of gas discharge
c. negative resistance region of its *I–V* characteristics
d. none of these.

7.3. With reference to simmer mode operation of flashlamps,
a. flashlamp life is enhanced
b. pulse-to-pulse reproducibility is improved
c. electromagnetic interference is less
d. all of the above.

7.4. In which method of flashlamp triggering will the trigger transformer be bulky?
a. external trigger
b. parallel trigger
c. series trigger
d. over-voltage trigger.

7.5. A pulse-forming network designed for critically damped current pulse has certain values of L and C. A reduced value of L will lead to
a. an under-damped current pulse

b. an over-damped pulse
c. no change in current pulse wave shape
d. a current pulse with larger pulse width.

7.6. In a single-mesh PFN of capacitor C and inductor L, the rise time of critically damped current pulse is given by
a. $\sqrt{(LC)}$
b. $\sqrt{(L/C)}$
c. $1.25 \times \sqrt{(LC)}$
d. none of these.

7.7. During flashlamp discharge after the peak current has been attained,
a. lamp voltage is proportional to lamp current
b. lamp voltage is proportional to square root of lamp current
c. lamp voltage is inversely proportional to lamp current
d. lamp voltage is inversely proportional to square root of lamp current.

7.8. Modulated CW and quasi-CW modes of operation are used in the case of
a. pulsed flashlamps
b. CW arc lamps
c. only xenon arc lamps
d. simmer mode of operation of linear flashlamps.

7.9. Flashlamp impedance parameter depends upon
a. bore diameter
b. arc length
c. gas-fill pressure
d. all of the above.

7.10. The capacitor-charging unit of a flashlamp-pumped pulsed solid-state laser delivers 10 J of electrical energy per pulse. If the laser has a pulse repetition frequency of 20 Hz, the output power delivery capability of the capacitor-charging unit is
a. 200 W
b. 2 W
c. 500 mW
d. indeterminate from given data.

Answers

1. (d) 2. (b) 3. (d) 4. (c) 5. (b) 6. (c) 7. (b) 8. (b) 9. (d) 10. (a)

Bibliography

1. *Optical Electronics*, 1990 by Amnon Yariv, Saunders College Publishing.
2. *Laser Electronics*, 1995 by Joseph T. Verdeyen, Prentice-Hall.
3. *Pulsed Power for Solid State Lasers*, 2007 by W. Gagnon, G. Albrecht, J. Trenholme and M. Newton, Lawrence Livermore National Laboratory LLNL.
4. *High Power Laser Handbook*, 2011 by Hagop Injeyan and Gregory D. Goodno, McGraw-Hill.
5. *Handbook of Laser Technology and Applications*, 2003 Volume-II by Collin E. Webb and Julian D. C. Jones, Institute of Physics Publishing.
6. *Solid State Laser Engineering*, 2006 by W. Koechner, Springer.
7. *Fundamentals of Light Sources and Lasers*, 2004 by Mark Csele, John Wiley & Sons.

8

Gas Laser Electronics

8.1 Introduction

Gas laser electronics primarily comprises the power supply used to initiate and subsequently sustain electrical discharge through the gas mixture contained in a sealed envelope. The active medium is usually excited either by passing an electric discharge current along the length of the tube, known as longitudinal excitation, or perpendicular to the length of the laser tube, known as transverse excitation. Frequency stabilization electronics used in the case of actively stabilized Doppler-broadened gas lasers such as helium-neon and carbon dioxide lasers is another area that relates to gas laser electronics. This chapter describes the fundamentals of gas laser power supplies in terms of requirement specifications, circuit configurations and design guidelines, with particular reference to the two most commonly used gas lasers: helium-neon and carbon dioxide lasers. Power supply configurations for metal vapour lasers and excimer lasers and, to a large extent, noble gas ion lasers are similar to those used for helium-neon and carbon dioxide lasers except for minor deviations, which are also discussed in this chapter. Frequency stabilization techniques used in the case of helium-neon and carbon dioxide lasers are also examined with the help of schematic diagrams to illustrate design issues, merits and demerits of each of the configurations.

8.2 Gas Discharge Characteristics

As outlined earlier in Chapter 4, most gas lasers are driven by electrical discharge through the laser gas mixture. The discharge is struck either longitudinally, by passing electrical current along the length of the discharge tube (longitudinal excitation), or perpendicular to the length of the tube, by passing electric current transverse to the length of the tube (transverse excitation). The two modes are reproduced schematically in Figure 8.1a and b for longitudinal excitation and transverse excitation, respectively. While the former is used in the case of low-power CW lasers, the latter is used in high-power CW or pulsed lasers.

The electrical discharge is initiated and subsequently sustained with the help of a suitably designed power supply. The electrons in the electrical discharge transfer their energy to the molecules that constitute the active species either directly (e.g. carbon dioxide lasers) or indirectly (e.g. helium-neon lasers), producing population inversion essential for laser action.

8.3 Gas Laser Power Supplies

The design of a power supply needed to initiate and sustain electrical discharge through the gas mixture is governed by the current–voltage characteristics of the electrical discharge. Although the nature of the current–voltage characteristics of electrical discharge is the same for different gas mixtures, the

Lasers and Optoelectronics: Fundamentals, Devices and Applications, First Edition. Anil K. Maini.

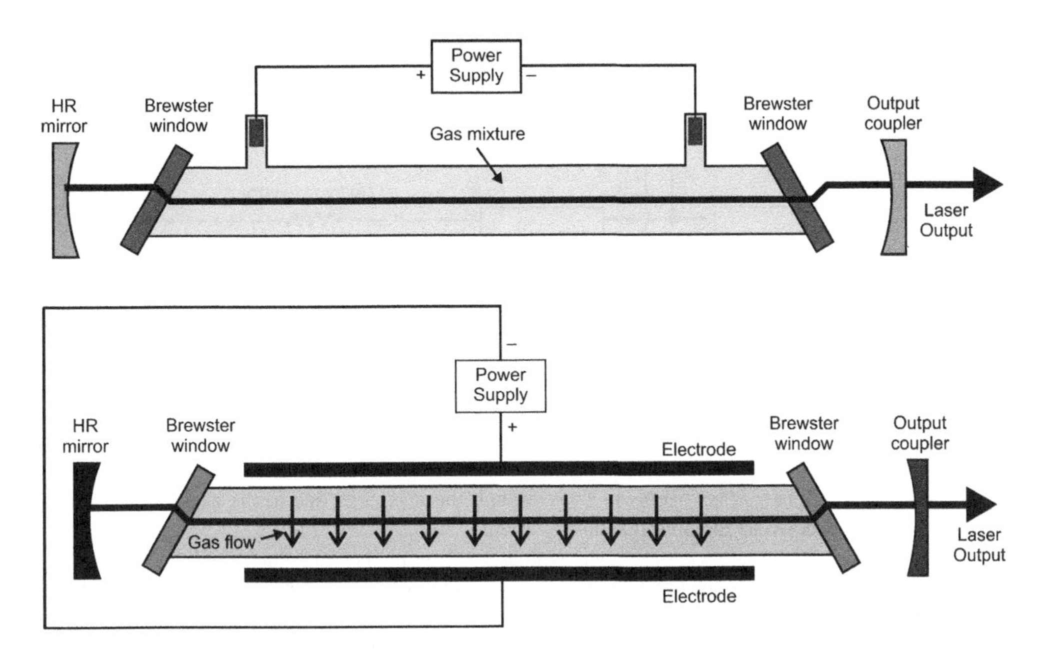

Figure 8.1 Gas laser excitation modes: (a) longitudinal excitation and (b) transverse excitation.

magnitudes of discharge current and corresponding voltage may be different. Consequently, the power supply topologies used for various gas lasers are similar but the design parameters vary. For example, the breakdown voltage of the gas fill depends not only on the nature of the gas but also on the gas pressure, size and shape of electrodes and also the inter-electrode spacing. Breakdown voltage is about 1.3 kV/cm of discharge length for neon at 1 atm and 3.4 kV/cm of discharge length for argon at 1 atm.

Figure 8.2 shows the generalized form of the current–voltage characteristics of the gas discharge. At low values of the applied voltage, the discharge current is zero and remains essentially zero until a certain minimum threshold voltage is reached. As this threshold voltage is exceeded, a small amount of current

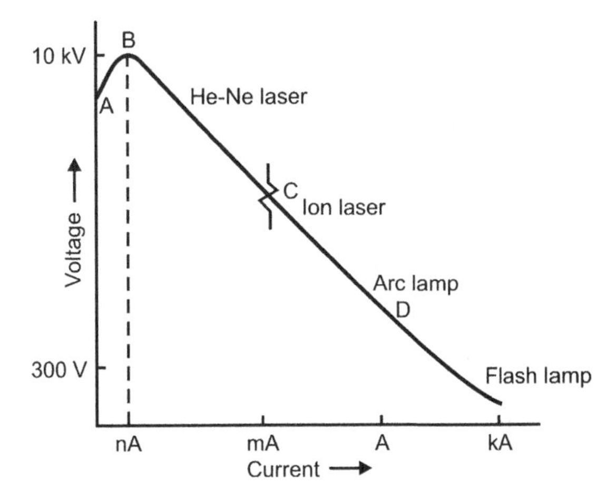

Figure 8.2 Gas-discharge current-voltage characteristics.

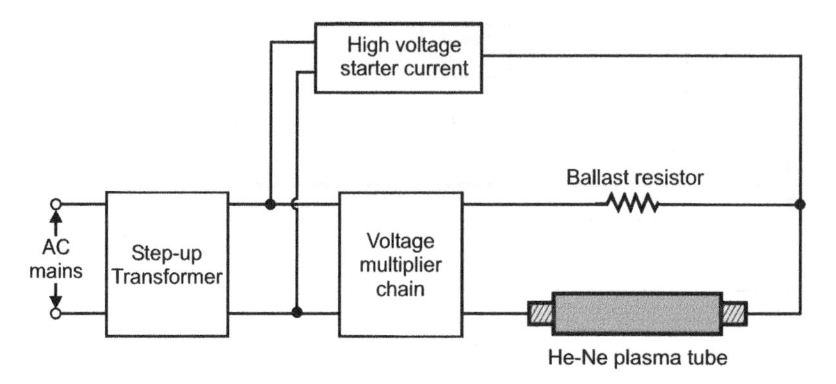

Figure 8.3 Generalized block schematic of power supply for a DC-excited gas discharge laser.

of the order of a few nanoamps begins to flow through the discharge due to the existence of some ionization. This is known as the pre-breakdown current.

The pre-breakdown current increases slowly with an increase in applied voltage until the breakdown point is reached (Point A). At the breakdown voltage, a large number of molecules in the gas mixture become ionized and conductivity of the gas mixture increases. Electrons are accelerated to velocities where they transfer energy to more molecules through collisions. An increase in discharge current causes further reduction in discharge resistance with the result that the voltage required to sustain the discharge actually decreases with increase in current. This gives rise to what is called negative resistance region in the gas-discharge current–voltage characteristics, as indicated by region B–C–D in Figure 8.2.

If the discharge current was allowed to increase without limit, catastrophe may occur at some value of discharge current. The discharge current therefore needs to be limited by connecting a positive resistance in series with the load. The series resistance used to limit the current to an acceptable value is known as the ballast resistance. In the gas discharge curve of Figure 8.2, current is shown to vary from nanoamps to kiloamps. This is intended to cover the whole range of gas discharge devices used for laser applications. The discharge current in the case of helium-neon lasers is of the order of a few milliamps. The discharge current would be of the order several amps in CW arc lamps and several kiloamps in the case of pulsed flash lamps.

The power supply essentially comprises a starter circuit and a power supply with current limiting feature. The starter circuit provides the high-voltage pulse to initiate the discharge. The peak amplitude of this high-voltage trigger pulse should necessarily be greater than the breakdown voltage of the gas mixture in question. The power supply delivers the steady-state current to maintain the discharge at the desired values of current and voltage. The current-limiting resistance or ballast resistance disallows unbounded increase in discharge current. Figure 8.3 depicts the generalized block schematic arrangement of a power supply for a DC-excited gas-discharge laser.

8.4 Helium-Neon Laser Power Supply

The task of designing a power supply for a helium-neon (He-Ne) laser could range from a routine exercise of designing a high-voltage low-current power supply with a moderate current regulation requirement to a highly complex task involving current stabilization of a very high order. At the lower end of the complexity scale, the designer would attempt to achieve the twin objectives of plasma excitation and sustenance in laboratory-use lasers; at the other end, design of a He-Ne-based system such as a ring laser gyroscope is a highly complex task.

The He-Ne laser uses a low-pressure helium-neon mixture (predominantly helium with about 10% neon) contained in a narrow bore glass tube. A mirror of almost 100% reflectance forms one end of the resonator and another mirror with about 1% transmission forms the other end. With an appropriate choice of gas mixture, processing technique and resonator optics, this laser can be made to emit at any of

Table 8.1 He-Ne power supply performance features

Specifications	Red lasers 05 LPL 340	Green lasers 05 LPL 370	IR lasers 05 LPL 344	IR lasers 05 LPL 343
Power range (mW)	1–5	7	0.5	1
Input voltage (V AC)	115/230 ± 10%	115/230 ± 10%	115/230 ± 10%	115/230 ± 10%
Input frequency (Hz)	50 or 60	50 or 60	50 or 60	50 or 60
Average input current (mA)	120/60	120/60	120/60	120/60
Recommended fusing (A)	1.0	1.0	1.0	1.0
Output current (mA)	6.5 ± 0.5	7.0 ± 0.5	3.7 ± 0.5	5.5 ± 0.5
Laser beam amplitude ripple (% RMS)	0.2	0.2	0.2	0.2
Operating voltage range (V DC)	1700–2450	2500–2700	1800	2400
Starting voltage (kV DC)	10	10	10	10
Conversion efficiency (%)	87	87	87	87
Heat sink	Not required	Not required	Not required	Not required
Required ballast	None	None	None	None
BRH time delay (%)	3–5	3–5	3–5	3–5

the four wavelengths: 0.6328 μm (red); 0.5435 μm (green); 1.523 μm (infrared); and 3.391 μm (infrared).

An He-Ne laser is pumped by an electric discharge through the gas mixture, thus creating He-Ne plasma. Commercial He-Ne lasers are almost invariably DC excited, although excitation by RF means, which is more popular in the case of CO_2 lasers, is also possible. The process of excitation in the case of a He-Ne laser involves application of a high voltage of the order of 10 kV initially across the electrodes of the plasma tube to initiate the plasma. Once the plasma is initiated, the voltage required to sustain the excitation is usually of the order 2500 V or so.

Both the excitation voltage as well as the sustaining voltage is dependent upon the intended laser power output of the tube, which in turn depends upon several He-Ne plasma tube parameters. Plasma current that needs to be supplied at the plasma tube sustaining voltage is in the range of 5–10 mA. Table 8.1 lists the major power supply performance features of different types of He-Ne lasers from Melles-Griot, a leading manufacturer of He-Ne lasers and He-Ne laser plasma tubes. These features are also broadly found in the power supplies of other leading He-Ne laser manufactures.

Another important requirement is that of providing an appropriate ballast resistance to be connected in series with the plasma discharge impedance, wired as close to the electrode as possible as shown in Figure 8.4. Figure 8.5 shows the current–voltage (*I–V*) characteristics of the He-Ne discharge. The plasma *I–V* characteristics exhibit a negative resistance as shown in the *I–V* curve of Figure 8.5. The ballast resistance ensures that the power supply encounters a positive resistance while delivering current to the plasma discharge. This is a necessary condition to initiate stable plasma discharge. The negative resistance offered by the plasma is given by the reciprocal of the slope of the tangent drawn at the operating point in the curve of Figure 8.6.

Example 8.1

In a certain He-Ne laser plasma tube, an operating current of 5 mA causes a voltage of 1250 V across the tube. A 1 mA change in plasma discharge current causes a 100 V change in voltage across the tube. Determine the minimum value of the required ballast resistance for stable discharge. Also determine the minimum value of the DC voltage at the output of power supply to sustain stable discharge.

Solution

1. Negative impedance offered by the laser plasma at the operating current = $100/(1 \times 10^{-3})\ \Omega = 100\ \text{k}\Omega$
2. Therefore, for stable operation, minimum value of ballast resistance $R_B = 100\ \text{k}\Omega$

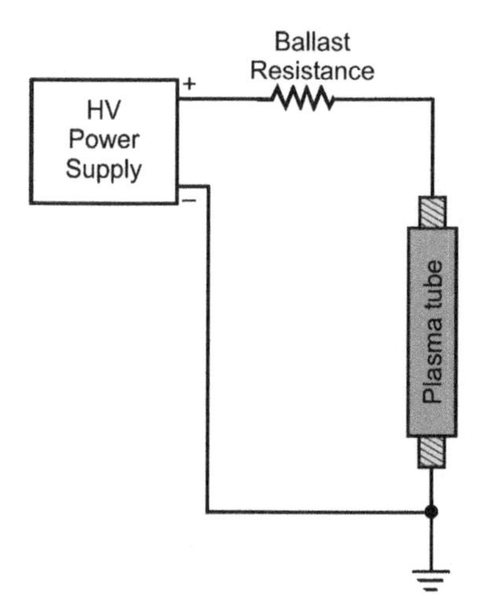

Figure 8.4 Use of ballast resistance.

3. Voltage drop across the ballast resistance at the operating current = $(100 \times 10^3) \times (5 \times 10^{-3})$ V = 500 V
4. Therefore, minimum value of DC voltage at the power supply output = 1250 + 500 V = 1750 V.

Example 8.2

In the case of the He-Ne laser plasma tube of Example 8.1, if the DC power supply output voltage is 2500 V what would be the value of the ballast resistance needed to operate the plasma tube at the same operating current?

Solution

1. At the operating current of 5 mA, voltage across the tube = 1250 V
2. Required voltage drop across the ballast resistance = (2500 − 1250) = 1250 V
3. Therefore, value of ballast resistance = $1250/(5 \times 10^{-3})$ Ω = 250 kΩ.

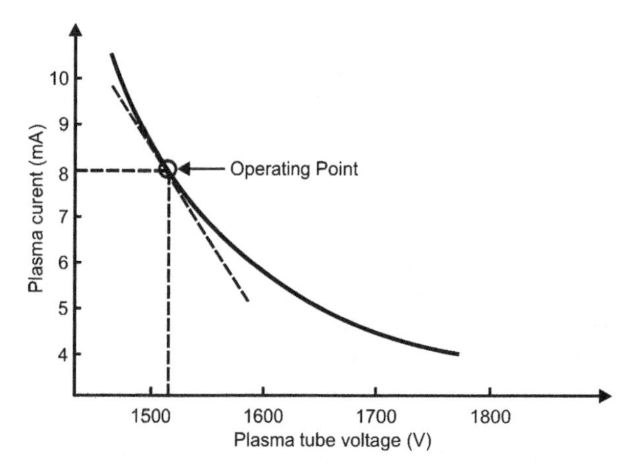

Figure 8.5 *I–V* characteristics of He-Ne plasma discharge.

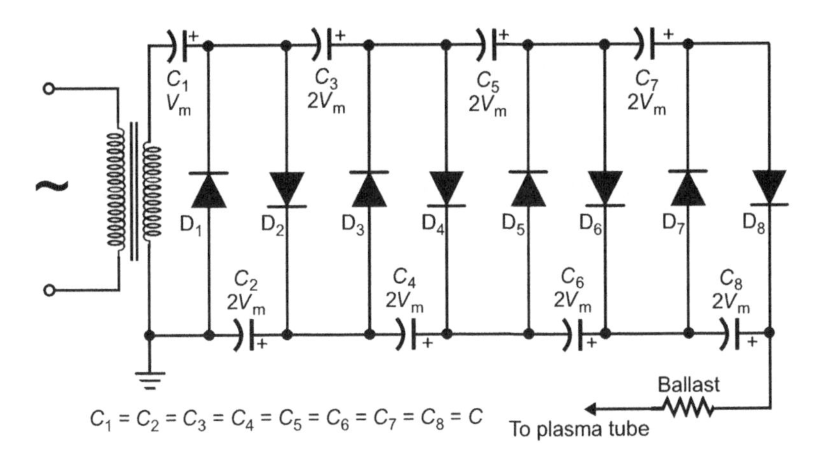

Figure 8.6 He-Ne power supply with a cascade multiplier chain having an even multiplication factor.

8.4.1 Power Supply Design

One design approach, which is also the most commonly used approach in the case of commercial He-Ne laser power supplies, is to use a voltage multiplier chain to generate the required high voltage for plasma excitation from a relatively much lower input voltage. The power supply in this case is usually designed for a power output capability equal to the product of the voltage (plasma tube voltage at the operating current plus the expected drop across the ballast resistance) and the plasma current. In a supply like this, the multiplier chain raises the output voltage to a magnitude needed for initiating the discharge. The moment plasma is struck and the current is drawn from the power supply, the output voltage falls to a lower value required to sustain the plasma.

One possible circuit configuration using a step-up power transformer followed by a cascade-type voltage multiplier chain is shown in Figure 8.6. The step-up ratio and the multiplying factor are chosen in order to produce an output DC voltage of about 10 kV required for initiating the plasma. A step-up transformer that produces an output voltage of 1000 V RMS followed by a cascade-type voltage multiplier chain with a multiplication factor of 8 could possibly be used for the circuit of Figure 8.6. Such a circuit would produce at its output a DC voltage equal to $8V_m$, where V_m is the peak AC voltage across the secondary of the transformer. For a 1000 V RMS across the transformer secondary, the no-load output voltage would be about 11 kV, which is just the right voltage for exciting the He-Ne plasma.

In the circuit of Figure 8.6, the multiplication factor of the cascade multiplier chain needs to be an even number. A similar circuit configuration with an odd multiplication factor is shown in Figure 8.7. The designer is free to exercise this flexibility in choosing the transformer step-up ratio and the multiplier's multiplication factor, depending upon what is optimum in a given situation. In the circuit configuration of Figure 8.7, the multiplication factor is 7.

Once the plasma is struck and current is drawn from the power supply, the output voltage falls due to voltage drops in the transformer windings and diodes in the multiplier chain. Ideally, this voltage should drop to a value equal to the required power supply voltage for plasma sustenance, which is the sum of the plasma tube voltage at the desired plasma current plus the voltage drop across the ballast resistance. If the voltage does not fall to the required value and is instead higher than that, then the excess voltage would have to be dropped across an additional resistance in series with the ballast resistance in order to obtain the desired plasma current. Otherwise, the plasma current would be greater than the desired operating value of the ballast resistor.

It may be mentioned here that most of the voltage drop occurs across the multiplier chain. The components of the chain are chosen in order to drop the required voltage across it. For a

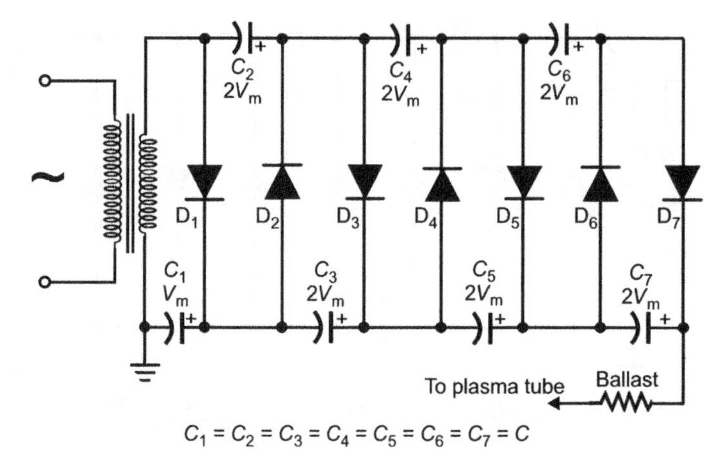

Figure 8.7 He-Ne power supply with a cascade multiplier chain having an odd multiplication factor.

multiplier chain with an even multiplication factor, the output voltage V_O can be expressed by Equation 8.1:

$$V_O = \frac{nV_p}{1 + [n(n^2/2 + 1)/6fCR_L]} \tag{8.1}$$

where n is the multiplication factor; V_p is the peak value of the voltage at the input to the multiplier chain; f is the frequency of the AC input; C is the capacitance of each capacitor in the chain; and R_L is the load resistance.

In the case where the multiplication factor is odd, the output voltage V_O can be expressed by Equation 8.2:

$$V_O = \frac{nV_p}{1 + [n(n^2 - 1)/12fCR_L]} \tag{8.2}$$

In both cases, as is clear from the above equations, the output voltage is n times the input voltage to the multiplier chain only when the factor within the brackets in the denominator is zero. For this to happen, the denominator of the bracketed term needs to be much greater than the numerator. For the given values of frequency and load resistance, this can be achieved by using large-value capacitors.

As the desired multiplication factor increases, it can be seen that the capacitance value becomes prohibitively large even for moderate values of load current. That is why the multiplier chain of this type is inherently poorly regulated. This can be used to advantage in the case of a He-Ne laser power supply design, however. For instance, for the given values of load resistance (which is determined by the required power supply output voltage for plasma sustenance and the plasma current) and the frequency of the input to the chain (50 Hz in the case of a chain operating from the AC mains), the value of C can be chosen in order to obtain an output voltage that equals the required output voltage for plasma sustenance.

For a multiplication factor of 8 and frequency equal to 50 Hz, the value of capacitance would be about 0.04 μF if the laser power supply output had to produce an output voltage of 2000 V for a plasma current of 5 mA, assuming that the input AC signal to the multiplier chain input is 1000 V RMS or about 1400 V peak. For the circuit of Figure 8.6, if the capacitance value is 0.04 μF the output voltage would automatically fall from an initial no-load value of about 11 kV to about 2000 V when a plasma current of 5 mA is drawn from it.

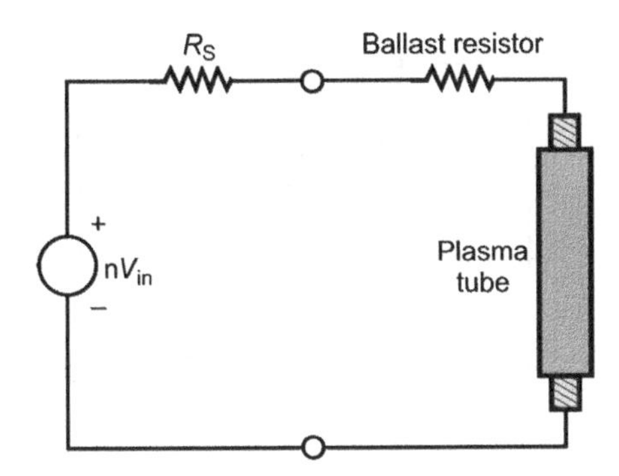

Figure 8.8 Equivalent circuit of cascade multiplier chain.

The multiplier chain as a whole offers a series resistance which drops a voltage across it when current is drawn from it. The multiplier chain may be represented by an equivalent circuit of the type shown in Figure 8.8. This equivalent series source resistance increases (to a very good approximation) with the cube of the multiplication factor. This implies that if the multiplication factor is doubled, then for a given operating frequency the capacitance will have to be increased by a factor of 8 if the multiplier circuit was to maintain the same output voltage on load.

Another multiplier circuit configuration that offers superior performance in terms of efficiency is shown in Figure 8.9. The configuration shown here is that of a multiplier with an even multiplication factor. The circuit works at an efficiency that is better than that of the circuit using equal value capacitances. The circuits shown in Figures 8.6–8.9 are half-wave circuits. A full-wave voltage multiplier that gives a far better ripple performance is depicted in Figure 8.10. The circuit shown is a quadrupler.

Example 8.3

A certain He-Ne laser power supply is designed using a cascade voltage multiplier chain of the type shown in Figure 8.6 and a multiplication factor of 8. If the peak value of the 50 Hz waveform at the input of the multiplier chain is 1500 V, determine the minimum capacitance value of the capacitors used in the multiplier chain to produce the desired value of 1000 V across the plasma tube at an operating current of 4 mA. Assume the value of ballast resistance to be 100 kΩ.

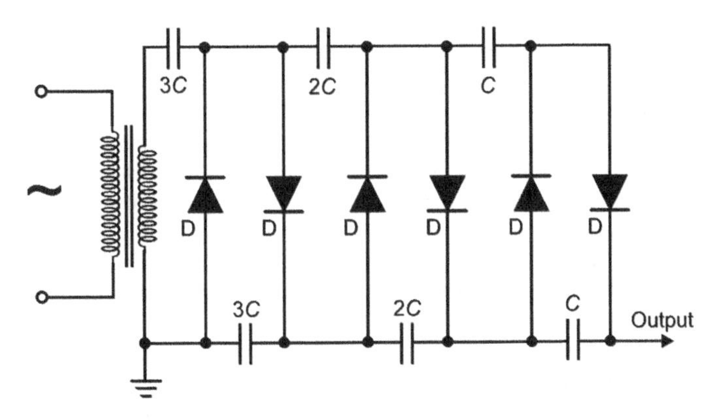

Figure 8.9 Multiplier chain with unequal capacitors.

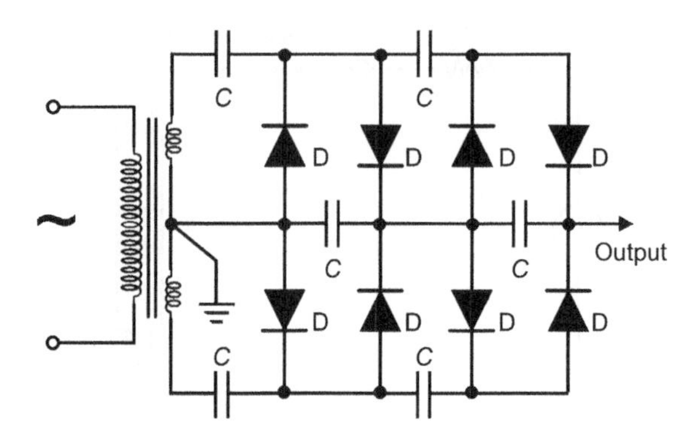

Figure 8.10 Full-wave voltage multiplier.

Solution

1. The output voltage in the case of cascade voltage multiplier chain with even multiplication factor is given by $V_O = \dfrac{nV_p}{1+[n(n^2/2+1)/6fCR_L]}$
2. The desired value of $V_O = (1000 + 4 \times 10^{-3} \times 100 \times 10^3)$ V = 1400 V

   Therefore, $R_L = 1400/(4 \times 10^{-3})\,\Omega = 350\,\text{k}\Omega$.

3. Substituting for R_L, f, n and V_P ($f = 50$ Hz, $n = 8$, $V_P = 1500$ V) in the above equation, we have:

$$1400 = \frac{8 \times 1500}{1+\left[8 \times (64/2) + 1/\left(6 \times 50 \times C \times 350 \times 10^3\right)\right]}$$

$$= \frac{12000}{1+\left(264/105000 \times 10^3 C\right)}$$

This gives $C = 0.332\,\mu$F.

8.4.2 Switched-mode Power Supply Configurations

The concept of using a multiplier chain fed at the input by an AC signal can be implemented far more effectively and efficiently by using a switched-mode power supply to drive the voltage multiplier chain. The most obvious SMPS configuration for such an application is the push-pull circuit as it generates a square-wave AC signal across the secondary of the switching transformer. Both self-oscillating as well as externally driven configurations can be used; the latter configuration has better control over various performance parameters.

A variety of switched-mode IC controllers are available that are suitable for driving the two transistors (bipolar junction transistor or BJT; metal-oxide semiconductor field-effect transistor or MOSFET) of the push-pull configuration; ICs LM/SG3524, TL494, TL594 and TL497 are a few examples. The peak value of the square-wave output across the transformer secondary equals the input DC voltage applied to the centre tap of the transformer primary, multiplied by the step-up ratio of the transformer. Again, the converter could operate off-line where a bridge circuit transforms the AC mains input into a proportional DC, which in turn feeds the centre tap of the push-pull transformer primary.

In the case where the converter is off-line, there is no isolation transformer for the mains. In such a case the feedback loop must have isolation so that the DC output is cut off from the AC line. This is usually accomplished by using an opto-isolator.

The other option is to operate the converter from a relatively lower DC voltage of, say, 24 V DC, which could be generated from a step-down transformer, rectifier and filter combination. The transformer here

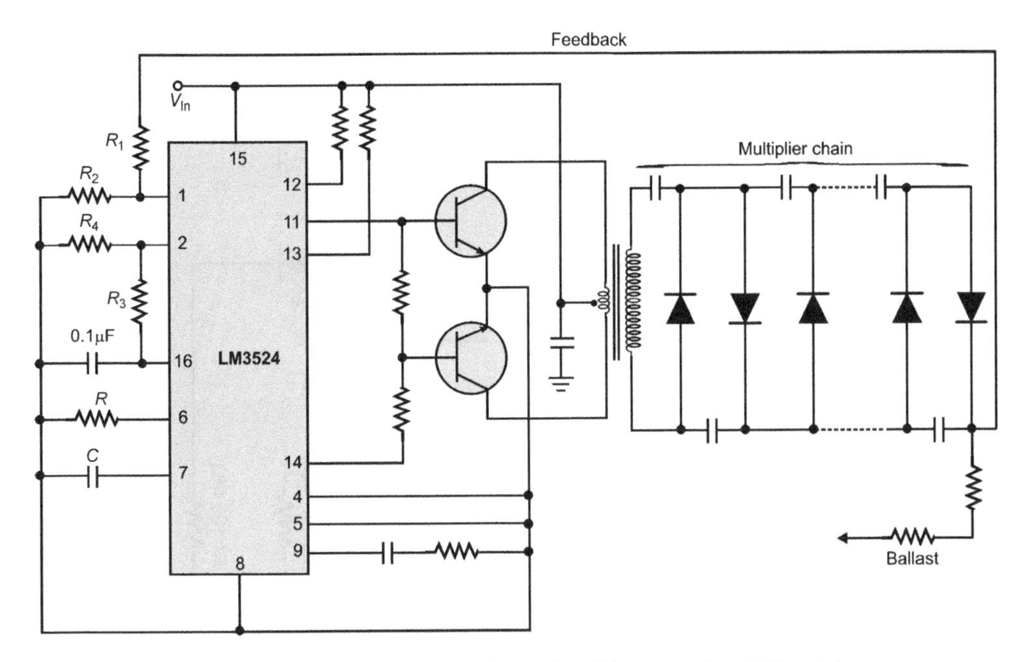

Figure 8.11 He-Ne power supply using push-pull inverter and multiplier chain.

provides isolation, which eliminates the need to use additional isolation on the output side. It may be mentioned that a typical He-Ne power supply needs to deliver about 10 W of power. The use of step-down transformer, rectifier and filter does not really add significantly to the size, while it allows the designer to use much lower-voltage switching devices.

The switched-mode version has all the advantages associated with switched-mode power supplies such as higher efficiency and small size. To use a push-pull converter with a voltage multiplier chain for designing a He-Ne power supply, the required capacitance value would be much smaller for a given multiplication factor due to high-frequency operation. On the other hand, if the operating frequency is say 50 kHz for a given value of the chosen capacitance, it would be possible to use a multiplication factor which is 10 times that possible in the case of a multiplier circuit operating at 50 Hz.

A typical He-Ne laser power supply configuration built around an externally driven push-pull inverter driving a voltage multiplier chain is shown in Figure 8.11. The controller IC used here is LM3524. Here the components R and C determine the switching frequency. R_1 and R_2 constitute the voltage sense loop whereas the R_3 and R_4 potential divider arrangement makes use of the internal reference of +5 V (available at pin 16 of the IC) to generate reference for the voltage sense loop. Any other suitable IC can also be used with associated minor changes in the external circuit. The data sheets of these ICs come with basic application circuits. In the absence of availability of the controller IC, the same configuration could be implemented using two 555 timer ICs or one 556 timer IC as shown in Figure 8.12.

A close examination of the circuit shown in Figure 8.12 reveals an interesting variation from the circuit of Figure 8.11 on the output side. As discussed earlier, the capacitance value is chosen in order to ensure that the power supply output voltage falls to the level of the sustaining voltage when plasma current is drawn from the power supply. Such a circuit inherently lacks good regulation. If we choose the capacitance value in such a way that the power supply output voltage on load is either equal to or slightly less than the open circuit voltage, the excess voltage would need to be dropped across an additional series resistance.

In the configuration of Figure 8.12, the plasma tube is initially fed from the output of the multiplier chain. The moment the plasma is struck, the tube draws its current from a point where the push-pull output is only doubled. The current drawn through the rest of the multiplier chain is almost negligible. In

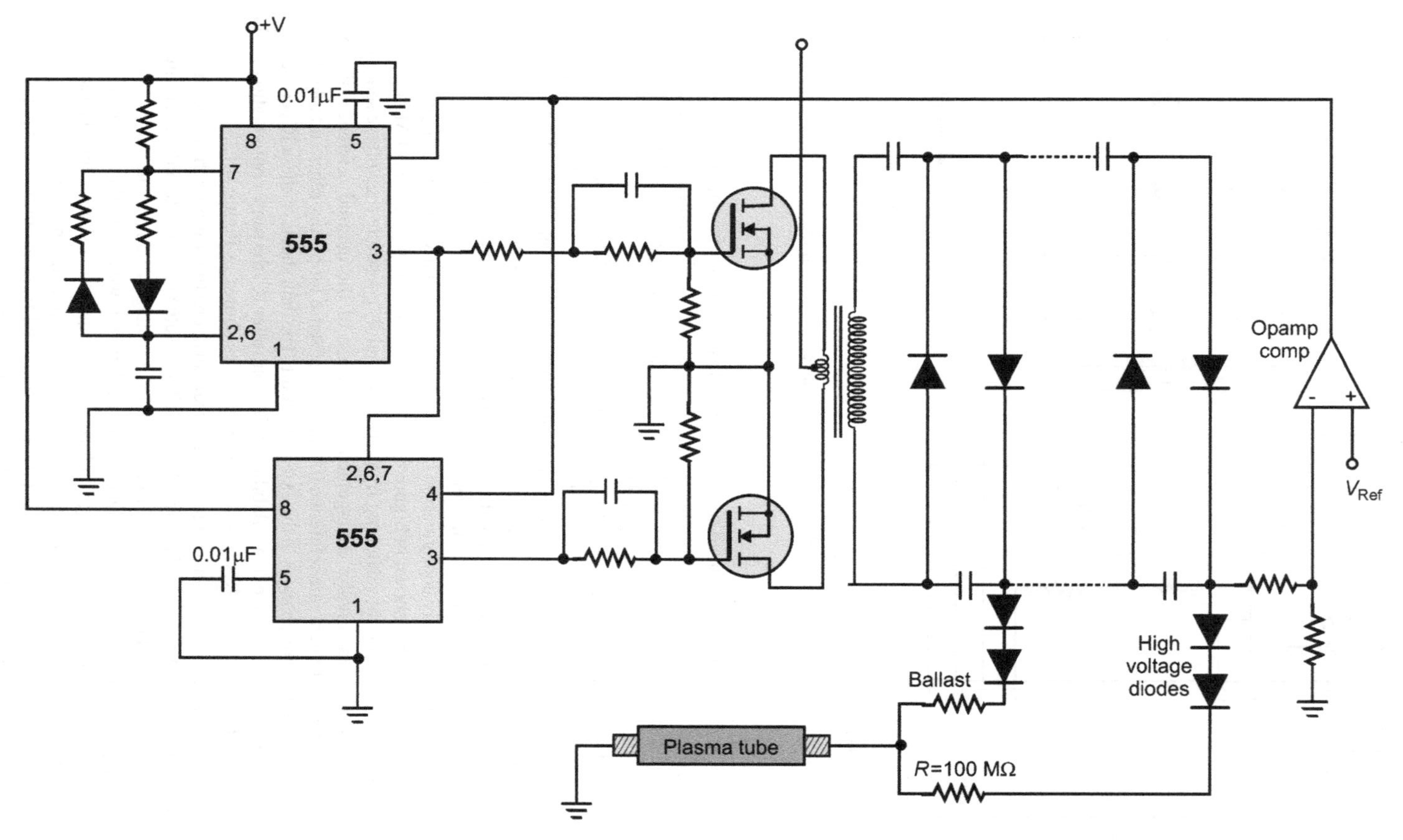

Figure 8.12 Push-pull converter with 555 timer IC-based drive circuit.

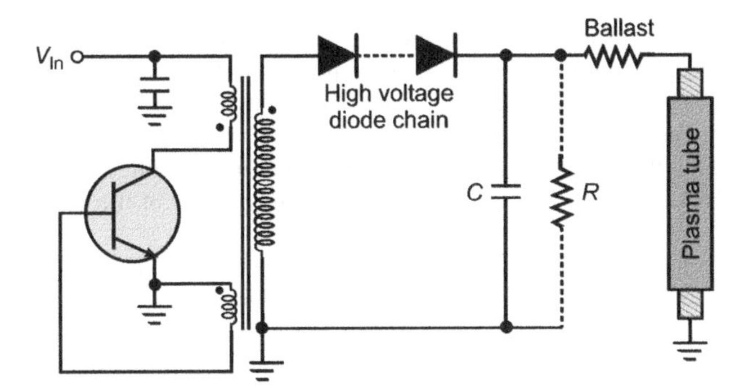

Figure 8.13 Constant output power self-oscillating flyback-converter-based power supply.

the circuit shown, this current is of the order of 80 μA for a multiplied output voltage of 10 kV, assuming a voltage of about 2 kV across the plasma tube.

8.4.3 Other Possible Configurations

A self-oscillating flyback converter, which is inherently a constant output power DC/DC converter, is another power supply configuration particularly attractive for use with He-Ne lasers; Figure 8.13 shows the basic circuit. The converter circuit in this case is designed to deliver an output power equal to the product of the required sustaining voltage at the power supply output and the plasma current. Since it is a constant output power converter, the output voltage would increase without limit in the ideal case in the event of zero current drawn from the power supply. When the output voltage exceeds the required initiating voltage, the plasma current drawn from the power supply forces the output voltage to fall to the sustaining voltage governed by the output power capability of the converter.

It is advisable to have a fixed-value high-voltage resistor across the output as shown in the diagram. The resistance of this resistor can be computed from $R = V^2/W$, where V is the required initiating voltage, typically 10 kV, and W is the power output capability of the converter. This resistance would ensure that the output voltage does not rise without limit.

Another configuration depicted in Figure 8.14 makes use of an externally driven flyback converter. It generates an output voltage equal to the required sustaining voltage. The power output capability is

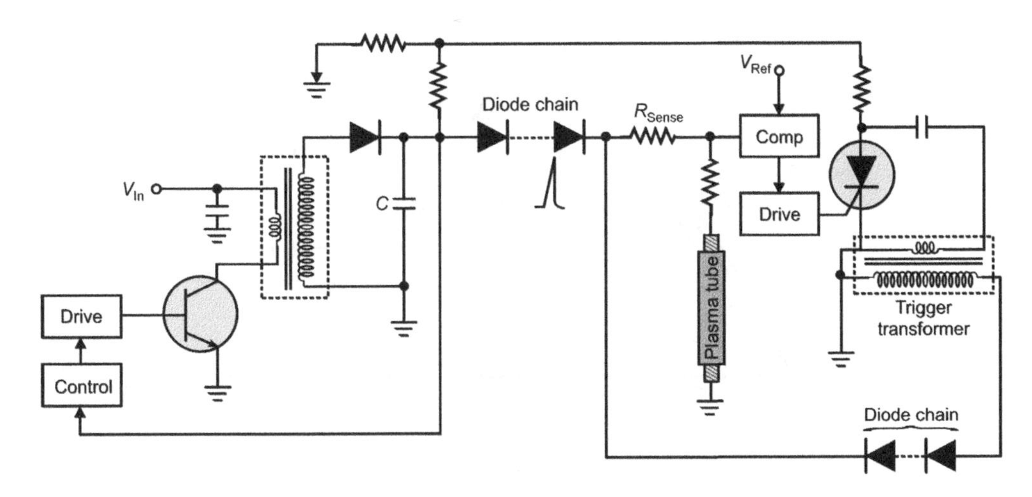

Figure 8.14 Externally driven flyback-converter-based power supply.

equal to the product of the sustaining voltage and the desired plasma current. A voltage feedback loop ensures the right voltage at output. The output of the power supply feeds the plasma tube through a ballast resistance. The ballast resistance here is split into two parts. The voltage present after the first resistance is used to sense the status of the plasma tube, to determine if the plasma has been struck. In the absence of plasma, this voltage equals the power supply output voltage. In the case where the plasma is struck, the voltage falls by an amount equal to the product of the plasma current and the resistance value.

A fraction of this voltage is compared with a reference voltage in a comparator whose output is used to enable or disable an astable multivibrator. The multivibrator circuit in turn triggers an SCR-based capacitive charge/discharge circuit. In the absence of plasma, the SCR circuit generates high-voltage pulses at a repetition rate of typically 20–30 Hz. The trigger pulses are withdrawn when the plasma is struck.

8.4.4 Configurations for Special Applications

The design of a power supply for a laboratory use He-Ne laser is not very critical when it is mainly used as an alignment tool. A slight variation or drift in the output power as a result of either the laser frequency drifting across the laser Doppler-broadened gain versus frequency curve or a drift in the discharge current makes no serious difference in such applications. However, these lasers also have applications in medical diagnostics, spectroscopy and, above all, in the laser of a ring laser gyroscope (Figure 8.15) where power supply design becomes highly critical.

The slow variation in the output power, particularly during the first 10–15 minutes after switch on due to the laser frequency varying across the Doppler gain profile, is not the subject of discussion here. Frequency-stabilized He-Ne lasers that feature an active stabilization module are also available for certain specific applications. The variation in the output power can be checked by designing a current-regulated power supply rather than a constant output voltage. A constant-output-voltage power supply would deliver varying plasma current to the plasma tube with a dynamically varying plasma impedance.

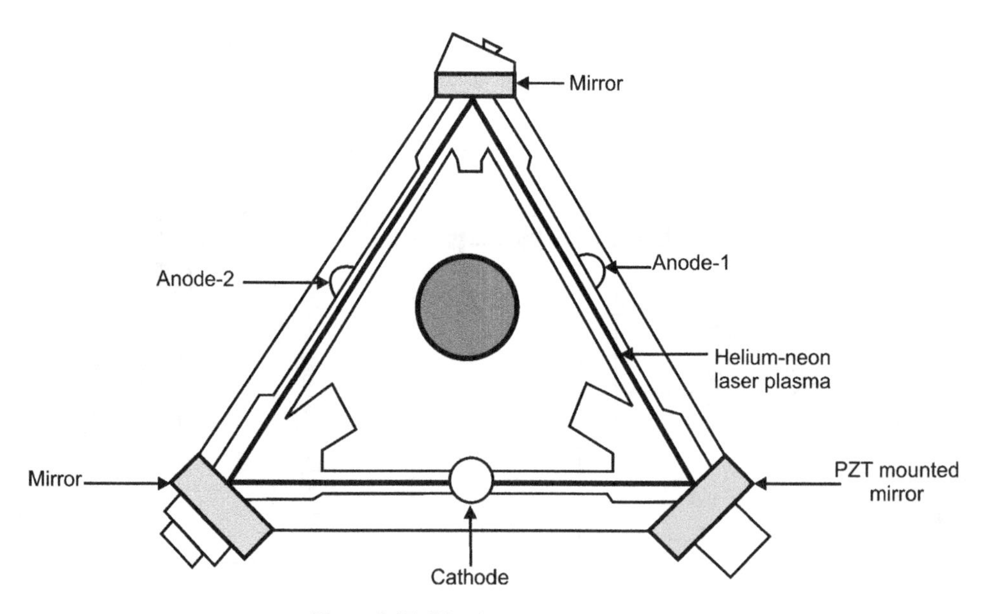

Figure 8.15 Ring laser gyroscope cavity.

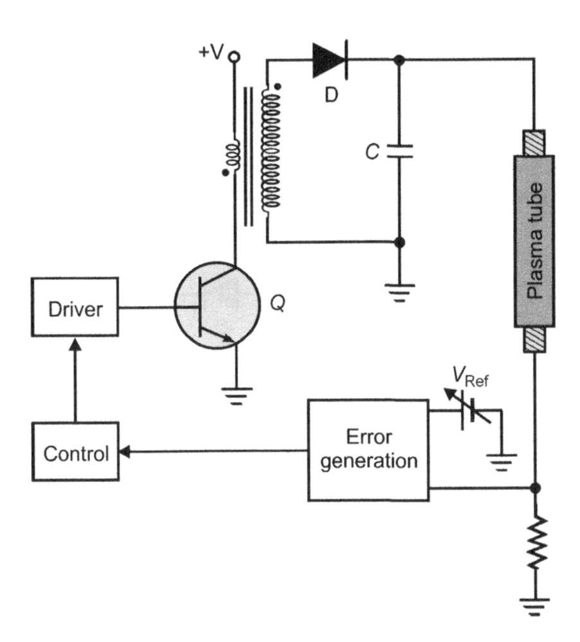

Figure 8.16 Current regulated He-Ne power supply.

The basic arrangement of the power supply where the output voltage would adjust itself in order to maintain a constant current though the plasma tube is shown in Figure 8.16. In this case, it is the plasma current and not a fraction of the output voltage that is compared with a reference to generate an error voltage. The error voltage in turn controls one of the parameters of the drive waveform such as pulse width or frequency to maintain a constant discharge current. The constant value of discharge current can be varied by varying the reference voltage. He-Ne lasers also have a tendency to degrade in output power with the passage of time. This can be checked by using an adjustable reference that determines the constant plasma current delivered by the power supply. This reference can be set for a little higher current in the case where the laser output power has deteriorated. However, current should not be increased indiscriminately as there is a well-defined plasma current range within which the plasma operates most optimally. Such a control should be used only for fine adjustment.

The He-Ne ring laser of a ring laser gyroscope is one application which probably demands the maximum in terms of stability specifications from the plasma discharge power supply. A ring laser gyroscope, which is at the heart of inertial navigation systems used in modern aircraft and missiles, makes use of a ring He-Ne laser in a closed cavity.

A ring laser basically has two counter-propagating helium-neon laser beams, one travelling in the clockwise and the other travelling in the anticlockwise direction. Such a laser has two plasma arms. For proper operation of this device, the two plasmas are operated with an exceptionally high current stability. Not only that, it is also important to maintain a precisely constant difference in discharge current of the two plasma arms (the reason for this is beyond the scope of the text).

Typically, a current stability in each arm of better than 1 μA in a nominal current of 5 mA and current difference stability of better than 100 nA may be achieved from the plasma power supplies in the inertial grade ring laser gyroscopes. The design objective of such a power supply would be to perform the twin jobs of plasma excitation and sustenance for both arms of the ring laser plasma with the required absolute current and current difference stability.

The other obvious objective in such an application is to achieve the highest possible conversion efficiency, small size and electromagnetic compatibility, which are as important as the electrical specifications. The preferred circuit topology for such an application is an externally driven pulse width modulation (PWM) controlled flyback converter generating a regulated output voltage equal to the sustaining voltage at the required plasma current. This is equal to the sum of the nominal plasma currents for the two arms and independent current stabilization loops for the two arms controlled by the same precision reference. Another possible approach is to use a precision reference for one arm and then use the actual controlled current flowing through that arm as the reference for controlling current in the other arm. Figure 8.17 shows one such circuit configuration.

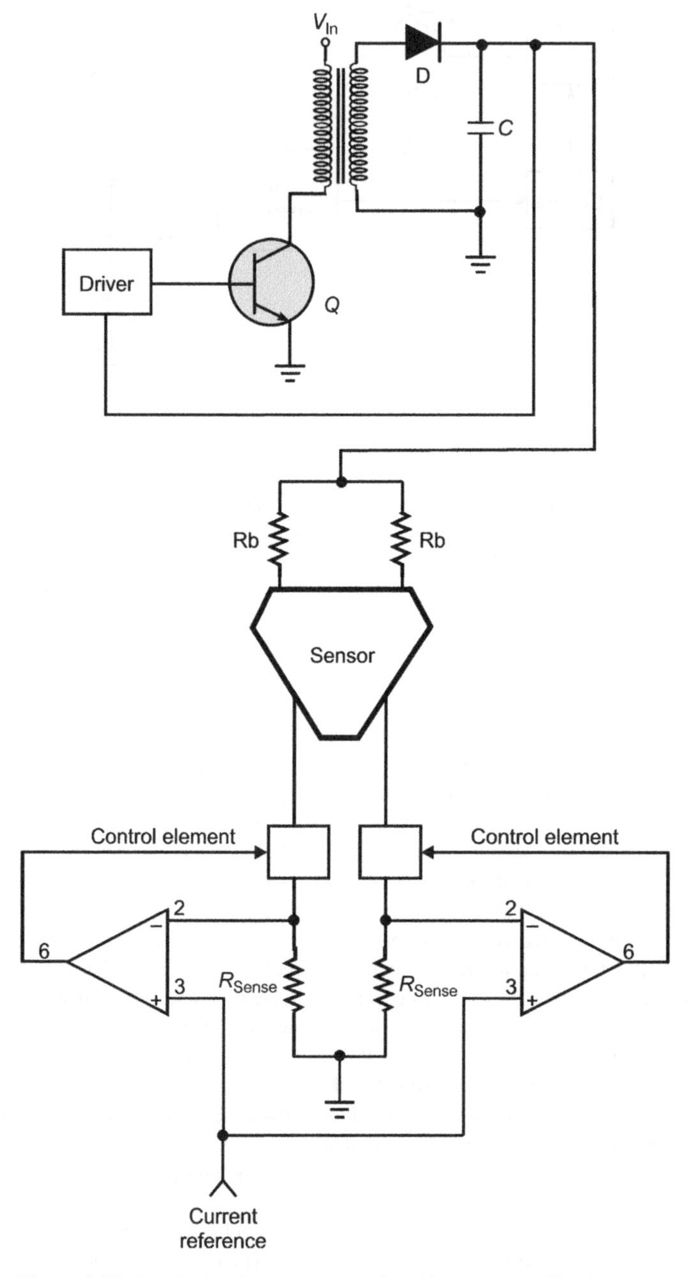

Figure 8.17 He-Ne ring laser power supply with current stabilization loops.

8.4.5 Ballast Resistance

As outlined in Section 8.3, the ballast resistance is a fixed resistance connected from the output of the power supply in series with the plasma tube to ensure that the power supply experiences an overall positive resistance while feeding the plasma tube whose *I–V* characteristics exhibit a negative resistance. Incorrect value of the ballast resistance leads to a highly unstable plasma that has a tendency to extinguish. The choice of ballast resistance is therefore governed by the plasma tube for which the power supply is being designed. The resistance typically varies over the range 50–100 kΩ. Commercial He-Ne lasers with associated power supplies take this aspect into account. Plasma tube manufactures specify the required value of the ballast resistance for the tubes manufactured by them. If designing a power supply for a given plasma tube, it is important to consult the manufacturer's catalogue for this information.

Another important aspect of the ballast resistance is to connect it as close to the plasma tube electrode as possible. Placement of the ballast resistance close to the electrode ensures that the stray capacitance is minimized. A stray parasitic element leads to noisy plasma. If the power supply is to be housed in a separate package, the ballast resistance is located in the plasma tube housing. The resistance should preferably be of carbon film or carbon composition type. If the required wattage specification dictates that a wire wound type is to be used, it should preferably be of the non-inductive variety and the rated wattage of the resistor should preferably be ten times the actual power dissipated in the tube.

8.5 Carbon Dioxide Laser Power Supplies

Carbon dioxide lasers mostly operate as continuous wave lasers with power levels ranging from modest values of a few watts, in the case of laboratory lasers and those used in biomedical applications, to several kilowatts for industrial applications such as cutting, welding and material processing. Carbon dioxide lasers are also used as pulsed lasers, the most common type being the TEA (transversely excited atmospheric pressure) CO_2 laser. The two most commonly used techniques for pumping carbon dioxide lasers are:

1. DC-excited CO_2 lasers
2. RF-excited CO_2 lasers.

8.5.1 DC-excited CW CO_2 Laser

The basic power supply for a DC-excited CW CO_2 laser consists of a high-DC-voltage power supply feeding the laser tube through a ballast resistor. The DC voltage required to produce plasma discharge depends upon the discharge length and fill pressure. The optimum magnitude of electric field is of the order of 30 V cm^{-1} per torr of gas pressure. This implies that an axial flow CO_2 laser with 1 m of discharge length and a gas pressure of 10 torr would need approximately 30 kV to produce plasma discharge. Similarly, a TEA CO_2 laser with discharge tube bore diameter of 1 cm would require about 23 kV ($30 \times 760 = 22\,800$ V). A high-voltage power supply with an output DC voltage rating of several tens of kilovolts and an output current specification of 100 mA meets the excitation requirement of most lasers. Figure 8.18 shows the typical set-up of a CW CO_2 laser excited by a commercial power supply. The ballast resistor limits the discharge current.

8.5.2 DC-excited Pulsed CO_2 Laser

The basic power supply arrangement employed in the case of a pulsed CO_2 laser such as a TEA laser is to charge an energy storage capacitor to a high voltage from a high-voltage (HV) DC power supply and then discharge the same through the gas mixture. The discharge process is controlled by a HV switch. Commonly used HV switches include spark gaps, thyratron tubes and solid-state switching devices. It may be mentioned here that the CO_2 laser's performance is not as badly affected by change in discharge impedance during pulsed discharge as for some of the other types of gas lasers such as ion lasers and metal vapour lasers.

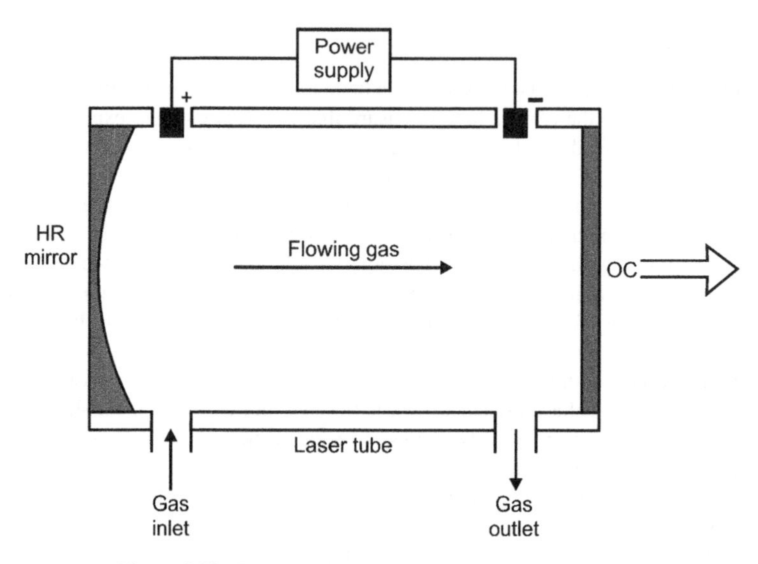

Figure 8.18 DC-excited CW CO_2 laser power supply set-up.

However, the discharge in the case of a TEA CO_2 laser does have a tendency to break down into bright line arcs. One way of solving this problem is to have cathode in the form of a large number of pins (Figure 8.19), each with its ballast resistor in series with it. The ballast resistors limit the current through their respective pins and therefore inhibit any arc formation.

Present-day TEA CO_2 lasers use Marx bank capacitor systems, where the arrangement of individual capacitors is such that they are in parallel during the charging process and in series during the discharge operation. This means if each capacitor in the bank were charged to a voltage V, then the total voltage available across the tube for discharge operation would be nV, n being the number of capacitors. Figure 8.20 shows one such Marx bank supply configuration.

In the arrangement of Figure 8.20, switches are open during the charging operation. They are triggered simultaneously to the closed position to initiate discharge. Although the Marx bank of Figure 8.20 uses four stages, Marx bank systems with a higher number of stages are also used. Marx banks are particularly suitable for producing fast-rising pulses.

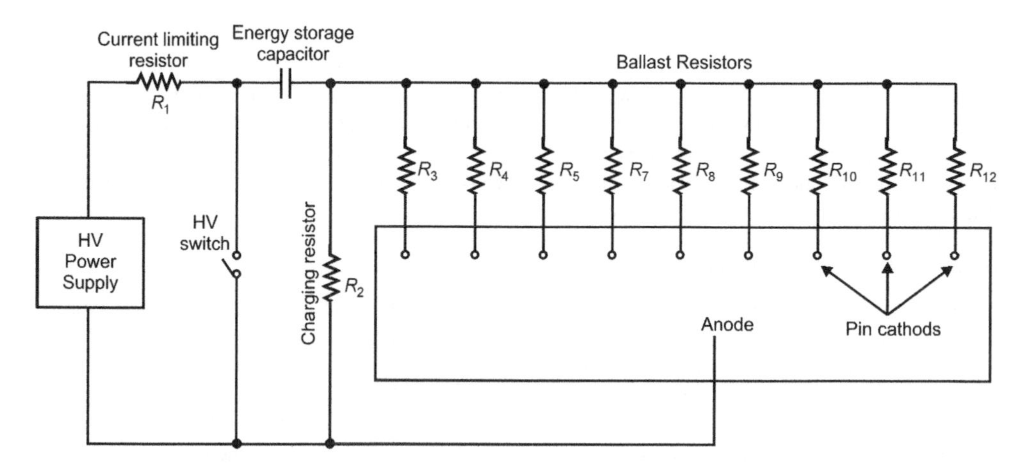

Figure 8.19 TEA CO_2 laser with cathode having large number of pins.

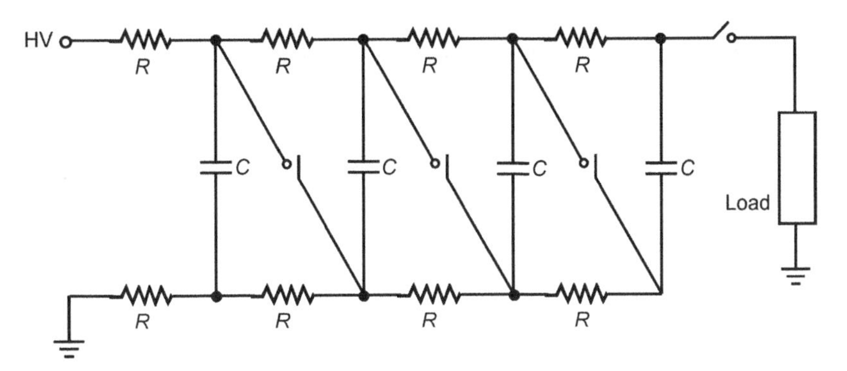

Figure 8.20 Marx bank power supply arrangement.

Modern TEA CO_2 lasers generally use a double-discharge approach. In this case, a low-energy discharge first preionises the gas mixture, which is then followed by main discharge. This method increases the probability of discharge plasma forming throughout the volume of the gas mixture and also improves the stability and uniformity of the discharge.

8.5.3 RF-excited CO_2 Lasers

Transverse RF excitation of CO_2 lasers offers some distinct advantages over DC excitation. These include the following.

1. RF excitation has enabled new laser device architectures due to the ability to control RF discharge over very large areas of all-metal electrodes at high power densities in relatively simple device structures. This has provided compact sealed-off high-average-power lasers.
2. RF discharges can be switched at much higher frequencies than DC discharges. This provides greater flexibility to the user in terms of temporal formats, which is a crucial requirement for many material processing applications.
3. It allows all-metal construction, which makes cooling more effective.
4. High electrode voltages are avoided, extending the life of the laser.

The typical excitation source comprises an RF source and an impedance-matching network. The RF source may be split into an RF oscillator and a cascade arrangement of RF amplifiers, depending upon the output power delivering capability. Frequency of operation is in the range of 50–150 MHz. RF power is fed to the laser cavity through an impedance-matching network. The laser cavity's mechanical structure combined with RF discharge can be considered as an electrical load comprising complex impedance. Power transfer efficiency from the RF source to the discharge is a key operational feature of the laser that is made more complicated by the fact that the discharge impedance is power dependent. Figure 8.21 depicts a typical RF-excited CO_2 laser.

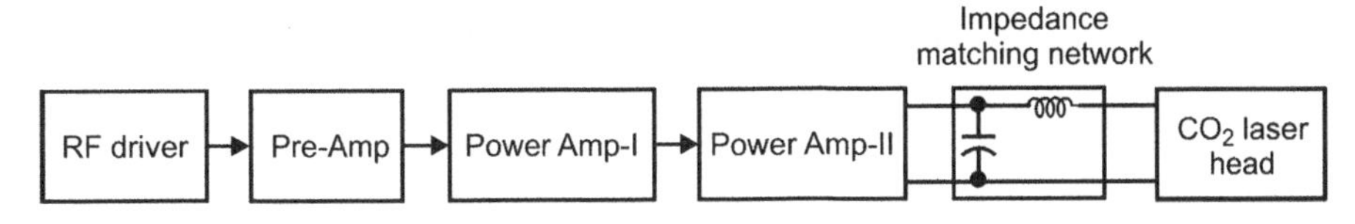

Figure 8.21 RF-excited CO_2 laser.

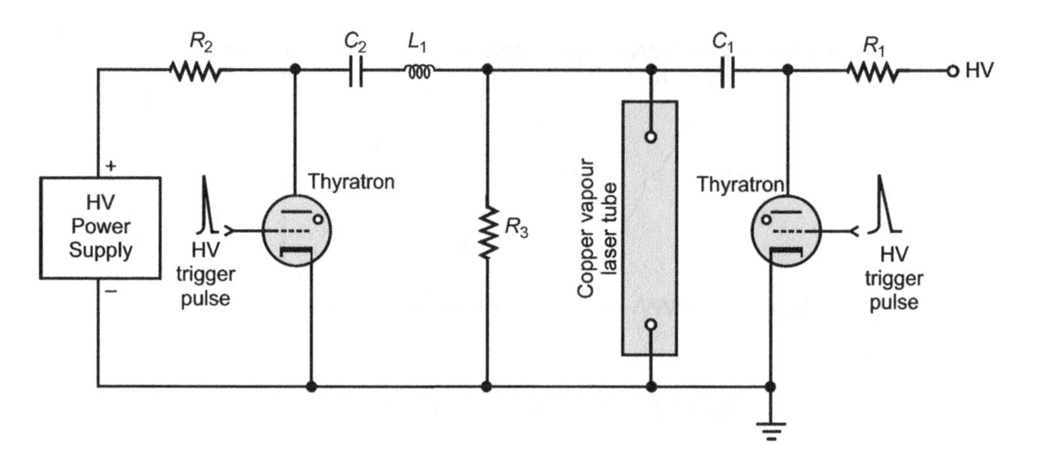

Figure 8.22 Basic power supply for pulsed metal vapour laser.

8.6 Power Supplies for Metal Vapour Lasers

Metal vapour lasers are categorized in two broad classes: the helium-cadmium family of lasers where the ionized cadmium atoms constitute the active species, and the second class where neutral atoms constitute the lasing species. Copper vapour and gold vapour lasers are the predominant candidates of the second category of metal vapour lasers. While the former type operates as CW lasers, the latter are essentially pulsed lasers.

The basic power supply arrangement for CW metal vapour lasers such as helium-cadmium lasers is similar to that discussed in the case of helium-neon lasers. In the case of pulsed metal vapour lasers such as copper vapour and gold vapour lasers, the basic power supply would be similar to that discussed in the case of TEA CO_2 lasers. Figure 8.22 shows the basic power supply for a pulsed metal vapour laser. The energy storage capacitor is charged through the inductor and then discharged through the laser tube by triggering the thyratron switch to the closed state.

Figure 8.23 shows the preferred power supply configuration for pulsed metal vapour lasers. In this case, the energy storage capacitor is charged to a relatively much lower voltage than is required across the discharge tube. The capacitor is made to discharge through the primary of the step-up pulse transformer by triggering the switch connected in series with the primary of the pulse transformer to the closed state. The discharge circuit is a multi-mesh pulse-forming network comprising saturable inductors. The pulse-forming network acts like a magnetic pulse-compression circuit. It shortens the

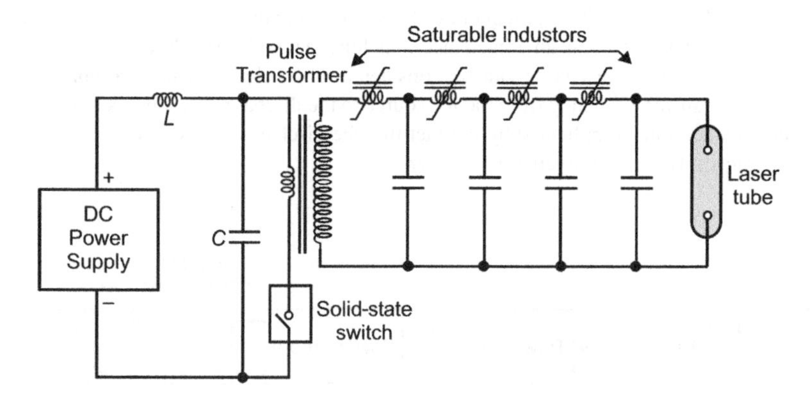

Figure 8.23 Preferred power supply configuration for pulsed metal vapour lasers.

discharge pulse duration from a few microseconds to about 100–150 ns. This configuration leads to improved conversion efficiency, as the reduced pulse duration is much better matched to the requirements of the discharge. Also, use of a pulse transformer allows the capacitor to be charged to relatively lower voltage, allowing the use of solid-state switching devices. Due to the reduced quantum of energy being stored in the capacitor, the lifetime of the switch also increases.

8.7 Power Supplies for Excimer Lasers

Excimer lasers are pumped either by pulsed electric discharge or by high-energy electron beams. Those excited by high-energy electron beams are capable of producing very high pulse energy of the order of tens of kilojoules. Such giant lasers find applications in areas such as laser-assisted thermonuclear fusion and defence research. Detailed discussion on this form of excitation is beyond the scope of the present text. On the other hand, electrically excited excimer lasers are relatively much smaller and far less expensive. These find use in a large number of industrial applications such as semiconductor fabrication, material processing and photochemistry.

Excimer lasers are excited by short-duration high-peak-current discharge pulses with pulse rise time of the order of tens of nanoseconds and a peak value of discharge current pulse in the range of a few kiloamps. Excitation of the active medium is usually a two-stage process. Energy is stored in a capacitor in a relatively slow process in one part of the circuit and then transferred to a peaking capacitor that enables a much faster discharge in another part of the circuit. Figure 8.24 depicts one such power supply circuit.

In the circuit of Figure 8.24, C is the charging capacitor. Stored energy is transferred to the peaking capacitor C_p across the pre-ionization spark gap. The basic circuit is adequate for excimer lasers with moderate values of pulse energy and repetition rate for the following reasons. When the discharge is initiated, voltage across the tube rises very rapidly. As it reaches the breakdown voltage, current begins to flow. Increasing current leads to a rapid fall in voltage due to the negative resistance characteristics of discharge. As a result, electrons in the discharge plasma cannot be accelerated to the velocities required to excite the lasing species. This forces the designer to store and switch more than the required energy, leading to a reduction in the efficiency and lifetime of the thyratron switch.

Commercial excimer laser power supplies employ a modified circuit where the charging part of the circuit is decoupled from the fast discharge part of the circuit. One such circuit, where the decoupling between charge and discharge parts of the circuit is provided by a saturable inductor, is shown in Figure 8.25.

Another circuit using a peaking capacitor that is internal to the laser enclosure close to the main electrodes is shown in Figure 8.26. When the peaking capacitor is located internal to the laser housing, it not only minimizes the number of feedthroughs into the enclosure but also allows the designer to have a discharge circuit with low inductance producing faster rise time discharge pulses.

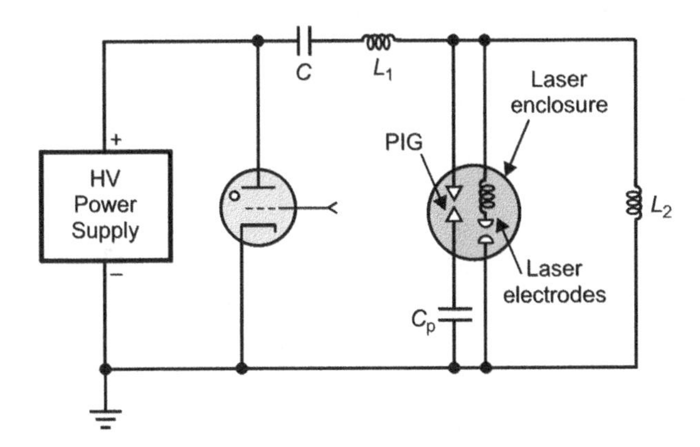

Figure 8.24 Basic power supply for excimer laser.

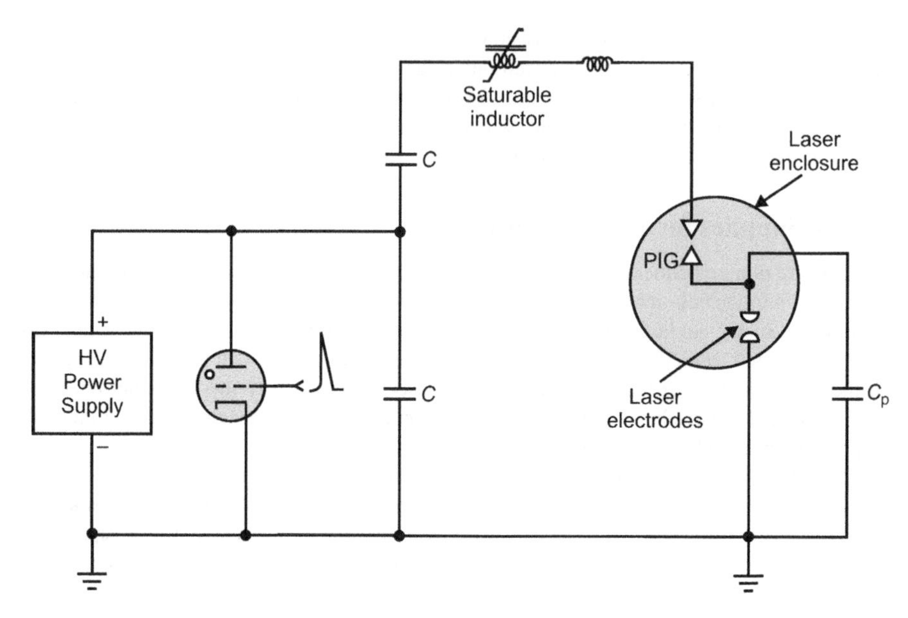

Figure 8.25 Modified power supply circuit for excimer laser.

8.8 Power Supplies for Ion Lasers

Gas lasers discussed so far in this chapter including helium-neon, carbon dioxide, excimer and metal vapour lasers which operate at high voltage and relatively lower current. In contrast, noble gas ion lasers operate at relatively lower voltages of the order of 200 V and much higher current values in the range 10–100 A. This is evident from the current–voltage characteristics of a typical gas discharge discussed earlier and shown in Figure 8.2. The excitation mechanism in ion lasers is electron impact in the gas in a DC discharge with high current density in the range of several hundreds of amps per squared centimetre. In some cases such as the argon ion laser, a high value of current density is maintained by using a magnetic field that confines the discharge. Operation at high current density means that there a suitable arrangement to transfer heat out of the tube is necessary and that the tube is fabricated from materials which can withstand such hostile operating conditions. Another requirement is that of maintaining uniform gas pressure in the tube, which is achieved by using a pressure sensor and a gas fill regulator. In addition, the power supply circuit needs to have the feedback control loops to operate the laser either in the constant-output-power or in the constant-discharge-current mode.

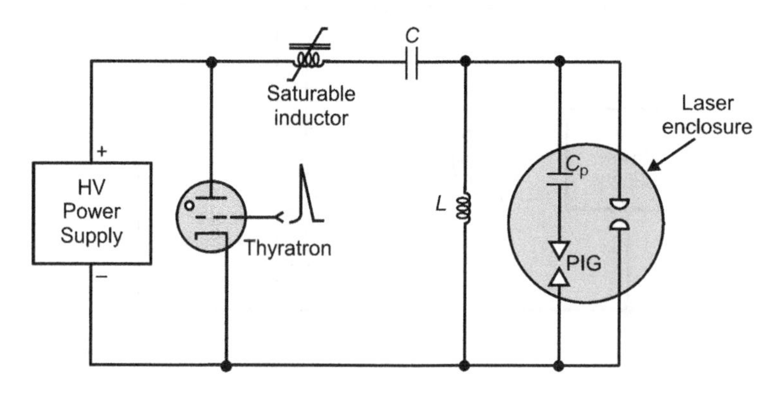

Figure 8.26 Modified power supply circuit for excimer laser with peaking capacitor internal to laser enclosure.

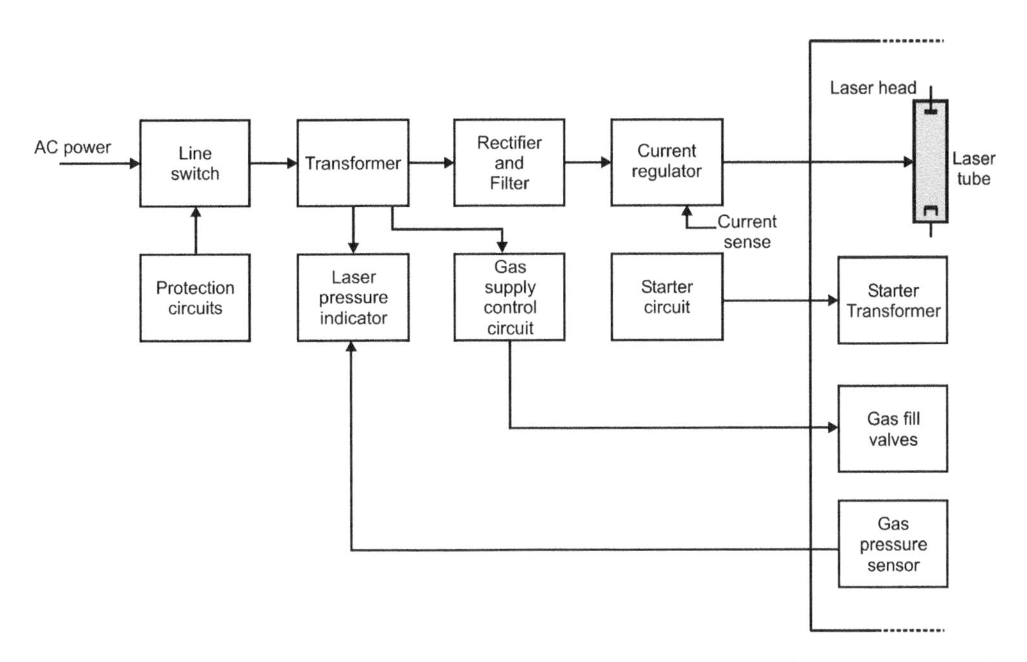

Figure 8.27 Generalized block schematic of ion laser power supply.

It may be mentioned here that it is the current regulator rather than voltage regulation that is implemented while designing the power supply for argon/krypton ion lasers. It is because of this reason that the incremental resistance of the discharge is very low of the order of a few ohms. This implies that a small change in voltage across the discharge will result in a relatively large change in the discharge current. Discharge current regulation therefore makes more sense.

Figure 8.27 shows the generalized block schematic arrangement of a typical ion laser power supply along with auxiliary circuits. Depending upon the output power level, discharge current in the case of ion lasers could be of the order of several tens of amps. The block schematic shown is that of a conventional current-regulated AC/DC power supply. While a DC voltage of the order of 200 V is needed to sustain a steady discharge in an ion laser, it needs a higher voltage of the order of a few kilovolts to initiate the discharge. This is provided by the starter circuit with the high-voltage trigger transformer located inside the laser head. The line switch is simply a kind of circuit breaker operated by protection circuits such as over current limit and over temperature cut-off.

In the case of modern ion lasers, the linearly current-regulated conventional AC/DC supply of Figure 8.27 may be replaced by a switched-mode power supply based on buck-boost topology. The inverted polarity available from the buck-boost topology allows easy connection to the cathode terminal. The anode terminal is generally used for applying the starting voltage. The ion laser power supplies also have auxiliary circuits for control of gas-fill valves and monitoring of gas pressure.

8.9 Frequency Stabilization of Gas Lasers

In this section we present a comparative study of the techniques available for frequency stabilization of carbon dioxide lasers. Different techniques commonly used for frequency stabilization of carbon dioxide lasers include:

1. dither stabilization
2. stark-cell stabilization

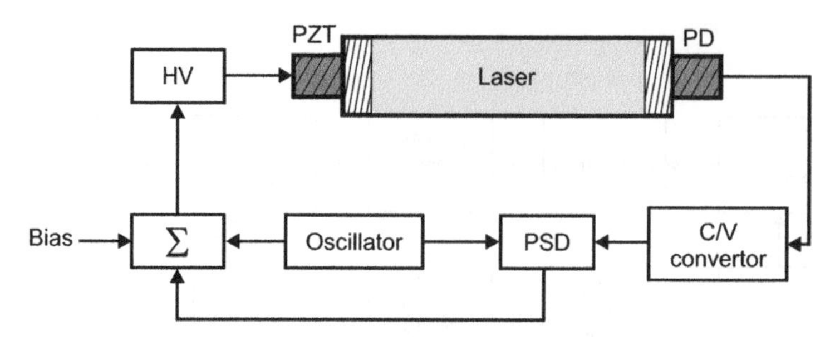

Figure 8.28 Dither frequency stabilization.

3. optogalvanic stabilization
4. stabilization by conventional saturation absorption dip
5. stabilization by stark saturation absorption cell.

8.9.1 Dither Stabilization

Figure 8.28 shows the typical experimental arrangement for a dither stabilization scheme. The grating allows selection of various lines within the laser tuning curve. The output of the low-frequency oscillator when applied to the lead zirconate titanate (PZT) moves the grating along the axis and hence varies the output frequency. This is called *dithering*. As a result, the laser output is detected and fed to a synchronous detector. The other input to the synchronous detector is the dithering signal that acts as the reference. The output of the synchronous detector is the error signal, which corresponds to the deviation in the laser output frequency from the line centre. In fact, the magnitude and phase of the error signal determine the location of operational point with respect to the centre of the Doppler-broadened gain curve. While the amplitude determines how far or close to the line centre the operational point is, the phase determines which side of the line centre it is located. As is evident from Figure 8.29, if the

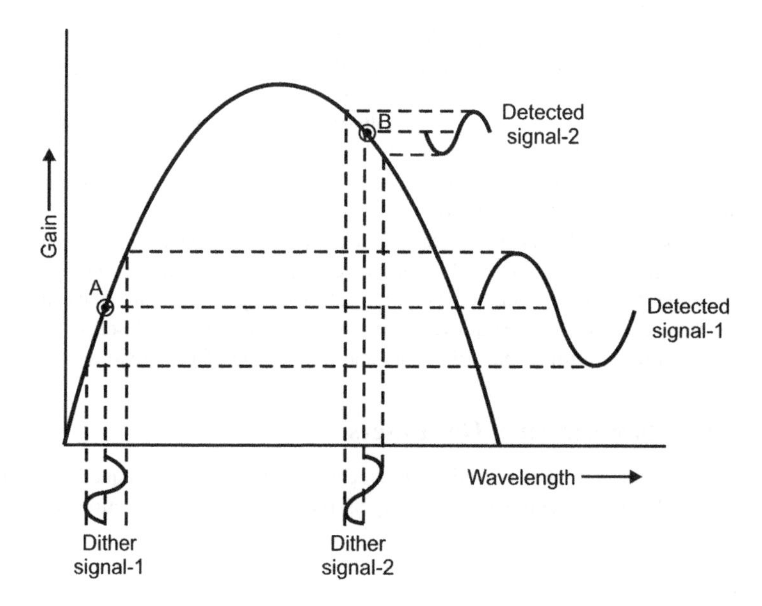

Figure 8.29 Principle of operation of dither stabilization.

instantaneous frequency or wavelength of the laser corresponds to point A located on the positive slope of the gain curve, the output of the photodiode corresponding to an applied sinusoidal dither signal (Dither signal 1) has the same phase. On the other hand, if the instantaneous frequency or wavelength of the laser corresponds to point B on the negative slope of the gain curve, the photodiode signal is 180° out of phase with the applied dither signal (Dither signal 2).

The amplitude of the photodiode signals in the two cases depends upon the slope of the gain curve at the point of interest. The closer the point is to the line centre, the smaller the amplitude of photodiode signal. The biggest disadvantage of this technique is the presence of frequency modulation in the output as a result of laser dithering.

There are however other techniques that overcome this shortcoming, where it is not the laser but an external gas-cell that is dithered to generate the error signal. As a result, the frequency modulation is absent from the output. These other techniques are described in the following sections.

8.9.2 Stark-cell Stabilization

Figure 8.30 shows the typical block schematic arrangement for stark-cell stabilization. A part of the output laser beam is made to pass through the stark cell. An appropriate value of DC voltage is applied to one of the stark plates to make the stark-shifted molecular transition of the gas contained in the cell precisely coincide with the CO_2 laser output line of interest. A slowly varying linear voltage ramp may be applied instead. An audio-frequency waveform (typically 5 kHz), either sinusoidal or square, is superimposed on this. The other stark plate is grounded. The beam at the output of the stark cell is thus modulated in both amplitude and frequency by the applied high-frequency signal (5 kHz or so) around the line of interest. The beam is detected and fed to a lock-in amplifier for phase-sensitive detection of the signal. This synchronous detection produces a frequency-discriminant error signal. The error signal is amplified and then fed back to the PZT through a suitable interface. One of the end-elements of the cavity is mounted on the PZT. The error signal adjusts the length of the cavity in the feedback mode to lock the laser frequency to the centre of the Doppler profile of the line of interest.

The biggest advantage of this technique stems from the fact that it is not the laser but the stark cell that has been dithered for generating the frequency-discriminant error signal. As a consequence, the troublesome frequency modulation present in the frequency-stabilized output in the case of the former is absent from the stark-cell stabilization scheme.

8.9.3 Optogalvanic Stabilization

The carbon dioxide laser is a high-efficiency device and the intensity of laser radiation field is often sufficiently large to cause significant variation in many of the laser and discharge macroscopic

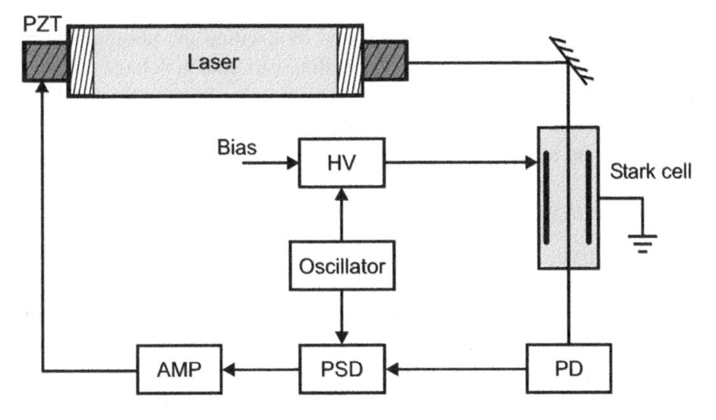

Figure 8.30 Stark-cell stabilization.

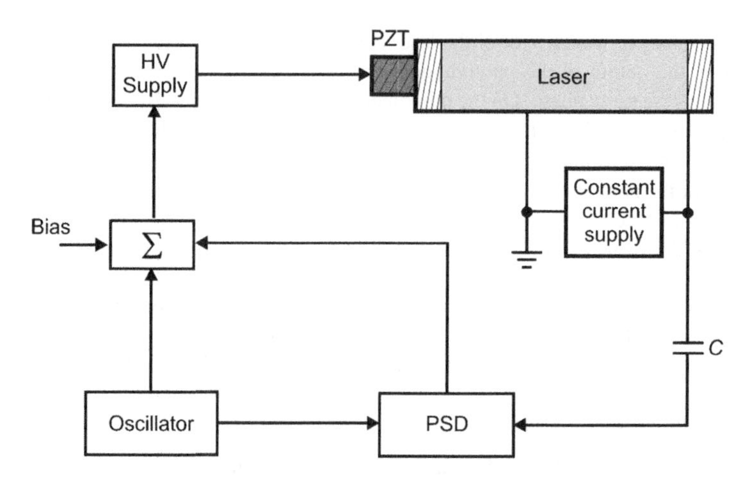

Figure 8.31 Optogalvanic stabilization.

parameters when there is a change in the field intensity. Parameters such as the spontaneous side light intensity and its spectral distribution, gas pressure, discharge current and discharge impedance all change as the internal laser radiation field intensity is altered, for instance by using an intra-cavity modulator or by changing the resonant cavity alignment.

In a typical set-up, the laser cavity length is modulated at some convenient frequency by oscillating the position of one mirror along the cavity axis using a piezoelectric mirror mount. The field intensity is therefore modulated at the same frequency as the cavity mode scans the transition line gain profile. An intensity variation of about 10% is sufficient to change the discharge impedance by about 0.1%. This impedance change can be detected as a current fluctuation (in which case the effect is called opto-galvanic) or as a voltage variation (opto-voltaic). If the mean frequency was lower than the line-centre frequency, the phase of impedance fluctuation is opposite to that if the mean frequency was higher than the line-centre frequency. The amplitude of impedance fluctuation increases with the increasing frequency offset from the line centre. If the impedance fluctuation is measured with a phase-sensitive detection system and its integrated output suitably amplified and applied to the cavity length transducer as a DC correction signal, the system will maintain the oscillating cavity mode at the line centre.

Figure 8.31 depicts a typical opto-galvanic frequency stabilization set-up. An AC signal is applied to the cavity length transducer, resulting in frequency modulation of the laser and therefore modulation of the plasma-tube impedance. The impedance variation which is proportional to the slope of the output power versus frequency curve of the laser is measured by exciting the plasma-tube by a high-speed current-regulated power supply and measuring the resulting variation in voltage drop across the plasma tube. This voltage is applied to the synchronous detector. The other input to the synchronous detector is the reference signal used to modulate the laser cavity length with the help of PZT. The error signal produced at the output of the synchronous detector is used to stabilize the laser frequency.

8.9.4 Stabilization using Saturation Absorption Dip

From the viewpoint of electronics set-up required for frequency stabilization of Doppler-broadened gas lasers, this approach is no different from that of the extra-cavity stark cell described in Section 8.9.2. In the present approach however, the absorption cell is placed in such a way that a saturation absorption dip (i.e. inverted lamb dip) is obtained on the Doppler-broadened gain versus frequency gain profile. The laser is then locked onto the centre of this inverted lamb dip. In the case of stark-cell stabilization, the laser is locked to the centre of the Doppler profile, taking advantage of the extreme narrowness of saturation absorption dip; an extremely high degree of frequency stabilization can then be achieved.

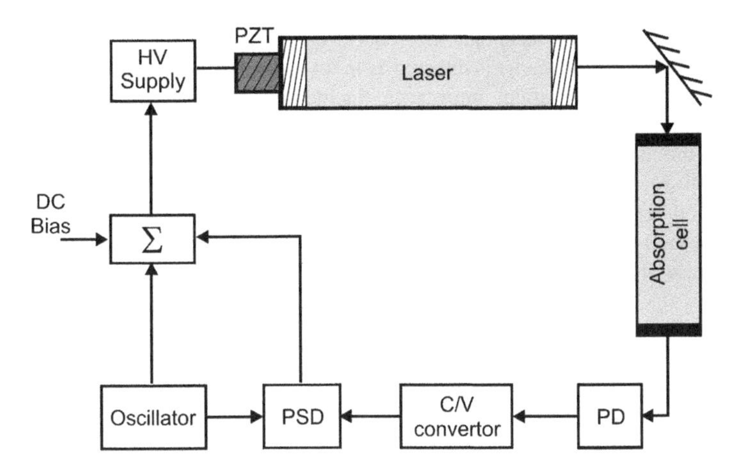

Figure 8.32 Frequency stabilization using saturation absorption dip.

Figure 8.32 shows the block schematic used for frequency stabilization of a carbon dioxide laser using an extra-cavity absorption cell. The arrangement is self-explanatory.

8.10 Summary

- Most gas lasers are driven by electrical discharge through the laser gas mixture. The discharge is struck either longitudinally by passing electrical current along the length of the discharge tube (longitudinal excitation) or perpendicular to the length of the tube by passing electric current transverse to the length of the tube (transverse excitation).
- The negative discharge impedance offered by the plasma is given by the reciprocal of the slope of the tangent drawn at the operating point in the *I*–*V* characteristics.
- Pre-breakdown current is the discharge current that flows before breakdown voltage is reached. It increases slowly with an increase in applied voltage until breakdown point is reached.
- At breakdown voltage, a large number of molecules in the gas mixture become ionized. The conductivity of the gas mixture increases. Electrons are accelerated to velocities where they transfer energy to more molecules through collisions. Increase in discharge current causes a further reduction in discharge resistance with the result that the voltage required to sustain the discharge actually decreases with increase in current.
- Ballast resistance is a fixed resistor connected from the output of the power supply in series with the plasma tube to ensure that the power supply experiences an overall positive resistance while feeding the plasma tube whose *I*–*V* characteristics exhibit a negative resistance. The resistance typically varies between 50 kΩ and 100 kΩ.
- Buck-boost topology produces an output voltage that can either be of higher magnitude (in the case of boost topology) or lower magnitude (in the case of buck topology) than the input voltage. The polarity of the output voltage is inverted from that of the input.
- TEA CO_2 laser is a carbon dioxide laser where the gas is filled at atmospheric pressure and the discharge is transversely excited.
- Dither stabilization is a technique for frequency stabilization of Doppler-broadened gas lasers. In this technique, one of the mirrors of the laser cavity is mounted on a PZT, which allows it to be dithered (or moved along the axis). The frequency modulation and consequent amplitude modulation of the output signal is used to generate an error discriminant. The error discriminant in the feedback loop is used to stabilize the laser at the line centre. The disadvantage of the technique is the presence of incidental frequency modulation (FM), even in the frequency-stabilized output.

- Optogalvanic stabilization is a technique for frequency stabilization of gas lasers. In this technique, one of the mirrors of the laser cavity is dithered as in the case of dither stabilization. However, in the case of optogalvanic stabilization, impedance fluctuations resulting from cavity dithering are measured by using a constant current source. The resultant voltage signal acts as a discriminant that is used in the feedback loop to stabilize the laser.
- Stark-cell stabilization is a technique for frequency stabilization of Doppler-broadened gas lasers. It is not the laser but the stark cell that is dithered to generate the discriminant. The troublesome frequency modulation present in the frequency-stabilized output in the case of Dither stabilization is therefore absent in the case of a stark-cell stabilization scheme.

Review Questions

8.1. With the help of suitable current-voltage characteristics relevant to electrical discharge in gas lasers, briefly describe the current and voltage requirements of the part of the power supply for helium-neon and ion lasers.

8.2. What problems does the negative discharge impedance pose to the power supply designers of gas laser power supplies? How are these problems usually overcome?

8.3. With the help of a suitable circuit diagram, briefly describe the operation of a helium-neon power supply constructed with a push-pull type of inverter followed by a cascade type of voltage multiplier chain.

8.4. Briefly describe the role of ballast resistor in a gas laser power supply. How is a suitable value for the ballast resistor chosen?

8.5. What is a Marx bank capacitor arrangement? Why is such an arrangement preferred in pulsed CO_2 laser power supplies?

8.6. How do the design requirements of an ion laser power supply differ from those of helium-neon or carbon dioxide gas laser supplies? With the help of block schematic arrangement, briefly describe the different building blocks of a typical argon/krypton laser power supply.

8.7. Why do we need to stabilize the frequency of gas lasers? What techniques are available for stabilizing the frequency of the laser to the centre of the Doppler-broadened gain curve? Briefly describe the principle of operation of each of them.

8.8. How is the problem of incidental FM present in the stabilized output in the case of dither frequency stabilization technique overcome in the case of Stark cell stabilization technique?

Problems

8.1. In a certain He-Ne laser plasma tube, 1 mA change in plasma discharge current causes a 150 V change in voltage across the tube. Determine the minimum value of the required ballast resistance for stable discharge. Also determine the minimum value of DC voltage at the output of the power supply to sustain stable discharge at an operating current of 4 mA, assuming that the voltage across the plasma tube at the operating current of 4 mA is 1000 V. If the DC power supply output voltage was 2000 V, what would be the value of ballast resistance needed to operate the plasma tube at the same operating current?
[150 kΩ, 1600 V, 250 kΩ]

8.2. Refer to the current-voltage characteristics of the electrical discharge in a helium-neon plasma as shown in Figure 8.33. Determine: (a) the value of ballast resistance; (b) the drop across ballast resistance; and (c) the minimum DC power supply voltage required to sustain the plasma discharge current of 8 mA
[(a) 100 kΩ, (b) 800 V (c) 2400 V]

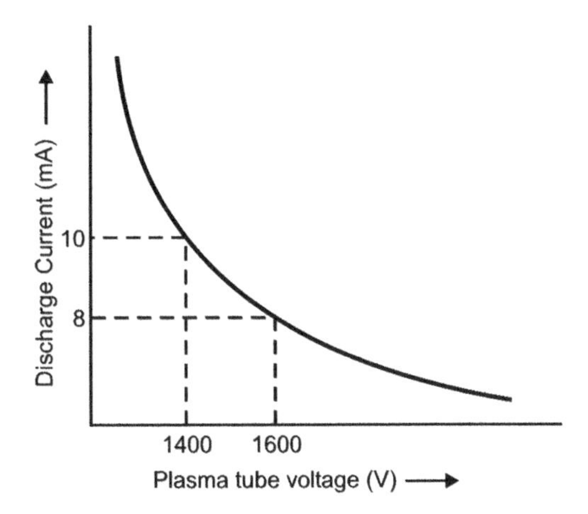

Figure 8.33 Problem 8.2.

Self-evaluation Exercise

Multiple-choice Questions

8.1. Incremental resistance of the gas discharge plasma can be determined by
  a. measuring discharge current and corresponding discharge voltage
  b. measuring change in discharge current and the corresponding change in discharge voltage
  c. measuring change in voltage drop across the ballast resistance for a known change in power supply voltage
  d. none of these.

8.2. When the operating current in a helium-neon laser plasma tube increases by 0.5 mA, it increases the voltage drop across the ballast resistance by 50 V. The incremental negative resistance of the plasma is
  a. 100 kΩ
  b. 200 kΩ
  c. 50 kΩ
  d. Indeterminate.

8.3. The incremental discharge resistance in the case of ion lasers is typically
  a. a few ohms
  b. tens of kilo-ohms
  c. hundreds of kilo-ohms
  d. none of these.

8.4. The purpose of connecting ballast resistance in series with the plasma tube in the case of DC-excited gas lasers is
  a. to limit the discharge current to the desired value
  b. to ensure a stable plasma discharge
  c. both (a) and (b)
  d. none of these.

8.5. The objective is to design a regulated DC power supply for an argon ion laser. The DC voltage is applied to the cathode terminal of the plasma tube while the starting voltage is applied to the anode terminal. Which of the following is the preferred topology for designing the DC power supply?
  a. boost topology
  b. buck topology

c. buck-boost topology
d. flyback-converter topology.

8.6. Non-isolating-type switching regulator topology similar to the flyback-converter topology is
a. buck-boost topology
b. buck topology
c. boost topology
d. none of these.

8.7. In which of the following gas laser power supplies would current rather than voltage regulation make more sense?
a. metal vapour laser power supplies
b. ion laser power supplies
c. helium-neon laser power supplies
d. carbon dioxide laser power supplies.

8.8. A Marx capacitor bank is used in power supplies for
a. DC-excited CW CO_2 lasers
b. DC-excited pulsed TEA CO_2 lasers
c. RF-excited CO_2 lasers
d. He-Ne lasers.

8.9. Which of the following frequency stabilization techniques does not suffer from the presence of incidental frequency modulation in the stabilized output?
a. dither frequency stabilization
b. stark cell frequency stabilization
c. optogalvanic frequency stabilization
d. optovoltaic frequency stabilization.

8.10. Which of the following techniques is capable of providing the highest level of frequency stability?
a. dither frequency stabilization
b. stabilization on the centre of lamb dip on the Doppler-broadened gain curve
c. stark cell frequency stabilization
d. stabilization on inverted lamb dip on the Doppler-broadened gain curve.

Answers

1. (b) 2. (a) 3. (a) 4. (c) 5. (c) 6. (a) 7. (b) 8. (b) 9. (b) 10. (d)

Bibliography

1. *Fundamentals of Light Sources and Lasers*, 2004 by Mark Csele, John Wiley & Sons.
2. *Handbook Of Laser Technology And Applications*, 2004 Volume-II by Collin E. Webb and Julian D. C. Jones, Institute of Physics Publishing.
3. *Switching Power Supply Design*, 1998 by A.I. Pressman, McGraw-Hill.
4. *Power Supply Cookbook*, 1990 by Marty Brown, Academic Press.

9

Laser Diode Electronics

9.1 Introduction

Laser diode electronics primarily comprises a source of drive current and a temperature controller. The wavelength of a semiconductor diode laser is strongly dependent on drive current and operating temperature. The stability of the drive current and operating temperature therefore dictates the output wavelength stability. The large significance of current and temperature control is such that many international companies offer a laser diode driver and temperature controller in a single package.

Not all laser diode applications have the stringent requirements of ultra-stable wavelength, necessitating a high degree of current and temperature stabilization. Applications such as data storage and retrieval, laser printing and laser pointing only require the conventional advantages of lasers. In the case laser diodes to be used for spectroscopy, biomedical diagnostics, pollution monitoring and characterization of materials however, a tight control on both the drive current and the diode temperature is essential.

In addition, laser diodes are highly prone to damage due to excessive drive current and transients. A laser diode drive circuit should not only provide a stabilized drive current, but it should also have inbuilt protection features. In this chapter we discuss the common mechanisms of laser damage and the precautions to be observed to protect them. This is followed by a discussion on different topologies commonly used for the design of laser diode drive circuits and temperature controllers to meet the requirements of different applications.

9.2 Laser Diode Protection

Laser diodes exhibit excellent reliability under ideal operating conditions, with lifetimes often exceeding 100 000 hours. However, they are highly susceptible to damage due to excessive drive current, electrostatic discharge (ESD) and transients. Laser diode damage manifests itself in the form of: reduced output power; a shift in threshold current; an increase in beam divergence; and failure to laser action, thus producing an LED-like output only. A laser diode protection strategy needs to be multi-pronged with desired focus on laser diode instrumentation, system set-up, power line conditioning and laser diode handling. Laser diode instrumentation includes the current source used to drive the laser diode and temperature controller. The relevant damage mechanisms are over-current, overheating, current spikes and power line surges. System set-up considerations include cables, proper grounding and shielding; improper set-up leads to radiated electrical transients. Improper power line conditioning also leads to severe fast transients. Electrostatic discharge caused by improper handling during storage, transport and mounting is considered as the single leading cause of premature laser diode failure. Note that typical laser diodes have rise times in the range of tens of picoseconds, and are therefore highly sensitive to fast events. The four areas of laser diode protection strategy are briefly described in the following sections.

Lasers and Optoelectronics: Fundamentals, Devices and Applications, First Edition. Anil K. Maini.

9.2.1 Laser Diode Drive and Control

Laser diode instrumentation includes the *current source* used to drive the laser diode and the *temperature controller* used to maintain a constant operating temperature regardless of the ambient conditions. The current source should be designed in order to provide an independent adjustable limit of the drive current, guard against the diode junction becoming reverse biased and suppress both power line and current transients. A temperature-control circuit should not allow the current driving thermoelectric module used to control the temperature to exceed its maximum rated value. The temperature-control feedback loop should guard against any thermal oscillations. Note that many important laser diode operating parameters such as operating wavelength, threshold current, electrical input to laser output efficiency and lifetime are strongly dependent upon operating temperature and temperature stability.

Protection features desirable for a laser diode current source include slow start, current limit, over-voltage protection and power line transient suppression. Commercial laser diode drivers maintain a short circuit across the output leads to which the laser diode is to be connected by using a shorting relay or a field effect transistor (FET) switch. The shorting is maintained until the output is turned on and also when power to the instrument is off. Shorted output provides protection against ESD damage. Figure 9.1 shows the basic topology of a laser diode current source with protection features.

Slow start protects the laser diode from turn-on transients by forcing the drive current through the laser diode to rise slowly to the final operating value. The start-up time is typically longer than 100 milliseconds. Start-up time ensures that the control circuits are fully active and circuit transients have died out before the drive current reaches its operating value.

The current source should have an inbuilt *adjustable current limit* that is effective irrespective of whether it is operated in DC, pulsed or constant-power-output modes. The current limit should preferably be independent which when set overrides any other condition that can lead to the drive current exceeding the maximum limit of the device. Excessive drive current leads to what is called catastrophic facet damage (CFD). Damage to the diode at the output facet usually occurs when the optical energy density exceeds 10 000 W mm^{-2}. In a typical laser diode, the area of emitting surface for a given output laser power is such that the power density at the output facet is of the order of 500–700 W mm^{-2}. Although the material of the active region is transparent in the pumped part of the active region, due to non-radiative combination it becomes absorbing at the facet surface. Temperature rises faster than normal and this temperature increase produces further increase in absorption at the

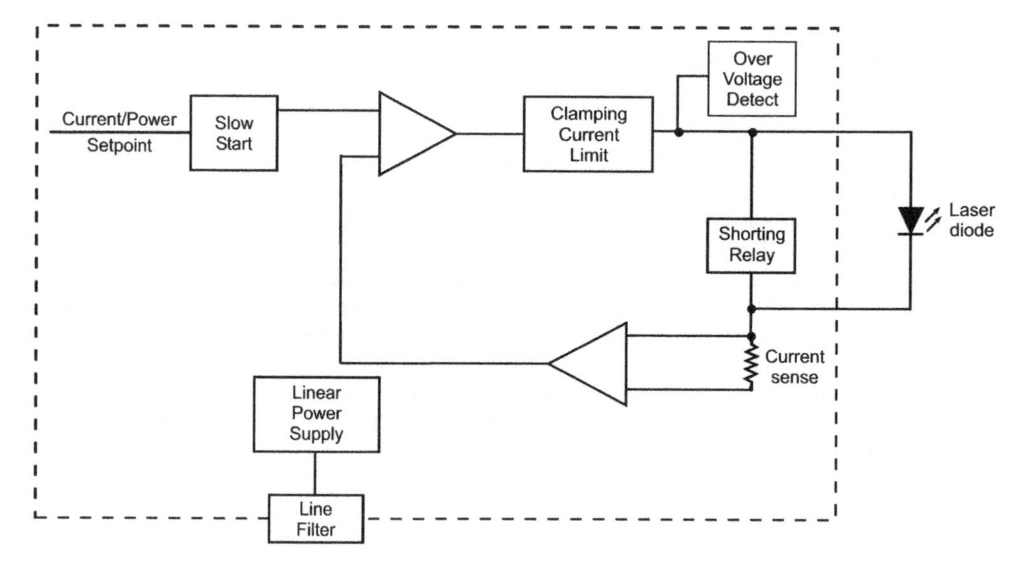

Figure 9.1 Laser diode current source with protection features.

surface, which leads to a further rise in temperature. This thermal runaway occurrence causes damage to the facet and/or facet coating.

Over-voltage protection is another important feature to be built into the current source designed to drive a laser diode. A constant current source can maintain a constant current through a load only up to a certain voltage across the load, known as compliance voltage. If the load impedance was to suddenly drop, the current control loop would drive the output current to short-circuit limit. The laser diode in this case will be overdriven for a brief period of time until the feedback control circuit adjusts itself to reduce the drive voltage. This short-term overdrive is generally sufficient to damage the device. Laser diode current source should be designed in order to guard against this condition.

Power line transients originate from a variety of sources, which include poor power conditioning, electronic equipment using switched-mode power supplies and computers. The laser diode driver and temperature controller should be designed in order to efficiently suppress transients that make their way to the power input stage. An appropriate combination of filters could do the job. However, it may not be possible to prevent transients that are radiatively coupled into the laser or drive cables.

The protection needs of the *temperature controller* are also very stringent. The laser diode lifetime is inversely proportional to the operating temperature. The lifetime has been observed to improve by an order of magnitude for every 30 °C reduction in the laser diode case temperature as shown in Figure 9.2. There is therefore a great motivation to operate laser diodes at as low a temperature as possible. The output wavelength is also a strong function of junction temperature which, in turn, depends on the environmental temperature, power dissipated at the junction and relevant thermal resistances. Wavelength shifts by 0.4 nm for each unit change in temperature. Heat removal could make use of passive

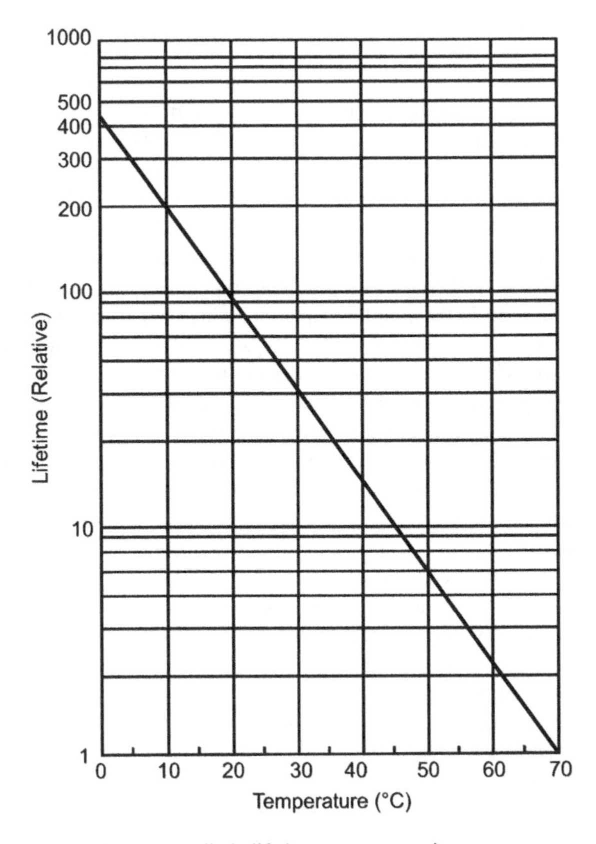

Figure 9.2 Laser diode lifetime versus operating temperature.

techniques where, at best, the case temperature can equal the ambient temperature or active techniques where there is no such limitation.

A thermoelectric (TE) module (also called Peltier cooler) temperature controller is used when the application demands a high degree of temperature stability. The current limit for the TE module should always be set below its maximum rating. Thermal oscillations should be avoided by setting the gain of the temperature controller's feedback loop appropriately. The chosen TE module should have a high current rating and the current limit for the TE module drive circuit should be set just above the operating value.

To summarize, a laser diode drive circuit should be a current source and not a voltage source. The current source should have inbuilt protection features such as shorting output, independently adjustable current limit, over-voltage protection, slow start and immunity to transients. Current limit should be set with due consideration to recommendations of laser diode manufacturer. In the absence of any specific guideline on the desirable current limit, it should be set just above the expected maximum operating current. A laser diode mount should have an adequate heat sinking capability and must not allow the laser diode to go to thermal runaway. The laser diode should be operated at the lowest temperature possible, dependent upon the type of laser diode and intended application.

9.2.2 Interconnection Cables and Grounding

Use of proper cabling and grounding methodology is crucial in determining the susceptibility of laser diode instrumentation (laser diode driver, temperature controller) to radiated noise. The main sources of radiated noise include transformers, fluorescent lights, switched-mode power supplies, high-speed data communication channels, motorized equipment and lightning discharge. Identification of all possible coupling mechanisms of the radiated transient noise, which in turn determines the radiated susceptibility, is usually a difficult proposition. Primary concerns however remain improper mounting, cabling and grounding methodologies. It may be noted here that the proper laser diode mounting, the correct choice of cable type and routing and grounding practice are also important.

The first and foremost precaution to be observed in the case of *cabling* is to make sure that the laser diode driver is turned off prior to connecting the laser diode. A loose cable connection could be very dangerous. An intermittent or bouncing cable connection is interpreted by the laser diode driver as an intermittent high impedance load, with the result that it tries to drive the laser diode harder.

Both twisted-pair and shielded cables can be used, depending upon the application requirement. A twisted-pair cable is preferred in the case of long cable runs due to the reduction in loop area and consequently the inductive term in the coupling equation, which helps in reducing the low frequency noise that would have inductively coupled into the system. Note that a twisted-pair cable offers practically no resistance to high-frequency noise, as the coupling mechanism in this case is capacitive. Use of a shielded cable helps in this case, but alone is not enough. Improper shield termination often limits its effectiveness. The shield on the cable provides a low-resistance path for the high-frequency noise to ground. As a result of this, it not only restricts radiation of energy from the internal conductor but more importantly it prevents high-frequency noise radiated by other sources of noise becoming coupled into the system. It therefore significantly improves both radiated emission as well as radiated susceptibility performance.

A cable shield should never be used as a current-carrying conductor. Both ends of the shield should preferably be terminated with a low-inductance ground. In the case of a ground loop formation leading to possible inductive coupling of low-frequency noise, one of the ends of the shield (preferable on the diode mount side) should be connected to ground through a capacitor of a few hundred picofarads. The capacitor serves the purpose of breaking the loop for low-frequency noise signals while maintaining a low-impedance path for high-frequency noise signals.

A *laser diode mount* also plays an important role in overall noise performance. If the mount is not properly grounded and/or the radiation environment is extremely harsh, the mount itself might act as an antenna and couple radiated noise signals from external sources into the system. A laser diode mount should be selected with due consideration to the type of laser diode package.

Grounding in the context of laser diode instrumentation serves two purposes. The first is to reference the laser diode with respect to ground and the second is to provide a return path for the device current. Ground loops should be avoided; ideally, a single-point ground connection should be used. If multiple ground nodes are unavoidable, the ground loop current should be minimized. With reference to the second requirement, the laser return path should not used as a return path for other circuits. The final implementation of grounding methodology is a compromise between the two conflicting requirements of avoiding ground loops and providing low-inductance high-frequency grounds to suppress fast transients. Multiple low-inductance ground connections are preferred when confronted mainly with high-frequency noise. If low-frequency noise is a bigger worry, a single-point ground connection is preferred.

To summarize, with reference to mounting, cabling and grounding requirements, the laser diode return path should be kept separate from other current paths as far as possible. Ground loops should be avoided and appropriate shielded cables that are properly fastened and terminated should be used to minimize radiated emission and susceptibility. Laser diode mounts should not couple the radiated noise into the system.

9.2.3 Transient Suppression

Laser diodes are highly prone to damage due to all types of electromagnetic interference including fast transients. These transients are caused by equipment such as soldering irons, motors and compressors which require large power line surge currents and systems such as computers using switched-mode power supplies. Although the use of shielded cables and proper grounding helps to accommodate radiated transients, fast transients often beat the best defences. Immunity to fast transients can be further improved by avoiding sharing power lines used for laser diode driver with other equipment. Isolation transformers or/and surge suppressors should be used on the power line. If forced to share the power line with other equipment, ensure that a separate surge protector is used for the laser diode driver.

9.2.4 Electrostatic Discharge

To ensure efficient and reliable operation of a laser diode throughout its lifetime, proper handling of laser diodes (e.g. during storage, transport and mounting) is no less important than the design requirements of laser diode instrumentation discussed in the previous sections. Electrostatic discharge is considered as the single-most predominant cause of premature device failure. AlGaAs and InGaAsP laser diodes may be highly prone to damage by ESD voltages as low as 1200 V. Figure 9.3 depicts laser power output versus drive current characteristics of a typical InGaAsP laser diode for different values of reverse-biased

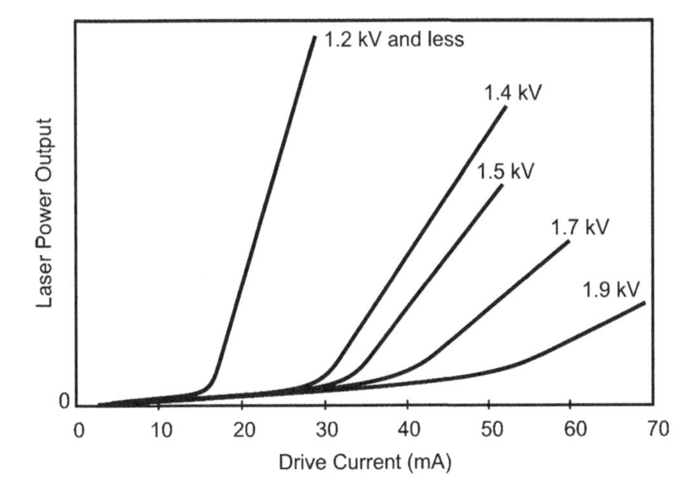

Figure 9.3 Laser power output versus drive current for different values of reverse-biased ESD stress.

ESD stress. As is evident from the family of curves shown in Figure 9.3, there is an onset of failure at 1400 V. More often than not, ESD causes latent damage to the laser diode without any immediate visible symptoms. Consequent performance degradation may appear long after the initial occurrence of the damage-causing ESD.

Laser diodes can be protected from ESD damage by taking certain precautions while handling laser diodes. The use of anti-static gloves, wrist straps and clothing and grounded soldering stations and other equipment are strongly recommended. Also recommended is the use of anti-static floor coverings, ionized air blowers and dissipative work surfaces. Charge-generating materials must be kept at least 30 cm away from the laser diode to prevent charge accumulation due to inductive coupling. When not in use and during transportation and storage, laser diodes should be fully enclosed in a conductive material. Laser diode pins also should be kept shorted when not in use by inserting them into conductive foam.

9.3 Operational Modes

Characteristics such as operating wavelength, threshold current, slope efficiency and lifetime of a laser diode, as outlined in Section 9.2.1, are strongly dependent on the temperature of the laser diode chip. In addition, operating wavelength is also dependent on the drive current. Optical power output is observed to be inversely proportional to the chip temperature. Operating wavelength for a typical AlGaAs laser diode varies directly at a rate of $0.25\,\text{nm}\,°\text{C}^{-1}$. Operating wavelength also changes at a rate of $0.025\,\text{nm}\,\text{mA}^{-1}$ of drive current. Further, in the case of a single-mode laser diode, change in wavelength leads to mode-hopping phenomenon.

Laser diodes are operated in either of the two operating modes: (a) constant-current mode and (b) constant-output-power mode. In the case of the former, the drive current to the laser diode is maintained constant irrespective of variations in the source of power input. Circuit topology used to achieve constancy of drive current varies from using a simple band-gap reference-controlled current source driven from a regulated DC voltage to using a feedback loop that continuously monitors the drive current and applies appropriate correction in the case of drive current deviating from the desired value. Requirements on the part of current stability are also governed by maximum acceptable frequency/ wavelength variation. For example, laser diodes used for pointing and printing applications may not require drive current stabilization to high precision. On the other hand, devices used for telecommunication or spectroscopy applications have very stringent requirements on drive current stabilization. Assuming a typical figure of $0.025\,\text{nm}\,\text{mA}^{-1}$ (which is equivalent to a frequency variation of $15\,\text{GHz}\,\text{mA}^{-1}$ at 700 nm output wavelength) and a nominal drive current of 100 mA (which is reasonable for laser diodes used for telecommunications and spectroscopy applications), 1 μA variation would cause a 15 MHz shift in frequency. A stability of ± 1 μA in 100 mA is equivalent to ± 10 ppm. Circuit design and the choice of components must therefore take such requirements into account. For instance, all circuit components should have drift specifications far better than the actual stability required.

9.3.1 Constant-current Mode

In view of important characteristic parameters of the laser diode being dependent on operating temperature and drive current, the preferred mode of operation of a laser diode is the constant drive current with precise control of operating temperature. Constant-current operation without temperature control is generally not desirable. Even at constant drive current, output power would increase with a decrease in operating temperature. With a significant decrease in temperature, output power could easily go past the absolute maximum value.

The circuit topology in the case of constant-current mode of operation is usually configured around a current sensing element that continuously senses the drive current and produces a proportional voltage. This is then compared to a reference voltage representing the desired value of drive current to generate the error signal. After suitable conditioning, the error signal is then fed back to restore the drive current to the desired value. Figure 9.4 shows the basic schematic arrangement of a constant-current drive circuit. Practical circuits with inbuilt protection features are far more complex. Several constant-current drive circuit topologies are discussed in Section 9.4.1.

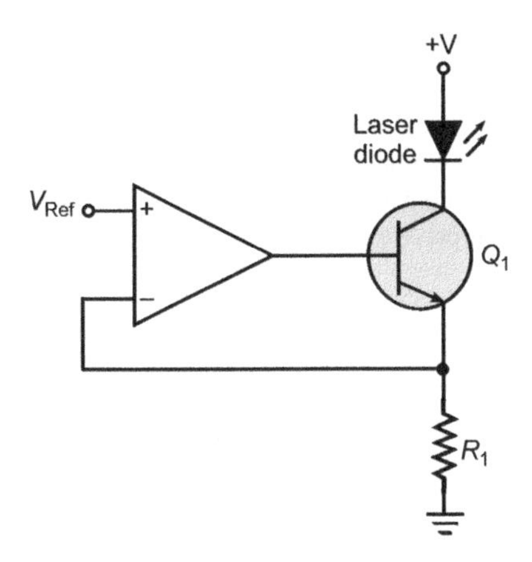

Figure 9.4 Constant current drive circuit schematic.

9.3.2 Constant-power Mode

There are applications where constant output power is more relevant. When a laser diode is driven by a constant current, the heat dissipated at the laser diode junction leads to a rise in temperature and hence a fall in output power. An increase in temperature could be because of an absence of any active temperature control mechanism or even inadequacy of heat sink in the presence of active temperature control. This reduction in output power can be compensated for by increasing the drive current. The current source can be designed in such a way that it adjusts the drive current in a feedback mode to maintain a constant output power instead of maintaining a constant drive current. Laser diode modules with an integral photodiode suit this operational mode well.

Constant-power mode also precludes the possibility of the output power increasing with a decrease in operating temperature. The basic circuit topology of a constant-output-power drive circuit, as shown in Figure 9.5, is similar to that of a constant-current drive circuit except for the nature of sense signal. In the case of constant-output-power mode, the sense signal is photocurrent, which is proportional to the output power. It may be mentioned here that constant-output-power mode may lead to the thermal-runaway situation, thus damaging the laser diode if operated without an absolute current limit. Noise intrinsic to

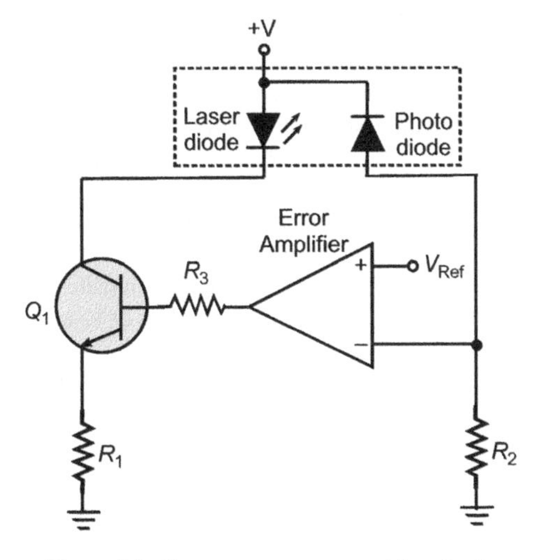

Figure 9.5 Constant output power drive circuit.

the integral photodiode also manifests itself in the form of noisy and unstable output in the case of constant-power mode of operation.

To summarize, operating a laser diode at a constant drive current with precise temperature control should always be the preferred option, more so when the laser diode is intended for applications such as spectroscopy, pollution monitoring, biomedical diagnostics and characterization of materials. Automatic power control may be resorted to if precise temperature control is not practical and constancy of output power is highly desirable.

9.4 Laser Diode Driver Circuits

In this section we describe some representative laser diode driver circuits for both constant-current and constant-power modes of operation. In the constant-current mode of operation, the primary objective is to provide a stable drive current to the laser diode. The drive current is continuously sensed and any deviation in the drive current is corrected for by changing the voltage controlling the drive current. In the case of constant-power mode, the drive current to the laser diode is continuously adjusted to maintain constant output power. As the output power increases or decreases, the drive current is decreased or increased to maintain a constant output power.

Irrespective of the operational mode however, certain protection features such as independently adjustable current limit, slow start, immunity to transients and other forms of interference need to be incorporated into laser diode driver design. As already outlined in Section 9.2, laser diodes are highly prone to damage or performance deterioration due to transient conditions, the current exceeding a specified upper limit and radiated noise present in the environment.

In the following sections we describe some basic laser diode drive circuits. Beginning with the most basic constant-current source, drive circuits with inbuilt feedback control for both constant-current as well as constant-power modes of operation are discussed. A large number of integrated circuits are also available that are capable of driving laser diodes in either of the two modes of operation. These integrated circuits, available from leading international manufacturers of digital and linear integrated circuits, have the necessary building blocks to offer the desired protection features in addition to providing the required drive current. These integrated circuits usually drive the relatively low-power laser diodes directly. In the case of high-power laser diodes, output of the integrated circuit drives an output stage which in turn provides the necessary drive to the laser diode.

9.4.1 Basic Constant-current Source

Figure 9.6 shows a simple circuit providing a constant current drive to the laser diode; this is one of the circuits which all electronics engineers are familiar with. The Zener diode ensures a constant voltage at the base of the PNP transistor with the result that the voltage at the emitter terminal $V_E = V_Z + 0.6V$ is also constant. The diode current $I_D = (V - V_E)/R_2$ which approximately equals the emitter current is constant, provided V and R_2 are constant. The choice of V_Z in the circuit is dictated by the desired value of V_E which, in turn, equals $(1.5 + V_{CE})$ V. The typical voltage drop across the laser diode when forward-biased as shown in the typical *I–V* characteristics of Figure 9.7 is 1.5 V. V_{CE} is chosen in order to operate the transistor in the centre of the active region of its output characteristics. Having determined V_E, V_Z is given by $(V_E - 0.6)$ V. R_2 can be computed from the known values of V_E and the desired drive current. R_2 can be varied to change the drive current.

The circuit shown in Figure 9.6 is seldom used in practice for the simple reason that it is probably too simple and inflexible to allow incorporation of other desired features such as current modulation, protection features and feedback mode of operation. One of the preferred configurations for driving laser diodes is shown in Figure 9.8.

This circuit has several advantages over the drive circuit of Figure 9.6. The drive current in this case is independent of any variation in transistor parameters. The circuit can be easily adapted to incorporate the modulation input if needed.

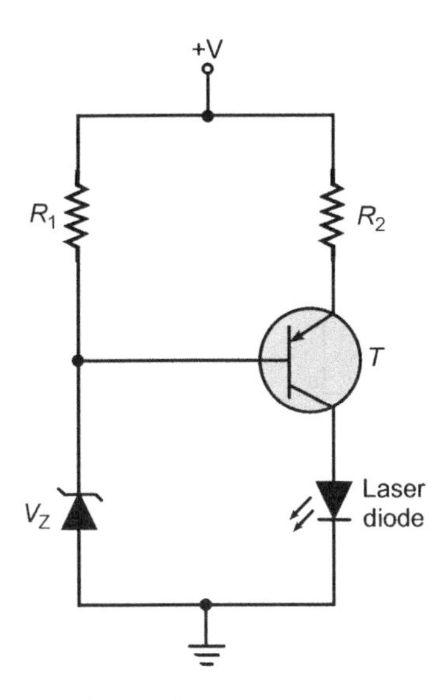

Figure 9.6 Basic constant current source.

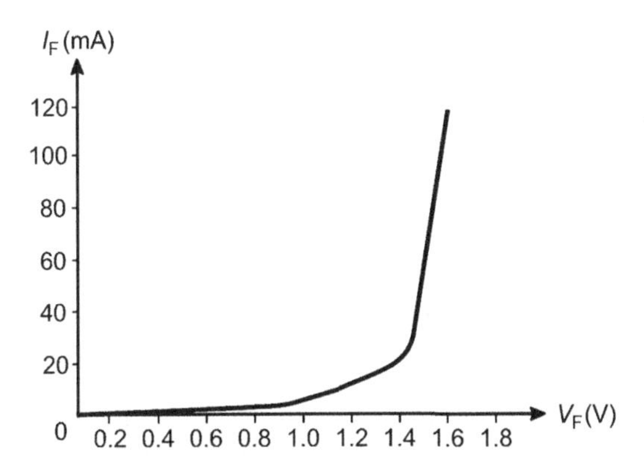

Figure 9.7 *I–V* characteristics of typical laser diode.

The voltage V_E at the emitter terminal of the transistor $V_E = V_{Ref} \times R_2/(R_1 + R_2)$ and hence the drive current $(= (V - V_E)/R_3)$ is held constant. V_{Ref} can be generated by using a band-gap reference and R_1, R_2 and R_3 can be precision metal-film resistors. Drive current can be varied by varying R_2.

Laser diode driver circuit of Figure 9.8 provides a constant-current drive to the laser diode only if the DC supply V and V_{Ref} are constant. If V_{Ref} is a band-gap reference and R_1, R_2 and R_3 are precision resistors, it would be safe to assume V_{Ref} as constant. The same might not be true for V; any variation in V manifests itself as a change in drive current.

9.4.2 *Laser Diode Driver with Feedback Control*

Figure 9.9 shows another voltage-controlled constant-current laser diode driver in which the drive current is independent of variation in the DC supply used for active devices. Laser diode current is sensed differentially by measuring voltage drop across R_{Sense} wired in series with the laser diode. The drive current is controlled by the input voltage V_{In}, which could be generated with a band-gap reference. If the

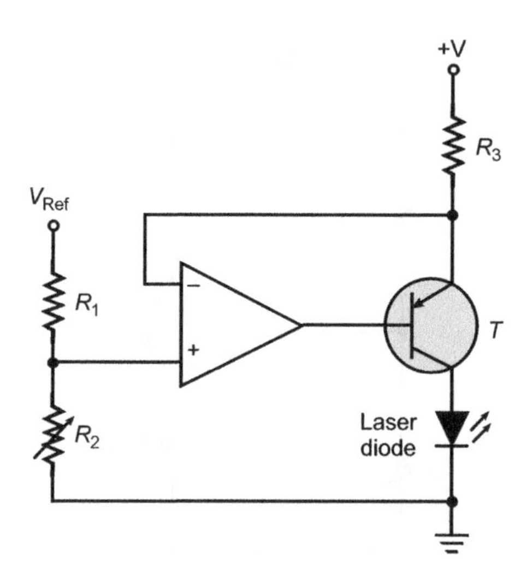

Figure 9.8 Improved basic constant current source.

drive circuit is to be controlled digitally, V_{In} could be provided by a voltage output digital-to-analogue converter. Drive current can be computed from Equations 9.1 and 9.2:

$$I_O \times R_{Sense} = V_{In} \times \frac{R_4}{R_3} \tag{9.1}$$

$$I_O = \frac{V_{In} \times R_4}{R_3 \times R_{Sense}} \tag{9.2}$$

$R_1 = R_3$ and $R_2 = R_4$

C_1 is the compensation capacitor to ensure stable operation. If the available control voltage V_{In} is of negative polarity, the circuit is then wired as an inverting amplifier by applying V_{In} to R_1 instead; R_3 in that case is grounded.

Another drive circuit with feedback control is shown in Figure 9.10. It is similar to the driver circuit of Figure 9.9, except for the fact that the laser diode in this case is connected as a floating load. A low-noise operational amplifier (opamp) feeds a current booster transistor. Laser diode current is sensed by measuring the voltage drop across a sense resistor R_{Sense} wired in series with the emitter terminal of bipolar transistor. Again, output current is controlled by the input voltage V_{In} and can be computed from Equations 9.3 and 9.4. C_1 is the compensation capacitor.

$$I_O \times R_{Sense} = V_{In} \times \frac{R_2}{R_1} \tag{9.3}$$

$$I_O = \frac{V_{In} \times R_2}{R_1 \times R_{Sense}} \tag{9.4}$$

The voltage-controlled constant-current sources discussed so far provide constant drive to the laser diode, provided that the DC supply voltage, the voltage at the emitter terminal and the resistance in the emitter lead are all constant. It would be a reasonably good assumption if the source voltage were derived from a precision band-gap reference and the resistors used had stability specifications equal to or better than the desired level of current stability.

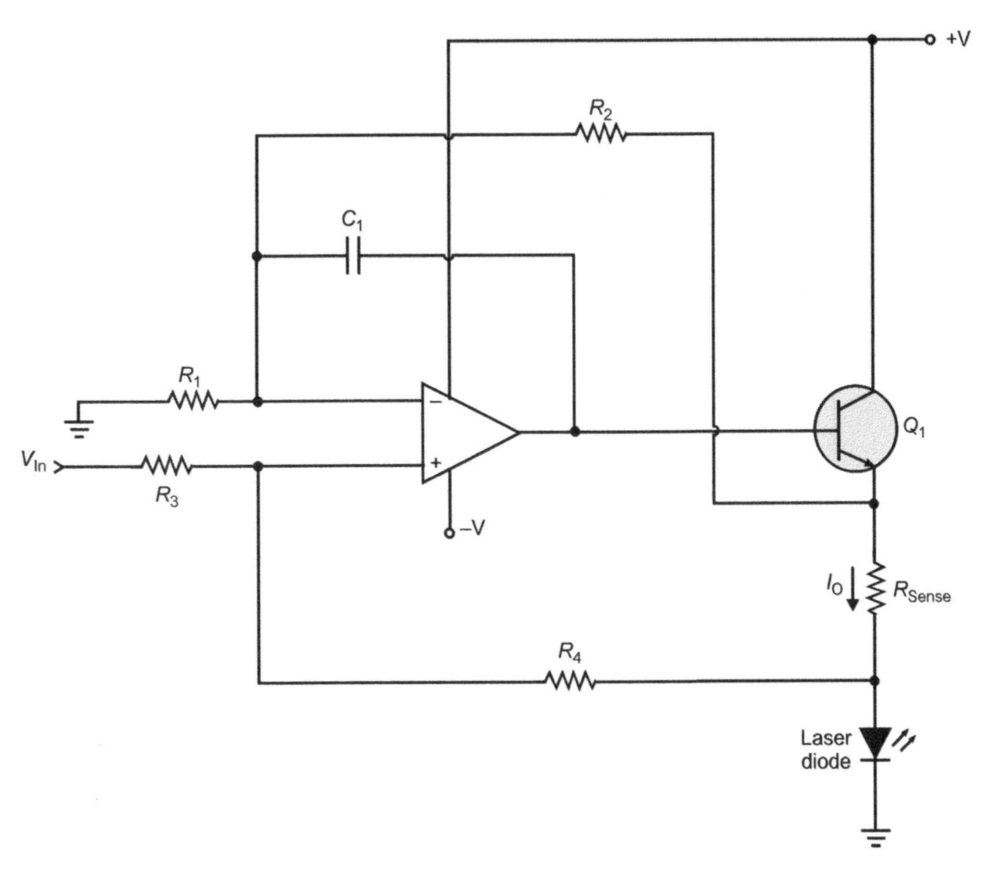

Figure 9.9 Laser diode driver circuit with feedback control (grounded load).

When it comes to achieving a current stability of ± 10 ppm or better (equivalent to ± 1 μA in 100 mA), desirable in laser-diode-based spectroscopy and other similar applications, a feedback loop that samples the diode current *in situ* and applies a correction in the case of drift in drive current becomes essential. A negative feedback loop of this kind would reduce the drift or error in the drive current by a factor that equals the loop gain.

One such circuit that achieves feedback-loop-based current regulation is shown in Figure 9.11. The circuit is self-explanatory. The basic circuit topology is similar to that used in driver circuits described in Figure 9.10. The only change is the inclusion of a junction field effect transistor (JFET) connected in series with the sense resistor. The JFET is wired here as a voltage-variable resistor. Having set the drive current at a certain value, any tendency on the part of the drive current to change is compensated by an appropriate variation in the drain-source resistance of the JFET. For instance, when the current tends to increase the JFET gate voltage is driven more negatively, increasing the drain-source resistance and restoring the current back to the set value. Initially, at the nominal value of drive current, circuit parameters are adjusted in order to ensure that the JFET with the feedback loop closed gets a negative gate voltage to keep it nearly in the middle of its variable-voltage resistor (VVR) characteristics. This is done to exploit fully the voltage-dependent resistance range.

Example 9.1

Refer to the laser diode drive circuit of Figure 9.12. Compute the values of R_1, R_2, R_3 and R_4 for a constant drive current of 500 mA.

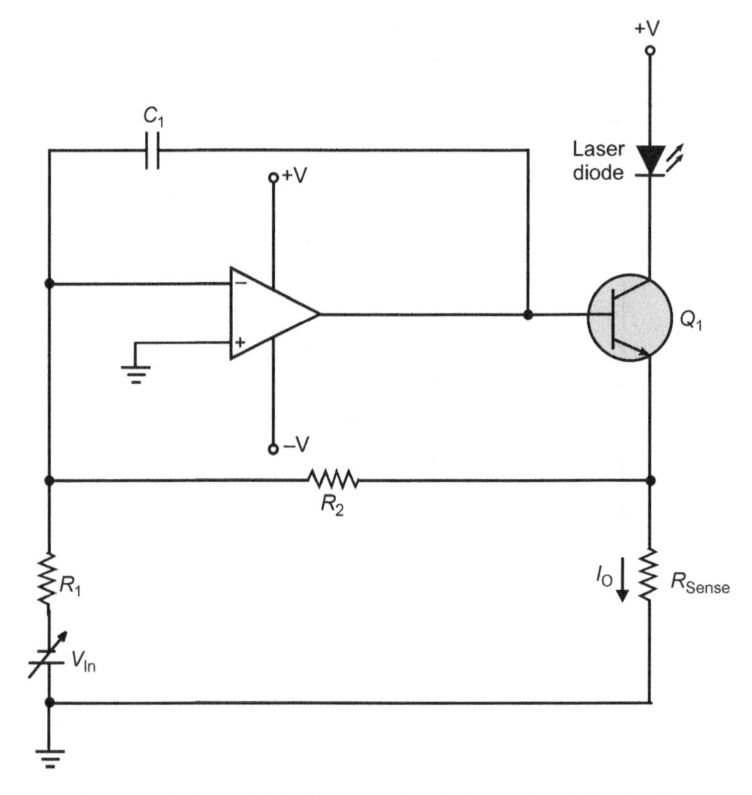

Figure 9.10 Laser diode driver with feedback control (floating load).

Solution

1. For the circuit of Figure 9.12, I_O can be computed via:

$$\begin{aligned} I_O \times R_{\text{Sense}} &= -V_{\text{In}} \times \frac{R_2}{R_1} \\ &= -(-2R_2/R_1) = 2R_2/R_1 \end{aligned}$$

   This gives $R_2/R_1 = I_O \times R_{\text{Sense}}/2$
2. Substituting values of I_O and R_{Sense}, $R_2/R_1 = (500 \times 10^{-3} \times 0.1)/2 = 25 \times 10^{-3}$
3. Assuming $R_1 = 100\ \text{k}\Omega$, we get $R_2 = 2.5\ \text{k}\Omega$
4. $R_3 = R_1 = 100\ \text{k}\Omega$ and $R_4 = R_2 = 2.5\ \text{k}\Omega$.

9.4.3 Laser Diode Driver with Modulation Input

An important functional requirement in many application circuits is the option to modulate the laser output. The laser diode driver circuit of Figure 9.13, a modification of the driver circuit of Figure 9.8, can be used to modulate the laser diode. The circuit is self-explanatory. The DC reference voltage that determines the unmodulated laser drive current and the modulating signal have been summed up in the inverting amplifier A_1. The polarity is restored in the unity gain inverting buffer A_2 following A_1.

Example 9.2

For the laser diode driver and input waveform of Figure 9.14, draw the laser diode current waveform.

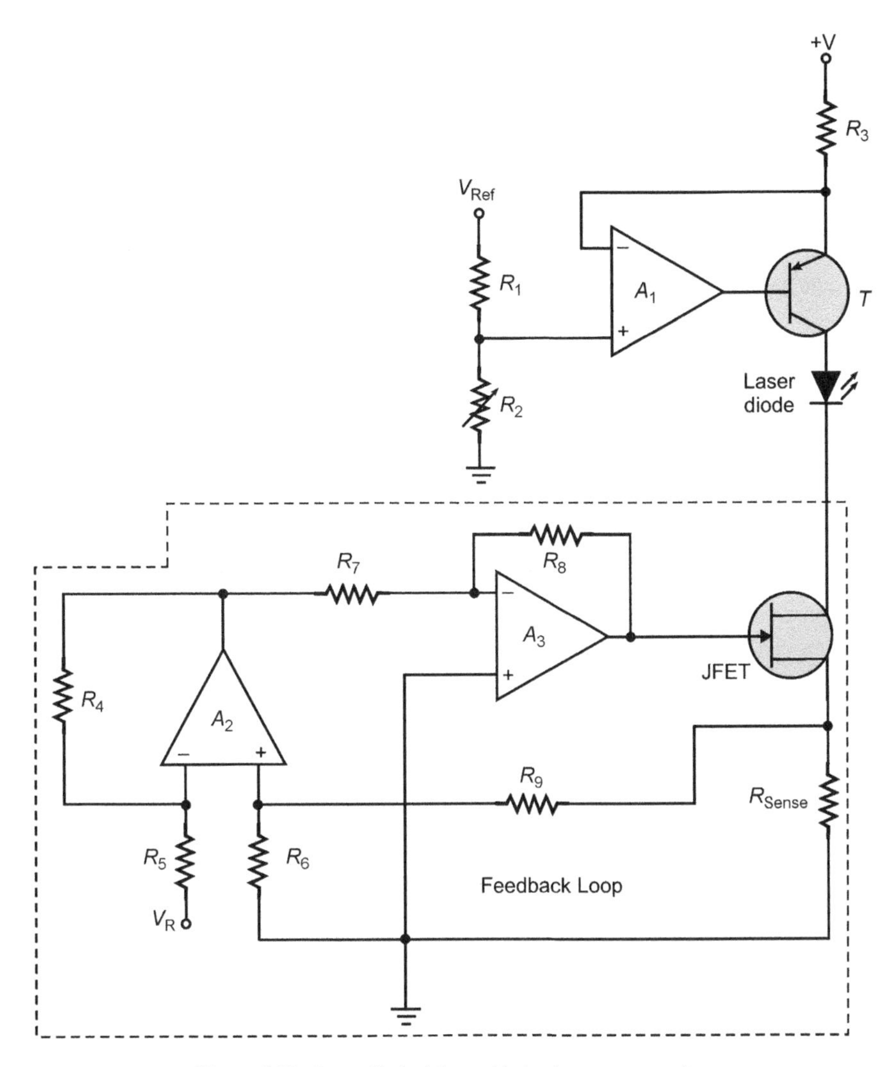

Figure 9.11 Laser diode driver with *in situ* current monitor.

Solution

1. Laser diode current = $(5 - V_E)/20$ A, where V_E is the voltage at the Q_1 emitter
2. V_E is equal to voltage at inverting input of amplifier A_3. A_1 is a unity gain inverting summer; A_2 is a unity gain inverting amplifier. Voltage at the non-inverting input of A_3 is therefore the sum of V_{In} and voltage appearing across 10 kΩ resistor at the input of A_1.
3. Voltage across 10 kΩ resistor = $(5 \times 10)/25$ V = 2 V
4. Therefore, at the positive peak of V_{In}, $V_E = 2 + 1$ V = 3 V
5. At the negative peak of V_{In}, $V_E = 2 - 1$ V = 1 V and for $V_{In} = 0$ V, $V_E = 2$ V.
6. Therefore, at the positive peak of V_{In}, laser diode current = $(5 - 3)/20$ A = 100 mA
7. At the negative peak of V_{In}, laser diode current = $(5 - 1)/20$ A = 200 mA
8. Laser diode current waveform is shown in Figure 9.15.

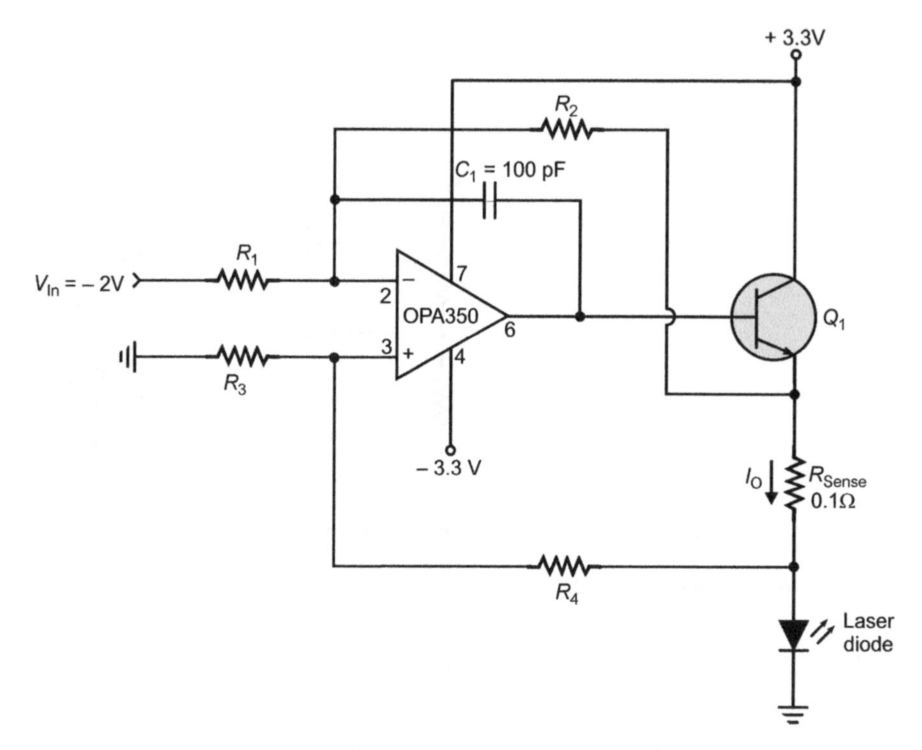

Figure 9.12 Example 9.1.

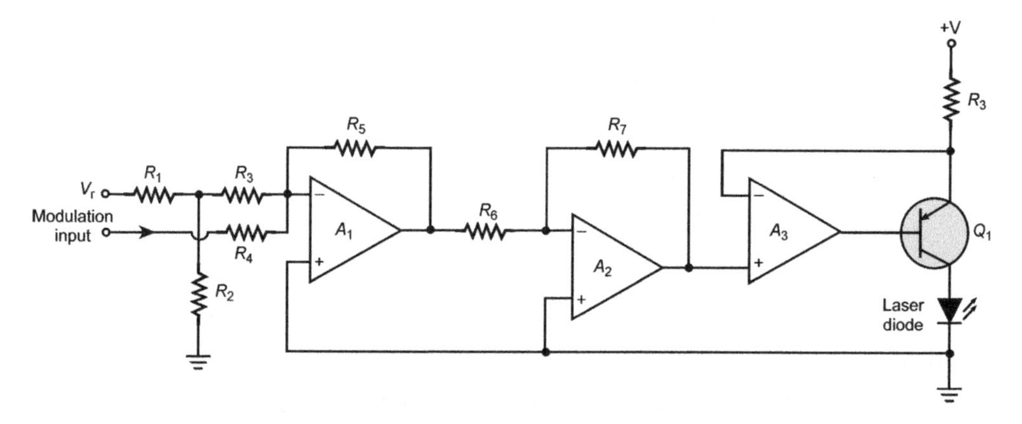

Figure 9.13 Laser diode driver with modulation input.

9.4.4 Laser Diode Driver with Protection Features

Laser diodes are highly prone to performance deterioration and permanent damage if drive current exceeds the maximum limit. Sudden spikes in the drive current could also be fatal to the laser diode. Protection features such as slow start, immunity to fast transients and over-current limit are essential requirements of every laser diode driver circuit.

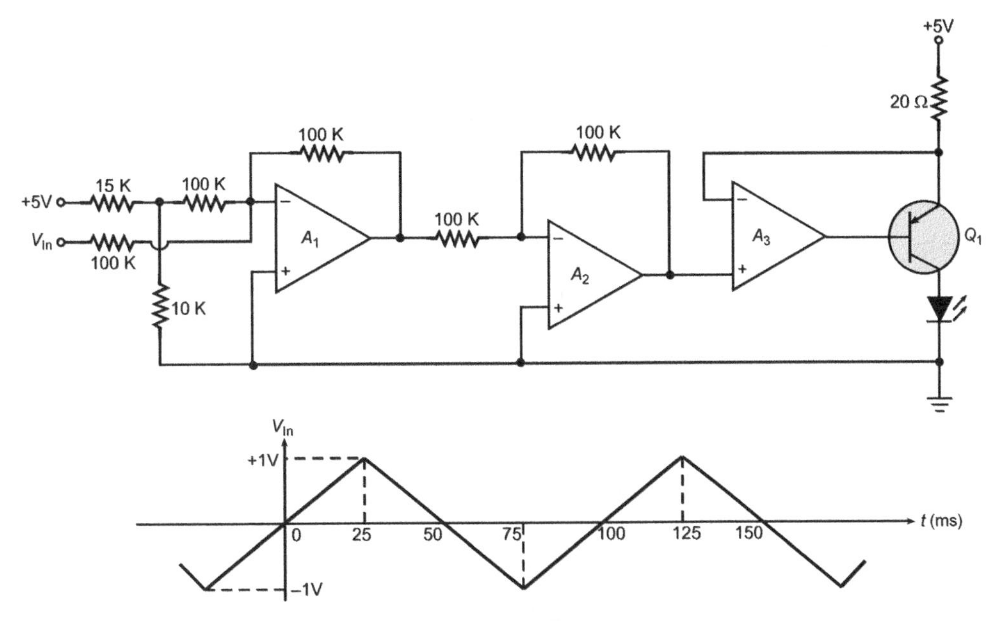

Figure 9.14 Example 9.2.

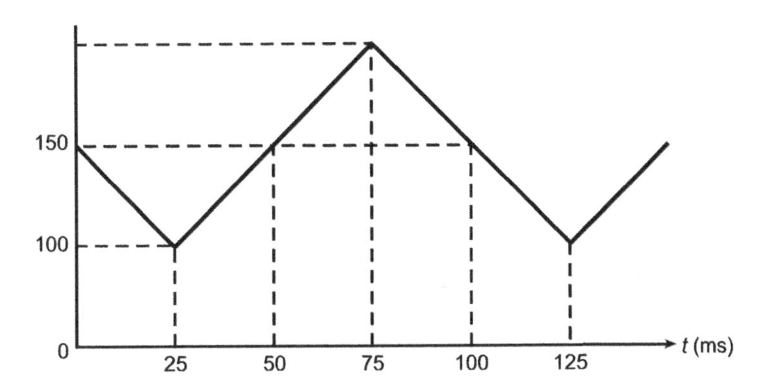

Figure 9.15 Answer to Example 9.2.

Figure 9.16 shows a laser diode driver with features of slow start, immunity to fast transients and current limit. The circuit operates as follows. Laser diode current is determined by the voltage appearing across non-inverting terminal of opamp A_4, resistor R_{10} and regulated DC supply voltage $(+V)$. The voltage at the non-inverting terminal of opamp A_4 depends upon the sum of voltage appearing across the R_3–R_4 junction and the modulation input. Opamps A_2 and A_3 are wired as unity gain inverting summing amplifier and unity gain inverting amplifier, respectively. The Schmitt comparator at the input provides a delay of the order of a few tens of milliseconds after the switch on to offer protection against switch-on transients. The time delay is determined by the R_1C_1 time constant. R_2 and C_2 provide the desired slow or soft start and decay during switch on and switch off. The R_2C_2 time constant is of the order of 1–2 s. R_{11} and C_3 provide additional protection against transients; D_1 protects the laser diode against reverse voltages. Opamp A_1 along with resistors R_{12} and R_{13} and diode D_2 provides the over-current limit. During normal operation, D_2 remains reverse-biased and therefore output of opamp A_1, which equals the voltage across the R_{12}–R_{13} junction, does not interfere with normal operation of the circuit. The laser diode drive

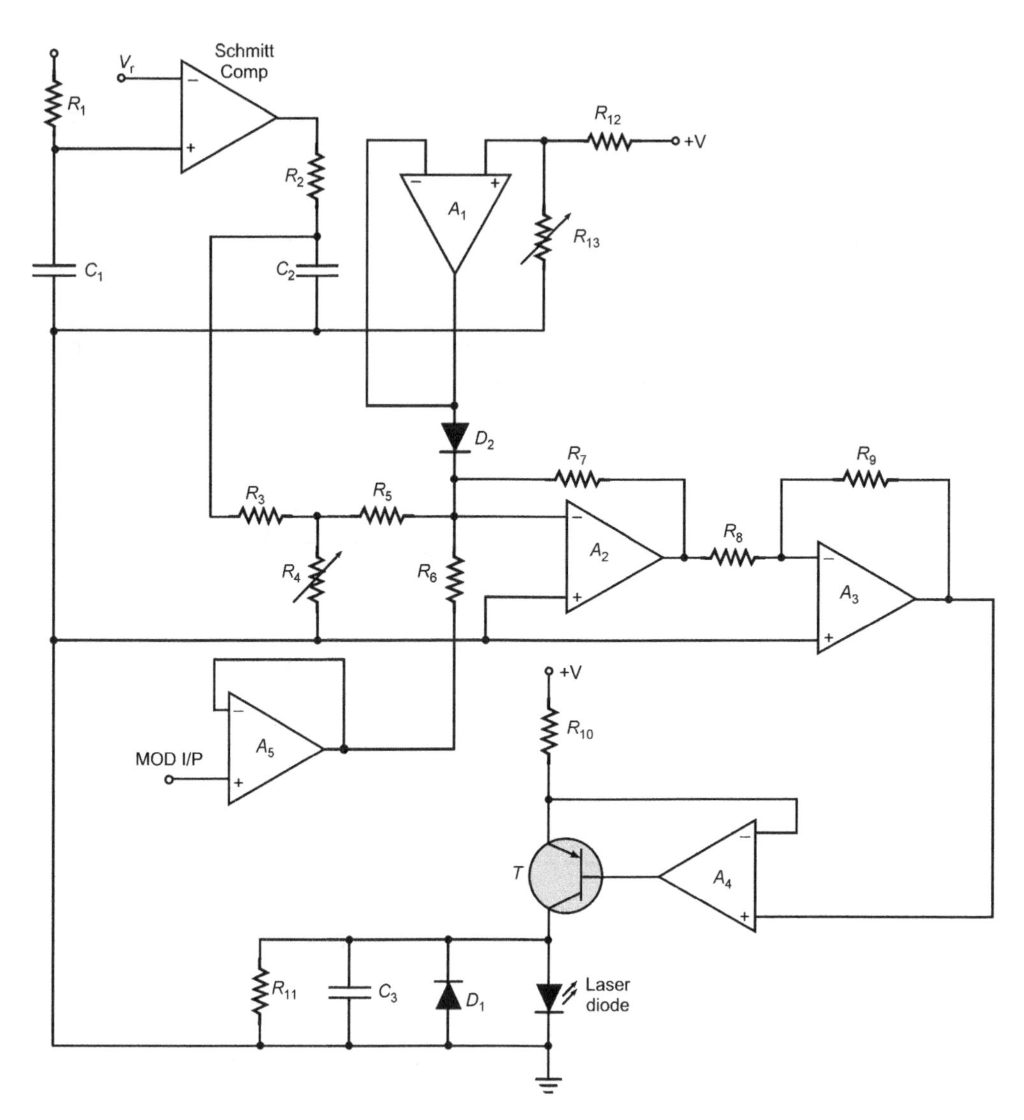

Figure 9.16 Laser diode driver with protection features.

current in this case continues to be dictated by the voltage equal to the sum of the modulation input voltage and the voltage appearing across the R_3–R_4 junction. The moment the upper current level tends to exceed the set limit, the diode becomes forward-biased. The voltage at the inverting input of opamp A_2, and hence its output, is clamped at a value equal to the voltage at R_{12}–R_{13} junction.

Figure 9.17 shows a typical laser diode driver circuit using a feedback loop for current regulation and having inbuilt protection features. The circuit is self explanatory; various building blocks shown in the figure have been described in sections 9.4.2 and 9.4.3.

9.4.5 Laser Diode Driver with Automatic Power Control

An alternative method of driving the laser diode is to vary the drive current using a feedback loop to maintain a constant output power. The laser diode with an integral photodiode is particularly suitable for this operational mode; Figure 9.18 shows a typical circuit configuration. The photodiode current is

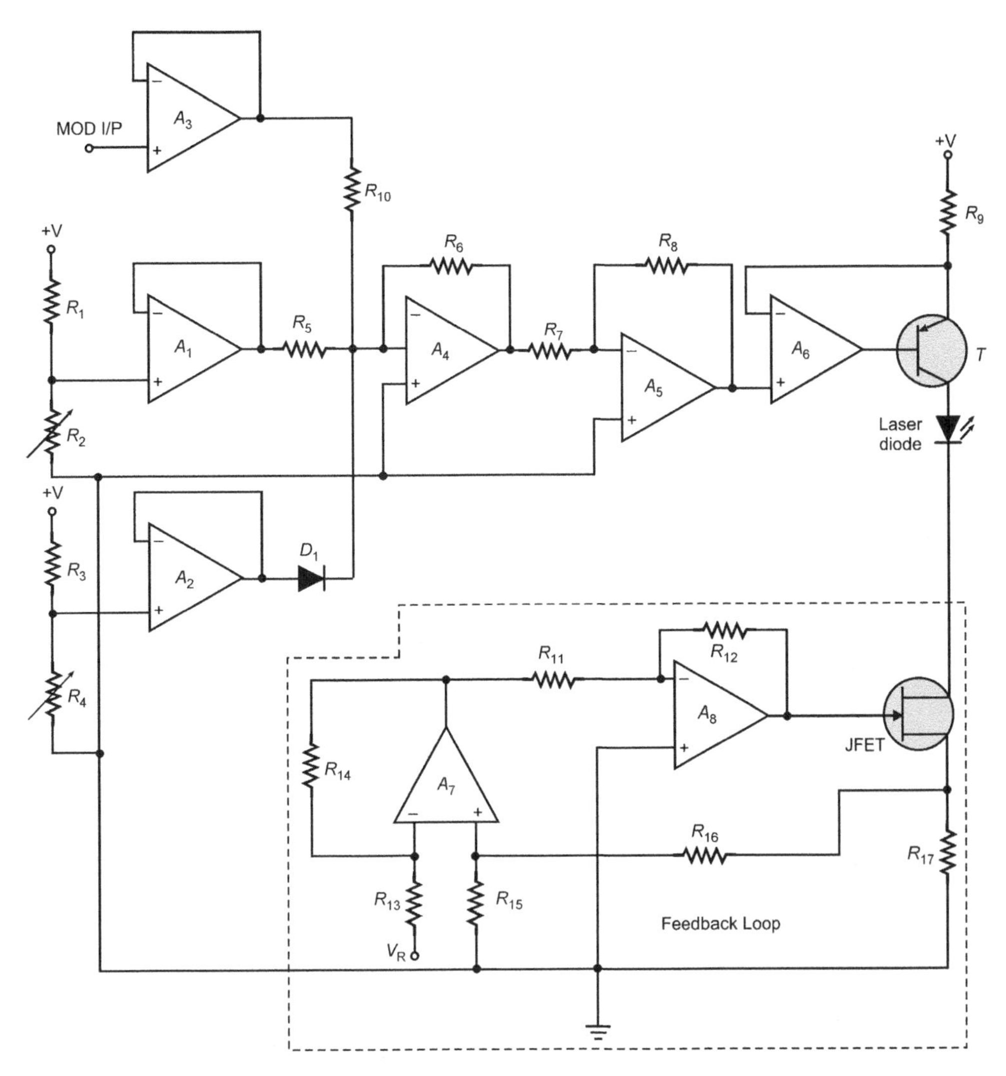

Figure 9.17 Laser diode driver with feedback control and protection features.

proportional to optical output power from the laser diode. This photocurrent is converted into a proportional voltage using a transimpedance amplifier configured around opamp A_4. The voltage representative of laser power is then summed up with a reference voltage representing the desired power level in an inverting summer configuration A_5. The output of the summer, which is null when the actual power level equals the desired power level, feeds an integrator A_6 that provides the correction signal even for infinitesimally small deviations from the desired power level. The integrator output summed up with a bias voltage feeds control element, which is a JFET in this case.

The circuit provides automatic power regulation as follows. A fall in the output power leads to a reduction in the photodiode current, which further results in a less-negative transimpedance amplifier output. As a result, the summing amplifier output that feeds the integrator becomes more negative. The integrator output becomes more positive and the FET gate becomes more negative than the nominal bias voltage. A more negative gate voltage means a larger drain-source resistance and a reduced gain for the amplifier A_2. Reduced gain means reduced voltage at the emitter terminal and an increased diode drive current. An increase in drive current increases the output power of the laser diode.

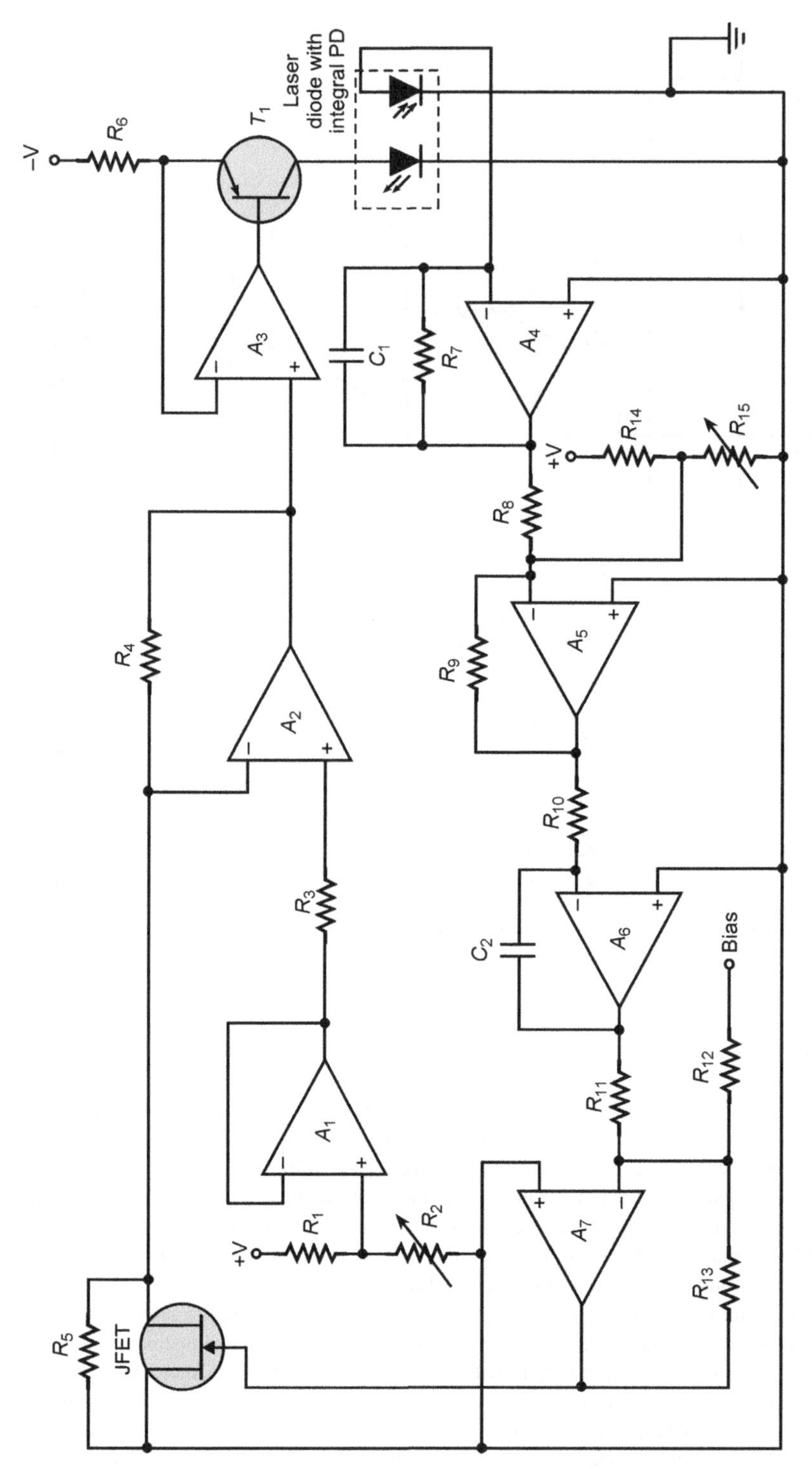

Figure 9.18 Laser diode driver with automatic power control.

When the output power from the laser tends to increase, photocurrent also tends to increase. An increase in photocurrent forces the transimpedance amplifier output to be more negative. As a result of this, the integrator output and hence the gate of JFET become less negative. The drain-source resistance of JFET decreases, thereby increasing the gain of amplifier A_2. An increase in gain of A_2 increases voltage at the non-inverting input of A_3, thereby reducing the drive current. This restores power to the original value. Power is stabilized at a point where the transimpedance amplifier output equals the voltage set by the power-setting potentiometer.

9.4.6 Quasi-CW Laser Diode Driver

Continuous wave (CW) with/without modulation is the most common mode of operation of laser diodes. Another possible mode of operation is known as quasi-CW mode of driving laser diodes. In the case of quasi-CW mode, the laser diode is driven by current pulses that are typically hundreds of microseconds to a few milliseconds wide as compared to a few tens to a few hundreds of nanoseconds wide in the case of conventional pulsed mode of operation. These diodes are operated at relatively low repetition rates of typically a few hertz to a few tens of hertz. The relatively high peak power and efficient operation of quasi-CW laser diodes have made them an ideal choice for pumping solid-state lasers. Higher peak powers are possible due to the low duty cycle of operation of quasi-CW devices, which keeps the average power low. The quasi-CW mode of operation can be a very powerful analytical tool for both pre-package/post-package testing and characterization of laser diodes. The pulsed mode of operation allows characterization with minimal thermal effects.

A laser diode driver for quasi-CW operation is configured around the same basic building blocks as those described in Sections 9.4.2–9.4.4 in the case of CW operation. Figure 9.19 shows the block schematic arrangement of a typical quasi-CW laser diode driver.

The circuit operates as follows. The oscillator, which is basically an astable multivibrator circuit, produces a train of pulses with pulse width and repetition rate at which the laser diode or laser diode array needs to be driven. A soft start inbuilt into the oscillator provides protection against damage due to instant power-on. The power stage contains the active device/s and associated circuitry capable of delivering the required current pulse amplitude. The peak current pulse amplitude typically varies from a few tens to a few hundreds of amps in the case of a diode array used for optical pumping of solid-state lasers.

The drive and control block is a preamplifier stage with some kind of control feature that allows the gain of the amplifier and hence the drive current to the power stage to be varied. The drive and control stage could be configured either around discrete bipolar transistors/MOSFETs or an opamp with JFET as

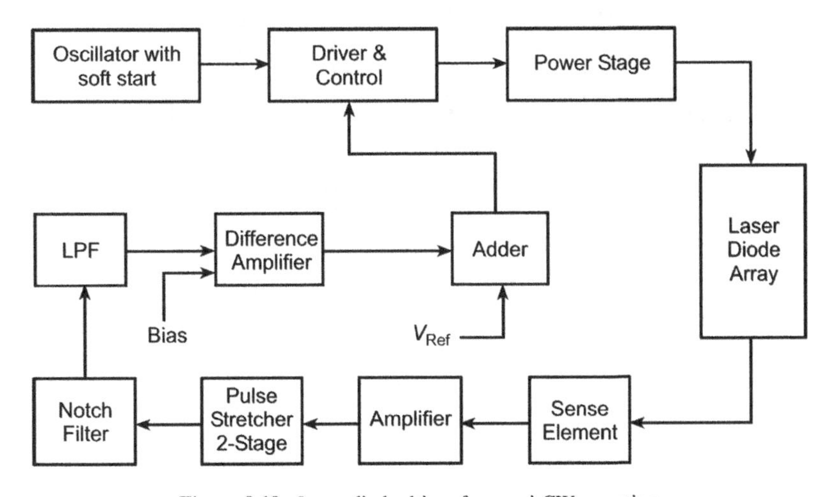

Figure 9.19 Laser diode driver for quasi-CW operation.

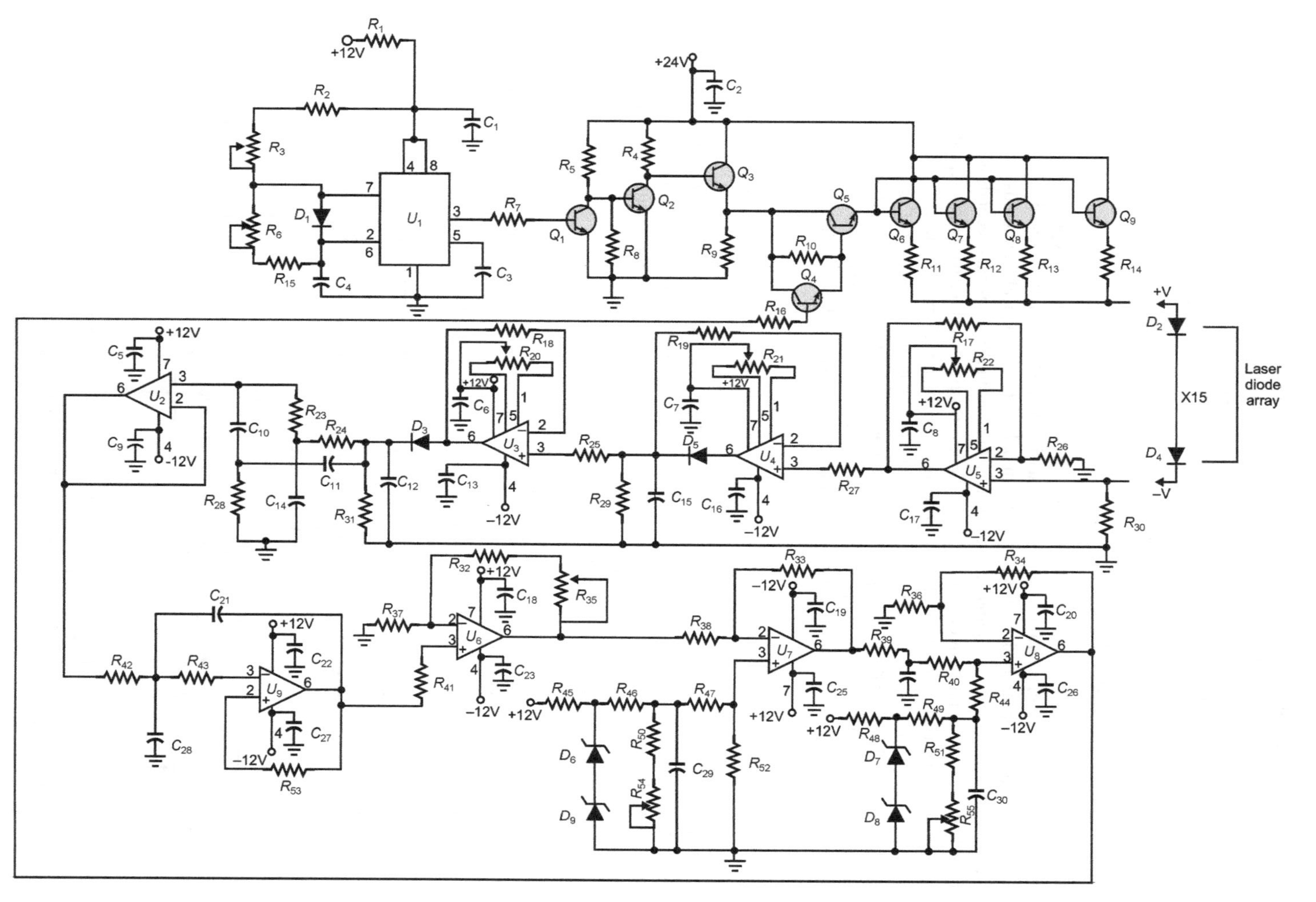

Figure 9.20 Circuit diagram of a typical quasi-CW laser diode drive circuit.

one of the gain-determining elements. The power stage could be configured around one or a parallel connection of more than one bipolar transistors/MOSFETs. The sensing element is usually a resistor connected in series with the diode or diode array. The peak pulse voltage across the sense resistor representing the peak amplitude of current pulse is amplified and then subsequently stretched. The stretched pulse train after filtering is fed to a differential amplifier stage where it is compared with a standard reference to produce an error voltage. The error voltage here represents the deviation of the peak amplitude of current pulse from the desired value. The error voltage is added to a fixed bias and fed to the control element in the drive stage to control the drive current to the power stage.

Figure 9.20 shows a detailed circuit diagram of a typical laser diode drive circuit for quasi-CW mode of operation; note that this circuit is also used to illustrate the building blocks of Figure 9.19 (other variations are also possible). In the circuit diagram of Figure 9.20, U_1 is a 555 timer wired as an astable rising-edge multivibrator to generate a train of rectangular pulses of desired pulse width and repetition rate. The R_1–C_1 combination constitutes a soft-start circuit. The soft-start circuit time constant is such that it takes about 3–4 current pulses before it attains nominal amplitude at the output of the timer IC. Bipolar transistors Q_6–Q_9 constitute the output stage. The number of bipolar transistors to be connected in parallel depends upon the desired peak current through the laser diode. Power MOSFETs can also be used in place of bipolar transistors. Resistors R_{11}–R_{14} are current-equalizing resistors and facilitate equal current sharing by the four transistors. The current-equalization resistors are not needed if power MOSFETs were used for the purpose, as they are voltage-controlled devices. In the circuit shown, transistors Q_1, Q_2 and Q_3 provide the base drive to the output stage. The higher the required value of peak laser diode current, the higher the required magnitude of base drive current. Again, a relatively simpler drive circuit would be required in the case of a power-MOSFET-based output stage.

R_{30} is the current sense resistor and is of the order of a few tens of milliohms. The voltage pulse across the sense resistor is amplified in U_5, which is then stretched in a two-stage pulse stretcher configured around U_3 and U_4. The stretching provided by U_3 and U_4 depends upon the R_{31}–C_{12} and R_{29}–C_{15} time constants, respectively. The notch filter configured around U_2 at the output of the pulse-stretching stage is tuned to the repetition rate of current pulses. The notch filter is followed by a low-pass filter configured around U_6. While the notch filter removes the repetition frequency component, a low-pass filter removes other high-frequency components arising out of sharp rising edges of current pulses. The output of a low-pass filter is a DC voltage representative of the peak value of drive current. After suitable conditioning in the non-inverting amplifier configured around U_6, this DC voltage is compared with a stable reference provided by a band-gap reference in a comparator circuit configured around U_7. The error voltage produced at the output of comparator circuit is added to a fixed-DC bias voltage in the adder circuit configured around U_8. This constitutes the control voltage that is used to control the base drive current for the output stage. In the circuit shown, transistors Q_4 and Q_5 constitute a variable-voltage resistance whose resistance value is dictated by the control voltage.

9.5 Laser Diode Temperature Control

The importance of temperature control in the operation of a laser diode has been highlighted in Section 9.2.1. Important characteristic parameters of a laser diode such as threshold current, slope efficiency, wavelength and lifetime are critically dependent on the operating temperature. Lower operating temperature leads to enhanced lifetime. A well-known technique to remove heat from the diode junction is to use an appropriate heat sink. The difference between the junction temperature and the ambient temperature depends on the junction-to-ambient thermal resistance and the power dissipated in the junction. Junction-to-ambient thermal resistance in turn depends on junction-to-case thermal resistance and case-to-ambient thermal resistance. Thermal resistance is measured in $°\mathrm{C\,W^{-1}}$. Junction-to-case thermal resistance depends on the device package. The heat sink reduces the case-to-ambient thermal resistance and therefore the overall junction-to-ambient thermal resistance. The junction-to-ambient thermal resistance can be computed from Equation 9.6:

$$1/\theta_{\mathrm{J-A}} = 1/\theta_{\mathrm{J-C}} + 1/\theta_{\mathrm{C-A}} \tag{9.6}$$

where θ_{J-A} is the junction-to-ambient thermal resistance; θ_{J-C} is the junction-to-case thermal resistance; and θ_{C-A} is the case-to-ambient thermal resistance.

The passive technique of heat removal using a heat sink becomes impractical for moderate and large-power diode lasers. The other limitation is that it cannot be used to operate the laser diode at a temperature lower than the ambient temperature, nor can it be used to stabilize the temperature. Use of an active temperature stabilization mechanism therefore becomes essential when working with moderate and high-power laser diodes and also in low-power laser diodes where the application demands precise temperature control. A thermoelectric cooling device based on the Peltier effect is the heart of such a system.

9.5.1 Thermoelectric Cooling Fundamentals

A common technique of active temperature stabilization of laser diodes makes use of a thermoelectric (TE) cooler as the control element, a temperature sensor and a properly designed feedback loop. The TE cooler is a reversible solid-state heat pump whose operation is based on the Peltier effect. According to the Peltier effect, when an electric current is passed through a junction of dissimilar metals heat is created or absorbed at the junction depending upon the direction of flow of current. The heat transfer according to the Peltier effect takes place in the direction of flow of charge carriers. For example, if we took an N-type semiconductor and connected it across a source of DC voltage, electrons (that are the charge carriers in this case) will flow from the contact connected to the negative terminal of the voltage source to the contact connected to the positive terminal of the voltage source. As a result, the negative terminal becomes cold due to absorption of heat and the positive terminal becomes hot due to the release of heat. In the case of a P-type semiconductor, the opposite occurs with the negative terminal becoming hot and the positive terminal becoming cold. Neither N-type or P-type pellets make practical thermoelectric devices. If we arranged N-type and P-type pellets in the form of a couple by connecting them electrically in series and thermally in parallel as shown in Figure 9.21, heat would be transferred in the same direction as shown in Figure 9.21. A cascaded arrangement of such couples as shown in Figure 9.22 can be used to increase the cooling/heating capacity of the device.

A practical TE module is usually a two-dimensional array of P-N couples connected electrically in series and sandwiched between two thermally conducting and electrically insulating faces as shown in Figure 9.23. Both single-stage and multiple-stage TE cooler modules are commercially available.

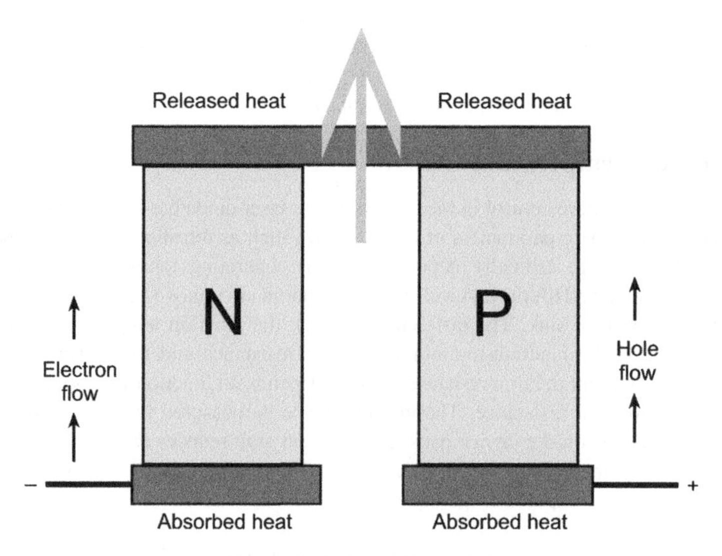

Figure 9.21 Basic TE couple.

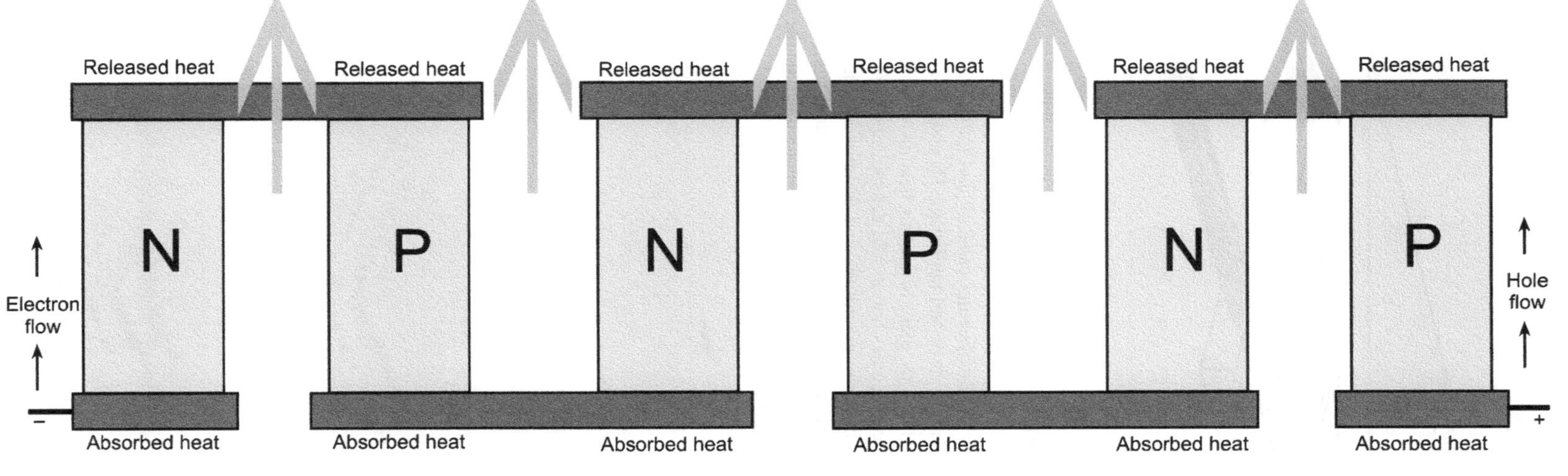

Figure 9.22 Cascaded arrangement of multiple TE couples.

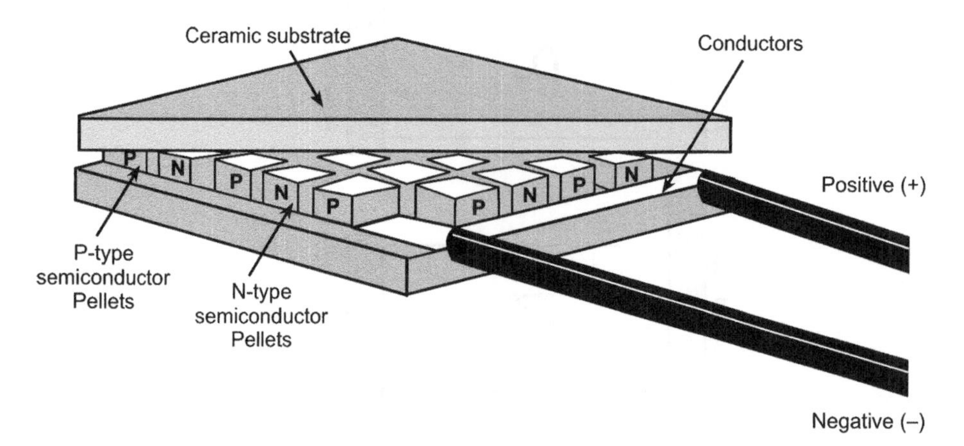

Figure 9.23 Practical TE module.

Multiple-stage modules offer higher cooling/heating capacity and maximum differential temperature specifications. Figure 9.24 depicts single-stage and multistage TE cooler modules.

A TE module literally pumps heat from one place to another, heating one face and cooling the other in the process. Whether a particular face becomes hot or cold depends upon the direction of current flow through the TE module. By controlling the magnitude and direction of the current drive to the TE module, it can therefore be used as a control element in a temperature-stabilization circuit. The TE module pumps the heat dissipated in the laser diode onto a heat sink mounted on the supposedly hot face of the TE module. The heat is transferred from the heat sink to the ambient.

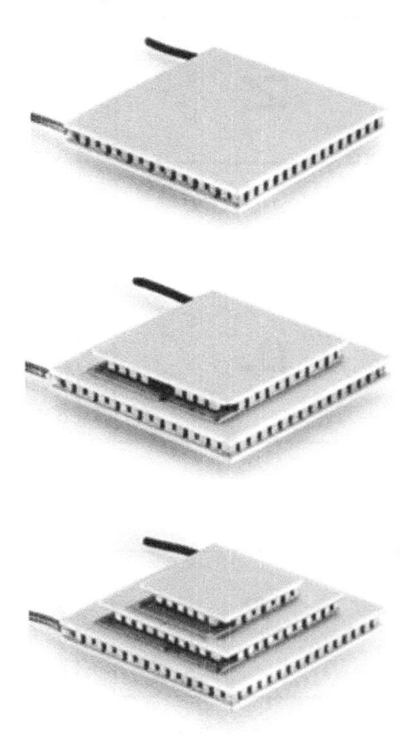

Figure 9.24 Photographs of single-stage and multistage TE modules.

9.5.2 Thermoelectric Cooler: Performance Characteristics

TE modules are characterized by maximal performance parameters, which include:

1. maximum temperature difference ΔT_{max} along the module at no heat load
2. cooling/heating capacity Q_{max} corresponding to zero temperature difference
3. device current I_{max} measured at ΔT_{max}
4. terminal voltage V_{max} corresponding to I_{max} with no heat load
5. coefficient of performance (COP)
6. AC resistance.

ΔT_{max} is the maximum possible differential temperature across the TE module for a given value of temperature at the hot side of the module. Maximum differential temperature always occurs when the heat load is zero.

Q_{max} is the maximum heat load for $\Delta T = 0$. Note that Q_{max} is not the maximum amount of heat the device can handle. If the heat load exceeds Q_{max}, the device still pumps heat with the difference that ΔT becomes negative. This implies that the cold side is now the hot side. The heat load the device can pump for a given value of ΔT decreases with increasing ΔT.

I_{max} is the DC current that will produce maximum possible ΔT (ΔT_{max}) across the TE module for a given value of hot side temperature. I_{max} is not the maximum current the device can withstand before it fails; if the drive current is less than I_{max}, it will be insufficient to produce the highest possible ΔT. If it is more than I_{max}, the power dissipated in the Peltier device will increase. This increases the temperature of the hot side and reduces the attainable ΔT.

V_{max} is the DC voltage that would deliver the maximum possible ΔT (ΔT_{max}) across the module. Again, it is not the maximum voltage that the device can withstand before failing. For terminal voltage greater than V_{max}, the power dissipated in the Peltier device goes up, increasing the hot side temperature, which reduces ΔT. If it is lower than V_{max}, there would be insufficient current to achieve maximum ΔT.

Coefficient of performance (COP) is defined as the ratio of pumped heat load to electrical power supplied to the device. Electrical power input to the TE module is the product of drive current and terminal voltage. COP is representative of a thermoelectric cooling/heating system.

AC resistance is the internal resistance of the TE module. It may be mentioned here that the parameter is dependent on the type of applied voltage. It refers to the fact that an AC voltage must be applied to the device to measure its resistance. If a DC voltage were applied, the resulting differential temperature developing across the device would give rise to a Seebeck voltage, which would oppose the applied DC voltage and make the device appear to be more resistive than it actually is. It may be mentioned here that special test equipment is used to measure this parameter.

Various performance parameters are interdependent. This interdependence is best understood with the help of performance curves, which are invariably given in the device data sheet provided by the manufacturer. The true analysis of a TE module operation can only be performed with the help of these performance curves. Figure 9.25a–d depicts performance curves of a typical TE module.

Refer to Figure 9.25a, which shows the variation of heat load pumping capability Q_C as a function of ΔT for different values of I_{max}. Two important inferences are evident from the family of curves: (1) heat load pumping capability is inversely proportional to ΔT for a given value of I_{max}; and (2) as I_{max} decreases, both heat load pumping capability for a given ΔT and attainable ΔT for a given heat load pumping capability decrease. The manufacturer's data sheet usually gives performance curves for two different hot side temperatures.

Figure 9.25b shows the family of curves depicting the relationship between terminal voltage V and differential temperature ΔT for different values of I_{max}. Again, two important inferences can be made: (1) terminal voltage is more or less independent of ΔT; and (2) for a given value of ΔT, terminal voltage V varies directly with I_{max}.

Figure 9.25c depicts the family of curves describing dependence of heat load Q_C on terminal voltage for different values of differential temperature ΔT. The following important conclusion can be drawn from these curves: for a given value of ΔT, heat load pumping capability reduces with a decrease in terminal voltage. In other words, for a given heat load pumping requirement, an increase in terminal voltage increases the achievable ΔT.

Figure 9.25d shows the family of curves depicting the dependence of COP on drive current for different values of ΔT. Two important inferences can be made: (1) for a given ΔT, COP peaks at a certain

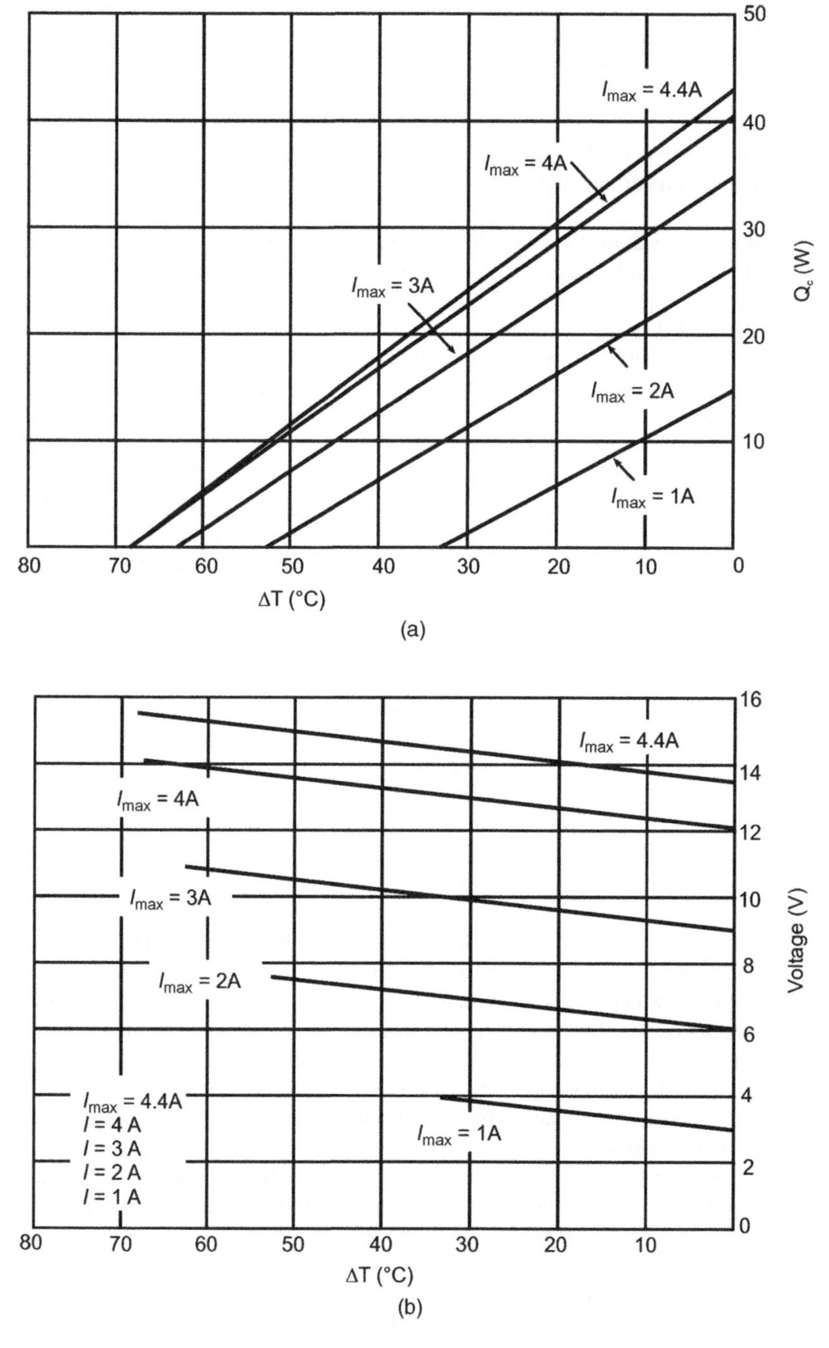

Figure 9.25 Performance curves of TE module.

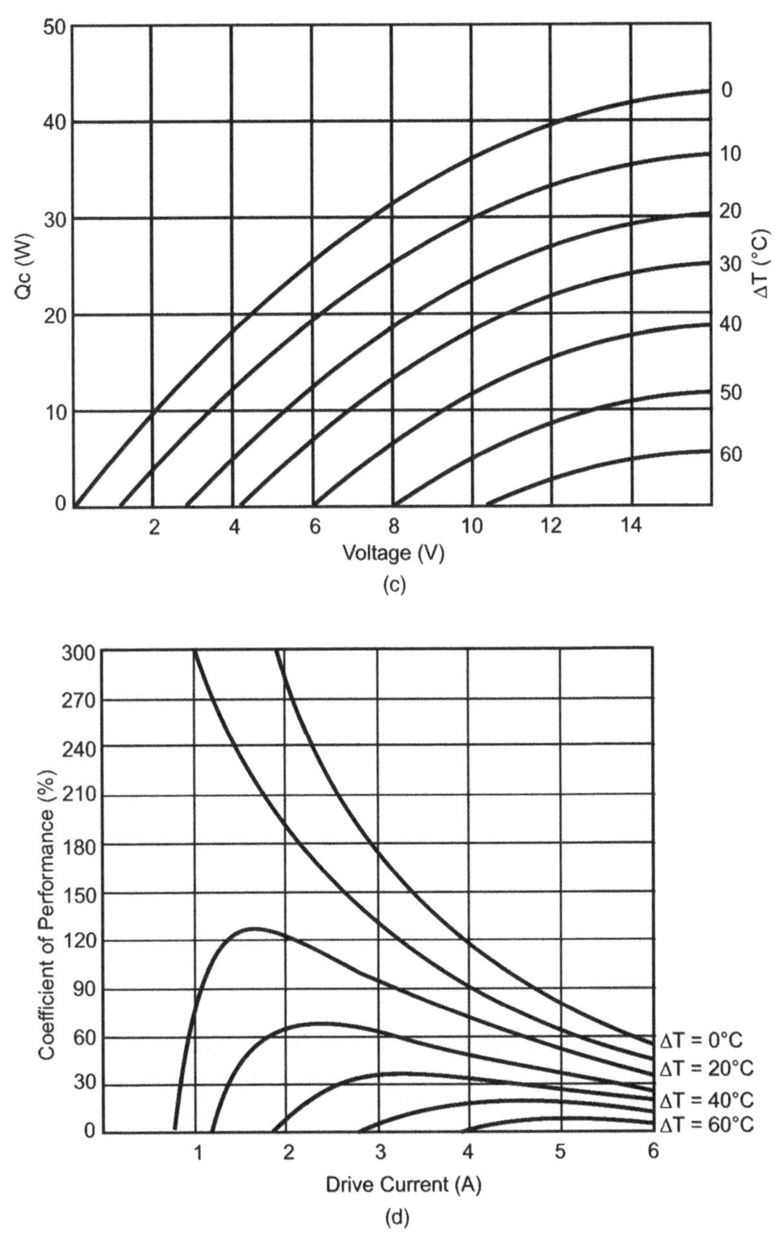

Figure 9.25 *(Continued)*

value of drive current; the curve rolls off on both sides; and (2) the peak value of COP also reduces with an increase in ΔT.

9.5.3 TE Module Selection

Major performance parameters that dictate selection criteria for the TE module for a given application include operating temperature difference ΔT between the cold side and the hot side, heat load to be pumped by the TE module, drive current and terminal voltage. In some cases, there may even be dimensional restrictions.

The heat load to be pumped by the TE module may consist of two types – active or passive – or a combination of the two. An active load is the heat dissipated by the device being cooled. In the case of a laser diode, it equals the total input electrical power minus the laser output power. Passive heat loads are parasitic in nature and may originate from conduction, convection or radiation phenomena.

Conductive heat transfer takes place by the direct impact of molecules moving from a high-temperature region to a low-temperature region. In most cases of thermoelectric cooling, TE modules provide the largest conductive path. Conductive heat loading on a system may occur through lead wires and mounting screws which form a thermal path from the device being cooled to the heat sink or ambient environment. Conductive loading is described by Equation 9.7:

$$Q_{\mathrm{cond}} = \frac{kA\Delta T}{L} \tag{9.7}$$

where

Q_{cond} = conductive heat load (W)
k = thermal conductivity of the material ($\mathrm{W\,m^{-1}\,{}^{\circ}C^{-1}}$)
A = cross-sectional area of the material ($\mathrm{m^2}$)
L = length of the heat path (m)
ΔT = temperature difference across the heat path (°C) equal to difference between hot side and cold side temperatures.

Convective heat loads on TE modules are generally a result of natural (or free) convection. Naturally occurring air currents cause heat transfer between the device to be cooled or heated and the ambient. Convective heat load may become a limiting factor for large temperature differences. The rate at which convective heat transfer takes place depends upon the geometry of the heat load and surroundings. Convective heat transfer load can be expressed by Equation 9.8:

$$Q_{\mathrm{conv}} = hA\Delta T \tag{9.8}$$

where

Q_{conv} = convective heat load
h = convective heat transfer coefficient ($\mathrm{W\,m^{-2}\,{}^{\circ}C^{-1}}$)
A = exposed surface area ($\mathrm{m^2}$)
ΔT = temperature difference.

Radiative heat transfer occurs through electromagnetic radiation emitted from one object and absorbed by the other. Radiation heat loads are usually considered insignificant when the system is operated in a gaseous environment, as the other forms of passive heat loads are relatively much stronger. Radiation loading is usually significant in systems with small active loads and large temperature differences, especially when operating in a vacuum environment. Radiative heat transfer varies with the fourth power of the temperature difference between the heat load and the surroundings. It is expressed by Equation 9.9:

$$Q_{\mathrm{rad}} = \sigma eA\left(T_1^4 - T_2^4\right) \tag{9.9}$$

where

Q_{rad} = radiative heat load
σ = Stephan–Boltzmann constant ($5.67 \times 10^{-8}\,\mathrm{W\,m^{-2}\,K^{-1}}$)
e = emissivity of enclosed body
A = area of enclosed body ($\mathrm{m^2}$)
T_1 = absolute temperature of enclosed body (K)
T_2 = absolute temperature of surroundings (K).

Having computed the heat load Q to be pumped by the TE module, the next obvious step is to compute the required value of differential temperature ΔT. Performance curves are used to pick the right TE module from a range of single-stage and multistage devices available from a host of manufacturers. For example, a TE module would give ΔT_{max} specification in the range of 65–70 K for single-stage devices, 90–100 K for two-stage devices, 110–120 K for three-stage devices and 120–130 K for four-stage device). Also, in both single-stage and multistage TE modules, devices with different heat load capacity are available to meet the designer's requirements. Equation 9.10 relates the pumped heat load Q to the performance parameters of the TE module:

$$Q = Q_{max} \times \left[1 - \left(\frac{\Delta T}{\Delta T_{max}}\right)\right] \tag{9.10}$$

$$\Delta T = \Delta T_{max} \times \left[1 - \left(\frac{Q}{Q_{max}}\right)\right] \tag{9.11}$$

Equation 9.10 or 9.11 may be used to choose the right TE module to meet the required heat load and differential temperature specifications, and are valid for single-stage or at the most two-stage TE modules.

9.5.4 Heat Sink Selection

Selecting the right heat sink from those available or designing one to meet application requirements is crucial to the overall operation of the thermoelectric system. Differential temperature ΔT between hot and cold sides of the TE module is an important specification of a thermoelectric cooling/heating system. The hot and cold side temperatures in turn depend upon the temperature of the surroundings and heat sink-to-ambient thermal resistance. The thermal resistance determines the temperature difference between that of the heat sink and the surroundings. The product of thermal resistance (measured in $°C\,W^{-1}$) and heat dissipated in the heat sink (measured in W) gives this temperature difference. An inadequate heat sink would cause a larger temperature difference and thus put more stringent demands on the TE module. A typical design parameter might be to limit the heat sink temperature rise above ambient in the range of 10–20 °C.

Heat power dissipated into the heat sink is equal to the sum of total pumped heat load, comprising active and passive loads, and the electrical power input to the TE module. The latter is the product of drive current and terminal voltage. The objective of the heat sink design is to minimize thermal resistance in order to keep the hot side temperature of the TE module as close to the ambient temperature as possible. This is achieved through increased exposed surface area and may require forced air or liquid circulation. The three basic types of heat sinks include natural convective, forced convective and liquid-cooled types. Typical values of thermal resistance are within the range 0.5–5 $°C\,W^{-1}$ for naturally convective and 0.02–0.5 $°C\,W^{-1}$ for forced convective heat sinks. Among the three types, liquid-cooled heat sinks offer the best performance with thermal resistance ranging from 0.005–0.15 $°C\,W^{-1}$. Most applications involving thermoelectric cooling require forced convective or liquid-cooled heat sinks. Having computed the required thermal resistance, the next step is to look for a heat sink or heat sink/fan combination that could deliver the intended performance. It may not always be possible to acquire one with the desired thermal resistance which fits the dimensional and other constraints; a practical approach would be to choose the best compromise.

Example 9.3

A typical lead selenide (PbSe) infrared detector with a resistance of 1 MΩ, operated at a bias voltage of 100 V, is to be thermoelectrically cooled to −23 °C from an ambient of 27 °C. If the total exposed surface area of the detector is $10\,\text{cm}^2$, determine the total heat load that would need to be pumped by the TE module. The Stefan-Boltzmann constant $\sigma = 5.67 \times 10^{-8}\,\text{W m}^{-2}\,\text{K}^{-4}$; emissivity $e = 0.9$; and convective heat transfer coefficient $h = 20\,\text{W m}^{-2}\,\text{K}^{-1}$.

Solution

1. We have: temperature $T_1 = 300\,\text{K}$, $T_2 = 250\,\text{K}$, emissivity $e = 0.9$, convective heat transfer coefficient $h = 20\,\text{W m}^{-2}\,\text{K}^{-1}$, Stefan–Boltzmann constant $\sigma = 5.67 \times 10^{-8}\,\text{W m}^{-2}\,\text{K}^{-4}$ and area $A = 10\,\text{cm}^2 = 0.001\,\text{m}^2$
2. Active load of the detector $= V^2/R = (100)^2/10^6 = 0.01\,\text{W}$
3. Heat load due to convective heat transfer:

   $$Q_{\text{conv}} = hA\Delta T = 20 \times 0.001 \times 50 = 1\,\text{W}$$

   Heat load due to radiative transfer:
   $$Q_{\text{rad}} = \sigma e A\left(T_1^4 - T_2^4\right) = 5.67 \times 10^{-8} \times 0.9 \times 0.001 \times \left[(300)^4 - (250)^4\right] = 0.214\,\text{W}$$
4. Total heat load $= 0.01 + 1.0 + 0.214\,\text{W} = 1.224\,\text{W}$.

Example 9.4

A thermoelectric cooling system for a laser diode is designed to handle a temperature difference of 30 °C between the hot and cold sides of the TE module. Total pumped heat load is 10 W and the TE module chosen for the purpose is driven by 1.5 A at a terminal voltage of 10 V. If the laser diode is cooled to 20 °C and the heat sink on the hot side has a thermal resistance of $0.4\,°\text{C W}^{-1}$, determine the maximum ambient temperature for which the system will provide the desired cooling.

Solution

1. Maximum hot side temperature that can be handled by the system $= 20 + 30 = 50\,°\text{C}$
2. Now, heat sink temperature $= 50\,°\text{C}$
3. Heat dissipated in the heat sink $= 10 + 1.5 \times 10\,\text{W} = 25\,\text{W}$
4. Thermal resistance of heat sink $= 0.4\,°\text{C W}^{-1}$
5. Therefore, temperature difference between heat sink and ambient $= 25 \times 0.4 = 10\,°\text{C}$
6. Maximum acceptable ambient temperature $= 50 - 10 = 40\,°\text{C}$.

Example 9.5

A 10 W laser diode with an electrical to optical conversion efficiency of 40% is to be operated at 10 °C against a worst-case ambient of 35 °C using a TE-cooler-based temperature controller. Heat load due to conduction and convection are estimated to be equal to 5 W. Radiation loss is negligible. The thermoelectric module chosen for the purpose requires electrical input power of 20 W to provide the desired cooling. If the temperature difference between the heat sink and ambient is not to exceed 5 °C, determine the following:

1. COP of the thermoelectric system
2. maximum allowable thermal resistance of heat sink
3. minimum value of $Q_{\max}$ of TE module if its $\Delta T_{\max}$ rating is 60 °C.

Solution

1. COP = pumped heat load/electrical input power to TE module
2. Electrical power input to laser diode $= 10/0.4\,\text{W} = 25\,\text{W}$. Therefore, active heat load of the laser diode $= 25 - 10\,\text{W} = 15\,\text{W}$
3. Passive heat load $= 5\,\text{W}$
4. Therefore, total pumped heat load $= 15 + 5 = 20\,\text{W}$
5. COP $= 20/20 = 1$

6. Worst-case ambient temperature = 35 °C; therefore maximum allowable temperature difference between heat sink and ambient = 5 °C
7. Total heat dissipated in heat sink = 20 + 20 W = 40 W
8. Therefore, maximum allowable thermal resistance of heat sink = 5/40 = 0.125 °C W^{-1}. Therefore, worst-case hot side temperature = 35 + 5 = 40 °C
9. Temperature difference between cold and hot sides of TE module, $\Delta T = 40 - 10 = 30$ °C
10. ΔT_{max} rating of TE module = 60 °C
11. Therefore, $\Delta T/\Delta T_{max} = 30/60 = 0.5$
12. The minimum value of Q_{max} can be computed:

$$Q = Q_{max} \times \left[1 - \left(\frac{\Delta T}{\Delta T_{max}}\right)\right] = Q_{max} \times (1 - 0.5) = 0.5 \times Q_{max}$$

13. Therefore, $Q_{max} = Q/0.5 = 20/0.5 = 40$ W.

9.5.5 Thermoelectric Cooler Drive and Control Circuits

Precise temperature control using thermoelectric modules is accomplished by continuously monitoring the temperature of the device to be cooled or heated with a temperature sensor such as thermistor and employing an electronic feedback loop to alter the drive current to the TE module in case the controlled temperature tends to deviate from the specified value. Figure 9.26 shows the basic block schematic of a TE module-based temperature control circuit. The key building blocks of the drive and control circuit include temperature sensor, error amplifier, error signal processor and a bipolar output drive circuit. A bipolar output driver provides the required power to drive the TE module. The output stage in most cases is designed to allow flow of drive current in either direction through the TE module. This enables both cooling and heating of the device to maintain its temperature at the specified value regardless of ambient temperature being higher or lower. Like laser diode drive circuits, leading international manufacturers offer integrated circuits with on-chip TE module drive and control functions.

9.5.5.1 Temperature Sensing Circuits

The temperature sensor in most cases is a negative temperature coefficient (NTC) thermistor. Other commonly used temperature sensors include semiconductor devices that employ temperature-dependent forward-biased junction voltage and resistance temperature detectors (RTD). A thermistor

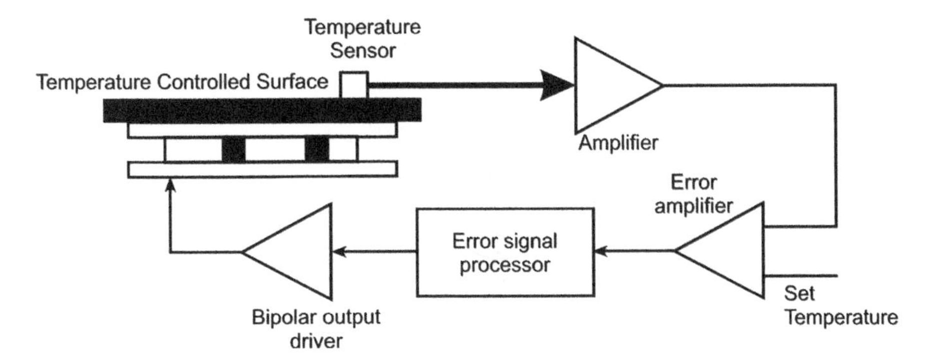

Figure 9.26 Block schematic of thermoelectric temperature drive and control circuit.

has the advantage of high sensitivity but suffers from non-linearity of resistance versus temperature characteristics. Semiconductor devices and RTD sensors are linear but are less sensitive than thermistor.

When using a thermistor as a temperature sensor, it is connected either in a simple voltage divider or a balanced bridge circuit as shown in Figure 9.27a and b. In both cases, the temperature-dependent resistance of the thermistor is compared to a known fixed resistance. The circuit of Figure 9.27a generates a single-ended temperature-dependent voltage. Resistance R in the voltage divider network is chosen to be equal to the resistance of the thermistor at the desired stabilized temperature in order to take full advantage of the available voltage range. If temperature is to be controlled over a specific range, R is taken to be equal to the resistance of the thermistor corresponding to the centre of the temperature range. For example, if the desired temperature range to be controlled is 0 to –40 °C, reference temperature would be −20 °C.

The balanced bridge arrangement of Figure 9.27b produces a differential output. Again, R is chosen to be equal to the resistance of the thermistor at the desired temperature. In this case, output voltage will be zero when the temperature is equal to the desired value. As the temperature deviates from the desired value, output voltage swings in a direction depending upon whether temperature is lower or higher than the desired value. An intrinsic advantage of this circuit is that voltage versus temperature characteristics is more linear than the resistance versus temperature characteristics of the thermistor.

Another commonly used temperature sensor is the integrated temperature sensor. AD590 from Analogue Devices is one such sensor. It produces an output current proportional to the absolute temperature. It has an operational range of −55 °C to +150 °C and a current to temperature coefficient of 1.0 μA K^{-1} over the supply voltage range of 4–30 V. Figure 9.28 shows the application circuit along with transfer characteristics.

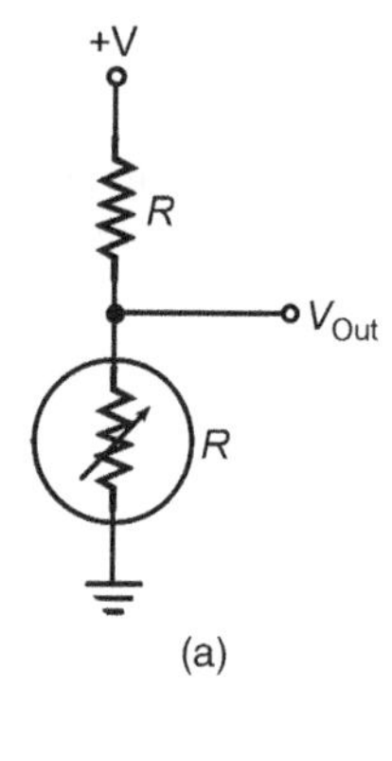

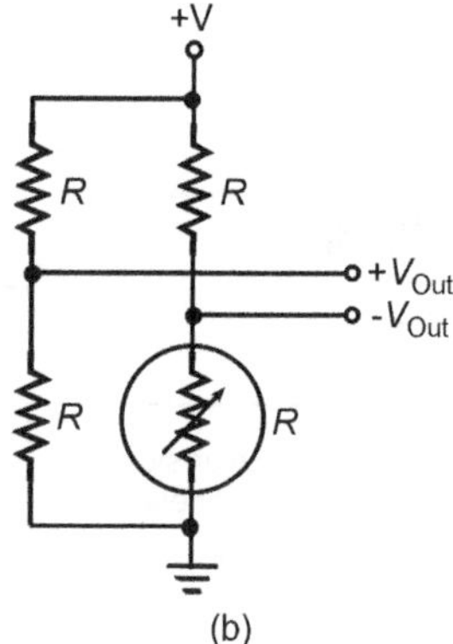

Figure 9.27 Temperature sensing circuits

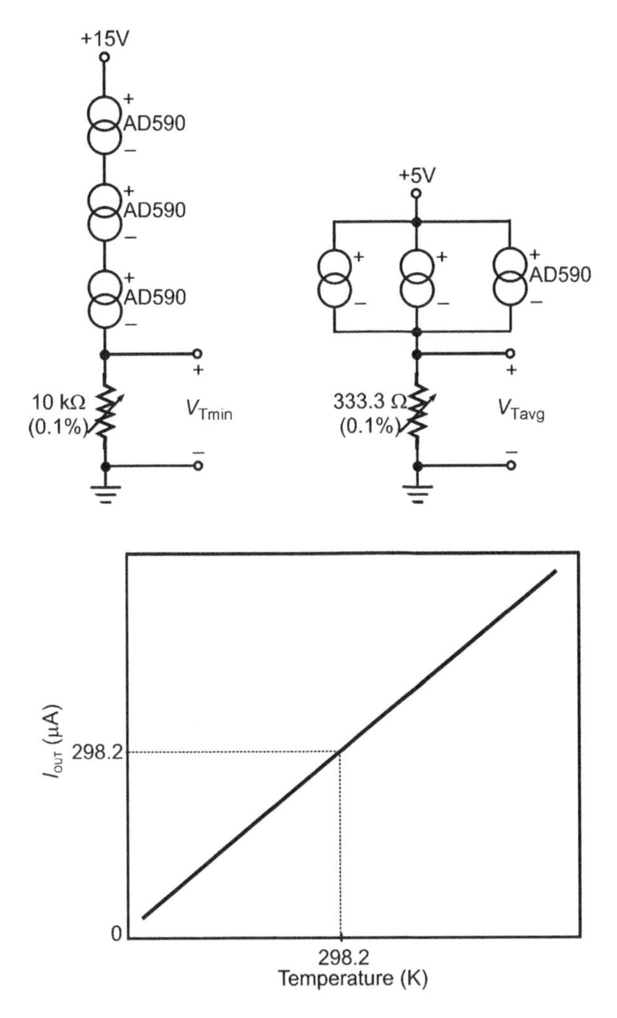

Figure 9.28 AD590 application circuit and transfer characteristics.

9.5.5.2 Error Amplifier

The error amplifier amplifies the difference between the two voltage signals representing the measured and the desired temperatures. Figure 9.29 shows a typical error amplifier interfaced with temperature-sensing and reference-voltage-generating circuits. In the circuit of Figure 9.29, a thermistor is used for temperature sensing and the output of the temperature-sensing circuit is compared with the set reference using an error amplifier.

9.5.5.3 Error Signal Processor

An error signal processor could be anything from a simple on-off controller to a proportional controller or proportional-integral (PI) or even fully digital proportional-integral-differential (PID) controller. *On-off control* is the simplest technique for controlling the temperature and is the least preferred. In this method, power to the TE module is switched from full on to full off, which degrades its life. A comparator circuit with the two inputs as reference voltage representing desired temperature and the voltage representing measured temperature makes an on-off controller.

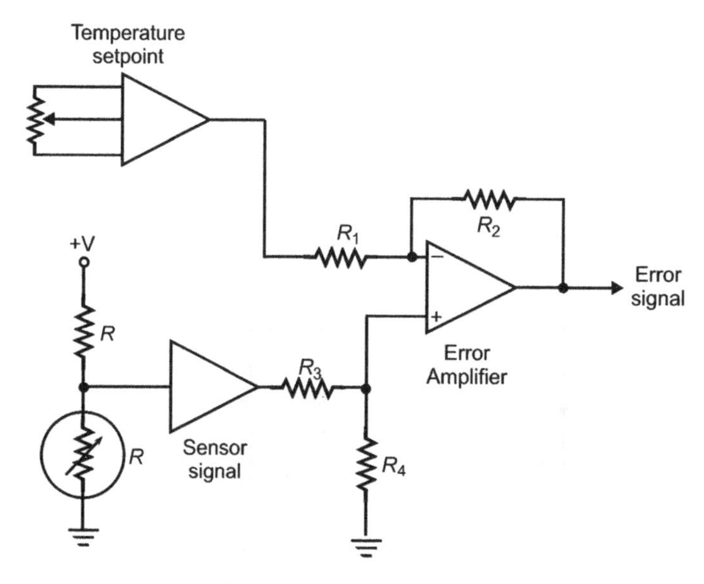

Figure 9.29 Error amplifier.

In *proportional control*, the drive signal to the TE module is proportional to the difference between the actual and desired temperatures. In a proportional controller, there is always a residual error even after the controller has settled to the final state. This error is proportional to the difference between the desired temperature and the actual temperature and is inversely proportional to the gain of the control loop. A simple inverting amplifier makes a proportional controller. In Figure 9.30, the inverting amplifier provides the desired value of the proportional gain. In a digital system, the proportional controller computes the error from the measured output and a user-defined input to a program. The error signal is multiplied by a proportionality constant to generate the control signal.

The problem of residual error of a proportional controller can be overcome by the addition of an integrator in the control loop. The result is a *proportional-integral controller*. One disadvantage of PI

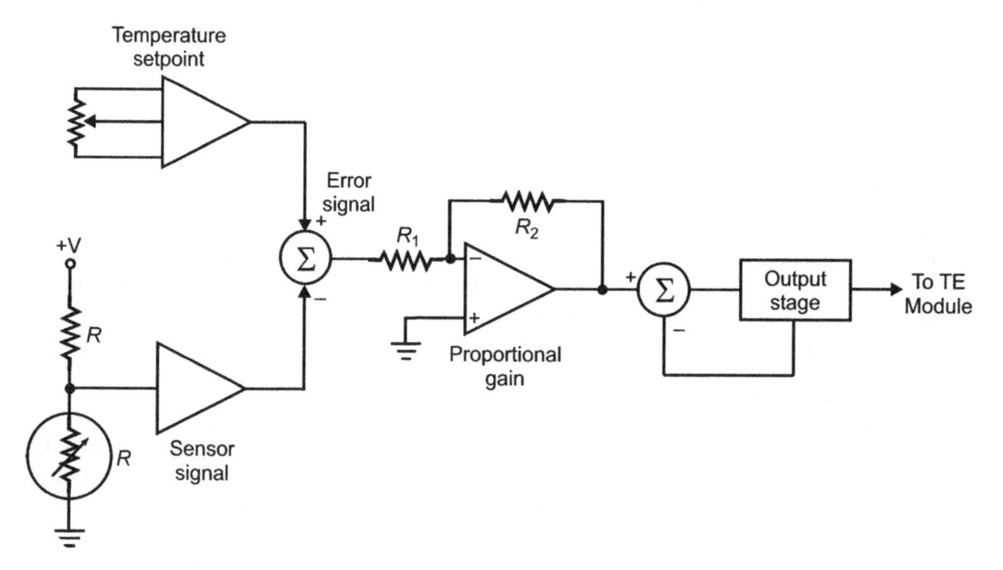

Figure 9.30 Error amplifier interfaced with temperature sensing and reference voltage generating circuits.

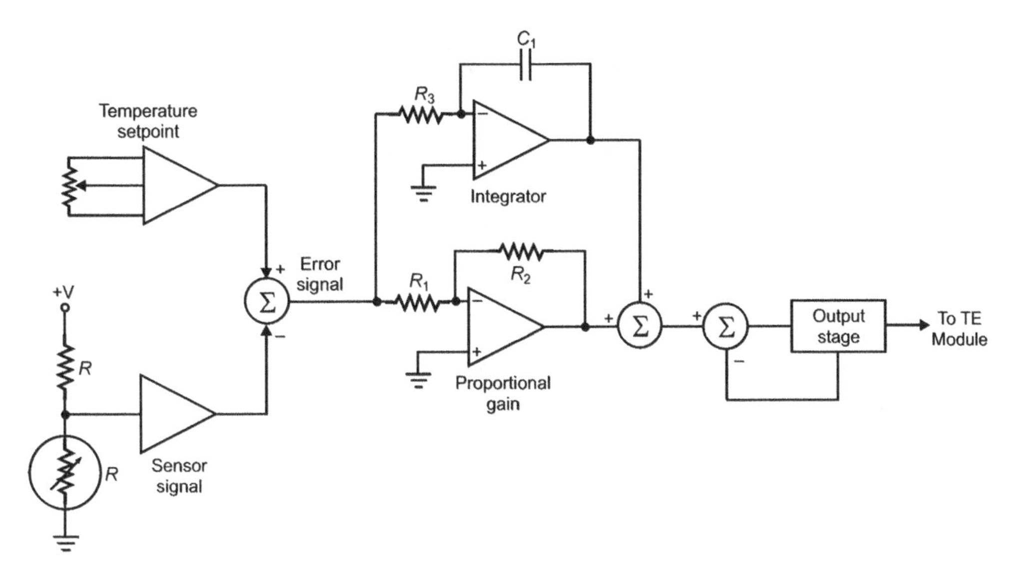

Figure 9.31 Proportional-integral controller.

control is that it would be slow to respond to large residual errors. Figure 9.31 shows one type of proportional-integral controller. The circuit is nothing but a parallel connection of a proportional controller and an integral controller.

A *proportional-integral-differential controller* (PID) overcomes the problem of slow response to large residual errors encountered in PI controllers. The addition of the derivative term improves the transient response of the loop. This type of controller is mainly used in applications where large thermal loads must be controlled rapidly and accurately. Figure 9.32 shows one type of PID controller, which is basically a parallel connection of proportional, integral and differential controllers.

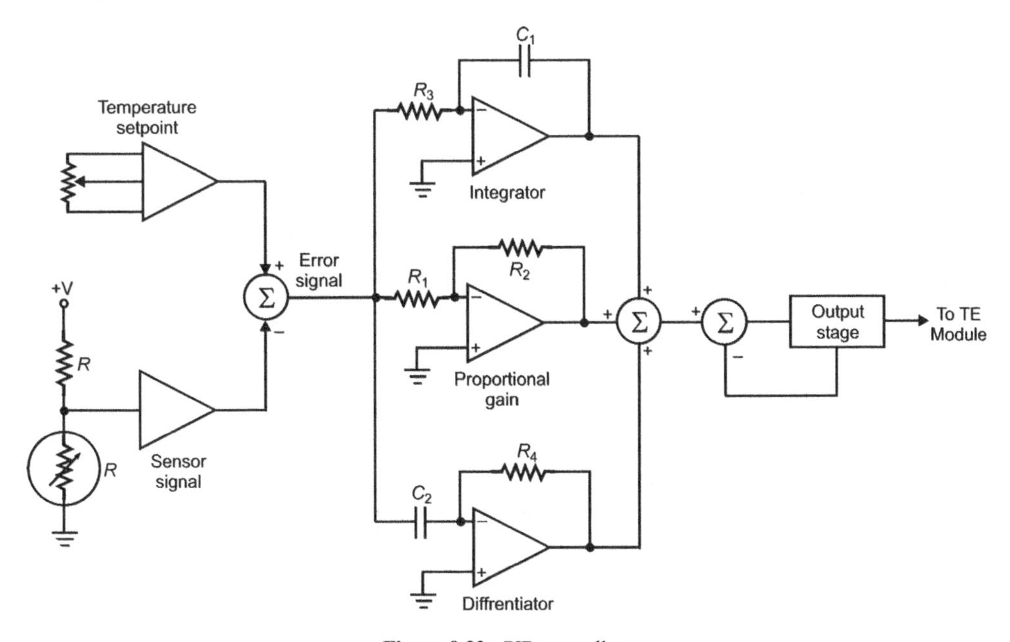

Figure 9.32 PID controller.

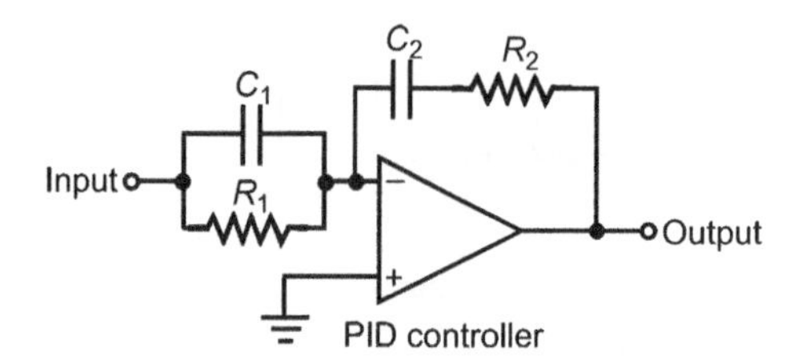

Figure 9.33 Single opamp PID controller.

Equations 9.12–9.14 define K_P, K_I and K_D (proportional, integral and differential gains, respectively):

$$K_P = R_2/R_1 \tag{9.12}$$

$$K_I = R_3 \times C_1 \tag{9.13}$$

$$K_D = 1/C_2R_4 \tag{9.14}$$

Figure 9.33 shows a common PID controller circuit configured around a single opamp. The expressions for proportional, integral and differential gains are given by Equations 9.15–9.17 respectively. A more practical circuit is shown in Figure 9.34, for which Equations 9.15–9.17 hold well.

$$K_P = R_2/R_1 + C_1/C_2 \tag{9.15}$$

$$K_I = R_1 \times C_2 \tag{9.16}$$

$$K_D = 1/C_1R_2 \tag{9.17}$$

9.5.5.4 Output Stage

The output stage provides the necessary drive power to the TE module. In most cases, electronic systems are designed to operate from a single positive DC voltage supply. TE modules also need to be driven in bipolar mode to cater for both heating and cooling operations. A commonly used circuit topology to

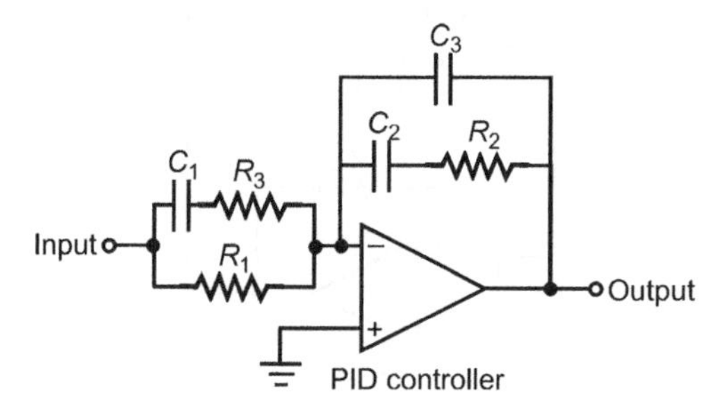

Figure 9.34 Single opamp PID controller: practical circuit.

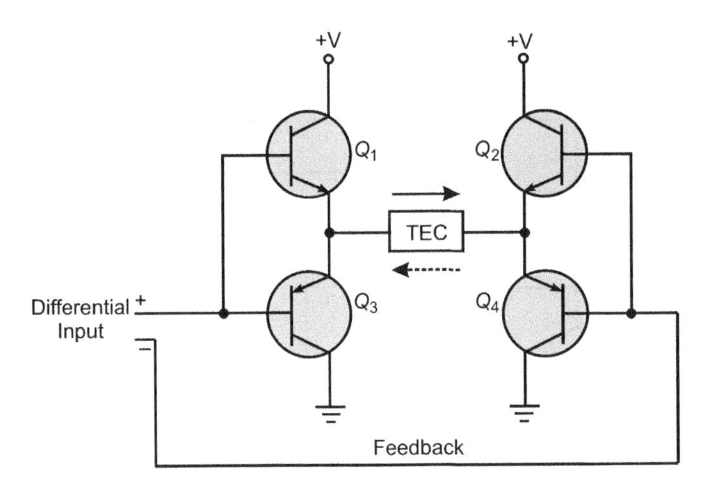

Figure 9.35 Output stage based on half-bridge circuit topology.

provide bipolar drive to TE modules while operating from a single DC voltage is the half-bridge circuit topology depicted by Figure 9.35. The circuit is usually driven at the input by a driver amplifier stage with a differential output.

The circuit functions as follows. When the differential input to the output stage swings positive ($V+ > V^{-}$), transistors Q_1 and Q_4 conduct while transistors Q_2 and Q_3 remain in cut-off. Current flows through the TE module in the direction as shown by the solid arrow. The voltage across the TE module and the corresponding current flowing through it is proportional to the input drive voltage. When the differential input drive swings negative, transistors Q_2 and Q_3 conduct while transistors Q_1 and Q_4 remain in cut-off. As a result, current flows through the TE module in the opposite direction as shown by the dashed arrow. Again, voltage across the module and the corresponding current are proportional to the input drive voltage. All transistors operate as emitter followers and, as a consequence of this, the maximum voltage available across the TE module is limited to the maximum differential voltage swing available at the input minus twice the base-emitter voltage drop of the transistor. Maximum current through the TE module is also limited to the maximum drive current supplied by the driver amplifier output multiplied by the DC current gain h_{FE} of the transistor.

The shortcoming of limited voltage and current drive capability of the half-bridge circuit of Figure 9.35 is overcome by having a cascaded arrangement of two half-bridge stages as shown in Figure 9.36. When the differential input drive swings positive, transistors Q_1, Q_4, Q_5 and Q_8 conduct while transistors Q_2, Q_3, Q_6 and Q_7 are non-conducting. Current flows through the TE module as shown by the solid arrow. When the differential input drive swings negative, transistors Q_2, Q_3, Q_6 and Q_7 conduct and transistors Q_1, Q_4, Q_5 and Q_8 go to cut-off. Current flows through the TE module as shown by the dashed arrow.

This circuit not only has a much higher current drive capability, but it also provides a rail-to-rail differential drive of nearly $\pm V$ across the TE module when operated from a DC supply of $+V$. The circuit of Figure 9.36 also provides current to the TE module while the circuit of Figure 9.35 applies a voltage. Since the heat pumped by a TE module is proportional to current flowing through it, current mode control allows better and more uniform control of the TE module than the voltage mode control. The DC transfer characteristics (I_{Out} versus V_{In}) of the half-bridge driver circuit of Figures 9.35 and 9.36 will have a dead zone of $\pm 2V_{BE}$. The dead zone may be reduced or eliminated by using negative feedback to control the gain of the driver amplifier, ensuring that the four transistors are not all switched on at the same time.

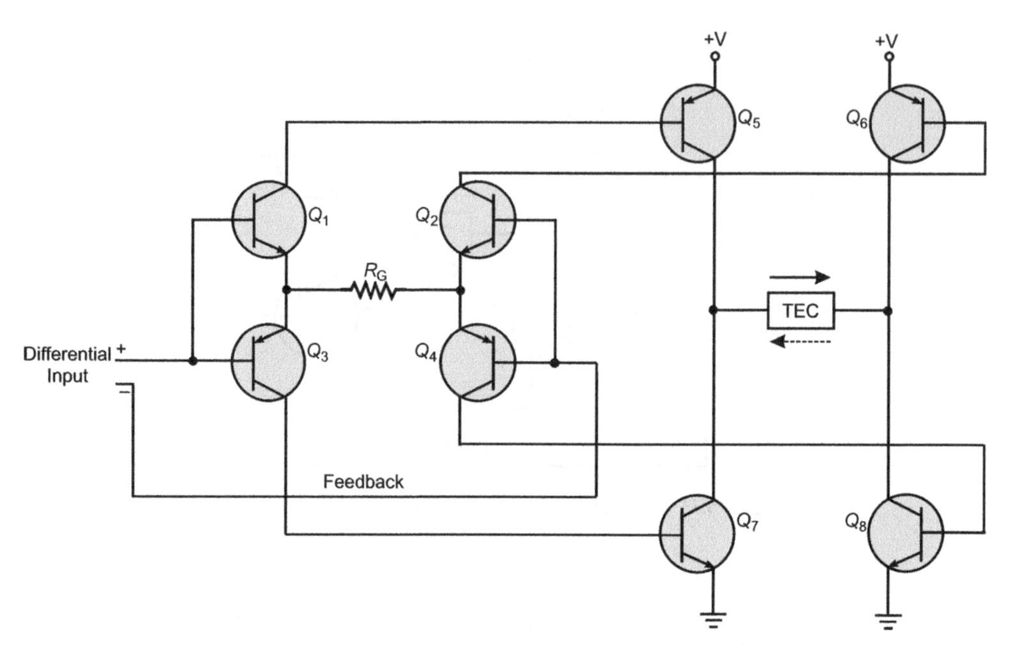

Figure 9.36 Output stage with enhanced voltage and current drive.

9.6 Summary

- Laser diode electronics primarily comprises a source of drive current and a temperature controller. Since the wavelength of a semiconductor diode laser is strongly dependent on drive current and operating temperature, the stability of drive current and operating temperature dictates the output wavelength stability.
- Laser diodes are highly susceptible to damage due to excessive drive current, electrostatic discharge (ESD) and transients. Laser diode damage manifests itself in the form of: reduced output power, shift in threshold current, increase in beam divergence and/or failure to laser action, thus producing an LED-like output only.
- A laser diode protection strategy needs to be multi-pronged with focus on laser diode instrumentation, system set-up, power line conditioning and laser diode handling. Laser diode instrumentation includes the current source used to drive the laser diode and temperature controller, and the relevant damage mechanisms are over-current, overheating, current spikes and power line surges.
- Protection features desirable for laser diode current sources include slow-start, current limit, over-voltage protection and power line transient suppression. Protection needs of the temperature controller are also very stringent. The laser diode lifetime is inversely proportional to operating temperature. The lifetime has been observed to improve by an order of magnitude for every 30 °C reduction in the laser diode case temperature. Laser diodes should be operated at as low a temperature as possible.
- Laser diodes are operated in either constant-current mode or constant-output-power mode. In the case of the former, the drive current to the laser diode is maintained constant irrespective of variations in the source of power input. The circuit topology is usually configured around a current-sensing element that continuously senses the drive current and produces a proportional voltage. This is then compared with a reference voltage representing the desired value of drive current to generate the error signal. After suitable conditioning, the error signal is then fed back to restore the drive current to the desired value.
- In the case of constant-power mode of operation, the current source is designed in such a way that it adjusts the drive current in a feedback mode to maintain a constant output power instead of maintaining a constant drive current. Laser diode modules with an integral photodiode suit this

operational mode well. Constant-power mode also precludes the possibility of the output power increasing with a decrease in operating temperature.

- In quasi-CW mode, a laser diode is driven by current pulses that are typically hundreds of microseconds to a few milliseconds wide (compared to a few tens to a few hundreds of nanoseconds in the case of conventional pulsed mode of operation). These diodes are operated at relatively low repetition rates typically of a few hertz to a few tens of hertz. The relatively high peak powers and efficient operation of quasi-CW laser diodes have made them an ideal choice for pumping solid-state lasers.
- A common technique of active temperature stabilization of laser diodes makes use of a thermoelectric (TE) cooler as the control element, a temperature sensor and a properly designed feedback loop. The TE cooler is a reversible solid-state heat pump whose operation is based on the Peltier effect.
- According to the Peltier effect, when electric current is passed through a junction of dissimilar metals heat is created or absorbed at the junction depending upon the direction of flow of current. A practical TE module is usually a two-dimensional array of P-N couples connected electrically in series and sandwiched between two thermally conducting and electrically insulating faces.
- A TE module literally pumps heat from one place to another, heating one face and cooling the other in the process. Whether a particular face becomes hot or cold depends upon the direction of current flow through the TE module. By controlling the magnitude and direction of the current drive to the TE module, it can be used as a control element in a temperature stabilization circuit. The TE module pumps the heat dissipated in the laser diode to a heat sink mounted on the supposedly hot face of the TE module. The heat is transferred from the heat sink to the ambient.
- Heat power dissipated into the heat sink is equal to the sum of total pumped heat load (comprising active and passive loads) and the electrical power input to the TE module. The electrical power input to the TE module is a product of drive current and terminal voltage. The objective of the heat sink design is to minimize thermal resistance in order to keep the hot side temperature of the TE module as close to the ambient temperature as possible.
- Key building blocks of the drive and control circuit for thermoelectric cooling include temperature sensor, error amplifier, error signal processor and a bipolar output drive circuit. The bipolar output driver provides the required power to drive the TE module. The output stage in most cases is designed to allow flow of drive current in either direction through the TE module. This enables both cooling and heating of the device to maintain its temperature at the specified value, regardless of ambient temperature being higher or lower.
- The temperature sensor in most cases is a negative temperature coefficient (NTC) thermistor. Other commonly used temperature sensors include semiconductor devices that employ temperature-dependent forward-biased junction voltage and resistance temperature detectors (RTD).
- The error amplifier amplifies the difference between the two voltage signals representing the measured and the desired temperatures. The error signal processor could be anything from a simple on-off controller to a proportional controller or PI or even fully digital PID controller. The output stage provides the necessary drive power to the TE module.
- In the case of on-off control, power to the TE module is switched from full on to full off. A comparator circuit with the two inputs as reference voltage (representing desired temperature) and the voltage representing measured temperature makes a typical on-off controller. In a proportional controller, the drive signal to the TE module is proportional to the difference between the actual and desired temperatures. In this case, there is always a residual error even after the controller has settled to the final state. This error is proportional to the difference between the desired temperature and the actual temperature and is inversely proportional to the gain of the control loop. The problem of residual error of a proportional controller can be overcome by the addition of an integrator in the control loop. The result is a PI controller. A PID controller overcomes the problem of slow response to large residual errors encountered in PI controllers by the addition of a derivative term.
- In most applications, TE modules need to be driven in bipolar mode to cater for both heating and cooling operations. A commonly used circuit topology to provide bipolar drive to TE modules while operating from a single DC voltage is the half-bridge circuit topology. The shortcoming of the limited

voltage and current drive capability of a half-bridge circuit is overcome by having a cascaded arrangement of two half-bridge stages.

Review Questions

9.1. Why is laser diode protection an important criterion while designing laser diode drive circuits? Briefly describe the protection features that should be built in by the designer, emphasizing the adverse effects if these features were not included.

9.2. Explain the following.

a. How is the performance of a laser diode affected by a change in drive current and operating temperature?
b. The importance of cabling and grounding in improving susceptibility of laser diode drive circuit to radiated noise.
c. The difference between proportional, proportional-integral and proportional-integral-differential controller.
d. Integrated circuits as temperature sensors.

9.3. With the help of basic circuit schematics, briefly describe the constant-current and constant-output-power modes of operation of laser diodes.

9.4. What is the most preferred mode of operation of laser diodes used for: (a) printing and pointing applications; and (b) spectroscopic studies? Give reasons in support of your answer.

9.5. What is the quasi-CW mode of operation of laser diodes? How does it differ from CW and conventional pulsed modes of operation? Name one common application where laser diodes are operated in quasi-CW mode.

9.6. Differentiate between passive and active cooling of laser diodes. What are the shortcomings of passive cooling and how they are overcome in the case of active cooling?

9.7. Briefly describe the principle of thermoelectric cooling. With the help of a suitable diagram, briefly describe the constructional and operational features of a practical single-stage thermoelectric cooler module.

9.8. Name the important performance parameters of a thermoelectric cooler module. Define each of these parameters with particular reference to their significance in the design of a thermoelectric-cooler-based laser diode temperature controller.

9.9. What are the basic building blocks of a laser diode temperature drive and control circuit? With the help of a block schematic arrangement, briefly describe the functioning of a thermoelectric-cooler-based laser diode temperature drive and control circuit, employing thermistor-based temperature sensing, proportional control and bipolar drive.

9.10. With the help of a circuit diagram, describe the operation of a two-stage half-bridge drive circuit to provide bipolar drive to a thermoelectric cooler module.

Problems

9.1. For the laser diode drive circuit of Figure 9.37, determine the values of resistors R_2 and R_4 to provide a constant drive current of 1.0 A to the laser diode.
[$R_2 = R_4 = 5.0\ \text{k}\Omega$]

9.2. Refer to the laser diode drive circuit of Figure 9.38. The laser diode is operated in the pulsed output mode with the two values of drive current corresponding to the two voltage levels of the pulsed signal applied at the input. Determine the laser diode current corresponding to low and high values of control signal.
[$I_{Low} = 0.5$ A, $I_{High} = 1.0$ A]

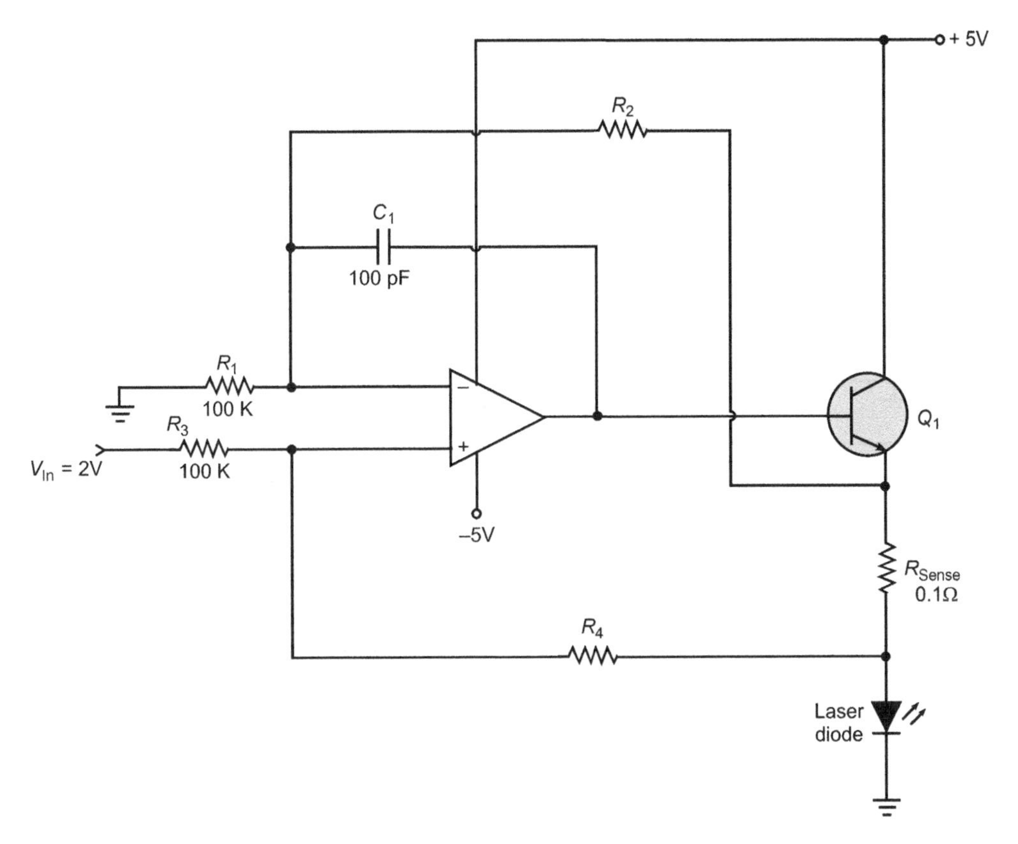

Figure 9.37 Laser diode driver circuit of Problem 9.1.

9.3. For the laser diode drive circuit of Figure 9.39, plot laser diode drive current I_D against control input voltage V_{In} for V_{In} varying from 0 to -2 V.
[Figure 9.40]

9.4. An infrared detector with a resistance of 1 MΩ is operated at a bias voltage of 200 V. The detector is to be thermoelectrically cooled to $-25\,°C$ from an ambient temperature of 25 °C. If the total exposed surface area of the detector is 5 cm^2, determine the active load of the detector and heat loads due to convective and radiative transfer mechanisms. The Stefan–Boltzmann constant is $\sigma = 5.67 \times 10^{-8}\,W\,m^{-2}\,K^{-4}$, emissivity is $e = 0.9$ and the convective heat transfer coefficient $h = 20\,W\,m^{-2}\,K^{-1}$.
[Active load of detector = 40 mW, $Q_{Conv} = 500$ mW, $Q_{Rad} = 175\,\mu W$]

9.5. A laser diode is to be operated at 20 °C. The thermoelectric module used for the cooling laser diode is designed to operate at ΔT of 40 °C, for which it is driven by 2.0 A at a terminal voltage of 12.5 V. If the maximum expected ambient temperature is 50 °C, determine the maximum allowable thermal resistance of heat sink on the hot side of the thermoelectric cooling module if the total heat load to be pumped is 20 W.
[$0.22\,°C\,W^{-1}$]

9.6. A laser diode is thermoelectrically cooled to the desired operating temperature by driving the thermoelectric cooler module with 2.5 A of current at a terminal voltage of 10 V. If the total pumped heat load including passive and active heat loads is 20 W, determine the coefficient of performance (COP).
[COP = 0.8]

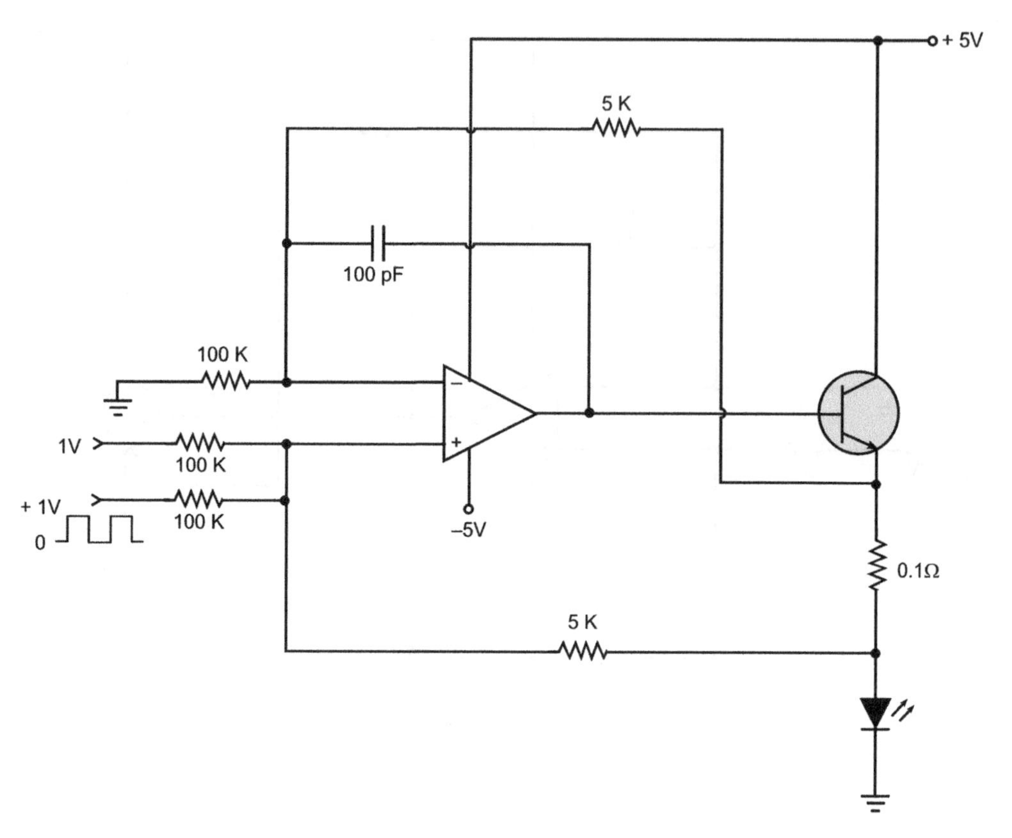

Figure 9.38 Laser diode driver circuit of Problem 9.2.

Self-evaluation Exercise

Multiple-choice Questions

9.1. A laser diode intended for carrying out spectroscopic studies should preferably be operated in
a. constant-current mode
b. constant-power mode
c. quasi-CW mode
d. either constant-current or constant-power mode.

9.2. In the case of thermoelectric cooling, ΔT_{max} specification can be achieved for:
a. pumped heat load $Q = Q_{max}$
b. pumped heat load $Q = 0$
c. pumped heat load in the range of 0–Q_{max}
d. none of these.

9.3. The coefficient of performance in the case of a thermoelectric cooling system is defined as:
a. the ratio of electrical power input to TE module to total pumped heat load
b. the ratio of total pumped heat load to heat dissipated in heat sink mounted on hot side of TE module
c. the ratio of total pumped heat load to electrical power input to TE module
d. none of these.

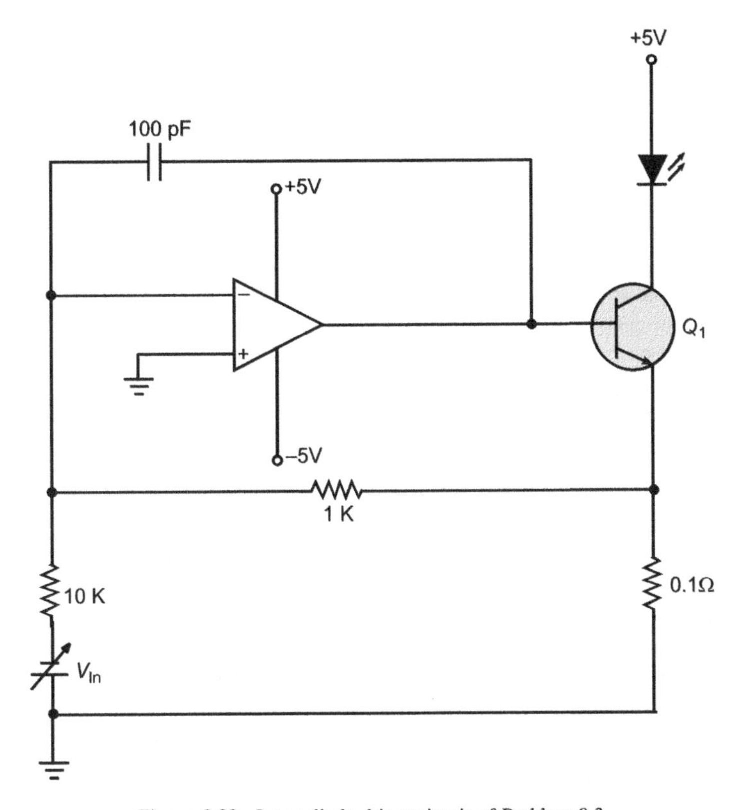

Figure 9.39 Laser diode driver circuit of Problem 9.3.

9.4. If the heat load pumped by a TE module is $Q_{c(max)}$, the achievable maximum ΔT is

a. 0

b. ΔT_{max}

c. 0–ΔT_{max}

d. $\Delta T_{max}/2$.

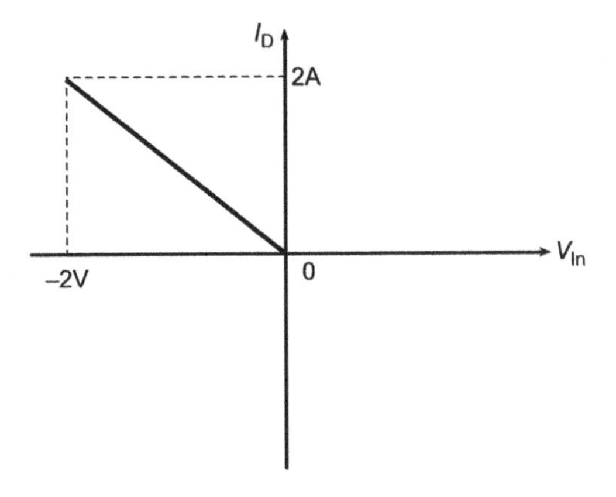

Figure 9.40 Solution to Problem 9.3.

9.5. To produce higher peak power output, laser diodes are
a. driven by larger current and terminal voltage
b. operated in pulsed mode
c. quasi-CW mode
d. connected in series.

9.6. In the case of laser diode arrays used in diode-pumped pulsed solid-state lasers, the laser diodes are operated in
a. conventional pulsed mode
b. quasi-CW mode
c. constant current mode
d. none of these.

9.7. In the case of a thermoelectric cooling system, the heat power dissipated in the heat sink is equal to
a. total heat load to be pumped by the system
b. electrical power input to the TE module
c. the sum of total heat load to be pumped and electrical power input to TE module
d. none of these.

9.8. In the case of thermoelectric cooling of laser diodes, the lower the thermal resistance of the heat sink,
a. the higher the heat sink-to-ambient temperature differential
b. the lower the heat sink-to-ambient temperature differential
c. the lower the maximum achievable ΔT for a given ambient temperature
d. none of these.

9.9. A commonly used circuit configuration for the output stage to drive a thermoelectric cooler module in bipolar mode is
a. a Darlington-transistor-based common-emitter stage
b. a multistage differential amplifier
c. a half-bridge circuit topology
d. none of these.

9.10. In a thermoelectric-cooler-based temperature controller for a laser diode, a residual error between the desired and actual temperatures was observed in the steady state. The residual error was observed to decrease with increase in loop gain. The controller is employing
a. a proportional-integral controller
b. a proportional-integral-differential controller
c. a proportional controller
d. an on-off controller.

Answers

1. (a) 2. (b) 3. (c) 4. (a) 5. (c) 6. (b) 7. (c) 8. (b) 9. (c) 10. (c)

Bibliography

1. *Fundamentals of Light Sources and Lasers*, 2004 by Mark Cselet, John Wiley & Sons.
2. *Handbook of Laser Technology and Applications*, 2004 by Collin E. Webb and Julian D.C. Jones, Institute of Physics Publishing.
3. *Laser Electronics*, 1995 by Joseph T. Verdeyen, Prentice-Hall.
4. *Laser Diode Beam Basics, Manipulations and Characterizations*, 2012 by Haiyin Sun, Springer Briefs in Physics, Springer.
5. *Fundamentals of Semiconductor Lasers*, 2004 by Takahiro Numai, Springer Series in Optical Sciences, Springer.
6. *Semiconductor Laser Engineering, Reliability and Diagnostics: A Practical Approach to High Power and Single Mode Devices*, 2013 by Peter W. Epperlein, Wiley.

10

Optoelectronic Devices and Circuits

10.1 Introduction

Optoelectronics is related to the study of electronic devices that emit, detect and control light. Commonly used optoelectronic devices include photoemitters, photosensors, displays and optocouplers. While photoemitters are electrical-to-optical transducers that are used to convert the electrical energy into output light, photosensors are optical-to-electrical transducers used for converting the incident light energy into electrical output. Photoconductors, photodiodes, phototransistors, photomultiplier tubes and image intensifiers are some commonly used photosensors. Photosensors have a significant role to play in the design and development of lasers and laser-based systems. Optocouplers are devices that use a short optical transmission path to transfer signals between elements of a circuit. Optoelectronic devices constitute the heart of a variety of systems ranging from simple gadgets such as light meters to the most complex of military systems such as precision-guided munitions, laser rangefinders and target trackers, from instrumentation, measurement and diagnostic systems to space-based weather forecasting and remote sensing systems, and from fibre-optic and laser-based communication applications to spectrophotometry and photometry applications.

This chapter discusses in detail the fundamentals and application circuits of different types of optoelectronic devices. We begin with a classification of optoelectronic devices, followed by the definition of various radiometric and photometric terms commonly used in the field of optoelectronics. This is followed by a detailed discussion on different types of photosensors in terms of principle of operation, characteristic parameters and application circuits. The chapter concludes with a brief description on optocouplers and different types of displays. The text is adequately illustrated with practical circuits and a large number of solved examples.

10.2 Classification of Photosensors

Photosensors are classified into two major categories: (1) photoelectric sensors and (2) thermal sensors.

10.2.1 Photoelectric Sensors

Photoelectric sensors can be classified further into two types: (a) devices that depend on the external photo effect for their operation and (b) devices that make use of some kind of internal photo effect. Photoelectric devices of types (a) and (b) include:

a. photoemissive sensors
b. photoconductors
c. junction-type photosensors.

Lasers and Optoelectronics: Fundamentals, Devices and Applications, First Edition. Anil K. Maini.

Of the three types mentioned above, photoemissive sensors belong to category (a) based on their external photo effect. The other two types of device (category (b)) use an internal photo effect. Common photoemissive sensors include non-imaging sensors such as vacuum photocells and photomultiplier tubes and imaging sensors such as image intensifier tubes. Junction-type photosensors are also classified into amplifying and non-amplifying types. The amplifying type of junction photosensors include phototransistors, photothyristors and photo field-effect transistors (FETs). Non-amplifying types of junction photosensors include photodiodes, solar cells and charge-coupled devices. A charge-coupled device is also a type of imaging sensor.

10.2.2 Thermal Sensors

Thermal sensors can be classified as either:

a. thermocouple (or thermopile) sensors
b. bolometric sensors
c. pyroelectric sensors.

Figure 10.1 depicts the classification of photosensors. While thermal sensors absorb incident radiation and operate on the resulting temperature rise, photoelectric sensors on the other hand are based on the quantum effect. Thermal sensors are relatively sluggish in their response to incident radiation than photoelectric sensors. Thermal sensors however offer a much wider operational wavelength band than photoelectric sensors.

10.3 Radiometry and Photometry

Radiometry is the study of properties and characteristics of electromagnetic radiation and the sources and receivers of electromagnetic radiation. Radiometry covers a wide frequency spectrum, but for the present chapter we will limit our discussion to frequencies from infrared to ultraviolet.

Photometry is the science that deals with visible light and its perception by human vision. The most important difference between radiometry and photometry is that in radiometry the measurements are made with objective electronic instruments and in photometry measurements are made with reference to the response of the human eye.

In this section, we define the commonly used radiometric and photometric quantities.

10.3.1 Radiometric and Photometric Flux

Flux is defined as a flow phenomenon or a field condition occurring in space. It is a measure of the total power emitted from a source or incident on a particular surface. Radiometric flux and photometric flux are denoted ϕ_R and ϕ_P, respectively. Radiometric flux or luminous flux is measured in watts (W) while photometric flux is measured in lumens (lm). A lumen is defined as the amount of photometric flux generated by 1/683 W of radiometric flux at 555 nm where the photopic vision sensitivity of eye is maximum.

Efficacy of a radiation source K (lm W^{-1}) is defined as the ratio of photometric flux to the total radiometric flux from the source. It is given by Equation 10.1:

$$K = \frac{\phi_P}{\phi_R} \tag{10.1}$$

10.3.2 Radiometric and Photometric Intensity

The intensity function describes the flux distribution in space. Radiometric intensity I_R is defined as the radiometric flux density per steradian. It is given by Equation 10.2 and is expressed in Watts per steradian (W sr^{-1}).

$$I_R = \frac{\phi_R}{\Omega} \tag{10.2}$$

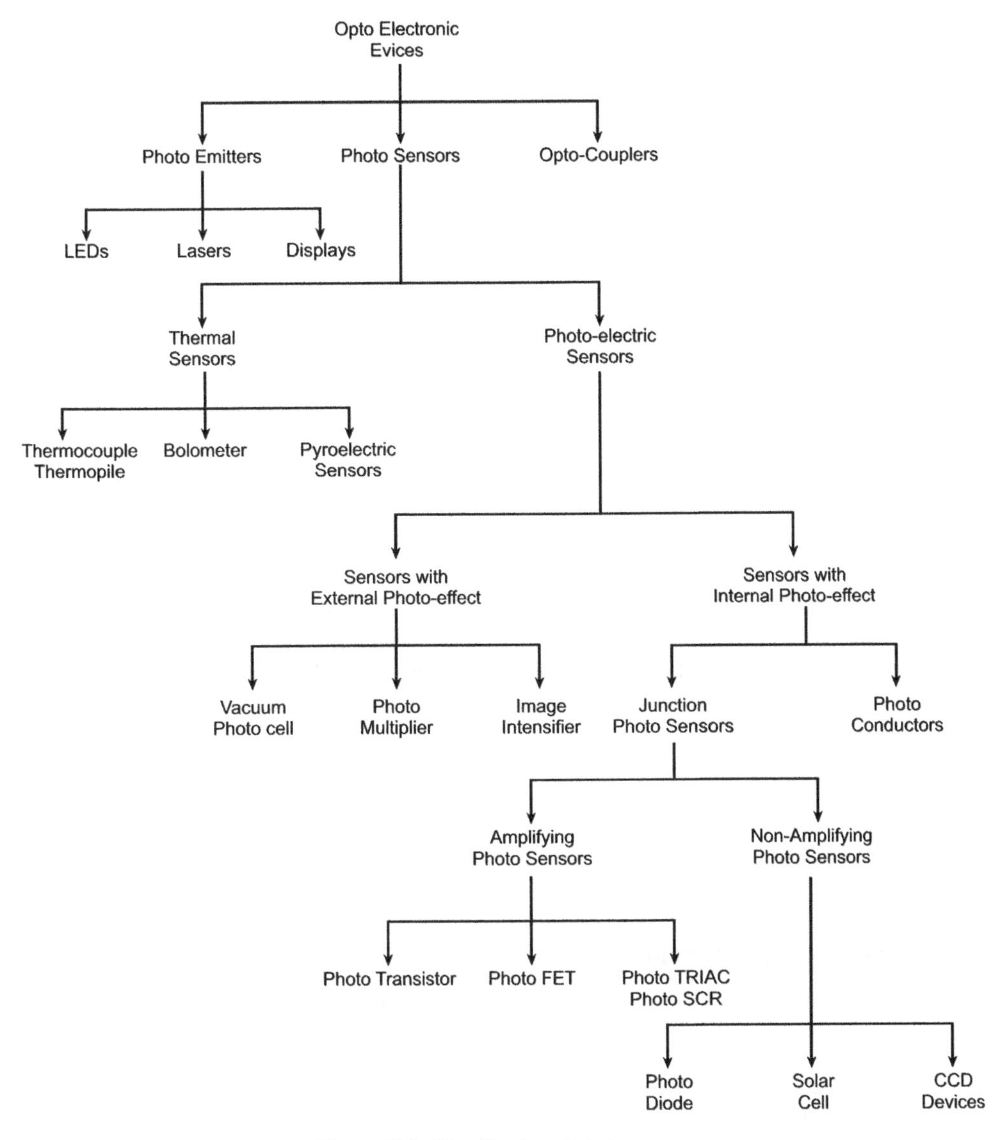

Figure 10.1 Classification of photosensors.

where

ϕ_R = radiometric flux (W)
Ω = solid angle (sr)

Photometric or luminous intensity I_P is defined as the ratio of luminous flux density per steradian. It is given by Equation 10.3:

$$I_P = \frac{\phi_P}{\Omega} \tag{10.3}$$

where

ϕ_P = photometric flux (lm)
Ω = solid angle (sr)

The unit of photometric flux density is the Candela (Cd), equivalent to luminous flux density of one lumen per steradian ($\mathrm{lm\,sr^{-1}}$).

10.3.3 Radiant Incidence (Irradiance) and Illuminance

Radiant incidence or irradiance E_R defines the radiometric flux distribution on a surface. It is expressed:

$$E_R = \frac{\phi_R}{A} \tag{10.4}$$

where A is the area of flux distribution (m^2).

Illuminance E_P defines the photometric flux distribution on a surface and is expressed:

$$E_P = \frac{\phi_P}{A} \tag{10.5}$$

Two very commonly used units to define illuminance are lux and foot-candle. Lux is defined as the illumination of one lumen of luminous flux evenly distributed over an area of one square metre. Foot-candle is an old English unit and is defined as an illumination of one lumen of luminous flux evenly distributed over an area of one square foot. One foot-candle is equivalent to 10.764 lux.

10.3.4 Radiant Sterance (Radiance) and Luminance

Radiant sterance (radiance) is defined as the ratio of radiometric flux per unit solid angle per unit area ($\mathrm{W\,sr^{-1}\,m^{-2}}$). Luminance is defined as the ratio of luminous flux per unit solid angle per unit area ($\mathrm{lm\,sr^{-1}\,m^{-2}}$).

10.4 Characteristic Parameters

Major characteristic parameters used to characterize the performance of photosensors include:

1. responsivity
2. noise equivalent power (NEP)
3. sensitivity, usually measured as detectivity D or D-star (D^*)
4. quantum efficiency
5. response time
6. noise
7. spectral response.

10.4.1 Responsivity

Responsivity is defined as the ratio of electrical output to radiant light input determined in the linear region of the response. It is measured in amps per watt ($\mathrm{A\,W^{-1}}$) or $\mathrm{V\,W^{-1}}$ if the photosensor produces a voltage output rather than a current output. Responsivity is a function of the wavelength of incident radiation and band-gap energy. Spectral response is a related parameter. It is the curve that shows variation of responsivity as a function of wavelength. Most photoelectric sensors have a narrow spectral response whereas most thermal sensors have a wide spectral response. As an example, the spectral response of silicon, germanium and indium gallium arsenide photodiodes are in the range of

200–1100 nm, 500–1900 nm and 700–1700 nm respectively, whereas the spectral response of thermistors is within the range 0.5–10 μm.

Silicon photodiodes exhibit a response from the ultraviolet through the visible and into the near-infrared part of the spectrum. With the band-gap energy of silicon being 1.12 eV at room temperature, its spectral response peaks in the near-infrared region between 800 nm and 950 nm. Peak responsivity figures for silicon PIN photodiodes are in the range of 0.4–0.6 AW^{-1} and for avalanche photodiodes in the range of 40–80 AW^{-1}. Thermal sensors have poorer responsivity as compared to photoelectric sensors. As an example, the responsivity figure for pyroelectric sensors is in the range of 0.5–5 μAW^{-1}.

The shape of the spectral response curve of silicon photosensors, particularly in the blue and UV part of the spectrum, can be altered by choosing an appropriate manufacturing process. Figure 10.2 shows typical spectral response curves of a silicon photodiode.

It is evident from the spectral response curves shown in Figure 10.2 that silicon becomes transparent to radiation of longer than 1100 nm wavelength. On the contrary, wavelengths in the ultraviolet region are absorbed in the first 100 nm thickness of the silicon. Even the most careful surface preparation leaves some surface damage, which reduces the collection efficiency for this wavelength. Anti-reflection coatings may be used to enhance the responsivity to about 25% at the required wavelength, at the cost of a reduction in efficiency at other wavelengths that they reflect. The package window also plays an important role in shaping the spectral response. The standard glass window absorbs wavelengths shorter than 300 nm. For UV detection, a fused silica or UV transmitting glass window is used. Various filter windows are also available to tailor the spectral response to suit the application. Optical filters can also be added to change the spectral response. A common example is the use of a specific filter to modify the normal silicon response to approximate the spectral response of the human eye.

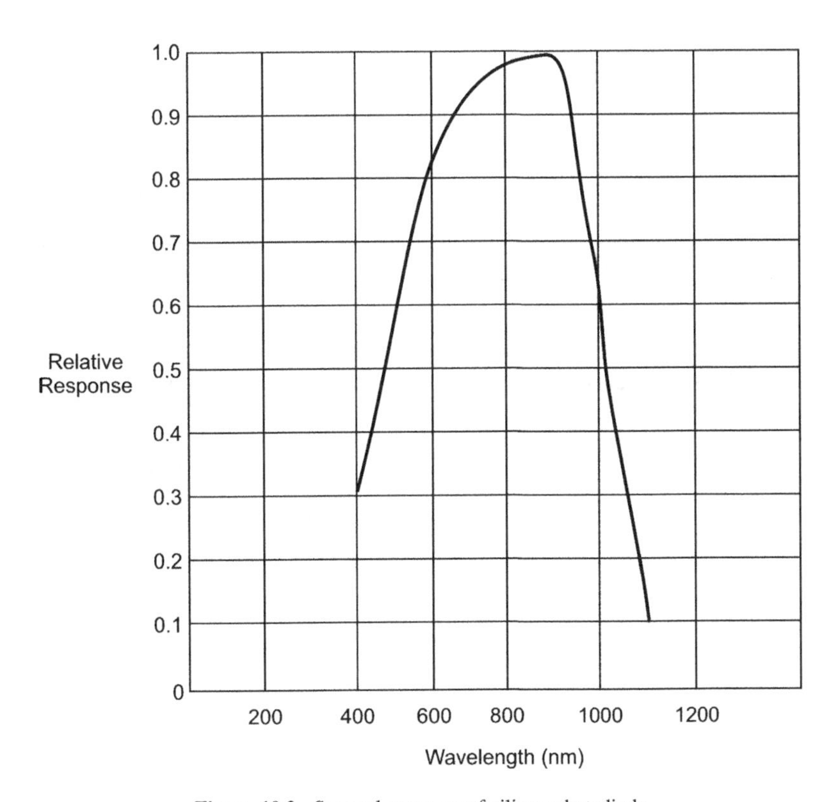

Figure 10.2 Spectral response of silicon photodiode.

In the case of photodiodes, the responsivity R_ν is typically highest in a region with photon energies slightly greater than the band-gap energy. It declines sharply for photon energies in the region of the band gap, where the absorption decreases. It can be calculated by using Equation 10.6:

$$R_\nu = \frac{\eta \times e}{h \times \upsilon} \tag{10.6}$$

where

h = Planck's constant ($=6.625 \times 10^{-34}$ J s)
ν = frequency of incident radiation (Hz)
η = quantum efficiency
e = electron charge ($=1.6 \times 10^{-16}$ C)

Responsivity increases slightly with applied reverse bias due to improved charge collection efficiency at the photodiode. Responsivity also exhibits dependence on temperature variations, due to the variation in band-gap energy (band-gap energy varies inversely with change in temperature). Figure 10.3 depicts the change in responsivity with temperature. As is evident from the family of curves, responsivity is more or less independent of temperature from 500 nm to 900 nm. For wavelengths less than 500 nm, responsivity decreases or increases gradually with an increase or decrease in temperature, respectively. For wavelengths greater than 900 nm, it increases or decreases rapidly with an increase or decrease in temperature, respectively. This can be explained from the nature of temperature dependence of band-gap energy.

The term *responsivity* should not be confused with *sensitivity*; the latter is the lowest detectable light level, which is typically determined by detection noise. It is also significantly influenced by the required detection bandwidth. A photosensor should ideally be operated in a spectral region where its responsivity is not far below the highest possible value, because this leads to the lowest possible detection noise and thus to a high signal-to-noise ratio and high sensitivity.

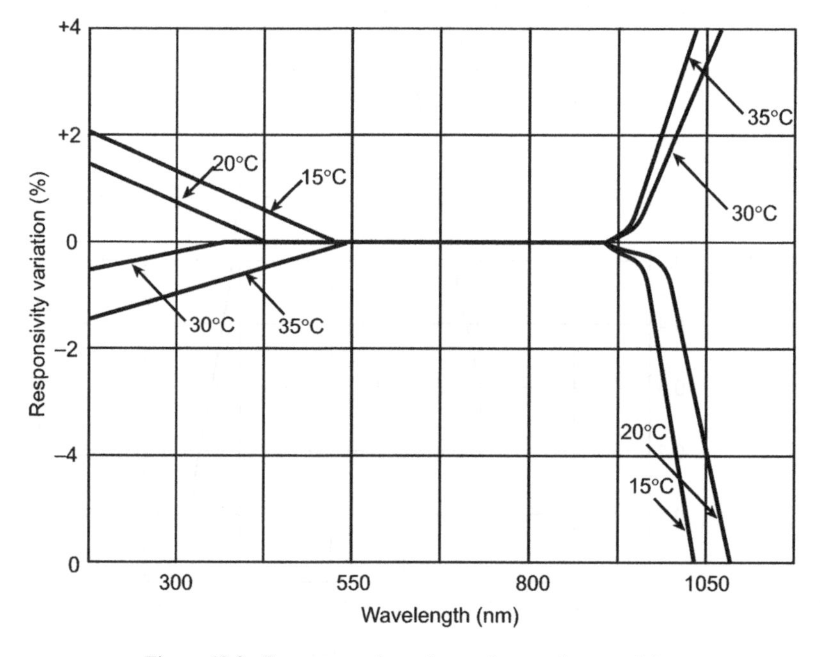

Figure 10.3 Temperature dependence of spectral responsivity.

10.4.2 Noise Equivalent Power (NEP)

The *noise equivalent power* (NEP) is the input power to a sensor which generates an output signal current equal to the total internal noise current of the device, which implies a signal-to-noise ratio of 1. In other words, it is the minimum detectable radiation level of the sensor. Obviously, a low NEP is desirable. NEP depends on the wavelength, since that influences the responsivity of the sensor. The lowest NEP is achieved for those wavelengths where the responsivity is the highest:

$$\text{NEP} = \frac{I_{\text{N}}}{R_{\nu}} \tag{10.7}$$

where

I_{N} = total noise current (A)
R_{ν} = responsivity ($\text{A}\,\text{W}^{-1}$)

The noise power and thus the noise-equivalent power depend on the assumed detection bandwidth. If the full-detection bandwidth of the device was used to compute NEP, then the NEP would not allow a fair comparison of sensors with different bandwidths. It is therefore common practice to assume a bandwidth of 1 Hz, which is usually far below the detection bandwidth. NEP is usually specified in units of $\text{W}\,\text{Hz}^{-0.5}$ rather than W. In that case, noise current is specified as $\text{A}\,\text{Hz}^{-0.5}$. Effectively, computation of NEP is based on power spectral density (PSD) rather than on power. Power is simply the power spectral density computed over a bandwidth of 1 Hz. Since the photodiode light power to current conversion depends on wavelength, the NEP is quoted at a particular wavelength. As for responsivity, NEP is non-linear over the wavelength range.

10.4.3 Detectivity and D-star

Detectivity D of a sensor is the reciprocal of its NEP. A sensor with a higher value of detectivity is more sensitive than a sensor with a lower detectivity value. Detectivity, as for NEP, depends upon noise bandwidth and sensor area. To eliminate these factors, a normalized figure of detectivity referred to as '*D*-star' or D^* is used. It is defined as the detectivity normalized to an area of 1 cm^2 and a noise bandwidth of 1 Hz. The value of D^* can be calculated using Equation 10.8:

$$D^* = D\sqrt{A\Delta f} \tag{10.8}$$

where

D = detectivity (W^{-1})
D^* = normalized detectivity ($\text{W}^{-1}\,\text{cm}\,\text{Hz}^{1/2}$)
A = sensor area (cm^2)
Δf = bandwidth (Hz)

10.4.4 Quantum Efficiency

An ideal photosensor should produce one photoelectron per incident photon of light. This is not true for practical sensors, however. The ratio of the number of photoelectrons released to the number of photons of incident light absorbed is referred to as the *quantum efficiency* of the sensor. This is the percentage of input radiation power converted into photocurrent. In other words, it is the intrinsic efficiency of the sensor. The value of quantum efficiency η is computed as the ratio of actual responsivity to the ideal responsivity as given by Equation 10.9:

$$\eta = \frac{R_{\text{Actual}}}{R_{\text{Ideal}}} \tag{10.9}$$

The ideal responsivity is defined

$$R_{\text{Ideal}} = \frac{\text{Electron charge}}{\text{Photon energy}} = \frac{e \times \lambda}{h \times c} \tag{10.10}$$

where c is the speed of light in vacumm (defined as $3 \times 10^8\,\text{m s}^{-1}$).
Substituting the values of h, c and e into Equation 10.10, we have:

$$R_{\text{Ideal}} = 8.044 \times 10^5 \lambda$$

where λ is wavelength in m. Substituting this definition for R_{Ideal} into Equation 10.9, and changing units of wavelength to nm,

$$\eta = \frac{1.24 \times 10^3 R_{\text{Actual}}}{\lambda} \tag{10.11}$$

10.4.5 Response Time

Response time is referred to as *rise/fall time* in photoelectric sensors and as the *time constant* parameter in thermal sensors. Rise and fall times are the time durations required by the output to change from 10% to 90% and from 90% to 10% of the final response, respectively. This determines the highest signal frequency to which a sensor can respond. *Time constant* is defined as the time required by the output to reach 63% of the final response from an initial value of zero.

Bandwidth BW (MHz) of photoelectric sensors is related to its rise time, defined:

$$\text{BW} = \frac{0.35}{t_r} \tag{10.12}$$

where t_r is rise time (μs).

Response time of a photodiode is governed by three parameters:

1. drift time t_{Drift}, which is the charge collection time of the carriers in the depletion region of the photodiode
2. diffusion time t_{Diff}, which is the charge collection time of the carriers in the undepleted region of the photodiode.
3. *RC* time t_{RC} is the time constant of the diode–external circuit combination.

RC time is defined as $t_{RC} = 2.2\,RC$, where R is the sum of the photodiode's series resistance and the load resistance and C is the sum of the photodiode's junction and the stray capacitances. Junction capacitance is directly proportional to the diffused area of the photodiode and inversely proportional to the applied reverse bias. As a consequence of this, photodiodes with smaller area and larger applied reverse bias have a faster rise time. In addition, stray capacitance can be minimized by using short leads and careful electronic circuit lay-out. The total rise time is given by Equation 10.13:

$$t_r = \sqrt{\left[t_{\text{Drift}}^2 + t_{\text{Diff}}^2 + t_{RC}^2\right]} \tag{10.13}$$

In photovoltaic mode of operation where there is no applied reverse bias, drift time can be considered to be negligible and the rise time is dominated by the diffusion time for diffused areas less than 5 mm^2 (smaller diffused areas offer very small junction capacitance) and by the RC time constant for larger diffused areas for all wavelengths. When operated in photoconductive mode where there is an applied reverse bias, the dominant factor is the drift time if the photodiode is fully depleted; all three factors are important if the photodiode is not fully depleted.

10.4.6 Noise

Noise is the most critical factor in designing sensitive radiation detection systems. Noise is generated in these systems in photosensors, radiation sources and post-detection circuitry. Photosensor noise mainly comprises:

1. Johnson noise
2. shot noise
3. generation-recombination noise
4. flicker noise.

Johnson noise, also known as *Nyquist noise* or *thermal noise*, is caused by the thermal motion of charged particles in a resistive element. The RMS value of the noise voltage depends on the resistance value, temperature and the system bandwidth, and is given by Equation 10.14:

$$V_{\mathrm{RMS}} = \sqrt{4kRT\Delta f} \qquad (10.14)$$

where

V_{RMS} = RMS noise voltage (V)
R = resistance value (Ω)
k = Boltzmann constant ($1.38 \times 10^{-23}\,\mathrm{J\,K^{-1}}$)
T = absolute temperature (K)
Δf = system bandwidth (Hz)

Shot noise in a photosensor is caused by the discrete nature of the photoelectrons generated. It is related to the statistical fluctuation of both dark current and the photocurrent. It depends on the average current through the photosensor and system bandwidth and is given by Equation 10.15:

$$I_{\mathrm{SRMS}} = \sqrt{2eI_{\mathrm{ave}}\Delta f} \qquad (10.15)$$

where

I_{SRMS} = RMS shot noise current (A)
I_{ave} = average current through the photosensor (A)
e = charge of an electron (= 1.60×10^{-19} C)
Δf = detection bandwidth (Hz)

Shot noise is the dominant source of noise for photodiodes operating in photoconductive mode.

Generation-recombination noise is caused by the fluctuation in current generation and the recombination rates in a photosensor. This type of noise is predominant in photoconductive sensors operating at infrared wavelengths. The generation-recombination noise can be calculated using Equation 10.16:

$$I_{\mathrm{GRMS}} = 2eG\sqrt{\eta EA\Delta f} \qquad (10.16)$$

where

I_{GRMS} = RMS generation-recombination noise current (A)
E = radiant intensity (W cm^{-2})
A = sensor receiving area (cm^2)
G = photoconductive gain
η = quantum efficiency

Flicker noise or 1/f noise occurs in all conductors where the conducting medium is not a metal and exists in all semiconductor devices that require bias current for their operation. Its amplitude is inversely proportional to the frequency. Flicker noise is usually predominant at frequencies below 100 Hz.

The total equivalent noise (I_{NEQ}) is calculated via Equation 10.17:

$$I_{NEQ} = \sqrt{(I^2_{JRMS} + I^2_{SRMS} + I^2_{GRMS} + I^2_{FRMS}} \quad (10.17)$$

Example 10.1

A photodiode has a noise current of 1 fA, a responsivity of 0.5 A W^{-1}, an active area of 1 mm^2 and a rise time of 3.5 ns. Determine its (a) NEP; (b) detectivity D; (c) D^*; and (d) quantum efficiency at 850 nm.

Solution

1. NEP = noise current/responsivity = $1 \times 10^{-15}/0.5 = 2 \times 10^{-15}$ W = 2 fW
2. Detectivity = 1/NEP = $1/2 \times 10^{-15}$ W^{-1} = 0.5×10^{15} W^{-1}
3. $$D* = D \times \sqrt{(A \times \Delta f)}$$

   where A = 1 mm^2 = 1×10^{-2} cm^2 and Δf (MHz) = $0.35/t_r$ (μs) = $0.35/(3.5 \times 10^{-3})$ = 100 MHz = 1×10^8 Hz. Therefore, $D^* = 0.5 \times 10^{15} \times \sqrt{(1 \times 10^{-2} \times 1 \times 10^8)}$ W^{-1} cm Hz$^{1/2}$ = 5×10^{17} W^{-1} cm Hz$^{1/2}$.
4. Quantum efficiency, $\eta = \left[\frac{(1240 \times R)}{\lambda}\right]$

   $\eta = (1240 \times 0.5)/850 = 0.729.$

Example 10.2

An oscilloscope is used to measure the output of a photodiode. Determine the rise time of the pulse as seen on the oscilloscope, given that the rise time of the photodiode is 1 ns, rise time of the light pulse is 5 ns and the bandwidth of the oscilloscope is 350 MHz.

Solution

1. Bandwidth of the oscilloscope = 350 MHz. Therefore, rise time of the oscilloscope is t_r (μs) = 0.35/ BW (MHz) = 0.35/350 μs = 1×10^{-3} μs = 1 ns.
2. Therefore, overall rise time = $\sqrt{[(5 \times 10^{-9})^2 + (1 \times 10^{-9})^2 + (1 \times 10^{-9})^2]} = 5.19 \times 10^{-9}$ s = 5.19 ns.

10.5 Photoconductors

Photoconductors, also referred to as photoresistors, light-dependent resistors (LDRs) and photocells, are semiconductor photosensors whose resistance decreases with increasing incident light intensity. They are bulk semiconductor devices with no P-N junction, and have a structure as shown in Figure 10.4a. When light is incident on the photoconductor, electrons jump from the valence band to the conduction band; the resistance of the semiconductor material therefore decreases. The resistance change in a photoconductor is of the order 10^6, ranging from a few tens of megaohms under dark conditions to a few tens or hundreds of ohms under bright light conditions. Other features include wide dynamic response,

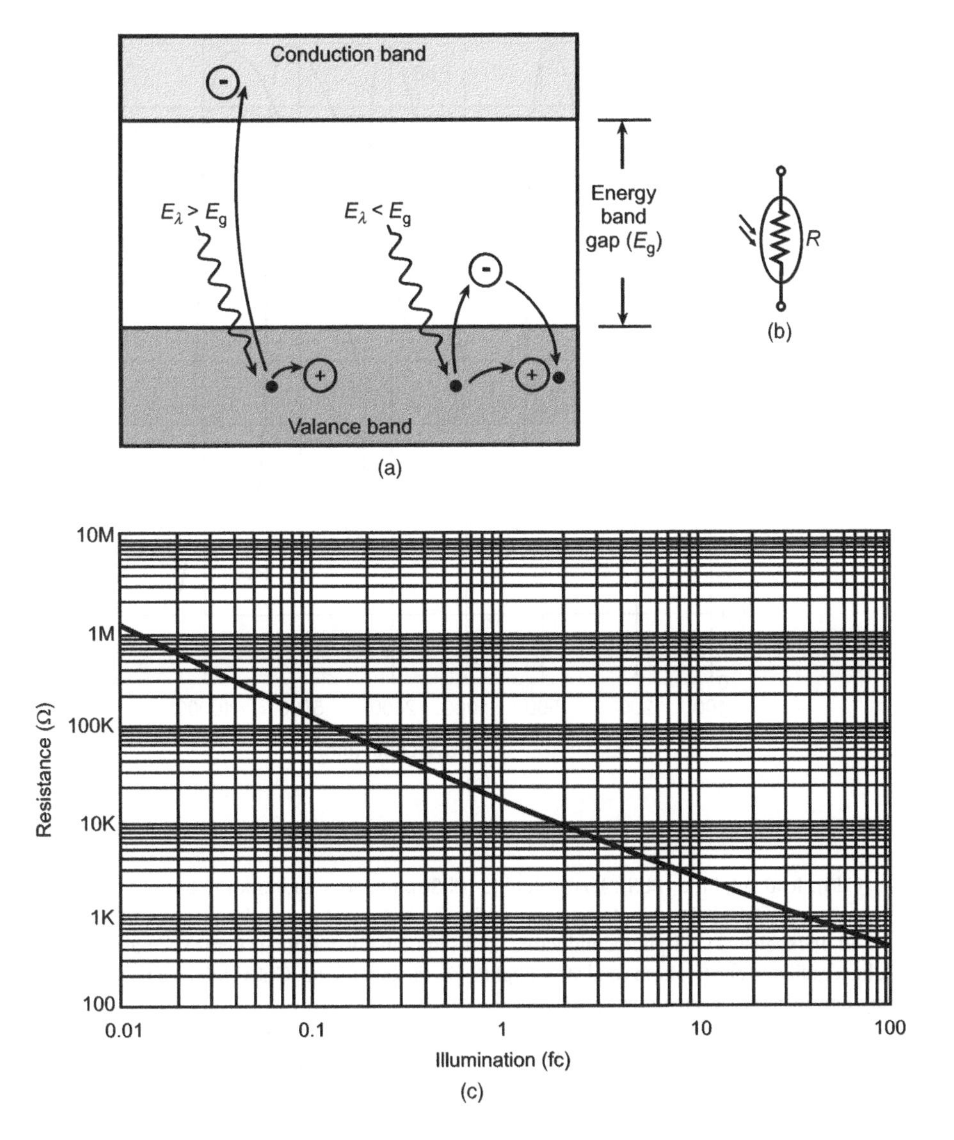

Figure 10.4 Photoconductors: (a) cross-section; (b) circuit symbol; (c) typical resistance–illumination curve; and (d) spectral response of commonly used photoconductor materials.

spectral coverage from ultraviolet to far-infrared and low cost. However, they are sluggish devices, having a response time of the order of hundreds of milliseconds.

The resistance–illuminance relation in photoconductors is described by Equation 10.18:

$$R_a = R_b \times \left(\frac{E_a}{E_b}\right)^{-\alpha} \tag{10.18}$$

where R_a and R_b are the resistances at illumination levels E_a and E_b, respectively; E_a and E_b are the illumination levels in lux or foot candles; and α is the characteristic slope of the resistance–illumination curve.

The value of α is in the range 0.55–0.9. Figure 10.4b depicts the circuit symbol of a photoconductor, which can be provided in various shapes and sizes. Figure 10.4c shows the typical resistance–illumination curve of a photoconductor. Commonly used materials in photoconductors are cadmium

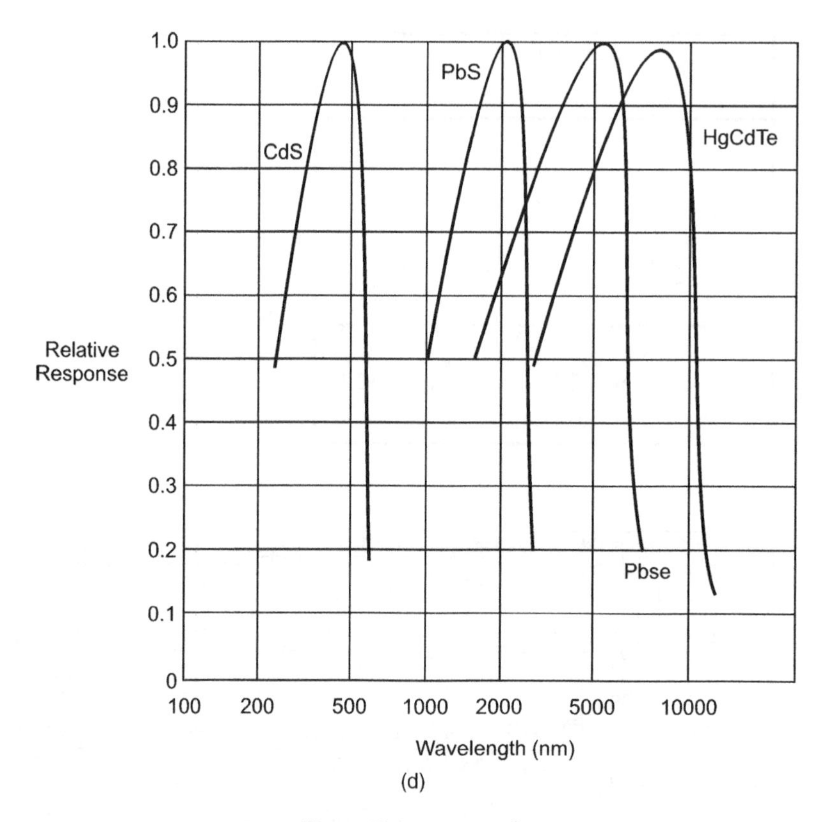

Figure 10.4 *(Continued)*

sulphide (CdS), lead sulphide (PbS), lead selenide (PbSe), mercury cadmium telluride (HgCdTe) and germanium copper (GeCu). The spectral response of some of these photoconductor materials is shown in Figure 10.4d. Inexpensive CdS photoconductors are used in many consumer items such as camera light meters, clock radios, security alarms and street lights. On the other hand, GeCu cells are used for infrared astronomy and infrared spectroscopy applications.

Photoconductors are further classified as intrinsic or extrinsic photoconductors depending upon whether an external dopant has been added or not to the semiconductor material. Intrinsic photoconductors operate at shorter wavelengths, as the electrons have to jump from the valence to the conduction band. Extrinsic photoconductors have a spectral response covering longer wavelengths.

10.5.1 Application Circuits

Photoconductors are usually used for detection of infrared radiation. When a bias is applied to the photoconductor in the absence of radiation, a current referred to as the dark current is generated. When light is incident on the photoconductor, its resistance decreases and the current flowing through it increases. Photosignal is the increase in the current caused by radiation. Generally this photosignal is much smaller (of the order of few parts per thousand) than the dark current. Extracting this small signal from the dark current is the primary task of the front-end circuit.

Figure 10.5a and b depict some simple circuits using photoconductors. Using photoconductors in these configurations however reduces the responsivity of the conductor as the relative change in the circuit resistance is smaller because of the load resistance R. The choice of R and R_{sen} also affects the output voltage from the circuit. For Figure 10.5a, the higher the value of R, the higher the output voltage but the relative responsivity is poorer. Similarly, in the case of circuit of Figure 10.5b, the higher the value of R, the lower the output voltage but the relative responsivity is better.

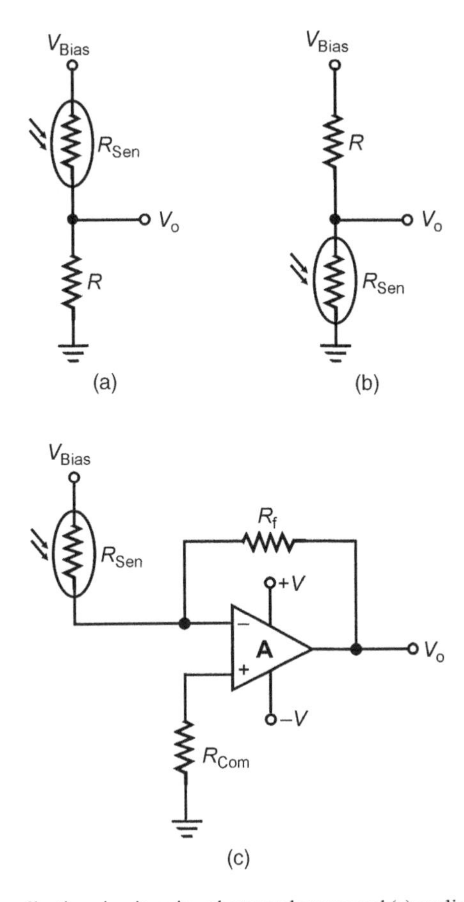

Figure 10.5 (a, b) Simplest application circuits using photoconductors; and (c) application circuit of photoconductor using opamp in the transimpedance mode.

To overcome these problems, photoconductors are used in conjunction with amplifiers to obtain both better responsivity and a high output voltage. There are two possible circuit configurations: voltage-mode amplifiers and current-mode or transimpedance amplifiers. The basic transimpedance amplifier is shown in Figure 10.5c. The non-inverting input of the operational amplifier (opamp) is connected to ground through resistance R_{com} to minimize the DC offset voltage.

The output voltage V_O is given by Equation 10.19:

$$V_O = -(R_f/R_{sen}) \times V_{bias} \tag{10.19}$$

The gain of the transimpedance amplifier should be set such that the amplifier does not saturate at the maximum expected radiation intensity. Also, if the bias voltage of the photoconductor is more than the maximum rated input voltage of the opamp, then a Zener diode should be connected between the inverting input of opamp and the ground terminal. Theoretically, the signal voltage can be obtained by subtracting the output voltage under dark conditions from the voltage signal in Equation 10.19, given by Equation 10.20:

$$V_O = -[(R_f/R_{sen}) - (R_f/R_{dark})] \times V_{bias} \tag{10.20}$$

where R_{dark} is the resistance value of the photoconductor in the absence of radiation.

Practically however, this is not a feasible solution as the dark resistance of the photoconductor is a strong function of temperature and even a slight increase in temperature decreases the value of dark resistance by a large amount and *vice versa*. The sensor temperature therefore has to be controlled to the

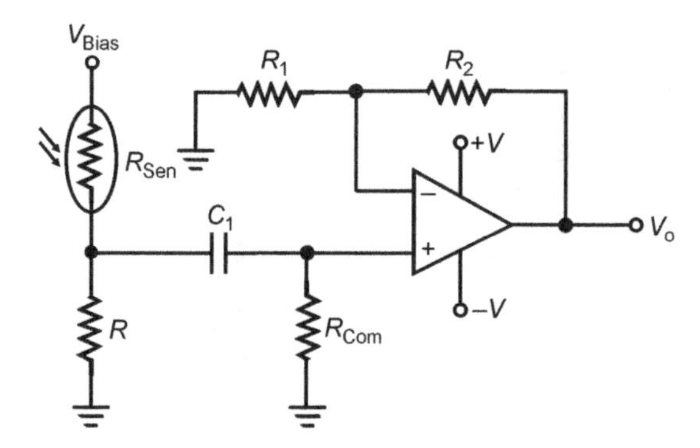

Figure 10.6 Application circuit of photoconductor using voltage mode amplifier with AC coupling.

order of 0.01 °C or better, which is often not feasible. The most common method used to extract the signal is to modulate the incident radiation at a specific frequency, either by placing a mechanical chopper in front of the sensor or by electrically modulating the radiation source. The signal generated due to radiation is now an AC signal while the dark current is a DC signal. The AC signal can be separated from the DC background signal using an AC coupled amplifier. A voltage-mode amplifier using AC coupling is shown in Figure 10.6.

Example 10.3

Design a circuit using a photoconductor that generates a logic HIGH voltage when the light incident on it is above 200 lux, given that the photoconductor has a resistance of 10 kΩ at a light level of 100 lux, $\alpha = 0.5$, power supply voltage $V_{CC} = 10$ V, reference voltage of the Zener diode is 2.5 V and the current flowing through the Zener diode can be taken as 10 mA.

Solution

1. The resistance of the photoconductor at 200 lux can be calculated using the expression $R_a = R_b \times \left(\frac{E_a}{E_b}\right)^{-\alpha}$ where $R_b = 10$ kΩ, $E_b = 100$ lux, $E_a = 200$ lux and $\alpha = 0.5$.

$$R_a = 10 \times 10^3 \times (200/100)^{-0.5} = 10 \times 10^3 \times (2)^{-0.5}\Omega = 7.07\,\text{k}\Omega$$

2. Figure 10.7 shows one of the possible circuit configurations that can be used for the given application.
3. The comparator output will go high when the voltage at the positive terminal exceeds 2.5 V.

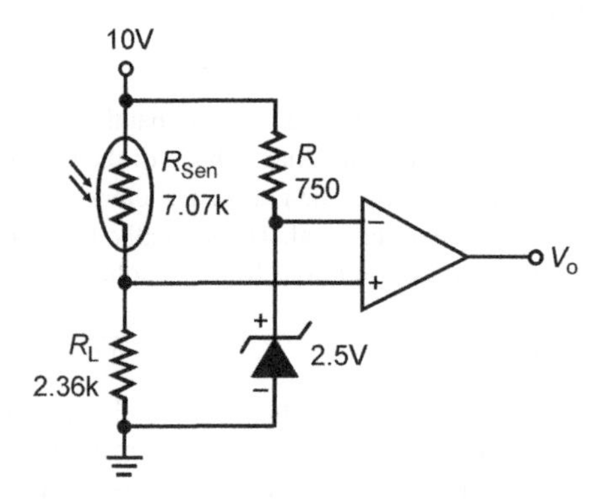

Figure 10.7 Example 10.3.

The value of load resistance R_L can be calculated using $(V_{CC} \times R_L)/(R_L + R_{sen}) = 2.5$ V. As $V_{CC} = 10$ V and R_{sen} at 200 lux is 7.07 kΩ,

$$(10 \times R_L)/(7.07 \times 10^3 + R_L) = 2.5$$

$$4R_L = 7.07 \times 10^3 + R_L \text{ and therefore } R_L = 2.36\ \text{k}\Omega.$$

4. For a 10 mA current through the Zener diode, the value of resistor R can be calculated using the expression $R = (V_{CC} - V_Z)/I_Z = (10-2.5)/(10 \times 10^{-3}) = 750$ kΩ.

10.6 Photodiodes

Photodiodes are junction-type semiconductor light sensors that generate current or voltage when the P-N junction in the semiconductor is illuminated by light of sufficient energy. The spectral response of the photodiode is a function of the band-gap energy of the material used in its construction. The upper cut-off wavelength λ_c (nm) of a photodiode is given by Equation 10.21:

$$\lambda_c = \frac{1240}{E_g} \tag{10.21}$$

where E_g is band-gap energy (eV).

Photodiodes are mostly constructed using silicon, germanium, indium gallium arsenide (InGaAs), lead sulphide (PbS) and mercury cadmium telluride (HgCdTe). Figure 10.8 shows the spectral characteristics of these photodiodes.

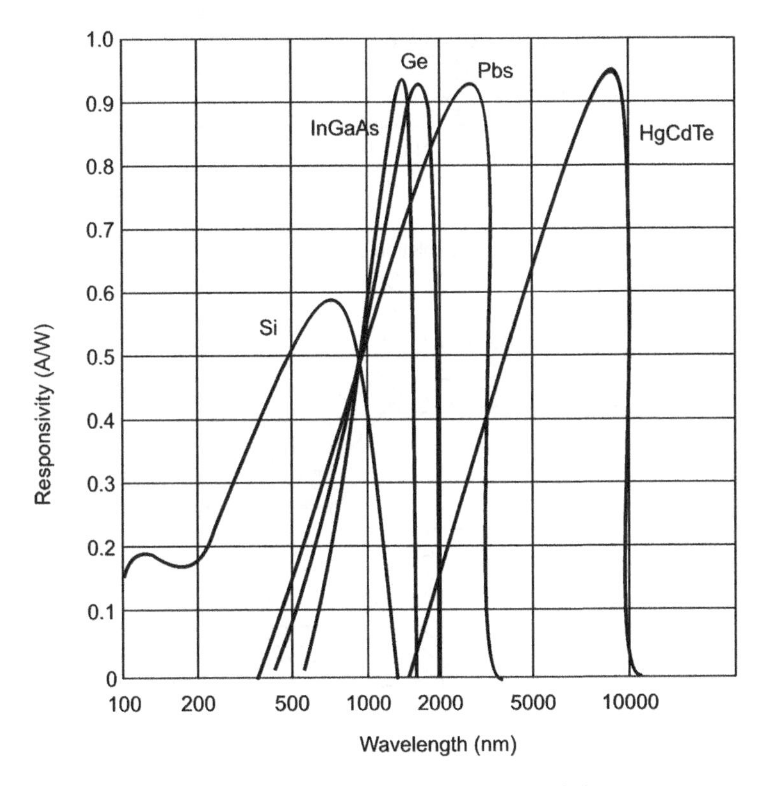

Figure 10.8 Spectral characteristics of photodiodes.

10.6.1 *Types of Photodiodes*

Depending upon their construction there are several types of photodiodes. These include PN photodiodes, PIN photodiodes, Schottky photodiodes and avalanche photodiodes (APDs).

10.6.1.1 PN Photodiodes

A PN photodiode comprises a PN junction as shown in Figure 10.9a. When light with sufficient energy strikes the photodiode the electrons are pulled up into the conduction band, leaving behind holes in the valence band. These electron-hole pairs occur throughout the P-layer, depletion layer and N-layer materials. When the photodiode is reverse biased, the photo-induced electrons will move down the potential hill from the P-side to the N-side. Similarly, the photo-induced holes will add to the current flow by moving across the junction to the P-side. Shorter wavelengths are absorbed at the surface while the longer wavelengths penetrate deep into the diode.

Figure 10.9b shows the mechanism of conversion of incident light photons into electric current in a PN photodiode. PN photodiodes are used for precision photometry applications such as medical instrumentation, analytical instruments, semiconductor tools and industrial measurement systems.

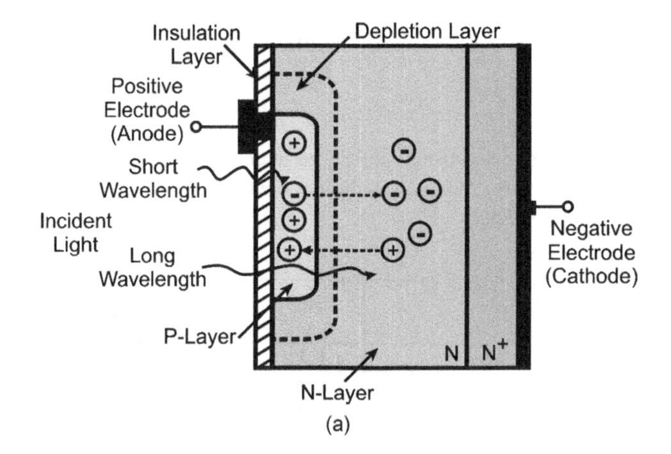

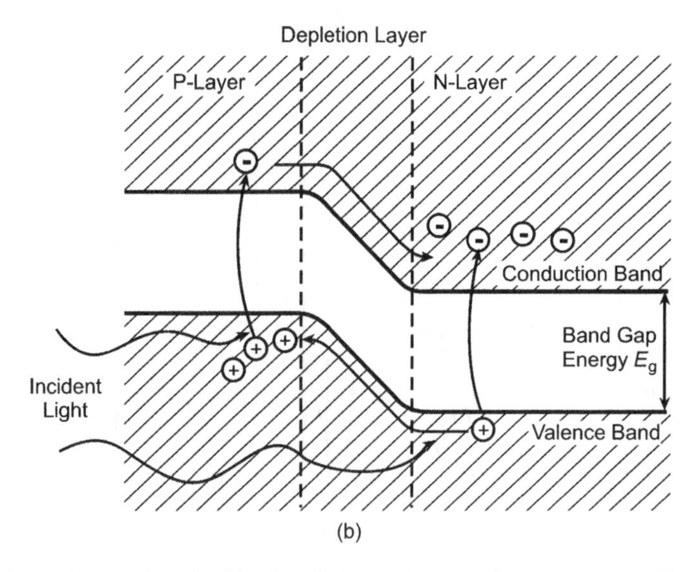

Figure 10.9 (a) Cross-section of a P-N photodiode; and (b) generation of current in a PN photodiode.

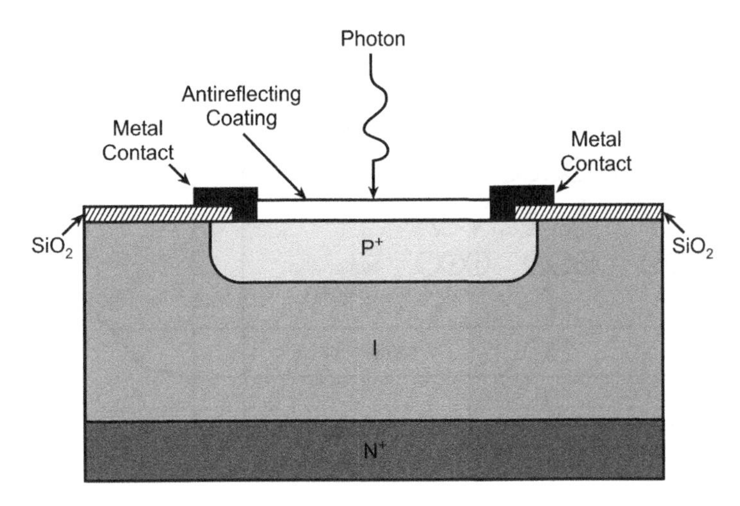

Figure 10.10 PIN photodiode.

10.6.1.2 PIN Photodiodes

In PIN photodiodes, an extra high-resistance intrinsic layer is added between the P- and the N-layers as shown in Figure 10.10. This has the effect of reducing the transit or diffusion time of the photo-induced electron-hole pairs, which in turn results in improved response time. PIN photodiodes feature low capacitance thereby offering high bandwidth, making them suitable for high-speed photometry as well as optical communication applications.

10.6.1.3 Schottky Photodiodes

A thin gold coating is sputtered onto the N-material to form a Schottky-effect P-N junction. Schottky photodiodes have enhanced UV response.

10.6.1.4 Avalanche Photodiodes

Avalanche photodiodes (APDs) are high-speed high-sensitivity photodiodes utilizing an internal gain mechanism that functions by applying a relatively higher reverse-bias voltage than that applied in the case of PIN photodiodes.

Figure 10.11 shows the cross-section of an avalanche photodiode. Avalanche photodiodes are constructed in order to provide a very uniform junction that exhibits the avalanche effect at reverse-bias voltages between 30 V and 200 V. The electron-hole pairs that are generated by incident photons are accelerated by the high electric field to force the new electrons to move from the valence band to the conduction band. In this way, multiplication of the order 50–100 is achieved. Avalanche photodiodes have fast response times, similar to that of PIN photodiodes. Responsivity figures for silicon PIN photodiodes are in the range 0.4–0.6 $A\,W^{-1}$ whereas for APDs they are 40–80 $A\,W^{-1}$, around 100 times greater than that of PIN photodiodes.

Moreover, APDs offer excellent signal-to-noise ratio of the order of that of photomultiplier tubes. They are therefore used in a variety of applications requiring high sensitivity such as long-distance optical communication and optical distance measurement.

10.6.2 Equivalent Circuit

A photodiode is electrically represented by a current source in parallel with an ideal diode. Figure 10.12 shows the equivalent circuit of a photodiode. The current source represents the photocurrent generated by incident radiation and the diode represents the PN junction. The current source is also shunted by a *junction capacitance* C_j and a *shunt resistance* R_{SH} across it. The parallel arrangement of four elements is then connected in series with a *series resistance* R_S.

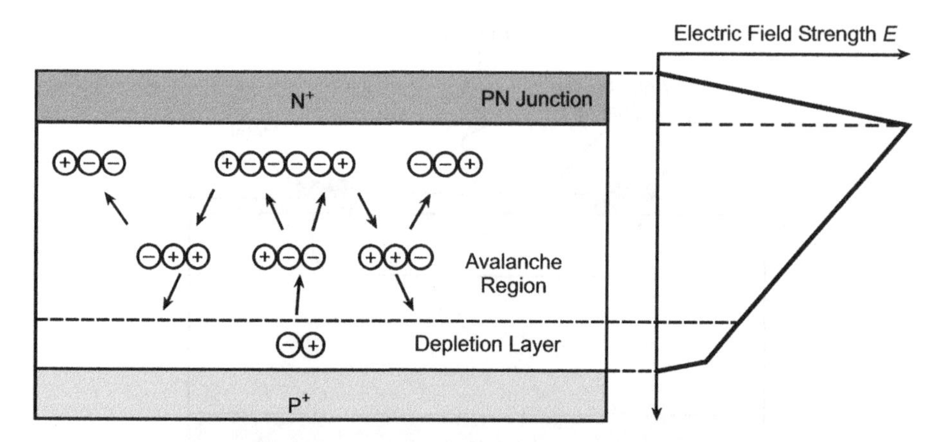

Figure 10.11 Cross-section of avalanche photodiode.

Shunt resistance is the slope of the current–voltage characteristic curve of the photodiode at the origin, where the applied bias voltage is zero. The value of R_{SH} in the case of an ideal photodiode is infinity. In the case of practical devices however, values range from tens to thousands of megaohms. A higher shunt resistance is always desirable. Shunt resistance is used to determine the noise current in the photodiode in the photovoltaic mode of operation where the applied bias voltage is zero.

Junction capacitance is formed by the boundaries of the depletion region acting as the plates of a parallel-plate capacitor and the depletion region acting as the dielectric medium. Consequently, it is directly proportional to the diffused area and inversely proportional to the width of the depletion region. Higher-resistivity substrates also offer lower junction capacitance. The value of junction capacitance C_j depends upon the applied reverse bias according to Equation 10.22:

$$C_j = \frac{\varepsilon_0 \varepsilon A}{\sqrt{[2\varepsilon\mu_0\rho(V_A + V_{bi})]}} \tag{10.22}$$

where

ε = dielectric constant of semiconductor material (= 11.9 for silicon)
ε_0 = permittivity of free space = $8.85 \times 10^{-12}\,\mathrm{F\,m^{-1}}$
μ_0 = mobility of electrons (= $1400\,\mathrm{cm^2\,V^{-1}\,s^{-1}}$ at 300 K)
ρ = resistivity of silicon (Ω)
V_A = applied reverse bias (V)
V_{bi} = built-in voltage of silicon (V)

Figure 10.13 depicts the variation of junction capacitance as a function of applied reverse bias for a silicon PIN photodiode, type BPX 65.

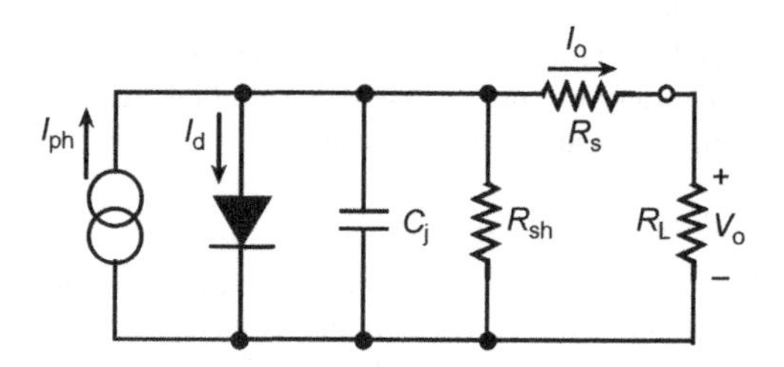

Figure 10.12 Equivalent circuit of a photodiode.

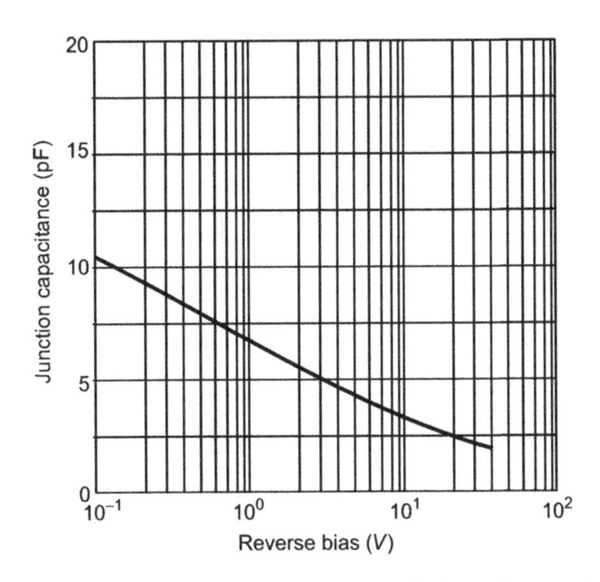

Figure 10.13 Variation of junction capacitance of a photodiode as a function of reverse bias.

Series resistance R_S of a photodiode arises from the resistance of the contacts and the resistance of the undepleted region. It is calculated from Equation 10.23:

$$R_S = \left(\frac{\rho W_S W_D}{A}\right) + R_C \tag{10.23}$$

where

W_S = substrate thickness (m)
W_D = depletion region width (m)
A = diffused area (m^2)
R_C = contact resistance (Ω)
ρ = resistivity of substrate material (Ω)

Series resistance in the case of an ideal photodiode is zero. In the case of practical devices, values range from 10 to 1000 ohms. It is used to determine the linearity of the photodiode response in photovoltaic mode of operation.

10.6.3 I–V Characteristics

Figure 10.14a and b show the circuit symbol and *I–V* characteristics of a photodiode. As we can see from the *I–V* characteristics of a photodiode, the curve in the dark state is similar to that of a conventional rectifier diode. When light strikes however, the curve shifts downwards with increasing intensity of light. If the photodiode terminals are shorted, a photocurrent proportional to the light intensity will flow in a direction from anode to cathode. If the circuit is open, then an open-circuit voltage will be generated with the positive polarity at the anode. It may be mentioned here that the short-circuit current is linearly proportional to light intensity while the open-circuit voltage has a logarithmic relationship with the light intensity.

Photodiodes can be operated in two modes: *photovoltaic* mode and *photoconductive* mode. No bias voltage is applied in the former mode of operation and, due to the incident light, a forward voltage is produced across the photodiode. In photoconductive operational mode, a reverse-bias voltage is applied

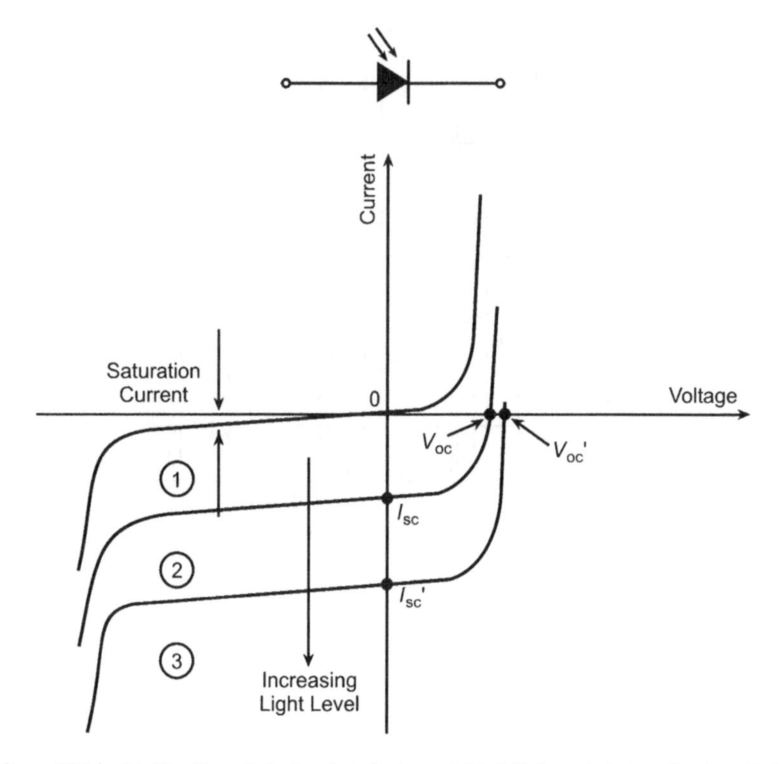

Figure 10.14 (a) Circuit symbol of a photodiode; and (b) *I–V* characteristics of a photodiode.

across the photodiode. This widens the depletion region, resulting in a higher speed of response. As a rule-of-thumb, all applications requiring bandwidth less than 10 kHz can use photodiodes in photovoltaic mode. For all other applications, photodiodes are operated in photoconductive mode. Moreover, the linearity of a photodiode is also improved when it is operated in the photoconductive mode. However, there is an increase in the noise current of the photodiode when it is operated in the photoconductive mode. This is due to the reverse saturation current referred to as the dark current flowing through the photodiode. The value of dark current is typically in the range 1–10 nA at a specified reverse-bias voltage. When the photodiode is operated in the photovoltaic mode, the value of dark current is zero.

10.6.4 Application Circuits

As discussed above, photodiodes can be operated in two modes (photovoltaic and photoconductive). In the photovoltaic mode, the photodiode is operated with zero external-bias voltage and is generally used for low-speed applications or for detecting low light levels. Figure 10.15a and b show two commonly used application circuits employing photodiodes in the photovoltaic mode. The output voltages for these circuits are given by $I_{det} \times R$ and $I_{det} \times R_f$, respectively, where I_{det} is the current through the photodiode. The circuit in Figure 10.15b offers better linearity than the circuit in Figure 10.13a as the equivalent input resistance across the photodiode in this case is R_f/A, where A is the open-loop gain of the operational amplifier. It is obvious that the value of R_f/A is much smaller as compared to that of R in the case of Figure 10.15a.

Figure 10.16 shows the load line analysis of a photodiode operating in the photovoltaic mode. The load line corresponding to the smaller load is closer to the current axis and the load line corresponding to a larger load is close to the voltage axis. As is evident from the figure, for a better linear response the equivalent resistance across the photodiode should be as small as possible.

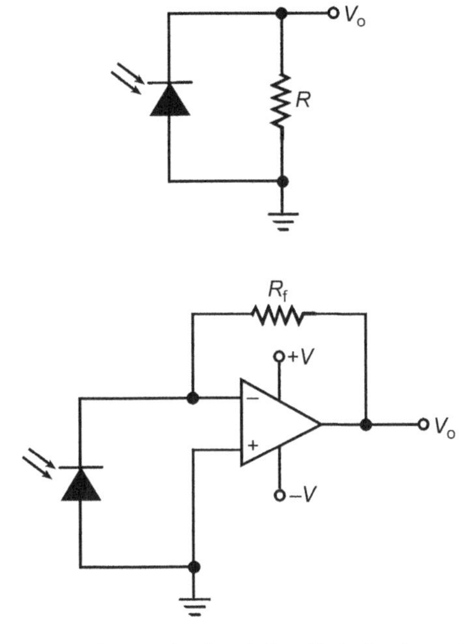

Figure 10.15 Application circuits of photodiodes in photovoltaic mode.

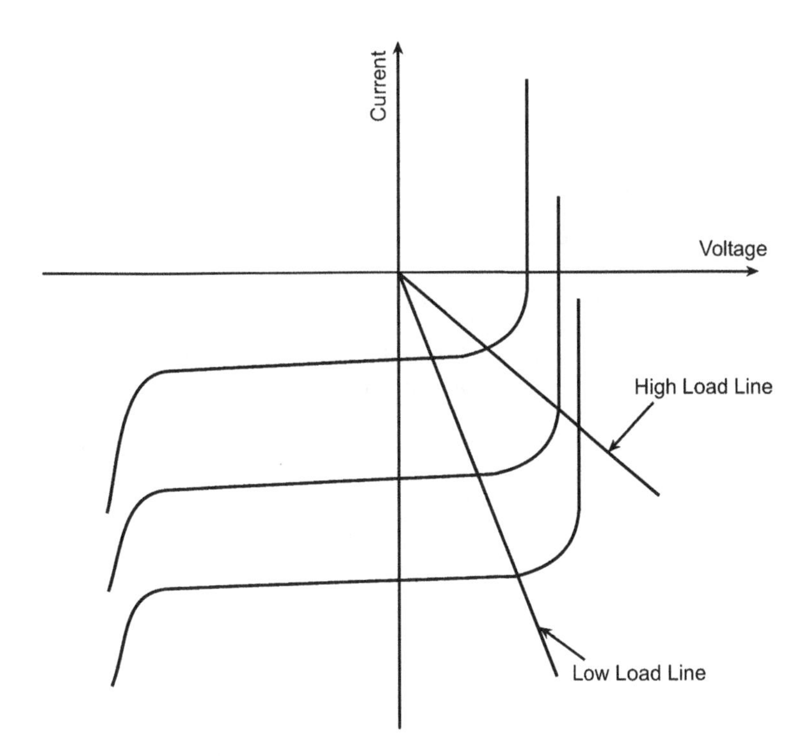

Figure 10.16 Load line analysis of photodiode in photovoltaic mode.

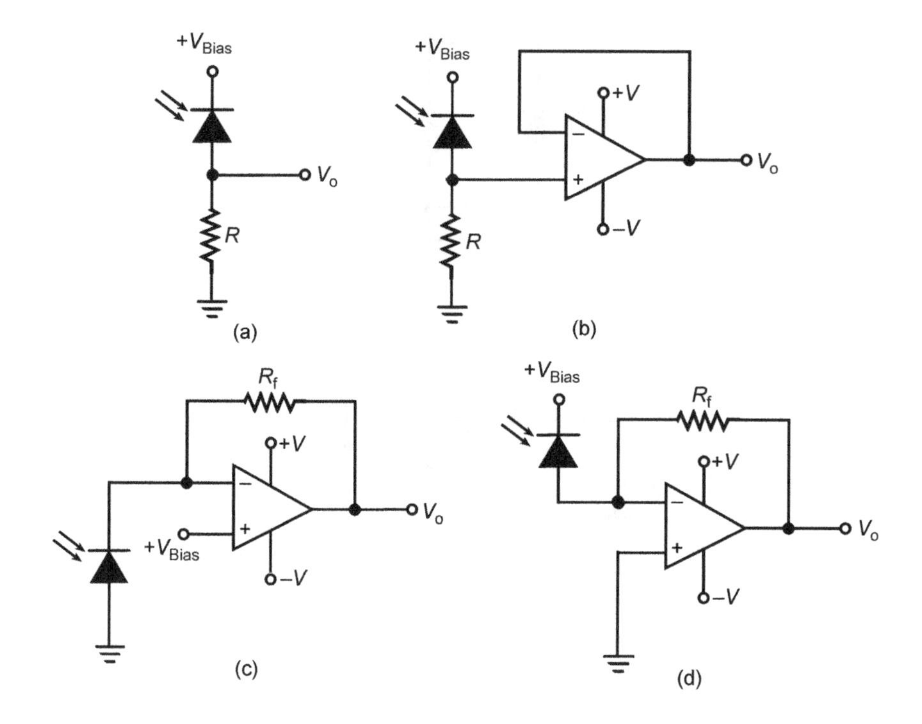

Figure 10.17 Application circuits of photodiodes operating in photoconductive mode.

Figure 10.17a–d show four possible circuits using photodiodes in the photoconductive mode. In Figure 10.17b, the operational amplifier is used as a voltage follower whereas in Figure 10.17c and d, the operational amplifier is used in the transimpedance mode. For the circuit in Figure 10.17b, the output voltage and the effective resistance across the photodiode are $I_{det} \times R$ and R, respectively. I_{det} is the current flowing through the photodiode. The output voltage and effective resistance across the photodiode in Figure 10.17c and d are $I_{det} \times R_f$ and R_f/A. The response of the photodiode for different loads operating in the photoconductive mode is shown in Figure 10.18. As we can see, circuits with lower-resistance load line offer better linearity.

Avalanche photodiodes (APDs) are also connected in a similar manner as discussed above, except that a much higher reverse-bias voltage is required for its operation. The power consumption of APDs during operation is much higher than that of PIN photodiodes and is given by the product of input signal, sensitivity and reverse-bias voltage. A protective resistor is therefore added to the bias circuit as shown in Figure 10.19, or a current-limiting circuit is used.

Since the gain of APDs changes with temperature, if they are operated over a wide temperature range a temperature-offset circuit has to be added; this changes the reverse-bias voltage with temperature in order to compensate for the change in gain with temperature. As an alternative, a temperature controller can be added to keep the temperature of APD constant. For detecting low signal levels, shot noise from the background light should be limited by using optical filters, source modulation and restricting the field-of-view.

10.6.5 Solar Cells

The operation of solar cells is very similar to that of a photodiode in the photovoltaic mode. Due to the photovoltaic effect, an open-circuit voltage is generated across a P-N junction when it is exposed to light, which is solar radiation in case of a solar cell. This open-circuit voltage leads to the flow of electric current through a load resistance connected across it, as shown in Figure 10.20.

As is evident from the figure, the impinging photon energy leads to the generation of electron-hole pairs. The electron-hole pairs either recombine and vanish or start drifting in opposite directions, with

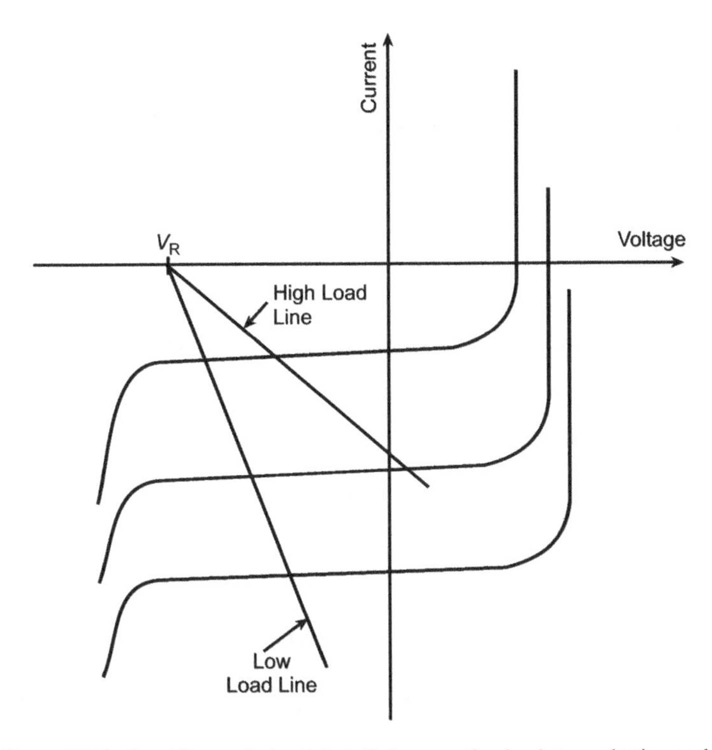

Figure 10.18 Load line analysis of photodiodes operating in photoconductive mode.

electrons moving towards the N-region and holes moving towards the P-region. This accumulation of positive and negative charge carriers constitutes the open-circuit voltage. This voltage can cause a current to flow through an external load or, when the junction is shorted, the result is a short-circuit current whose magnitude is proportional to input light intensity.

The voltage output and the current-delivering capability of an individual solar cell are very small when considered as possible electrical power input to any system. As an example, a typical solar cell would produce

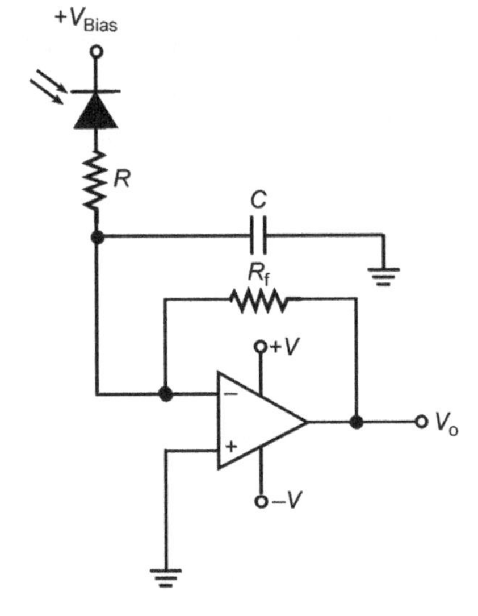

Figure 10.19 Application circuit using avalanche photodiode.

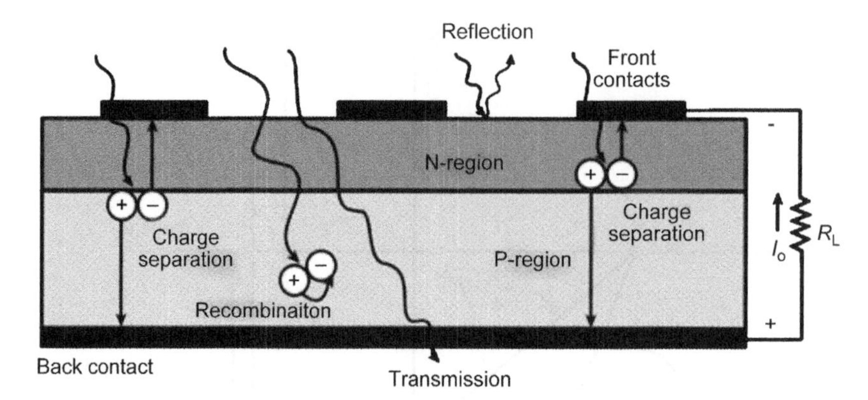

Figure 10.20 Principle of operation of a solar cell.

500 mV output with a load current capability of about 150 mA. Solar cells are therefore arranged in series-parallel to obtain the desired output voltage with required power delivery capability. The series combination is used to enhance the output voltage while the parallel combination is used to enhance the current rating.

Figure 10.21 shows the current–voltage and power–voltage characteristics of a solar cell. As is evident from the figure, a solar cell generates its maximum power at a certain voltage. The power–voltage curve has a point of maximum power called maximum power point (MPP). The cell voltage and the

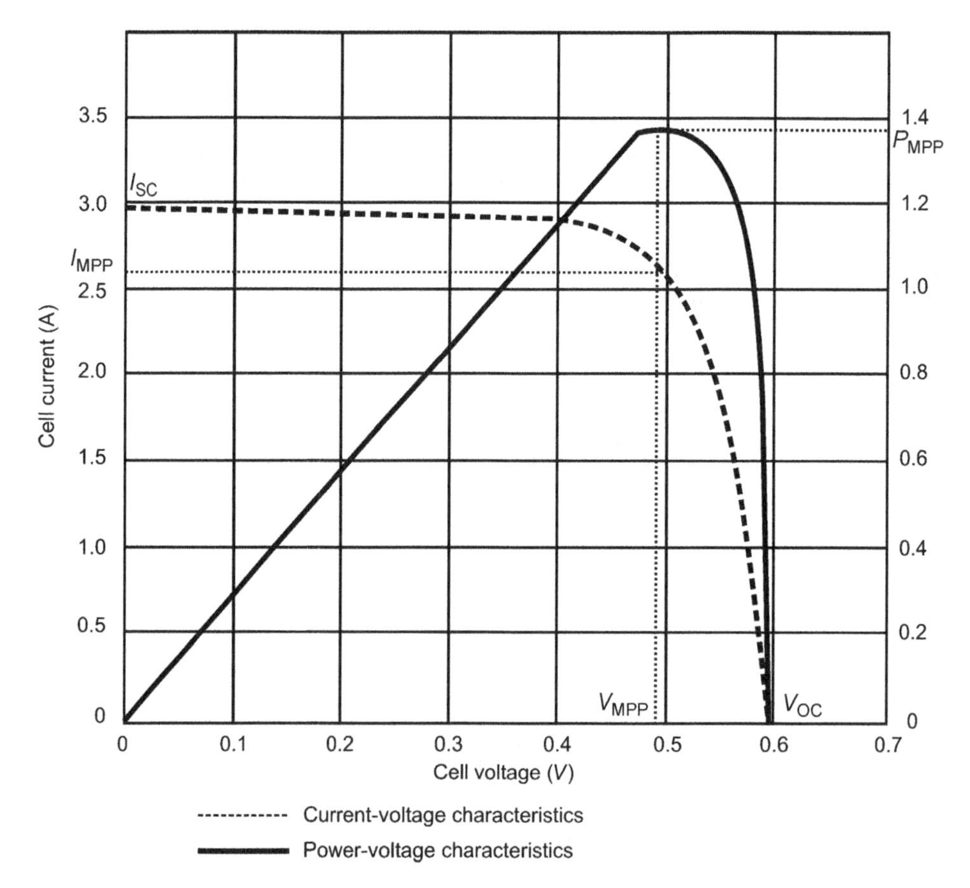

Figure 10.21 Current–voltage and power–voltage characteristics of a solar cell.

corresponding current are less than the open circuit voltage V_{OC} and the short-circuit current I_{SC}, respectively, at the maximum power point.

Solar cell efficiency is the ratio of maximum electrical solar cell power to the radiant light power on the solar cell area. The efficiency figure for some crystalline solar cells is in excess of 20%. The most commonly used semiconductor material for making solar cells is silicon. Both crystalline and amorphous forms of silicon are used for the purpose. Another promising material for making solar cells is gallium arsenide (GaAs). When perfected, GaAs solar cells will be lightweight and more efficient.

Example 10.4

Determine the cut-off wavelengths for silicon and germanium photodiodes, given that their band-gap energies are 1.1 eV and 0.72 eV, respectively, at 25 °C. How will the cut-off wavelength change when the operating temperature changes from 25 °C to 200 °C?

Solution

1. The cut-off wavelength is given by the formula $\lambda_c = \dfrac{1240}{E_g}$
2. For silicon photodiode, at 25 °C $E_g = 1.1$ eV; therefore $\lambda_c = 1240/1.1$ nm = 1127.27 nm.
3. The temperature variation of the band-gap energy of silicon semiconductor is given by $E_g(T) = 1.21 - 3.60 \times 10^{-4}T$, where T is the temperature in Kelvin. Band-gap energy of the silicon photodiode at 200 °C is therefore given by $E_g = 1.21 - 3.60 \times 10^{-4} \times 473$ eV = 1.21 – 0.17 eV = 1.04 eV.
4. The cut-off wavelength of the silicon photodiode at 200 °C is given by $\lambda_c = 1240/1.04$ nm = 1192.31 nm.
5. At 25 °C for the germanium photodiode, $E_g = 0.72$ eV; therefore $\lambda_c = 1240/0.72$ nm = 1722.22 nm.
6. The temperature variation of the band-gap energy of germanium semiconductor is given by $E_g(T) = 0.785 - 2.23 \times 10^{-4}T$, where T is the temperature in Kelvin. Band-gap energy of the germanium photodiode at 200 °C is therefore $E_g = 0.785 - 2.23 \times 10^{-4} \times 473$ eV = 0.785 – 0.105 eV = 0.68 eV.
7. The cut-off wavelength of germanium photodiodes at 200 °C is $\lambda_c = 1240/0.68$ nm = 1823.53 nm.

Example 10.5

For the circuit shown in Figure 10.22, determine the amplitude of the output voltage pulse when the light pulse (wavelength 1000 nm, pulse width 1 s and energy 10 mJ) is incident on the active area of the photodiode. The responsivity of the photodiode is 0.5 A W^{-1} at 1000 nm.

Solution

1. The incident light pulse has an energy of 10 mJ and a pulse width of 1 s. The input peak power therefore = $10 \times 10^{-3}/1$ W = 10 mW

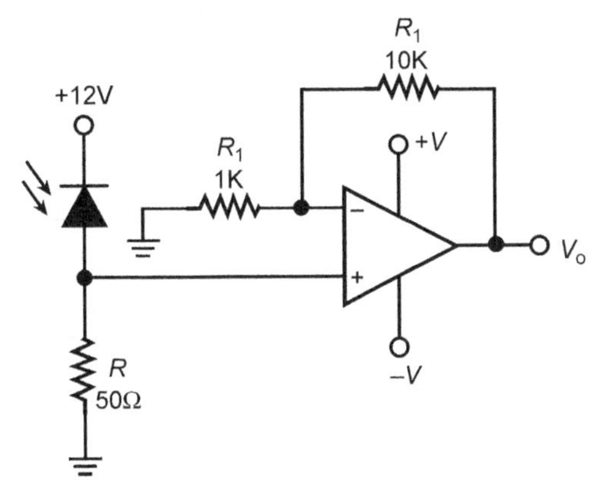

Figure 10.22 Example 10.5.

2. Output current from the photodiode $= 0.5 \times 10 \times 10^{-3}$ A $=$ 5mA
3. Voltage across the resistance $R = 50 \times 5 \times 10^{-3}$ V $= 250$ mV
4. Gain of the amplifier $= 1 + R_2/R_1 = 1 + (10 \times 10^3/1 \times 10^3) = 11$.
5. Voltage amplitude of the output pulse $= 250 \times 10^{-3} \times 11$ V $= 2.75$ V.

10.7 Phototransistors

Figure 10.23 depicts the construction of a phototransistor. Phototransistors are usually connected in a common emitter configuration with base open and the radiation is concentrated on the region near the collector-base junction. Figure 10.24a shows the circuit symbol of the phototransistor and Figure 10.24b shows the typical I–V characteristics of a phototransistor. When there is no radiation incident on the

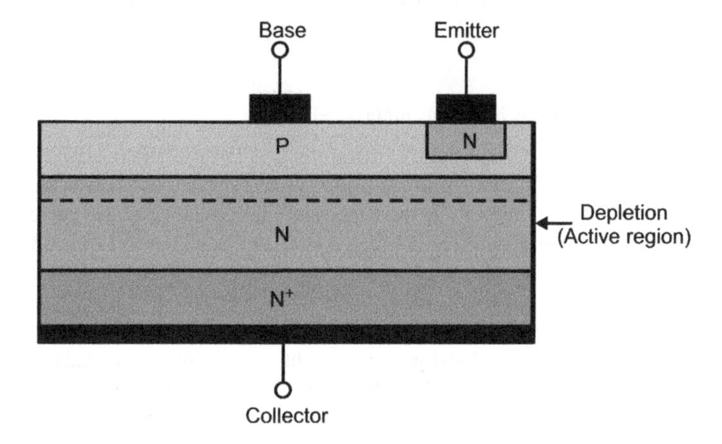

Figure 10.23 Cross-section of a phototransistor.

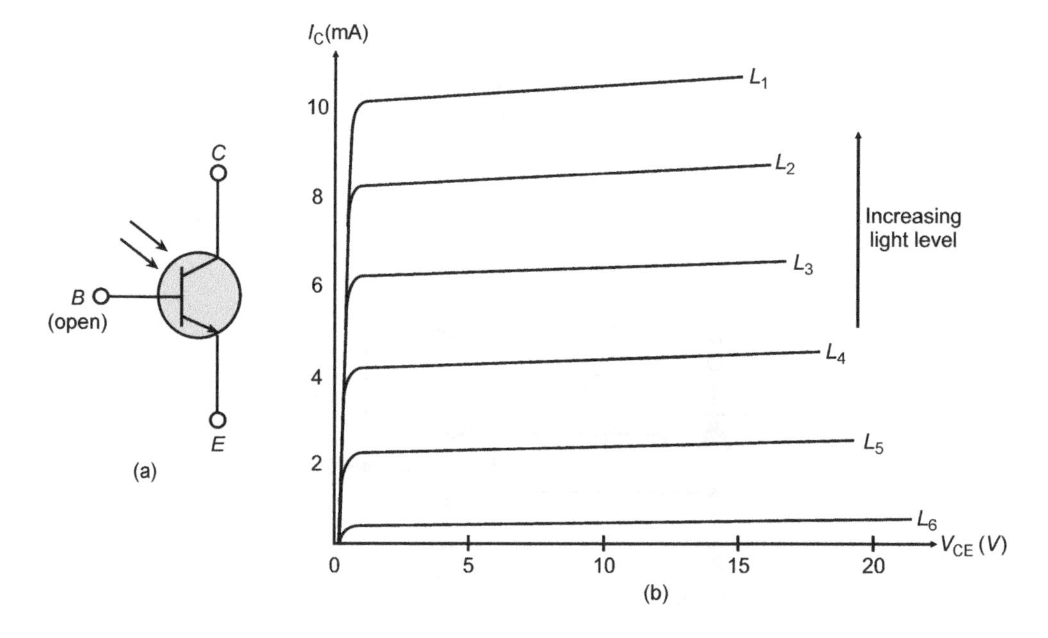

Figure 10.24 (a) Circuit symbol of a phototransistor (B is base); and (b) I–V characteristics of a phototransistor.

phototransistor, the collector current I_C is due to the thermally generated carriers and is given by Equation 10.24:

$$I_C = (\beta + 1)I_{CO} \tag{10.24}$$

where I_{CO} is the reverse saturation current and β is the transistor current gain in the common emitter configuration.

In phototransistors, this current is referred to as the dark current. When light is incident on the phototransistor, photocurrent is generated and the magnitude of the collector current increases. The expression for the collector current I_C is given by Equation 10.25:

$$I_C = (\beta + 1)(I_{CO} + I_\lambda) \tag{10.25}$$

where I_λ is the current generated due to incident light photons.

10.7.1 Application Circuits

Phototransistors can be used in two configurations: the common emitter configuration (Figure 10.25a) and the common collector configuration (Figure 10.25b). In the common emitter configuration, the output is high and goes low when light is incident on the phototransistor. In common collector configuration, the output goes from low to high when light is incident on the phototransistor. The transistor in both the configurations can act in two modes: active mode and switched mode. In active mode, the transistor operates in the active region of its characteristics and the output is proportional to input light intensity. In switched mode, the phototransistor is switched between cut-off and saturation and output is in the high and low states, respectively. The modes are controlled by the value of the resistor R. The output of the phototransistor can be amplified using an opamp or a transistor-based amplifier circuit.

Example 10.6

Determine the output voltage of the phototransistor circuit shown in Figure 10.26a when a CW light radiation of 0.1 mW is incident on the active area of the phototransistor. The active area of the transistor is $10\,\text{mm}^2$ and its output characteristics are shown in Figure 10.26b. The base-emitter voltage of the transistor $Q_2 = 0.7$ V, values of resistors R_1, R_C and R_E are 1 kΩ, 2.2 kΩ and 1 kΩ, respectively, and the supply voltage V_{CC} is 12 V.

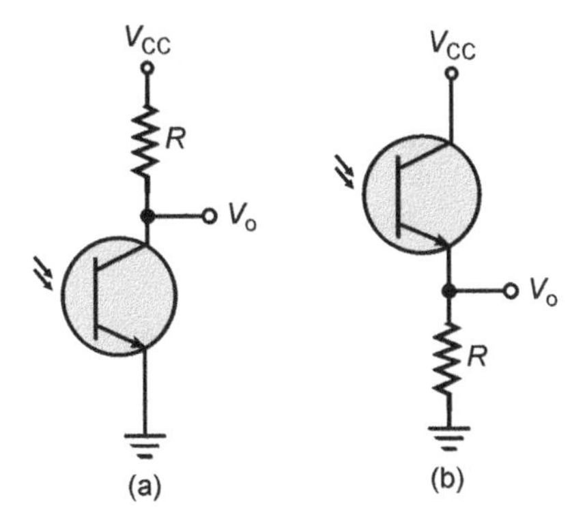

Figure 10.25 Application circuits of phototransistors.

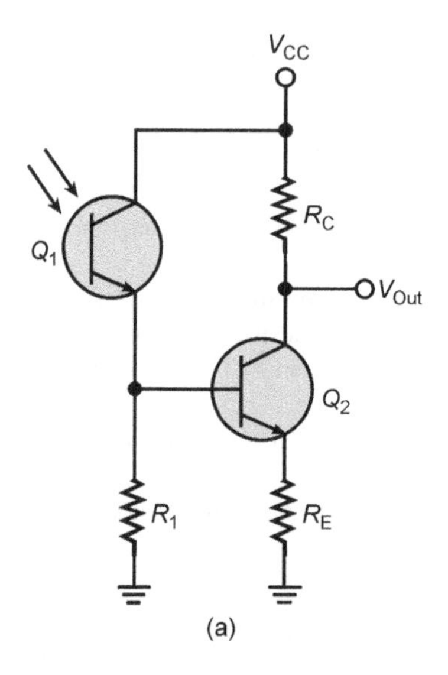

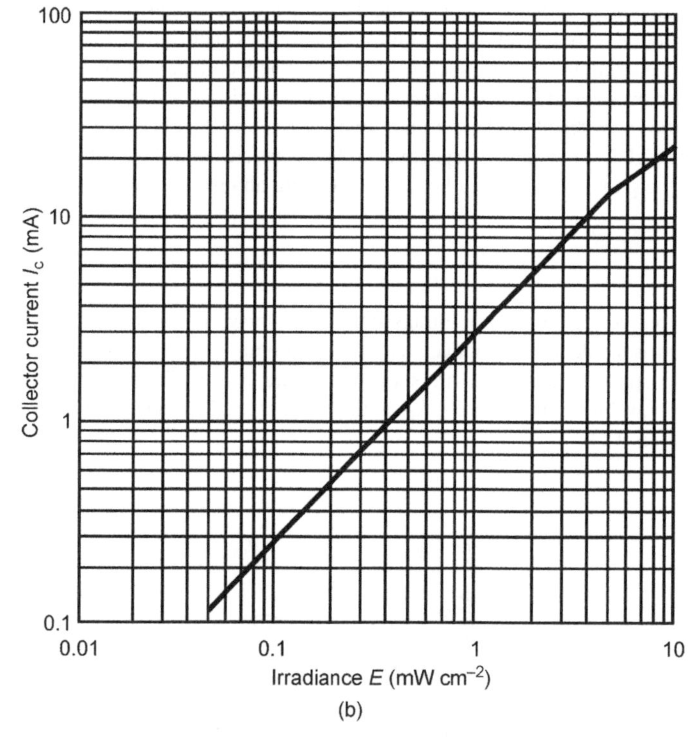

Figure 10.26 Example 10.6.

Solution

1. The incident irradiance on the phototransistor $= (0.1 \times 10^{-3})/(10 \times 10^{-2})$ W cm^{-2} = 1 mW cm^{-2}
2. From output characteristics of the transistor, collector current generated is 3 mA.
3. The voltage generated across the resistor R_1 is $3 \times 10^{-3} \times 1 \times 10^3$ V = 3 V. The voltage applied to the base of the transistor Q_2 is therefore 3 V and the transistor Q_2 goes into the conducting mode.

4. The voltage across resistor $R_E = 3-0.7$ V $= 2.3$ V.
5. The value of emitter current I_E is given by $I_E = 2.3/1 \times 10^3$ A $= 2.3$ mA. The collector current I_C is approximately equal to the emitter current, therefore $I_C = 2.3$ mA.
6. The output voltage $V_O = V_{CC} - I_C \times R_C = 12 - 2.3 \times 10^{-3} \times 2.2 \times 10^3 = 12 - 5.06$ V $= 6.94$ V.

10.8 Photo- FET, SCR and TRIAC

In this section we shall discuss the three other important photosensors, namely photo-FETs, photo-SCRs and photo-TRIACs. While photo-SCRs and photo-TRIACs are latching types of photosensors, photo-FETs are non-latching photosensors, such as photodiodes and phototransistors.

10.8.1 Photo-FET

Photo-FETs are light-sensitive FET devices where the diode formed by the reverse-biased gate-channel junction acts as a photodiode. Incident light generates additional photocarriers, resulting in increased conductivity level. Gate current flows if the gate is connected to an external resistor. Figure 10.27a shows the circuit symbol of a photo-FET. When no light is incident on the photo-FET, the gate impedance is very high. When light is incident on the photo-FET, the value of gate impedance decreases. Figure 10.27b shows the typical application circuit using a photo-FET. When no light is incident, the gate voltage is approximately equal to the voltage $-V_{GG}$. When light is incident the negative gate voltage decreases, resulting in an increase in the value of drain current I_D and a decrease in the output voltage V_O.

10.8.2 Photo-SCR

Photo-SCRs, generally referred to as light-activated SCRs (LASCRs), are essentially the same as conventional SCRs except that they are triggered by light incident on the gate junction area. They comprise a window and lens to focus more light on the gate junction area, more specifically on the middle junction J_2 of the SCR. They conduct current in one direction when activated by sufficient light, and continue to conduct until the current falls below a specified value. In other words, photo-SCRs act as a latch that can be triggered on by the light incident on the gate junction but they do not turn off when the light source is removed. They can be turned off by reducing the current below its threshold value.

Photo-SCRs can handle large amounts of current as compared to a photodiode or a phototransistor. They have a high rate of change of voltage with time, that is, high dV/dt rating, which is important for triggering the SCR on application of light input. Photo-SCRs are most sensitive to light when their gate terminal is open. They are generally used in the receiving channel of optocouplers.

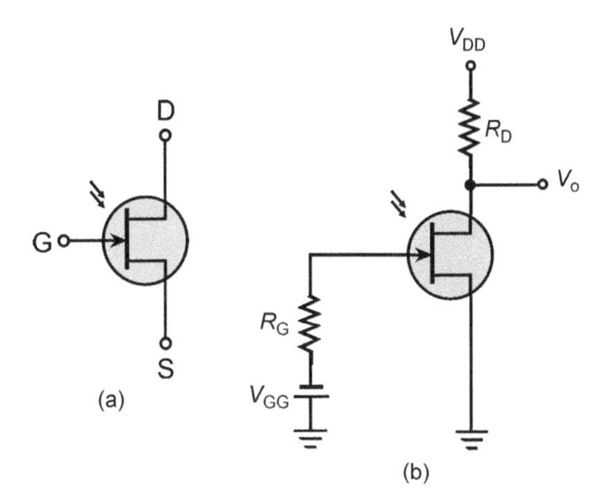

Figure 10.27 (a) Circuit symbol of photo-FET (where D, G and S are drain, gate and source); and (b) a simple application circuit using photo FET.

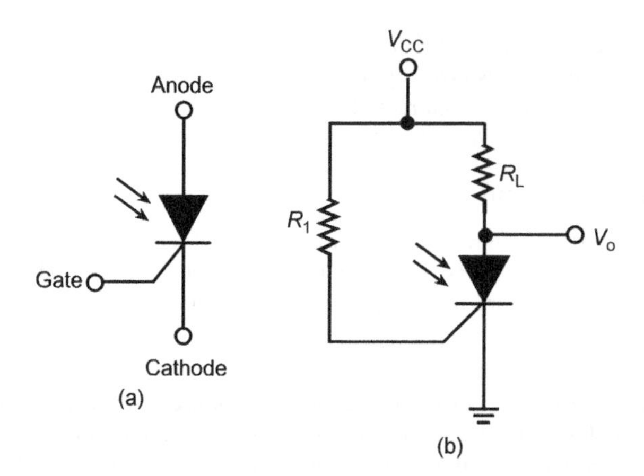

Figure 10.28 (a) Circuit symbol of photo-SCR; and (b) simple application circuit using photo SCR.

Figure 10.28a shows the circuit symbol of a photo-SCR and Figure 10.28b shows a simple application circuit built around it. When no light is incident on the photo-SCR, it is off and no current flows through the load resistor R_L. When the photo-SCR is illuminated, it turns on and hence allows current to flow through load resistance R_L.

10.8.3 Photo-TRIAC

Photo-TRIACs, also referred to as light-activated TRIACs, are bidirectional thyristors that are designed to conduct current in both directions when the incident light radiation exceeds a specified threshold value. Photo-TRIACs are generally used as solid-state AC switches and as photosensors in optocouplers to provide isolation from the driving source to the load. Figure 10.29 shows the circuit symbol of a photo-TRIAC. The operation of a photo-TRIAC is similar to a standard TRIAC, except that the trigger current is generated indirectly for a photo-TRIAC by the light incident on it and is supplied directly to a standard TRIAC. One of the most important parameters for describing the performance of a photo-TRIAC is the output dV/dt rating. Other important parameters are the breakdown voltage and the power rating of the device.

There are two different types of photo-TRIACs available: non-zero-crossing and zero-crossing. Non-zero-crossing photo-TRIACs are used for applications that require fine control involving small time constants. Zero-crossing photo-TRIACs are used in applications where the control time constant is fairly large.

Example 10.7

The circuit shown in Figure 10.30 is kept in a dark room. Determine the output voltage V_O. Also determine the output voltage when a bright light is flashed on the circuit.

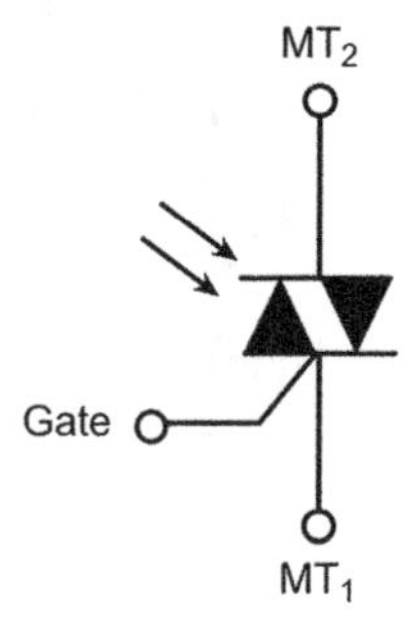

Figure 10.29 Circuit symbol of a photo-TRIAC.

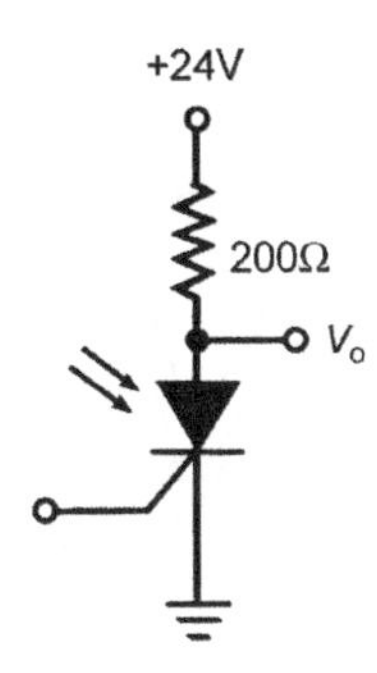

Figure 10.30 Example 10.7.

Solution

1. When the circuit is kept in the dark room with no incident light, the photo-SCR is in the non-conducting state. Therefore, the output voltage V_O is equal to the supply voltage, that is, $V_O = +24$ V.
2. When a bright light is incident on the photo-SCR, it goes into the conducting state and ideally the output voltage is zero, that is, $V_O = 0$ V.

10.9 Photoemissive Sensors

The photosensors discussed so far are based on the internal photoeffect where the photoelectrons generated by the incident radiation remain within the semiconductor material. The other category of photosensors relies on the *external* photoeffect, where the photo-generated electrons travel beyond the physical boundaries of the material. These sensors are also referred to as photoemissive sensors. In this section we discuss some of the commonly used photoemissive photosensors such as vacuum photodiodes, photomultiplier tubes (PMTs) and image-intensifier tubes.

10.9.1 Vacuum Photodiodes

The vacuum photodiode is the oldest photosensor. It comprises an anode and a cathode placed in a vacuum envelope. When irradiated, the cathode releases electrons that are attracted by the positively charged anode, thus producing a photocurrent proportional to the light intensity.

10.9.2 Photomultiplier Tubes

Photomultiplier tubes are extremely sensitive photosensors operating in the ultraviolet, visible and near-infrared spectrum. PMTs have an internal gain of the order of 10^8 and can even detect single photon of light. They are constructed from a glass vacuum tube which houses a photocathode, several dynodes and an anode. When the incident photons strike the photocathode, electrons are produced as a result of the photoelectric effect. These electrons accelerate towards the anode and, in the process, electron multiplication takes place due to secondary emission processes from the dynodes. PMTs require a few kilovolts of biasing voltages for proper operation. Figure 10.31 shows the cross-section of a photomultiplier tube. As we can see from the figure, the dynodes are given progressively increasing positive voltages with the dynode nearest to the cathode having the lowest voltage and the dynode nearest to the anode having the maximum voltage.

Important features of photomultiplier tubes include low noise, high frequency response and large active area. By virtue of these features, PMTs are used in nuclear and particle physics, astronomy, medical imaging and motion picture film scanning. Avalanche photodiodes have replaced PMTs in some applications, but PMTs are still frequently used.

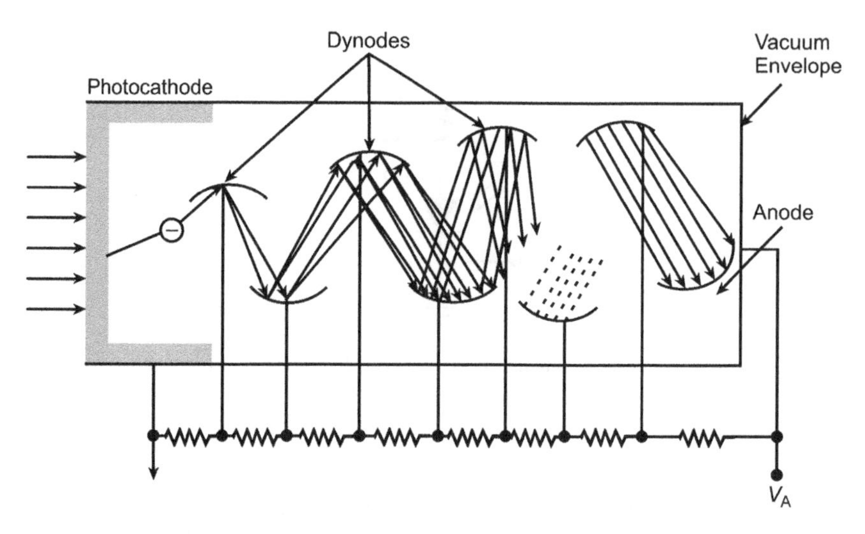

Figure 10.31 Cross-section of a photomultiplier tube.

10.9.3 Image Intensifiers

Image intensifiers are devices that amplify visible and near-infrared light from an image so that a dimly lit scene can be viewed by a camera or by human eye. Contemporary image intensifiers comprise an objective lens, vacuum tube with photocathode at one end, tilted micro-channel plate (MCP) and a phosphor screen (Figure 10.32). An objective lens focuses the image onto the photocathode. When the photons strike the photocathode, electrons are released due to the photoelectric effect. These photoelectrons are accelerated through around 4–5 kV into a tilted MCP where secondary electron multiplication takes place. The electrons all move together due to the potential difference across the tube and, for each photoelectron, hundreds or even thousands of electrons are created. All these electrons hit the phosphor screen at the other end, releasing one photon for every electron. The screen therefore converts the high-energy electrons into photons, which corresponds to the input image radiation but with the incident flux being amplified many times.

Image intensifiers are classified into the following categories: generation 0, generation 1, generation 2 and generation 3. Generation 0 and generation 1 devices did not have an MCP and the stream of electrons generated by the photocathode was accelerated towards the phosphor screen by the applied potential. Generation 1 devices were a tremendous improvement upon generation 0 devices, and had three times the photosensitivity of generation 0 devices. Generation 2 devices introduced the concept of MCPs.

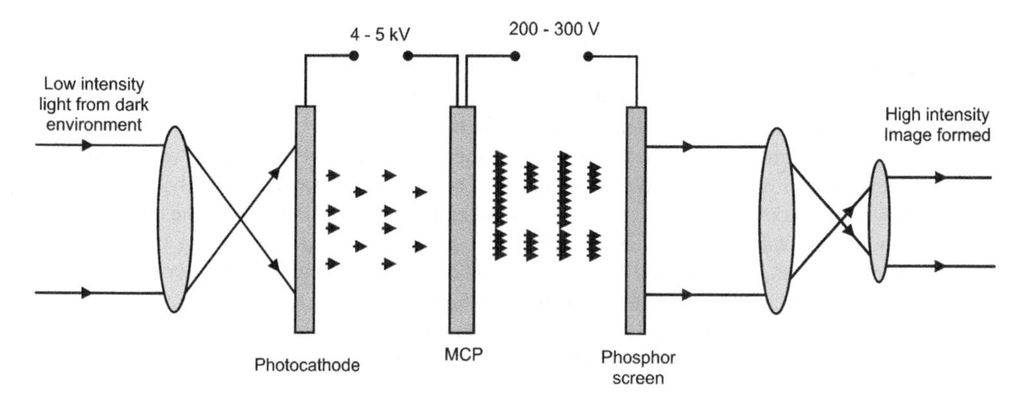

Figure 10.32 Image intensifier tube.

Generation 3 devices are the same as generation 2 devices, except that the photocathode material in generation 3 devices is gallium arsenide (GaAs) and for generation 2 devices is S-25. Generation 3 devices also have a better MCP. Generation 3 ultra and generation 4 tubes are also available, which offer slight improvement over generation 3 devices. Image intensifiers are used in night-vision devices for military applications.

10.10 Thermal Sensors

Thermal sensors absorb radiation, producing a temperature change that in turn causes a change in the physical or the electrical property of the sensor. In other words, thermal sensors respond to a change in their bulk temperature caused by the incident radiation. Thermocouple, thermopile, bolometer and pyroelectric sensors belong to the category of thermal sensors. Thermal sensors lack the sensitivity of photoelectric sensors and are generally slow in response, but have a wide spectral response. Most of these sensors are passive devices, requiring no bias. In this section, we discuss the different types of thermal sensors and their application circuits.

10.10.1 Thermocouple and Thermopile

Thermocouple sensors are based on the Seebeck effect, that is, the temperature change at the junction of two dissimilar metals generates an electro-motive force (EMF) proportional to the temperature change. The commonly used thermocouple materials are bismuth-antimony, iron-constantan and copper-constantan, which have temperature coefficients 100 $\mu V\,°C^{-1}$, 54 $\mu V\,°C^{-1}$ and 39 $\mu V\,°C^{-1}$ respectively. To compensate for the changes in the ambient temperature, thermocouples generally have two junctions: a measuring junction and a reference junction.

The responsivity of a single thermocouple is very low; to increase this responsivity, several junctions are connected in series to form a thermopile. Thermopiles are series combinations of around 20–200 thermocouples. The spectral response of thermocouples and thermopiles extends into the far-infrared band up to 40 μm. They are suitable for making measurements over a large temperature range up to 1800 K. However, thermocouples are less suitable for applications where smaller temperature differences need to be measured with great accuracy, e.g. 0–100 °C measurement with 0.1 °C accuracy. For such applications, thermistors and resistance temperature detectors (RTDs) are more suitable.

The responsivity of thermopiles is of the order 10–100 $V\,W^{-1}$ and the typical signal output varies from a few tens of microvolt to a few millivolts. They therefore need a low-noise and very-low-offset operational amplifier for providing the gain. The gain required varies from as little as 10 to as large as 10 000 or more. Generally, for gain <1000, a single-stage amplifier is used. For gain values >1000, two stages are used. Figure 10.33 shows the application circuit where two amplifier stages are used. As we can see from the figure, thermopiles require no bias voltage.

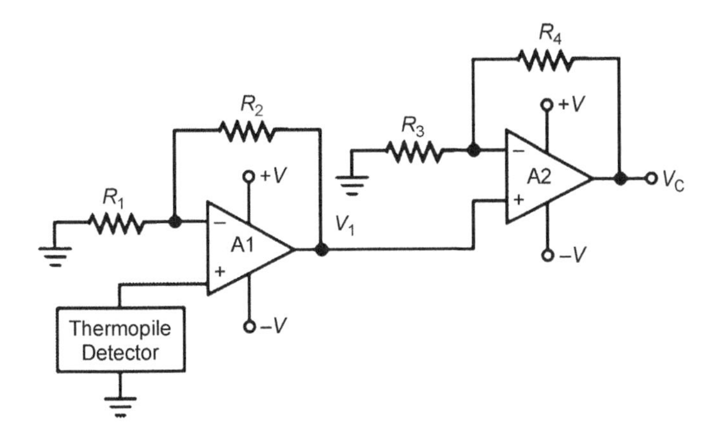

Figure 10.33 Application circuit using thermopiles.

The thermopile signal is positive or negative depending upon whether the temperature of the object filling the field-of-view of the thermopile is greater than or less than that of the thermopile. The output of the circuit varies with change in the ambient temperature which must be compensated for; many thermopile modules therefore have an inbuilt thermistor to compensate for the ambient temperature variations.

10.10.2 Bolometer

The bolometer is the most popular type of thermal sensor. The sensing element in a bolometer is a resistor with a high temperature coefficient. A bolometer is different from a photoconductor in that a direct photon–electron interaction causes a change in the conductivity of the material in the former, whereas the increased temperature and the temperature coefficient of the element causes the resistance change in the bolometer. Bolometers can be further categorized as metal bolometers, thermistor bolometers and low-temperature germanium bolometers.

A metal bolometer uses metals such as bismuth, nickel or platinum, having a temperature coefficient in the range 0.3–0.5% $°C^{-1}$. A thermistor bolometer is the most popular and has applications in burglar alarms, smoke sensors and other similar devices. The sensor in this case is a thermistor, an element made of manganese, cobalt and nickel oxide. They have high temperature coefficients up to 5% $°C^{-1}$ which vary with temperature as T^{-2}. They are classified as negative temperature coefficient (NTC) and positive temperature coefficient (PTC) thermistors, depending whether their temperature coefficient of resistance is negative or positive.

Figure 10.34a shows the circuit symbol of a thermistor. The spectral response of thermistors extends from 0.5 to 10 μm. More sensitive thermistors typically have NEP and response time of the order of 10^{-10} W and 100 ms. Less-sensitive thermistors have NEP and response time values of 10^{-8} W and 5 ms, respectively. Figure 10.34b and c show the simplest possible configurations in which a thermistor can be used for measurement of light intensity. The output of the circuits in Figure 10.34b and c can be fed to an operational amplifier or to a comparator for linear light control or light on-off control, respectively.

A low-temperature germanium bolometer is a sensitive laboratory bolometer that uses germanium as the sensor. It has the highest responsivity when operated at a few degrees above absolute zero.

10.10.3 Pyroelectric Sensors

Pyroelectric sensors are characterized by spontaneous electric polarization, which is altered by temperature changes as light illuminates these sensors. Pyroelectric sensors are low-cost high-sensitivity devices that are stable against temperature variations and electromagnetic interference. Pyroelectric sensors only respond to modulating light radiation, and there is no output for a CW incident radiation. Figure 10.35a shows the circuit symbol of pyroelectric sensors.

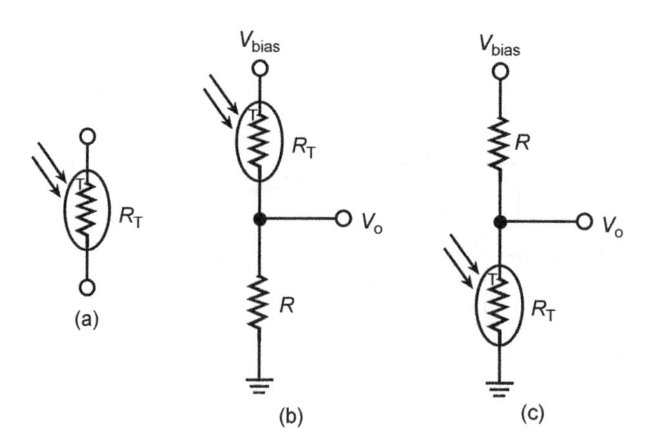

Figure 10.34 (a) Circuit symbol of a thermistor; and (b, c) application circuits of a thermistor.

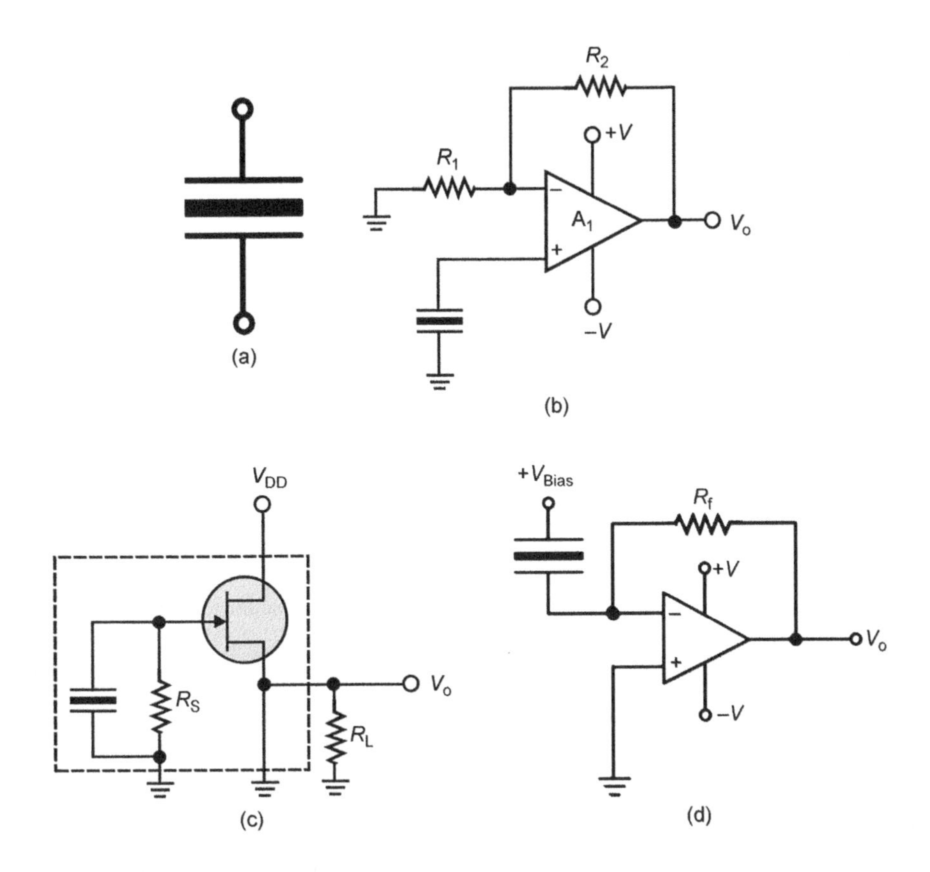

Figure 10.35 (a) Circuit symbol of pyroelectric sensors; (b, c) voltage-mode pyroelectric sensor application circuits; and (d) current-mode pyroelectric sensor application circuit.

Pyroelectric sensors operate in two modes: voltage mode and current mode. In the voltage mode, the voltage generated across the entire pyroelectric crystal is detected. In the current mode of operation, current flowing on and off the electrode on the exposed face of the crystal is detected. Voltage mode is more commonly used than current mode.

The circuit for voltage mode is shown in Figure 10.35b. The operational amplifier chosen should have very high input impedance of the order of 10^{12}–10^{14} ohms. The circuit is sensitive to ambient temperature variations, however; ambient temperature variations can be compensated for by employing AC coupling between the amplifier stages or by adding a compensation crystal in opposition (either in series or parallel). One crystal is exposed to radiation and the other is shielded from radiation. As the ambient temperature changes, the surface charge generated on one crystal is cancelled by the equal and opposite charge generated on the other crystal. The incident radiation however generates charge only on one crystal and is not cancelled.

Voltage-mode pyroelectric sensors are generally integrated with a FET. A shunt resistor R_S in the range of 10^{10}–10^{11} ohms is added to provide thermal stabilization. External connections include a power supply and load resistor R_L as shown in Figure 10.35c. The output voltage appears across R_L.

The circuit for current-mode operation is shown in Figure 10.35d. The modulation frequency can be much higher in case of current-mode operation than for voltage-mode operation. It is therefore much easier to separate the signal from the ambient temperature drift.

10.11 Displays

Displays are output devices that are used for visual presentation of information. Displays form an interface between the machine and the human. In this section, we shall discuss different types of displays and the characteristic parameters used to define the quality of displays.

10.11.1 Display Characteristics

Three factors are critical for a good visual display: legibility, brightness and contrast.

Legibility is the property of a display by virtue of which the characters are easy to read with speed and accuracy. The factors which contribute to the legibility of the display are its style, size, character sharpness and shape. *Brightness* refers to the perception of luminance by the visual world. The *contrast* of a display depends on the background luminance and the source luminance. The readability of the display depends upon the contrast parameter. It is defined in different ways for passive and active displays. In the case of the former such as liquid-crystal displays (LCDs), contrast C is defined by Equation 10.26:

$$C = (L_{\mathrm{O}} - L_{\mathrm{B}})/L_{\mathrm{O}} \tag{10.26}$$

where L_{O} is the object or source luminance and L_{B} is the background luminance (both measured in Cd m^{-2}. Contrast can have values between 0 and 1, 0 being the case when the object and the background luminance are the same and 1 when the background has zero luminance.

For active displays such as light-emitting diodes (LEDs), the contrast parameter is defined in terms of contrast ratio CR, defined by Equation 10.27:

$$\mathrm{CR} = L_{\mathrm{O}}/L_{\mathrm{B}} \tag{10.27}$$

The contrast ratio can have values between 1 and infinity. For contrast ratio equal to 1, the object and the background have the same luminance and the displayed characters are not visible at all. The background luminance is zero and the display has best visibility at contrast value equal to infinity ($\mathrm{CR} = \infty$).

10.11.2 Types of Displays

Displays can be categorized into different types depending upon the manner in which they display information. These include bar graph displays, segmented displays, dot-matrix displays and large displays.

Bar graph displays are composed of several bar elements as shown in Figure 10.36. These are replacing analogue instruments as indicators due to their simplicity and cost-effectiveness.

Segmented displays are available in two configurations: 7-segment displays (Figure 10.37a) and 16-segment displays (Figure 10.37b).

A 7-segment display comprises 7 bars and 1 or 2 decimal points and is an industry standard for numeric displays. These displays are used for displaying numerals and limited alphabet. A 16-segment display can present the entire upper case alphabet and numerals. Segmented displays can present only limited information.

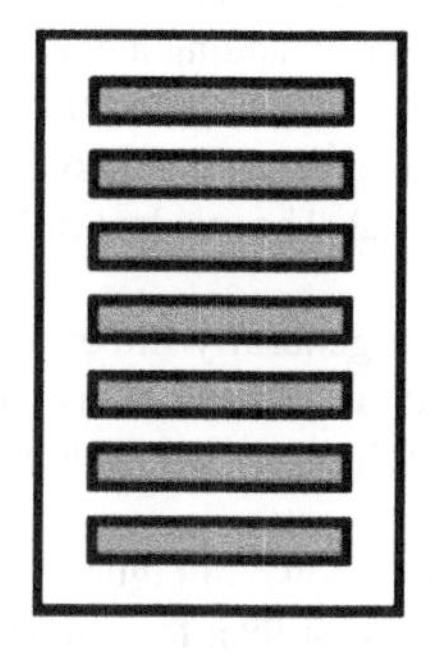

Figure 10.36 Bar graph display.

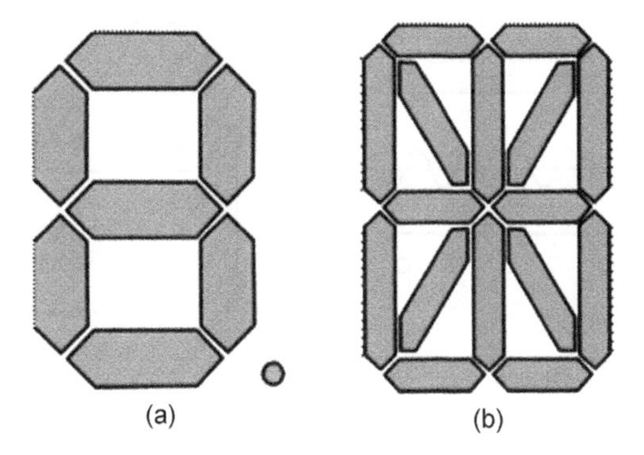

Figure 10.37 (a) 7-segment display; and (b) 16-segment display.

Dot matrix displays are the simplest displays that represent the set of lower-case and upper-case alphabet and numerals at reasonable cost and complexity. The most commonly used dot matrix display is the 5×7 display which comprises 35 display elements set in a pattern of 5 rows and 7 columns (Figure 10.38). Each element is addressed selecting the relevant row and column. Bar, segmented and 5×7 dot-matrix displays can be constructed using LEDs or LCDs.

Large-scale displays include cathode ray tube (CRT) displays, plasma displays and LCD thin-film transistor (TFT) displays.

LEDs and LCDs are discussed in Sections 10.12 and 10.13, respectively, followed by CRT displays and emerging trends in display technology in Sections 10.14 and 10.15, respectively.

10.12 Light-emitting Diodes

An LED is a semiconductor P-N junction diode designed to emit light when forward biased. It is one of the most popular optoelectronics source. LEDs consume very little power and are inexpensive.

When a P-N junction is forward-biased, the electrons in the N-type material and the holes in the P-type material travel towards the junction. Some of these holes and electrons recombine with each other and radiate energy in the process. The energy will be released either in the form of photons of light or in the form of heat. In silicon and germanium diodes, most of the energy is released as heat and the emitted light is insignificant. However, in some materials such as gallium phosphide (GaP), gallium arsenide (GaAs) and gallium arsenide phosphide (GaAsP), substantial photons of light are emitted. These materials are therefore used in the construction of light-emitting diodes.

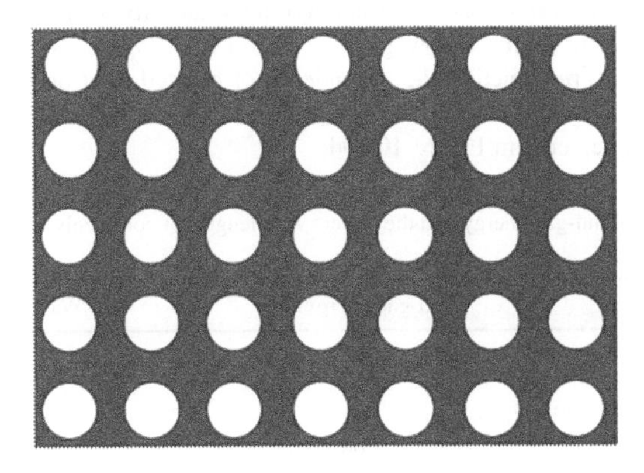

Figure 10.38 5×7 dot-matrix display.

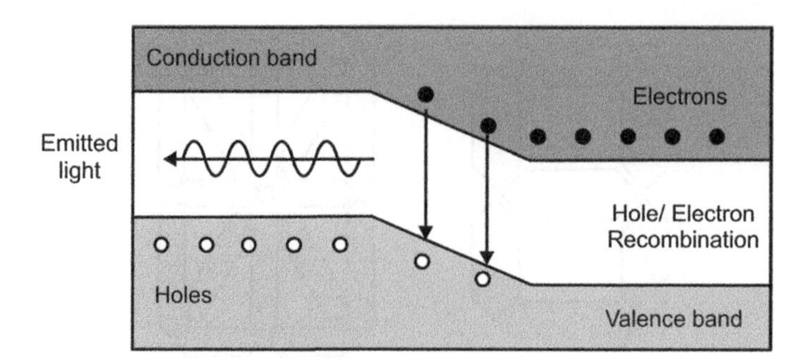

Figure 10.39 P-N junction of a LED.

In the absence of an externally applied voltage, the N-type material contains electrons while the P-type material contains holes that can act as current carriers. When the diode is forward biased, the energy levels shift and there is a significant increase in the concentration of electrons in the conduction band on the N-side and of holes in valence band on the P-side. These electrons and holes combine near the junction to release energy in the form of photons (Figure 10.39). It may be mentioned here that the process of light emission in a LED is that of spontaneous emission, that is, the photons emitted are not in phase and travel in different directions.

The energy of the photon resulting from this recombination is equal to the band-gap energy of the semiconductor material E_g (eV) and is defined by the empirical formula given in Equation 10.28:

$$E_g = \frac{1240}{\lambda} \tag{10.28}$$

where λ is wavelength (nm).

Some of the commonly used semiconductor materials used for fabricating LEDs are GaAs, GaP, GaAsP, aluminium antimonide (AlSb) and indium antimonide (InSb). Table 10.1 enlists the band-gap energies and the typical wavelengths emitted by these materials.

10.12.1 Characteristic Curves

The parameters of interest in a LED are the *I–V* characteristics, spectral distribution curve, light output vs input current and the directional characteristics.

Figure 10.40a shows the *I–V* characteristics of LEDs of different colours. As the LED is operated in the forward-biased mode, the *I–V* characteristics in the forward-biased region are shown. *I–V* characteristics of LEDs are similar to that of conventional P-N junction diodes, except that the cut-in voltage in the case of LEDs is in the range 1.3–3 V compared to 0.7 V for silicon diodes and 0.3 V for germanium diodes.

The *spectral distribution curve* shows the variation of light intensity with wavelength. Figure 10.40b shows the typical spectral curves for yellow, green and red LEDs. Figure 10.40c shows a typical *light output vs input current curve* depicting the dependence of emitted light on forward current flowing through the LED. Finally, *directional characteristics* refer to the variation in the light output with change in the viewing angle, as depicted in Figure 10.40d.

Table 10.1 Band-gap energy and the typical wavelengths of commonly used LED materials

Material	Band-gap energy (eV)	Wavelength (nm)
GaAs	1.43	910
GaP	2.24	560
$GaAs_{60}P_{40}$	1.91	650
AlSb	1.60	775
InSb	0.18	6900

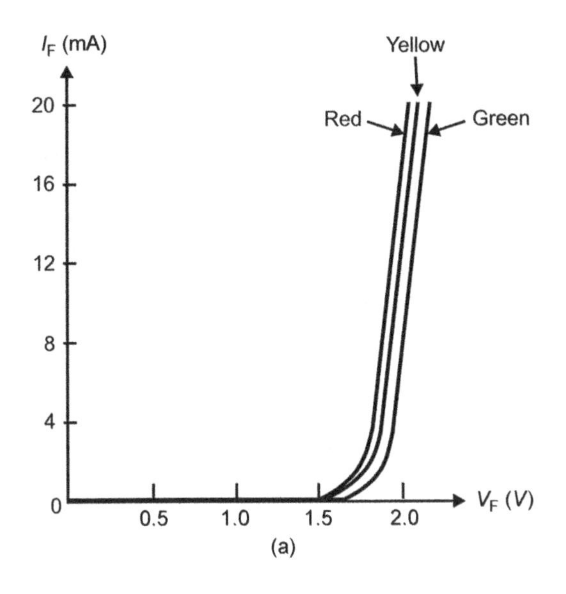

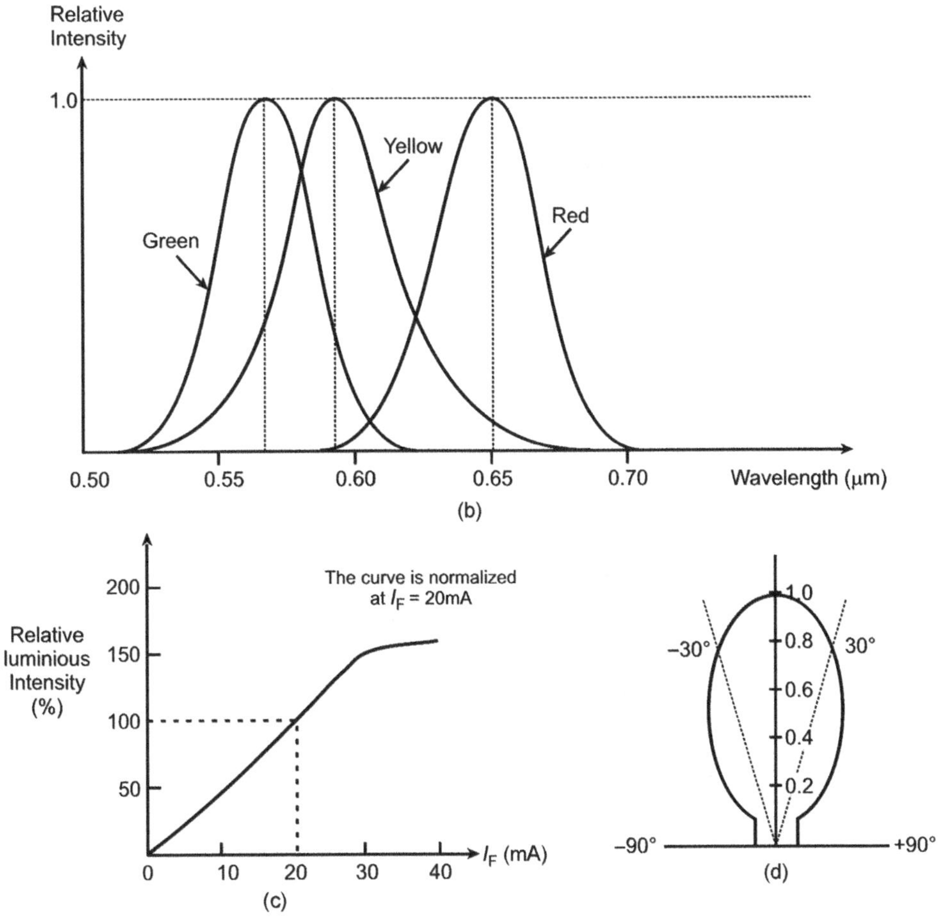

Figure 10.40 LED characteristics: (a) *I–V*; (b) spectral; (c) light output vs. input current; and (d) directional.

10.12.2 Parameters

The parameters of interest in the case of LEDs are forward voltage V_F, candle power (CP), radiant power output P_O, peak spectral emission λ_P and spectral bandwidth.

Forward voltage V_F is the DC voltage across the LED when it is on. Typical values for V_F for LEDs are in the range 1.3–3.0 V. As can be seen from Figure 10.40a, V_F is near 1.5 V for yellow, green and red LEDs. *Candle power* (CP) is a measure of the luminous intensity or the brightness of the light emitted by the LED. It is the most important parameter of a LED. It is a non-linear function of LED current, and the value of CP increases with an increase in the current flowing through the LED. *Radiant power output* P_O is the light power output of the LED. *Peak spectral emission* λ_P is the wavelength where the intensity of light emitted by the LED is a maximum. *Spectral bandwidth* gives an indication as to how concentrated the brightest colour is around its nominal wavelength.

10.12.3 Drive Circuits

LEDs are operated in the forward-biased mode. As the current through the LED changes very rapidly with change in the forward voltage above the threshold voltage, LEDs are current-driven devices. Figure 10.41a shows a simple circuit for driving an LED. The resistor R is the current-limiting resistor

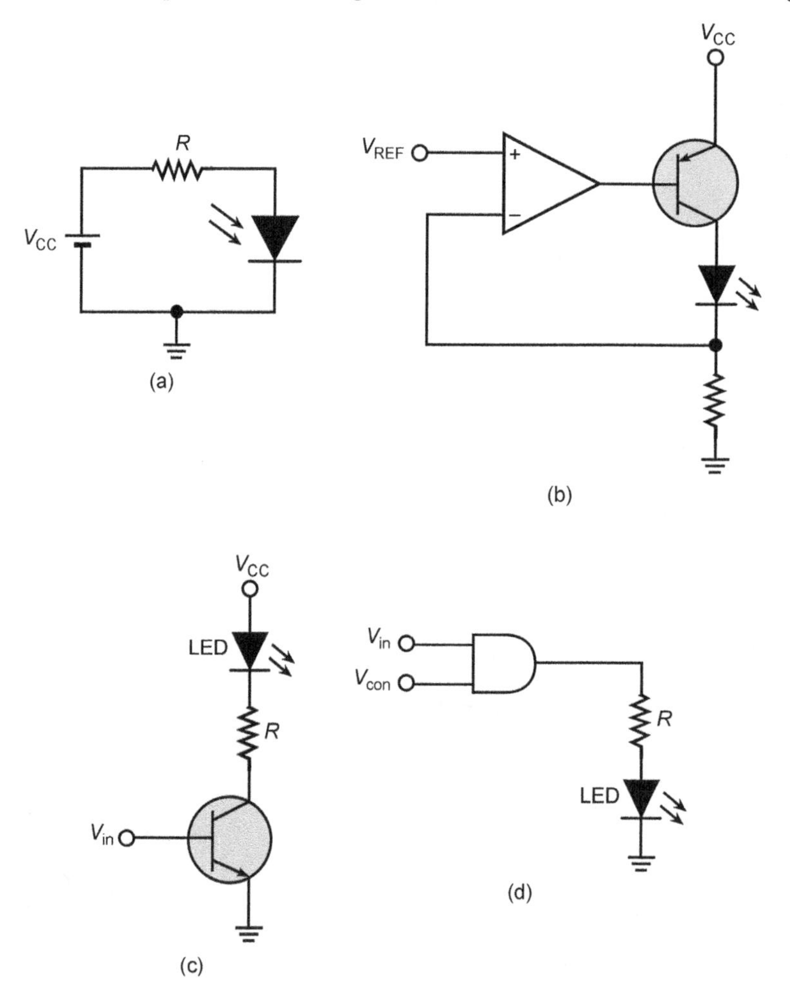

Figure 10.41 (a) Simple LED circuit; (b) constant-current LED driver circuit; and (c, d) logic circuits for driving LEDs.

used to limit the current flowing through the LED. In this case, the I–V characteristics of the LED are used to determine the voltage that needs to be applied to the LED to generate the desired forward current. A silicon diode can be placed inversely parallel to the LED for reverse-polarity voltage protection. The current that flows through the LED is given by Equation 10.29:

$$I = (V_{CC} - V_F)/R \tag{10.29}$$

where V_{CC} is the supply voltage.

Any change in the forward voltage of the LED due to temperature changes or variation from device to device causes a change in the LED current. Moreover, there is power dissipation across the series resistor R, resulting in reduced efficiency. A better drive circuit configuration is one that employs a constant current source as shown in Figure 10.41b. The current flowing through the LED is determined by the reference voltage V_{Ref} and the resistor R.

LEDs can also be used to display the logic output states. Figure 10.41c and d show typical logic circuits that can be used to drive LEDs. Figure 10.41c uses a transistor-based switch while Figure 10.41d employs a logic gate/buffer. In both these circuits, the LED glows when the voltage V_{in} is in the logic HIGH state.

When the light emitted by one LED is not sufficient, several LEDs can be connected in series to enhance the light level to the desired value. LEDs can be connected in series as shown in Figure 10.42a. In a series connection, the current flowing through each LED is the same. The value of the supply voltage V_{CC} should be sufficiently large to drive the desired number of LEDs. It should also be checked that the series current flowing in the circuit through each LED is within their operating range. The value of the resistor R to be connected is given by Equation 10.30:

$$R = [V_{CC} - (V_{F1} + V_{F2} + V_{F3.} \ldots\ldots + V_{Fn})]/I \tag{10.30}$$

where V_{F1}, V_{F2}, . . . , V_{Fn} are the forward voltages across LED_1, LED_2, . . . , LED_n respectively.

LEDs can also be connected in parallel to enhance the output light level. However, more care is needed in the case of parallel connection. Figure 10.42b shows the parallel connection of LEDs. The resistors R_1, R_2, . . . , R_n are used to protect the diodes. If these resistors are not used, then the LED with the lowest forward voltage will draw excess current and is likely to be damaged. The LED with the next-smallest forward voltage will then be damaged, and this process will continue until all the LEDs are damaged. The values of these resistors determine the current flowing through individual LEDs.

Example 10.8

For the circuit shown in Figure 10.43, determine the current through the LED when (a) $V_B = 0$ V and (b) $V_B = 10$ V. (V_{BE} of both the transistors is 0.7 V and $V_{CE(sat)}$ of both the transistors is 0.2 V).

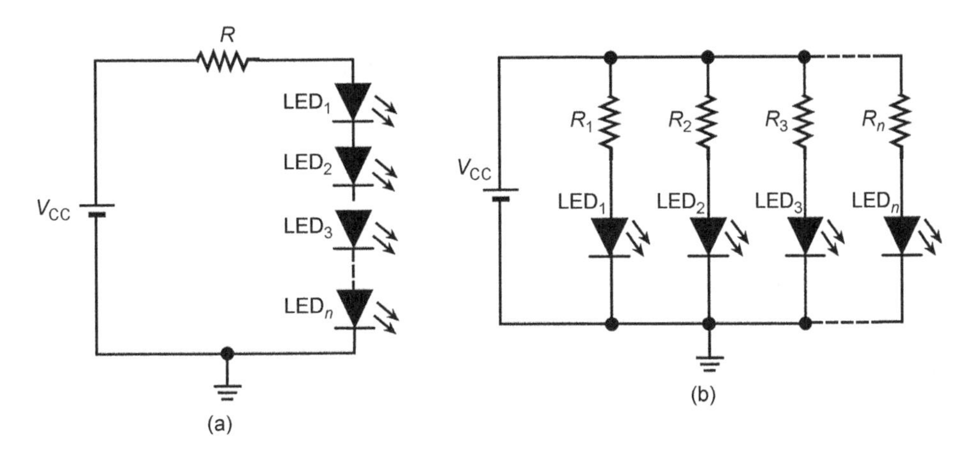

Figure 10.42 Connecting LEDs in (a) series; and (b) parallel.

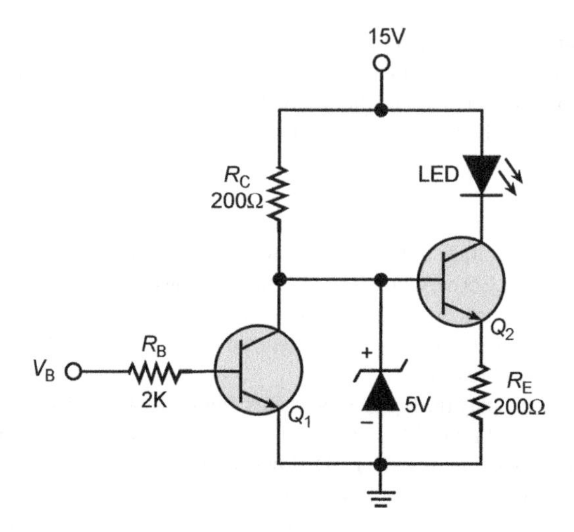

Figure 10.43 Example 10.8.

Solution

1. When $V_B = 0\,V$, the transistor Q_1 is in the cut-off region and there is no collector current flowing through the transistor. The base voltage of transistor Q_2 is therefore equal to the reverse voltage of Zener diode Z_1. Therefore, the base voltage of transistor Q_2 is $V_{B2} = 5\,V$.
2. The transistor Q_2 is in the conducting mode and the voltage at the emitter terminal of transistor $Q_2 = V_{B2} - 0.7\,V = 5 - 0.7\,V = 4.3\,V$.
3. The emitter current flowing through transistor $Q_2 = V_{E2}/R_E = 4.3/200\,A = 21.5\,mA$, where V_{E2} is the emitter voltage of transistor Q_2. As the collector current is approximately equal to the emitter current, the value of $I_C = 21.5\,mA$. The current flowing through the LED is the same as the collector current of the transistor. Therefore, the current flowing through the LED is 21.5 mA.
4. When $V_B = 10\,V$, transistor Q_1 is in saturation. Therefore, the collector-emitter voltage of transistor Q_1 is $V_{CE1} = 0.2\,V$. The base voltage of transistor Q_2, $V_{B2} = 0.2\,V$.
5. The transistor Q_2 is in cut-off and the value of collector current is zero. Therefore, the current flowing through the LED is also zero.

10.13 Liquid-crystal Displays

Liquid crystals are materials that exhibit properties of both solids and liquids, that is, they are an intermediate phase of matter. They can be classified into three different groups: nematic, smectic and cholestric. Nematic liquid crystals are generally used in LCD fabrication, with twisted-nematic material being the most common.

10.13.1 Construction

Figure 10.44a shows the construction of a twisted-nematic LCD display. As can be seen from the figure, it comprises a cell of liquid-crystal fluid, conductive electrodes, a set of polarizers and a glass casing.

Polarizers are components that polarize light in one plane. The polarizer attached to the front glass is referred to as the front polarizer, while that attached to the rear glass is the rear polarizer. On the inner surface of the glass casing, transparent electrodes are placed in the shape of the desired image. The electrode attached to the front glass is referred to as the segment electrode while that attached to the rear glass is the backplane or the common electrode. The patterns of the backplane and segment electrodes form numbers, letters and symbols. The liquid crystal is sandwiched between the two electrodes. The basic principle of operation of the LCD is to control the transmission of light by changing the polarization of the light passing through the liquid crystal with the help of an externally applied

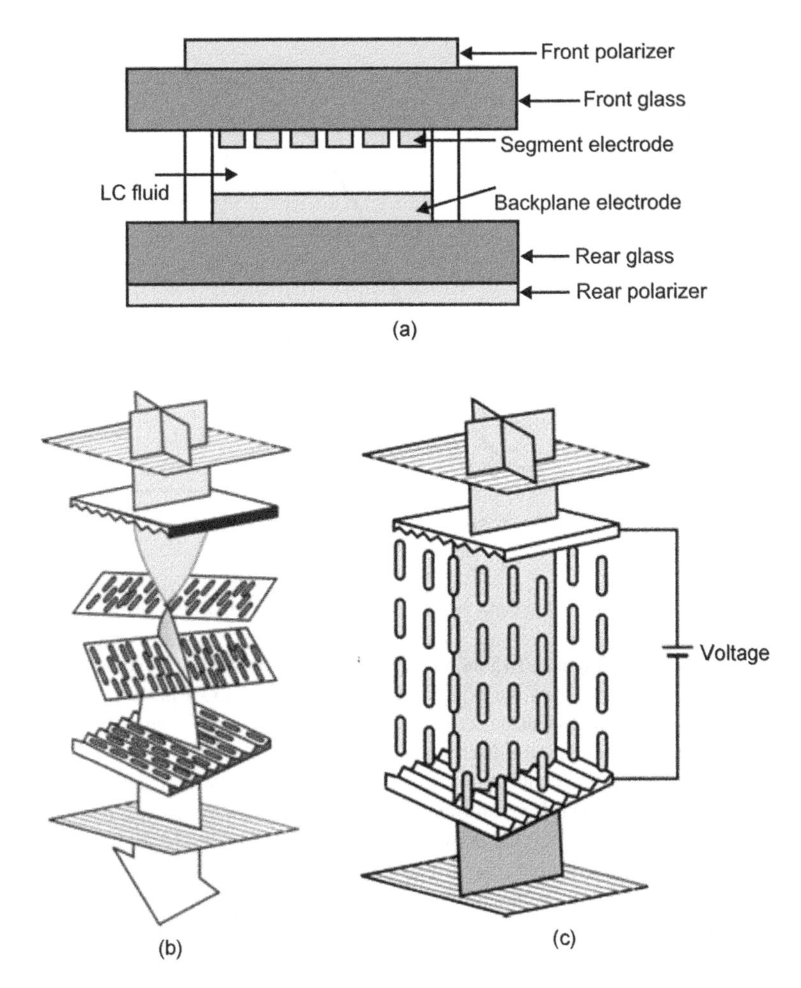

Figure 10.44 (a) Cross-section of a twisted-nematic LCD display; (b, c) twisted-nematic LCD operation.

voltage. As LCDs do not emit their own light, backlighting is used to enhance the legibility of the display in dark conditions. A variety of methods exist for backlighting LCD panels, such as use of incandescent lamps, LEDs and electroluminescent lamps.

LCDs have the ability to produce both positive as well as negative images. A positive image is defined as a dark image on a light background. In a positive image display, the front and the rear polarizers are perpendicular to each other. Light entering the display is guided by the orientation of the liquid-crystal molecules that are twisted by 90° from the front glass plate to the rear glass plate. This twist allows the incoming light to pass through the second polarizer as shown in Figure 10.44b. When voltage is applied, the liquid-crystal molecules straighten out and stop redirecting light. As a result, light travels straight through and is filtered out by the second polarizer. No light can pass through, making this region darker compared to the rest of the screen as shown in Figure 10.44c. In order to display characters or graphics, voltage is applied to the desired regions making them dark and visible to the eye.

A negative image is a light image on a dark background. In negative image displays, the front and the rear polarizer are aligned to each other.

10.13.2 Driving LCD

The LCD driver waveforms are designed to create a zero DC potential across all the pixels, as a DC voltage deteriorates the LC fluid such that it cannot be energized. LCDs are driven with symmetrical waveforms with

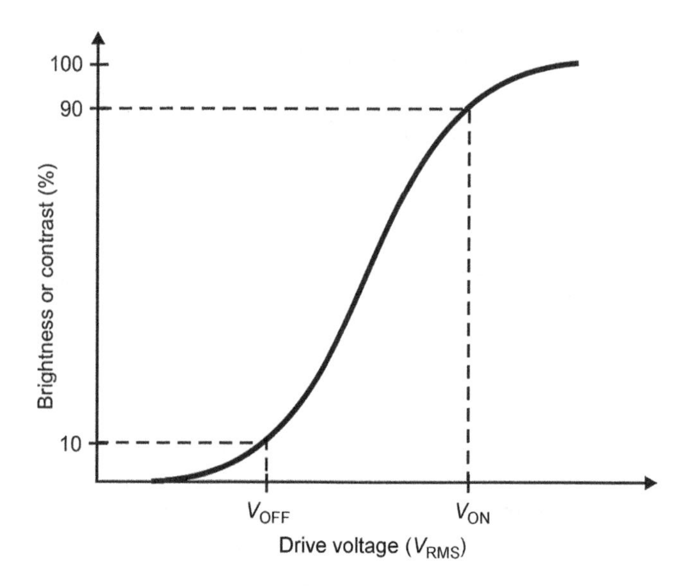

Figure 10.45 Brightness vs drive voltage curve for LCD displays.

less than 50 mV DC component. Figure 10.45 shows the brightness vs the RMS drive voltage curve for LCDs. V_{ON} is the RMS voltage applied across the liquid crystal that creates an ON pixel, typically at the 90% contrast level. V_{OFF} or V_{Th} is the RMS voltage across the liquid crystal when the contrast voltage reaches 10% level. Another important specification is the discrimination ratio, which is defined as the ratio of V_{ON} to V_{OFF}. The discrimination ratio defines the contrast levels which the LCD panel will achieve.

LCDs can be classified as direct-drive and multiplex-drive displays depending upon the technique used to drive them. Direct-drive displays, also known as static-drive displays, have an independent driver for each pixel. The drive voltage in this case is a square waveform having two voltage levels, namely ground and V_{CC} as shown in Figure 10.46a. In the figure, segment 0 is the ON segment whereas segment 1 is the OFF segment. Direct-drive displays offer the best contrast ratios over a wide operating temperature range. As the display size increases, the drive circuitry becomes very complex.

Multiplex-drive circuits are used for larger-size displays. These displays reduce the total number of interconnections between the LCD and driver. They have more than one backplane and the driver produces an amplitude-varying time-synchronized waveform for both the segments and backplanes. Figure 10.46b shows a typical multiplexed segment and the backplane-drive waveforms. The segment 0 is inactive whereas segment 1 is active. Segment 0 is inactive since the voltage across the LCD never crosses its activation threshold voltage.

10.13.3 Response Time

The LCD response time is defined by the ON and OFF response times. ON time refers to the time required by an OFF pixel to become visible after the application of proper drive voltage. The OFF time is defined as the time required by the ON pixel to turn OFF after the application of proper drive voltage. The response time of LCDs varies widely with temperature and increases rapidly at low operating temperatures. LCDs can therefore only operate at low temperatures when used along with temperature controllers. At high temperatures, the liquid-crystal molecules begin to assume random orientations, resulting in the pixels on the positive image display becoming completely dark while the pixels on the negative image become completely transparent. Figure 10.47 shows the typical variation of the ON and the OFF times of a LCD display with temperature.

10.13.4 Types of LCD Displays

LCDs are non-emissive devices, that is, they do not generate light on their own. Depending upon the mode of transmission of light in a LCD, they are classified as reflective, transmissive and transreflective LCD displays.

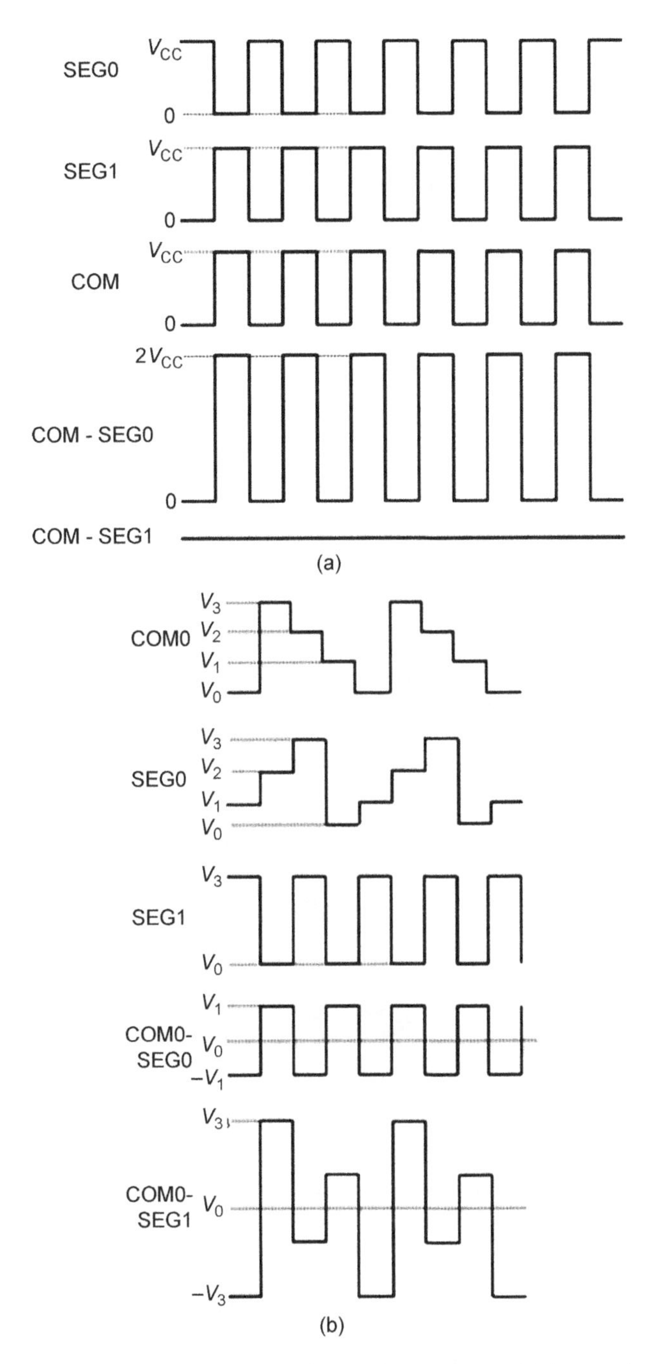

Figure 10.46 (a) Direct drive waveform; and (b) multiplexed drive waveform.

Reflective LCD displays have a reflector attached to the rear polarizer which reflects incoming light evenly back into the display. Figure 10.48 shows the principle of operation of reflective LCD displays. These displays rely on the ambient light to operate and do not work in dark conditions; they produce only positive images. The front and the rear polarizers are perpendicular to each other. These types of displays are commonly used in calculators and digital wristwatches.

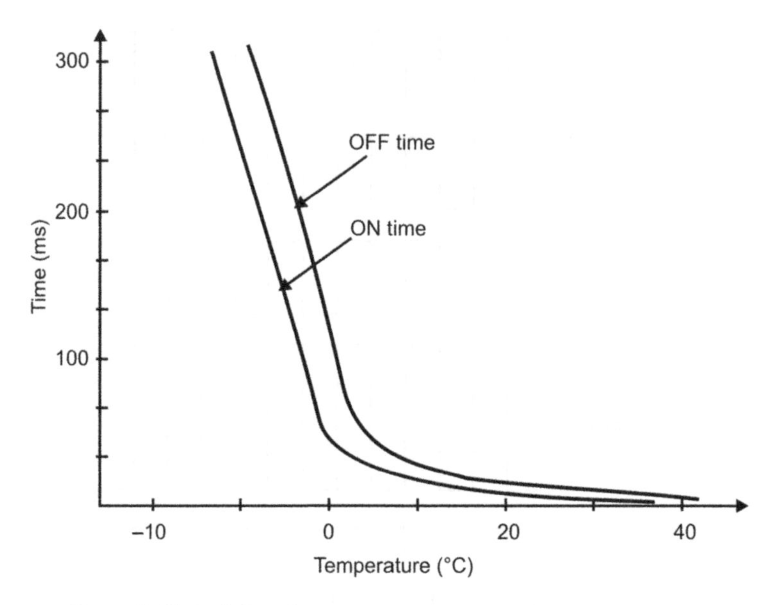

Figure 10.47 Variation of response time of LCD display with temperature.

In *transmissive LCD displays*, back light is used as the light source. Most of these displays operate in the negative mode, that is, the text will be displayed in light colour and the background is a dark colour. Figure 10.49 shows the basic construction of a transmissive display. Negative transmissive displays have front and rear polarizers in parallel with each other whereas positive transmissive displays have the front and the rear

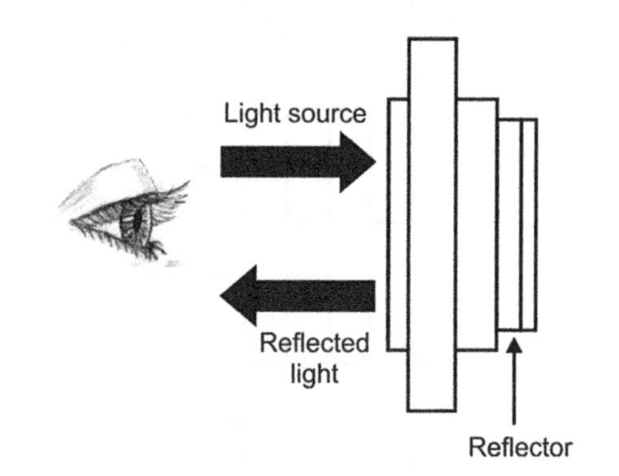

Figure 10.48 Reflective LCD display.

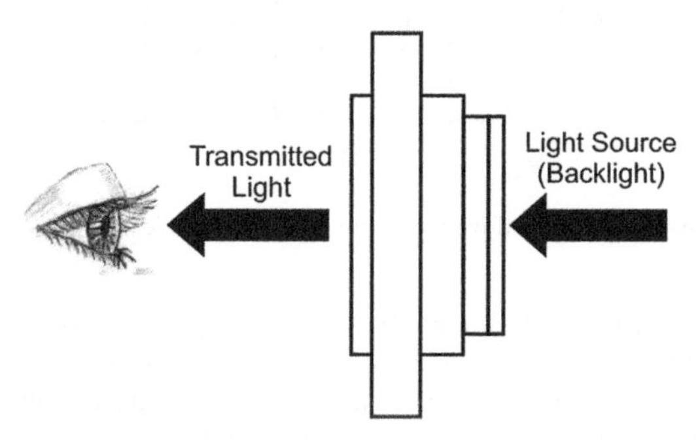

Figure 10.49 Transmissive LCD displays.

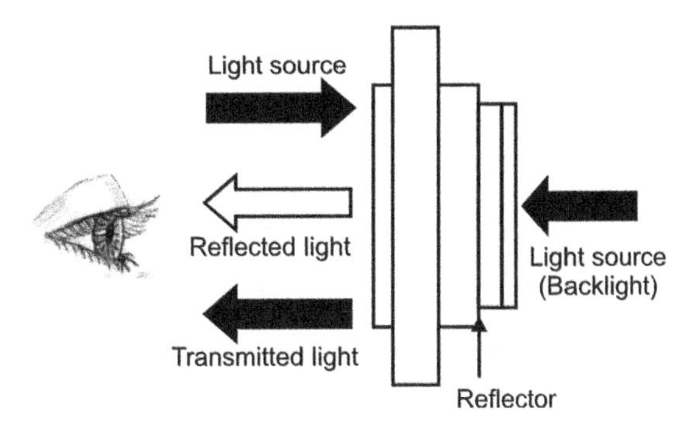

Figure 10.50 Transreflective LCD displays.

polarizers perpendicular to each other. Transmissive displays are good for very low light level conditions. They offer very poor contrast when used in direct sunlight, because sunlight swamps out the backlighting. These displays cannot be used in natural sunlight, but provide good- quality pictures indoors. They are generally used in medical devices, electronics testing and measuring equipment and in laptops.

Transreflective displays are a combination of reflective and transmissive displays (Figure 10.50). A white or silver translucent material is applied to the rear of the display, which reflects some of the ambient light back to the observer. It also allows the backlight to pass through. They are good for displays operating in varying light conditions. However, they offer poorer contrast ratios than reflective displays.

LCD displays can also be classified as passive or active LCD displays, depending upon the nature of the activation circuit. Passive displays use components that do not supply their own energy to turn the desired pixels ON or OFF. They are made up of a set of multiplexed transparent electrodes arranged in a row/column pattern. To address a pixel, the column containing the pixel is sent a charge and the corresponding row is connected to ground. Passive displays can have either direct-drive or multiplexed-drive circuitry. For larger displays however, it is not possible or economical to have separate connections for each segment. Also, as the number of multiplexed lines increase, the contrast ratio decreases due to the cross-talk phenomenon where a voltage applied to the desired pixel causes the liquid-crystal molecules in the adjacent pixels to partially untwist.

These inherent problems of passive displays are absent from active displays. Active displays use an active device such as a transistor or a diode into each pixel which acts like a switch that precisely controls the voltage that each pixel receives. Active displays are further classified as thin-film transistor (TFT) displays and thin-film diode (TFD) displays, depending upon whether the active device used is a transistor or a diode. In both these devices a common electrode is placed above the liquid-crystal matrix. Below the liquid crystal is a conductive grid connected to each pixel through a TFT or a TFD. The gate of each TFT is connected to the row electrode, the drain to the column electrode and the source to the liquid crystal. The display is activated by applying the display voltage to each row electrode line by line. One of the major advantages of active displays is that nearly all effects of cross-talk are eliminated.

10.13.5 Advantages and Disadvantages

As LCD displays are not active sources of light, they offer considerable advantages such as very low power consumption, low operating voltages and good flexibility. However, their response time is too slow for many applications, they offer limited viewing angles and they are sensitive to temperature.

10.14 Cathode Ray Tube Displays

Cathode ray tube (CRT) displays are used in a wide range of systems ranging from consumer electronic systems (e.g. televisions and computer monitors) to measuring instruments (e.g. oscilloscopes) to military systems (e.g. radar). A CRT display is a specialized vacuum tube in which the images are produced when the electron beam strikes the fluorescent screen. CRT displays can be monochrome displays as well as coloured displays.

Monochrome CRT displays comprise a single electron gun, a fluorescent screen and an internal or external mechanism to accelerate and deflect the electron beam. Figure 10.51 shows the cross-sectional

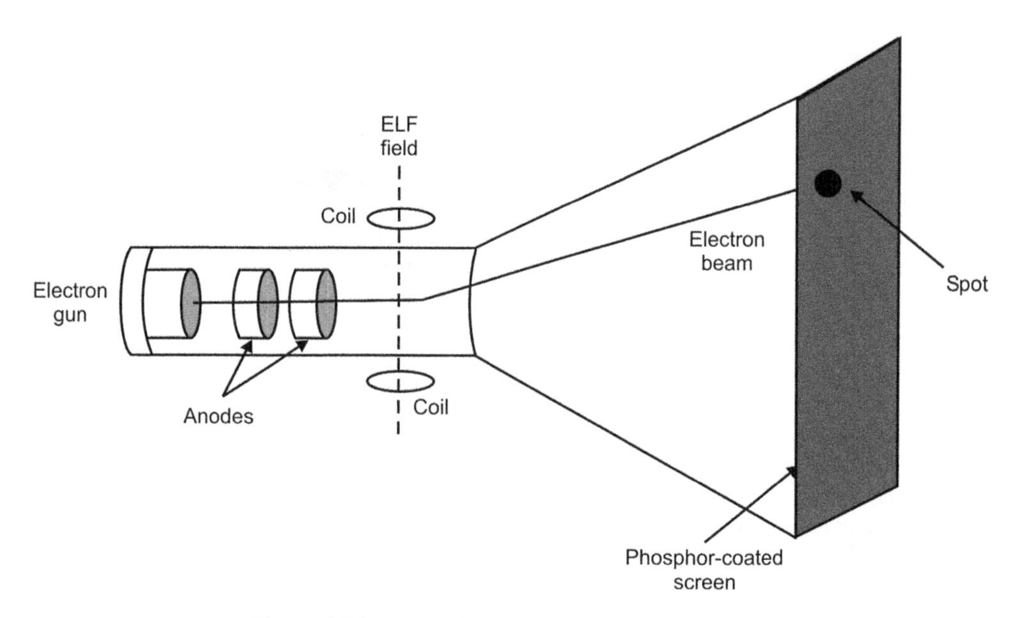

Figure 10.51 Cross-section of a monochrome CRT display.

view of a CRT display. The electron gun produces a narrow beam of electrons which are accelerated by the anodes. There are two sets of deflecting coils, namely the horizontal coil and the vertical coil. These coils produce an extremely-low-frequency electromagnetic field in the horizontal and vertical directions to adjust the direction of the electron beam. CRT tubes also have a mechanism to vary the intensity of the electron beam. In order to produce moving pictures in natural colours on the screen, complex signals are applied to the deflecting coils and to the circuitry responsible for controlling the intensity of the electron beam. This results in movement of the spot from right to left and top to bottom in a sequence of horizontal lines referred to as a raster. The speed of the spot movement is so fast that the person viewing the screen sees a constant image on the entire screen.

Coloured CRT displays comprise three electron guns, one each for the three primary colours (red, blue and green). The CRT produces three overlapping images, one in each colour. This is referred to as the RGB colour model.

CRT displays offer very high resolution and, as these displays emit their own light, very high values of peak luminance. Moreover, these displays offer wide viewing angles of the order of 180°. CRT display technology is more mature as compared to alternative display technologies and is also cheaper.

Despite the significant advantages offered by CRT displays as mentioned above, alternative display technologies are slowly replacing CRT displays due to the drawbacks of the CRT displays: they are bulky and consume significant power. Moreover, they require high voltages to operate and cause fatigue and strain to the human eye.

10.15 Emerging Display Technologies

This section provides an introduction to emerging display technologies including organic light-emitting diodes (OLEDs), digital light-processing (DLP) technology, plasma displays, field-emission displays (FEDs) and electronic ink displays. All these display technologies are explained in brief in the following sections (detailed descriptions are beyond the scope of the book).

10.15.1 Organic Light-emitting Diodes (OLEDs)

OLEDs are composed of a light-emitting organic material sandwiched between two conducting plates, one of N-type material and the other of P-type material. When an electric potential is applied between these plates, holes are ejected from the P-type plate and electrons are ejected from the N-type plate. Due to the recombination of these holes and electrons, energy is released in the form of light photons. The wavelength of light emitted depends upon the band-gap energy of the semiconductor material used. In order to produce visible light, band-gap energy of the semiconductor material is of the order 1.5–3.5 eV.

Depending upon their basic structure, OLEDs can be classified into three types: small-molecule OLEDs (SMOLEDs), polymer LEDs (PLEDs) and dendrimer OLEDs. OLEDs can be driven using passive as well as active matrix driver circuits.

As OLEDs are emissive devices, they offer significant advantages compared to LCD displays, including faster switching speeds, higher refresh rates, lower operating voltages and larger viewing angles.

10.15.2 Digital Light-processing (DLP) Technology

DLP technology makes use of an optical semiconductor device referred to as a digital micro-mirror device (DMD) which is basically a precise light switch that can digitally modulate light through a large number of microscopic mirrors arranged in a rectangular array. These mirrors are mounted on tiny hinges and can be tilted away or towards the light source with the help of the DMD chip, thus projecting a light or a dark pixel onto the screen. Use of DLP systems is currently limited to large projection systems.

10.15.3 Plasma Display Panels (PDPs)

Plasma displays are composed of millions of cells sandwiched between two panels of glass. Two types of electrode, namely the address electrodes and display electrodes, are also placed between the two glass plates covering the entire screen. The address electrodes are printed on the rear glass plate and the transparent display electrodes are located above the cells along the front glass plate. These electrodes are perpendicular to each other, forming a grid network.

Each cell is filled with a xenon and neon gas mixture. The electrodes intersecting at a specific cell are charged to excite the gas mixture in each cell. When the gas mixture is excited, plasma is created releasing ultraviolet light which then excites the phosphor electrons located on the sides of the cells. These electrons in turn release visible light and return to their lower energy state. Each pixel is composed of three cells containing red, green and blue phosphors.

Plasma displays offer several advantages such as: each pixel generates its own light, offering large viewing angles; superior image quality; and the image quality is not affected by the area of the display. However, these displays are fragile in nature and are susceptible to burn-out from static images.

10.15.4 Field Emission Displays (FEDs)

Field emission displays (FEDs) function much like the CRT displays; the main difference is that the former use millions of small electron guns to emit electrons at the screen, instead of just one electron gun in the case of the latter. The extraction of electrons in FEDs is based on the tunnelling effect. FEDs produce the same quality of image as produced by a CRT display, but without being bulky. These displays can be as thin as LCD displays and as large as plasma displays.

10.15.5 Electronic Ink Displays

Electronic ink displays, also referred to as electronic paper, are active matrix displays making use of pigments that resemble the ink used in print.

10.16 Optocouplers

An optocoupler, also referred to as an optoisolator, is a device that uses a short optical transmission path to transfer signals between the elements of a circuit. Optocouplers are sealed units that house an optical transmitting device and a photosensitive device that are coupled together optically. The optical path may be air or a dielectric waveguide. Figure 10.52 shows the basic structure of an optocoupler.

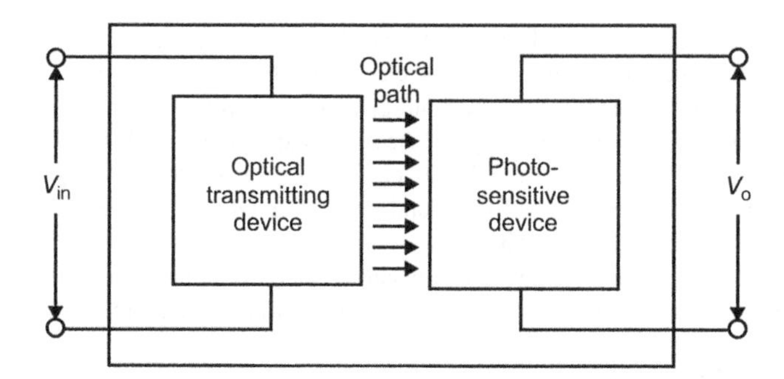

Figure 10.52 Basic structure of an optocoupler.

The transmitter unit contains either a lamp, an LED or, in some cases, a laser diode. The receiver unit may be a photodiode, a phototransistor, a photo-FET, a photo-SCR, a photo-DIAC, a photo-TRIAC, a photoconductor or any other photosensitive material. As the coupling between the source and the photosensor is optical, high isolation exists between the input and the output circuitry. Hence, low-power sensitive circuits can be used to actuate high-power devices using optocouplers. In other words, optocouplers are used in applications that require isolation between input and output signals. Optocouplers have also replaced low-power relays and pulse transformers in many applications. They can also be used for applications such as the detection of objects, liquid level detection, smoke and fog detection and end of tape detection.

Optocouplers with photodiodes (Figure 10.53a), phototransistors (Figure 10.53b), photo Darlington transistors (Figure 10.53c), photoconductor (Figure 10.53d) and photo-FET (Figure 10.53e) sensors at the receiving end are referred to as non-latching optocouplers. Optocouplers with photo-SCR (Figure 10.54a), photo-DIAC (Figure 10.54b) and photo-TRIAC (Figure 10.54c) at the receiving side are referred to as latching optocouplers.

10.16.1 Characteristic Parameters

The important parameters that define the performance of an optocoupler include forward optocoupling efficiency, input to output isolation voltage, input current and bandwidth.

The *forward optocoupling efficiency* is specified in terms of the current transfer ratio (CTR). CTR is the ratio of the output current to the input current. For logic output optocouplers, the coupling efficiency is defined as the input current to the LED that would cause a change in the logic state of the optocoupler's output. To ensure high coupling efficiency, the wavelength response of the receiver is matched to the emission spectrum of the phototransmitter. Typical values of CTR range over 10–50% for optocouplers having phototransistors as the photosensor and can even be as high as 200% for optocouplers having photo Darlington transistors at the output.

Isolation voltage is the maximum permissible DC potential that can be allowed to exist between the input and the output circuits. Typical values of isolation voltages offered by optocouplers are in the range 500–4000 V.

Input current is the maximum permissible forward current that is allowed to flow into the transmitting LED. Typical values of forward current fall within the range 10–100 mA.

The maximum collector-emitter voltage $V_{CE(max)}$ is applicable for optocouplers having phototransistors at the output. It is defined as the transistor's maximum collector–emitter voltage rating. It limits the supply voltage that can be applied to the output circuit.

Bandwidth determines the maximum signal frequency that can be successfully passed through the optocoupler. The bandwidth of an optocoupler depends upon its switching speed. Optocouplers offer bandwidths in the range 10 kHz–1 MHz, depending upon the device construction. To achieve faster operating speeds, integrated photodiode-transistor sensors or Schottky transistors are used.

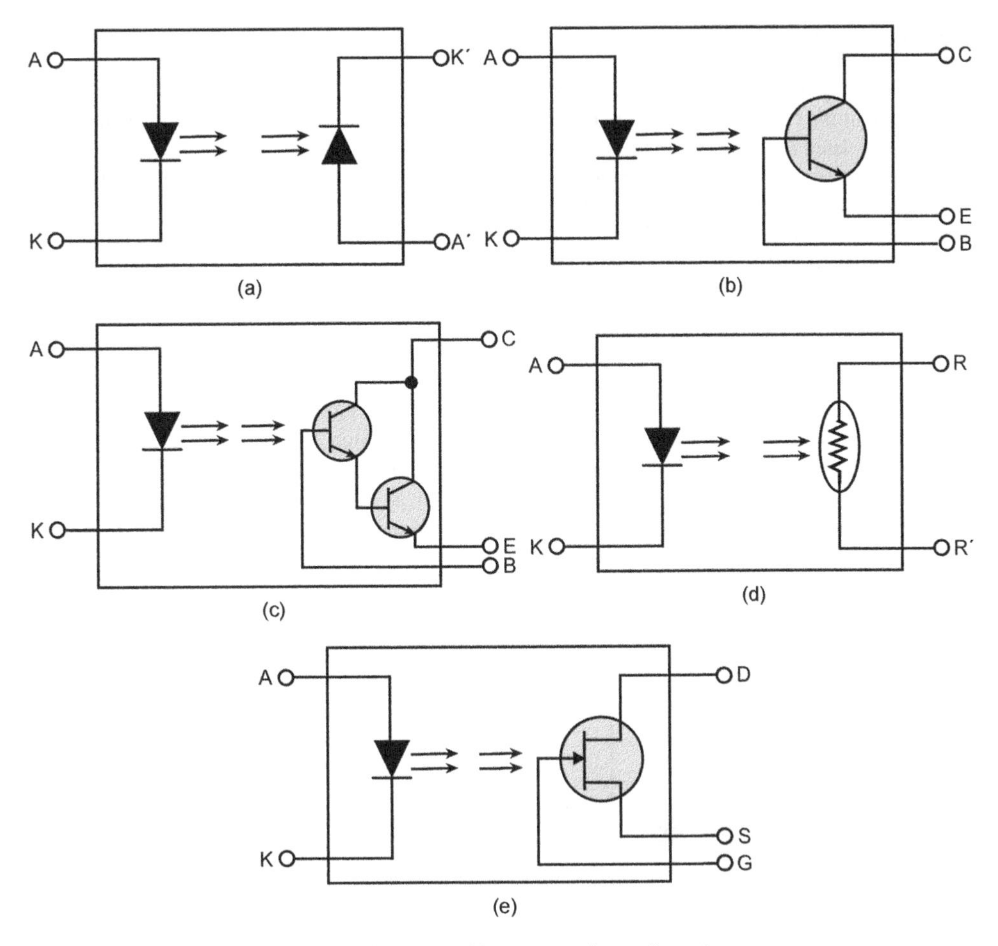

Figure 10.53 Non-latching optocoupler configurations.

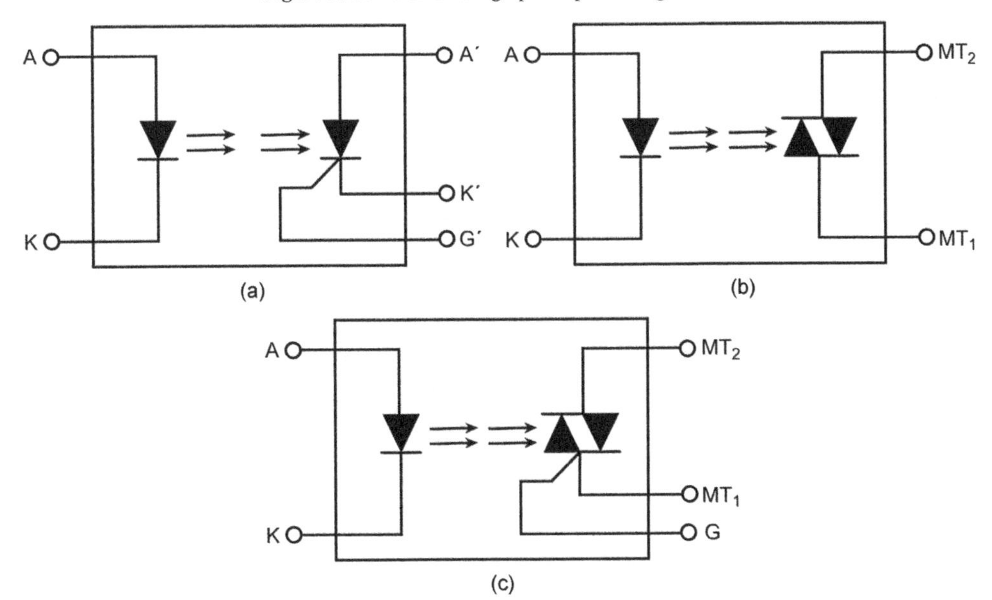

Figure 10.54 Latching optocoupler configurations.

10.16.2 Application Circuits

The simplest way to visualize an optocoupler is in terms of its two important components: the input optical transmitting element and the output photosensor. The transmitting and the receiving elements are electrically isolated in an optocoupler, offering a lot of flexibility in connecting them.

The transmitting element most commonly used in an optocoupler is an LED. The LED in an optocoupler can be driven in a manner similar to that for a discrete LED. Figure 10.55 shows the various configurations in which the LEDs in an optocoupler can be connected. These circuits are similar to those discussed in Section 10.12 on LEDs. Figure 10.55a shows a conventional circuit for driving the LED. The diode across the LED is used for protecting the LED against reverse polarity voltages. Figure 10.55b and c show how to drive an LED using a transistor and a logic buffer, respectively. For both the circuits, the LED is ON for logic HIGH input and is OFF for logic LOW input.

At the receiving side of the optocoupler, the photosensor used may be a photodiode, phototransistor, photo Darlington transistor, photo-FET, photo-DIAC or a photo-TRIAC. Circuits for driving these photosensors are similar to that used for driving discrete sensors. It may be mentioned here that how the output circuit is configured is dependent upon the intended application.

In many cases, the phototransistors are simply connected as a light-operated switch as shown in Figure 10.56a and b. The phototransistor is used in the pull-up mode in Figure 10.56a and in the pull-down

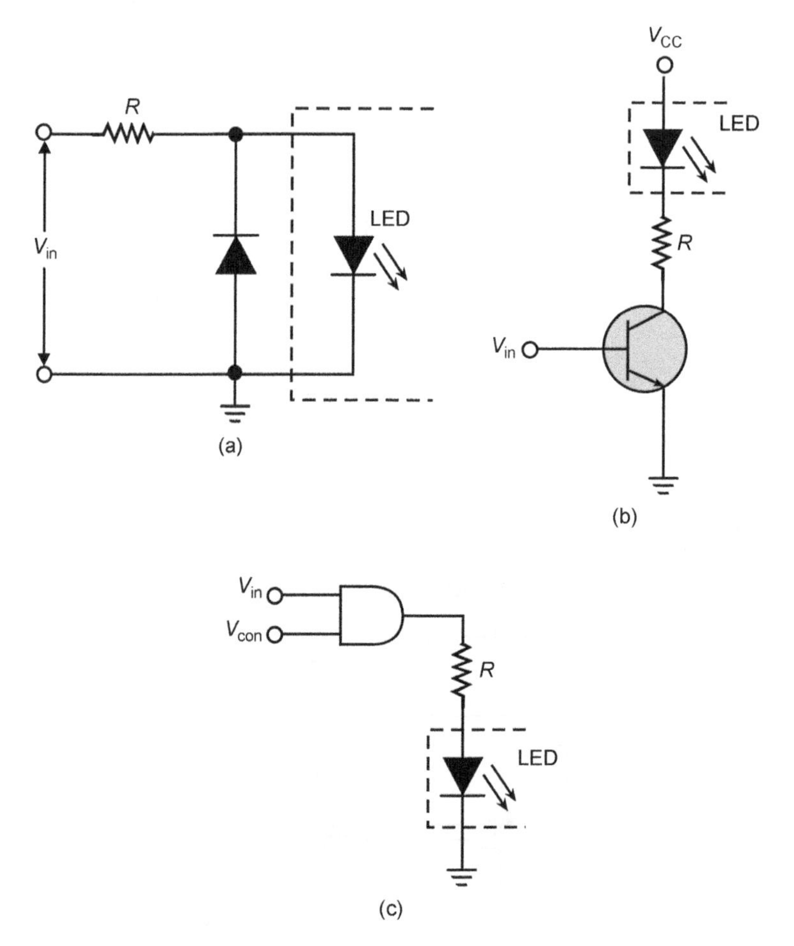

Figure 10.55 Driving LEDs of an opt coupler.

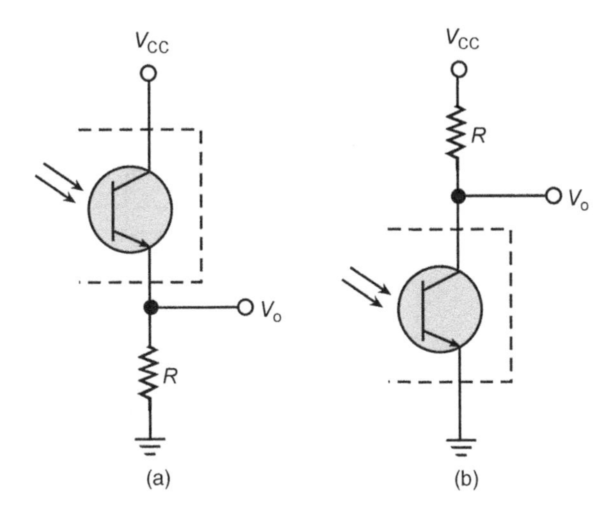

Figure 10.56 Driving the phototransistors of an opto coupler.

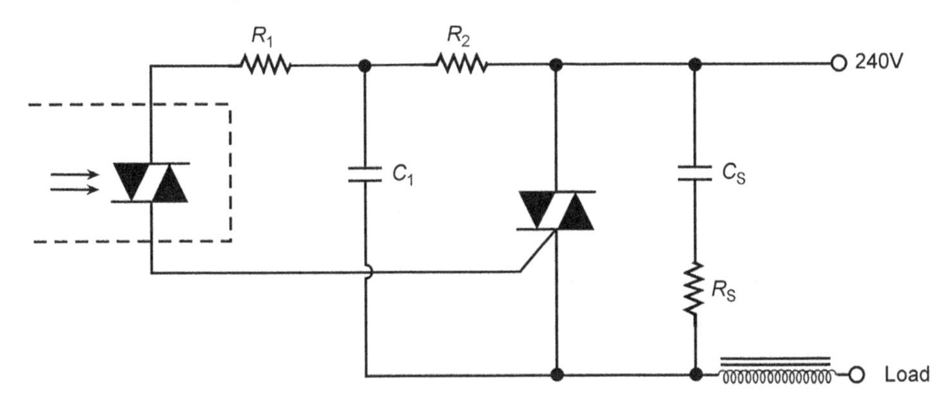

Figure 10.57 Application circuit to drive the photo-DIAC sensor of an opto coupler.

mode in Figure 10.56b. The output of the phototransistor can be connected to a logic gate, a transistor or an operational amplifier. An optocoupler with a photo Darlington transistor can also be driven in a similar manner.

Figure 10.57 shows the application circuit of an optocoupler having a photo-DIAC as a photosensing element. In the circuit, the photo-DIAC is used to trigger a TRIAC. The circuit employs a filter/delay circuit comprising resistors R_1 and R_2 and capacitor C_1 and also a snubber circuit across the TRIAC comprising resistor R_S and capacitor C_S to ensure correct triggering with inductive loads.

Example 10.9

For the optocoupler circuit of Figure 10.58a, the voltage across the resistor R is 4 V. Determine the value of the voltage V_B applied to the base of the transistor Q_1. The relationship between the input current flowing through the LED and the output phototransistor collector current is shown in Figure 10.58b. The base-emitter voltage of transistor Q_1 is equal to 0.7 V.

Solution

1. The current flowing across the resistor $R = 4/(4 \times 10^3)$ A $= 1 \times 10^{-3}$ A $= 1$ mA.
2. The collector current of the phototransistor I_C is equal to the current across resistor R; therefore, $I_C = 1$ mA.

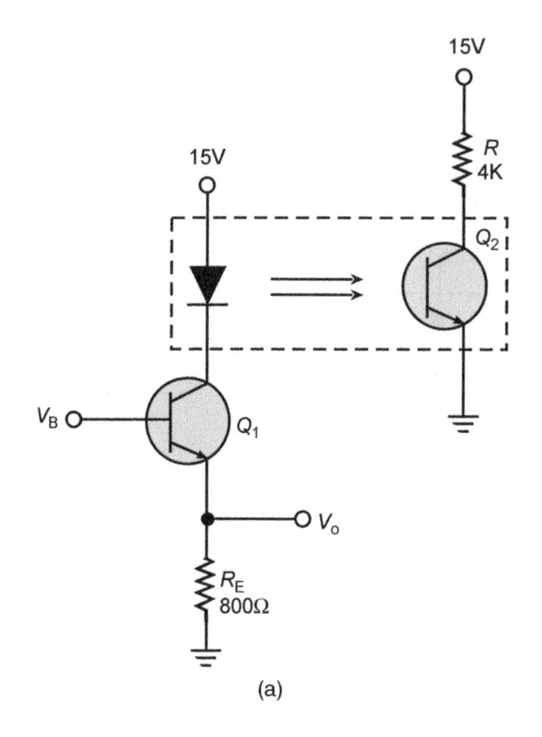

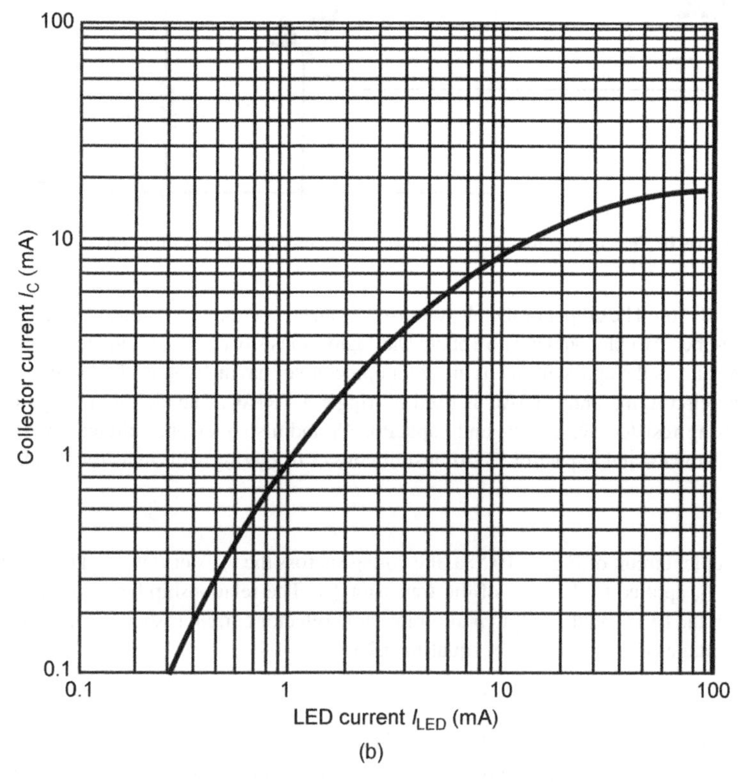

Figure 10.58 Example 10.9.

3. From the curve of Figure 10.58b, the input current flowing through the LED is approximately equal to 1.1 mA.
4. The current flowing through the LED is equal to the collector current of transistor Q_1.
5. As the collector current of the transistor is approximately the same as its emitter current, the emitter current is equal to 1.1 mA.
6. The voltage drop across resistor $R_E = 800 \times 1.1 \times 10^{-3} = 0.88$ V.
7. The voltage $V_B = V_E + V_{BE} = 0.88 + 0.7$ V $= 1.58$ V.

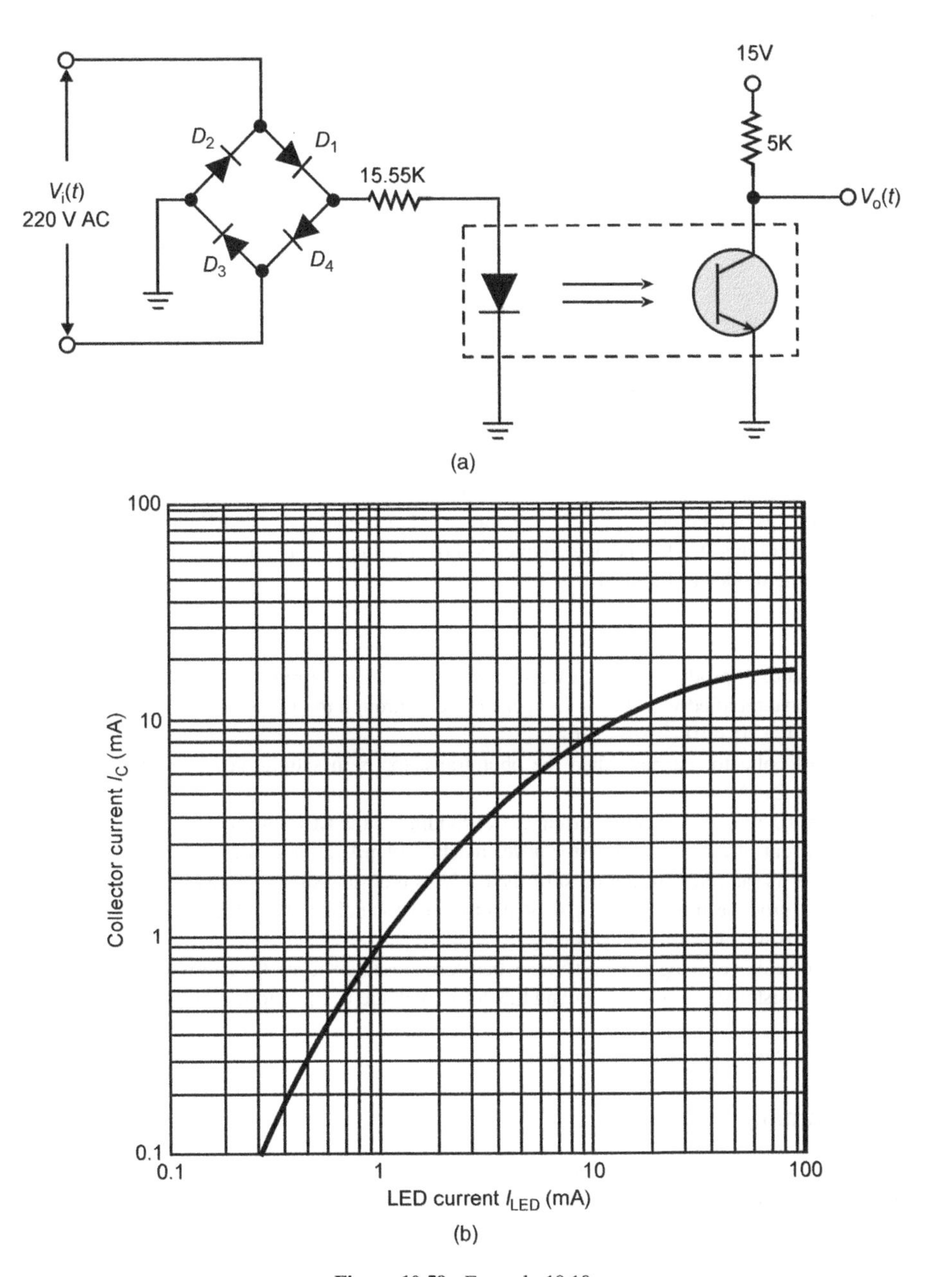

Figure 10.59 Example 10.10.

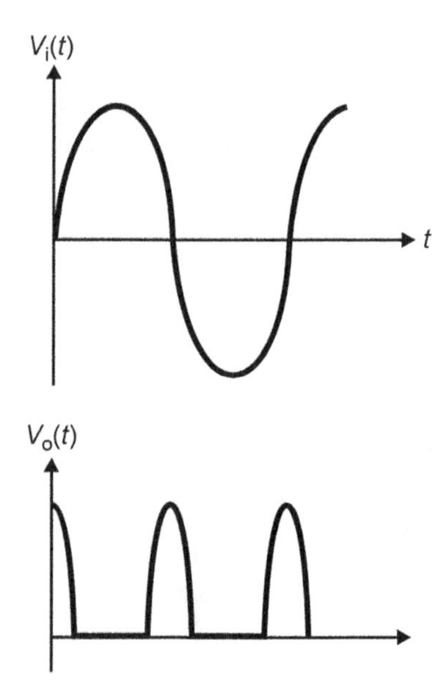

Figure 10.60 Solution to Example 10.10.

Example 10.10

The optocoupler-based circuit of Figure 10.59a is used to provide isolation from the power line and detect zero crossings of the line voltage. The relationship between the input LED current and the output current of the phototransistor for the optocoupler is shown in Figure 10.59b. Draw the output waveform $V_O(t)$. Assume the diodes to be ideal.

Solution

1. The bridge rectifier produces a full-wave rectified output. The current through the LED is therefore also a full-wave rectified waveform. The peak current through the LED is $I_{LEDpeak} = (1.414 \times 220)/15.55 \times 10^3$ A $= 20 \times 10^{-3}$ A $= 20$ mA.
2. The value of collector current when the phototransistor is in saturation $I_{Csat} = 10/(5 \times 10^3)$ A $= 2 \times 10^{-3}$ A $= 2$ mA.
3. From Figure 10.59b, the collector current corresponding to the LED current of 20 mA is 14 mA. The collector current of the phototransistor never reaches 14 mA, as it saturates at 2 mA.
4. The transistor is saturated during most of the cycle as the transistor saturates for all values of LED currents that produce more than 2 mA of phototransistor collector current. Therefore, the output voltage is zero for most of the cycle.
5. When the line voltage changes polarity, i.e. when zero crossings occur, the LED current drops to zero. The phototransistor stops conducting and the output voltage is equal to the supply voltage 10 V, i.e. $V_O(t) = 10$ V.
6. Figure 10.60 shows the output waveform.

10.17 Summary

- Photosensors are classified as either photoelectric sensors or thermal sensors. Photoelectric sensors – which can either depend on the external photo effect for their operation or make use of some kind of internal photo effect –include the photoemissive sensors, photoconductors and junction-type photosensors. Thermal sensors include thermocouples (or thermopiles), bolometric sensors and pyroelectric sensors.

- *Radiometry* is the study of the properties and characteristics of electromagnetic radiation and the sources and receivers of electromagnetic radiation. *Photometry* is the science that deals with visible light and its perception by human vision.
- Important photometry and radiometry parameters include radiometric and photometric flux, radiometric and photometric intensity, radiant incidence (irradiance) and illuminance and radiant sterance (radiance) and luminance.
- Parameters used to characterize the performance of photosensors include responsivity, noise equivalent power (NEP), sensitivity (usually measured as detectivity D and D^*), quantum efficiency, response time, noise and spectral response.
- Responsivity is defined as the ratio of electrical output to radiant light input determined in the linear region of the response. It is measured in $\mathrm{A\,W^{-1}}$ or $\mathrm{V\,W^{-1}}$ if the photosensor produces a voltage output rather than a current output.
- NEP is the input power to a sensor which generates an output signal current equal to the total internal noise current of the device, which implies a signal to noise ratio of 1. Detectivity D of a sensor is the reciprocal of its NEP. D^* is defined as detectivity normalized to an area of $1\,\mathrm{cm}^2$ and a noise bandwidth of 1 Hz.
- Quantum efficiency is defined as the ratio of the number of photoelectrons released to the number of photons of incident light absorbed.
- Response time is defined as rise/fall time parameter in photoelectric sensors and as time constant parameter in thermal sensors. Rise and fall times are the time durations required by the output to change from 10% to 90% and from 90% to 10% of the final response respectively. It determines the highest signal frequency to which a sensor can respond. Time constant is defined as the time required by the output to reach 63% of the final response from zero initial value.
- Sensor noise mainly comprises Johnson noise, shot noise, generation-recombination noise and flicker noise. Johnson noise, also known as Nyquist or thermal noise, is caused by the thermal motion of charged particles in a resistive element. Shot noise in a photosensor is caused by the discrete nature of the photoelectrons generated. Generation-recombination noise is caused by the fluctuation in current generation and the recombination rates in a photosensor. Flicker noise or $1/f$ noise occurs in all conductors where the conducting medium is not a metal and exists in all semiconductor devices that require bias current for their operation.
- Photoconductors, also referred to as photoresistors, light-dependent resistors (LDRs) and photocells, are semiconductor photosensors whose resistance decreases with increasing incident light intensity. They are bulk semiconductor devices with no P-N junction.
- Photodiodes are junction-type semiconductor light sensors that generate current or voltage when the P-N junction in the semiconductor is illuminated by light of sufficient energy. The spectral response of the photodiode is a function of the band-gap energy of the material used in its construction.
- Depending upon their construction, photodiodes can either be PN, PIN, Schottky or avalanche photodiodes.
- PN photodiodes comprise a PN junction and are used for precision photometry applications. In PIN photodiodes, an extra high resistance intrinsic layer is added between the P and the N layers to improve the response time. A thin layer of gold coating is sputtered on to the N-type material to form a Schottky effect P-N junction, which has enhanced ultraviolet response. An avalanche photodiode (APD) is a high-speed high-sensitivity photodiode which utilizes an internal gain mechanism that functions by applying a high reverse-bias voltage of the order 30–200 V.
- Solar cells are devices whose operation is based on the photovoltaic effect; when exposed to light (solar radiation), an open-circuit voltage is generated across a P-N junction. This open-circuit voltage leads to the flow of electric current through a load resistance connected across it.
- Photo-FETs are light-sensitive FET devices where the diode formed by the reverse-biased gate-channel junction acts as a photodiode. Incident light generates additional photocarriers, resulting in increased conductivity. Gate current flows if the gate is connected to an external resistor.
- Photo-SCRs, generally referred to as light-activated SCRs (LASCRs) are essentially the same as conventional SCRs except that they are triggered by light incident on the gate junction area.

Photo-TRIACs, also referred to as light-activated TRIACs, are bidirectional thyristors that are designed to conduct current in both directions when the incident light radiation exceeds a specified threshold value.

- Photoemissive sensors rely on the external photo effect where the photo-generated electrons travel beyond the physical boundaries of the material. Some of the commonly used photoemissive photosensors include vacuum photodiodes, photomultiplier tubes (PMTs) and image intensifier tubes.
- A vacuum photodiode comprises an anode and a cathode placed in a vacuum envelope. Photomultiplier tubes are extremely sensitive photosensors operating in the ultraviolet, visible and near-infrared spectrum. PMTs have internal gain of the order of 10^8 and can even detect single photon of light. Image intensifiers are devices that amplify visible and near-infrared light from an image so that a dimly lit scene can be viewed by a camera or by a human eye. Contemporary image intensifiers comprise an objective lens, vacuum tube with photocathode at one end, tilted micro-channel plate and a phosphor screen.
- Displays are output devices that are used for visual presentation of information. The three factors critical for good visual display are: legibility, brightness and contrast.
- Displays can be categorized into different types depending upon the manner in which they display information. These include bar graph displays, segmented displays, dot-matrix displays and large displays. Bar graph displays are composed of several bar elements. A 7-segment display comprises 7 bars and 1 or 2 decimal points and is an industry standard for numeric displays.
- A light-emitting diode (LED) is a semiconductor P-N junction diode designed to emit light when forward biased.
- Liquid-crystals are materials that exhibit properties of both solids and liquids, that is, they are an intermediate phase of matter.
- A CRT display is a specialized vacuum tube in which the images are produced when the electron beam strikes the fluorescent screen. CRT displays can be monochrome displays as well as coloured displays.
- An organic light-emitting diode (OLED) is composed of light-emitting organic materials sandwiched between two conducting plates, one of N-type material and the other of P-type material.
- Digital light processing (DLP) technology makes use of an optical semiconductor precise light switch named digital micro-mirror device (DMD) that can digitally modulate light through a large number of microscopic mirrors arranged in a rectangular array.
- Field emission displays (FED) function much like CRT displays, with the main difference being that these displays use millions of small electron guns to emit electrons at the screen instead of just one in case of CRT displays.
- Plasma displays are composed of millions of cells sandwiched between two panels of glass. Two types of electrodes (address and display electrodes) are also placed between the two glass plates covering the entire screen.
- Electronic ink displays, also referred to as electronic paper, are active matrix displays making use of pigments that resemble the ink used in print.
- An optocoupler is a device that uses a short optical transmission path to transfer a signal between the elements of a circuit. Optocouplers are sealed units that house an optical transmitter and a receiver that are coupled together optically.

Review Questions

10.1. Name the two operating modes of a photodiode. Describe the difference between these two modes using schematic diagrams to highlight the operation of the photodiode in both the modes.

10.2. What is a phototransistor? Draw the schematic symbol of a phototransistor and state any two application areas of phototransistors.

10.3. Explain in detail the principle of operation of an LED. Also explain the factors that determine the output wavelength of LEDs.

10.4. Name the most sensitive photosensor and explain in brief its principle of operation.

10.5. What are optocouplers? Explain the important characteristic parameters of optocouplers.

10.6. Explain the following terms
  a. NEP
  b. illuminance
  c. thermal noise
  d. current transfer ratio.

10.7. Compare and contrast the following
  a. radiometry and photometry
  b. detectivity D and D^*
  c. thermal sensors and photoelectric sensors
  d. radiant intensity and illuminance.

10.8. How does the response time of an LCD display vary with temperature?

10.9. Explain the different modes of operation of an LCD display? State the advantages and disadvantages of each of these operating modes.

10.10. State any two important applications of optocouplers. Draw the basic circuits that can be used for these applications.

10.11. Explain in detail the difference between bolometers and photoconductors.

10.12. Which photosensor is used in night-vision devices? Explain in brief its principle of operation.

10.13. Which parameters of a photosensor are of utmost importance while designing a receiver for detecting weak optical signals?

10.14. How does the wavelength of an LED vary with change in the band-gap energy of the semiconductor material?

10.15. What are the design considerations to be kept in mind while connecting LEDs in series and in parallel?

Problems

10.1. For the circuit shown in Figure 10.61, determine the current flowing through the LED. The base-emitter voltage of the PNP transistor is −0.6 V and the forward voltage drop of the LED is 1.4 V. [10 mA]

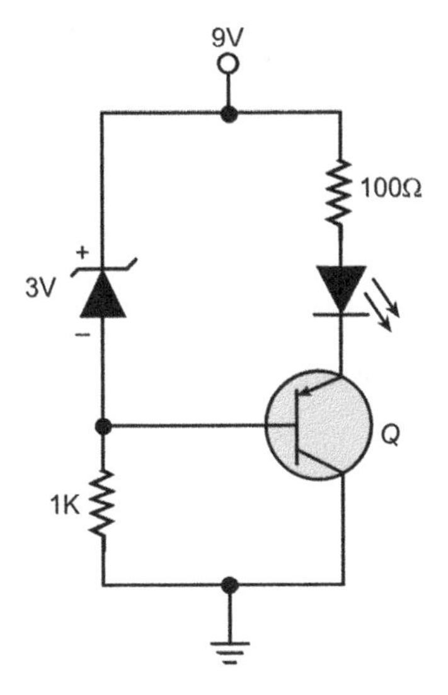

Figure 10.61 Problem 10.1.

10.2. Design a panel of solar cells capable of delivering an output voltage of 5 V with a load delivering capability of 1A. Each of the solar cells is capable of generating an output voltage of 0.45 V with an output current of 150 mA.
[Figure 10.62]

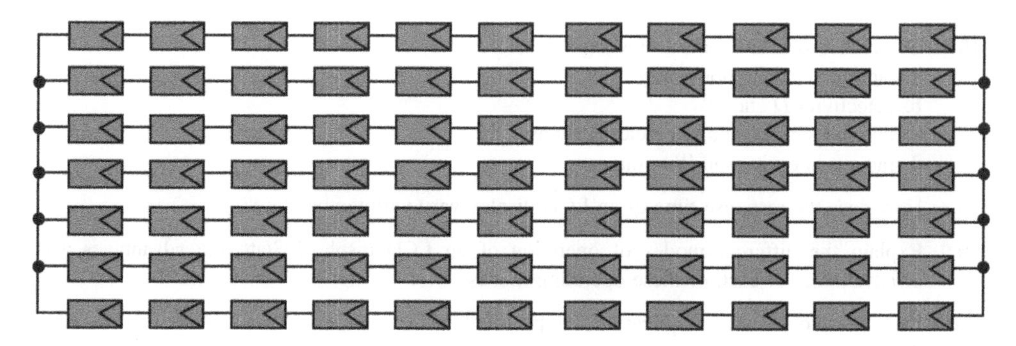

Figure 10.62 Solution to Problem 10.2.

10.3. A laser pulse with a rise time of 10 ns is incident on a PIN photodiode. The electrical pulse is seen on an oscilloscope having a bandwidth of 35 MHz and the rise time is measured to be 20 ns. Determine the rise time of the photodiode.
[14.14 ns]

10.4. A photosensor has total noise current of 100 pA, responsivity of 1 $\mathrm{A W^{-1}}$, active area of 2 $\mathrm{mm^2}$ and rise time of 3.5 ns. Determine its (a) NEP, (b) detectivity D and (c) D^*.
[(a) 0.1 nW, (b) $10^{10}\ \mathrm{W^{-1}}$ and (c) $1.414 \times 10^{13}\ \mathrm{W^{-1}\,cm\,Hz^{0.5}}$]

10.5. Calculate the (a) NEP, (b) detectivity D and (c) D^* parameters of a sensor with the same responsivity and rise time as that in problem 10.4, with an active area of 5 $\mathrm{mm^2}$ and noise current of 10 nA.
[(a) 10 nW, (b) $10^8\ \mathrm{W^{-1}}$ and (c) $2.236 \times 10^{11}\ \mathrm{W^{-1}\,cm\,Hz^{0.5}}$]

10.6. For the optocoupler circuit of Figure 10.63a, determine the voltage across the resistor R given that the value of the voltage V_B applied to the base of the transistor Q_1 is 1 V. The relationship between the input current flowing through the LED and the output phototransistor collector current is shown in Figure 10.63b. The base-emitter voltage of transistor Q_1 is equal to 0.7 V.
[1.24 V]

Self-evaluation Exercise

Multiple-choice Questions

10.1. What factors determine the type of photosensor you select?
a. speed of response
b. minimum and maximum light levels
c. spectral response
d. (a) and (b)
e. all of the above.

10.2. A photocell's resistance vs light can vary by as much as
a. 10:1
b. 20:1
c. 50:1
d. several orders of magnitude.

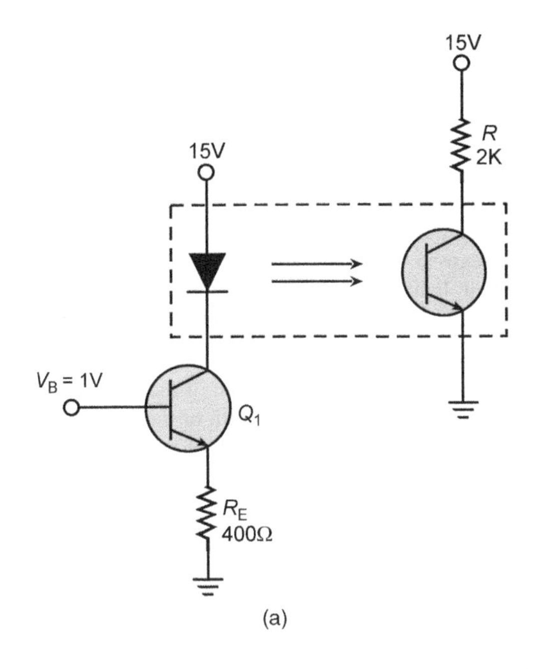

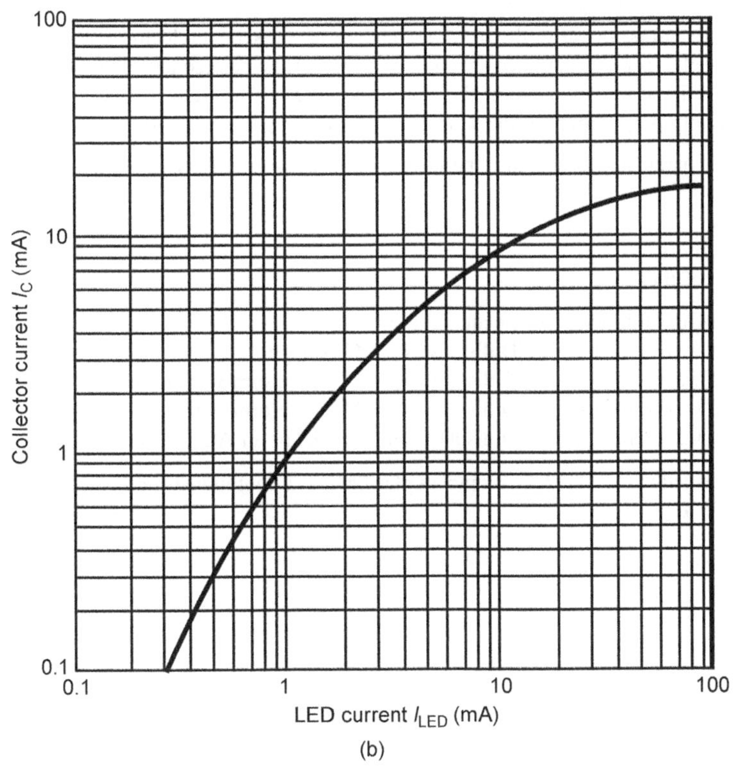

Figure 10.63 Problem 10.6.

10.3. The efficiency of an LED is a measure of the electrical energy required to produce a certain
a. current
b. resistance
c. photometric efficiency
d. light output.

10.4. The resistance of a photoconductor
a. decreases with increase in light intensity
b. increases with increase in light intensity
c. does not change with light intensity
d. can increase or decrease with an increase in light intensity.

10.5. An increase in reverse-bias voltage across the photodiode
a. increases the rise time and decreases the width of the depletion region of the photodiode
b. decreases the rise time and decreases the width of the depletion region of the photodiode
c. decreases the rise time and increases the width of the depletion region of the photodiode
d. increases the rise time and increases the width of the depletion region of the photodiode.

10.6. The noise current and responsivity of a sensor are 10 nA and 1 AW^{-1} respectively. Its NEP is:
a. 10 nW
b. 1 nW
c. 100 nW
d. 0.1 nW.

10.7. The output wavelength of a LED is
a. directly proportional to the band-gap energy of the semiconductor material
b. inversely proportional to the band-gap energy of the semiconductor material
c. proportional to the square of the band-gap energy of the semiconductor material
d. inversely proportional to the square of the band-gap energy of the semiconductor material.

10.8. The best parameter to compare the noise performance of any two sensors is
a. NEP
b. *D*-star
c. detectivity
d. none of these.

10.9. Which of the following statements is true?
a. drive waveform for a LCD display is a DC voltage
b. drive waveform for a LCD display is an AC waveform with zero RMS value
c. drive waveform for a LCD display is an AC waveform with zero DC value
d. drive waveform for a LCD display can be a DC or an AC waveform

10.10. A typical value of CTR for an optocoupler having a phototransistor at the output is
a. 10–50%
b. 1–10%
c. 60–90%
d. 10–70%.

Fill in the Blanks

10.1. Dark current in a photodiode is a measure of its ________________ current.

10.2. ________________ photosensor can be used to count even a single photon.

10.3. The width of the depletion region of a photodiode is ________________ proportional to the applied reverse-bias voltage.

10.4. ________________ is the most important photosensor parameter to be considered for low-noise applications.

10.5. When you are looking for an ultra-low-noise performance and the frequency of operation is less than 1 kHz, you operate the photodiode in ________________ mode.

10.6. LCD displays offer ________________ viewing angle as compared to the LED displays.

10.7. The main component of a CRT display is an ________________.

10.8. LEDs are ________________ biased for their normal operation.

10.9. Optocouplers are mostly used in applications to provide ________________ between the input and output circuits.

10.10. Phototransistors offer ________________ response time as compared to photodiodes.

Answers

Multiple-choice Questions

1. (e) 2. (d) 3. (d) 4. (a) 5. (c) 6. (a) 7. (b) 8. (b) 9. (c) 10. (a)

Fill in the Blanks

1. reverse saturation current; 2. photomultiplier tube; 3. directly; 4. NEP; 5. photovoltaic; 6. smaller; 7. electron gun; 8. forward; 9. isolation; 10. larger

Bibliography

1. *Optoelectronics*, 1995 by Endel Uiga, Prentice Hall.
2. *Optoelectronics & Photonics: Principles & Practices*, 2012 by Safa O. Kasap, Prentice Hall.
3. *Optoelectronics*, 2002 by Emmanuel Rosencher, Borge Vinter and P. G. Piva, Cambridge University Press.
4. *Applied Electro Optics*, 1997 by Louis Desmarais, Prentice Hall.
5. *Electro-Optics Handbook*, 2000 by Ronald Waynant and Marwood Ediger, McGraw-Hill Professional.

Part IV
Laser Applications

11

Lasers in Industry

11.1 Introduction

Although the laser was invented in 1960, new applications are still being found. Over the years, lasers have either completely replaced existing processes and techniques or enhanced their performance efficacy. As a result, lasers have a use in every conceivable area of application. One of the most widespread uses of laser technology has been in industry, as was evident during the period immediately after the invention of this magical device. The majority of industrial applications of lasers relate to material processing, and we discuss in detail the key industrial applications in this chapter. Material-processing applications include cutting, welding, drilling, marking, rapid prototyping and photo-lithography. Laser printing is another common application of lasers in industry. We focus on the principles behind the application, the merits and limitations and the different types of laser system configurations.

11.2 Material-processing Applications

Laser-based material processing is a versatile technology that has the required flexibility to process a wide range of materials for various applications, such as cutting, welding, drilling, marking, rapid manufacturing, ablation and so on. Due to its advantages of non-contact processing, high positioning accuracy, a narrow heat-affected zone and a high processing speed, the laser is the preferred choice for a wide range of material-processing applications. Such applications range from routine operations to more complex processes such as cutting fine intricate cardiovascular stents, drilling guide vanes in the aerospace industry and welding thick steel in the ship-building industry. In the following sections we discuss the classification of laser material-processing applications followed by important processing considerations and the advantages and limitations of the technology.

11.2.1 Classification

Laser-based material-processing applications can be classified into one of the two major groups as follows:

1. those requiring limited laser power/energy and causing no significant change in phase or state of the material before and after processing; and
2. those involving phase changes and requiring higher power/energy to induce those changes from solid to liquid or solid to vapour phase.

Lasers and Optoelectronics: Fundamentals, Devices and Applications, First Edition. Anil K. Maini.

Operations such as curing, annealing and marking fall within the first category. Common applications belonging to the second category include cutting, drilling and welding. A more practical way of classifying material-processing applications is:

1. forming (e.g. rapid manufacturing, deposition, bending/straightening);
2. joining (e.g. welding, brazing, soldering);
3. machining (e.g. cutting, drilling, marking); and
4. surface engineering (e.g. surface hardening, surface alloying, amorphization).

11.2.2 Important Considerations

All laser-based material-processing applications involve the use of a pulsed or continuous wave (CW) laser beam, which is directed to strike the surface of the work piece for the intended operation. When the laser beam strikes the surface of the work piece, only three possible processes can occur: transmission, reflection and absorption. That is, part of the incident laser power/energy is transmitted, part is reflected and a part of it is absorbed. The sum of these three fractions is 1. Different material-processing applications can be better understood by analyzing the three phenomena. In most applications, a significant proportion of the laser energy is absorbed. The quantum of laser energy absorbed is of primary interest in material-processing applications, and is mainly dependent upon the properties of the actual material and on the laser parameters.

11.2.2.1 Material Properties: Absorption Length and Diffusion Length

Light is absorbed in the form of electronic and vibration excitation of the atoms. The absorbed laser energy is converted into heat, leading to an increase in temperature of the material which increases the fraction of laser energy absorbed by the material. This is a cumulative process, resulting in a rapid rise in temperature in a very short time (e.g. a millisecond or so for welding applications). The rate of temperature rise depends on the rates of energy absorption and energy dissipation by the material. Laser energy is absorbed up to a depth called *absorption length*, which is the depth where incident laser intensity drops to 37% of the original value. Absorbed energy is converted into thermal energy that is diffused into the material. The *diffusion length* is directly proportional to the square root of the product of thermal diffusivity and laser pulse width. The temperature rise at the laser spot depends upon the relative proportion of absorption length and diffusion length. The temperature rise is very rapid if the diffusion length is much smaller than the absorption length. If the diffusion length is much longer than the absorption length, there is a limited temperature rise.

11.2.2.2 Laser Parameters

Laser parameters of importance in material-processing applications include: wavelength, pulse width, pulse energy, power density (also called intensity), repetition rate and average power for pulsed lasers, and laser beam quality defined by the M^2 value.

Wavelength plays an important part in determing what fraction of incident laser energy is absorbed by a given material and also the reaction of the material. The energy quantum of each photon of light is a function of wavelength. If photon energy is high enough, it may even break an atomic bond under certain special conditions. The majority of laser processes are however thermal in nature, governed by well-controlled melting and vaporization processes. The wavelength absorption spectrum of the material is a very important factor in laser-based material processing. For example, in the case of metals laser energy is partially absorbed and partially reflected with practically no transmission. Metals are also poor absorbers of CO_2 laser energy at 10.6 μm, meaning that CO_2 lasers are not very suitable for working with metals. CO_2 lasers can however be used for processing metals for cutting and welding operations if the energy density is very high, compensating for the loss due to inadequate absorption. The absorption fraction increases as wavelength decreases. On the other hand, ceramics and glasses have good absorption at both lower and higher wavelengths. Ceramics are relatively more difficult to process due to their poorer thermal shock characteristics and higher melting points. Due to their poor thermal conductivity, glasses are easily melted. Plastics absorb laser energy even better than ceramics and glasses, especially in the UV and CO_2 regions. Certain specific wavelengths in the UV range have sufficient energy to break atomic bonds, and these wavelengths can be used to selectively alter the surface properties of a material and even the properties below the surface if the material has a certain transparency.

Pulse width is another very important laser parameter affecting the quantum of laser energy absorbed by the material and subsequent rise in its temperature. Pulsed lasers are used in most material-processing applications. CW lasers are mainly used for cutting, welding and heat-treatment applications. Since temperature rise is determined by relative values of absorption and diffusion lengths, for a given class of materials with similar absorption length the optimum pulse width depends on the thermal diffusivity of the material. For example, longer pulse widths can be used for micromachining operations on stainless steel compared to nickel, due to the poorer thermal conductivity of nickel. Similarly, silicon requires shorter pulse widths to produce ablation compared to nickel, due to the high thermal conductivity of silicon. Due to a higher power density and shorter time frame, pulse widths of the order of femtoseconds are capable of knocking off atoms from the surface of the material without affecting adjacent atoms. This makes them highly suitable for micromachining operations.

Pulse energy and *spot diameter* together determine the *power density*. Although pulse width determines the rise in temperature, it is not sufficient to determine the efficacy for a given material-processing application. Pulse energy needs to be sufficiently high to heat up a useful volume of the material to be processed in each pulse. After selecting a suitable wavelength for a given application, pulse energy and pulse width together determine the quality of processing. For enhanced performance, pulse energy can be modified to allow a ramp-up of energy at the beginning followed by a gradual cool-down towards the end of pulse.

The *repetition rate* of pulsed lasers is also important. For a given value of pulse energy, it is the repetition rate that determines the *average power*. While wavelength and pulse width determine the initial reaction, average power controls the process rate.

Beam quality or M^2 value also plays an important role in determining the laser spot diameter for a given pulse energy and therefore the power density. The closer the value of M^2 to 1, the tighter the focus. An M^2 value close to 1 is preferred for micromachining. For welding and heat-treatment applications, a larger value of M^2 in the range 30–100 is generally used. A higher beam quality associated with a lower M^2 value offers finer-feature machining, improved process robustness and overall superior process results. In addition to allowing for small spot sizes, a low M^2 value results in an increased depth of focus when the beam is focused to a particular spot size. A common misperception is that, regardless of the value of M^2, suitable optics will always give the desired laser spot size for the targeted feature; however, if high- and low-M^2 beams are focused to the same spot size, the low-M^2 beam will have a longer Rayleigh range and hence a better tolerance for system defocusing.

11.2.3 Common Material-processing Applications

Although all materials can be processed by laser, the processes for some materials are better-established than others. The list includes metals, plastics, ceramics, cloth, diamonds and even human tissue. Common material-processing applications of lasers include cutting, welding, drilling, marking, photo-lithography, micromachining, rapid manufacturing, photolithography and laser printing.

The *laser cutting* process involves a high-intensity infrared laser beam focused onto the surface of the work piece with the help of optics. This heats up the material and creates a localized melt throughout the depth of the work piece. A gas jet acting coaxially with the laser beam removes the molten material. The laser beam is moved across the surface to produce the cut. Cutting by laser is a fast method of obtaining a narrow cut width with reduced material waste. The cutting process is far safer than its mechanical counterpart and it can be a fully computer-numerical-controlled (CNC) non-contact process.

In the case of *laser welding*, a high-energy pulsed laser beam is focused onto the surface of the work piece with the help of either bulk optics or a fibre-optics module. The high energy density in the focused laser beam penetrates, vaporizes and melts some of the metal; joining of the two parts is established as the common melt solidifies. The advantages of laser welding include low heat input and a small heat-affected zone, which makes it ideal for products that require welding near electronics and glass-to-metal seals. Pulsed lasers are used to weld thin materials and CW lasers are used for deep welds.

Laser drilling generally occurs through the processes of melting and vaporization; it is also referred to as ablation when the energy of a focused laser beam is absorbed. The drilling operation largely depends upon pulse energy and duration. Use of laser drilling has been common since the 1970s for drilling of gemstones in wire dies and watch jewels. Present-day applications include drilling of cooling holes in aircraft engine blades and combustors, inkjet nozzles, fuel injector nozzles, cigarette filter paper and vias in printed circuit boards (PCBs). Laser drilling is the preferred choice in these applications due to its ability to drill high-quality small holes at high speed in most materials.

Laser marking is one of the most widely exploited industrial applications of lasers. Laser marking could be in the form of an alphanumeric code imprinted on a label or surface of a product carrying information such as manufacturing data, expiry date or serial/part number, a machine-readable barcode or a two-dimensional symbol. In addition to the above-mentioned processes (classified as laser coding), commercially available laser-marking machines are also equipped to perform functional and decorative markings. Laser marking is primarily a surface process; the pulsed laser light is absorbed by the work piece and is converted into heat. The short pulse width (of the order tens of nanoseconds to several microseconds) causes an instantaneous rise in temperature, leaving a resultant mark. Commonly used types of lasers for laser-marking equipment include Nd:YAG, CO_2 and excimer lasers.

Laser micromachining has applications in many high-technology areas where high-precision microfabrication is an enabling technology. Laser micromachining offers many distinct advantages over conventional mechanical, chemical, ion or particle beam processes, including: high-resolution precision non-contact processing and unlimited material coverage; the ability to machine flat and multicontoured surfaces; and multiprocess compatibility. Commonly used lasers for micromachining applications include excimer, Nd:YAG, CO_2 and metal vapour lasers.

Laser rapid manufacturing is a new class of technology used to fabricate engineering components directly from computer-aided design data. This is also known as additive manufacturing, layered manufacturing, solid freeform manufacturing, e-manufacturing or digital manufacturing. Unlike subtractive machining (e.g. CNC machining which removes material to form the final part), this is an additive process that deposits layers of material to form the final part. The laser is used here as a heat source. Rapid manufacturing has widespread use in the fabrication of models and prototypes, patterns and functional components. Rapid manufacturing is well established for the fabrication of large products from metals, plastics and composite materials in the military and aerospace industry, but also of small products and microsystem applications in medical, consumer electronics, diagnostics and sensor technologies.

Photolithography is an important process in the semiconductor industry for the fabrication of semiconductor devices and integrated circuits to selectively remove parts of a thin film or bulk of a substrate. Photolithography makes use of light to transfer a geometric pattern from a photomask onto a photoresist on the substrate. Chemical treatment enables deposition of a new material in the desired pattern on the substrate material. The resolution at which the pattern can be transferred to the photoresist is an important parameter in the fabrication of integrated circuits, more so for ultra-large-scale integrated (ULSI) circuits. Higher resolution is desirable to produce denser and faster chips. Gas-discharge lamp-based lithography processes operating at relatively larger wavelengths cannot meet this requirement. Excimer lasers emitting in the ultraviolet wavelength band are the preferred laser sources in photolithography, and have largely replaced conventional gas-discharge mercury lamps emitting in the ultraviolet band. Low wavelengths, such as the 193 nm of an argon fluoride laser, offer greater resolution.

Laser printing uses a non-impact photocopier technology. The process is very similar to xerography, with the printing mechanism involving the same basic steps of charging, exposure, development, transfer and fusing. Laser printers have higher print capacity and print speed, higher resolution and lower operational costs (but higher initial costs) compared to inkjet printers.

11.2.4 Advantages

The general advantages of laser processing, applicable to all processes, include the following:

1. contact-less material processing (offering a tool that never becomes 'blind');
2. high positioning accuracy, allowing laser beam positioning on areas that are difficult to reach or access;
3. a small heat-affected zone, so only the desired area of the work piece is processed with no affect on neighbouring areas;
4. high-speed material processing since light is not affected by inertia;
5. flexibility, versatility and easy adaptation to automation and computer control (particularly advantageous in rapid manufacturing); and
6. reduced processing times and cost (through reduced wages and tool costs) and increased production and quality.

In addition to the important general features of laser material processing outlined above, it also offers specific advantages for various applications:

1. *Laser cutting*: the cut width can be extremely small with the result that evaporated material volume is minimal; in most cases, there is no requirement for post-cut finishing; difficult materials (e.g. asbestos and glass fibre) can be cut without leaving dust; equally effective for cutting of soft materials (e.g. rubber) without deforming or damaging the material; 2D and 3D cutting of metals with a thickness of greater than 30 mm is possible.
2. *Laser drilling*: drilling of very small diameter holes; high-quality drilling of cooling holes in aircraft engine blades and combustors, inkjet nozzles and fuel injector nozzles.
3. *Laser welding*: electron-beam-like welding operation without a vacuum; welding without any additional material; different materials can be easily welded; point welding on miniature and sensitive parts; welding of ceramics.
4. *Laser heat-treatment processing*: produces hardened surfaces by a metallurgical transformation of specific areas without the need for bulk heating of the entire part; allows re-melting of special alloys to enhance certain features such resistance-to-abrasion and corrosion stability.

Laser processing provides possibilities for many non-conventional industrial applications, material-processing applications in particular. It can be used for: precise control of chemical processes through localized heating or selective excitation of molecules; precise drilling of deep holes or making a molecular sieve from an opaque sheet material; instantaneously sterilizing of precise areas for food or pharmaceutical industry needs; cutting, welding or fusing of metals and hard-to-work materials such as heat-resistant refractory materials; and separating selected isotopes from mixtures, split friable materials and degas materials in high-vacuum systems.

11.3 Laser Cutting

Laser technology has made possible numerous industrial processes not previously possible with conventional mechanical operations. It has redefined the speed of production line manufacturing and the range of industrial applications. Using the directed energy of laser radiation to cut materials is one of the most widely used industrial applications of lasers today. Being a non-contact process, laser cutting does not contaminate the material being cut; a small focus spot generates a very fine cut that does not require any post-cut processing.

11.3.1 Basic Principle

Laser cutting refers to the use of a directed high-energy infrared-emitting high-repetition-rate pulsed laser or high-power CW laser to cut the material. The different steps involved in laser cutting are briefly outlined in this section.

The high-power infrared laser beam pulsed or CW is tightly focused onto the intended location on the surface of the work piece with the help of focusing optics. The focus spot diameter is in the range of 25–100 μm for fine cutting and 100–300 μm for cutting thicker sections. The laser energy is transformed into heat energy, which produces a localized melt throughout the depth of the material; this localized melt diameter is generally less than 0.5 mm. The molten material is blown away from the area by a pressurized gas jet acting coaxially with the laser beam to accelerate the cutting process. Oxygen and air are generally used to promote the cutting of ferrous alloys and cellulose materials. Air is often used with plastics and fabrics. In the case of some metals requiring an unoxidized cutting edge with little dross (re-solidified metal at the lower edge of a deep cut), a high-pressure inert gas is used. The laser beam is moved across the surface of the material producing the cut. Movement of the laser spot is achieved either by laser beam scanning or by moving the sheet of material in the *x*–*y* direction. Figure 11.1 depicts the schematic arrangement of the laser cutting process.

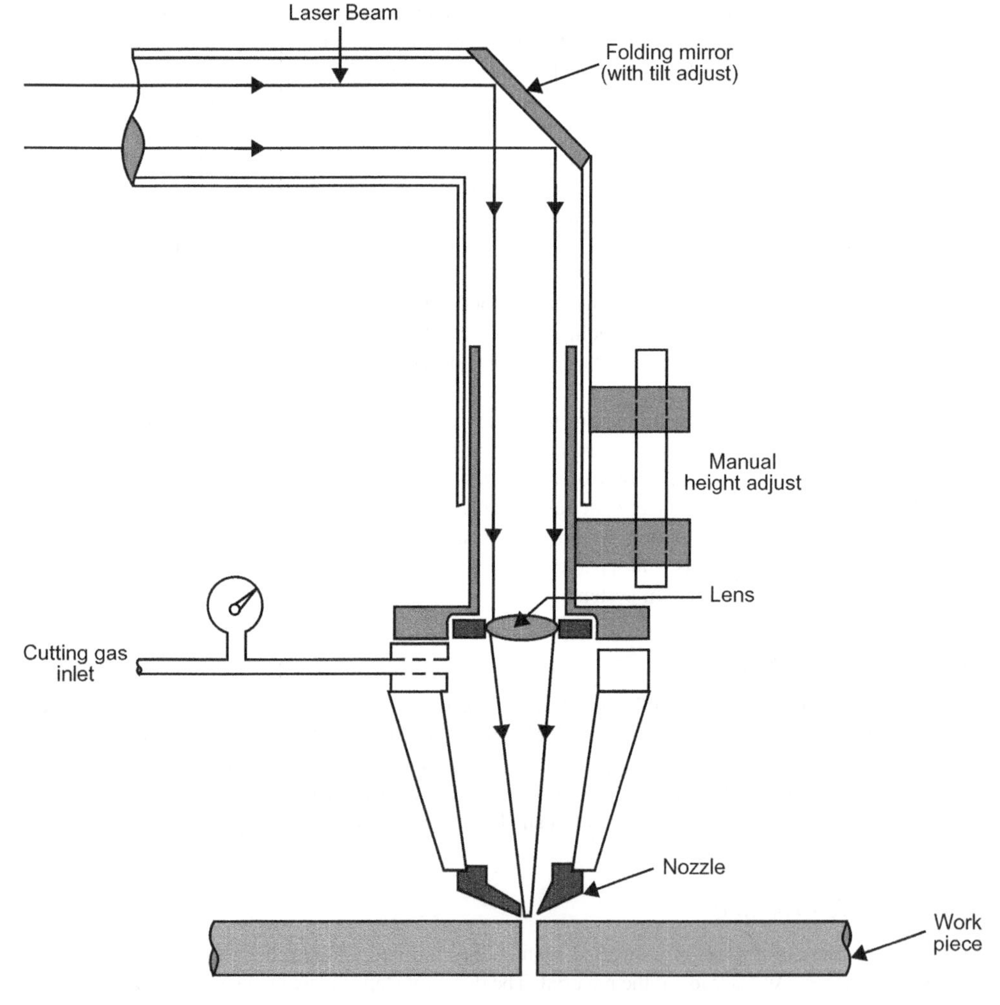

Figure 11.1 Laser cutting process.

11.3.2 Laser Cutting vs Plasma Cutting

A large number of technologies are available for cutting metals and other materials, such as laser cutting, plasma cutting, mechanical and thermal machining, flame cutting, electro-discharge machining and abrasive jet cutting. Laser and plasma cutting are the most widely used processes in metal cutting in industry. While laser cutting makes use of the directed energy of a high-power laser for cutting, plasma cutting (also called plasma arc cutting) uses electrically conducting gases such as argon, hydrogen, nitrogen, oxygen and air to generate a plasma arc from input electrical energy. Plasma cutting equipment typically comprises a source of electrical energy, an arc initiation circuit and a plasma gun that generates a directed flow of plasma. In the following we present a comparison of the merits and demerits of laser cutting and plasma arc cutting techniques.

Laser cutting yields higher precision and accuracy and causes negligible surface cracking. It is faster; for example, a 1500 W CO_2 laser cutting machine can cut a 2.5-mm-thick mild steel sheet at a rate of 7.5 m min^{-1} or a 5-mm-thick acrylic sheet at a rate of 12 m min^{-1}. Laser cutting is suitable for all metals except copper and aluminium (which reflect a large fraction of the incident energy). It can also be used to cut plastics, wood and glasses effectively. It is hard to cut materials thicker than 15 mm using laser cutting; laser cutting is less efficient and more expensive than plasma cutting in this case. Laser cutting is not effective for cutting materials of different melting points; plasma cutting has no such limitation. From the viewpoint of safety considerations, laser cutting is better than plasma cutting (but it is recommended to use safety glasses while working with a laser cutting machine).

The plasma cutting process can cut any metal, including copper and aluminium, up to a thickness of 150 mm. It is less expensive, but adversely affects the area surrounding the cut (laser cutting has a very narrow heat-affected zone). Plasma cutting generally leaves a dross or re-solidified metal at the lower edge of the cut. Safety glasses are essential while working with plasma cutting equipment.

11.3.3 Laser Cutting Processes

Laser cutting technology today is well established and widely used in industry for cutting both ferrous and non-ferrous metals and non-metals such as polymers, ceramics, glasses, wood and fabric. A number of laser cutting processes are in use, each suited to a specific class of materials. In some cases, a combination of more than one process is used for better results. Commonly used laser cutting processes:

1. melt shearing or fusion cutting;
2. vaporization cutting or boiling;
3. chemical degradation or burning;
4. oxidation cutting;
5. thermal stress cracking; and
6. scribing.

In the case of *melt shearing* or *fusion cutting* mechanism, the focused laser energy heats up the material to the melting point. A high-pressure gas jet is used to blow the molten material out of the cutting area. Due to the removal of the molten material by gas jet, the temperature of the material does not need to be raised any further. The resulting cut edge is of good quality, but may be covered with some microscopic ripples. The lower edge of the cut generally has a residue of re-solidified melt (dross) since not all melt is blown away from the cutting area. This is particularly true for larger-thickness materials. This technique is common for cutting metals such as mild and carbon steels and thermoplastics including polyethylene, polystyrene, polypropylene and polyamide. Polyvinyl chloride (PVC) and polymethyl methacrylate such as plexiglas, acrylic, and so on cannot be cut by this technique. The former degrades chemically when heated to produce corrosive and highly toxic hydrogen chloride fumes, and the latter vaporizes rather than melts. PVC may be cut using a melt-shearing mechanism, providing adequate ventilation arrangements are made.

A CO_2 laser with power output in the range 100–1500 W is generally used to cut thermoplastics. To a first approximation, maximum material thickness that can be cut using this technique varies linearly with power in this range. Typically, a 500 W laser can cut a 10 mm thick sheet of a thermoplastic material at a rate of 0.5–1.5 m min^{-1} depending upon the actual material.

In the case of *vaporization cutting*, the focused laser beam heats the material surface to the boiling point and generates a keyhole. This increases absorptivity, thus deepening the keyhole. The boiling vapours are blown away from the cut zone by a gas jet. As the vapours are blown away, they leave behind a thin liquid layer on the cut edges. If the velocity of the gas jet is kept less than a certain maximum value, the liquid layer dries up producing a glossy-edged cut. For velocities higher than this, the solidifying liquid layer becomes frosted. Thermoplastics such as plexiglas and acrylic are cut by this technique.

The *chemical degradation* or *burning* technique of cutting is common for cutting thermoset plastics such as epoxy and phenolic resins, kevlar and natural rubber products. The focused laser energy burns the work piece, reducing it to a smoke containing carbon and other constituents of the material. The cut edge is generally flat, smooth and covered with a thin layer of carbon. The maximum material thickness that can be cut and cutting speed are lower compared to those achievable with the melt shearing technique. This is due to the process requiring a higher input energy than the melting process. Wood-based products are also cut by a similar mechanism.

The *oxidation cutting* mechanism, also known as *flame cutting* or *reactive cutting*, is basically a combination of the processes of melt shearing and chemical reaction with the cutting gas jet, which is oxygen. The cutting equipment is like an oxygen torch with laser as the source of ignition. This technique is very common for cutting mild and carbon steels. It can cut thick sheets of steel at high speed, producing high-quality dross-free cut edges. A 1500 W CO_2 laser cutting machine using oxidation cutting can achieve a cutting speed of about 1 m min^{-1} for a 10-mm-thick sheet of mild steel. Oxygen in the gas jet reacts with iron in steel to produce iron oxides. The chemical reaction produces heat that speeds up the cutting process. It also produces a liquid that is oxidized and has a lower melting point than steel. The oxidized liquid does not adhere well to the walls of solid steel, and is therefore blown away by the oxygen gas jet. This produces a dross-free cut edge.

Thermal stress cracking is primarily suitable for brittle materials that are particularly sensitive to thermal fracture. The focused laser beam causes localized heating of the surface, producing thermal expansion. This leads to the formation of a crack that can be guided by moving the laser beam across the surface. Glass is cut using this technique.

Scribing is commonly used for cutting silicon wafers into chips and for cutting thin sheets of ceramics such as alumina used to produce microelectronic substrates. Scribing involves drilling lines of small blind holes. Pulsed Nd:YAG lasers are generally used for this purpose, with each laser pulse producing a hole. The lines of holes allow cutting of the work piece along the lines by snap action. The technique achieves a cutting speed that is an order of magnitude faster than full-penetration cutting, which may be needed if curved shapes rather than rectangles are to be cut. In scribing, cutting speeds of 20 m min^{-1} are common.

11.3.4 Machine Configurations

Based on the mechanism of moving the laser beam across the surface of the work piece to produce the desired cut, laser cutting machines are categorized as:

1. moving-platform-bed machine configuration
2. flying optics configuration
3. hybrid configuration.

For a *moving-platform-bed machine configuration*, the laser cutting edge is stationary and the material is moved under it in the x–y direction to produce the cut. This machine configuration is characterized by a constant laser-head–work-piece distance, single point removal of cutting effluent and relatively simple beam delivery optics. The disadvantages include a cutting speed which is the slowest of the three configurations and machine dynamics being affected by the variations in the size of the work piece. Figure 11.2 shows one such laser cutting machine from M/s Coherent Inc. The META E-series

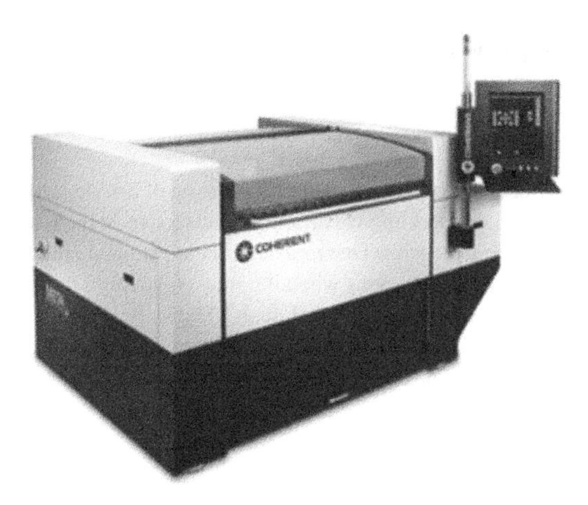

Figure 11.2 META E-series laser cutting machine (Reproduced by permission of Coherent Inc.).

machine is interfaceable with 150 W, 400 W and 1000 W average power pulsed CO_2 lasers. When interfaced with a 1000 W laser, the machine is capable of cutting material thicknesses of 3 mm for aluminium and stainless steel, 10 mm for mild steel and 25 mm of acrylic, plastics and wood products. The machine offers 1.25 m × 1.25 m of cutting area and a deep moving platform bed with 300 mm of vertical travel.

For a *flying optics configuration*, the laser head moves in the x–y direction. The work piece is stationary. This machine configuration is characterized by high-speed cutting (the highest of the three configurations). Since the moving mass (i.e. the laser head) is constant, equipment dynamics are not affected by a variation in the size of the work piece. Additionally, the work piece need not be clamped. There is however a changing beam path length due to the movement of the laser head, which can be taken into account by the use of suitable optics. Figure 11.3 depicts a laser cutting machine of Omni-Beam series from Coherent Inc. The machine features a high peak power CO_2 laser, a cutting area of 1.23 m × 1.23 m and a flying optics beam delivery. The machine is primarily used for cutting a wide range of plastics and organics as well as metals; it can cut non-metallic materials of thickness up to 25 mm.

A *hybrid configuration* is a combination of the two above-mentioned configurations. In this case, the table housing the work piece is moved in one horizontal direction, usually the x-direction, and the laser cutting head is moved in the shorter y-axis direction. This configuration boasts the advantages of both the moving platform bed and flying optics configurations, and is characterized by simpler beam delivery

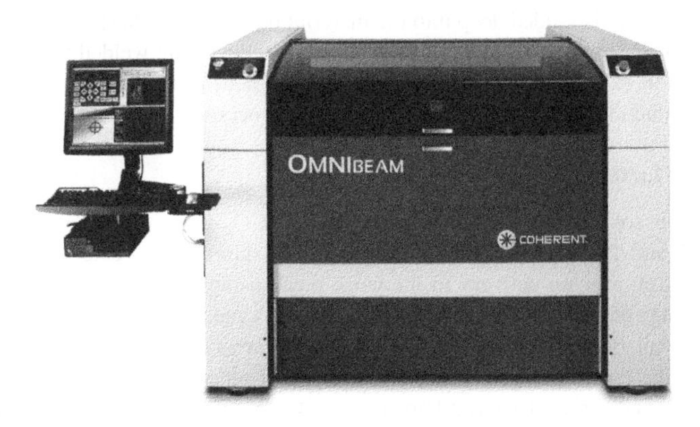

Figure 11.3 Laser cutting machine of OMNIBEAM series (Reproduced by permission of Coherent Inc.).

optics and higher capacity per watt (due to lower associated loss) compared to the flying optics configuration.

11.4 Laser Welding

Laser welding is extensively used for both spot and seam welding of a wide range of materials including metals, alloys and non-metals. Laser welding is particularly attractive in high-volume applications such as the electronics and automotive industries. Due to a small heat-affected zone and minimal heat input, laser welding is also distinctively advantageous when it comes to welding thin sections or welding near sensitive areas such as electronics or glass-to-metal seals. The only disadvantage preventing it from being more widely used is its high investment cost. In the following sections we briefly describe the basic principles of laser beam welding, the different welding processes, the types of lasers used for welding and its key advantages.

11.4.1 Laser Welding Processes

A laser beam is focused onto the surface of the work piece to be welded. The concentrated light energy produces high energy density at the focal spot. The light energy is absorbed by the material and is converted into heat energy and the work piece begins to melt. The laser beam energy is maintained such that the resultant temperature rise is below the vaporization temperature of the material of the work piece (note that vaporization temperature must be attained in cutting and drilling operations). As the laser energy is terminated, the common melt re-solidifies producing a spot welded joint. Laser welding is a type of fusion welding that can be used to produce selective spot welds or linear continuous seam welds. There are two types of laser welding processes in use: heat conduction and deep penetration.

In the case of *heat conduction welding*, the materials to be joined are melted by absorption of laser energy at the material surface and the solidified melt joins the materials. Welding depths in this case are typically less than 2 mm. The laser beam is focused onto a specified location on the material surface. The heat energy produced as a result of absorbed laser energy conducts to the joint area to be welded by virtue of close proximity. The two materials change state from solid phase to liquid phase producing a common melt. When the laser energy is terminated, the common melt is re-solidified producing a spot welded joint at that location. Laser conduction welding is mainly used for spot welding, continuous and partial seam welding.

The *deep penetration welding* process dominates for power densities of 10^6 W cm^{-2} or higher. Higher power density causes localized heating to raise the temperature to the vaporization temperature creating, a vapour capillary inside the material. The capillary diameter is about 1.5 times the diameter of the focused spot. The capillary moves through the material following the contour to be welded. Capillary formation is also referred to as the keyholing effect. The equilibrium achieved due to hydrostatic pressure, vapour pressure in the capillary and surface tension of the melt prevent the capillary from collapsing. The laser beam is guided deep into the material by multiple reflections in the capillary. The re-solidification of melt after the removal of laser energy produces the welded joint. It is possible to achieve weld depths in excess of 25 mm in steel using deep penetration welding. Figure 11.4a and b depicts the heat conduction and deep penetration welding processes.

11.4.2 Welding Lasers

Commonly used laser sources for welding applications are CO_2 lasers and neodymium-doped lasers including Nd:YAG and Nd:Glass lasers. Welding by CO_2 laser is characterized by an operating wavelength of 10.6 μm, a beam delivery to the work piece using bulk optics and not fibre optics, a laser power in the range of a few hundreds of watts to kilowatts and a high-speed operation. Both pulsed and CW lasers are available for the purpose. CO_2 lasers are capable of welding both metallic and non-metallic materials. The far-infrared wavelength has an initial reflectance of 80–90% for most metals. Because of their higher power level and capillary action, CO_2 lasers overcome the limitations due to high initial reflectivity. Once the material surface temperature at the focal point approaches melting point,

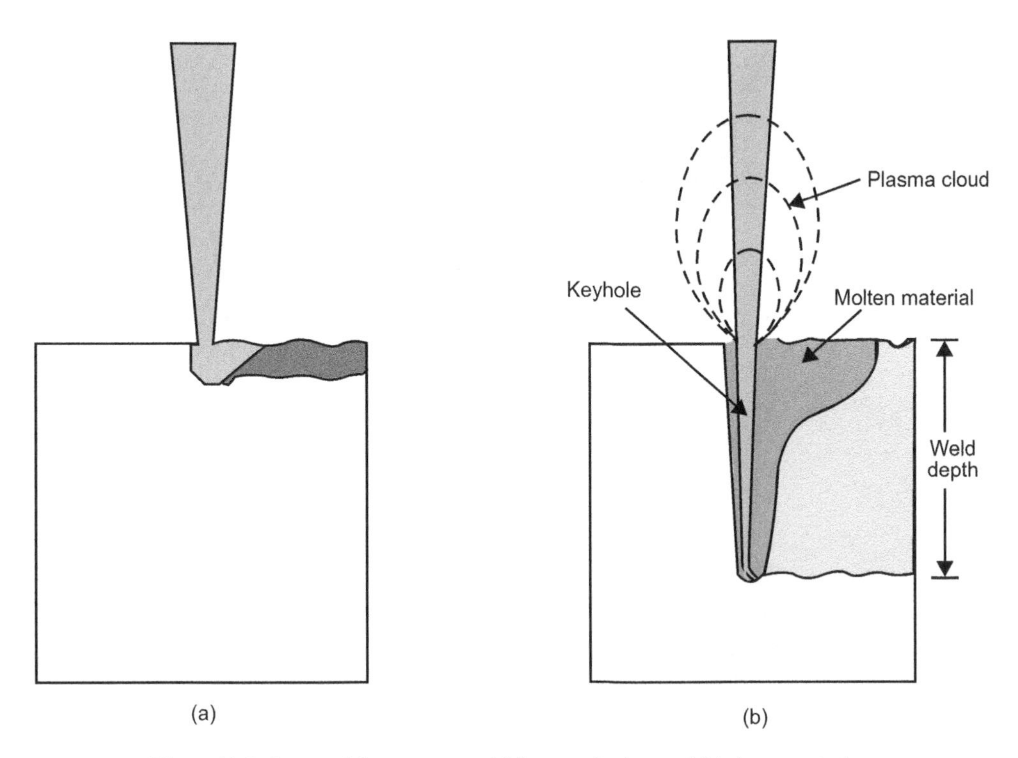

Figure 11.4 Laser welding processes: (a) heat conduction; and (b) deep penetration.

reflectivity drops within microseconds. With advances in laser technology, fast spiral flow lasers are available, producing fundamental mode output at the kilowatt level. The higher energy density available from these lasers makes them suitable for welding thermally sensitive alloys where thermal distortion is a problem. For applications requiring rapid energy coupling and low heat input, slow axial flow lasers with enhanced pulse capability are preferred over fast axial flow systems. Higher peak power and lower average power overcome the surface reflectivity problem and minimize damage to the surrounding material.

Nd:YAG laser welding is characterized by an operating wavelength of 1.064 μm, a beam delivery by either bulk optics or fibre optics and a typical power level in the range of tens of watts to a few hundreds of watts. The key advantage of using Nd:YAG laser welding stems from the high absorptivity of metallic materials, including aluminium and copper, at 1.064 μm wavelength. The near-infrared wavelength also permits the use of standard optics to achieve focused spot sizes of a few tenths of a millimetre. Nd:YAG and Nd:Glass lasers are used for low to moderate power applications. These lasers are preferred in the electronics industry for the fabrication of semiconductor devices and integrated circuits. Another advantage of Nd:YAG and Nd:Glass laser welding is the coaxial viewing optics offered by these machines, which is a big advantage in delicate welding applications such as lamp filaments. The operating wavelength allows the use of standard optics to transmit laser as well as image work piece. Nd:YAG laser welding machines are also relatively smaller in size compared to CO_2 laser welding machines.

Figure 11.5 depicts an Nd:YAG laser spot welding machine from M/s Sintec Optronics Pte Ltd. The major specifications of the system include maximum average power of 200 W, pulse width and pulse repetition frequency of less than or equal to 20 ms and 15 Hz respectively and a focused spot diameter of 0.1–0.3 mm. The machine is capable of spot welding a range of materials including cold-work alloy steel, hot-work alloy steel, nickel tool steel, high-grade steel and high-tenacity aluminium alloy.

Figure 11.5 Nd-YAG laser spot welding machine (Reproduced by permission of Sintec Optronics Technology Pte Ltd).

11.4.3 Advantages

The key advantages of laser welding compared to conventional techniques such as gas metal arc welding, submerged arc welding, resistance welding and electron beam welding are as follows. Laser welding:

1. is a non-contact process, which eliminates the problems of tool wear and tear and debris build-up;
2. can be used for a wide range of materials including metals, alloys and plastics;
3. is easily adapted to automation;
4. is a high-speed operation, faster welding rates and high-quality welding, eliminating the need for any post-welding processing;
5. has adjustable input energy to best suit different materials;
6. has higher operational safety with proven beam delivery systems;
7. offers the possibility of simultaneous spot welds by using either different machines or the same machine with beam splitters;
8. has higher reliability and repeatability;
9. is associated with low thermal material influences;
10. can do single-pass two-side welding;
11. facilitates access to inaccessible locations; and
12. does not require operation in vacuum or magnetic shielding (as for electron beam welding) and can be used to weld magnetic materials.

11.5 Laser Drilling

Laser drilling is a thermal process by which holes are drilled using a tightly focused laser beam to produce a high energy density at the focal spot. The thermal removal process is by melting and evaporation and the melt is ejected out by vaporization with or without the assist gas pressure. Laser drilling has enabled the drilling of small holes with extraordinary precision in different shapes and orientations in a wide range of materials, including difficult-to-machine aerospace alloys. The drilling of hundreds and thousands of extremely fine cooling holes in aerospace turbine engine parts at unusual angles has significantly enhanced cooling efficiencies, by controlling the magnitude and direction of flow of cooling fluids. Laser technology is extensively employed in the fabrication of parts of turbine engines in aircraft and land-based power generation systems, in the automotive industry and in medicine.

11.5.1 Basic Principle

A beam from a high-power laser is focused on the surface of the work piece, producing high power density in the range of megawatts to hundreds of megawatts per squared centimetre. The light energy is absorbed and converted to heat energy, leading to a localized increase in temperature. The material melts, vaporizes and drills into the material thickness. The material is thrown out by a pressure gradient created by the vaporization pressure, with or without the pressure exerted by an assist gas. In the melt expulsion without an assist gas, expulsion is the result of a rapid build-up of gas pressure created by vaporization within a cavity. For melt expulsion due to vaporization pressure alone, there must be a molten layer and the pressure gradient acting on the surface must be sufficiently large to overcome the force due to surface tension. Vaporization pressure is a strong function of peak temperature. The desired pressure gradient occurs only for a temperature greater than a certain critical temperature at which gas pressure and the surface tension of molten material are equal. Figure 11.6 illustrates the laser drilling operation.

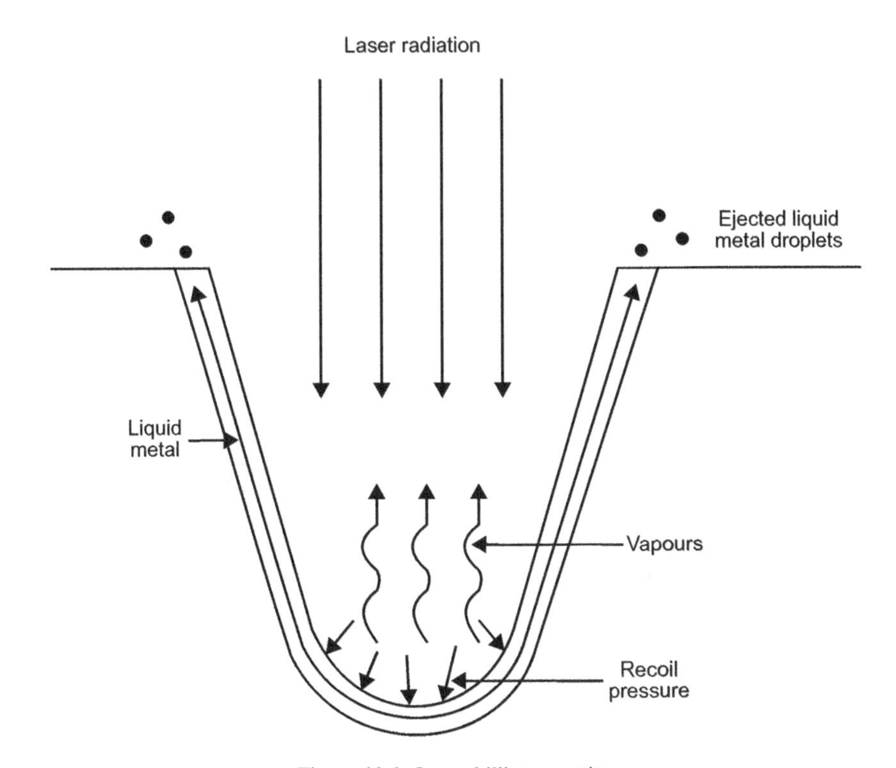

Figure 11.6 Laser drilling operation.

11.5.2 Laser Drilling Processes

Commonly used drilling processes include: single-pulse drilling; percussion drilling and its variants; trepanning and its variants; and imaged drilling.

Single-pulse drilling uses a single pulse to drill each hole as shown in Figure 11.7 a, and is used for high-speed drilling blind and through holes. The process is commonly used in the automotive industry to create a scribed guideline for breaking off a connecting rod in diesel engines, by drilling blind holes close enough to produce a notch. Another example from the automotive industry is the manufacture of filters.

Percussion drilling, illustrated in Figure 11.7b, is used to drill holes with diameters in the range 20–1200 μm. The aspect ratio (defined by the ratio of depth to diameter of the hole) can be as high as 200:1. Percussion drilling can be accomplished with or without an assist gas. If no assist gas is used for the expulsion of the molten material, an alternative method for the protection of the focusing optics needs to be employed. In percussion drilling operation, both laser and work piece are usually stationary. More than one laser pulse is required to drill each hole, with the number of pulses depending upon the material thickness. Percussion drilling is a relatively slow process however, and this limitation is overcome by

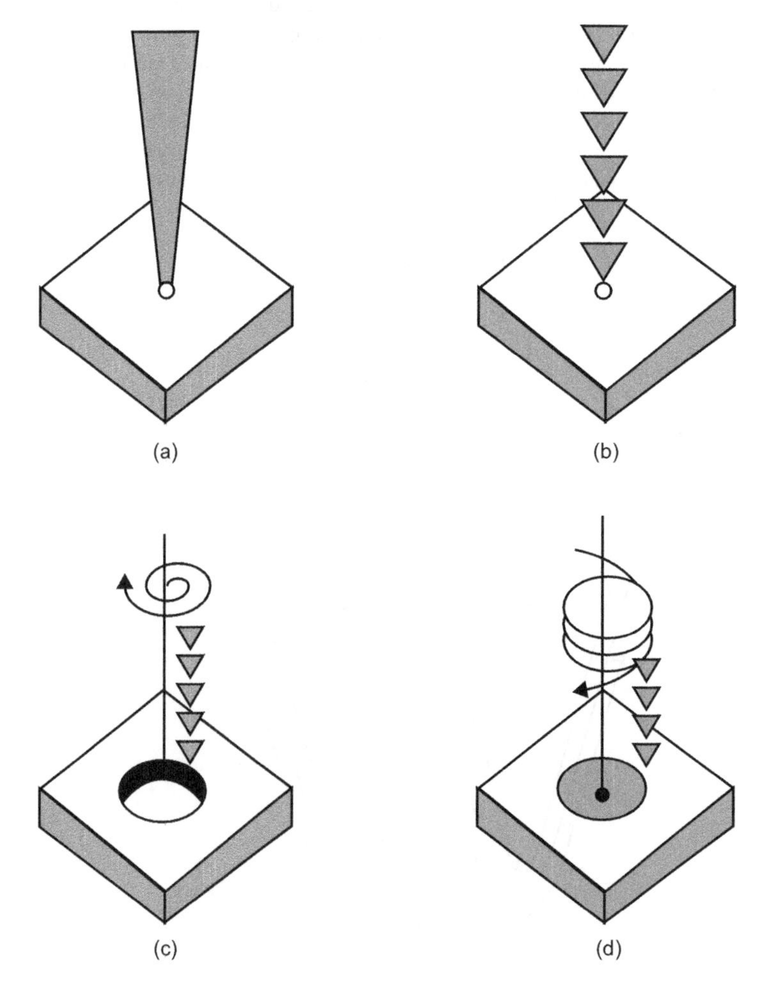

Figure 11.7 Drilling processes: (a) single-pulse; (b) percussion; (c) trepanning; and (d) imaged.

parallel percussion drilling. In this case, the laser beam is split into an array of smaller beams which allows the simultaneous drilling of holes (typically, 100 holes may be drilled simultaneously). The drilling rate either approaches or exceeds the rate achievable by percussion drilling-on-the-fly (also called single-pulse serial drilling) without compromising the quality of the percussion drilling.

In *drilling-on-the-fly operation*, a stationary laser beam and a moving work piece configuration is employed. Air or an inert gas may be used as an assist gas. Oxygen is also used in what is referred to as exothermal percussion drilling. Oxygen causes an exothermic reaction with many metals, and the consequent rise in temperature helps in efficient removal of molten material.

Trepanning is used when the laser beam diameter is much smaller than the diameter of the hole to be drilled. It is also used to cut features other than drilling circular holes. Note that the term 'drilling' refers to operations where the diameter of the hole or cut feature is less than or equal to the thickness of the material. The trepanning process employs a moving laser beam from a rotating optical head with or without a moving work piece, as shown in Figure 11.7c. The availability of high-precision and repeatable laser positioning systems has made many unique and tight-tolerance trepanned features possible. Shaped-hole drilling is an emerging variation of the trepanning process. It has enabled designers to try new cooling system designs for turbine components of aerospace engines and land-based power generation systems.

A variation of trepanning process is *helical trepanning*. The process is essentially trepanning combined with percussion drilling. In this process, the laser beam is rotated around the circumference of the hole many times until full penetration is obtained (Figure 11.7d). This leads to a significant improvement in the quality achieved from drill-on-the-fly processes. However, it necessitates the use of a very high-repetition-rate copper vapour laser.

Imaged drilling is a highly specialized method producing high-quality drilling. In this, an image of a mask is projected onto the surface of the work piece. Percussion drilling is then used to drill holes, with the hole shape being exactly that of the mask. It requires short-pulse lasers with uniform beam profile. It is generally used for drilling a small number of highly accurate holes. Figure 11.8 illustrates imaged drilling process.

11.5.3 Lasers for Drilling

Commonly used lasers for drilling operation are the Nd:YAG lasers and CO_2 lasers. Pulsed Nd:YAG lasers are far more popular than CO_2 lasers for drilling operation. Both free-running output and Q-switched Nd:YAG lasers are used for the purpose. Free-running lasers produce pulse widths of hundreds of microseconds to a millisecond and power density up to the megawatt per squared centimetre level. In contrast, Q-switched lasers used for drilling produce pulse widths of a few nanoseconds to a few tens of nanoseconds and a power density in the range of tens to hundreds of megawatts per squared centimetre.

Power density achieved at the work piece depends on both power level and focused spot diameter; focusing optics therefore plays an important part. The focused spot diameter is also correlated to the hole

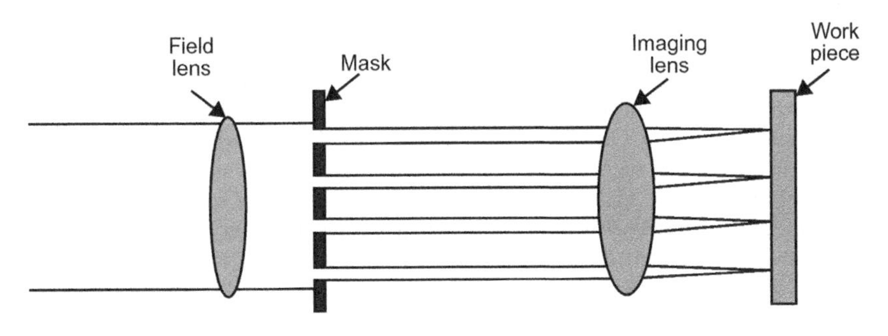

Figure 11.8 Imaged drilling.

diameter. For thin materials with thickness less than or equal to 6 mm, focused spot diameter is generally the same as the hole diameter. Focal spot position is also kept slightly above or below the material surface to achieve best results. In most cases, focal spot is below the surface by a length equal to 5–15% of the material thickness.

Pulse repetition frequency is determined on the basis of required throughput and quality: 5–200 Hz in the case of percussion drilling with Nd:YAG lasers, and up to 1 kHz with CO_2 lasers. CO_2 lasers are used for non-metallic materials.

11.5.4 Advantages of Laser Drilling

There are many advantages of laser drilling as listed below. Laser drilling:

1. is a non-contact process free from the problems of tool wear and tear associated with conventional drilling techniques;
2. has a small heat-affected zone and can be used on a wide range of materials including alloy steels, precious metals, non-ferrous materials, tungsten, molybdenum, tantalum, nickel, beryllium, aluminium, silicon and non-metallic materials;
3. enables drilling in difficult-to-drill materials including super alloys, composites and ceramics;
4. is capable of much higher drilling rates than conventional mechanical drilling;
5. allows operation at high angles, which is extremely difficult with conventional mechanical drilling;
6. is particularly suited for drilling operations in non-conductive substrates or metallic substrates coated with non-conducting materials where electric discharge machining poses substantial problems; it is also suited to drilling in the thermal-barrier-coated superalloys and multilayer carbon-fibre composites used in aerospace applications;
7. works very well for drilling high aspect ratios at shallow angles to the surface;
8. adapts very well to real-time characterization during drilling operations.

11.6 Laser Marking and Engraving

Laser marking is a surface process; the laser beam is focused on the work piece and laser energy is converted into heat energy. This causes an instantaneous rise in temperature over a time period which is much shorter than the thermal conductive time constant of the material. Laser marking is a non-contact process, unlike other marking techniques that either use ink or bit heads. Laser marking is extensively used in a variety of applications such as: adding part numbers, manufacturing and expiry dates on food packages and medicines; adding traceable information for quality control; printing bar codes, logos and other information on products; and marking PCBs and electronic components and modules. Laser marking is used on a wide variety of materials including metals, acrylic, paper, wood, cardboard, plastic materials and leather. Figure 11.9 shows some laser marked/engraved samples.

11.6.1 Principle of Operation

The laser beam from a pulsed laser source is focused on the surface of the work piece to be marked. The laser beam modifies the optical appearance of the surface at the point where the laser beam hits the surface. The modification of optical appearance occurs as a result of the following processes: *material ablation* (also referred to as *engraving*), which removes a coloured surface layer; *melting*, which modifies the surface structure; *slight burning* (also called *carbonization*), usually employed on paper, wood, cardboard or polymers; transformation or *bleaching of pigments* in plastics; *expansion of a polymer* and generation of surface structures such as small bubbles. The laser beam is then scanned using a vector scan or raster scan to mark the object in two dimensions. An alternative approach is to use a mask that is imaged onto the work piece. The process is known as mask marking or projection marking. Although less flexible than scanning process, mask marking is much faster and can even be used on moving objects.

Figure 11.9 Laser engraved samples.

11.6.2 Laser Marking Processes

Commonly used laser marking processes include: laser bonding; laser engraving/vaporization; laser etching/melting; laser ablation; laser coating and marking; laser coloration/annealing; laser engraving and melting; and 3D engraving.

In the *laser bonding process*, pigments and other materials applied to the material surface are bonded by the heat energy generated by the absorption of laser energy. Marking by laser bonding is resistant to heat and is unaffected by fluids and salt. It is particularly attractive for safety-critical parts.

In *laser engraving*, heat energy generated by absorption of laser energy causes the material to vaporize, leaving a mark with very little damage to the material itself. It is not preferred for safety-critical parts.

In *laser etching*, the heat energy causes the material surface to melt rather than vaporize. The melted material expands, producing a raised mark. The laser etching process however changes the surface finish. Also, penetration depth is limited to 2–3 μm. Its key advantage is its speed.

In the *laser ablation process*, a coated material surface is engraved. The technique works well with coatings and paints and other surface treatments. It produces excellent contrast without affecting the underlying material. This technique is popular for marking anodized aluminium and painted steels.

In *laser coating and marking*, the material surface is first coated and etched. *Laser colouration* uses a low-power laser beam moving across the surface, discolouring the material in the process to leave behind high-contrast marks. It is also called charring in plastics and annealing in metals.

Engraving and melting is characterized by material removal and surface melt. *3D engraving* creates very high-quality 3D marks at significant depths.

The efficacy of the above-mentioned laser marking techniques depends on one or more of the physical properties of the material to be marked. The *colour of the material* marginally affects the absorption of laser energy, although dark colours absorb slightly more than lighter colours. Painted colours have little effect, as the paint is usually vaporized away by laser energy.

The *surface finish* has a significant bearing on the contrast of laser mark. A smooth surface produces excellent readability even with shallow engraving. On the other hand, a rough surface requires deeper engraving for good contrast (assuming that the laser does not induce a colour change). *Reflectivity* should be small, as laser energy needs to be absorbed for marking. Higher reflectivity materials require larger laser power. Variation of *absorptivity* as a function of temperature is another parameter that affects performance. While on one hand it may force the material to go to runaway condition, a reduced power may not be able to achieve the desired initial temperature rise. High *thermal conductivity* is also undesirable as it tends to conduct heat energy away from the point where the temperature is required to rise in order to leave a mark. For all practical purposes, material hardness does not affect marking efficacy.

11.6.3 Lasers for Marking and Engraving

Commonly used lasers for marking applications include flash- and diode-pumped pulsed Nd:YAG lasers operating at 1.064 μm and their harmonics at 532 nm, 355 nm and 266 nm; ytterbium-doped pulsed fibre lasers operating at 1.07 μm; pulsed CO_2 lasers operating at 10.6 μm; and excimer lasers operating in the ultraviolet wavelength band. CO_2 lasers are less suitable for marking metals due to the small absorptivity of metals at the CO_2 laser operating wavelength of 10.6 μm. CO_2 lasers can be used to mark all materials except metals and are widely used to mark electronic components and packaged goods using both imaged mask (Figure 11.10) and dot matrix formats. They are also used to mark anodized aluminium and other materials that show good absorption at larger wavelengths, and in applications demanding higher average power laser.

Nd:YAG and fibre lasers are more popular for marking due to the strong absorption characteristics of most materials at 1.0 μm. For precious metals such as gold, which exhibits low absorption at 1.0 μm, a frequency-doubled Nd:YAG laser emitting at 532 nm is used. Q-switched Nd:YVO_4 lasers may be used where a high repetition rate of greater than 100 kHz is required. Figure 11.11 depicts a laser marking station, Model No. LWS 780 from M/s Trotec Laser Inc. Depending upon the intended application, the marker station is configurable with both the CO_2- as well as the fibre-laser-based marking heads of the SpeedMarker series from the same company. The two types of laser marking heads are shown in Figure 11.12 . The laser marking station can be used for marking components in machine and tool construction, in medicine and in the electronics industry. State-of-the-art laser marking machines are usually equipped with the required hardware and software to automate all operations of the machine. Typically, the design to be marked or engraved may be stored in the computer that controls all operations to mark the design on the object.

Excimer lasers such as KrF lasers emitting at 248 nm are used in a few niche areas where a UV wavelength is necessary to make a mark. They are generally used for marking part numbers on small ceramic chip capacitors, medical catheters and fluoropolymer-coated aircraft wire by virtue of titanium dioxide additives using imaged mask technique. KrF lasers are used to inscribe gratings inside doped optical fibres. In this case, UV light is imaged inside the fibre using a diffractive mask.

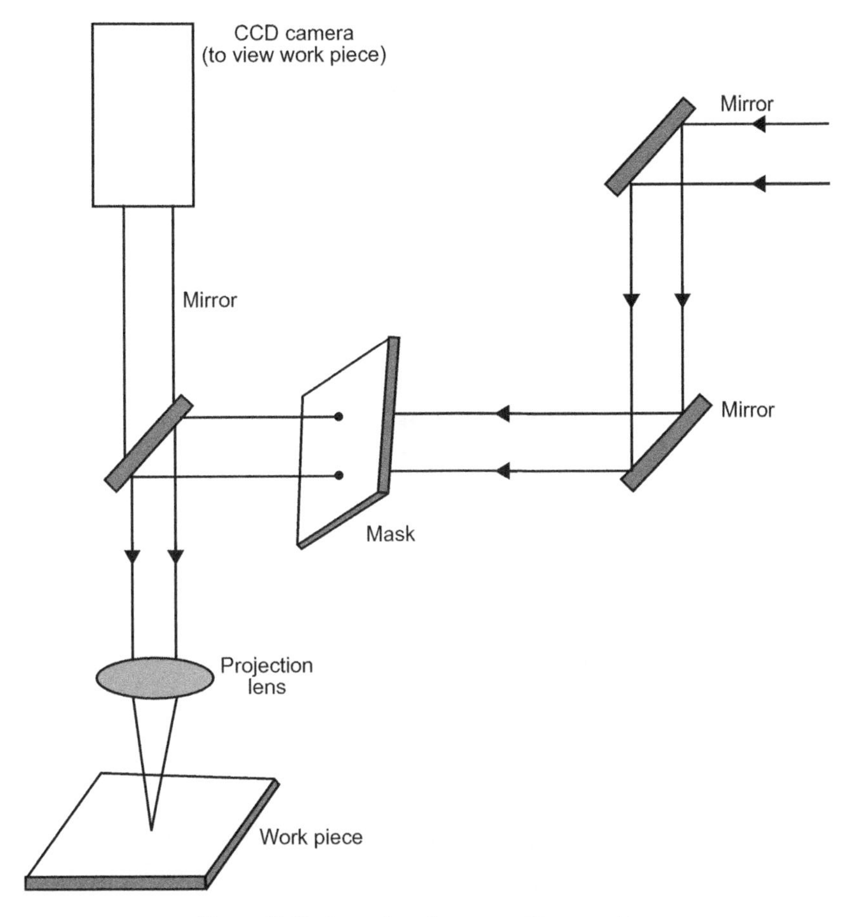

Figure 11.10 Imaged mask process of laser marking.

11.6.4 Advantages

Laser marking and engraving offer a large number of advantages over conventional marking and engraving techniques, as listed below:

1. Conventional marking/engraving employs toxic chemicals, leaving behind a chemical residue. In laser engraving, no toxic chemicals are used and there is no chemical residue left after engraving.
2. Laser marking is a non-contact and a quiet process and there is no requirement for any special environment.
3. Since the laser marking machine has no direct contact with the surface of the material to be marked or engraved, it is easier to adapt to uneven surfaces.
4. In conventional engraving, tool tips need to be replaced depending upon the type of surface to be marked or engraved. There is no such associated problem in case of laser marking.
5. Laser marking produces high-quality permanent marks at low operational cost.
6. Laser marking is effective even on difficult-to-access areas and adapts very well to automation.
7. Due to precise beam positioning and localized energy transfer, the heat-affected zone is very small; this practically eliminates any damage to adjacent areas.
8. Laser marking is characterized by high reproducibility, high speed and high throughput.
9. Laser marking can be used effectively on a wide variety of metallic and non-metallic materials of various sizes, shapes and orientations.

Figure 11.11 Laser marking machine Type LWS 780 (Courtesy of Trotec Laser Inc.).

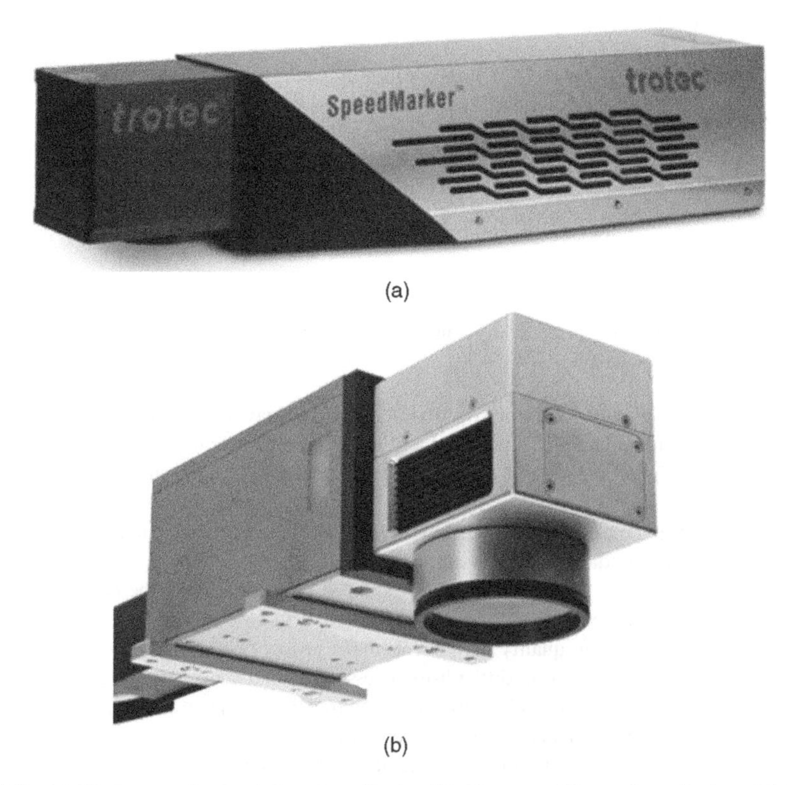

Figure 11.12 (a) CO_2 laser marker head, Type LaserMarker CL (Courtesy of Trotec Laser Inc.); and (b) fibre laser marker head, Type LaserMarker FL (Courtesy of Trotec Laser Inc.).

11.7 Laser Micromachining

Laser micromachining is an established technology for high-precision microfabrication. It is a very powerful technique that allows new device designs to be realized and evaluated quickly, perhaps in a matter of days, in a cost-effective manner. Most industrial applications are centred on the processes of material ablation and melting, which includes cutting, drilling, welding and marking. It is the ability of the ultrafast lasers to etch out extremely small features on a micrometer scale that has spurred on the development in micromachining. Mechanical micromachining processes mainly include microcutting, microdrilling, microwelding, microturning and micromilling.

Figure 11.13 depicts some micromachined samples, covering a variety of applications. The samples shown in Figure 11.13 include a micromachined blackbody emitter (Figure 11.13a), a microlens array

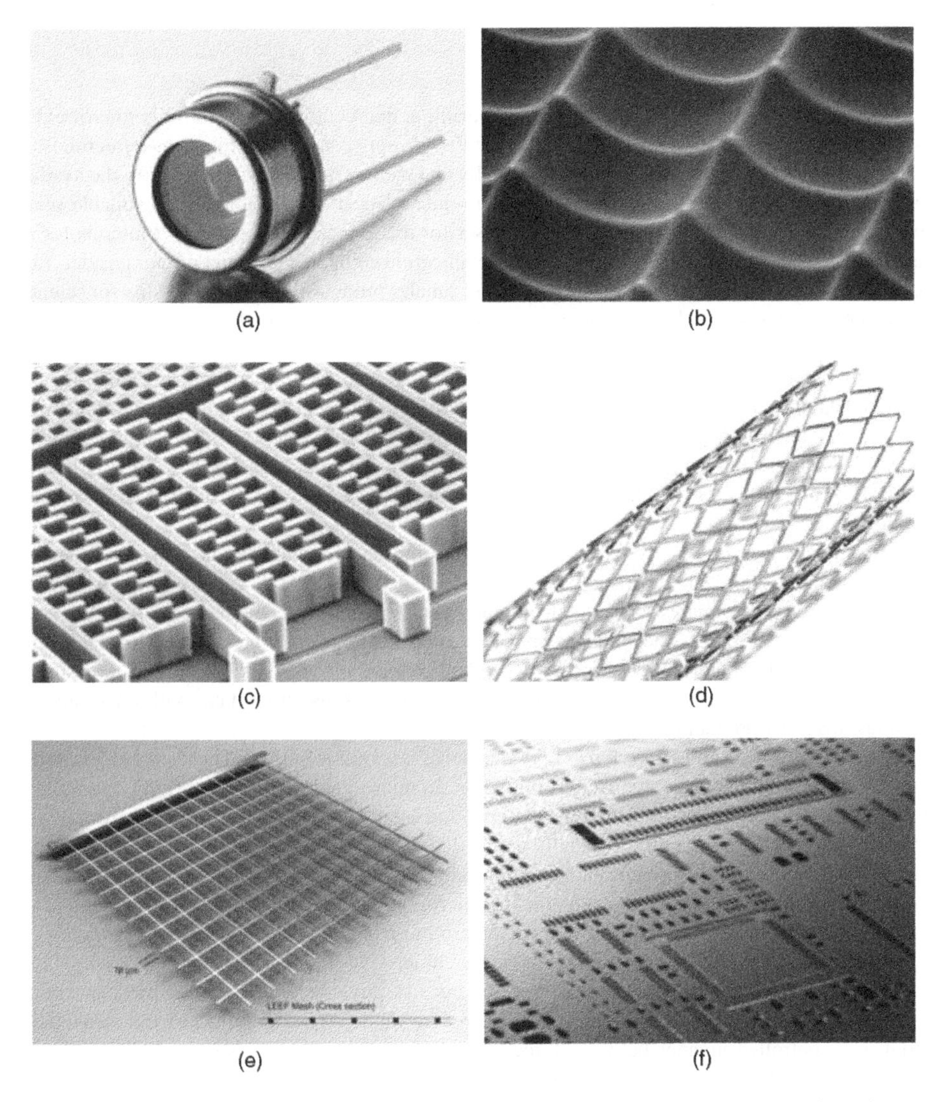

Figure 11.13 Laser micromachined samples: (a) blackbody emitter; (b) microlens array; (c) MEMS accelerometer; (d) stent; (e) wire mesh; and (f) laser-cut stencil.

(Figure 11.13b), a microelectromechanical system (MEMS) accelerometer (Figure 11.13c), a stent (Figure 11.13d), a wire mesh for RFI shielding application (Figure 11.13e) and a laser-cut stencil (Figure 11.13f).

Micromachining processes can be applied to a wide range of materials. These include *metals* such as aluminium, steel and stainless steel, molybdenum, copper, titanium, platinum, gold, silver, nickel, tantalum, chromium and tungsten; *semiconductors* such as silicon, germanium, gallium arsenide and gallium nitride; *ceramics* such as alumina, bulk and thin-film silicon nitrides, silicon carbide, lithium niobate, zirconium and metalized ceramics; *dielectrics* such as glasses, borosilicate, soda lime, sapphire, silicon dioxide, synthetic diamond and fibres; *polymers* such as polyester, polyethylene, polyamides, polyimides, polypropylene, polystyrene, polyurethanes, silicone resin, epoxy resins and elastomers; and other materials such as sol-gel, carbon composites and bio-materials such as cornea and protein. Micromachining processes find extensive application in the automotive, space, biomedical, electronics and micromechanics industries.

11.7.1 Laser Micromachining Operations

Common micromachining operations include microdrilling, microcutting, micromilling, microwelding, marking and engraving, scribing and dicing. *Laser microdrilling* can be used to produce microholes of pinholes, nozzles, orifices, vias and photovoltaic cells in a wide variety of materials. With the available technology, position and diameter tolerances of sub-micron level can be achieved. Available micro-drilling techniques, coupled with an appropriate laser for the job, can be used to drill microholes in a wide range of components and devices including fuel injection components, inkjet printer nozzles, fibre-optic interconnects, vertical probe cards, metred dose inhaler products, pin holes and slits for scientific instrumentation, sensors, fuel cells and medical devices with a very high diameter accuracy and controlled taper.

Laser microcutting is an attractive micromachining technique for most materials, which includes metals (e.g. stainless steel, hardened steel, copper, aluminium, tungsten, brass and titanium); ceramics (e.g. alumina, silicon nitride, tungsten carbide, PZT and zirconium); glasses and crystalline materials (namely BK7, fused silica and sapphire); hard materials (e.g. diamond); and plastics (e.g. polyimide, PTFE, PMMA and ABS). Laser microcutting allows features as small as $<1.0\,\mu m$ wide and 2 mm deep to be machined.

Laser micromilling is ideal for a wide variety of applications including the production of micromoulds and microstamping tools and the creation of microfluidic devices. Laser micromilling is simple in concept. Each laser pulse ablates a very small amount of material. The laser is scanned across the surface to produce the desired feature. The number of pulses at each position determines the depth of the feature. In practice, the quality of the result is strongly dependent on the choice of laser and the scanning strategy adopted. In many cases the morphology of the starting material is also important, with fine-grained or amorphous materials producing the best results.

Laser *micro-engraving* and *micromarking* is a subset of laser micromilling. These processes are used to engrave patterns and shapes on different materials. In micro-engraving and marking operations, a laser beam is scanned over the surface of the material and small amount of material is removed by each laser pulse. The depth of the mark is usually controlled by the number of laser pulses at each point. Common applications of micro-engraving and micromarking include surface microtexturing, decorative engraving and 2D bar codes. State-of-the-art micro-engraving and micromarking machines can engrave microfeatures down to submicron level. MicroStar from M/s Ohio Gravure Technologies Inc (Figure 11.14) is one such laser micro-engraving machine.

Laser microscribing can be used to scribe fine features in many materials including metals, ceramics and glasses. It is commonly used as the first step before dicing. The scribing process weakens the material in a controlled manner before a force is applied to fracture it. It is commonly used in the electronics industry for dicing silicon wafers and ceramic sheets and also in the solar power industry for cutting photovoltaic cells for solar concentrators.

Figure 11.14 Laser micro-engraving machine model MicroStar (Reproduced by permission of Ohio Gravure Technologies, Inc.).

11.7.2 Lasers for Micromachining

The most commonly used laser types for micromachining applications include diode-pumped solid-state lasers, fibre lasers and excimer gas lasers producing pulsed outputs. Although pulsed-output CO_2 lasers are also widely used for cutting applications, particularly for cutting plastic materials, they have limitations in the materials they can machine and the precision of micromachining features they are capable of producing.

Micromachining lasers are almost exclusively pulsed-output lasers as they provide better control of heat input than continuous wave output lasers, which can easily melt small features. Also, the ability to adjust pulse characteristics such as pulse repetition frequency and pulse energy provides process flexibility. It may be mentioned here that shorter wavelengths are key to the production of small feature sizes. Commonly used solid-state lasers include diode-pumped Q-switched Nd:YAG lasers operating at 1064 nm (near IR) and their harmonics at 532 nm (visible), 355 nm (near UV) and 266 nm (UV). A commonly used fibre laser is a pulsed Yb-doped fibre laser operating at 1070 nm. Excimer lasers operate in the UV region; commonly used excimer lasers include ArF lasers operating at 193 nm, KrF lasers operating at 248 nm, XeCl lasers operating at 308 nm and XeF lasers operating at 351 nm. There exists a wide choice of operating wavelengths from the UV to near-IR region.

When choosing an appropriate laser for a given micromachining application, factors such as the nature of job to be performed (e.g. whether surface modification or ablation), whether 2D or 3D, resolution in terms of required minimum feature sizes, area of the sample to be worked upon, material characteristics, process scalability, desired processing speed and operational costs all need to be considered. The nature of the job determines the processing approach to be adopted. Resolution determines laser wavelength, processing technique and optical system. Material absorption characteristics dictate the operational wavelength. For example, ultraviolet wavelength band is preferred for micromachining silicon as silicon absorbs ultraviolet wavelengths better than visible and infrared wavelengths. Laser matter interaction is also influenced by pulse width and energy density, also called fluence. Although there may be more than one type of laser broadly meeting the application requirements, there is always an optimum choice of laser.

The majority of lasers intended for micromachining applications have pulse energy in the range 1–500 mJ, pulse width of 10 ns–10 μs, pulse repetition frequency in the range 10 Hz–500 kHz and average power of 1.0–100 W. If the intended micromachining operation relates to the fabrication of a medical laser, it is important to consider regulatory requirements before finalizing a particular laser for a micromachining application.

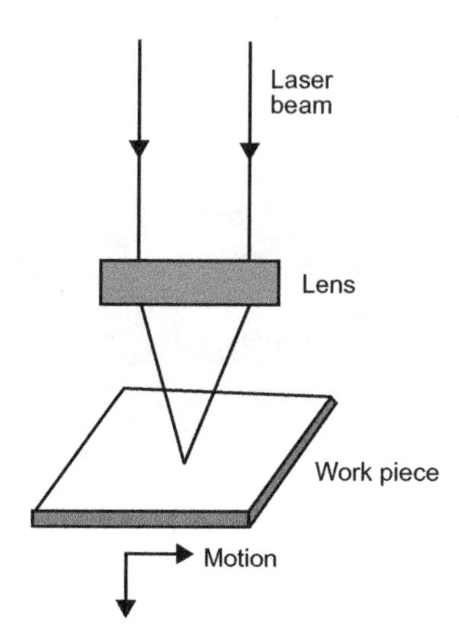

Figure 11.15 Direct-writing micromachining concept.

11.7.3 Laser Micromachining Techniques

There are two broad categories of laser micromachining techniques: *beam focusing* (also called *direct writing*) and *mask projection*. In the case of the former, the laser beam is scanned across the work piece by moving either the work piece or the laser beam or a combination of the two. In the case of the latter, the laser beam illuminates a mask which is then imaged or projected onto the work piece to perform the desired micromachining operation. The two techniques are briefly described in the following.

In the case of the *beam-focusing* (or *direct-writing*) technique, the laser beam is tightly focused to a small spot on the work piece and then scanned across it to perform the desired machining operation. The focal spot diameter here defines the smallest feature dimension. Figure 11.15 depicts the direct-writing laser micromachining system. In view of the requirement of a small focal spot in the direct-writing technique, the lasers capable of producing high beam quality output with $M^2 < 1.3$ are generally used. Focal spot sizes in the range 5–10 μm are common. Commonly used lasers for this technique include solid-state lasers such as Nd:YAG, Nd:YVO_4 and Ti:Sapphire lasers and single-mode RF-excited CO_2 lasers.

There are three system variants of this technique. These include: (a) fixed laser beam position and work piece on CNC-controlled *x–y* stages; (b) static work piece and CNC-controlled galvanometer scanner mirrors to steer the laser beam; and (c) a combination of (a) and (b). Figure 11.15 represents system (a). In the case of beam scanning system (b), the laser beam is rapidly scanned over a small area on the work piece by galvanometer *x–y* scanning mirrors. Desired features are produced by suitably controlling the scanning mirrors with the help of CAD data files. Figure 11.16 shows the schematic representation of beam scanning system. Flat field lenses are used for focusing to ensure that the laser beam maintains a good beam focus while scanning over the area. Scan field sizes of several tens of millimetres are commonly available.

The third type of micromachining system (c) uses a combination of galvanometer scanning mirrors for laser beam steering and *x–y* stages for motion of the work piece. This technique combines the advantages of ultra-high-speed processing of beam scanning and the ability to process large dimensions of greater than 200 mm in the case of direct writing. In the case of beam scanning, the processing dimensions are limited to less than 50 mm by the size of the flat field image size. This allows large samples to be processed at high speed. This technique is usually used when it is feasible to stitch together the separate scan fields. One such example is processing of PCBs, where individual hole positions can be easily defined in each scan field. State-of-the art systems allow use of this technique even for continuous micromachining without it being constrained by the existence of discrete scan fields.

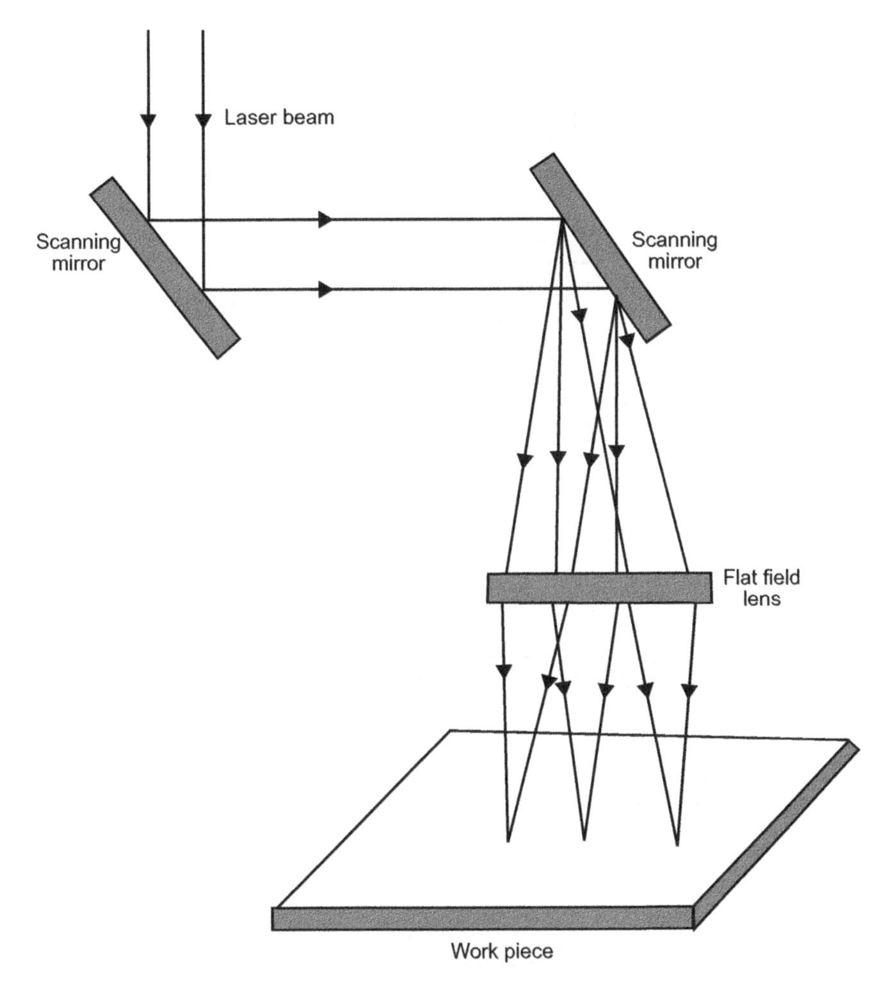

Figure 11.16 Beam-scanning micromachining concept.

Key advantages of the direct-writing micromachining technique include large wavelength coverage from visible to long-infrared, high repetition rates from tens to hundreds of kilohertz, different pulse widths from a few nanoseconds to a few microseconds, high resolution and low operational costs. The variability of laser parameters gives process flexibility to this technique.

The *mask-projection technique* is almost invariably used in all micromachining applications demanding the use of multimode excimer lasers to achieve ultra-high feature resolution. The multimode nature of the excimer laser output lends itself very well to the mask-projection technique. Since the multimode output cannot be tightly focused, the excimer laser is not a feasible option for micromachining by the direct-writing technique. Key attributes of the mask-projection excimer laser include high feature resolution, fine depth control, excellent reproducibility and large sample areas.

Figure 11.17 illustrates the basic concept of the mask-projection technique. The laser beam from the excimer laser is shaped using an expansion telescope and then homogenized with the help of a single or double fly's eye lens array. The purpose of the homogenizer is to produce a laser beam with uniform energy density distribution illuminating the mask. The plane of uniformity of the beam is made coincident with the plane of positioning of the mask. The pattern on the mask is imaged or projected onto the work piece.

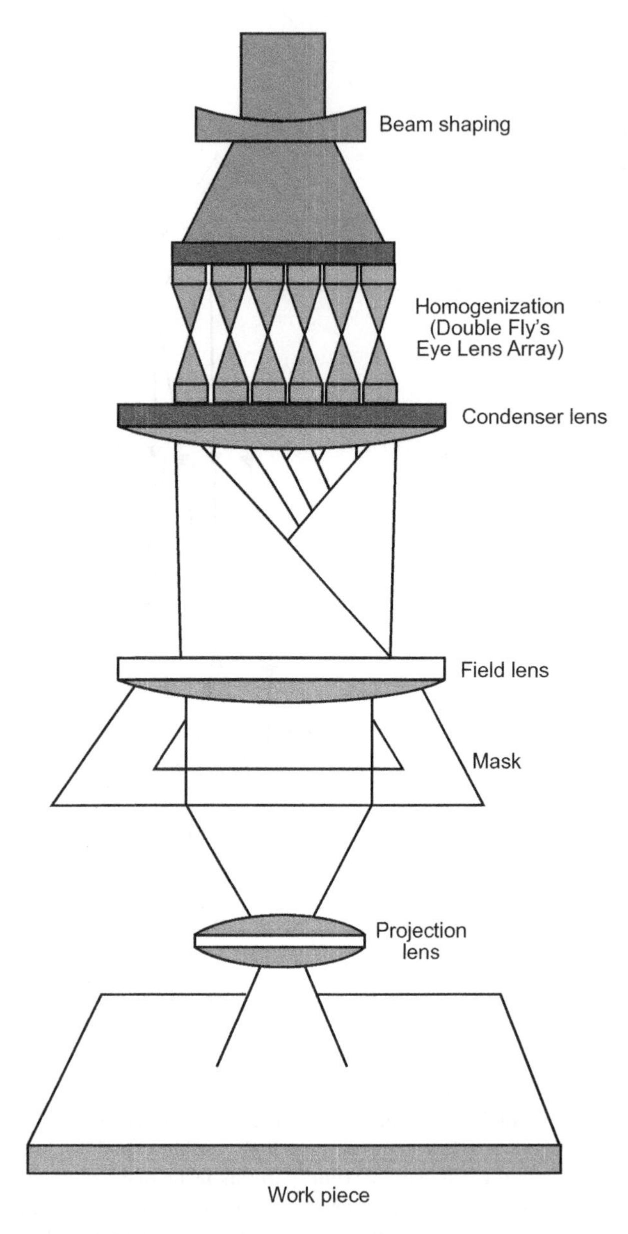

Figure 11.17 Mask-projection technique concept.

Two types of mask-projection techniques are in common use: *standard mask-projection technique* and *synchronized overlay scanning technique*. In *standard mask-projection* systems, while feature resolution is determined by the mask and optical projection system, the depth of the microstructures is controlled by the number of laser pulses fired at a given location. The entire sample area is usually machined under the same laser conditions, which means that all microstructures are machined to the same depth. Although uniformity of depth of different microstructures may be important in most applications, there are many emerging applications demanding tailoring of depth profile of the microstructures across the sample size. Some of these applications include printing devices, multifunctional

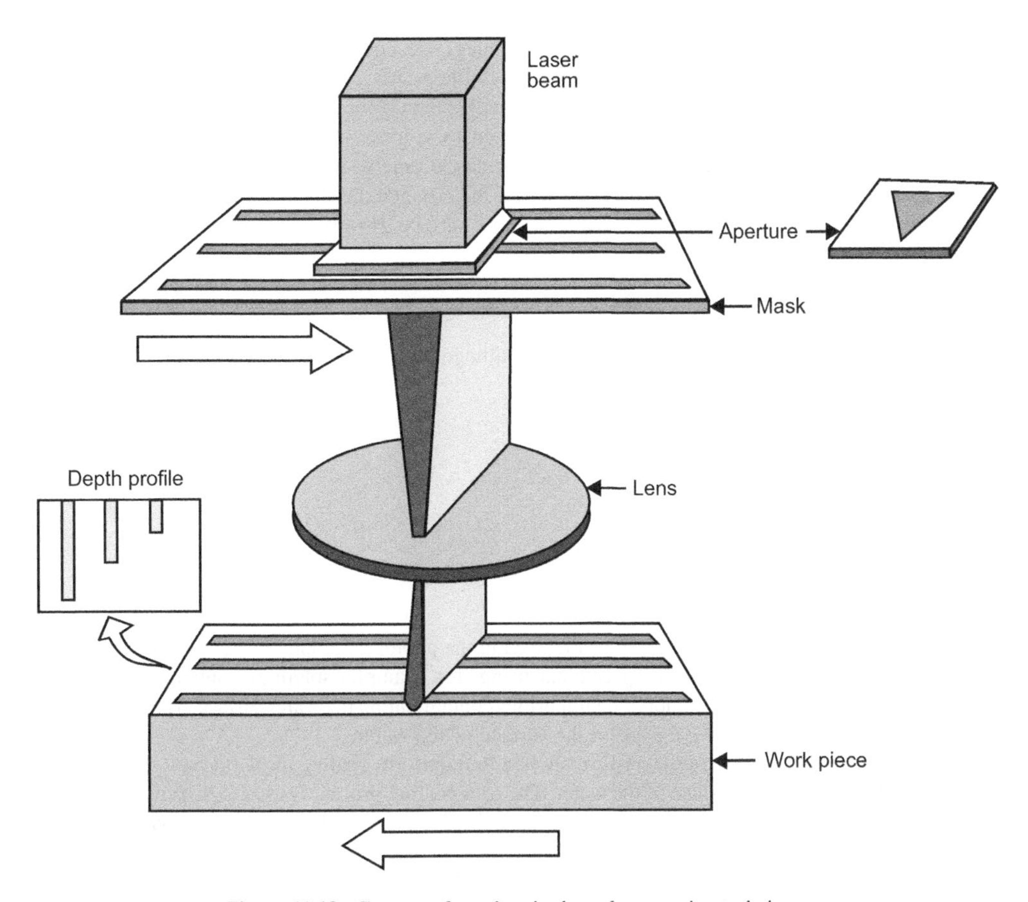

Figure 11.18 Concept of synchronized overlay scanning technique.

units using micro-optical-electromechanical systems (MOEMS) and microfluidic systems. Although the standard mask-projection technique is capable of producing the desired depth profiles by appropriately synchronizing sample position and laser firing sequence, the level of depth profile control needed in many complex jobs of the type outlined above may not be optimally achieved.

The synchronized overlay scanning (SOS) mask-projection technique offers a solution to the limitations of the standard mask-projection technique. For the SOS technique, the mask and work piece are moved in unison in the same manner as for the standard mask-projection approach. In addition, an aperture is used above the mask to additionally shape the beam. The shape of the aperture determines the shape of the beam and hence the depth profile of microstructures. Figure 11.18 illustrates the SOS concept. In the diagram shown, the aperture gives a triagular shape to the beam, resulting in a triangular depth profile. The feature-determining mask can be chosen independent of the aperture shape that determines depth profile. An appropriate combination of mask and aperture can therefore be used to meet the specific requirements of the intended application.

11.8 Photolithography

Photolithography is a type of microfabrication process that uses light to transfer a geometric pattern on a photomask to a light-sensitive material called a photoresist on a part of thin film or bulk substrate. The photoresist material is then developed in a developer solution to form resist patterns. The principle is somewhat similar to that of photography. Photolithography is extensively used in the semiconductor industry for fabrication of complex integrated circuits, in particular the ultra-large-scale integrated circuits (ULSI). In contrast to the photolithography systems of the early 1960s to mid-1980s which employed gas-discharge lamps filled with mercury or a combination of mercury and xenon as the source of ultraviolet light, modern systems use ultraviolet output from excimer lasers such as KrF and ArF lasers. Photolithography can create extremely small feature sizes down to a few tens of a nanometre, can exercise precise control over the size and the shape of features created and is cost effective when used to create patterns over an entire surface. However, it requires extremely clean conditions and a flat substrate.

11.8.1 Basic Process

The various steps involved in one iteration of photolithography include:

1. wafer cleaning;
2. barrier layer formation;
3. photoresist application;
4. soft baking;
5. mask alignment;
6. exposure to UV light and development; and
7. hard baking.

Figure 11.19 illustrates the different steps involved in the photolithography process. In the first step, the wafer is *cleaned* by wet chemical treatment using, for example, solutions containing hydrogen peroxide to remove any organic or inorganic impurities contaminating the wafer surface. The next step is to deposit a silicon dioxide (SiO_2) layer on the surface of the wafer.

After wafer cleaning and barrier layer formation, a thin uniform coating of *photoresist* of thickness 0.5–2.5 μm is applied to the surface of the wafer. The spin-coating process is used to dispense a viscous liquid solution of the photoresist onto the wafer by spinning it at 1200–4800 rpm for a period of 30–60 s. The resulting photoresist layer has a uniformity of better than 10 μm. When exposed to ultraviolet light, the photoresist coating material undergoes a change in its chemical structure and becomes either soluble in a developer solution or is polymerized and becomes very difficult to dissolve.

There are two types of photoresist: positive and negative. A positive photoresist is used whenever the parts of it exposed to ultraviolet light are to be removed. If a negative photoresist is used, the developer solution removes the unexposed parts and the exposed portion remains on the surface. Positive photoresists are more widely used than negative photoresists as they offer relatively better process control for small feature sizes; positive photoresists are therefore the dominant type in the fabrication of very-large-scale integrated (VLSI) and ULSI circuit chips. Figure 11.19a and b depicts the sequence of steps involved in the photolithography process for both positive and negative photoresists, respectively.

The photoresist-coated wafer is then pre-baked (also referred to as *soft baked*), typically at 90–100 °C for 30–60 s to drive off the excess photoresist solvent. The photoresist coating only achieves its photosensitive property after soft baking. Over-soft-baking can degrade the photosensitivity of the photoresist by either reducing developer solubility or destroying a portion of the sensitizer. On the other hand, under-baking can prevent light from reaching the sensitizer.

Mask alignment follows the soft-baking process. A photomask is basically a square glass plate with a patterned emulsion of a metal film on one side. The geometric pattern is transferred onto the wafer surface for subsequent selective etching. The mask is aligned to the wafer. Note that the masking operation is usually carried out several times to complete the entire photolithography process. Each mask after the first must be aligned with respect to the previous pattern.

The next step after mask alignment is exposure of the wafer to intense ultraviolet radiation through the patterned mask. Commonly employed methods of exposure are contact printing, proximity exposure and projection printing. In contact printing, the wafer with the photoresist layer is in physical contact with the photomask during its exposure to ultraviolet light. Contact printing is characterized by very high

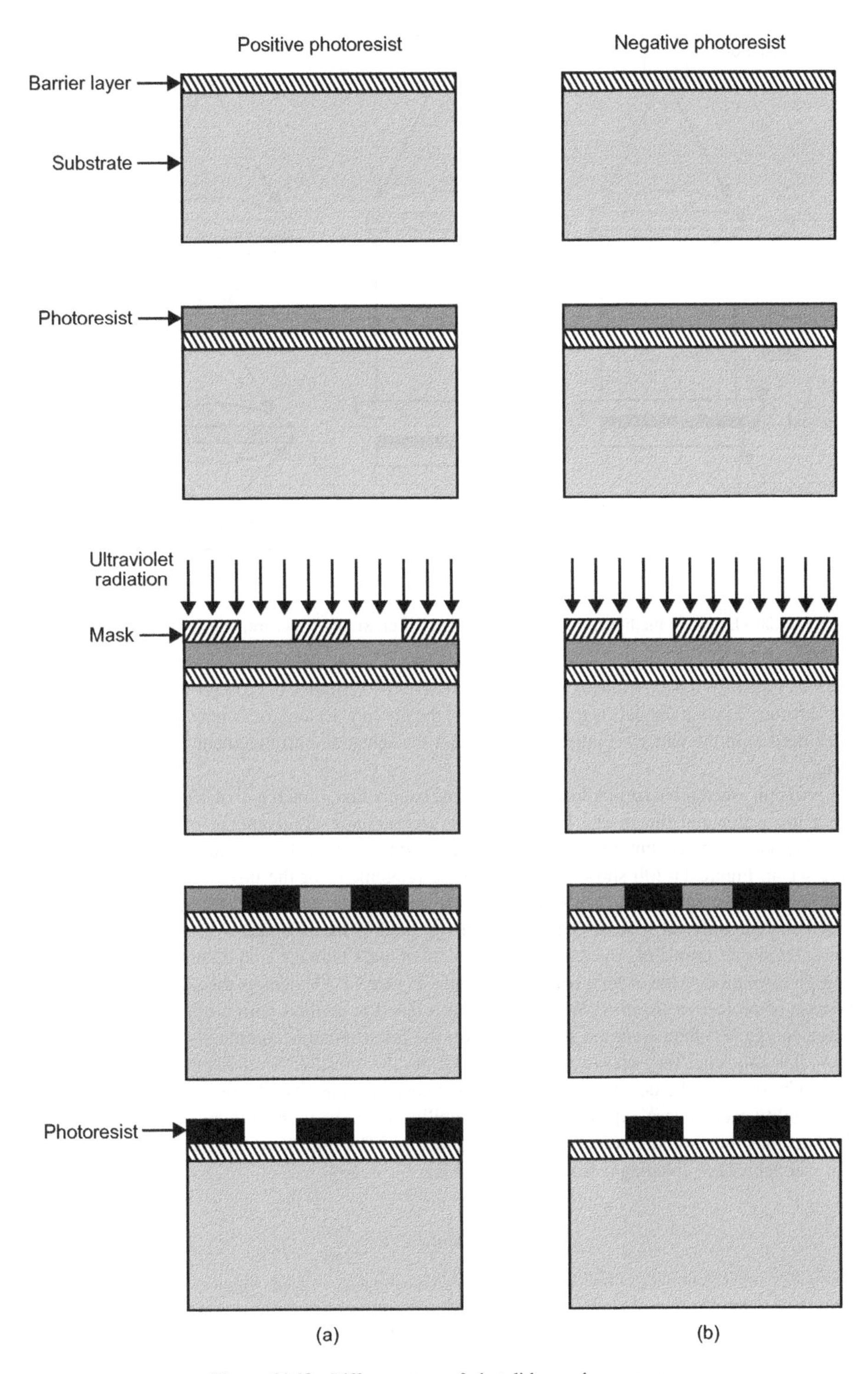

Figure 11.19 Different steps of photolithography process.

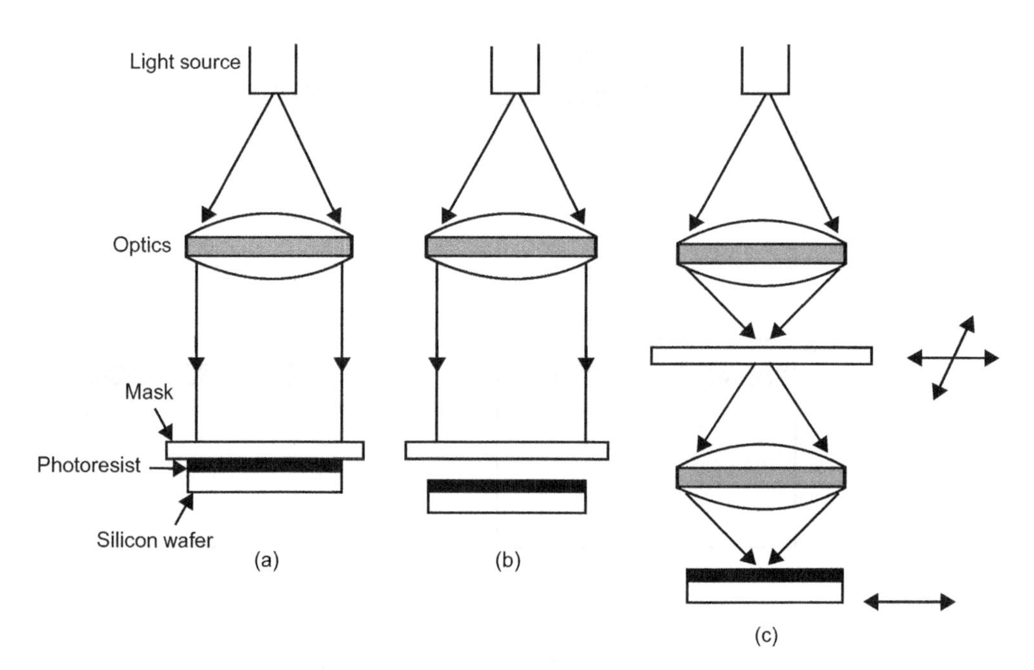

Figure 11.20 Exposure methods: (a) contact printing; (b) proximity exposure; and (c) projection printing.

resolution. Typically, it is possible to achieve feature sizes of 1 μm in a 0.5 μm positive photoresist layer. The disadvantage is that the debris trapped between the photoresist and mask may damage the mask and produce defects in the pattern. Figure 11.20a shows the schematic arrangement in the case of contact printing.

The proximity method is similar to contact printing except that a small gap of 10–25 μm is maintained between the wafer and the mask. This largely overcomes the disadvantage of contact printing by minimizing mask damage, but only at the cost of reduced resolution. The resolution in this case is of the order 2–4 μm. Figure 11.20b shows the schematic arrangement for the proximity method.

In projection printing, wafer is centimetres away from the mask and only a small portion of the mask is imaged onto the wafer at a time. This small image field is the scanned over the surface of wafer. Projection printing completely overcomes the problem of mask damage, and its small image field means that a high resolution of the order 1 μm is achievable. Figure 11.20c depicts the schematic arrangement in the case of projection printing. While using the projection method with a negative photoresist, for exposure energy less than a certain threshold level the resist remains completely soluble in developer solution. It is only when the exposure energy is 2–3 times the threshold level that the photoresist remains mostly undissolved in the developer solution. In the case of a positive photoresist, the resist has some finite solubility even at zero exposure energy. Solubility increases with an increase in exposure energy until it becomes fully soluble at a certain exposure energy. The minimum feature size that a projection system can achieve is calculated from Equation 11.1:

$$\text{Minimum feature size} = \frac{k_1 \lambda}{\text{NA}} \tag{11.1}$$

where λ is wavelength, k_1 is a coefficient dependent upon process parameters (typically 0.4) and NA is numerical aperture of the projection lens as seen from the wafer.

Hard baking is the last step in the photolithography process. The wafer is hard baked at 120–180 °C for a period of 20–30 min. It solidifies the remaining photoresist and improves its adhesion to the wafer surface, which helps in subsequent operations of etching or ion implantation.

Depth of focus is another important requirement, which is calculated from Equation 11.2:

$$\text{Depth of focus} = \frac{k_2 \lambda}{\text{NA}^2} \tag{11.2}$$

where k_2 is another process-related coefficient.

As is evident from Equations 11.1 and 11.2, the two parameters of minimum feature size and depth of focus compete against each other. For a given minimum feature size requirement, depth of focus restricts the thickness of the photoresist and also the depth of topography on the wafer.

11.8.2 Lasers for Photolithography

During the 1960s–mid-1980s, photolithography equipment included gas-discharge lamps filled with either mercury or a combination of mercury and a noble gas such as xenon as the source of ultraviolet radiation. The usually broadband radiation from these lamps has some strong peaks in the ultraviolet band, which can be filtered to produce a single line. Spectral lines from mercury lamps include the g-line at 436 nm, h-line at 405 nm and i-line at 365 nm. In the mid-1980s, mercury-lamp-based lithography equipment began to be replaced by excimer-laser-based equipment. The use of excimer lasers significantly enhanced resolution to produce faster and denser chips by operation at lower ultraviolet wavelengths, and reduced operational costs by increased throughput. Excimer laser lithography has enabled feature sizes as small as 25 nm to be achieved, which is expected to approach 10 nm by 2020.

Commonly used excimer lasers for lithography equipment include ArF and KrF lasers emitting at 193 nm and 248 nm, respectively. The F_2 excimer laser emitting at 157 nm was previously considered to achieve higher resolution than that from ArF and KrF lasers. Due to persistent problems with 157 nm technology and also economic considerations, the industry has continued with the better-established 193 nm KrF laser lithography for high-resolution requirement. The introduction of immersion lithography has also enabled higher-resolution performance due to increased numerical aperture. Note that the minimum achievable feature dimension is directly proportional to operating wavelength and inversely proportional to numerical aperture. In immersion lithography, the usual air gap between the projection lens and the wafer is replaced by a higher-refractive-index liquid, enabling the use of optics with a numerical aperture exceeding 1.0. Use of immersion lithography with the 193 nm KrF laser has replaced the 157 nm F_2 laser lithography equipment for higher-resolution requirements.

Other excimer lasers such as XeCl and KrCl lasers are not used for lithography. While the former is not used due to insufficient resolution capability, use of the latter would necessitate the development of new photoresist materials. The choice of photoresist and optics material have a strong dependence on the operating wavelength. Materials for photoresists and optics are available in the case of ArF, KrF and F_2 lasers.

Figure 11.21 illustrates the excimer laser IndyStar from Coherent Inc., suited to photolithography applications. The IndyStar is available in four variants: two are ArF lasers emitting at 193 nm with a stabilized output pulse energy of 8 mJ at 1 kHz and 4 mJ at 2 kHz; the other two are KrF lasers emitting at 248 nm with stabilized output energy of 12 mJ at 1 kHz and 6 mJ at 2 kHz.

11.9 Rapid Manufacturing

Rapid manufacturing is a collection of technologies developed to shorten the design and production cycle of a product. This is also known by other names such as additive manufacturing, layered manufacturing, solid freeform manufacturing, e-manufacturing or digital manufacturing. These technologies allow rapid prototyping of a new product design, an exercise that is usually carried out before full-scale production with specialized moulds, tools or jigs. With conventional subtractive toolroom fabrication techniques, prototyping is a tedious process involving considerable skilled hand labour in various fabrication processes such as cutting, bending and shaping. Further, prototyping process requires several iterations before the product design is finalized and declared ready for full-scale production. As a consequence, prototyping with conventional technologies is cumbersome and often very expensive.

Figure 11.21 ArF excimer laser Type IndyStar (Reproduced by permission of Coherent Inc.).

Rapid manufacturing is also used for the fast fabrication of tools required for mass production. Rapid prototyping and manufacturing technologies offer the advantages of reduced time to market, reduced fabrication cost and enhaced capability to fabricate complex features that otherwise would not have been possible with traditional machining. These are extensively used for the fabrication of both models and prototypes as well as functional components from metals, plastics and composite materials in the military and the aerospace industry. The technologies are also well established for the fabrication of small products and microsystem applications in medicine and in the consumer electronics industry.

11.9.1 Additive Versus Subtractive Manufacturing

Manufacturing technologies are broadly categorized as *additive* and *subtractive*. In the case of additive manufacturing, the technology used in rapid manufacturing, a 3D CAD model of the product is transformed into thin horizontal virtual cross-sections. The machine reads in the transformed CAD drawing data of the product to be manufactured and deposits successive layers of liquid, powder or sheet metal to build the final model. These different layers correspond to different horizontal virtual cross-sections.

The term subtractive manufacturing has been coined only recently to distinguish it from additive manufacturing. It refers to traditional CNC (computer numerical control) machining methods such as cutting, milling and shaping, involving the removal of material rather than addition of it. In the case of subtractive manufacturing, a cutting machine such as a lathe or milling machine is commanded by a computer to cut a specific shape. The process usually involves a large number of different steps and several changes of cutting tool. In additive manufacturing, the final product is built by adding material instead of cutting it away, enabling a much wider range of shapes including cavities or intricate geometries that would otherwise be impossible to machine.

11.9.2 Rapid Manufacturing Technologies

The various types of rapid manufacturing technologies differ mainly in usable materials and in the way different layers are deposited to create the final product. Commonly used rapid manufacturing technologies include the following:

1. stereo lithography;
2. selective laser sintering;

3. shape deposition manufacturing;
4. laminated object manufacturing;
5. 3D printing.

In *stereo lithography*, the base material used is photosensitive polymer resin. A UV laser beam is used to scan the liquid surface of a resin bath. The laser radiation causes the resin to cure in the shape of a layer of the part to be manufactured. Different layers combine to form the desired 3D shape of the final part.

In *selective laser sintering*, material in the powdered form is deposited on a platform. A CO_2 laser is used to selectively melt or sinter the powder into the shape of a layer. As successive layers are formed, these are lowered onto the platform. The loose powder material around the growing structure acts as a support for the top powder layer. Selective laser sintering can be employed to a wider range of base materials including thermoplastics and metal and ceramic powders. Laser power and scanning speed can be varied to control porosity and strength of the material. The technique is commonly used for the fabrication of turbine rotors and medical inserts.

Shape deposition manufacturing combines the deposition process with CNC machining. In this technique, each successive layer is deposited and machined. After every deposition and machining operation, new material is added and part machined again before the next layer is deposited. Incremental machining allows the use of thick layers while maintaining a smooth finish. Use of support material enables layers with overhanging, undercut and separated features to be supported during fabrication.

In the *laminated object manufacturing* process, parts are made from a sheet material bonded together in layers to form a laminated structure. Commonly used base materials include plastic, water-repellent paper, ceramic and metal powder tapes. The technique is used for fabrication of casting dies for automotive parts.

3D printing is similar to inkjet-printing technology. A group of print heads is scanned across a powdered material bed. During the scanning operation, the print heads distribute a liquid binder that binds the material to form the desired shape of a layer. The process is repeated to form successive layers lowering the part and adding additional powder before each successive layer is formed. After all layers are formed, the part is removed from the powder bed and cleaned. 3D printing technology is extensively used in the fabrication of functional metal parts and mould inserts.

11.9.3 Lasers for Rapid Manufacturing

Commonly used laser types for rapid manufacturing include pulsed and CW CO_2 and Nd:YAG lasers emitting at 10.6 μm and 1.064 μm, respectively. Use of semiconductor diode lasers emitting at 810 nm have also been experimented with. The key disadvantage with a diode laser is its low beam quality.

Most commercial rapid manufacturing machines are equipped with CO_2 lasers as they have a higher efficiency and low operational and maintenance costs compared to Nd:YAG lasers. Output laser power is typically in the range 50–500 W. Higher-power CO_2 lasers with output power level in the kilowatt range are also used.

Pulsed and CW Nd:YAG lasers are also widely used for rapid manufacturing, particularly parts made of metals and metal alloys, due to much higher absorptivity and better coupling efficiency of Nd:YAG wavelength of 1.064 μm compared to 10.6 μm of CO_2 laser. This leads to a higher melting depth for the same power density of Nd:YAG laser compared to CO_2 laser. Another advantage of the Nd:YAG laser is the possibility of using optical fibre for output beam delivery. Studies have shown that, for the same output laser energy of Nd:YAG and CO_2 lasers, the former produces higher energy density, deeper sintering depth and a larger processing window. Compared to CW Nd:YAG lasers, pulsed lasers offer better metallurgical bonding of tracks and layers by using high-pulse-energy and narrow-pulse-width output. Q-switched Nd:YAG lasers producing nanosecond pulses at repetition rates of the order of tens of kilohertz have been used.

Figure 11.22 depicts a laser prototyping machine, Model EOS P760 from EOS GmbH. The machine employs selective laser sintering technology and uses twin CO_2 lasers to fabricate single-piece large parts with dimensions up to 700 mm × 380 mm × 580 mm.

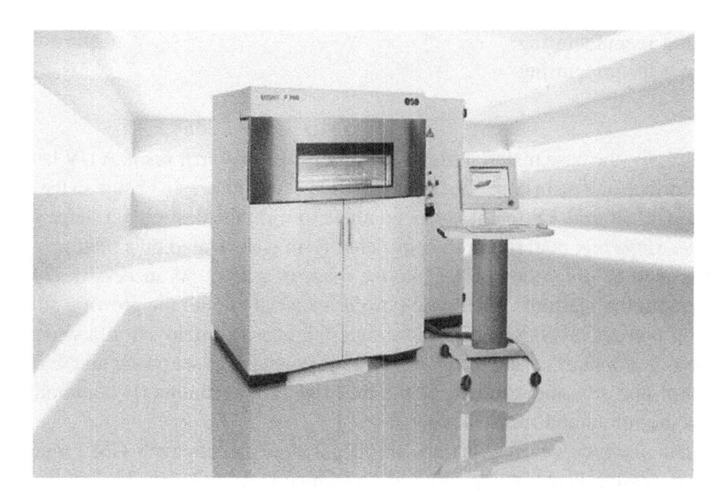

Figure 11.22 Laser prototyping machine, Model EOS P760 (Reproduced by permission of EOS GmbH).

11.9.4 Advantages

There are many advantages of rapid manufacturing, which is characterized by its high accuracy, low cost and reduced time to market. Rapid manufacturing:

- can be used to process a wide range of materials that can be melted and fused by a laser into a useful component in a single step;
- offers process flexibility due to the possibility of adjustment of process parameters such as laser power, feed rate and layer thickness;
- allows the same equipment to be used in the processing of diverse materials;
- can process different materials (or gradations of materials) within a single component to enhance certain properties; and
- allows component features that are either impossible or difficult to machine or required joined assemblies of subcomponents to be fabricated in a single step.

11.10 Lasers in Printing

Laser printer is a common type of computer printer that uses a non-impact photocopier technology to produce high-quality printing of both text and graphics on paper. The process is very similar to xerography (also known as electrography) with the printing mechanism involving the same basic steps: charging, exposure, development and transfer and fusing.

Laser printing technology was invented in 1969 at Xerox Corporation, an American multinational company engaged in the manufacture of a wide range of monochrome and colour printers, copier machines, document management systems, multifunction systems and digital production presses. The first prototype was created by modifying a photocopier machine in 1971 at Xerox. IBM introduced the first commercial laser printer in 1975 for use with its mainframe computers, followed by the first LaserJet printer for use with personal computers in 1984 by Hewlett-Packard.

Early laser printers were restricted to printing letters, office memos, spreadsheets, white papers and other routine business requirements, but today's laser printers provide advanced graphics functions such as desk top publishing, the creation of promotional materials and customized designs. This has become possible due to the simplicity and user-friendliness of laser printers, the availability of multiple functions such as printing, copying and scanning in a single integrated machine, the improvement in the specifications of existing features such as resolution and print speed, the addition of new features and their reduced price tag.

11.10.1 Laser Printing Process

The complete laser printing process can be described as a set of operations, including: charging, data transfer, drum writing, paper feed, toner pick-up, image transfer, fusing and drum preparation. These steps are briefly described in the following.

In the first step, the computer generates a query to make sure that a printer is attached to the output port and is ready. The printer sends a signal back indicating it is ready for data. The computer sends the print job to the printer, which is stored in the printer's memory. In most laser printers, some memory is built into the system board but they also allow external memory modules to be attached. Additional memory allows larger print jobs to be stored in the printer's memory.

In the next step, the photosensitive drum is charged. The photosensitive coating on the drum is usually of an organic photoconducting material. The organic photoconductor has insular-like properties when in the dark, and is often negatively charged due to its photoconductive properties. In earlier machines, electrostatic charging was carried out by a contact-free corona discharge method; in recent times, it has become common practice to use contact-charging methods involving charging by a roller or a brush, producing significantly lower ozone. The electrostatically charged drum is then ready to receive the image stored in the form of data in the printer's memory.

In the third step, the data stored in the printer's memory is written onto the drum using a laser. The data stored in the printer's memory is used to control the on and off operation of the laser by selectively hitting the charged drum at certain spots, thereby changing the electrical charge of those spots. This leads to the formation of a latent image on the drum, which is a negative of the print image.

In the fourth step, the photoconducting surface with the latent image is exposed to the toner. The toner particles are given a negative charge, so are electrostatically attracted to areas on photoconductor's latent image left neutralized by the laser. The negatively charged toner does not attract areas on the drum that have retained negative charge.

In the fifth step, paper is fed in with the help of a set of rollers. The paper is charged before it comes into contact with the photoconductor drum. As the sheet of paper completes a rotation, it magnetically attracts the toner particles from the drum, thereby transferring the image to the paper. Charged areas of the drum do not attract toner, and result in white areas on the paper.

At this stage, the toner is loosely held onto the paper by gravity and a weak electrostatic charge. For permanent adherence, it must be fused. The fusing operation is basically the process of melting the toner's plastic particles so they stick or fuse to the fibres in the paper. The fusing operation is performed with a non-stick roller with a high-power lamp that heats the paper to about 160–180 °C. A pressure roller in the fusing assembly, which is a rubber roller, presses against the fuser roller. The paper feeds between the pressure and fuser rollers on its way through the printer.

The last step involves cleaning the drum to remove any traces of previous print and prepare for the next printing operation. First, a rubber blade wipes the excess toner from the drum. This is followed by electrostatically cleaning by discharge lamps, which neutralize any residual electrical charges on it.

Figure 11.23 depicts the laser printer hardware, illustrating the different operations involved in laser printing.

11.10.2 Anatomy of Laser Printer

A laser printer has four subparts: paper feeder unit, processing unit, print head and fuser unit. The *paper feeder unit* feeds the paper into the print mechanism and comprises paper tray, paper take-up roll and paper take-up switch. The *processing unit* comprises toner cartridge, developer and drum. The *toner* is a mixture of fine particles of dry plastic powder and carbon black or colouring agents. The colouring agent is blended into fine particles of plastic powder. The plastic particles melt during the fusing operation due to heating, thereby binding the colouring agent to the paper. The *developer* supplies toner to the drum and makes sure that it is uniformly spread over the photoconductor drum. The *drum* consists of an aluminium cylinder coated with a photosensitive material such as selenium (used in earlier printers) or organic

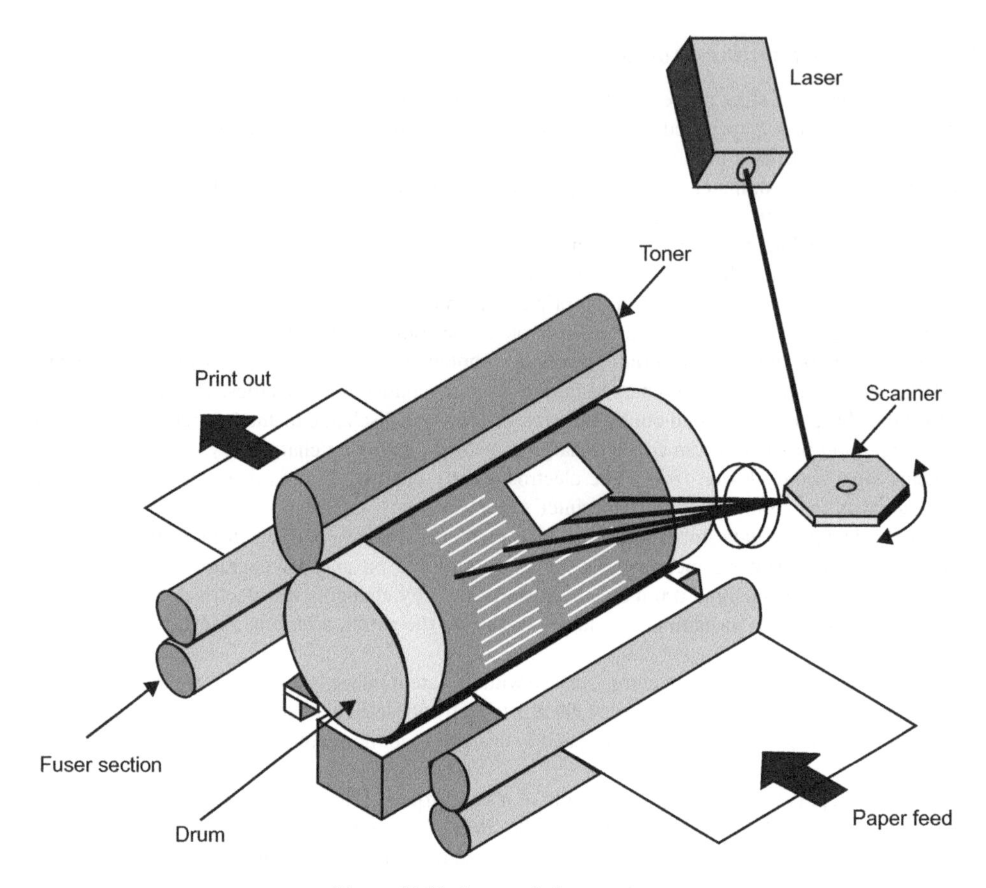

Figure 11.23 Laser printing process.

photoconductors (used in modern printers). Commanded by data representing the document to be printed, a laser beam produces a latent image on the drum. The drum subsequently transfers this image to the paper. The *print head unit* consists of the laser and beam delivery optics, and is used to create an electrostatic image on the drum. The *fuser unit* performs the function of binding the loosely attached toner to the paper.

11.10.3 Choice Criteria

Important laser printer features that should be considered when choosing a printer for your application include resolution, print capacity and speed, paper handling, FPOT (first-paper-out time) and warm-up time, network compatibility and printer languages.

Printer *resolution* describes the number of dots that can be put onto the paper within a square inch. It is usually measured as dpi (dots per inch). Some printers have different resolution in the vertical and horizontal directions. A resolution of 720×360 implies 720 dots per inch in the vertical direction and 360 dots per inch in the horizontal direction. The standard resolution in most laser printers today is 600 dots per inch. This resolution is sufficient for normal everyday printing, including small desktop publishing jobs. A high-end printer intended for heavy-duty production jobs would usually have a resolution of 2400 dpi. Resolution of 300 dpi or lower causes uneven lines to appear on the outer edge of an image. Some printer manufacturers use resolution enhancement technology. If you must buy a printer with a resolution of 300 dpi, make sure that it is equipped with this technology.

Print speed and *capacity* is another important feature. Low-end printers print up to 8 ppm (pages per minute). Home and small office printers print up to 24 ppm and are required to print up to 1000 ppm. Production printers used by large commercial houses may be required to print 50 000 pages or more per week and work 24 hours a day, 7 days a week. These printers are capable of printing up to 1000 ppm or even more.

Paper handling capability is an important consideration when choosing a printer. Most laser printers can only print on letter-size or A4-size paper. High-end printers can print on various paper sizes including legal and ledger paper, A3 and other larger paper sizes, transparencies, adhesive labels and lightweight cards. Most laser printers print on one side of paper with the facility for manual duplex printing. Manual duplex printing is achieved by changing the print options in the printer's properties or printing one side and taking that same paper and reinserting it into the printer to print on the other side. Some printers have a duplex printing facility, in which the printer can print on both sides of paper without any manual intervention.

Other considerations when choosing a printer are FPOT, warm-up time (the time for the fuser to warm up to the operating temperature) and the time taken by the printer to prepare itself to print a new job (typically 5–30 s). If the printer has a standby mode or is turned off between printing jobs, the warm-up time becomes even more important. Large production printers take up to several minutes to warm up, which can slow down the printing process.

Laser printers can either have a toner and drum integrated within a single printer unit, or be provided with separate toner and drum units. The latter have the advantage of reduced operational costs, as the toner would usually run out much earlier than the drum. It is a lot easier to replace integrated package. Laser printers can also be shared in a network (although some of the low-end computers can only be used by a single computer). The replacement cost of consumables such as toner cartridge and drum is also a consideration.

11.10.4 Laser Printers vs Inkjet Printers

Laser and inkjet printers can be compared and contrasted on the basis of initial and operating costs, print speed and quality and networking compatibility.

Inkjet printers have low initial but high operating costs, compared to the high initial and low operating costs of laser printers. The purchase cost of a low-end inkjet colour printer is in the range of US$30–40 as opposed to US$300–350 in the case of colour laser printers. Laser printers are a better option for heavy-duty business applications as they are designed to handle a high volume work. The toner cartridge offers higher capacity in terms of number of pages it can print before needing replacement. In contrast, ink cartridges require to be replaced quite often when in regular use for printing jobs. On the other hand, although toner cartridges are more expensive than ink cartridges, they hardly need to be replaced, lowering the total cost per page.

Laser printers are far more superior to inkjet printers when it comes to printing speed for a specified print quality. Inkjet print speed is usually specified for draft mode or quick-print mode. The speed drops significantly for high-quality printing. Laser printers are preferred over inkjet printers when required to handle large volumes of print jobs, and are more suited to general office use than inkjet printers. Inkjet printers are generally preferred by individual users for lesser printing needs.

Laser printers offer higher resolution than inkjet printers and, as a result of this, print a better-quality text. High resolution also helps the laser printers create precise fonts without fuzzy edges, an effect which is more pronounced when printing text in smaller font sizes. However, relatively lower-cost inkjet printers are usually recommended for inexpensive colour printing. Laser printers can produce good-quality prints on all kinds of printing paper, whereas inkjet printers will require special inkjet paper to produce good-quality prints.

High-end laser printers usually come with networking facilities and can therefore be directly connected to a computer network. On the other hand, inkjet printers cannot be directly connected to a network; they are networked via a computer.

11.11 Summary

- Laser-based material-processing applications can be classified into (1) applications requiring limited laser power/energy and causing no significant change in phase or state of the material in solid state and (2) those involving phase changes and requiring higher power/energy to induce those changes from solid to liquid or solid to vapour phase.
- Different material-processing applications can be better understood by analyzing the three basic phenomena of transmission, reflection and absorption. The quantum of laser energy absorbed is of primary interest in material-processing applications and mainly depends upon some laser parameters such as wavelength, pulse energy, pulse width, repetition rate, M^2 value and material properties.
- Some common material-processing applications of lasers include cutting, welding, drilling, marking, photolithography, micromachining, rapid manufacturing, photolithography and laser printing.
- Laser material processing is a non-contact operation. It offers high processing speed and positioning accuracy. It is flexible and versatile. It is easily adaptable to automation. It has a narrow heat-affected zone. It can be used to machine a wide range of materials and also difficult-to-work materials.
- Laser cutting involves use of a high-intensity infrared laser beam focused on the surface of the work piece with the help of optics. This in turn heats up the material and creates a localized melt throughout the depth of the work piece. A gas jet acting coaxially with the laser beam removes the molten material. The laser beam is moved across the surface to produce the cut. Laser cutting yields high-speed cutting, a narrow cut width and reduced material waste. The cutting process is far safer than its mechanical counterpart, and it can be a fully CNC non-contact process.
- Laser welding involves use of high-energy pulsed laser beam focused on the surface of the work piece with the help of either bulk optics or fibre-optics module. The high energy density in the focused laser beam drills into the metal, vaporizes and melts some of the metal; the joining of the two parts is established as the common melt solidifies. The advantages of laser welding include low heat input and small heat-affected zone, which makes it ideal for products that require welding near electronics, glass-to-metal seals and so on.
- Laser drilling generally occurs through the processes of melting and vaporization (also called ablation) of the work piece when the energy of a focused laser beam is absorbed. The drilling operation largely depends upon pulse energy and duration. Present-day applications include drilling of cooling holes in aircraft engine blades and combustors, inkjet nozzles, fuel injector nozzles, cigarette filter paper and vias in printed circuit boards.
- Laser marking is one of the most widely exploited industrial applications of lasers. Laser marking has various forms such as an alphanumeric code imprinted on a label or surface of product carrying information (e.g. manufacturing data, expiry date, serial/part number), a machine-readable barcode or a 2D symbol. It is primarily a surface process; the pulsed laser light is absorbed by the work piece and is converted into heat. The short pulse width in the range of tens of nanoseconds to several microseconds causes an instantaneous rise in temperature, leaving a resultant mark.
- Laser micromachining is an established technology that finds application in high-precision micro-fabrication. Laser micromachining offers many distinct advantages over conventional mechanical, chemical, ion or particle beam processes, including high-resolution non-contact processing, unlimited material coverage, the ability to machine flat and multicontoured surfaces and multiprocess compatibility. Commonly used lasers for micromachining applications are excimer, Nd:YAG, CO_2 and metal vapour lasers.
- Laser rapid manufacturing is also known as additive manufacturing, layered manufacturing, solid freeform manufacturing, e-manufacturing or digital manufacturing. Unlike subtractive machining that removes material to form the final part (e.g. CNC machining), this is an additive process that deposits layers of material to form the final part. The laser is used as a heat source.

- Photolithography is an important process used in the semiconductor industry for the fabrication of semiconductor devices and integrated circuits to selectively remove parts of a thin film or bulk of a substrate. Photolithography makes use of light to transfer a geometric pattern from a photomask onto a photoresist on the substrate. Chemical treatment is then used to enable deposition of a new material in the desired pattern on the substrate material. Excimer lasers emitting in the ultraviolet wavelength band are the preferred laser sources in photolithography, which have largely replaced conventional gas-discharge mercury lamps emitting in ultraviolet band.
- Laser printing uses a non-impact photocopier technology very similar to xerography with the printing mechanism involving the same basic steps of charging, exposure, development and transfer and fusing. Laser printers have higher print capacity and print speed, higher resolution and higher initial but lower operational cost compared to inkjet printers.

Review Questions

11.1. Outline the important considerations in terms of laser parameters and material properties when it comes to the use of lasers for material-processing applications. Briefly describe the laser parameters which are particularly important for material-processing applications.

11.2. What are the common laser material-processing applications? Briefly describe the basic process behind each of these applications.

11.3. What are the available laser cutting processes? Briefly compare the different cutting processes in terms of their suitability for various types of materials.

11.4. Briefly describe the heat conduction and deep penetration methods of laser welding. What are the advantages offered by laser welding over conventional methods of welding?

11.5. What are the advantages of laser drilling? Describe the basic drilling operation. Name at least three application areas where laser drilling has made a significant difference.

11.6. Briefly describe commonly used laser marking processes. Name the types of lasers along with major parameters generally used for laser marking operation.

11.7. What is laser micromachining? Name several different micromachining operations and the types of lasers suitable for performing these operations

11.8. Briefly discuss the following:
a. laser cutting versus plasma cutting
b. additive machining versus subtractive machining
c. excimer lasers and photolithography
d. rapid prototyping and its advantages.

11.9. With the help of a schematic diagram showing different parts of a laser printer, describe the different steps of the laser printing process from paper feed to print out.

11.10. Name different types of lasers commonly used for laser material-processing applications and the applications they are mainly suited to. Briefly describe different laser parameters that would play a significant role in different applications.

11.11. Name at least five material-processing applications that would be extremely difficult to execute using conventional machining techniques. Briefly describe the laser-based machining method used to perform these applications.

11.12. Name different materials that would be extremely difficult to process using conventional machining techniques. Name the lasers suitable for working on these materials, with particular reference to the laser parameters.

Self-evaluation Exercise

Multiple-Choice Questions

11.1. When the laser beam falls on the work piece during a material-processing operation, the dominant phenomenon is:
a. transmission
b. reflection
c. absorption
d. all of the above.

11.2. Laser-based material processing is characterized by:
a. high throughput
b. minimized collateral damage
c. non-contact operation
d. all the above.

11.3. Rapid manufacturing is also known as:
a. additive manufacturing
b. e-manufacturing
c. layered manufacturing
d. all of the above.

11.4. Commonly used lasers for material-processing applications include:
a. Nd:YAG
b. Nd:YAG, CO_2 and excimer lasers
c. Nd:YAG and He-Ne lasers
d. CO_2 and metal vapour lasers.

11.5. A laser cutting technique that makes use of the combination of melt shearing and chemical reaction is called:
a. fusion cutting
b. vaporization cutting
c. oxidation cutting
d. none of these.

11.6. When the laser beam diameter is much smaller than the diameter of the hole to be drilled, the drilling technique used is called:
a. percussion drilling
b. trepanning
c. imaged drilling
d. drilling-on-the-fly.

11.7. Processes such as material ablation, bleaching of pigments and carbonization are common in:
a. laser cutting operations
b. laser welding operations
c. laser drilling operations
d. laser marking and engraving operations.

11.8. The mask-projection technique is used in:
a. laser cutting operations
b. laser micromachining operations
c. laser marking and engraving operations
d. none of these.

11.9. Which of the following lasers is likely to give better resolution in the photolithography process?

a. F_2 laser
b. ArF laser
c. KrF laser
d. they will provide the same resolution.

11.10. Laser printer resolution and speed are respectively measured as:
a. dpi (dots per inch) and ppm (pages per minute)
b. cpi (characters per inch) and pps (pages per second)
c. dpi (dots per square inch) and ppm (pages per minute)
d. microseconds and ppm (pages per minute).

Answers

1. (d) 2. (d) 3. (d) 4. (b) 5. (c) 6. (b) 7. (d) 8. (b) 9. (a) 10. (a)

Bibliography

1. *Industrial Applications of Lasers*, 1997 by John F. Ready, Elsevier Science.
2. *Laser Material Processing*, 2010 by W.M. Steen, Springer.
3. *CO_2 Laser Cutting*, 1993 by John Powell, Springer.
4. *Laser Machining: Theory and Practice*, 1991 by George Chryssolouris, Springer-Verlag.
5. *Laser Cutting Guide for Manufacturing*, 2003 by Charles L. Caristan, Society of Manufacturing Engineers.
6. *Rapid Manufacturing: Technologies and Applications of Rapid Prototyping and Rapid Tooling*, 2011 by D.T. Pham and S.S. Dimov, Springer.
7. *Additive Manufacturing Technologies*, 2009 by Ian Gibson and Brent Stucker, Springer.
8. *Fundamental Principles of Optical Lithography: The Science of Microfabrication*, 2007 by Chris A. Mack, John Wiley & Sons.

12

Lasers in Medicine

12.1 Introduction

Due to their unique properties, lasers find diverse applications in both diagnostic as well as therapeutic medicine. Laser radiation can be easily controlled, manipulated, focused and conveniently delivered with the help of fibre optics. The possibility of varying the laser power or energy density over a large range and the number of available laser wavelengths covering a wide spectrum from ultraviolet to far-infrared mean that a laser treatment can be selected to match optimum tissue absorption and intended interaction. Consequently, lasers can be used to diagnose and treat many diseases.

Commonly used laser types for medical applications mainly include CO_2 lasers, solid-state lasers, semiconductor diode lasers, fibre lasers, dye lasers and excimer lasers. Some of the established medical disciplines where lasers have applications include angioplasty, cancer diagnosis and treatment, lithotripsy, dermatology, ophthalmology including laser-assisted *in situ* keratomileusis (LASIK) and photocoagulation, orthopaedics, cosmetic applications such as hair and tattoo removal, mammography, prostatectomy, optical coherence tomography and surgery.

The chapter begins with the fundamentals of light–tissue interaction mechanisms before discussing major medical applications of lasers, including both diagnostic and therapeutic applications. The focus is on the basic operational concept, the type of laser device or equipment used for the purpose and key advantages of using lasers as compared to alternative techniques.

12.2 Light–tissue Interaction

Medical applications of lasers as outlined in the introduction are classified as diagnostic and therapeutic applications. In the case of diagnostic applications, namely light–tissue interaction, the objective is to understand the pathology and physiology of the tissue through its interaction with light photons. For example, chemical composition finger-printing of the tissue could be performed by analyzing the fluorescence spectra emitted by the tissue. The chemical composition in turn may indicate the severity of a certain disease. Although optical diagnostics usually depend upon the interaction of light with endogenous agent constituents, it could also be performed by administering an exogenous agent. In the case of therapeutic applications, the objective is permanent modification of the tissue in question for treatment of the disease. The optical energy in this case is absorbed by the tissue or an exogenous agent, which further leads to a photochemical, photomechanical or thermal process depending upon the rate at which photon energy is absorbed per unit volume of the tissue. In the following sections we discuss the mechanisms of light–tissue interaction for therapeutic and diagnostic applications.

Lasers and Optoelectronics: Fundamentals, Devices and Applications, First Edition. Anil K. Maini.

12.2.1 Light–tissue Interaction for Diagnostic Applications

For diagnostic applications, relevant topics of light–tissue interaction include interaction mechanisms, optical properties of tissues and the fluence rate distribution in tissues. Each of these is briefly described in the following sections.

12.2.1.1 Fundamental Interaction Mechanisms

The fundamental mechanisms of light–tissue interaction include *absorption*, *luminescence* including fluorescence and phosphorescence and *scattering* including elastic and inelastic scattering and some others.

When a photon interacts with an atom or a molecule, *absorption* takes place if the photon energy ($= h\nu$) matches the energy difference between the discrete quantum states of the atom or molecule. The number of photons absorbed per unit volume per unit time is given by product of fluence rate H (photons/cm^2/s) and linear absorption coefficient μ_a (cm^{-1}). The absorption coefficient equals the product of molecular concentration N (cm^{-3}) and molecular cross-section σ (cm^2). The processes that follow absorption are non-radiative transitions, fluorescence and phosphorescence. Multi-photon absorption is also possible if the fluence rate is sufficiently high. For example, two photons each having energy of 0.6 eV may be absorbed to a transition energy of 1.2 eV. For further information on the absorption mechanism, see Chapter 1.

Luminescence is the emission of light from excited electronic states and includes processes of fluorescence and phosphorescence. Fluorescence emission is associated with electronic transitions allowed by the spin conservation rule. These occur over a timescale of nanoseconds. Fluorescence transitions can occur to various vibrational levels associated with the ground state; the fluorescence emission spectra therefore show several peaks. While absorption occurs from the ground state to any of the vibrational levels associated with the excited state, fluorescence emission occurs between excited states to any of the vibrational levels associated with the ground state. In fluorescence emission, a low-wavelength (higher-energy) photon is absorbed and is almost immediately followed by emission of a photon at a longer wavelength (lower energy). Since fluorescence emission occurs from same excited state, the shape of the emission profile is independent of the wavelength of photon used for excitation.

In the case of phosphorescence emission, the overall spin of the electron pair in the excited state is different from that in the ground state. These excited states are the result of rearrangement of the electron spin in the excited state in a process called intersystem crossing. Phosphorescence emission occurs between excited states resulting from intersystem crossing. These excited states are referred to as triplet states. Transition from an excited state to the ground state is forbidden, and therefore triplet state lifetime can be very long (of the order microseconds). An oxygen molecule has an unusual triplet ground state, which allows energy transfer from the triplet to molecular oxygen. This is significant from a medical diagnostics viewpoint as it allows *in vivo* measurement of oxygen concentration in tissue.

The processes of absorption, fluorescence and phosphorescence are best illustrated in a *Jablonski diagram*. Figure 12.1 shows a generalized Jablonski diagram indicating the three types of processes and different types of possible transitions. A Jablonski diagram illustrates the electronic states of a molecule and associated transitions. Different states are arranged vertically by energy and are grouped horizontally by spin multiplicity. In Figure 12.1, radiative and non-radiative transitions are represented by straight and dotted lines with arrows, respectively. Vibrational levels of the lowest electronic state are represented by thick lines, whereas higher vibrational levels are indicated by relatively thinner lines.

Radiative transitions include absorption and emission. Non-radiative transitions occur through several different mechanisms. These include transitions involving relaxation of the excited state to its lowest vibrational level with the molecule dissipating energy to its surroundings. A second type of non-radiative transition is the internal conversion, involving transition from the vibrational state of an electronically excited state to a vibrational state of a lower electronic excited state. Another type of non-radiative transition is intersystem crossing (mentioned above). In this case, transition occurs to a state with a different spin multiplicity.

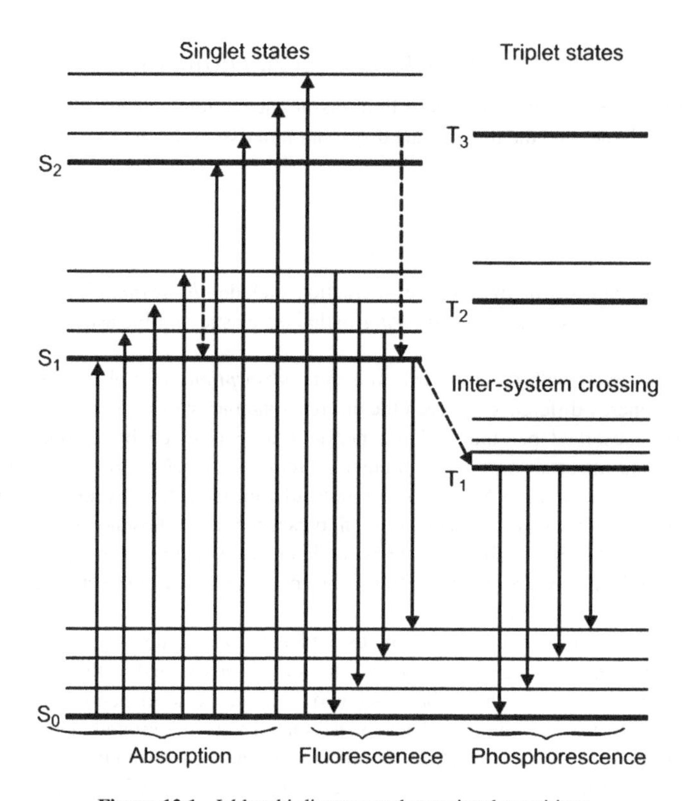

Figure 12.1 Jablonski diagram and associated transitions.

There are two broad categories of *scattering* mechanisms: elastic scattering and inelastic scattering. In the case of elastic scattering, the frequency of the scattered radiation is the same as that of the incident radiation. In the case of inelastic scattering, the frequency of the scattered radiation is different from that of incident radiation.

The origin of *elastic scattering* lies in the acceleration of electrons in the tissue caused by forces exerted by incident electromagnetic radiation. The scattered radiation in this case is of the same frequency as the incident radiation. Two major forms of elastic scattering are *Rayleigh scattering*, which occurs from particles much smaller than the wavelength, and *Mie scattering*, which occurs from particles of almost the same size as the wavelength. In a uniform medium, only the incident radiation is observable. In the case of media where refractive index varies with position, it is possible to detect scattered radiation. In tissue, the refractive index changes over different length scales extending from cellular organelles of submicron scale to multicellular organization on a millimetre scale.

Inelastic scattering is produced either by scattering particles in motion or when the incident electromagnetic radiation excites molecular vibrations. In the case of scattering from a moving particle, the scattered radiation is Doppler shifted. For a group of moving scattering particles, the resultant Doppler-shifted radiation spectrum is related to the distribution of particle velocities. One common medical application of this type of scattering mechanism is the measurement of blood flow in tissues.

Inelastic scattering due to molecular vibrations can be of two types: Stokes Raman scattering and anti-Stokes Raman scattering. Stokes Raman scattering occurs when the incident photon loses some of its energy to the scattering particle, with the result that the emitted photon has less energy (or a longer wavelength) than that of the incident photon. In the case of anti-Stokes Raman scattering, the incident photon gains energy from the scattering molecule and the emitted photon therefore has higher energy (or a lower wavelength) than that of the incident photon.

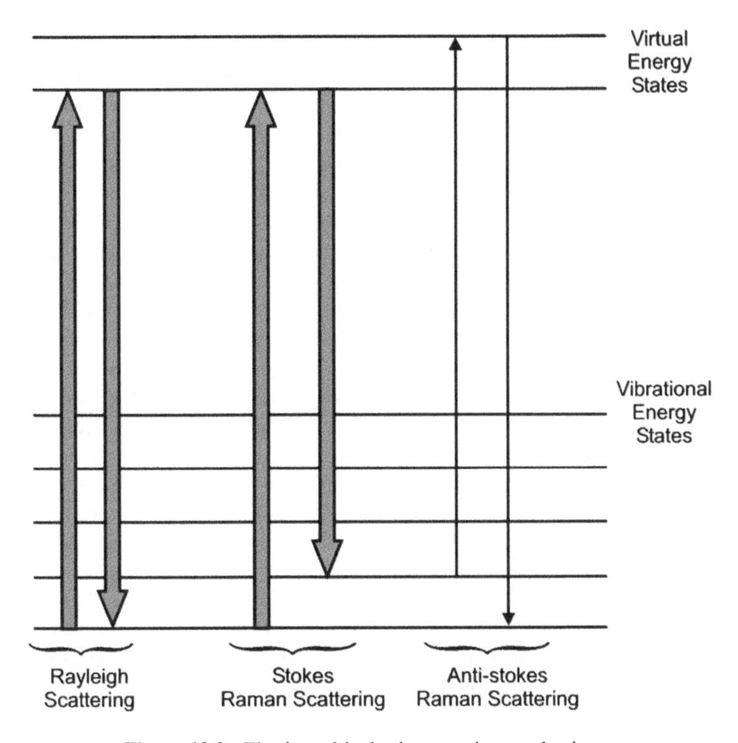

Figure 12.2 Elastic and inelastic scattering mechanisms.

Figure 12.2 illustrates the processes of elastic scattering and the two types of inelastic scattering. It may be mentioned here that the probability of Raman scattering is several orders of magnitude lower than that of elastic scattering. In Figure 12.2, the intensity of emission is indicated by the thickness of the transition lines. *In vivo* applications of Raman spectroscopy, which is otherwise a highly potent tool for analytical chemistry, evolved only after technological advances in detectors, laser sources and fibre-optic delivery systems.

Raman spectroscopy has been used for the detection and identification of even single bacteria and fungi cells due to its unique finger-printing feature, which allows detection and identification of biochemical components to a high degree of selectivity. The choice of excitation wavelength plays an important role. Different excitation wavelengths allow different molecular components inside the cell to be focused on. While visible and near-infrared excitation wavelengths probe the entire set of cell components and lead to a phenotypic characterization, UV wavelength excitation causes relative enhancement of DNA (deoxyribonucleic acid), RNA (ribonucleic acid) and aromatic amino acid vibrations. These vibrations are very specific taxonomic markers of microorganisms. A Raman spectrum captures the sum of all Raman active cell components, therefore characterizing metabolic cellular status. Raman spectroscopy allows analysis of change in cellular metabolism, and the data can be used to diagnose cancer and other pathological anomalies.

12.2.1.2 Optical Properties of Tissues

Tissue optical properties of interest include the absorption coefficient, scattering coefficient and scattering phase function. Another important consideration is the propagation of light in the tissue. From the viewpoint of biomedical diagnostics, it is important to measure the magnitude of absorption and scattering coefficients at the wavelength of interest and also their variation as a function of wavelength. Knowledge of these parameters is important as they are the determinant factors, in addition to the scattering phase function and light propagation mechanisms that govern the fluence rate distribution in tissue, during interpretation of results of a diagnostic measurement or execution of a therapeutic intervention. Optical properties such as birefringence are not important here as they do not impact fluence rate as much as the previously mentioned parameters.

The *absorption coefficient* for a given incident wavelength can be determined from the known values of concentration and molecular cross-section of all chemical species present in the tissue. Since absorption has a strong dependence upon incident wavelength, the absorption spectrum varies for different tissues. For example, proteins have strong absorption in the ultraviolet region of electromagnetic radiation. Amino acids phenylalanine, tyrosine and tryptophan are also efficient absorbers of ultraviolet radiation in the 250–300 nm range. Protein absorption rapidly decreases with increase in wavelength. At increased wavelength, dominant absorption occurs in other constituents such as haemoglobin. The absorption spectrum changes with oxygenation. At infrared wavelengths, absorption in tissue is predominantly due to water content. To summarize, the three important constituents of a tissue – namely protein, oxyhemogobin and water – are individually responsible for absorption of radiation wavelengths in ultraviolet, visible and infrared bands, although the absorption coefficient varies from one wavelength band to other. The absorption coefficient also depends on the concentration of constituents.

The *scattering coefficient* in tissues is not as simple and straightforward to predict as for the case of absorption. In the case of tissue, the refractive index varies in a complex manner over different length scales in different tissues. Several models have been developed for the prediction of tissue scattering, taking into consideration the fractal distribution of scattering particles of different size ranges. Results obtained from these models closely match the experimental data. Based on these models and experimental results, the following general observations can be made about tissue scattering.

1. The scattering coefficient is very high for optical bands compared to low-energy X-rays. At visible wavelengths, typical value is in the range 10–50 mm^{-1}, corresponding to a mean free path of 20–100 μm.
2. The scattering coefficient is observed to decrease monotonically with and increase in wavelength.
3. The scattering coefficient is observed to be orders of magnitude larger than absorption coefficient, except for ultraviolet and infrared wavelength bands.

Tissue structures that are comparable to or larger than the optical wavelength dominate the scattering phase function. This leads to scattering that is forward peaked.

12.2.1.3 Fluence Rate Distribution

Fluence rate distribution in tissue plays an important role during therapeutic procedures; it is therefore important to understand and examine fluence rate distribution issue as a function of wavelength. For a given irradiance and irradiance geometry, fluence rate distribution mainly depends upon the absorption and scattering coefficients of the tissue. In general, the absorption coefficient varies by several orders of magnitude across a wide range of biomedical applications. In contrast, the scattering coefficient decreases slowly as a function of wavelength. Scattering coefficient has an insignificant impact on determining fluence rate distribution in ultraviolet and infrared wavelength bands; its influence is more predominant in the visible and near-infrared wavelength bands.

Light propagation in tissue also plays an important role in the prediction of fluence rate distribution during therapeutic procedures, the interpretation of results of diagnostic measurements and modelling for optimization of diagnostic and therapeutic applications. Propagation models such as the radiative transport equation (RTE) and Monte Carlo simulation are commonly used. A detailed description of these models is beyond the scope of the present text.

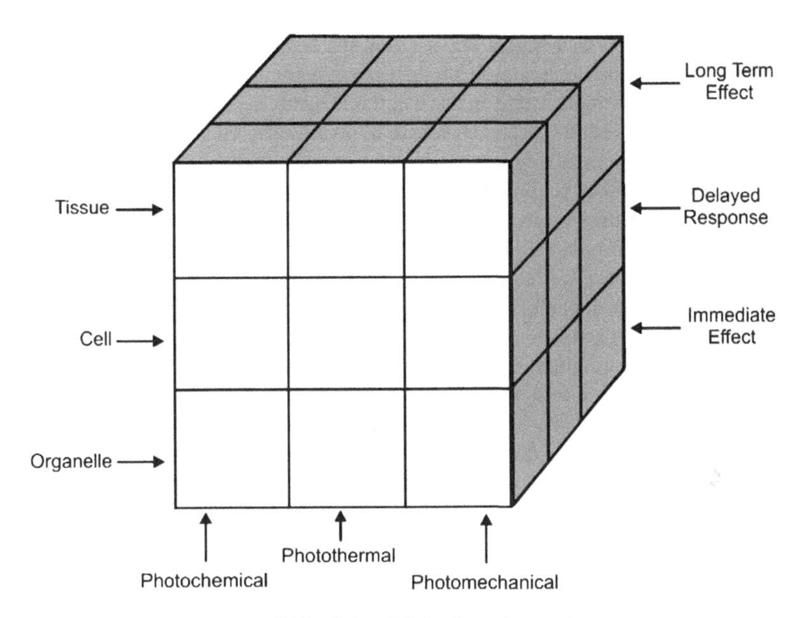

Figure 12.3 Cube of light–tissue interaction.

12.2.2 *Light–tissue Interaction for Therapeutic Applications*

In light–tissue interaction for diagnostic applications, it is the behaviour of photons that is affected by the presence of cells or tissues. Cells or tissues are not affected during the process. Here we describe the effect of light photons on cells or tissues. It is only when tissue is affected by the photons that a therapeutic function is possible. Although both lasers and conventional light sources such as arc lamps, incandescent lamps and light-emitting diodes (LEDs) can be used for the purpose, the laser offers distinctive advantages over conventional sources due to its monochromaticity, directionality and ability to generate short intense pulses. The laser effect on tissue is mainly determined by its CW power/pulse energy, wavelength and pulse duration. From the viewpoint of therapeutic applications, the nature of the light–tissue interaction can be either photochemical, photothermal or photomechanical. The biological structure that interacts with light could be an organelle, a cell or a tissue or an organ. A third factor that determines the nature of the interaction is the time duration.

Light–tissue interaction for therapeutic applications can therefore be explained in terms of three parameters: the mechanism, the biological structure and the time course. This is diagrammatically represented by the cube of laser–tissue interaction as shown in Figure 12.3. The three axes of the cube represent these three parameters. The x-axis represents the mechanism of interaction (chemical, thermal and mechanical), the y-axis the level of biological structure that absorbs photons (organelle, tissue and organ) and the z-axis the time course (immediate effect, delayed effect and long-term effect). Time course here signifies any physical effect or a variation in biological response with time. In the following sections we briefly describe the three mechanisms of interaction (photochemical, photothermal and photomechanical).

12.2.2.1 Photochemical Effects

Photochemical reactions have a wide range of applications in industry and medicine. The phenomenon is a common occurrence in nature; one such example is the photochemical reaction produced by the absorption of sunlight by chlorophyll in plants to yield excited molecular states. The energy of excited molecular states is then used to bind water and carbon dioxide as sugars. Light-activated adhesives are routinely used in industry. Photochemical effects form the basis of treatment of various diseases in different branches of medicine including oncology, dermatology and ophthalmology.

Laser wavelengths in ultraviolet, visible and near-infrared bands are used for photochemical reactions involving changes in electronic states. Laser wavelengths in mid-infrared and far-infrared bands are used for photochemical reactions involving vibrations in molecular bonds. The rate of production of photochemical reaction product is directly proportional to the incident irradiance, the absorption coefficient of the photosensitive molecule and the quantum efficiency that specifies the fraction of absorbed photons causing conversion. The absorption coefficient is given by the product of extinction coefficient and the concentration of the photosensitive molecule.

Some common applications of photochemical effects in medicine include treatment of psoriasis of the skin involving ultraviolet light activation of psoralen (a type of photoreactant) to achieve cross-linking of the DNA in order to control pathological proliferation. Another application is in photodynamic therapy (PDT), involving the use of visible or near-infrared light in the presence of singlet oxygen at ambient levels. Light is absorbed by the photosensitive drug, which passes its activation energy to oxygen thus oxidizing and killing the tumour. The therapy is also in use for the treatment of macular degeneration of the eye.

12.2.2.2 Photothermal Effects

Photothermal effects produced in the tissue on absorption of laser energy are the basis of a wide range of therapeutic applications. The treatment of a detached retina through coagulation of tissues is a common example. Laser coagulation is also an effective tool for treatment of tumours.

When laser energy is absorbed in a tissue, it is converted into heat which causes a rise in temperature. Laser heating density Q ($\mathrm{Jcm^{-3}}$) and the associated surface temperature rise ΔT are given by Equations 12.1 and 12.2:

$$Q = \mu_a E t \tag{12.1}$$

$$\Delta T = \frac{Q}{\rho C_p} \tag{12.2}$$

where

μ_a = absorption coefficient ($\mathrm{cm^{-1}}$)
E = laser power density ($\mathrm{W\,cm^{-2}}$)
T = exposure time (s)
ρ = tissue density ($\mathrm{g\,cm^{-3}}$)
C_p = specific heat ($\mathrm{J\,g^{-1}\,°C^{-1}}$)

Temperature rise varies as a function of depth z as given by Equation 12.3:

$$\Delta T(z) = \Delta T e^{-\mu_a z} \tag{12.3}$$

For a given laser energy density, the surface temperature rise is proportional to the absorption coefficient. The temperature rise below the surface is a function of absorption coefficient and depth. Laser heating therefore depends not only on laser energy delivered to the tissue but also on the absorption coefficient and the penetration depth. Penetration depth is the distance below the surface where the temperature rise falls to 37% ($=1/e$) of its maximum value at the surface.

Thermal diffusion is another parameter involved in the heating of the tissue. Thermal diffusion causes heat energy to dissipate, thus lowering the temperature. A small heat zone cools faster than a large heat zone by thermal diffusion. If the energy is deposited over a diameter d, then the time in which the temperature rise drops to 37% of the initial value, also called relaxation time t_r, is given by Equation 12.4:

$$t_r = \frac{d^2}{4\alpha} \tag{12.4}$$

where α is thermal diffusivity ($m^2 s^{-1}$).

Thermal damage caused to the tissue is an important requirement when it comes to the use of photothermal effects for therapeutic applications. In the case of diagnostics, there is no permanent change to the tissue. The objective of therapeutic use of lasers is to cause some kind of controlled damage to the tissue. Prediction of thermal damage to the tissue due to laser irradiation involves modelling light propagation and distribution in the tissue and predicting the resultant thermal damage. Damage results when the tissue is exposed to a high temperature for a prolonged period of time. If the temperature is less than a certain threshold temperature, called critical temperature, the rate of damage accumulation is negligible. In diagnostic and photochemical therapeutic applications, temperature is kept below the critical temperature.

As tissue temperature exceeds the critical temperature, the first form of thermal damage observed in the tissue is coagulation, caused primarily by denaturation of cellular and tissue proteins. Denaturation increases the scattering coefficient with the result that light is scattered out of the tissue making it appear white. Cooked egg white is a good example of coagulation. Collagen-rich tissues such as tendon and skin become transparent on coagulation as the scattering coefficient decreases. Coagulation falls more within the regime of non-ablative heating. If the temperature continues to rise beyond the range for coagulation, the result is vaporization. Most tissues have a high water content. Water vaporizes around 100 °C and, for still higher fluence rates, more vapour is generated than is possible to escape by diffusion. Excess vapours are trapped between tissue layers and subsequently become superheated to create vapour bubbles or expanding vacuoles. This further leads to rupturing of the walls of the vacuoles, creating holes in the tissue.

In addition to the use of photocoagulation to treat a detached retina and small tumours, another common application of photothermal effects is in dermatology where the thermal damage effect is used for the treatment of vascular lesions such as portwine stains. In the process, laser energy is selectively deposited in blood vessels helped by the strong absorption of light by haemoglobin. The transient temperature rise causes thermal damage to the vessel wall, leading to clotting called thrombosis of the vessel. The clotted vessel is removed by the body's wound-healing response, removing the portwine stain lesion.

12.2.2.3 Photomechanical Effects

Mechanical effects produced as a result of deposition of laser energy as thermal energy include thermoelastic expansion, cavitation, vaporization and plasma formation.

Thermoelastic strain is produced due to the temperature rise caused by deposition of thermal energy. Strain produces stress or pressure; the pressure P exerted by the thermoelastic strain is given by Equation 12.5:

$$P = M\varepsilon = M\beta\Delta T \tag{12.5}$$

where

β = thermal expansivity (°C^{-1})
M = bulk modulus ($J\,cm^{-3}$)
ΔT = temperature rise (°C)

Pressure waves tend to propagate away from the tissue as sound waves. In order to produce condition of stress confinement, which is desirable, it is important that the energy is deposited in a target tissue volume at a rate faster than it can dissipate as propagating pressure waves. To achieve this, the laser pulse width should be less than the ratio of the size of tissue volume to the velocity of sound in tissue. Laser-induced pressure waves are not used for any therapeutic application, but they are important as they form the basis of cavitation.

Cavitation is the result of stress caused by thermoelastic strain exceeding a certain limit. When the limit is exceeded, the tissue breaks. In the case of solids, it leads to the formation of a void, which subsequently grows usually along the plane of the fracture; this process is called spallation. In liquids, it leads to the initiation and subsequent growth of a bubble; this process is known as cavitation. Cavitation can have possible applications in the removal of thin layers of tissue without causing any significant thermal damage or surface modification of a biomaterial.

Vaporization of water in tissue is another effect produced by temperature rise. The vaporization process can take three forms: surface vaporization, subsurface explosive vaporization and explosive expansion of superheated fluid. *Surface vaporization* takes place when a vapour is formed at the surface. The phenomenon can be used to desiccate superficial layers of a tissue. In the case of *subsurface explosive vaporization*, vapour is formed below the surface. Vapour in this case is confined and not able to escape. With a further rise in temperature, vapour pressure builds up. When vapour pressure overcomes tissue strength, rapid vapour expansion disrupts the tissue.

Another situation is that of *superheated fluid expansion*, which occurs when temperature rise is significantly large to supply the entire heat for vaporization. In this case, water in tissue is explosively ejected. Vaporization process is used to remove large tumours such as obstructing lung or brain tumours. The process is particularly suitable for removing difficult-to-access tumours. The tumour in this case breaks into gas-borne particles, which are then removed by vacuum dump. A CO_2 laser emitting at 10.6 μm is the preferred laser due to its high absorption by water at 10.6 μm. Reshaping corneal collagen is another application based on the vaporization process. An ArF excimer laser emitting at 193 nm is the laser of choice due to its high absorption by proteins.

When the laser irradiance on the tissue exceeds approximately $10\,\mathrm{MW\,cm^{-2}}$, there is an optical breakdown leading to the formation of plasma. Plasma formation essentially involves electrons being stripped off by intense electric fields, leading to the ionization of molecules. A bubble is formed, which expands to its maximum diameter and then collapses to generate extreme pressures producing tissue disruption. It can be created by using nanosecond pulses such as those from Q-switched Nd:YAG lasers focused at the target point. One common application of the use of plasma is in post-cataract surgery treatment where the laser is focused into the eye near the implanted artificial lens to disrupt an opaque film interfering with vision. The mechanism is also used to disrupt large kidney stones.

12.3 Laser Diagnostics

Laser diagnostics and optical diagnostics in general are based on changes in one or more measurable properties of light in ultraviolet, visible or near-infrared wavelength bands as it transmits through and/or reflects from tissues after light photons interact with tissue structure and molecules. There are two broad categories of diagnostic techniques: *in vivo* and *in vitro*.

In vivo diagnostic techniques are either non-invasive or minimally invasive and are performed on a living organism, as opposed to *in vitro* techniques that are used outside a living organism in a controlled environment. Some common examples of *in vivo* diagnostic applications include medical imaging such as X-ray imaging, magnetic resonance imaging (MRI) and computed tomography (CT) and monitoring techniques such as electrocardiography (ECG) and electroencephalography (EEG).

In vitro diagnostics on the other hand involves the removal of tissue such as blood, saliva and biopsy samples from a living organism for examination in the controlled environment of a laboratory. Some examples of *in vitro* diagnostic equipment and tools range from over-the-counter pregnancy testing kits to highly sophisticated molecular or nucleic acid tests. In addition to the test elements, that is, the reagents, *in vitro* diagnostics also includes equipment used to analyze results.

In the following sections we discuss the fundamentals of optical diagnostics, their comparison with other medical diagnostic techniques and the advantages and limitations of optical diagnostic techniques. Some common *in vivo* diagnostic techniques are also described briefly.

12.3.1 Basic Principle

The underlying basic principle of all optical diagnostic techniques is derived from two fundamental characteristics of light–tissue interaction: one related to the magnitude of photon energy and the to the wavelength of light, namely size or dimension of cellular structures. The photon energies corresponding to light in ultraviolet, visible and near-infrared wavelength bands is of the same order as the excitation or vibrational energy levels of the molecule, which makes the process attractive to probe molecular states. Predominantly, visible and near-infrared wavelengths in the range of 700–1100 nm are used due to their high molecular specificity and greater penetration of tissue in this wavelength band. The ultraviolet wavelength band has limited application due to significantly lower tissue penetration and also due to its potential mutagenicity. Wavelengths above 1300 nm are also of no use for diagnostics due to their absorption by water content in tissue being too high for any meaningful measurement.

Another feature of optical diagnostic techniques underlying its functioning is that the optical wavelength region, that is 700–1100 nm, matches very well the dimensions of the cellular structures of interest, which also have dimensions of the order of a micron.

Optical diagnostic techniques are based on the fact that the optical properties of biological soft tissues are a function of physiological, functional and pathological condition of tissues and organs. Different spectral optical properties such as absorptance, reflectance, fluorescence and scattering of tissues and liquids, blood in particular, are different for functional and pathological cases, enabling knowledge of the tissue or organ clinical state to be obtained by an *in situ* measurement of these parameters.

In any *in vivo* optical diagnostic process, the tissue area is illuminated by a low-intensity light flux. The light flux interacts with the tissue and the backscattered and/or transmitted light flux is registered by a photosensor. Light–tissue interaction is a complex phenomenon and leads to many linear and non-linear physical phenomenon such as absorption, scattering, fluorescence and Doppler frequency shift. The spectral and spatial distribution of measured radiation carries a lot of information about the biochemical and/or structure composition of the subject tissue. Analysis of the registered physical data is used to extract the desired diagnostic information. For example, absorption spectroscopy is used to determine information on water content, lipids and melanin in a tissue. Elastic scattering spectroscopy is used to extract information on the morphological structure of the tissue. Tissue reflectance oxymetry is used to calculate the level of oxygenated fraction of haemoglobin in blood. Doppler flowmetry can be used for measuring the dynamics of blood flow.

12.3.2 Comparison with Other Techniques

Optical diagnostic techniques either eliminate or minimize invasiveness as well as pharmaceutical, chemical and other physiologically undesirable effects on a human organism. This is one of the distinguishing features that characterize optical diagnostics from other techniques.

The sensitivity of the molecular structure of biological tissues to optical radiation in the ultraviolet, visible and near-infrared wavelength bands, and also the fact that optical properties of the tissues are indicators of physiological, functional and pathological state of tissues is another feature of optical diagnostics that distinguishes it from other diagnostic methods. Note that other techniques such as X-ray imaging or ultrasound imaging do not offer this sensitivity to molecular structure. In the case of X-rays, interactions are at an atomic level; in ultrasound imaging, the diagnostic information resides in the mechanical properties of tissues.

The spatial resolution offered by optical diagnostic techniques is far superior to that achievable from any other *in vivo* diagnostic technique. This is due to the wavelength of the optical radiation being of the same order as the dimension of the cellular structures of interest.

Another salient feature of optical diagnostics is the possibility to use optical labelling to provide additional molecular, physiological or genetic information not easily obtainable with other methods. As an example, it is possible to conjugate fluorescent dyes with antibodies or DNA fragments and follow these by fluorescence imaging as the biologically active agents bind themselves to specific cellular targets. Although radio isotope labelling is also possible, the range of possible optical compounds is very large. Optical labelling will play an increasingly important role in tailoring treatment for individual patients, supported by knowledge of the molecular/genetic basis of the disease.

Another significant advantage of optical diagnostics comes from its compatibility with other diagnostic techniques, which allows it to be used simultaneously with other treatments. This has been made possible by fibre-optic delivery and collection of optical radiation. As an example, it is possible to perform optical monitoring of the brain function while the patient is inside an MRI unit. A fibre-optic Raman spectroscopy probe may be combined with ultrasound devices, providing high complementarity.

One of the major limitations is the need to have a trade-off between tissue depth and spatial resolution, two conflicting requirements. Depth in tissue is often limited by high optical attenuation levels, even in the optical window of transmission in the tissue. Spatial resolution is severely affected by elastic scattering. Optical measurements up to tissue depths of a few centimetres may reduce spatial resolution to a millimetre or so. On the other hand, micron-level imaging resolution may be achieved *in vivo* for tissue depths limited to a few hundreds of microns below the tissue surface.

Another contrasting feature of optical and other diagnostic methods is that the instrumentation for the latter category is far more generic and broad based. In the case of optical diagnostics, there may be a large number of distinctly different optical instruments for a single clinical application. Diagnostic instruments such as MRI and CT are very generic with multiple capabilities. On the other hand, optical instruments are application specific due to the need for tailoring to a specific purpose and due to the fact that optical technologies are relatively less expensive.

12.3.3 In Vivo Optical Diagnostic Techniques

All *in vivo* optical diagnostic techniques make use of one or more types of interactions that take place when optical radiation is incident on tissue area. After encountering interaction mechanisms, the light photons exit the tissue either as a transmitted or reflected radiation. It is this exiting light that carries the diagnostic information. The exiting radiation is measured by a detector system, which may range from a single photodetector or a highly sophisticated imaging system along with associated optics. The radiation is comprehensively analyzed to extract the desired diagnostic information.

Since there are a number of different interaction mechanisms, this has led to the introduction of several *in vivo* imaging techniques exploiting one or more of the interaction mechanisms and having the ability to be applied to different clinical applications. Some common diagnostic applications include white light imaging, diffuse optical spectroscopy, elastic scattering spectroscopy, optical coherent tomography, Raman spectroscopy, confocal imaging and fluorescence spectroscopy. In addition to these diagnostic techniques, there are several other optical techniques for clinical diagnostic application that are under development. These include acousto-optic spectroscopy, photothermal spectroscopy, fluorescence correlation spectroscopy, speckle interferometry and laser Doppler blood flow measurement. The more common and established techniques are described in the following sections.

12.3.3.1 White Light Imaging

White light imaging is a type of reflectance imaging in which diagnosis is based on the spectral and spatial appearance of the tissue in the visible range. Optical diagnostic techniques previously depended heavily upon the physician's interpretation, expertise and experience. Today, the ability of the physician's eye has been extended by use of optical instruments that enhance spectral range and sensitivity and also provide access to inner structures that otherwise hidden from view.

Examination of a retina by a widefield retinal scope is one such example; a typical handheld ophthalmoscope is depicted in Figure 12.4. The instrument uses xenon halogen technology and a concentrated white light for perfect illumination and crisp retinal detail. Figure 12.5 illustrates a retina with irreversible retinal damage, obtained by white light imaging. White light endoscopy for examination of the inner lining of hollow organs such as gastrointestinal tract, bladder, cervix or bronchus is another common and well-established clinical application of white light imaging.

White light endoscopy has undergone much improvement in terms of instrumentation and associated capabilities. Hollow tubes with associated lenses and other optical elements and a camera placed external

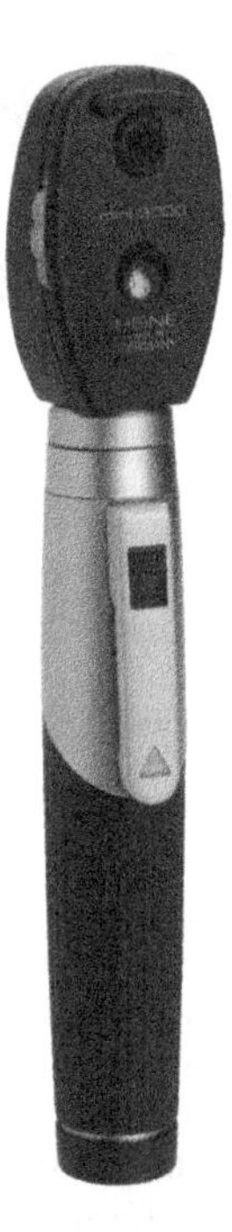

Figure 12.4 Ophthalmoscope Type Heine 3000 (Reproduced by permission of HEINE OPTOTECHNIK GmbH & Co. KG).

to the body have been replaced by fibre-optic delivery and a miniature RGB camera placed at the working end of what is today called a fibre-optic endoscope. Figure 12.6 depicts one such fibre-optic endoscope comprising a white light source, a fibre-optic cable and a camera fitted to the tip of the edge. This allows transmission of high-resolution image information electronically for the purpose of viewing. The use of contrast agents such as absorbing dyes enhances the features of abnormal structures. This feature in combination with high-magnification endoscopy can be used to see fine surface features of the tissue. An emerging trend in *in vivo* optical diagnostics is use of hyperspectral imaging which allows imaging at different wavelengths for the purposes of enhanced pathological diagnosis.

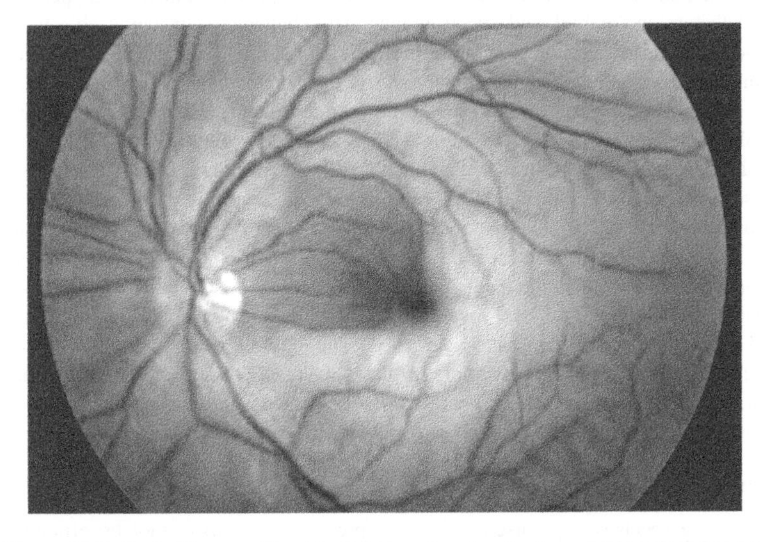

Figure 12.5 Retinal image with white light imaging showing irreversible retinal damage.

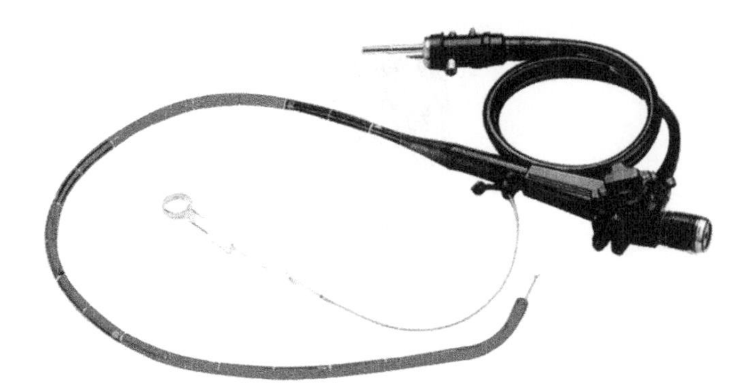

Figure 12.6 Fibre-optic endoscope.

12.3.3.2 Diffuse Optical Spectroscopy

Diffuse optical spectroscopic imaging is a non-invasive optical diagnostic technique that can quantify the absorption and scattering coefficients of tissues as deep as several centimetres. By measuring these optical properties, quantifiable and qualitative information about the target tissue can be ascertained. Elastic scattering and absorption are the basis of light–tissue interaction mechanisms responsible for most of the diagnostic applications. The elastic scattering coefficient is very high, of the order 10–100 cm^{-1}, at visible and near-infrared wavelengths. Elastic scattering is usually forward peaked with a high anisotropy factor of about 0.9, leading to an effective transport scattering coefficient of 1–10 cm^{-1}. This is the scattering coefficient in the case of isotropic scattering and quite relevant to diffuse optical spectroscopy measurement. The measurement of the spectral features of diffusely reflected or transmitted light at multiple wavelengths yields far greater diagnostic information than is attainable by the simple qualitative nature of white light imaging discussed in the previous section. The spectral content of collection optics comprising measured radiation that is diffusely reflected or transmitted due to multiple scattering depends upon the absorption spectrum of the tissues.

Major naturally occurring chromophores in the visible and near-infrared wavelength range include oxygenated and deoxygenated forms of haemoglobin (HbO_2 and Hb, respectively), water, lipids, melanin and various pigments. Wavelengths that are absorbed more are reflected or transmitted weakly. The diagnostic equipment could range from a simple instrument comprising a white light source, light delivery and collection optics and a photosensor with a set of filters, to complex instruments equipped with multiple sources and detectors and/or high-end spectrographs. The challenge is in being able to identify the absorption and elastic scattering components from the measured total reflectance or transmittance.

One common application of diffused optical spectral measurements is in continuous monitoring of health status of premature infants during birth or under intensive care. This is usually carried out by monitoring blood content ($Hb + HbO_2$) and/or oxygen saturation ($HbO_2/(Hb + HbO_2)$) of tissues (in particular brain tissues) through spectral measurements at discrete visible and near-infrared wavelengths. Hb has an absorption peak at 550 nm and HbO_2 exhibits a double absorption peak at 540 nm and 585 nm.

Another common diagnostic application of diffused optical spectroscopy is as a potential non-invasive alternative to blood sampling for identification and quantification of glucose levels in tissues in diabetic patients. Another application is in the measurement of changes in breast tissue that may be associated with breast cancer. In both glucose level measurement and monitoring changes in breast tissue, highly sophisticated spectral analysis techniques are used though the instrumentation needed for these applications is relatively simple. For example, in case of breast cancer detection, the challenge lies in the detection of subtle changes observed in the form of changed water, fat and blood content of breast with structural and/or physiological changes associated with cancer risk.

12.3.3.3 Elastic Scattering Spectroscopy

Diffused reflectance spectroscopy (discussed in the previous section) cannot be efficiently used to determine cellular information about surface layers because the strength of the backscattered light from

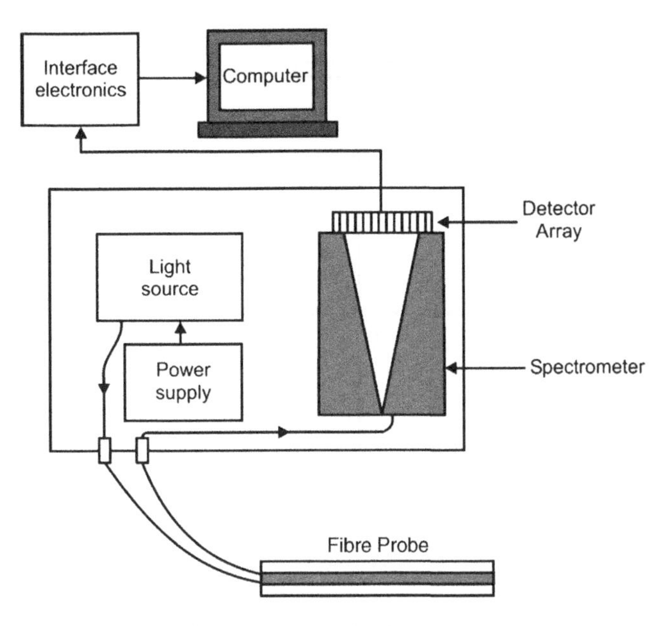

Figure 12.7 Elastic scattering spectroscopy set-up.

the surface layer (e.g. the mucosal layer in hollow organs) is completely swamped by a strong diffuse reflection from a deeper structure. This serious limitation can be overcome by using a multi-layer model of tissues with known approximate values of absorption and scattering coefficients and employing numerical subtraction. Another solution to the problem is by employing a polarized light beam that would allow discrimination between elastic scattered light from the surface layer and that received from deeper tissues after a large number of scatters. Backscattered light is analyzed to determine scattering particle size distribution, which is indicative of the size distribution of cell nuclei. It may be mentioned here that the size of cell nuclei changes when malignant. Elastic scattering spectroscopy has been found to be accurate in the identification of abnormalities of soft tissues, including ischaemia and associated inflammation and pre-cancer and cancer. Figure 12.7 depicts the elastic scattering spectroscopy set-up.

12.3.3.4 Optical Coherence Tomography

When light interacts with tissues, there is a near 180° backscattering that occurs at the microscale interfaces within tissues due to localized refractive index change. This forms the basis of optical coherence tomography (OCT), which enables the generation of high-resolution structural images up to a depth of a few millimetres. The operational principle of OCT imaging is similar to high-resolution ultrasound imaging, except for the fact that it makes use of a Michelson interferometry to provide spatial localization rather than echo-ranging by sound waves to map the depth distribution of micro-inhomogeneities in the tissue. Echo-ranging is not practical in optical coherence tomography at the micron resolution scale due to the much higher speed of light compared to that of sound waves.

Figure 12.8a depicts the basic principle of operation of OCT. It is basically a Michelson interferometer with an axially scanned reference mirror in one arm and the tissue in the other. The beam splitter splits the light beam almost equally into the two paths. The light beam specularly reflected (near-180° backscattering) from the internal optical interfaces of the tissue mixes with the reference beam to produce an interference pattern at the detector. Constructive interference and a measurable signal at the detector output are only obtained when the path lengths of the probe beam and the reference beam are the same or nearly the same, with a path length difference of less than the coherence length of the light source. Figure 12.8b and c show typical detected signals from a single site and a single lateral scan, respectively. A 2D or 3D cross-sectional image of the tissue may be generated with appropriate scanning. OCT synthesizes cross-sectional images from a series of laterally adjacent depth scans.

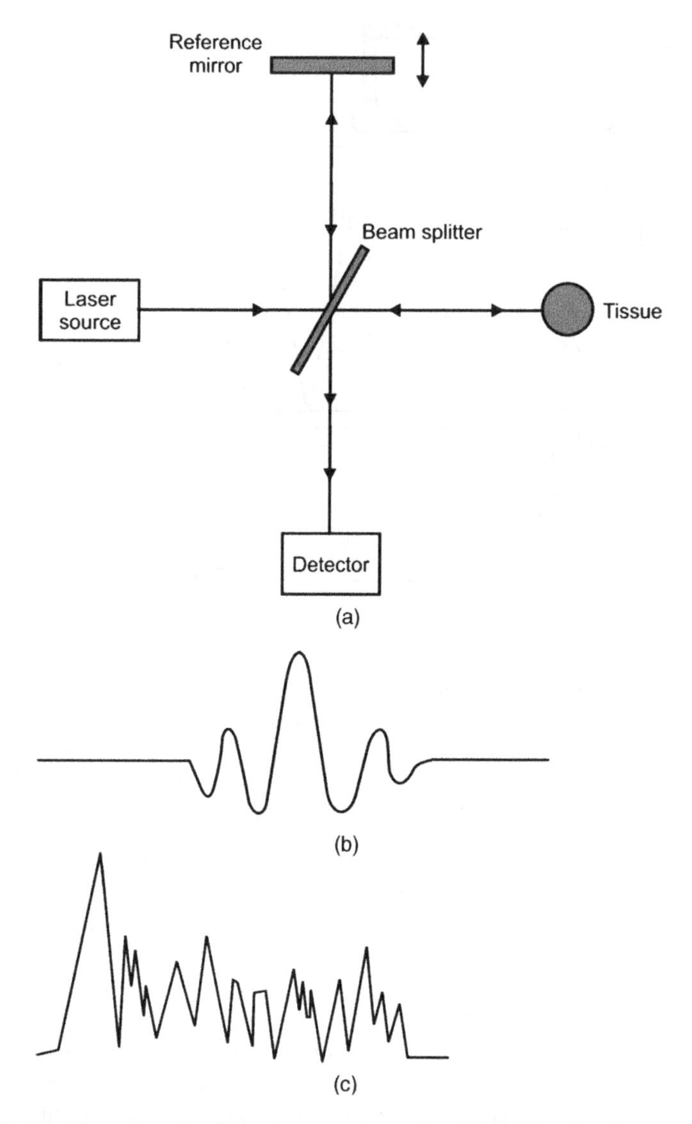

Figure 12.8 Principle of operation of optical coherence tomography: (a) Michelson interferometer; (b) signal from single site; and (c) signal from a single lateral scan.

Axial resolution achievable in OCT is a function of coherence length which, for a given wavelength λ and refractive index of the medium n, is inversely proportional to line width $\Delta\lambda$ and is given by $\lambda^2/(n \times \Delta\lambda)$. The operational wavelength range is 800–1500 nm with 800–850 nm and 1300 nm being the preferred wavelengths (the latter being particularly important for fibre-optic implementation of OCT). The chosen wavelength needs to be different from the water absorption peak at 970 nm. Figure 12.9 depicts the fibre-optic implementation of OCT via a Michelson interferometer.

A wavelength of 1300 nm and a line width in the range of 60–70 nm has an axial resolution better than 10 μm in a medium with a refractive index of about 1.4. Typical sources are super-luminescent diodes emitting in the range of 800–1500 nm and Ti-sapphire lasers pumped by frequency-doubled Nd:YAG lasers and emitting at around 800 nm. Emerging light sources include photonic crystal fibre and super-fluorescent fibre sources.

Lateral resolution depends upon the numerical aperture (NA) of the probe beam. High NA provides a tight focus but a limited depth of focus. Low NA offers a high depth of focus. Lateral resolution typically achievable is 2–3 times poorer than achievable axial resolution.

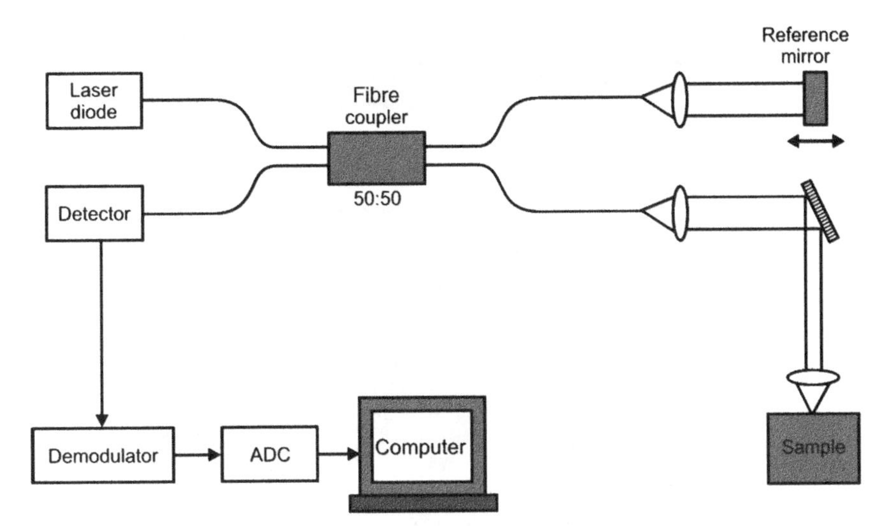

Figure 12.9 Fibre-optic implementation of OCT.

OCT is presently used in three different fields of optical imaging: macroscopic imaging of structures using weak magnification; microscopic imaging using very high magnification; and endoscopic imaging using low and medium magnifications. One major application area is in ophthalmology due to the transparent nature of ocular structures. Figure 12.10 depicts an OCT eye fundus examination instrument, a Canon OCT-HS100, which is mainly intended for the diagnosis of retinal disorders including the leading causes of eyesight loss, age-related macular degeneration and glaucoma. Use of near-infrared wavelength has made infrared imaging possible for strongly scattering molecules. Optical *in vivo* biopsy is one of the most challenging emerging applications of OCT. High resolution and high penetration depth with associated capability of functional imaging allow a high-quality optical biopsy, enabling *in situ* assessment of tissue and cell function and morphology.

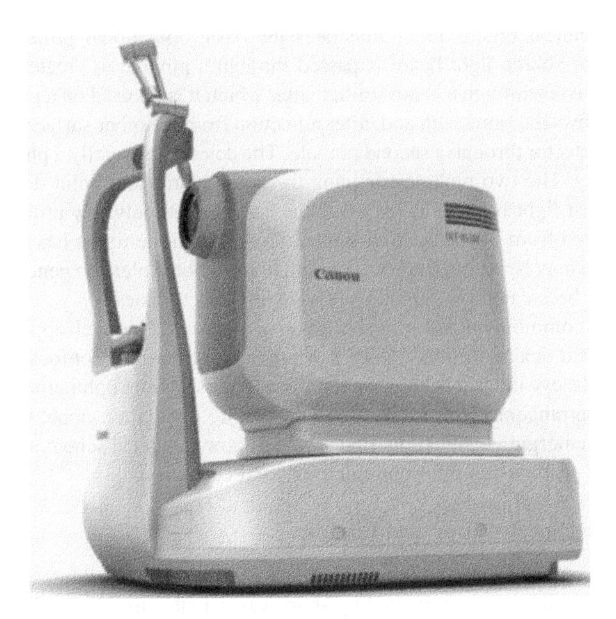

Figure 12.10 OCT fundus examination ophthalmoscope (Courtesy of Canon Europa NV).

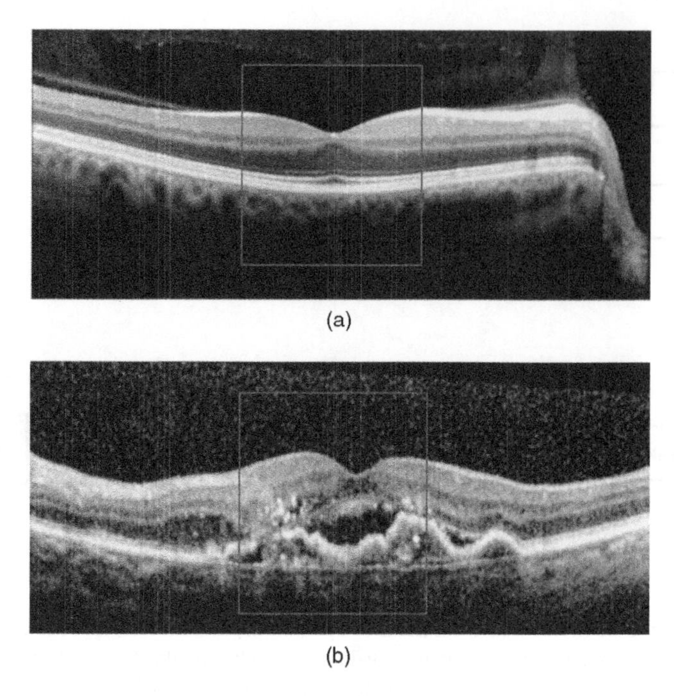

Figure 12.11 Retinal images.

Figure 12.11 shows 2D images generated by such an instrument during a retinal examination. Figure 12.11a shows a retinal image of a normal eye whereas the image shown in Figure 12.11b illustrates age-related macular degeneration.

12.3.3.5 Confocal Imaging

As for optical coherence tomography, confocal imaging is also used for high-resolution imaging of subsurface tissue structures. Figure 12.12 illustrates the basic operational principle of the confocal imaging technique. The source light beam is passed through a pinhole to create a point source. The expanded light beam passes through a beam splitter, after which it is focused on a point inside the tissue. The reflected light follows the same path and, after reflection from the other surface of the beam splitter, is focused on a photodetector through a second pinhole. The detector is usually a photomultiplier tube to achieve high sensitivity. The two pinholes defining the point source and point detector are conjugate points. This implies that light back-reflected off-focus has an extremely low probability of detection. Only the signal collected from the diffraction-limited point within the tissue has a high probability of detection. A laser beam may be scanned across the sample in the focal plane to generate a 2D x–y image. A 3D image may also be created by successively scanning the z-planes.

Confocal imaging is commonly used for *in vivo* imaging of tissues. One such application is in imaging skin tumours to guide surgical resection. Another common application of confocal imaging is in high-resolution imaging of the eye in what is known as confocal laser scanning ophthalmoscopy. Figure 12.13 depicts the schematic arrangement of a typical laser scanning ophthalmoscope. Endoscopic confocal imaging is yet another emerging application. Figure 12.14 shows a typical schematic arrangement of an endoscopic fibre-optic probe using fibre-optic delivery and collection.

12.3.3.6 Fluorescence Spectroscopy and Imaging

Fluorescence spectroscopy and imaging-based diagnostics is based on the fluorescence emission from specific chromophores in tissue when excited by short wavelengths, usually in the blue or ultraviolet

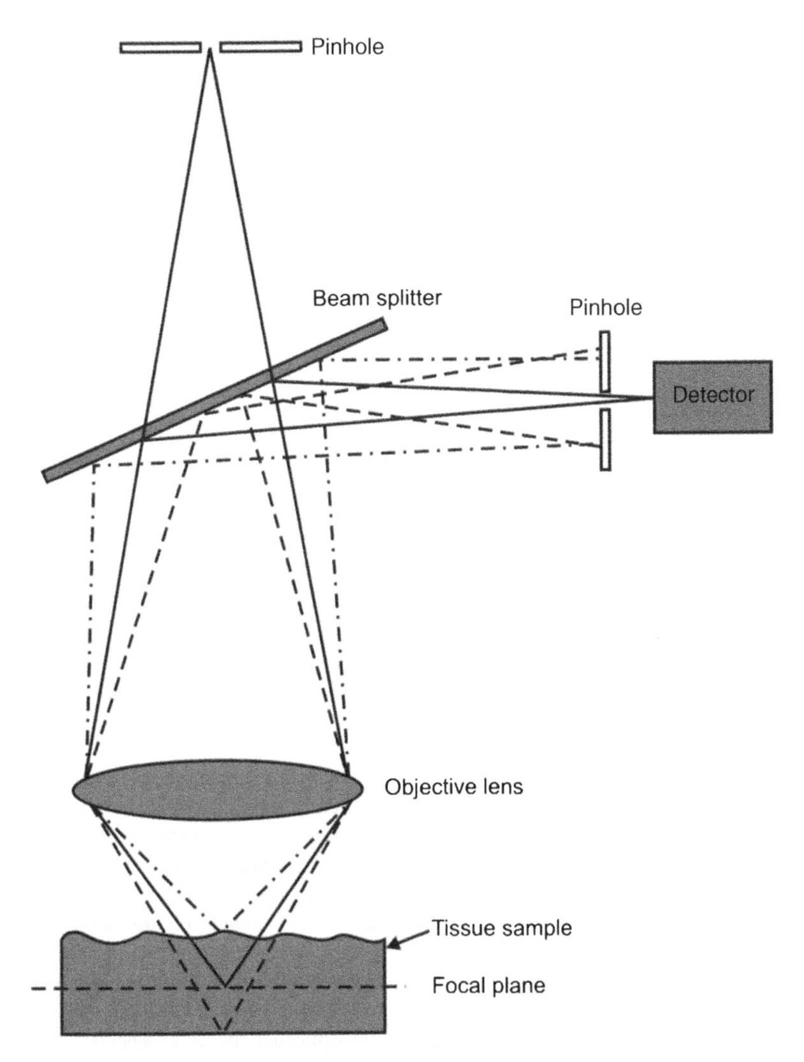

Figure 12.12 Confocal imaging.

range. The emission is in the longer wavelength region of 500–800 nm. Fluorescence-based diagnosis is implemented in two ways: (1) fluorescence is excited in and detected from naturally occurring molecules in the tissue, known as auto-fluorescence; and (2) a known fluorophore is administered either systemically (orally or intravenously) or topically, for example on to skin lesions. In the case of auto-fluorescence, contributing molecules include structural proteins such as collagen and elastin and various metabolic compounds. The onset of disease changes the concentration or spatial distribution of these fluorophores. In the case of exogenous fluorophores, diagnostic information depends upon uptake and/or retention of the administered drug by abnormal tissues.

Although the fluorescence signal is about three orders of magnitude weaker than the diffused reflectance signal obtained from the tissue due to elastic scattering, it is easily measurable by high-sensitivity photodetectors making both point spectroscopy as well as fluorescence imaging possible. Fluorescence-based diagnostic technique is particularly suitable for *in vivo* investigation of biochemical, physiological and structural profile of surface layers in tissues, such as the lining of hollow organs such as lungs, gastrointestinal tract, bladder, oral cavity and cervix. This is important as these are prominent

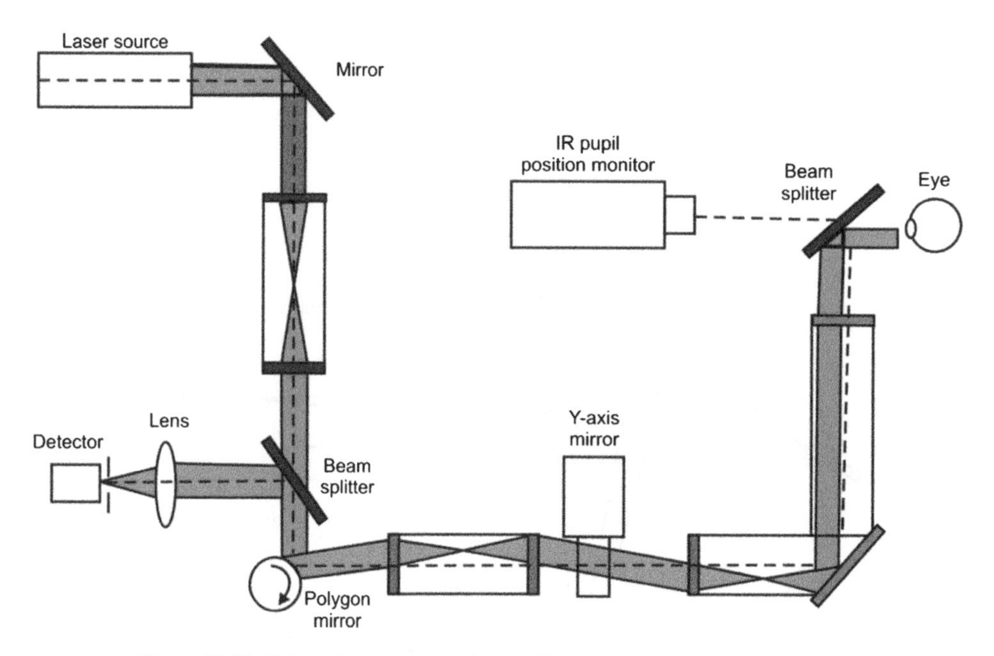

Figure 12.13 Schematic arrangement in a confocal laser scanning ophthalmoscope.

sites for early cancer and pre-malignant changes in tissues. In addition to the use of fluorescence spectroscopy and imaging for the detection of cancers in early stages, it has also been used to guide surgical operations particularly in determining tumour margins with precision. Another possible application is in assisting intravascular operations such as angioplasty used to clear blocked arteries.

There can be various possible fluorescence spectrometer configurations, either using free space optics or fibre optics or a combination of the two for transmit and receive channels. Figure 12.15 shows a typical schematic arrangement of a fibre-optic fluorescence spectroscopy diagnostic set-up. Output from the light source of the desired output wavelength is fed to the double-clad fibre (DCF) coupler through output optics (usually comprising a combination of filters and a focusing lens) and a single-mode fibre (SMF) with the same core diameter as the DCF. The SMF is fusion-spliced to the DCF. A neutral density filter is

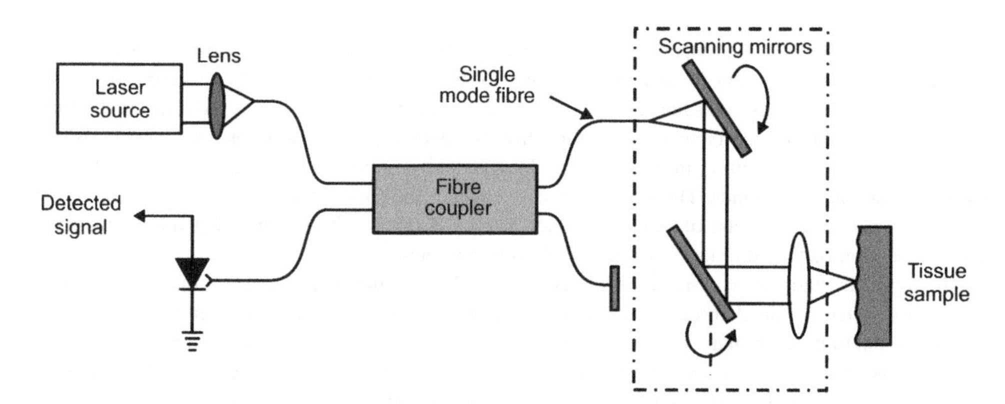

Figure 12.14 Schematic arrangement of endoscopic fibre-optic probe.

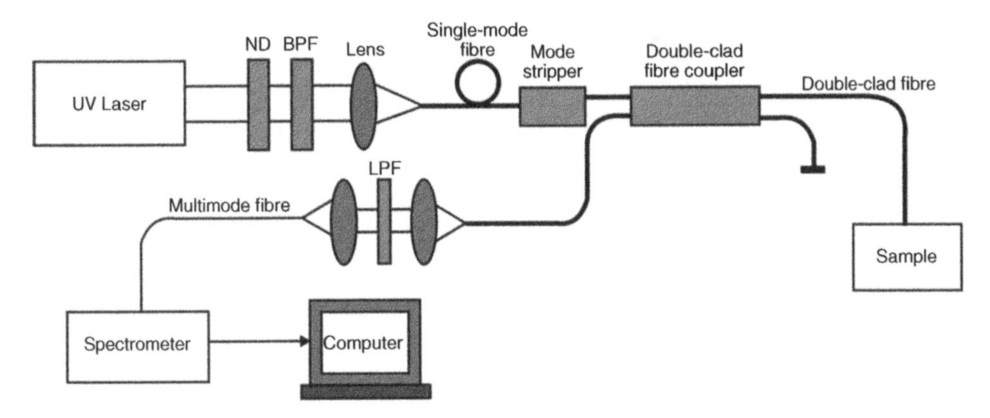

Figure 12.15 Schematic arrangement of fibre-optic fluorescence spectroscopy set-up.

used to adjust the intensity of laser radiation input to the sample and a bandpass filter centred on the laser wavelength is used to eliminate laser background noise. The DCF core is used to deliver laser radiation to the sample, while the inner cladding of DCF is used to collect fluorescence emission from the sample. The receiving channel has a long-pass filter, a beam collimator and a spectrometer as shown. The long-pass filter is used to block fundamental excitation wavelength elastically backscattered from the sample.

Fluorescence spectroscopy and imaging however have some limitations. First, discrimination between normal and abnormal tissues sometimes becomes difficult due to a lack of prior information on optimal excitation and emission wavelengths for various organs and diseases. This limitation is largely overcome in exogenous fluorophores as in that case optimal excitation and emission wavelengths are known. The fluorescence signal is also relatively stronger, allowing use of less-sensitive detectors. Secondly, the detected spectra in fluorescence spectroscopy and imaging is difficult to correctly interpret due to (1) variation in fluorophore concentration with depth; (2) distortion caused by wavelength-dependent absorption by the tissue; and (3) point-to-point and patient-to-patient variation in the auto-fluorescence signal. A fluorescence spectrometer is also relatively more expensive than its reflectance counterpart, which may restrict widespread usage for clinical applications.

12.3.3.7 Raman Spectroscopy

The Raman spectroscopy technique of medical diagnostics depends upon inelastic scattering occurring as a result of light–tissue interaction. There are two forms of inelastic scattering: Stokes scattering, where the photon loses energy on interaction and the scattered photon has a longer wavelength, and anti-Stokes scattering, where the photon gains energy and is scattered at a shorter wavelength. This interaction causes a corresponding change in the vibrational and rotational state of the molecule. Different scattering mechanisms including elastic scattering and two types of inelastic scattering mechanisms were described in Section 12.2.1.1 and illustrated by Figure 12.2. The Stokes signal is relatively stronger than the anti-Stokes signal and, despite the fact that the Stokes signal suffers from the presence of fluorescence background, it is this mechanism that is normally used for diagnostics. The Raman signal in this case is characteristic of the stretching and bending modes of the molecule, thus providing molecular finger-printing. Compared to auto-fluorescence spectroscopy, the Raman signal is very weak. The weak Raman signals prevent Raman spectroscopy from being a practical proposition as an imaging technique for *in vivo* diagnostics, due to the prohibitively large data collection time required for an acceptable signal-to-noise ratio. This is not an issue in the case of microscopy of tissues and cells, however.

Figure 12.16 shows the schematic arrangement of a typical fibre-optic Raman spectrometer. In addition to the well-established use of Raman spectroscopy for analysis of excited tissue samples such as biopsies using bench-top spectrometers, *in vivo* Raman spectrometers are also now available. This is due

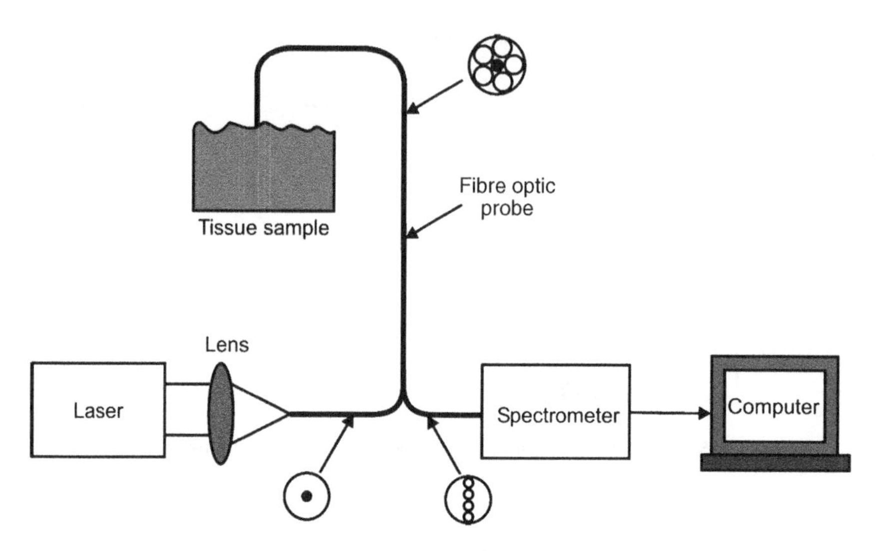

Figure 12.16 Schematic arrangement of fibre-optic Raman spectrometer.

to the advances in near-infrared diode lasers which produce several hundreds of milliwatts of power, high throughput holographic spectrographs and cooled CCD array detectors.

Raman spectroscopy for *in vivo* diagnostics has several serious limitations however, which include the following.

1. The Raman signal as outlined earlier is a weak signal. Signal-to-noise ratio further suffers due to the presence of fluorescence background. Although the Raman signal strength could be increased by increasing sample irradiance, it cannot be increased beyond a few hundreds of milliwatts because of safety concerns. Higher irradiance may increase tissue heating beyond an acceptable limit. Raman signal strength increases with a decrease in wavelength as λ^{-4}. This too is not a practical solution, as a decrease in wavelength towards the ultraviolet is associated with an increase in fluorescence background.
2. The detector often needs to be liquid-nitrogen cooled to reduce detector noise and improve signal-to-noise ratio.
3. Increasing wavelength in an attempt to reduce the fluorescence background causes a higher reduction in the Raman signal due to the λ^{-4} dependence and reduced quantum efficiency of the detector. The fluorescence background can be removed to some extent by taking measurements at two different wavelengths and subtracting the measured signals.
4. Spectral analysis to extract the desired diagnostic information is very complex in the case of Raman spectroscopy, as changes in the tissue due to a disease do not generally produce any marked spectral changes in the form of an increase or decrease in specific signal peaks. The effect for example may be in the form of changes in line widths. As a result, simple algorithms such as taking ratio of signal peaks do not work with Raman diagnostics. More complex spectral analysis techniques, the description of which is beyond the scope of this book, are required.

12.4 Therapeutic Techniques: Application Areas

Light–tissue interactions of relevance to diagnostic and therapeutic applications of lasers were discussed Section 12.2. Interaction mechanisms for therapeutic applications include photothermal, photomechanical and photochemical effects. Applications based on photothermal effects depend upon a temperature rise caused by the absorption of laser energy and are further categorized as photocoagulation and photo-ablation of tissues. While retinal photocoagulation is a routine clinical procedure, photo-ablation enables precise surgical cuts. The photomechanical effect causes explosive ablation of tissue surface with a relatively smaller thermal damage zone and higher ablation efficiency compared to thermal ablation. Photodisruption of tissues is also a routine clinical procedure in ophthalmology for treatment of post-cataract membranes. The photochemical effect is mainly used in photodynamic therapy for selective destruction of pathological tissues such as tumours, pre-cancerous or neovascular tissues. Common therapeutic applications of lasers are briefly described in the following sections.

Common therapeutic applications of lasers include their use in clinical procedures in ophthalmology, dermatology, dentistry, vascular surgery, photodynamic therapy and thermal therapy.

Some of the most widely used clinical procedures in *ophthalmology* (that incidentally have no other alternative therapy) include retinal photocoagulation, photodisruption of post-cataract membranes and photodynamic therapy in age-related macular degeneration. Photorefractive laser surgery of the cornea used to correct for refractive errors such as myopia, hyperopia and astigmatism is another routinely performed clinical procedure in ophthalmic centres.

Photothermal is the most important light–tissue interaction mechanism in *dermatology*. Most clinical applications of dermatology are based on the concept of photothermolysis. With the appropriate choice of laser wavelength, pulse duration and energy, photothermolysis enables selective destruction of targeted biological tissues. Common applications include the treatment of cutaneous vascular lesions, pigmented lesions and tattoos and the removal of unwanted hair.

Laser *dentistry* has rendered many painful and uncomfortable dental operations far less painful. Common applications include caries prevention, cavity preparation, root canal preparation and disinfection (endodontics), calculus removal (periodontics), various types of soft tissue surgery and sterilization, which is killing bacteria without excessive loss of tissue.

A common example of the use of laser therapeutics in *vascular surgery* is in angioplasty, which includes peripheral angioplasty (PA) and percutaneous coronary intervention (PCI), commonly known as coronary angioplasty. PA refers to the use of a balloon to open a blood vessel outside the coronary arteries, usually carried out to treat atherosclerotic narrowing of the abdomen, leg and renal arteries. PCI is a therapeutic procedure to treat the narrowed coronary arteries of the heart.

Photodynamic therapy mainly utilizes the photochemical effect and finds numerous applications in oncology such as in the treatment of small tumours in the wall of hollow organs, areas of pre-cancerous change in hollow organs, localized tumours in solid organs, skin tumours and as an aid in surgical operation to eliminate local residual disease.

Thermal therapy uses the temperature rise caused by the absorption of heat energy for the purpose of treatment of benign and malignant tumours. The ultimate effect depends upon the quantum of heat energy delivered to the target tissue, time duration over which it is delivered and the volume of tissue in which it is absorbed. In addition to the treatment of tumours, another common application of thermal therapy is prevention of bleeding from the edges of surgical incisions.

12.5 Ophthalmology

The most widespread use of lasers in medicine is in the field of ophthalmology for treatment of a wide range of eye disorders. The coherence of laser light allows it to be transported to a specific point without adversely affecting the adjacent areas, and the fact that many of the eye's structures are transparent to the visible spectrum of light makes it an excellent tool for the ophthalmologist. Different interaction mechanisms responsible for various treatment procedures include photocoagulation, photo-ablation, photodisruption and photodynamic therapy.

In *photocoagulation* for the treatment of tumours, laser light in the visible green wavelength is selectively absorbed by haemoglobin, the pigment in red blood cells, to seal bleeding blood vessels. It finds diverse therapeutic applications in ophthalmology for the treatment of a detached retina, the destruction of abnormal blood vessels in the retina and for the treatment of tumours. *Photo-ablation* is volatilization of tissue by laser light in the deep ultraviolet wavelength band. It is mainly used in corrective eye surgery such as LASIK, where it is used to remove tissues by vaporization by transferring

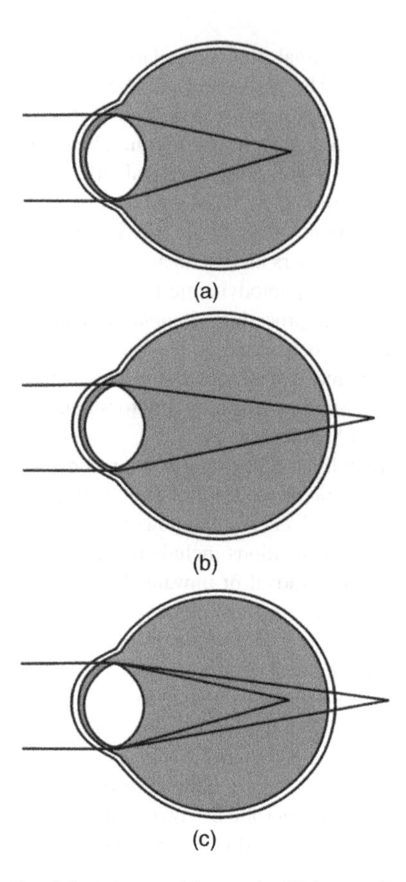

Figure 12.17 Refractive defects in eye: (a) myopia; (b) hyperopia; and (c) astigmatism.

the energy to the target area. Corneal shaping is used to correct the refractive power of the eye. *Photodisruption* is disruption of tissues by a laser-induced rapid ionization of molecules. It is an established method for treatment of post-cataract membranes. *Photodynamic therapy* is used for the treatment of wet age-related macular degeneration. PDT is a form of treatment that makes use of a photosensitive agent administered orally or intravenously. The photosensitive agent becomes concentrated selectively in certain cells. The affected tissues are exposed to ultraviolet light to destroy abnormal tissues.

12.5.1 Refractive Surgery

Refractive defects in human eyes cause blurred vision such as myopia, hyperopia and astigmatism. In the case of a normal eye, the light that enters the eye is focused on the retina to form the image, which is transmitted to the brain. In the case of an eye suffering from refractive defects, the image is not formed on the retina, causing blurred vision. There are three types of refractive defects: myopia, hyperopia and astigmatism. In a myopic eye, an image is formed in front of the retina as shown in Figure 12.17a causing distant vision to be blurred. In the case of hyperopia, the image is formed behind the retina as shown in Figure 12.17b causing blurring of near vision and, with time, also the distant vision. In the case of astigmatism, illustrated in Figure 12.17c, vision is distorted for both near and distant images. For clear vision the images must be formed on the retina, and this can be achieved with appropriate glasses, contact lenses or refractive surgery.

The use of lasers for refractive surgery is one of the most widely used applications of ophthalmology. The primary objective of the treatment is to achieve the correct relationship between the length of the

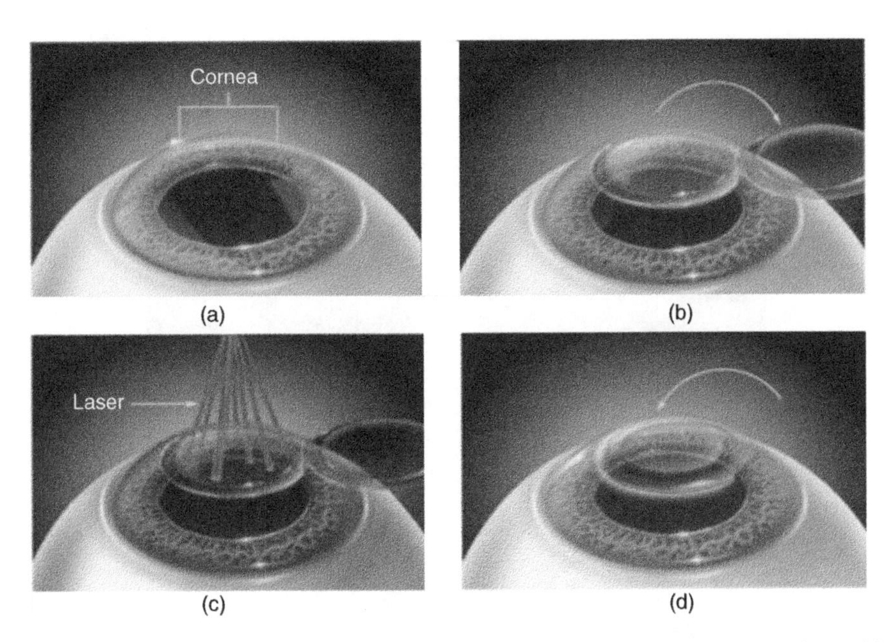

Figure 12.18 LASIK eye surgery: (a) eye before surgery; (b) flap cutting; (c) laser treatment; and (d) repositioning the flap.

eyeball and the power of its optical centres. An excimer laser emitting at 193 nm is commonly used for the purpose. In a commonly used refractive method, pulsed laser radiation with a pulse width in the range of 10–20 ns penetrates the cornea to a depth of 1 μm, leading to photo-ablation of the tissues. Excimer laser ablation is carried out under a partial thickness lamellar corneal flap. Excimer laser refractive surgery is not recommended in patients suffering from glaucoma, diabetes, uncontrolled cardiovascular disease and autoimmune disease. It is also not advisable for pregnant women or people with certain eye diseases.

Laser *in situ* keratomileusis (LASIK), laser subepithelial keratomileusis (LASEK) and photorefractive keratectomy (PRK) are the three established processes of refractive eye surgery. In the case of LASIK, a thin hinged flap (thin slice of cornea) is cut open as shown in Figure 12.18b. The flap is cut by either a microkeratome that involves a blade or a femtosecond laser. Once the flap is cut open, an excimer laser is used to reshape the cornea by removing tissue from the stromal layer of the cornea, thereby correcting the refractive error (Figure 12.18c). The corneal flap is repositioned after refractive correction and allowed to heal (Figure 12.18d). The corneal flap bonds to the cut edge of the cornea naturally and quickly. Figure 12.19 depicts LASIK surgery in action. The whole process takes about 15 minutes.

In LASEK eye surgery, the microkeratome used to make the incision in the cornea is not used. Instead of cutting a flap, the outer layer of the cornea, called the epithelium, is lifted. This is carried out by using a relatively thinner blade called a trephine. The epithelium is treated with 20% diluted alcohol solution to loosen it and detach it from the underlying tissue. A laser is used to reshape the cornea, thereby avoiding the complications associated with cutting and reattaching the corneal flap. At the end of the procedure, a contact lens is inserted to help the epithelium regenerate itself within the next 3–4 days. The procedure is carried out under topical anaesthesia. Figure 12.20 illustrates the different steps involved in the LASEK procedure.

As for the LASIK and LASEK procedures, PRK is based on the reshaping of the cornea by excimer laser, allowing light entering the eye to be properly focused onto the retina. In a PRK procedure, the entire epithelial layer is removed to expose the corneal area to be treated. An excimer laser works on the stromal layer of the cornea to correct the refractive error. In PRK the epithelial layer is lifted, preserved

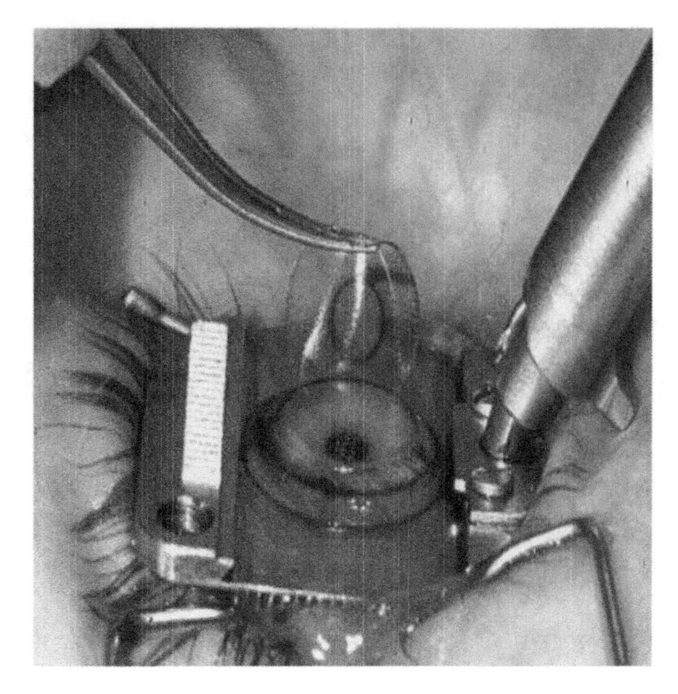

Figure 12.19 LASIK surgery in action.

Figure 12.20 Different steps in LASEK procedure.

during surgery and repositioned at the end of the procedure (as opposed to it being discarded in LASEK surgery to allow a new epithelial layer to grow). The healing of the repositioned layer in PRK takes longer than the growth of a new epithelial layer in LASEK, however.

12.5.2 Treatment of Glaucoma

Glaucoma is an eye disease in which the optic nerve suffers damage. It is caused by increased intraocular pressure, which causes an atrophy of the optical nerve resulting in a reduced field of vision. It may cause irreversible blindness in the patient if not controlled. The most common form is known as simple chronic glaucoma. Its incidence increases from 40 years of age. In the case of glaucoma, vision loss is slow, progressive and irreversible. Early diagnosis and treatment of this disease are very important to avoid serious damage to the vision. The cases in which there is a constant increase of intraocular pressure without any associated optic nerve damage are referred to as ocular hypertension, in which there is no glaucoma damage. If the ocular pressure is normal or low with typical glaucoma, the effect is called normal or low-tension glaucoma. However, raised intraocular pressure is still the most significant risk factor for glaucoma progression. Glaucoma can also be classified as open-angle glaucoma and closed-angle glaucoma; progression of the disease and associated vision loss is much slower in the case of former, while symptoms of the disease can appear rapidly in the latter.

Both laser and conventional surgeries are performed to treat glaucoma. Selective laser trabeculoplasty (SLT) is an efficient (and one of the newest) methods for the treatment of glaucoma. It is a form of laser surgery used to lower intraocular pressure in patients suffering from open-angle glaucoma. In the SLT procedure, a frequency-doubled Q- switched Nd:YAG laser emitting at 532 nm is used. This laser affects the cells of the trabecular meshwork without causing any destruction or coagulation (Figure 12.21). The 532 nm wavelength is highly absorbed by melanin and can selectively target pigmented cells while preserving the surrounding tissue.

The second method of laser treatment in glaucoma is argon laser trabeculoplasty (ALT). This method is focused on drainage correction of fluid from the eye. In open-angle glaucoma the drainage site of the eye does not function normally and the iris is encircled by the trabecular meshwork of an eye. The increase of pressure is due to inadequate drainage. In ALT therapy an argon laser beam is directed at the trabecular meshwork, thereby enhancing fluid drainage efficacy. While the SLT procedure can be repeated on the same area of the eye, the ALT procedure can only be carried out once on a given area. Laser iridotomy is another procedure used to reduce intraocular pressure, performed by drilling a hole in the iris with a laser in order to improve the outflow of aqueous humour towards the drainage angle.

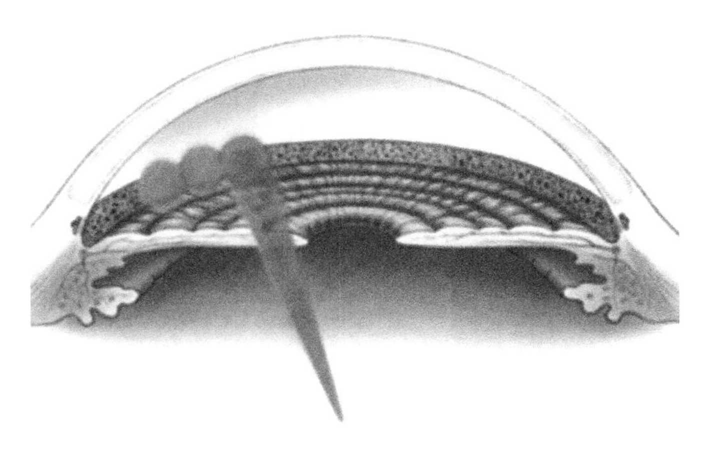

Figure 12.21 Selective laser trabeculoplasty (SLT).

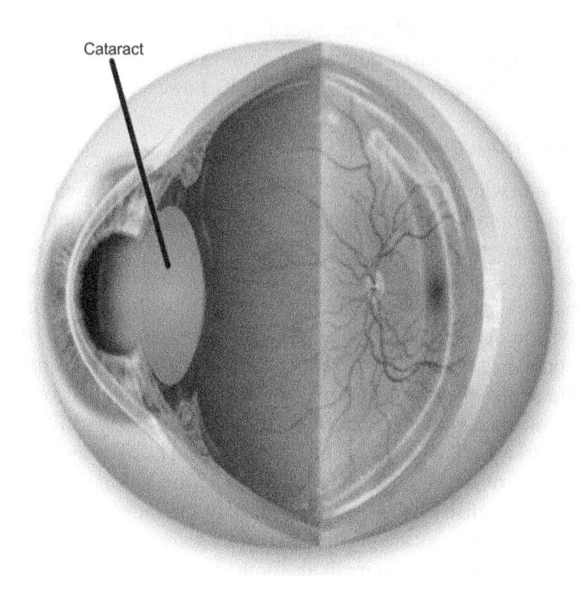

Figure 12.22 Eye affected by cataract.

12.5.3 Cataract Surgery

Cataracts are one of the most frequent causes of loss of vision, which occurs due to opacity of the crystalline lens of the eye as shown in Figure 12.22. Although it is more common in those over the age of 65 years, it may also be present in younger people. Although the cataract is not an ophthalmological emergency, the surgery must be undertaken as and when recommended by the ophthalmologist rather than waiting until the cataract matures. Figure 12.23a and b illustrate the mechanism of loss of vision due to cataracts, depicting a normal eye (Figure 12.23a) and a cataract-affected eye (Figure 12.23b).

The most popular method of cataract surgery is phacoemulsification in which the nucleus is cracked or chopped into smaller pieces to make emulsification easier. Two commonly employed techniques for emulsification include extra-capsular cataract extraction (ECCE) and intra-capsular cataract

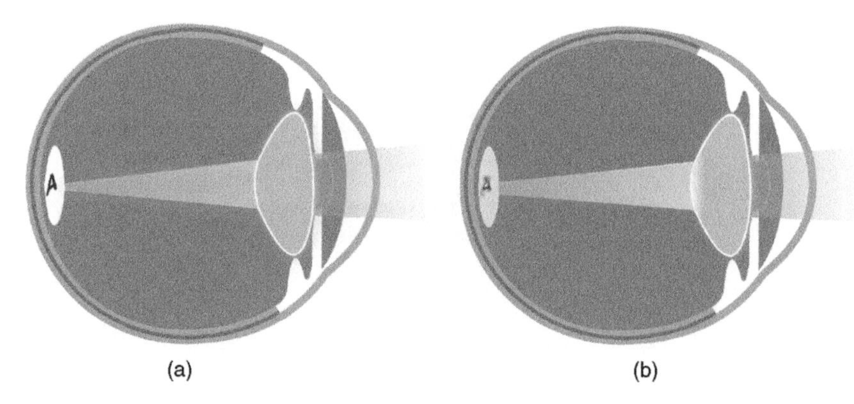

Figure 12.23 Effect of cataract: (a) normal eye; and (b) eye affected by cataract.

extraction (ICCE). ECCE involves the removal of almost the entire natural lens, keeping the elastic lens capsule intact to allow implantation of an intraocular lens. This method is used in patients with very hard cataracts. ICCE on the other hand involves removal of the lens and surrounding lens capsule in one piece. After removal of the lens, an artificial plastic lens is implanted.

Laser surgery is one of the most popular cataract treatments available today. It uses light to dissolve the cataract. Emulsification of the lens using the Er:YAG laser is very effective for performing small-incision cataract surgery in eyes with soft and medium nuclei. The small ablation zones created during treatment help prevent damage to surrounding ocular structures. The Er:YAG technique causes low ablation energy and does not result in thermal injury. A Q-switched Nd:YAG laser with a pulse width of a few nanoseconds and output energy of the order of a millijoule is used to remove post-cataract membranes.

12.5.4 Treatment of Retinal Detachment (Retinopexy)

The retina, a thin membrane covering the inside of the eye, is adhered to the internal surface of the eyeball; it is analogous to the film in a photographic camera. The cornea and the crystalline lens focus the light entering the eye onto the retina, which converts it into a visual impression and transmits to the brain through the optical nerve. The retina receives its nutrition from choroid, the layer behind the retina. In the case of retinal detachment, the retinal membrane is separated from the choroid and floats in the vitreous located in the centre of the eye. Detachment of partial or full retinal membrane leads to loss of vision and complete blinding, respectively. In most cases, retinal detachments are caused either by a tear or perforation in the retina or traction on the retina by vitreous filling most of the ocular cavity. Other factors responsible for retinal detachment could be shrinkage of vitreous and thinning of the retina with age, high myopia, trauma or hereditary reasons.

A number of surgical procedures are in use for treatment of detached retina. These include laser photocoagulation, cryopexy, sceral buckle surgery, pneumatic retinopexy and virectomy. Surgical procedures other than laser photocoagulation will not be discussed here. In the case of laser retinal photocoagulation, the laser beam combined with ophthalmic equipment and optics is focused on the retina. The light absorbed in melanin of the retinal pigmented epithelium (RPE) and choroid is used to cause a controlled damage thereby producing a scar during the healing process. This forms a mechanical connection between neural retina and the choroid preventing detachment of retina. Argon Excimer laser emitting at 193 nm and producing output power in the range of 100–400 mW and retinal spot diameter of about 0.5 mm is commonly used for the treatment. The treatment is usually given at multiple spots.

12.5.5 Treatment of Proliferative Diabetic Retinopathy

Laser photocoagulation is also extensively used to treat proliferative diabetic retinopathy. Diabetic retinopathy is a disease of the retina caused by diabetes that involves damage to the tiny blood vessels at the back of the eye. As the disease enters its proliferative stage, fragile new blood vessels grow along the retina and in the transparent gel-like vitreous that fills the inside of the eye. If untreated, these new blood vessels can bleed, cloud vision and destroy the retina. In order to treat the proliferative stage of diabetic retinopathy, pan-retinal photocoagulation is used. Unlike retinopexy, a much larger number of laser spots (1000–2000) occupying 20–30% of the retinal area in the middle and outer periphery is used. Figure 12.24a and b show photographs illustrating laser photocoagulation treatments in retinopexy and diabetic retinopathy, respectively.

12.6 Dermatology

Dermatological applications of laser therapy make use of the photothermal effect. When absorbed by the tissue, laser energy causes a rise in temperature. Tissue heating causes structural changes in complex molecules such as proteins, DNA and RNA. The structural changes further lead to functional impairment, called denaturation of tissue or grosser structural disorder called coagulation. Coagulation leads to necrosis and vaporization, which further results in tissue ablation and carbonization. The exact photothermal effect produced depends upon the magnitude and the duration of the peak temperature

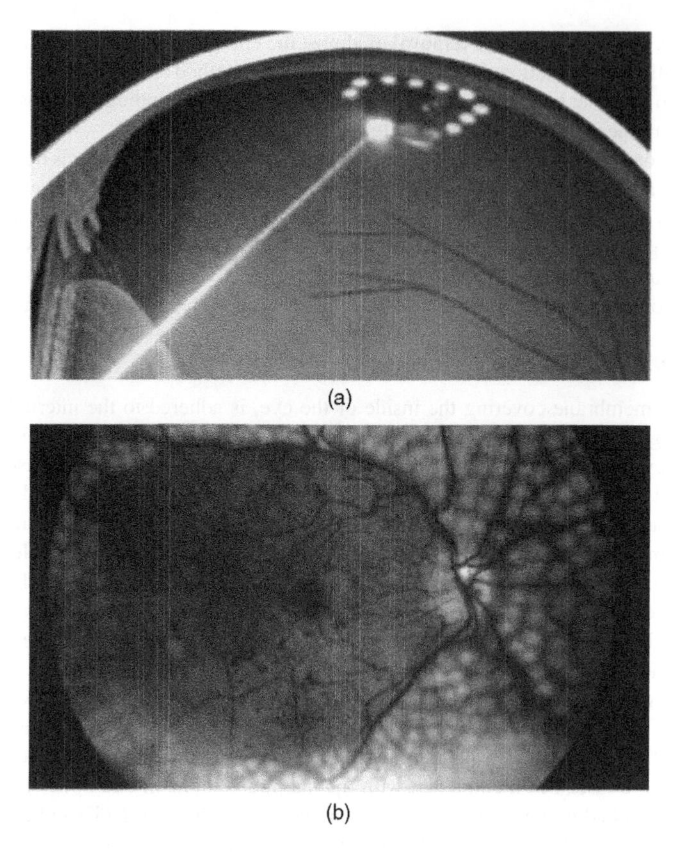
(a)

(b)

Figure 12.24 Laser treatment for retinal detachment: (a) retinopexy; and (b) diabetic retinopathy.

attained by the target tissue on absorption of laser energy. Peak temperature and its duration in turn depend upon the magnitude of laser fluence absorbed by the tissue and duration of the laser pulse. The absorption of laser energy is a function of wavelength. Wavelength also determines the depth to which laser penetrates the tissue. Laser fluence, wavelength and duration are therefore key parameters that determine the photothermal effect produced in the tissue.

This forms the basis of what is known as selective photothermolysis. The concept of selective photothermolysis allows use of an appropriate wavelength, duration and energy fluence to thermally precisely damage the targeted biological tissue without adversely affecting the neighbouring tissue. Appropriate selection of pulse duration can be used to confine the heat energy to the targeted tissue by keeping it less than the thermal time constant of the target. It is this precision that is largely responsible for extensive use of laser therapy for dermatological applications. Having chosen the appropriate value of wavelength and duration, laser fluence is then selected to produce the desired temperature rise in the tissue.

Some common applications of laser therapy in dermatology include laser treatment of cutaneous vascular lesions such as portwine stains (PWS), treatment of pigmented lesions and tattoos and hair removal. Each of these treatment procedures is briefly described in the following sections.

12.6.1 Portwine Stains

A 'portwine' stain is a mark on the skin, usually on the face, resembling port wine in its rich ruby-red colour. It is a type of vascular malformation due to an abnormal aggregation of capillaries. The colour of

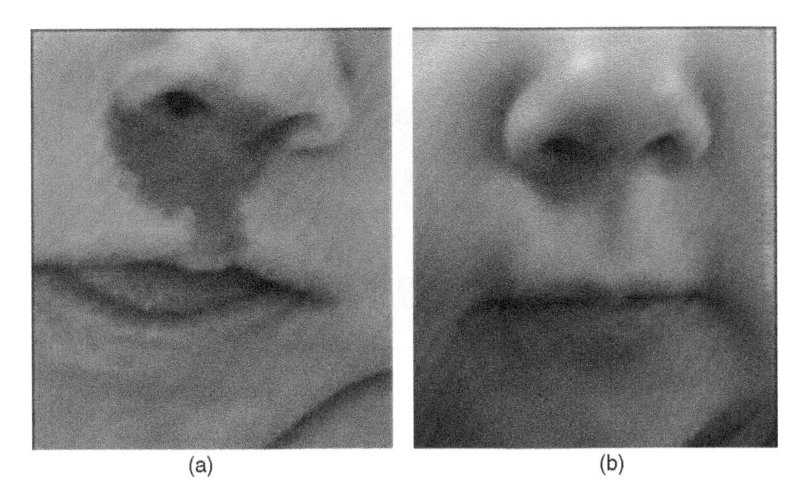

Figure 12.25 Laser treament of portwine stai: (a) before treatment; and (b) after treatment.

the stain partly depends upon the erythrocyte content of these capillaries. Laser treatment of PWS is based on the principle of selective photothermolysis, aiming to reduce the number, size and erythrocyte content of the capillaries resulting in lightening of the stain. A light beam at 577 nm is selectively absorbed by haemoglobin compared to other cutaneous structures. The light penetrates the dermal vessels. Depending upon the laser pulse duration in comparison with thermal relaxation time, there may be non-selective thermal damage of connective tissues resulting in scarring. It is balance between keeping the pulse duration short enough to allow selective absorption, but long enough in order to cause permanent damage to the blood vessels with the consequent lightening of the stain. In the case of patients with dark skin, a relatively higher laser energy needs to be used as a part of it is absorbed within pigmented epidermis and less energy reaches the blood vessels, causing an increased incidence of unwanted post-inflammatory pigmentary changes. This shortcoming is partly addressed by dynamic cooling of epidermis.

Figure 12.25 shows photographs of a child having undergone laser treatment for a portwine stain in the region between upper lip and nose. Figure 12.25a shows the condition before treatment while Figure 12.25b shows the condition after treatment. As is evident from the photographs, the stain has almost vanished, illustrating the efficacy of the treatment.

12.6.2 Pigmented Lesions and Tattoos

Use of lasers for the treatment of pigmented lesions and tattoos is another common application of laser therapy in dermatology. The treatment procedure makes use of the principles of selective photothermolysis. While the target chromophore in the case of pigmented lesions is melanin present within cells as melanosome, tattoos use intradermal, intracellular insoluble ink particles. Following the principles of selective photothermolysis, the laser pulse width is chosen to be less than the thermal relaxation time of melanosome, which is in the range of 100 ns–1 μs. Although melanosomes show significant absorption in the 200–1200 nm wavelength band, operational wavelength is chosen to be within the range 630–1100 nm. This is done to eliminate absorption by competing chromophores such as haemoglobin in the visible band and tissue water in the infrared. Also, longer wavelengths penetrate better with the result that ultraviolet wavelengths are not an optimal choice. The chosen pulse width is less than 1 μs, suitable for selective photothermolysis of melanosomes. Damage is caused both due to thermal effects as well as the shock waves due to rapid thermally induced tissue expansion.

For the treatment of tattoos and pigmented lesions, lasers such as Nd:YAG lasers are generally used. Figure 12.26 shows photographs depicting the efficacy of laser treatment for tattoo removal. However,

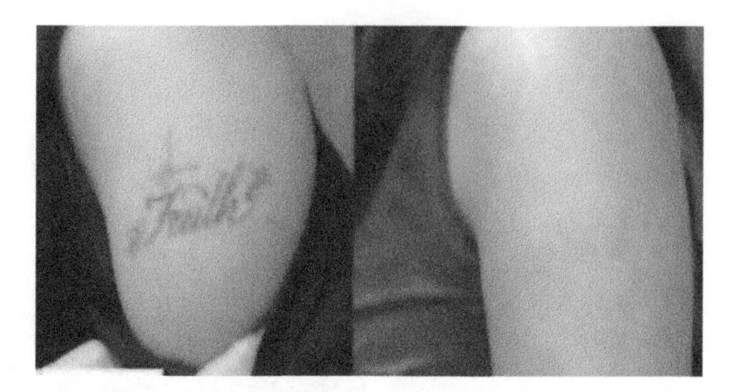

Figure 12.26 Tattoo removal by laser treatment.

different tattoo inks may absorb the laser energy differently, with the result that laser treatment may not completely remove the tattoo in some cases.

12.6.3 Hair Removal

In the case of laser treatment for hair removal by selective photothermolysis, melanin in the hair follicles is the chosen chromophore. Since the base of hair follicles is 2–7 mm below the surface of the skin, red and infrared wavelengths are used for treatment. Diode lasers emitting at around 800 nm and Nd:YAG lasers emitting at 1064 nm are commonly used for the purpose. Since the thermal relaxation time of hair follicles is in the range of several milliseconds to about 100 ms, the chosen pulse width normally falls within the range 1–50 ms.

An important factor to consider in selective photothermolysis of hair follicles is that the epidermis also contains melanin. One of the treatment approaches can be to make use of differential amounts of high melanin content in hair follicles as in dark-coloured hair and low melanin content in pale skin. The other approach is through appropriate selection of pulse duration, making use of different thermal relaxation times of hair follicles (up to 100 ms) and the epidermis (up to 50 ms). Figure 12.27 shows a photograph illustrating the efficacy of laser treatment for the removal of unwanted hair.

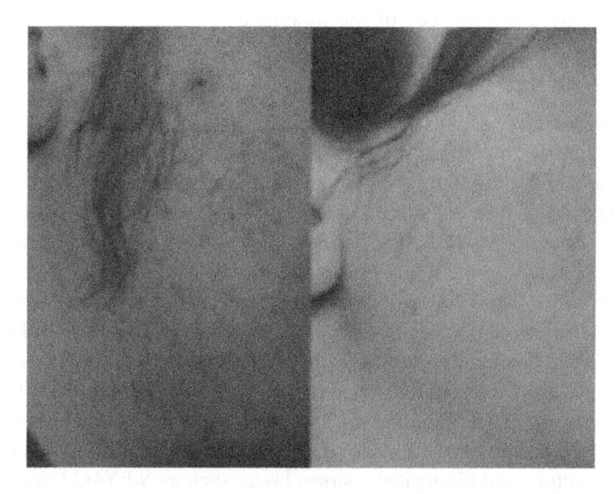

Figure 12.27 Laser treatment for hair removal.

12.6.4 Lasers for Dermatology

Possible types of lasers that have found applications in dermatology include CO_2 lasers, Nd:YAG and Er:YAG lasers, diode lasers, argon lasers, Alexandrite lasers and pulsed dye lasers.

CO_2 laser energy is absorbed by water, and it can be used as a focused or defocused laser. The focused laser is used for treating acne scars, chickenpox scars, fine lines, wrinkles and sun damage. The defocused mode is used to remove skin tags, dermatosis papulosa nigra, warts, syringomas, moles, seborrhoeic keratosis, milia and sebaceous hyperplasia.

Q-switched Nd:YAG lasers are used for blue tattoos, deep tattoos and dermal pigmented lesions. Frequency-doubled Nd:YAG lasers emitting at 532 nm are used for superficial brown lesions, solar lentigenes, freckles and red/orange tattoos. Er:YAG lasers emitting at 2940 nm are used for the treatment of wrinkles, acne scars and pre-cancerous lesions.

Short-pulse dye lasers emitting in the visible band are used to treat vascular lesions such as like portwine stains, naevus flemmus, hemangiomas, keloids and hypertrophic scars. Long-pulse dye lasers are used to treat fine veins, telangiectasia and blushing.

Diode lasers are usually preferred for the removal of brown and black hair. GaAlAs diode lasers emitting at 810 nm are found be very effective on dark skin.

Alexandrite lasers emitting at 755 nm are used for the removal of tattoos and benign lesions. Long-pulse Alexandrite lasers are used for hair removal.

Argon lasers emitting at 488 nm and 514 nm due to their absorption in pigments like haemoglobin and melanin are used to treat skin diseases like A-V malformations, Hemangiomas, fine veins, spider nevi and acne rosacea.

12.7 Laser Dentistry

Although the dental drill continues to be the most widely used mechanism of caries removal and the related processes, the laser holds tremendous potential to make cavity preparation, caries removal and other associated operations a less painful treatment. In addition, laser treatment offers the potential of minimally invasive caries removal and is also a methodology to treat inaccessible areas without adversely affecting the healthy tissue. Common present and potential future applications of laser dentistry include caries removal and the related operations of sterilization, conditioning and curing of fillings. Other areas where laser finds application in dentistry include endodontics involving root canal preparation and disinfection, periodontics involving removal of calculi, disinfection of implants therapy and various types of soft tissue surgery. Lasers offer the possibility of using selective ablation for the removal of caries from inaccessible areas without affecting surrounding tissue.

12.7.1 Considerations in Laser Dentistry

The process of caries removal involves the removal of affected tissue and its replacement by a filling material. During the removal of affected tissue, a geometrical cavity is prepared to ensure better adhesion of the filling material to the cavity. The conventional method of cavity preparation and caries removal involves use of a high-speed turbine. This is a source of discomfort due to vibrations, noise and pain; treatment by laser largely overcomes these problems. Laser dentistry also eliminates many of the more serious problems of high-speed drills, including cracks in the teeth from the vibration caused by the rapidly turning heat-producing drill bit. This can weaken the teeth and cause even more problems. With laser treatment, an extremely narrow focused stream of laser energy gently and precisely removes a wide range of damaged tissue much like a drill, but without the noise, vibration, heat and even anaesthesia in most cases.

However, to make laser treatment procedure as painless as possible, thermal and mechanical side effects have to be minimized. For this, laser absorption should be high enough to confine the heating to the surface. Laser irradiance also needs to be higher than the micro-explosive ablation threshold at an acceptable ablation rate and lower than the value that causes shock waves and interfering repulsion. Laser energy is generally combined with a spray of cool water to gently cut teeth and even bone without

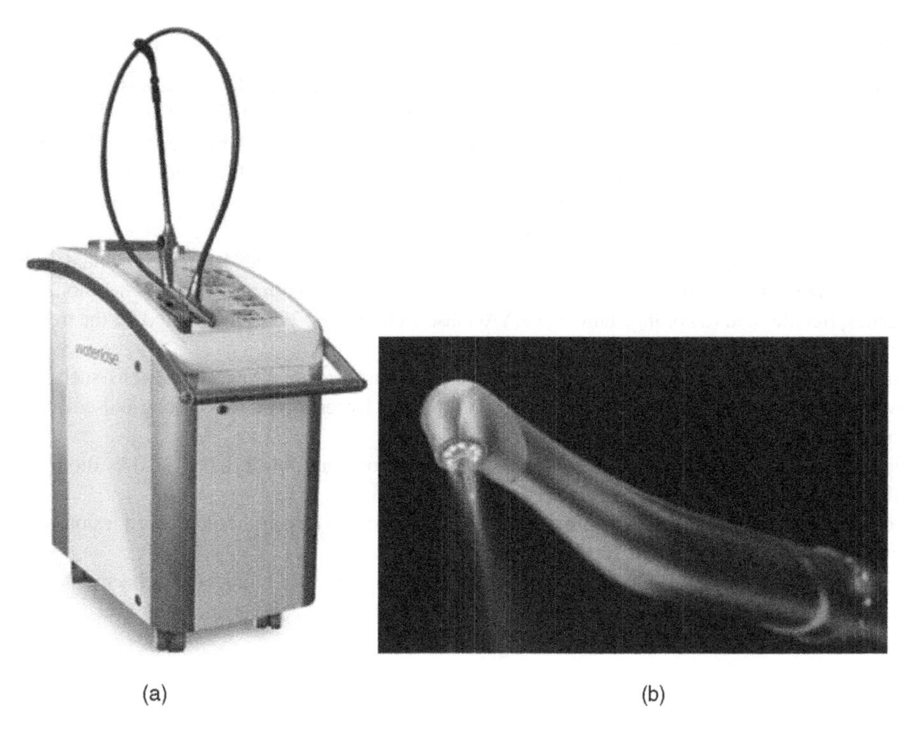

Figure 12.28 Laser dentistry system by Waterlase (12.28 (a) Courtesy of Biolase Technology Inc.).

generating heat, vibration or shock. Figure 12.28 illustrates one such system with its laser/water stream delivery probe.

Laser-assisted caries removal opens up the possibility of treating inaccessible areas. To enable this, laser therapy relies on the intrinsic selective ablation and feedback control. Intrinsic selective ablation is possible if ablation threshold of caries were lower than that of surrounding healthy tissue. A Q-switched frequency-doubled Alexandrite laser has been used to clinically demonstrate this concept. The associated disadvantages however are lower ablation rates and the need to fire the laser even if there is no caries to ablate, resulting in heating of tissue.

Use of feedback control is the other option which can overcome some of the problems mentioned above. In this approach, fluorescence spectroscopy is used for detection of caries and calculi. After detection, a thin burr is used to create an opening for inserting a thin optical fibre to guide the laser radiation to the treatment site. The fluorescence detection is used to control laser irradiance. The laser is made to fire pulses only when the fluorescence feedback signal exceeds a certain pre-determined threshold indicating decay. This method therefore eliminates the need for visual information about the treatment site.

The Key-3 laser dentistry system configured around a pulsed Er:YAG laser from M/s KaVo, USA uses feedback control, which ensures that the laser energy is used exactly as much as is necessary for the treatment. Figure 12.29 illustrates the system.

12.7.2 Lasers for Dentistry

Commonly used laser types for dentistry are Er:YAG laser (2.94 μm), Er:YSGG laser (erbium-doped yttrium scandium gallium garnet, 2.97 μm), Ho:YAG laser (holmium-doped yttrium aluminium garnet, 2.1 μm), CO_2 laser (9.6 μm) and excimer laser (193 nm).

Figure 12.29 Key-3 laser dentistry system (Courtesy of KaVo Dental).

An Er:YAG laser is important as its emission wavelength exactly matches the absorption peak of water molecules. A free-running Er:YAG laser with pulse duration and pulse energy of the order of 0.3 ms and 100 mJ respectively is used. Thermomechanical ablation is the relevant effect; the process leaves a roughness in the cavity walls without affecting the dentine underlying the enamel. This is due to the relatively greater strength and reduced water content of the dentine. An Er:YSGG laser also produces comparable results.

A CO_2 laser with its emission wavelength tuned to 9.6 μm is also suitable for caries removal due to its spectral match with the mid-infrared absorption spectrum of inorganic microcrystals, mainly hydroxy-apatite (HAP). It may be mentioned here that HAP determines the stability and strength of hard tissue bone, dentine and enamel. HP and water determine the absorption characteristics of tissue. An excimer laser emitting at 193 nm is another potential candidate, as the wavelength is highly absorbed by HAP.

Picosecond and femtosecond lasers are other prospective lasers. Ultra-short-pulse lasers have associated thermal and mechanical side effects. Of the above-mentioned laser types, Er:YAG continues to be the most preferred laser for dentistry.

12.8 Vascular Surgery

12.8.1 Conventional Treatment of Angioplasty: PTA

Laser angioplasty is better understood if we are familiar with the background of conventional angioplasty, also known as percutaneous transluminal angioplasty (PTA). PTA is a well-established clinical procedure used for treatment of atheromatous occlusive disease, a condition that refers to the narrowing or occlusion of arterial lumens by plaque made up of free cholesterol, a type of inflammatory cell filled with fat and fibrous tissue. The narrowing or occlusion of coronary and carotid arteries due to accumulation of these plaques manifests as angina. Rupturing of the plaque coupled with thrombosis can lead to myocardial infarcts and strokes.

In PTA, a fine wire is inserted into the arterial lumen through a needle inserted into the artery, usually the femoral artery in the groin. The wire is guided to the point in the artery where it is narrowed by plaque under X-ray guidance. The wire is passed through the narrowed segment of the artery. A thin flexible plastic tube called canula is passed over the guide wire and an angiogram is taken to confirm the extent of

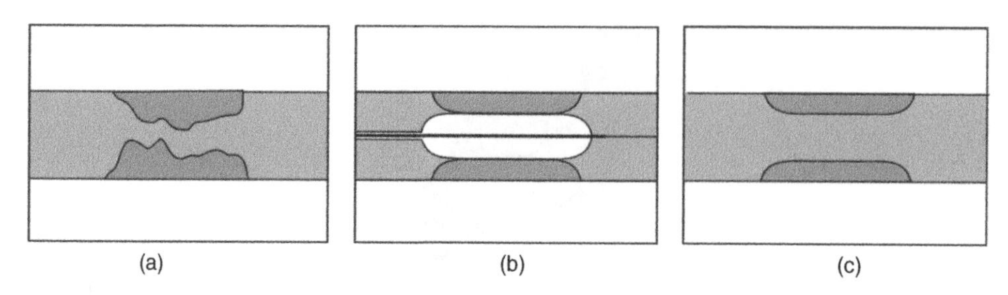

Figure 12.30 Percutaneous transluminal angioplasty: (a) narrowed artery; (b) insertion of guide wire and balloon catheter; and (c) withdrawal of guide wire with balloon catheter.

the narrowing of the artery. After the angiography is performed, the canula is removed and replaced by an undilated balloon catheter slid over the guide wire. Dilatation of the balloon pushes the atheroma into the wall of the artery, thereby opening the lumen to blood flow. Figure 12.30 illustrates the different steps involved in the PTA procedure.

Although very effective for a partial blockage of the arteries, it is not suitable for complete occlusions mainly due to the inability to pass the guide wire through the occlusion. A good percentage of patients having undergone PTA treatment suffer from recurrent arterial blockage, a condition known as restenosis, even after a successful PTA. This is overcome by placing a stent, a slotted metal mesh tube, inside the artery during the PTA procedure. An unexpanded stent mounted on the angioplasty balloon is introduced using a guide wire and aligned with the lesion. The stent expands and becomes embedded into the lesion on inflation of the balloon. The balloon is deflated and subsequently withdrawn along with the guide wire. The diameter of the vessel increases. Stenting significantly reduces the probability of restenosis. The PTA and stenting procedure is diagrammatically shown in Figure 12.31.

12.8.2 Laser Angioplasty

Laser angioplasty is the procedure by which an occluded artery is opened with laser energy delivered through a fibre-optic probe. Laser angioplasty offers a non-surgical alternative to cleansing of clogged arteries with extraordinary precision. The PTA and stenting procedure described in the previous section is best suited for treatment of discrete atherosclerotic stenoses. The success rate of the PTA procedure is however much lower in patients suffering from diffuse atherosclerotic disease and total occlusions, and PTA is also plagued by a high restenosis rate. Laser angioplasty addresses these limitations. In contrast to the PTA procedure where plaque is either compressed or displaced, laser angioplasty ablates the plaque. This allows treatment of previously untreatable lesions and also significantly reduces restenosis rate. With laser angioplasty, it is feasible to pass a fibre-optic catheter through the entire length of coronary circulation and vaporize all atherosclerotic plaque along the arterial wall.

Laser angioplasty is similar to balloon angioplasty with the balloon-tipped catheter replaced by a laser-tipped catheter. A flexible catheter is inserted into an artery in the groin. The catheter is directed under X-ray guidance to thread into the blocked coronary artery. Laser light is delivered through multiple fibre-optic bundles within the catheter. The area of plaque build-up is vaporized by firing short bursts of laser radiation using an excimer laser. Ultraviolet light is much cooler than infrared laser radiation, and so reduces the risk of damaging surrounding tissue. Figure 12.32 illustrates the use of a laser for angioplasty procedure.

12.9 Photodynamic Therapy

Photodynamic therapy (PDT) involves the use of photosensitizer drugs which, on activation by light of appropriate wavelength and intensity, produce photoproducts that can destroy diseased cells or tissues in the presence of oxygen. The drugs are usually administered intravenously. In the case of skin diseases,

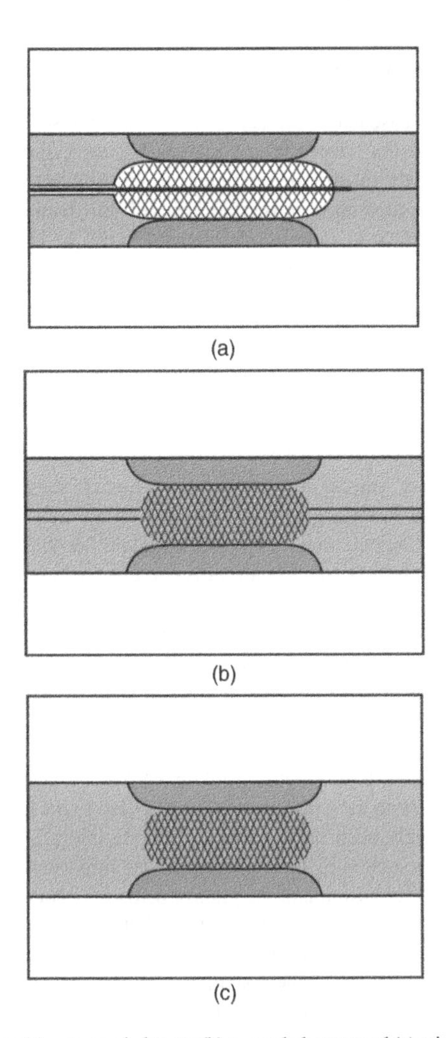

Figure 12.31 PTA and stenting: (a) unexpanded stent; (b) expanded stent; and (c) withdrawal of stent and guide wire.

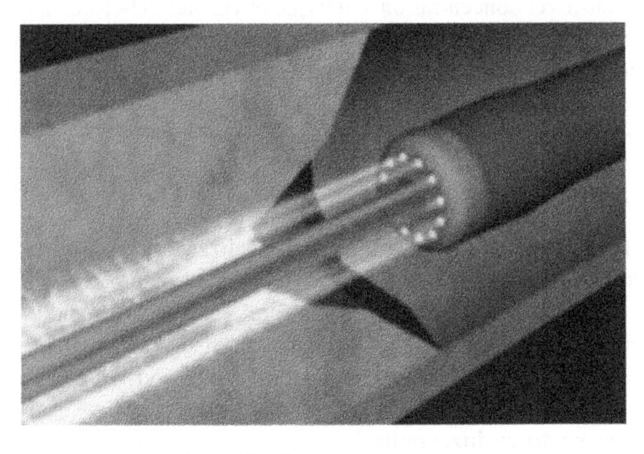

Figure 12.32 Laser angioplasty.

drug administration is by topical application. The photosensitizer is allowed to accumulate in the target tissue or organism over a period of time that could range from minutes to days. The light radiation irradiates the local site of the disease leading to activation of the photosensitizer, which is then followed by generation of photoproducts that destroy the targeted cells or tissue. Clinically approved photosensitizer drugs that are presently in use for treatment of solid tumours, tumours of hollow organs, premalignant tissues, macular degeneration and certain types of skin damage include photofrin, mTHPC, $AlPcS_4$, 5-aminolaevulinic acid (ALA) and protoporphyrin IX (PPIX). For most photosensitizer agents in clinical use or under investigation, the active photoproduct is mainly the excited singlet state of molecular oxygen ($^1O_2$).

12.9.1 Important Considerations

The efficacy of PDT depends upon many factors including the properties of the photosensitizer, the characteristics of the light source and its delivery mechanism, the photobiological response of the cells and tissue, and the optical technologies used for treatment monitoring.

With reference to light source, important considerations include intensity, wavelength and spatial profile. While the intensity and wavelength should be appropriate in order to cause efficient photosensitizer activation in an acceptable treatment time, the spatial profile should match the tissue geometry. Different types of sources of light that are currently in use include high-brightness lamps with appropriate filters, LED arrays and CW and high-repetition-rate pulsed lasers. Lasers are of particular significance as they can be efficiently coupled into fibre-optic cables. Fibre-optic delivery facilitates light delivery into the body either endoscopically or interstitially with the fibre-optic cable inserted directly into tissues through needles. The wavelength band of interest is 630–800 nm. Output power levels used for treatment are in the range 1–5 W.

Argon ion lasers were used earlier up until the 1980s. They were replaced by frequency-doubled Nd:YAG lasers in the late 1980s. A high-repetition-rate Nd:YAG short-pulse laser is photobiologically equivalent to a CW laser power with the same average power. Due to the high cost and relatively bigger size and weight, they have largely been replaced by CW diode lasers.

While using non-laser sources such as LEDs and lamps, a very important consideration is the degree of overlap between the light source emission and the photosensitizer activation spectra. Since the spectral width of a non-laser source is relatively much larger than that of a laser source, the non-laser source would necessarily be of a higher output power rating than a laser source for a given treatment job.

An important requirement of a light delivery system is to efficiently transport light to the target tissue and to distribute it to match the 2D or 3D geometry of the tissue. A multimode fibre is generally used with a simple lens system attached to a fibre tip to achieve the desired spatial profile. A scanning laser beam is used in the case of treatment of retinal diseases by PDT.

Monitoring of PDT treatment is yet another critical requirement. The parameters of interest include tissue irradiance, photosensitizer concentration and tissue oxygenation before, during and after the PDT. Tissue irradiance may be measured with the help of fibre-optic probes with a modified tip to offer a near-isotropic response. Fluorescence or diffused reflectance spectroscopy is generally used for non-invasive *in vivo* measurement of photosensitizer concentration. Both optical and non-optical probes are used for measurement of tissue oxygenation.

12.9.2 Applications of PTD

As outlined earlier, PDT is a method of treatment of localized tumours by prior administration of photosensitizer drug in the presence of oxygen. One of the key features of this treatment is that the connective elements such as collagen and elastin are largely unaffected, which ensures that the mechanical integrity of a hollow organ is maintained. Also, PDT-treated areas heal with more regeneration and less scarring. It is a localized treatment and can be considered to be safe despite the fact that, in the case of intravenous administration, the photosensitizer reaches every part of the body. If acting alone, neither of the photosensitizer or the light dose produce any effect. This allows the boundary of the tissue to be treated to be defined with an acceptable degree of selectivity. PDT is currently in use for treatment of tumours in hollow organs, including areas of pre-cancerous change in hollow organs, and treatment of tumours in solid organs including skin tumours.

PDT is particularly suited to treatment of localized cancers of hollow organs that have not spread beyond the wall of the organ. For example, localized cancers of the mouth have been effectively treated by PDT with porfimer sodium, a purified form of HpD, and mTHPC used as photosensitizer agents.

PDT is particularly attractive for treatment of such cancers as the possibility of cosmetic or functional impairment resulting from other forms of treatment may be unacceptable to most patients. It also suits patients who are otherwise not fit for surgery. Since normal bones are highly resistant to PDT, treatment of mouth cancers that have invaded the upper or lower jaw is also possible without adversely affecting the normal bone.

PDT is also an attractive alternative to surgery for patients suffering from localized cancers of the gastrointestinal tract and major airways if found unfit for surgery. Superficial skin tumours such as basal cell carcinomas can be effectively treated by PDT using ALA photosensitizer. ALA is a naturally occurring substance that leads to the generation of protoporphyrin IX (PPIX), which is a photosensitizer. It is converted to haem, which is an essential substance in all nucleated cells. ALA can be administered through the mouth, produces clinically useful levels of PPIX in 3–6 hours and is cleared from the body in about 24 hours.

Although better established for the treatment of hollow organs and skin cancers, PDT has also been used for the treatment of lesions of solid organs. Because of the fact that connective tissues are not adversely affected by PDT, it is being considered for treatment of some common cancers such as those of lungs, pancreas and prostrate. A flexible endoscope may be used for light delivery in the case of small tumours of major airways. Where airways are too small for endoscopic access, treatment is by interstitial PDT. In this case, the fibre is inserted through the skin up to the cancer site under image guidance. Prostrate and pancreas cancers have reportedly been treated using the mTHPC photosensitizer agent. PDT has also been exploited as additional therapy to conventional surgery to destroy small tumour deposits that may be invisible to the surgeon or those involving vital structures not desirable to be cut or removed.

12.10 Thermal Therapy

Photochemical and photothermal are the two most widely exploited light–tissue interaction mechanisms for the treatment of both benign and malignant cancers. While photodynamic therapy discussed in the previous section is based on the photochemical effects of light–tissue interaction, thermal therapy makes use of the photothermal effect. The efficacy of thermal therapy depends upon the amount of thermal energy delivered to the target tissue, the time over which it is delivered and the volume of the tissue in which it is absorbed.

Thermal therapy finds extensive use in general surgery for bloodless cutting, treatment of haemorrhages in peptic ulcers and cancers of the gastrointestinal tract, lungs and oesophagus. It is also used for treatment of lesions of external genitalia.

Commonly used lasers in thermal therapy include CO_2 lasers, Nd:YAG lasers and diode lasers, emitting at 10 600 nm, 1064 nm and 630–900 nm, respectively. Both Nd:YAG and diode lasers are used in conjunction with fibre-optic delivery. In the following sections we discuss common therapeutic applications of thermal therapy.

12.10.1 Treatment of Haemorrhages of Peptic Ulcers

Peptic ulcer condition can be best managed using a combination of pharmacologic, endoscopic and surgical treatment procedures. The management of peptic ulcer bleeding in the last couple of decades has benefited immensely from the advent of endoscopic haemostasis and potent acid suppressing agents. A combination of pharmacologic and endoscopic therapy is the best treatment for patients suffering from active bleeding ulcers. Although the best way to control bleeding, surgical interventions should only be considered after other treatment procedures have failed.

Endoscopic treatment therapy including injection therapy, thermal coagulation and mechanical haemostasis are widely established treatment procedures for upper gastrointestinal bleeding. Most clinical trials have demonstrated the efficacy of endoscopic therapy in reducing both recurrent bleeding in peptic ulcers and also the need for surgical intervention. Injection therapy involves injections with solutions of diluted epinephrine (1:10 000) and a sclera therapy needle. Solutions of other agents such as poliodocanol, saline and dextrose are equally effective. Mechanical therapy is mainly used for the treatment of variceal haemorrhage and rarely in the treatment of peptic ulcers.

Thermal therapy is the most widely used treatment procedure for endoscopic haemostasis. Thermal therapy is further classified as contact and non-contact. Contact thermal therapy may use a heater probe or electro-coagulation. Non-contact therapy is mainly laser based; a relatively recent non-contact technique is that of argon plasma coagulation. The haemostatic effects of contact therapy and laser-based non-contact therapy are clinically well established. Laser-induced coagulation is based on the absorption of laser energy by tissue proteins leading to rapid heating. In the case of a contact heat probe, heating is through diffusion; in electro-coagulation, heat is produced by the passage of a high-frequency electrical current through the tissue.

Lasers are very effective for direct coagulation of 0.25 mm arteries, but become less effective for larger arteries. Laser light is focused on a bleeding point to induce rapid heating of tissue, which causes blood coagulation and tissue necrosis. A Nd:YAG laser is widely used for the treatment as it offers higher coagulation capacity and increased perforation potential.

12.10.2 Treatment of Cancer

The symptoms of cancers of the oesophagus and gastrointestinal tract usually manifest themselves in the advanced stage of the disease, with the result that patients suffering from these cancers have practically very little hope of cure. One of the major objectives of the therapy is to relieve the patient of the pain and discomfort during the remaining life. Laser endoscopy therapy is a simple procedure to help the patients achieve this objective, and an Nd:YAG laser is generally used for the purpose. Endoscopic Nd:YAG laser canalization can be used in oesophageal cancers wherever there are visible lumps of tumours. Laser therapy can be used to reduce the bulk of these lumps, particularly those causing major obstruction. It can also be used to vaporize prominent or coagulate smaller nodules. The deep penetration of laser prevents bleeding from underlying tissues. Radiotherapy can be used to prolong a period of palliation.

Nd:YAG laser therapy plays a useful palliating role in patients suffering from lung cancer with tumours obstructing major airways, especially when surgical intervention is not possible. The procedure is more complicated and hazardous than in the case of gastrointestinal cancers; while temporary total occlusion of the oesophagus is tolerable, this is not the case for the trachea. There is also a convenient route for debris removal in the case of the gastrointestinal tract; again, this is not so for major airways. In the case of laser therapy for treatment of lung cancer, a flexible endoscope is inserted through a rigid endoscope. As for oesophageal cancers, radiotherapy can be used as a follow-up treatment for recanalization of major airways to extend the palliation period.

CO_2 and Nd:YAG lasers have also been effectively used to treat lesions of external genitalia such as warts and superficial penile cancer. Endoscopy treatment has also been used for the treatment of bladder tumours. Both laser coagulation and electro-coagulation can be used, although laser therapy offers higher precision and better control. It may be mentioned here that treatment of bladder tumours can be managed in the early stage of the disease; bladder tumours usually bleed into urine allowing early detection unlike cancers of the oesophagus and lungs that are detected quite late.

12.11 Summary

- Medical applications of lasers are classified as diagnostic and therapeutic applications. In the case of diagnostic applications, the objective is to know about the pathology and physiology of the tissue through its interaction with light photons. In the case of therapeutic applications, the objective is permanent modification of the tissue in question for treatment of the disease. The optical energy in this case is absorbed by the tissue or an exogenous agent, which further leads to a photochemical, photomechanical or photothermal process depending upon the rate at which the photon energy is absorbed per unit volume of tissue.

- Fundamental mechanisms of light–tissue interaction include absorption, luminescence including fluorescence and phosphorescence and scattering including elastic and inelastic scattering.
- The processes of absorption, fluorescence and phosphorescence are best illustrated in a Jablonski diagram. A Jablonski diagram illustrates the electronic states of a molecule and its associated transitions.
- Tissue optical properties of interest include absorption coefficient, scattering coefficient and scattering phase function. Another important consideration is the propagation of light in the tissue. From the viewpoint of biomedical diagnostics, it is important to measure the magnitude of absorption and scattering coefficients at the wavelength of interest and also their variation as a function of wavelength.
- From the viewpoint of therapeutic applications, the nature of light–tissue interaction can be either photochemical, photothermal or photomechanical.
- Laser diagnostics or optical diagnostics in general is based on changes in one or more measurable properties of light in the ultraviolet, visible or near-infrared wavelength bands as it transmits through and/or reflects from tissues as a result of the interaction of light photons with tissue structure and molecules.
- There are two broad categories of diagnostic techniques: *in vivo* and *in vitro*. *In vivo* diagnostic techniques are either non-invasive or minimally invasive and are performed on a living organism, as opposed to *in vitro* techniques that are used outside a living organism in a controlled environment.
- White light imaging is a type of reflectance imaging in which diagnosis is based on the spectral and spatial appearance of the tissue in the visible range.
- Diffuse optical spectroscopic imaging is a non-invasive optical diagnostic technique that can quantify the absorption and scattering coefficients of tissues. Elastic scattering and absorption are the basis of light–tissue interaction mechanisms responsible for most of the diagnostic applications.
- When light interacts with tissues, there is a near-180° backscattering that occurs at microscale interfaces within tissues due to localized refractive index change. This forms the basis of optical coherence tomography (OCT), enabling generation of high-resolution structural images. Confocal imaging such as OCT is also used for high-resolution imaging of subsurface tissue structures.
- Fluorescence spectroscopy and imaging diagnostics is based on fluorescence emission from specific chromophores in tissue when excited by short wavelengths, usually in the blue or ultraviolet range.
- Raman spectroscopy technique of medical diagnostics depends upon inelastic scattering occurring as a result of light–tissue interaction. There are two forms of inelastic scattering: Stokes (where the photon loses energy on interaction and the scattered photon has a longer wavelength) and anti-Stokes (where the photon gains energy and is scattered at a shorter wavelength) scattering.
- Common therapeutic applications of lasers include their use in clinical procedures in ophthalmology, dermatology, dentistry, vascular surgery, photodynamic therapy and thermal therapy.
- Different interaction mechanisms responsible for various treatment procedures in ophthalmology include photocoagulation, photo-ablation, photodisruption and photodynamic therapy (PDT).
- In photocoagulation for the treatment of tumours, laser light in visible green wavelength is selectively absorbed by haemoglobin, the pigment in red blood cells, to seal bleeding blood vessels. Applications in ophthalmology include treatment of detached retina, destruction of abnormal blood vessels in retina and treatment of tumours.
- Photo-ablation is volatilization of tissue by laser light in the deep ultraviolet wavelength band. It is mainly used in corrective eye surgery such as LASIK, where it is used to remove tissues by vaporization by transferring the energy to the target area. Corneal shaping is used to correct the refractive power of the eye.
- Photodisruption is the disruption of tissues by laser-induced rapid ionization of molecules. It is an established method for treatment of post-cataract membranes.

- PDT involves the use of photosensitizer drugs which, on activation by light of appropriate wavelength and intensity, produce photoproducts that can destroy diseased cells or tissue in the presence of oxygen. The drugs are usually administered intravenously. In the case of skin diseases, drug administration is by topical application. PDT is particularly suited to the treatment of localized cancers of hollow organs that have not spread beyond the wall of the organ.
- Thermal therapy exploits the photothermal effect. The efficacy of thermal therapy depends upon the amount of thermal energy delivered to the target tissue, the time for which it is delivered and the volume of the tissue in which it is absorbed. Thermal therapy finds extensive use in general surgery for bloodless cutting, treatment of haemorrhages in peptic ulcers and cancers of gastrointestinal tract, lungs and oesophagus. It is also used for treatment of lesions of external genitalia.
- Dermatological applications of laser therapy make use of the photothermal effect. Some common applications of laser therapy in dermatology include laser treatment of cutaneous vascular lesions such as portwine stains (PWS), treatment of pigmented lesions and tattoos and hair removal.
- Common applications of laser dentistry include caries removal and the related operations of sterilization, conditioning and curing of fillings. Other areas where the laser finds application in dentistry include endodontics involving root canal preparation and disinfection, periodontics involving removal of calculi, disinfection of implants therapy and various types of soft tissue surgery.
- Laser angioplasty is the procedure by which an occluded artery is opened with laser energy delivered through a fibre-optic probe. Laser angioplasty offers a non-surgical alternative to cleansing of clogged arteries with extraordinary precision.

Review Questions

12.1. Briefly describe light–tissue interaction mechanisms of relevance to laser diagnostics and laser therapeutics. Give examples of diagnostic applications and therapeutic procedures that make use of these interaction mechanisms.

12.2. Distinguish between *in vivo* and *in vitro* diagnostic techniques. Briefly describe the operational principles and applications of elastic scattering spectroscopy and confocal imaging types of *in vivo* optical diagnostic techniques.

12.3. Name common ophthalmological applications of lasers. Briefly describe the use of laser technology for refractive surgery.

12.4. What do you understand by retinopexy? How is laser photocoagulation used for treatment of this disease?

12.5. What is the underlying principle of use of thermal therapy for treatment of tumours? Briefly describe the procedure for use of thermal therapy for treatment of peptic ulcers and tumours of hollow organs.

12.6. Which photo effect is mainly responsible for use of laser therapy in dermatology? Briefly describe the procedure for treatment of a portwine stain. Which laser(s) is (are) usually preferred for the treatment and why?

12.7. How does laser angioplasty essentially differ from conventional percutaneous transluminal angioplasty (PTA)? What are the advantages of laser angioplasty over PTA?

12.8. Briefly describe the treatment procedure for laser angioplasty.

12.9. Name the possible treatment procedures where laser technology can be used with advantage over conventional techniques in dentistry. Compare and contrast laser-assisted caries removal with the conventional drill-based procedure.

12.10. Name any five types of lasers that find applications in therapeutic and diagnostic procedures along with their major specifications and application areas.

12.11. Write short notes on the following:
a. photodynamic therapy
b. selective photothermolysis
c. photocoagulation
d. LASIK

Self-evaluation Exercise

Multiple-choice Questions

12.1. If the wavelength of the scattered radiation is the same as that of incident wavelength, then it is
a. elastic scattering
b. inelastic scattering
c. Raman scattering
d. none of these.

12.2. If the scattering particle size is much smaller than the incident wavelength, then the scattering is referred to as
a. Mie scattering
b. Rayleigh scattering
c. Stokes Raman scattering
d. anti-Stokes Raman scattering.

12.3. If the scattering particle size is comparable to the incident wavelength, then the scattering is referred to as
a. Rayleigh scattering
b. Stokes Raman scattering
c. Mie scattering
d. anti-Stokes Raman scattering.

12.4. Refer to Figure 12.33a–d. The diagrams show part of a possible Jablonski diagram indication absorption, fluorescence and non-radiative transitions. Which of these diagrams is correct?
a. Figure 12.33a
b. Figure 12.33b
c. Figure 12.33c
d. Figure 12.33d

12.5. The principle of Michelson interferometry is put to use in
a. diffused reflectance spectroscopy
b. elastic scattering spectroscopy
c. optical coherence tomography
d. confocal imaging.

12.6. One of the most important applications of photochemical effect in laser therapeutics is in
a. photothermal therapy
b. photodynamic therapy
c. photomechanical therapy
d. photocoagulation.

12.7. One of the most important applications of the photothermal effect in laser therapeutics is in
a. laser coagulation
b. photodynamic therapy
c. laser dentistry
d. none of these.

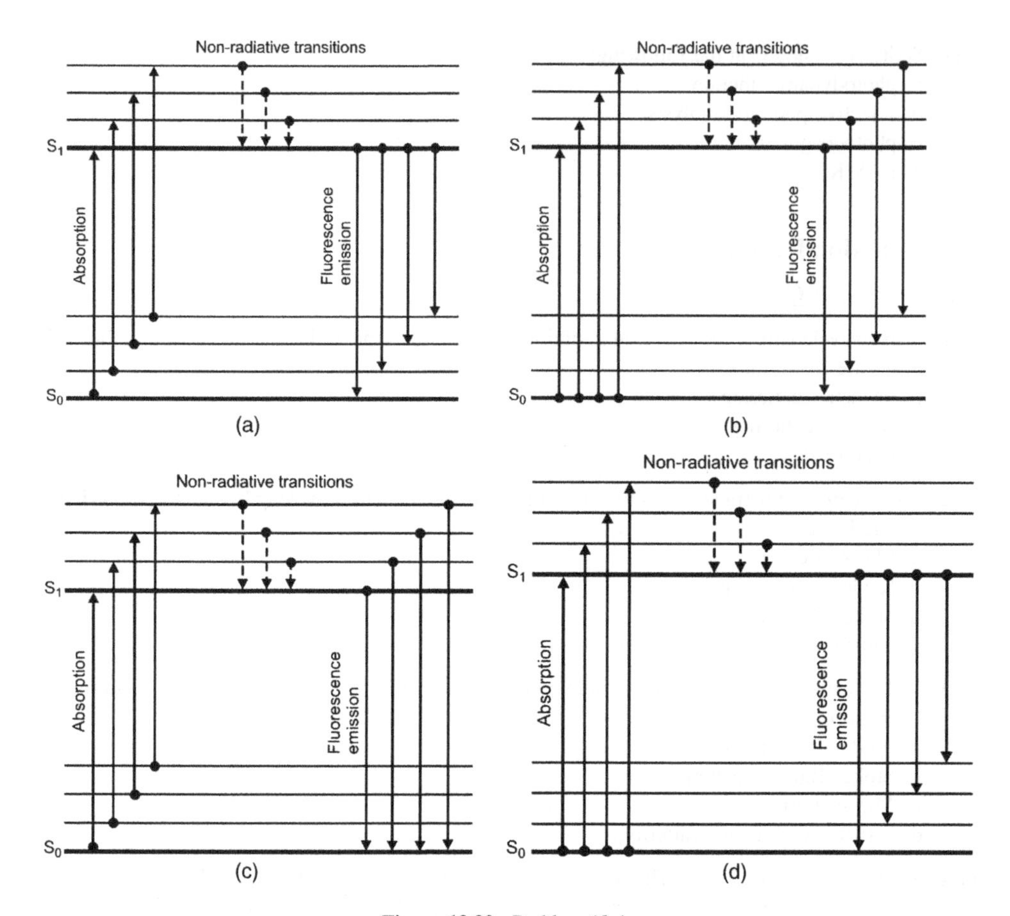

Figure 12.33 Problem 12.4.

12.8. The preferred laser in laser dentistry applications is the
   a. Nd:YAG laser
   b. Er:YAG laser
   c. CO_2 laser
   d. argon laser.

12.9. The preferred laser for LASIK surgery is the
   a. excimer laser
   b. argon laser
   c. frequency-doubled Nd:YAG laser
   d. Er:YAG laser.

12.10. Which wavelength band would have deeper tissue penetration?
   a. ultraviolet wavelength band
   b. visible wavelength band
   c. infrared wavelength band
   d. none of these.

12.11. The light output from which of the following lasers is not amenable to transmission through fibre-optic cables?
a. diode laser
b. Nd:YAG laser
c. CO_2 laser
d. Er:YAG laser.

12.12. Which of the following lasers is not popular for laser dentistry?
a. Er:YAG laser
b. frequency-doubled Nd:YAG laser
c. Er:YSGG laser
d. excimer laser.

12.13. The wavelength band usually preferred for tattoo removal applications is
a. 630–1100 nm
b. 200–1200 nm
c. 400–1300 nm
d. 600–800 nm.

12.14. The concept of use of appropriate laser wavelength, pulse duration and energy fluence to thermally damage the targeted biological tissue with precision without adversely affecting neighbouring tissue is called
a. photocoagulation
b. photosynthesis
c. selective photothermolysis
d. photo-ablation.

12.15. Oxy-haemoglobin and water show absorption peaks at
a. 418 nm and 2940 nm, respectively
b. 418 nm and 1064 nm, respectively
c. 577 nm and 1064 nm, respectively
d. 2940 nm and 418 nm, respectively.

Answers

1. (a) 2. (b) 3. (c) 4. (d) 5. (c) 6. (b) 7. (a) 8. (b) 9. (a) 10. (c) 11. (b) 12. (b) 13. (a) 14. (c) 15. (a)

Bibliography

1. *Handbook of Laser Technology and Applications: Volume III*, 2003 by COLIN E. WEBB and Julian DC Jones, Institute of Physics Publishing.
2. *Lasers in Medicine*, 2001 by Ronald W. Waynant, CRC Press.
3. *Medical Applications of Lasers*, 2002 by D.R. Vij and K. Mahesh, Springer.
4. *Laser Tissue Interactions: Fundamentals and Applications*, 2007 by Markolf H. Niemz, Springer.
5. *Lasers in Dermatology and Medicine*, 2011 by Keyvan Nouri, Springer.
6. *Applied Laser Medicine*, 2003 by Hans-Peter Berlien and Gerhard Muller, Springer-Verlag.

13

Lasers in Science and Technology

13.1 Introduction

Since their inception in the 1960s, lasers have been used in industry (Chapter 11), in medicine (Chapter 12) and in defence for a host of applications from target ranging and designation to tracking and imaging and from electro-optic countermeasures to their use as directed-energy weapons. In addition to these more visible applications of lasers, there are many important areas where lasers have been used in pursuit of science and technology. Some of the important areas include their use in *optical metrology*, mainly exploiting laser interferometry for making precise engineering measurements such as length, velocity, vibration, stress and strain; in Earth and environmental sciences for the remote sensing of Earth's environment and *satellite laser ranging*, contributing to geodetic studies such as determining the variation of Earth's centre of mass, detection of Earth and ocean tides, modelling temporal and spatial variation of Earth's gravity field and related studies; in *astronomy* such as in the use of the laser guide star technique in astronomical telescopes; and *laser holography*, which has a wide range of applications in industry, medical diagnostics and optical computing. We discuss some of these topics in detail in this chapter.

13.2 Optical Metrology

As a source of electromagnetic radiation with exceptionally high frequency stability, a laser is the basis of length/distance metrology. Optical frequency metrology allows high-precision measurements of optical frequencies, the atomic clock being an example. Laser interferometry is at the core of length metrology, measurement of time-of-flight of laser pulses being the other commonly used technique. Interferometric techniques such as self-heterodyne interferometers are used for the measurement of wavelength and line width. The measurement of temperature is another common application. In the following sections, beginning with an overview of interferometers (which form the basis of many engineering measurements), the concepts of measurement of length/distance, frequency, wavelength and line width and finally temperature are briefly described. Other related topics such as laser velocimetry, laser vibrometry and electron speckle pattern interferometry are discussed in Sections 13.3–13.5.

13.2.1 Interferometers

An interferometer is an optical device that makes use of the interference phenomenon. An optical beam is split into two separate beams travelling two different paths before they are recombined to produce an interference pattern. Beam splitters/combiners are used for the purpose. Certain interferometers use a common beam path but different polarizations, which makes them immune to measurement errors caused by geometric path length fluctuations. In the more common types of interferometers, while

Lasers and Optoelectronics: Fundamentals, Devices and Applications, First Edition. Anil K. Maini.

travelling through their designated paths the two beams are subjected to some external influences such as a variation in refractive index of the medium and path length differences caused by the parameter under measurement. The interference pattern is then analyzed for the intended measurement.

A number of interferometers have been developed over the years for various applications. Some common types include the Michelson interferometer, the Twyman–Green interferometer, the Mach–Zehnder interferometer, the Fabry–Pérot interferometer and the Sagnac interferometer. All these interferometers can be implemented with both free-space optics as well as fibre optics. In fibre optic interferometers, fibre couplers replace beam splitters. With fibre optic interferometers, the potential difficulties posed by a change in polarization state while propagating through the fibre and optical phase shifts caused by temperature fluctuations have to be addressed. The different interferometers mentioned above are briefly described in the following sections.

13.2.1.1 Michelson Interferometer

The Michelson interferometer, invented by Albert Abraham Michelson, uses a single beam splitter for separating and combining the beams as shown in the basic schematic arrangement of Figure 13.1a.

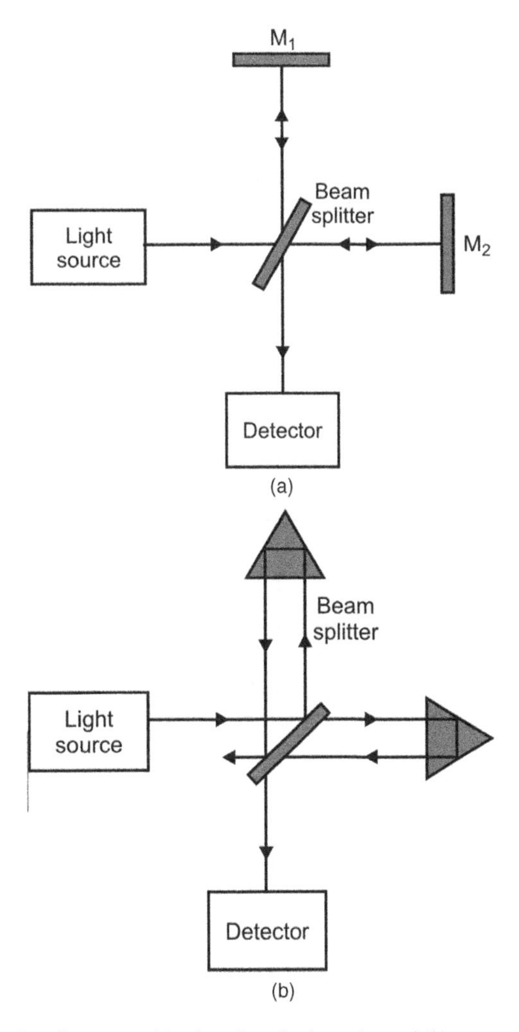

Figure 13.1 Michelson interferometer: (a) mirror-based schematic; and (b) retro-reflector-based schematic.

In this basic arrangement, if two mirrors were aligned for exact perpendicular incidence, only output from mirror M_2 would be accessible and the other output (from mirror M_1) would go back to the light source. If optical feedback to the source is undesirable and/or this output was also required, simple mirrors with slightly non-normal incidence or retro-reflectors instead of simple mirrors as shown in Figure 13.1b can be used. The use of retro-reflectors has the added advantage that the interferometer becomes insensitive to slight misalignments.

A constructive or destructive interference pattern results from a non-zero path length difference. If a white light source is used in the interferometer, it is necessary that the path length difference between the two beams is precisely zero. In the case of a monochromatic source of light such as a laser, the path length difference needs only to be less than the coherence length of the light source. A Michelson interferometer is a very useful optical tool for the measurement of wavelength, distance, refractive index and coherence length of optical beams.

13.2.1.2 Twyman–Green Interferometer

The Twyman–Green interferometer introduced by Frank Twyman and Arthur Green is essentially an adaptation of the Michelson interferometer. While the Michelson interferometer may use a broadband white light source or a laser, the Twyman–Green interferometer essentially uses a monochromatic point source of light. It is used for characterizing optical elements such as lenses, prisms and flats. Figure 13.2 shows the basic optical schematic of the interferometer to be used for characterizing lenses and other curved surfaces. As shown in the diagram, light from the point laser source is first collimated into a parallel beam. The collimated beam is then split into two paths with the help of a beam splitter. The optical component such as a lens under test is placed in one of the arms.

If testing a lens component, a convex spherical mirror with its centre of curvature coinciding with the focus of the lens being tested constitutes the other end component. The emergent beam is recorded by an imaging system for analysis. The fringe pattern is analyzed to test the quality of the optical component. To test optical flats, the component under test is the end component and a reference flat is used in the reference arm. To test prisms, the prism under test is placed in the path of the probe beam.

13.2.1.3 Mach–Zehnder Interferometer

The Mach–Zehnder interferometer developed by Ludwig Mach and Ludwig Zehnder is an optical tool used to determine the relative phase shift between two collimated beams from a coherent light source.

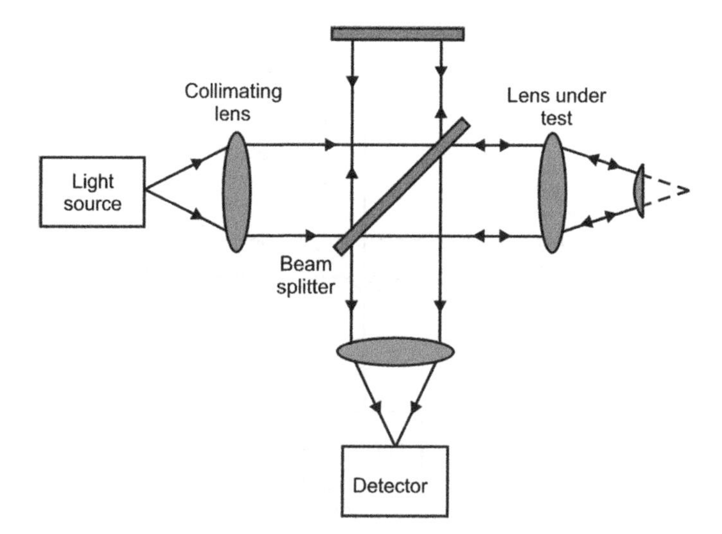

Figure 13.2 Twyman–Green interferometer for testing lenses or curved surfaces.

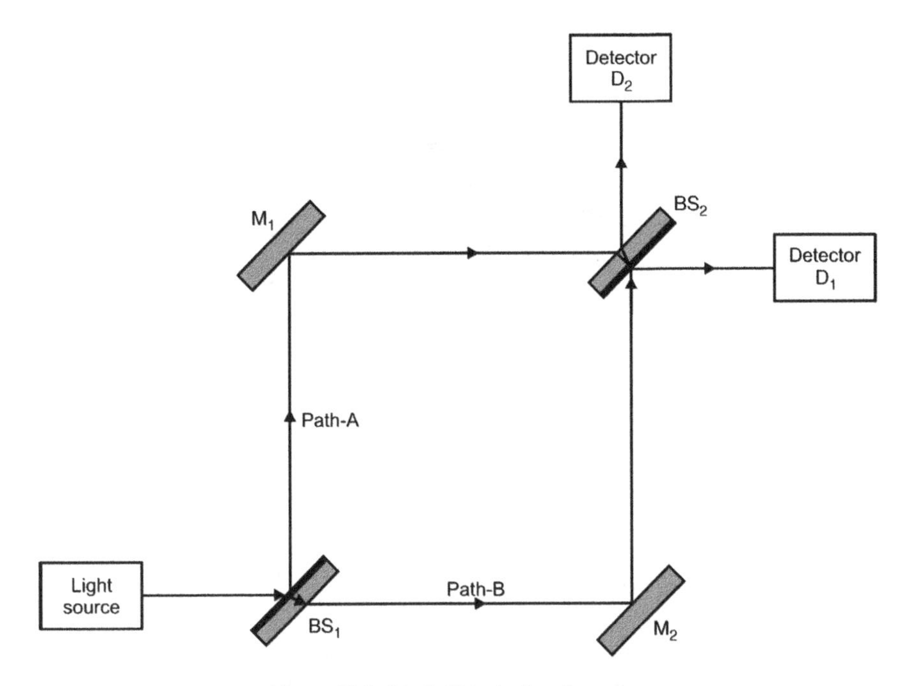

Figure 13.3 Mach–Zehnder interferometer.

Figure 13.3 shows the optical schematic of the interferometer. Compared to the Michelson interferometer which employs a single beam splitter for splitting and recombination of the beams, a Mach–Zehnder interferometer uses two separate beam splitters: one for splitting the light beam from the light source into two different paths and the other to combine the two beams.

The beam splitter BS_1 splits the light beam usually equally into two paths. Part of the beam reflected towards the fully reflecting mirror M_1 (path A) undergoes a phase shift corresponding to $\lambda/2$ at the reflecting surface of mirror M_1. It then travels towards beam splitter BS_2 and is further divided into two parts, one reaching detector D_1 and the other reaching detector D_2 after reflection from the reflecting surface of beam splitter BS_2. Beam splitter BS_2 does not introduce any phase shift at its reflecting surface. Also, part of the beam of path A reaching detector D_2 undergoes a phase shift of twice the path length in beam splitter BS_2 and that reaching detector D_1 corresponding to the single path length in beam splitter BS_2.

Part of the main beam transmitted towards the fully reflecting mirror M_2 (path B) undergoes a phase shift corresponding to $\lambda/2$ at the reflecting surface of mirror M_2 and also at the reflecting surface of beam splitter BS_2 before reaching detector D_1. The part of the path B beam reaching detector D_2 undergoes a phase shift corresponding to a single path length in beam splitter BS_2.

The combined beam reaching detector D_1 therefore has contributions from path A and path B and the components are in phase. On the other hand, the combined beam reaching detector D_2 also has two components which arrive at detector D_2 out of phase. This leads to constructive interference at detector D_1 and destructive interference at detector D_2. This will be true regardless of wavelength. If a sample is introduced in either of the two paths, the phase shift introduced by the sample alters the phase relationships between the two beams. Measuring the relative amount of light entering detector D_1 and detector D_2 allows computation of the phase shift introduced by the sample.

13.2.1.4 Fabry–Pérot Interferometer

The Fabry–Pérot interferometer, developed by Charles Fabry and Alfred Pérot, consists of two parallel optical flats with partially reflecting mirror surfaces facing each other. The flats in an interferometer have

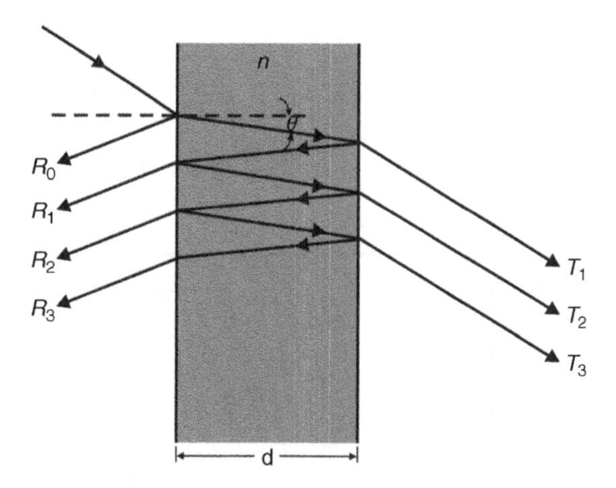

Figure 13.4 Fabry–Pérot interferometer.

an anti-reflection coating and are often made in a wedge shape to prevent the rear surfaces from producing interference fringes. The monolithic version of the interferometer, also referred to as an etalon, consists of a transparent plate with two reflecting surfaces. The transmission spectrum of the interferometer exhibits peaks corresponding to certain wavelengths. A Fabry–Pérot interferometer differs from a Fabry–Pérot etalon in that it allows the distance d between the plates to be fine tuned to make transmission peaks correspond to the desired wavelength. Since transmission is angle dependent, the peaks can also be shifted by rotating the etalon with respect to the beam. Figure 13.4 illustrates the operational principle of a Fabry–Pérot interferometer or etalon.

In both cases, whether interferometer or etalon, the changing transmission function is produced by interference between the multiple reflections that take place between the two reflecting surfaces. There is constructive interference if the transmitted beams are in phase, leading to a transmission peak or maxima. If the transmitted beams are out of phase, destructive interference produces a transmission minima. The phase relationship between the transmitted beams that ultimately determines the spectral response of the interferometer or etalon depends upon wavelength λ; refractive index n of the material between the two reflecting surfaces; distance d between the two reflecting surfaces; and the angle θ at which light travels in the space between the reflecting surfaces. The condition of maxima is defined by $m\lambda = 2nd\cos\theta$, where m is the order of interference.

Important parameters characterizing the interferometer or etalon include free spectral range $\Delta\lambda$, full-width half-maximum (FWHM), spectral width $\delta\lambda$ of transmission peak and finesse F. The free spectral range is defined as the wavelength separation between two adjacent peaks, and is given by:

$$\Delta\lambda = \frac{\lambda_0^2}{2nd\cos\theta + \lambda_0} \cong \frac{\lambda_0^2}{2nd\cos\theta}.$$

where λ_0 is the centre wavelength of the nearest transmission peak. Finesse F is defined as the ratio of free spectral range to FWHM spectral width, and is approximated by

$$F = \frac{m\pi R^{1/2}}{1-R}$$

for $R > 0.5$ where R is the reflectance of the two reflecting surfaces. FWHM spectral width can be computed from known values of finesse and free spectral range. Figure 13.5 illustrates the important parameters with the help of a typical transmission spectrum plotted for two different values of finesse. As is evident from the curves, higher finesse shows sharper transmission peaks with lower minimum transmission coefficients. For a given value of finesse, the wavelength resolution can be improved by increasing distance d, but only at the cost of reducing free spectral range.

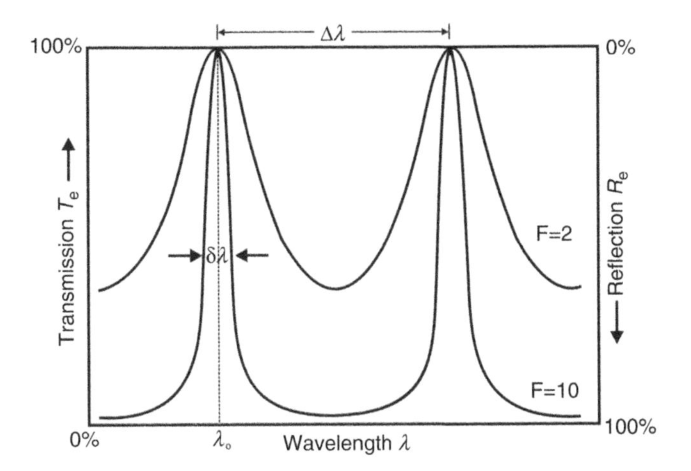

Figure 13.5 Transmission spectrum of Fabry–Pérot interferometer.

Some of the common applications of the Fabry–Pérot interferometer include: laser resonator design, where an etalon with appropriate values of finesse and free spectral range can be used to suppress all cavity modes except one hence transforming it from a multimode laser into a single-mode laser; laser absorption spectrometry where Fabry–Pérot etalons are used to increase the interaction length; dichroic filters, sources, cameras and astronomical equipment; telecommunications using wavelength division multiplexing; and optical wave meters.

13.2.1.5 Sagnac Interferometer

The Sagnac interferometer, named after the French physicist George Sagnac, is based on what is known as the Sagnac effect. According to the Sagnac effect, when a ring laser comprising two counter-propagating laser beams is rotated around an axis perpendicular to the plane of the ring laser, the light beam travelling in the same direction as the direction of rotation experiences a longer path length while the light beam travelling in the opposite direction experiences a shorter path length in one round-trip. This produces a relative phase shift between the two counter-propagating beams due to the two light beams travelling different path lengths with the magnitude of phase shift being proportional to the rotation rate.

Figure 13.6 shows the optical schematic of a Sagnac interferometer using a square geometry. Triangular geometry is the widely used ring laser geometry. The beam splitter splits the beam into two counter-propagating laser beams. After recombination, the two beams produce a fringe pattern that can be analyzed to compute the rotation rate. One of the most widely exploited applications of Sagnac interferometers is in a ring laser gyroscope used in inertial navigation systems. The Sagnac effect and ring laser gyro are discussed in detail in Chapter 14 on military applications.

13.2.2 Length Metrology

Lasers offer a non-contact methodology for precise measurement of distances and displacements. Applications range from architecture to scientific studies and from forensics to defence. There are a number of concepts that have been put to use for determining distances, each with its merits and shortcomings, for a given application or set of applications. Most laser-based methods of determining distances and displacements include the time-of-flight method, the phase shift method, the triangulation method, the frequency modulation method and interferometry.

In the *time-of-flight technique*, the distance to an object is determined by measuring time-of-flight of a narrow laser pulse to travel to the targeted object and return after reflection. This method is particularly

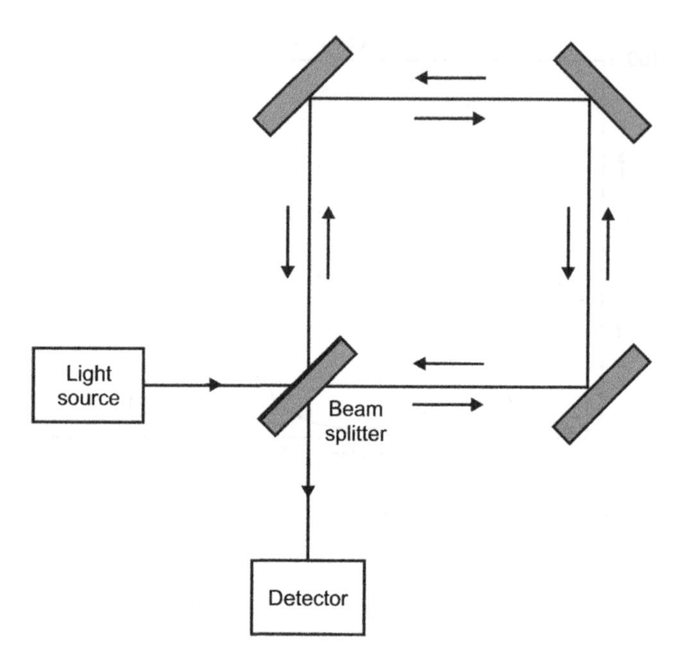

Figure 13.6 Sagnac interferometer.

suited to measurement of relatively larger distances ranging from hundreds of metres to tens of kilometres, with a typical range accuracy of a few metres.

The *phase shift technique* of range finding describes how the laser beam with sinusoidal power modulation is transmitted towards the target and the diffused or specular reflection from the target is received. The target distance is determined from the phase of the received laser beam compared to the transmitted laser beam. The phase shift is 2π times the product of time-of-flight and modulation frequency. This allows us to compute the time-of-flight and hence the distance to the target from known values of phase shift and modulation frequency.

In the *triangulation method*, the laser transmitter, laser receiver and the intended object form a triangle. The distance to the target is determined from known values of the distance between the transmitter and receiver and measured angles between the lines joining the transmitter and the receiver, the transmitter and the object and the receiver and the object. This method is capable of measuring distances from millimetres to kilometres.

For the *frequency modulation method*, the frequency of a narrow-line-width laser is modulated with a ramp or sinusoidal signal and then transmitted towards to the target. The received signal corresponding to reflected laser beam, specular or diffused, is mixed with the reference signal representing the transmitted laser beam. The beat frequency produced as a result of homodyne detection is used to derive the distance information.

The *interferometry method* is particularly suited to the measurement of changes in distance rather than absolute value. It is capable of measuring distances to an accuracy better than the wavelength of light used. Interferometers have been discussed in detail in Section 13.2.1.

These methods are discussed at length in Chapter 14 on military applications.

13.2.3 Time and Frequency Metrology

It is important for a variety of applications to determine both the frequency difference between two optical frequencies and also the absolute value of optical frequencies. Measurement of the absolute value of an optical frequency by first measuring the wavelength (e.g. by a wavelength meter) and then converting it to the corresponding frequency (by multiplying the reciprocal of wavelength by the velocity of light in vacuum c) often does not meet the precision requirement. This is due to the fact that the accuracy of measurement of wavelength is limited by effects such as wavelength distortions. For high-accuracy frequency measurements, it is important to measure optical frequencies either directly or as a difference between two optical frequencies related to a microwave reference. It may be mentioned here that the base unit of time, that is, the second in the SI system of units, has also been defined in terms of a microwave frequency. According to the standard, 1 s = 9 192 631 770 periods of radiation corresponding to the transition between two hyperfine levels of the ground state of caesium-133 atom, thereby defining time standard with respect to 9.192 631 770 GHz reference.

The difference between two optical frequencies can be measured by making them beat on a photodetector and comparing it to a microwave reference, provided the beat signal frequency is less than a few tens of GHz. The measurement of the absolute value of optical frequencies is a relatively far more complex task. One of the approaches experimented with earlier was to begin with a stable microwave frequency reference linked to a caesium atomic clock and then generate higher-frequency references through a frequency chain of oscillators. Higher-frequency references were connected to the lower frequencies by recording the beat signal of higher frequencies with the harmonics of lower frequencies, and then using the constancy of the beat signal to adjust the higher frequency in case of drift. Devices such as Schottky diodes, non-linear crystals, metal-insulator-metal diodes and so on were used to generate harmonics in different spectral regions.

One of the major limitations of the traditional frequency chain approach to generate frequency references is that it allows frequency measurement around a single optical frequency, unlike the frequency comb techniques that allow measurement over a wide spectral range. An optical frequency comb technique has revolutionized the optical frequency metrology. An optical frequency comb is simply an optical spectrum consisting of equidistant lines. Figure 13.7 shows the spectrum of a typical optical frequency comb. One of the methods of generating a frequency comb is to use a single-frequency continuous wave laser strongly driving an electro-optic modulator. A more commonly used technique for the generation of broadband frequency combs is by using ultra-short-pulse mode-locked lasers. The optical spectrum of the output of a mode-locked laser consists of discrete lines with an exactly constant frequency spacing equal to the pulse repetition frequency. Each frequency in the pulse train is an integer multiple of the pulse repetition frequency. It may be mentioned here that the generation of a frequency comb requires that the periodicity applies not only to the pulse envelopes, but that coherence between the pulses is also required. Techniques such as self-referencing, commonly used to achieve this objective, are not described here.

Such a frequency comb acts like an optical ruler. If the location of comb frequencies are precisely known, unknown frequencies could be determined by measuring the frequency difference between the

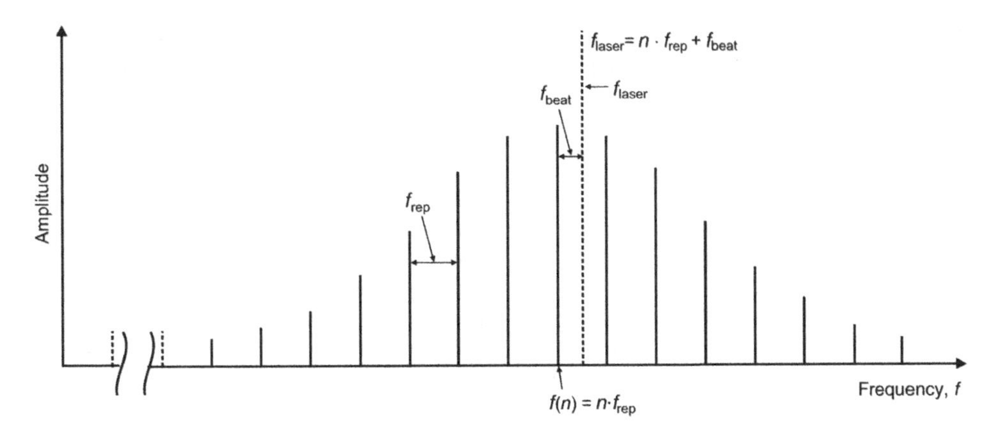

Figure 13.7 Typical optical frequency comb.

unknown frequency and the nearest comb frequency. As mentioned above, the optical spectrum of the mode-locked laser output is a comb of exactly equidistant lines. Such a frequency comb can be defined by two parameters: (1) the frequency spacing, which equals the pulse repetition frequency of the laser f_{rep}; and (2) the absolute position specified in terms of carrier-envelope offset frequency, which is the rate at which the peak of the carrier frequency slips from the peak of the pulse envelope on a pulse-to-pulse basis. If the two parameters were related to a microwave frequency reference, all frequencies in the comb would be precisely known.

The dotted line in the frequency spectrum of Figure 13.7 highlights the location of an unknown frequency. As shown in the figure, absolute frequency can be measured by measuring the frequency difference from either of the two adjoining comb frequencies. For measurement of unknown frequencies over a wide range, the frequency comb needs to have as wide a bandwidth as possible.

13.2.3.1 Optical Clock

An optical clock derives its output from an ultra-stable optical frequency standard, offering an exceptionally high precision that is far superior to the best-available caesium atomic clocks. Timing precision of the order of 10^{-15} can be achieved by averaging time of few seconds with optical frequency standards, whereas the same precision with microwave clocks would only be achievable by averaging a few hundreds of hours. An optical clock is predicted to replace the caesium clock as the fundamental timing reference in the not-too-distant future.

Optical clockwork phase coherently relates a high optical frequency standard (typically in the range of hundreds of terahertz) to a lower frequency (typically in the microwave frequency region between 1 and 100 GHz), thereby allowing processing by fast electronics. The optical clockwork relates an optical frequency standard to an electronic standard such as that based on the caesium atomic clock.

As outlined in Section 13.2 on frequency metrology, an earlier technique of relating high optical frequency to a lower microwave frequency involved use of frequency chains that were extremely cumbersome and expensive to set up. The use of frequency combs generated by broadband mode-locked lasers has made it possible to realize simpler optical clockworks. As outlined earlier, different frequencies in a frequency comb are defined by pulse repetition frequency and carrier-envelope offset frequency. An optical frequency from an optical frequency standard can then be expressed as the sum of certain integer multiple of pulse repetition frequency, carrier-envelope offset frequency and beat frequency between the optical standard frequency and the adjacent frequency comb frequency. These three components (pulse repetition frequency, beat frequency and carrier-envelope offset frequency) can all be measured and processed with fast electronics. This makes it possible to phase-coherently compare the optical frequency standard and the microwave frequency standard caesium clock, and therefore correct the timing signal of the caesium atomic clock with the much higher stability of the optical frequency standard.

Figure 13.8 shows the schematic arrangement of an optical clock with a carrier-envelope offset frequency stabilization by an $f - 2f$ interferometer-based feedback loop. Generation of the three signals that constitute an optical frequency from the optical frequency standard are evident in the schematic diagram.

13.2.4 *Measurement of Line Width*

Line width of a laser is an indicator of the degree of its monochromaticity and is defined as the width of its optical spectrum measured typically as the full-width half-maximum (FWHM). Line width is strongly related to temporal coherence of the laser, characterized by coherence length or coherence time. Lasers with narrow line width find use in a wide range of applications including fibre optic communications, spectroscopy, fibre optic sensors and testing and measurement purposes. Laser line width can be measured by a number of different methods, with their efficacy depending upon the order of line width to be measured.

In the case of multimode lasers with line widths of the order of 10 GHz or more, the conventional diffraction-grating-based technique of optical spectrum analysis may be used. Another possible

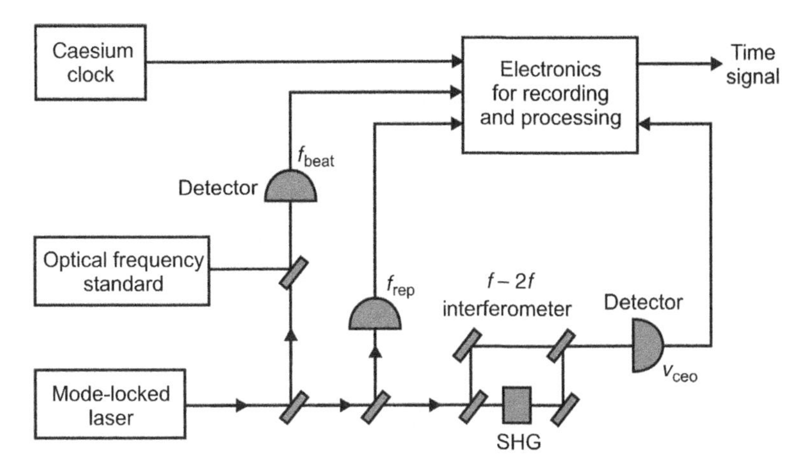

Figure 13.8 Schematic arrangement of optical clock.

technique is to convert frequency fluctuations to intensity fluctuations, using a frequency discriminator such as an unbalanced interferometer or a high-finesse reference cavity.

For single-frequency lasers with narrower line width, the self-heterodyne technique is often used. Line width in this case is measured as a beat frequency between the frequency-shifted laser output and a delayed version of it. Figure 13.9 shows the basic set-up. One portion of the laser beam is sent through a long optical fibre which provides the desired time delay. For self-heterodyne action, the two laser beams being superimposed to generate the beat note should be essentially uncorrelated so that the output spectrum becomes a simple self-convolution of the laser output spectrum. For this to happen, the induced time delay in one of the paths should be larger than the coherence length of the laser. Another portion of the laser beam is frequency-shifted by some tens of megahertz with an acousto-optic modulator (AOM). Both beams are finally superimposed on a beam splitter, and the resulting beat note (centred at the AOM frequency) is recorded with a photodetector. The laser line width is then retrieved from this beat note. In

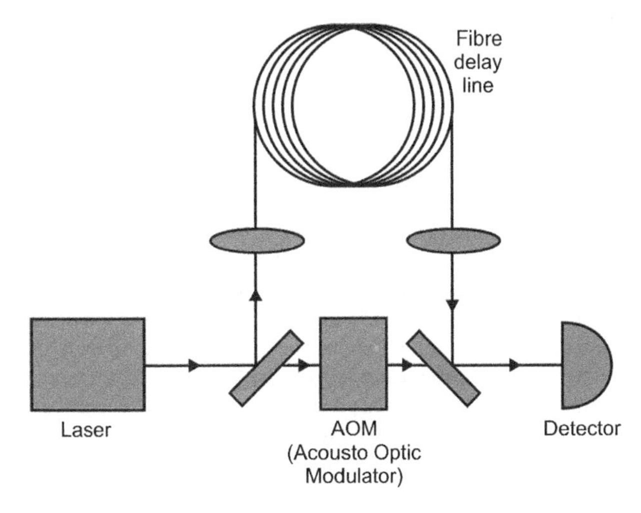

Figure 13.9 Set-up for self-heterodyne measurement of a laser line width.

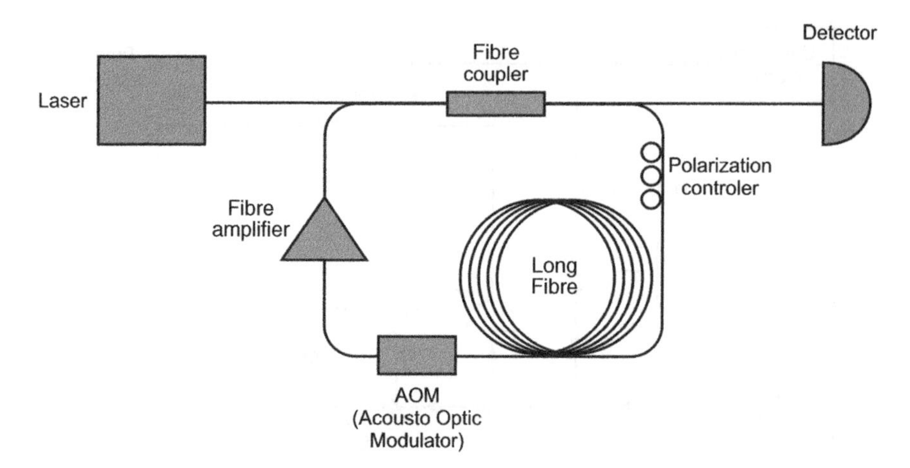

Figure 13.10 Schematic set-up of a recirculating fibre loop.

the case of lasers with relatively much narrower line widths of the order of kilohertz or sub-kilohertz, necessitating long time delays which are impractical to implement with single-pass fibre loop, a shorter time delay with sophisticated data processing or a recirculating fibre loop can be used instead.

The method of recirculating fibre loops for measurement of the laser line width of the order of sub-kilohertz is an extension of the self-heterodyne line width measurement technique. The long time delay needed at line widths in the sub-kilohertz range is provided by a moderately long recirculating fibre loop. Figure 13.10 shows the basic measurement set-up.

In order to keep the light from different round-trips well separated, an AOM in the loop shifts the optical frequency by about 100 MHz in each round-trip. On account of the fact that the frequency shift is much larger than the line width, the components corresponding to different numbers of round trips are well separated in the frequency domain. The beat notes of the original laser light with different frequency-shifted components can be used to measure their line widths. The amplifying element in the loop compensates for the losses in the fibre and the AOM, which otherwise would limit the number of round-trips possible. The sensitivity of the recirculating fibre loop is limited by noise from the fibre amplifier.

Yet another approach of measuring line width is to measure the beat frequency note between the laser under test and a reference laser operating at a nearby frequency. This method offers very high resolution. Although conceptually very simple and reliable, the availability of a second laser operating at a nearby frequency with the desired noise performance could be a difficult proposition. Use of a frequency comb source could facilitate measurement of line width over a wide spectral range.

13.2.5 Infrared Thermometer

Infrared thermometers allow non-contact measurement of an object's temperature, therefore allowing users to measure temperature in situation where use of conventional sensors would be either impossible or highly impractical. An infrared thermometer in its most basic form consists of focusing optics that focuses the received infrared energy onto a detector. It measures the temperature by detecting the infrared energy emitted by the object and converting it into an equivalent electrical signal which, after compensation for the ambient temperature, can be displayed in the units of temperature. Temperature is computed from known values of the emitted infrared energy and the emissivity of the object.

It may be mentioned here that all materials at temperature above absolute zero emit infrared energy, and the total energy emitted per unit surface area per unit time is directly proportional to fourth power of absolute temperature, as per Stefan's law. According to Stefan's law (also known as the Stefan–Boltzmann law), emitted infrared power P is given by

$$P = A\varepsilon\sigma T^4$$

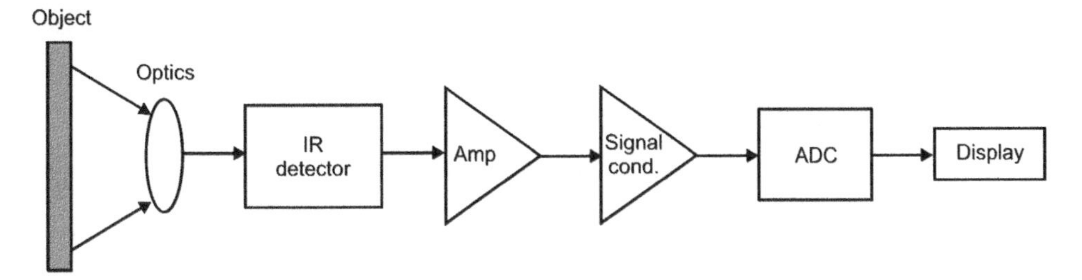

Figure 13.11 Block schematic of infrared thermometer.

where A, ε, σ and T are surface area (m^2), emissivity of the material, Stefan's constant and absolute temperature (K), respectively. The peak emitted wavelength from the hot body is also inversely proportional to the absolute temperature as per Wien's displacement law, that is,

$$\lambda_{\max} \propto 1/T$$

Figure 13.11 shows the basic block schematic of an infrared thermometer.

Some of the common application scenarios where non-contact infrared temperature measurement can be particularly useful include moving objects such as rollers and conveyor belts, chemically or electrically hazardous situations such as high voltages and applications where the temperatures are too high to be measured by thermocouples and other contact sensors. Infrared thermometry is also useful where the object is contained in vacuum or controlled atmosphere or where the object is surrounded by an electromagnetic field, as is the case in induction heating.

Important parameters to be considered when choosing an infrared thermometer for a given application include: field-of-view, target emissivity, spectral response, temperature range, response time and mounting considerations. Infrared thermometers are available today in a variety of shapes, form factors and mounting configurations, including those designed for flexible and handheld usage and those designed for fixed-mount applications. Figure 13.12 depicts a handheld high-temperature infrared thermometer, type OS 523E/524E series from Omega Engineering. This infrared thermometer is capable

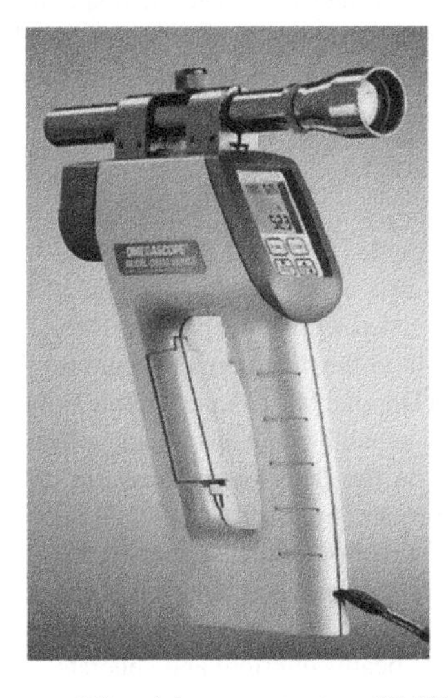

Figure 13.12 Handheld high-temperature infrared thermometer, type OS 523E/524E series (Courtesy of OMEGA Engineering INC.).

of measuring temperature up to 4500 °F, has a distance measuring feature and is also equipped with laser sight.

13.3 Laser Velocimetry

Laser velocimetry is extensively used for the measurement of one or more velocity components over one or more dimensions. Although the most widely exploited application of laser velocimetry continues to be in the measurement of transparent and semi-transparent fluid flow, it has also been used for the measurement of linear or vibratory motion of opaque reflecting surfaces. Different techniques of laser velocimetry include: laser Doppler velocimetry (LDV) and its variant phase Doppler velocimetry (PDV), particle image velocimetry (PIV), particle transit velocimetry (PTV) and global Doppler velocimetry (GDV) and its variant global phase Doppler (GPD). While LDV is an interferometric technique, PIV and PTV are imaging techniques. As for PIV, GDV is a planar technique but directly utilizes the Doppler shift of light scattered from moving particles. Laser velocimetry techniques can be further extended in phase Doppler and global phase Doppler techniques to be used for the measurement of particle size in addition to fluid flow velocity, assuming spherical and homogeneous particles. One of the major advantages of laser velocimetry techniques is their non-intrusiveness. These techniques allow measurement of fluid flow properties without interfering fluid flow.

All laser velocimetry techniques mentioned above are tracer-based methods depending upon elastic light scattering from tracer particles. Tracer particles are added to the fluid flow under study and the velocity of fluid flow is measured in terms of the velocity of these particles, making the assumption that the particle velocity is representative of the fluid flow velocity. For this assumption to hold, the size and density of tracer particles have to be carefully chosen for a given fluid flow. Factors such as tracer particle generation, insertion into the flow and their light scattering properties significantly affect the accuracy of measurement. The randomness of the flow field sample inherent in tracer-based techniques either in space (using planar techniques such as PIV and GDV) or in time (using point techniques such as LDV) makes signal processing for estimation of fluid flow properties an extremely complex task. Two of the above-mentioned techniques – LDV and PIV, interferometric and imaging techniques, respectively – are briefly described in the following sections.

13.3.1 Laser Doppler Velocimetry

Laser Doppler velocimetry (LDV) is an interferometric method that can be used to remotely and non-intrusively measure a single velocity component at a highly localized point in space. More than one optical system can be combined to simultaneously measure up to three velocity components of a fluid flow.

13.3.1.1 Operational Principle

Two possible approaches have been used to build an LDV set-up. In one of the approaches used in earlier devices, the laser beam is split into two collimated beams. One of the beams (probe or measurement beam) is focused into the fluid flow seeded with tracer particles, while the second beam (reference beam) passes outside the fluid flow. The light of the measurement beam passing through the fluid and scattered by the tracer particles experiences a Doppler shift. A part of this Doppler-shifted scattered light is made to fall on the photodetector. The reference beam is also made to fall on the photodetector. The optical heterodyne detection taking place on the detector produces an electrical signal proportional to the Doppler shift. The Doppler shift allows computation of the particle velocity component and hence the fluid flow velocity component in a direction perpendicular to the plane of the two light beams.

The two beams are derived from a single monochromatic and coherent laser beam to ensure coherence between the two beams. The transmitting optics focuses the two laser beams in order to intersect inside the fluid flow at their waists. The two beams interfere and generate a set of straight fringes. As the fluid entrained with appropriate tracer particles flows past the fringes, the particles scatter light elastically. The scatter could be forward scatter, backscatter or even side scatter. (The LDV can be set up to receive any of

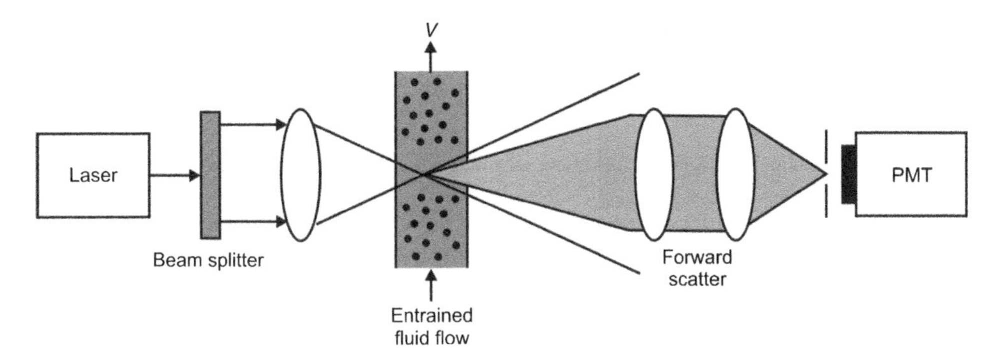

Figure 13.13 LDV set-up with forward scatter detection.

the three scattered lights.) The scattered light as collected by the receiving optics is focused onto a photodetector, which is usually a photomultiplier tube (PMT) or an avalanche photodiode (APD). The intensity of the scattered light fluctuates or modulates at a frequency equal to the Doppler shift between the incident and the scattered light. The Doppler shift can then be used to compute the velocity component in the direction of the plane of the two laser beams. Figure 13.13 shows the LDV set-up that makes use of forward scatter. The Doppler frequency f_D in this case is given by Equation 13.1:

$$f_D = \frac{(2V_x \sin\theta)}{\lambda} \tag{13.1}$$

where V_x is fluid flow velocity, θ is the angle between the direction of fluid flow and the scattered signal and λ is wavelength.

Figure 13.14 depicts the LDV set-up that makes use of backscatter. Lasers with output wavelength in the visible spectrum in the range of 390–750 nm are generally used for LDV applications, allowing the

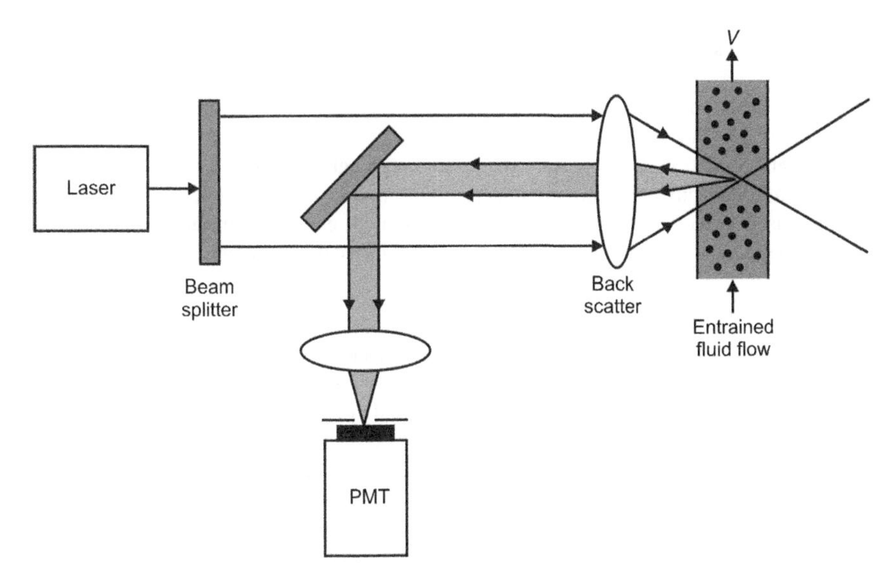

Figure 13.14 LDV set-up with back scatter detection.

beam path to be observed. Compared to infrared wavelengths, visible wavelengths produce a stronger Doppler signal.

13.3.1.2 Advantages

Laser Doppler velocimetry offers the following advantages:

1. It can be effectively used for different types of fluid flows including gas and liquid flows.
2. It is a non-intrusive method, that is, no physical probe is required with the result that there is no obstruction to the flow.
3. It offers excellent spatial resolution with the measuring volume being of the order of a very small fraction of a cubic millimetre.
4. The output electrical signal is linearly related to the Doppler frequency which in turn is linearly related to flow velocity.
5. It offers a fast response to fluctuating velocities. For a Doppler frequency of 1 MHz, the typical frequency response goes up to 50 kHz.

13.3.1.3 Applications

Laser Doppler velocimetry has been widely used in a variety of applications in flow research, medical diagnostics and process and quality control in industrial production processes. Common applications related to flow research include: the testing of aerodynamics of aircraft, missiles, automobiles, buildings and other structures by wind tunnel velocity experiments; hydrodynamics research, ship hull design and pipe and channel flows through velocity measurements in water flows; fuel injection and spray research by measuring velocities through nozzles and inside the engines; and environmental research including combustion research, tidal modelling, wave dynamics, river hydrology and so on.

There are also many applications in the field of medicine; for example, laser Doppler flowmetry is used to partially quantify blood flow in human tissues such as skin, which is further used to monitor the effect of exercise, environmental or physical manipulations and drug treatment of targeted micro-sized vascular areas. The laser light in this case is scattered with a Doppler shift from red blood cells. It is also used in clinical otology, which refers to the study and medical care of the ear. LDV is used for the measurement of displacement of the tympanic membrane (a thin cone-shaped membrane that separates the external ear from the middle ear), malleus (hammer-shaped small bone of the middle ear which connects with the incus or anvil) or prosthesis head in response to sound inputs of 80–100 dB sound pressure level.

Common applications of LDV in industry relate to process control, for example measuring the velocity and length of moving surfaces is the basis of all industrial process and quality control applications. Laser surface velocimetry is also used to measure the speed and length of moving surfaces on coils, strips, tubes, paper, film, foil or any other moving material. Different tasks that can be performed on moving surfaces using laser surface velocimetry include cut-to-length control, path length and spool length measurement, speed measurement and speed control, encoder calibration and inkjet marker control.

13.3.2 Particle Image Velocimetry

Particle image velocimetry is a planar and an imaging technique. While LDV depends upon the coherence property of the laser for its operation, PIV does not require coherence of the laser. PIV measures whole velocity fields by taking two images captured by a charged-couple device (CCD) camera within a short time differential and calculating the distance travelled by individual particles within this time. The velocity of tracer particles and hence the flow velocity is then computed from the known time difference and the measured displacement values. The displacement is commonly measured by dividing the image plane into small interrogation spots and then cross-correlating the images from the two time exposures. The spatial displacement producing the maximum cross-correlation statistically approximates the average displacement of the particles in the interrogation cell. The velocity component associated with each interrogation is given by the ratio of displacement to the time between the two pulses.

A Q-switched frequency-doubled Nd:YAG laser operating at 532 nm with pulse energy of the order of a few hundreds of millijoules and pulse width in the range 5–10 ns is generally used. A short pulse width of less than 10 ns is used to ensure that the tracer particle motion is frozen during each exposure. Although coherence is not a requirement in PIV, a laser is used as only laser light can be focused into a thin enough sheet to image particles only in a given plane. For a typical CCD camera of resolution 1000×1000 pixels, up to 4000 interrogation cells with an equal number of velocity vectors per image pair can be obtained. Framing rates of most PIV cameras are of the order 10–20 Hz, compatible with pulse repetition rates of Q-switched Nd:YAG lasers.

A variant of PIV is particle tracking velocimetry (PTV). The two techniques are exactly the same in terms of operational principle and measurement hardware, but the latter is used for sparsely seeded flows or for dispersed two-phase flows. In PTV, individual particle movements are examined.

Although pulsed lasers are commonly used for both PIV and PTV techniques, CW lasers can also be used in principle. In this case, the functions of pulse duration and pulse interval can be obtained by control of the camera shutter. CW lasers cannot compare with pulsed lasers in terms of peak power level; as a result of this, their use is limited to low-speed water flows where large tracer particles with good scattering properties can be used.

Figure 13.15 shows the optical set-up of particle image velocimetry. A dual-cavity laser is used in order to illuminate a plane within the flow twice in quick succession with a small time differential.

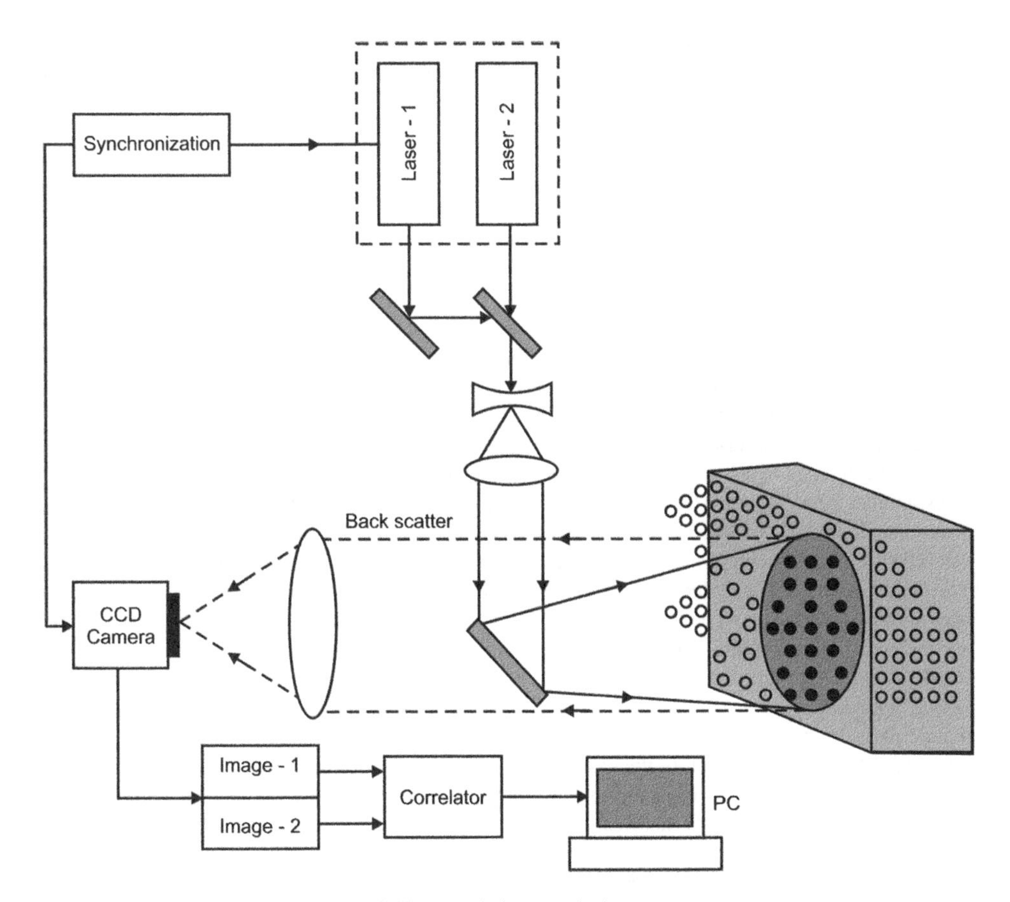

Figure 13.15 Particle image velocimetry set-up.

This time differential is matched to the local flow velocity and hence to the expected displacement of tracer particles between successive pulses. One or two CCD cameras record the scattered light. One camera is sufficient to record the two in-plane velocity components. A third camera is required only when the third out-of-plane velocity component is also to be measured.

13.4 Laser Vibrometry

Laser vibrometry is a non-contact technique used for measurement of surface vibrations. It is operationally similar to laser velocimetry, extracting information on the vibration amplitude and frequency from the Doppler shift undergone by the laser beam frequency due to vibratory motion of the surface of interest. Compared to electromechanical devices such as an accelerometer, used for measurement of vibratory motion, laser Doppler vibrometry offers all the advantages of being a non-contact method. For example, it does not mass load the target surface. It can be used on surfaces that are hard to access and also those surfaces that are either too small or too hot for attachment of a physical transducer.

13.4.1 Operational Principle

Laser Doppler vibrometry is also an interferometric technique that utilizes the coherence property of the laser beam. A two-beam laser interferometer extracts the vibration data from the interference pattern resulting from a probe beam reflected from the vibrating surface and a reference beam.

When two coherent light beams interfere, the resulting intensity is not simply the sum of the single intensities; is modulated by a third term resulting from the interference phenomenon. This interference term is a function of the path length difference between the two beams. The resultant intensity I is given by Equation 13.2:

$$I = I_1 + I_2 + 2\sqrt{I_1 \times I_2 \times \cos\left(\frac{2\pi(L_1 - L_2)}{\lambda}\right)} \tag{13.2}$$

where I_1 and I_2 are the intensities of the two interfering laser beams; L_1 and L_2 are the corresponding path lengths traversed by them before they recombine; and λ is the operational wavelength.

As is evident from Equation 13.2, the interference term depends upon the path length difference between the two laser beams. If this path length difference is an integer multiple of the laser wavelength, the overall intensity is four times the single intensity. The overall intensity is zero if the two beams had a path length difference of half of one wavelength. This gives rise to a fringe pattern.

Figure 13.16 shows the laser Doppler vibrometry set-up. The laser beam is split by a beam splitter (BS_1) into a reference beam and a probe beam. The reference beam after reflection from mirror M_1 reaches another beam splitter (BS_2). The probe beam is frequency shifted by about 40 MHz with an acousto-optic modulator such as a Bragg cell before it reaches another beam splitter (BS_3). After passing through BS_3, the measurement beam is focused onto the object under investigation and reflected by it. This reflected beam is now deflected upwards by BS_3 before it recombines with the reference beam in BS_2 as shown in the figure. The combined beam is then directed onto the photodetector.

As the path length of the reference beam is constant over time (except for negligibly small thermally induced variations), a vibratory motion of the target surface causes a variation in path length L_1 with respect to time. This produces an interference pattern of dark and bright fringes. One complete dark-bright cycle on the detector corresponds to an object displacement of exactly half of the wavelength of the light used. In the case of the helium-neon laser, which is the most widely used laser type for laser vibrometry applications, this corresponds to a displacement of 316.4 nm. The rate of change of optical path length manifests itself as the Doppler frequency shift of the probe beam. This means that the modulation frequency of the interferometer pattern determined is directly proportional to the velocity of the object. Measurement of both velocity as well as displacement is possible with the interferometer.

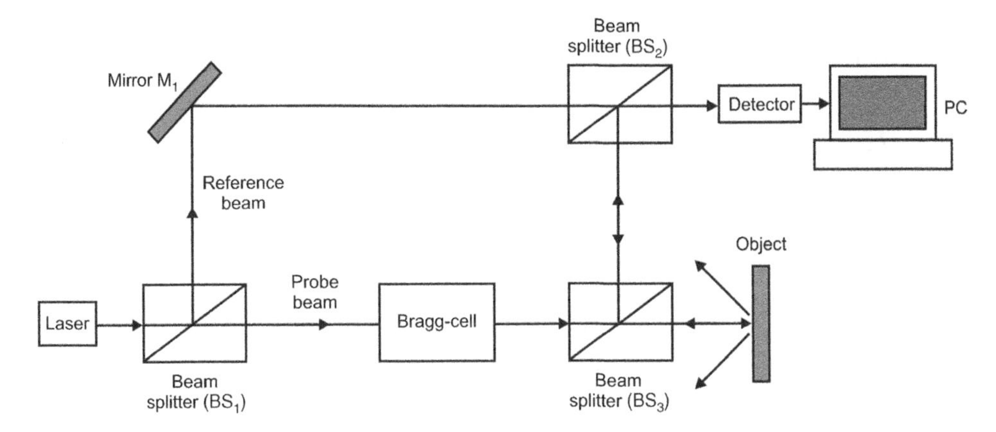

Figure 13.16 Laser Doppler vibrometry set-up.

The introduction of the Bragg cell in the path of the probe beam to shift the laser beam frequency of the probe beam by a known magnitude is intended to enable determination of direction of motion. In the absence of this frequency shift introduced by the Bragg cell, the interferometer will generate the same frequency shift or interference pattern irrespective of whether the object was moving towards or away from the interferometer. In the case of frequency-shifted operation, there is a modulation frequency of the fringe pattern of 40 MHz when the object is stationary. During object movement, the frequency is either greater or less than 40 MHz, depending upon direction of motion. This allows measurement of both amplitude and direction of motion.

Figure 13.17 depicts a laser Doppler vibrometer, Model no LVS-01 from Sunny Instruments Singapore Pte Ltd. The instrument uses a class 2 He-Ne laser operating at 632.8 nm wavelength and is designed for precise measurement of vibration, speed and displacement. Speed, displacement and acceleration measurements can be made at a resolution of 1 μm s^{-1}, 0.3 nm (at 500 Hz) and 0.0003 *g* (at 500 Hz), respectively.

Applications include motor dynamics and rotation accuracy measurement; machine tool dynamics and accuracy testing; diagnostics in life sciences, medicine and zoology; dynamic testing of computer

Figure 13.17 Laser Doppler vibrometer, Model No LVS-01 (Reproduced from Sunny Instruments Singapore Pte Ltd).

peripherals such as hard disk drives, CD and tape machines; bridge and structure vibration testing; MEMS system dynamic testing; measurement of roughness and thickness; the detection of mechanical vibration; and fault diagnosis.

13.4.2 Types of Laser Doppler Vibrometers

Laser Doppler vibrometers are made in various configurations. Some common variants include single-point vibrometers, scanning vibrometers, 3D vibrometers, rotational vibrometers, differential vibrometers, multibeam vibrometers, self-mixing vibrometers and continuous-scan laser vibrometers.

A *single-point vibrometer* is the most common type of vibrometer and is used to measure displacement or velocity at a single chosen point on the vibrating object at a time. This is the type described in the previous section. A *scanning vibrometer* allows the single laser beam to be moved across the surface of interest with the help of x–y scanning mirrors.

A *3D vibrometer* is designed to measure all three components of the velocity. It may be mentioned here that a standard laser vibrometer measures the velocity of the target along the direction of the laser beam. To measure all three velocity components at a point on the target object, the 3D vibrometer uses three independent laser beams which strike the target from three different directions. This allows determination of the two in-plane components and one out-of-plane component of velocity.

Rotational vibrometers are used to measure rotational or angular velocity. *Differential vibrometers* measure the out-of-plane velocity difference between two different locations on the target. *Multibeam vibrometers* are capable of simultaneously measuring the target velocity at several locations. *Self-mixing vibrometers* are very rugged and compact instruments. They are generally configured around a laser diode with a built-in photodetector. A *continuous-scan vibrometer* is a modified form of a laser Doppler vibrometer, in which the laser beam is continuously swept across the surface of the test specimen to capture the motion of a surface at different points.

13.4.3 Applications

Laser Doppler vibrometers are used in a wide variety of applications in science, industry, aerospace, defence and medicine. Laser vibrometry is extensively used in aerospace industry for applications such as the non-destructive inspection of aircraft components and the automotive industry for structural dynamics, quantification of noise and vibration and measurement of speed accuracy.

Common biological applications of laser vibrometry include ear drum diagnostics and insect communication. Laser vibrometry is used in the music industry for speaker design and for diagnosing the performance of musical instruments, and by the computer industry for dynamic testing of computer peripherals such as hard disk drives.

An emerging application of laser vibrometry is in the detection of buried land mines. The operational concept is based on exciting the ground by using an audio source, inducing ground vibration. The difference in the vibration characteristics of the ground above and away from the buried land mine enables its detection. The ground surface above the location of a buried mine shows enhanced vibration at the resonance frequency of the land mine and soil combination. A laser vibrometer measures the vibration amplitude and extracts information on land mine location. The concept has been successfully demonstrated using single-beam scanning type and array laser vibrometers. Figure 13.18 illustrates the concept.

13.5 Electronic Speckle Pattern Interferometry

Electronic speckle pattern interferometry (ESPI) is an optical interferometric technique enabling measurements of surface displacements or deformations on material surfaces caused by some kind of loading, which could be mechanical, thermal, vibrational or pressure. It is a non-contact, full-field measurement technique that allows computation of the 3D distribution of displacement and strain/stress of the object under test in response to loading, with a sensitivity equal to half of the operating wavelength. ESPI has been successfully used for the study of material properties, fatigue testing, fracture

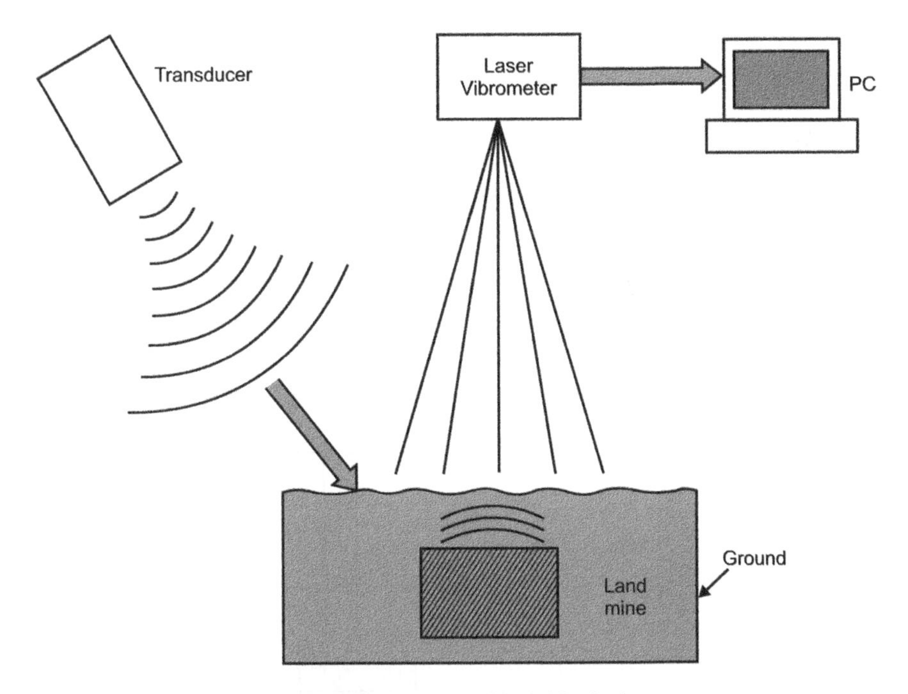

Figure 13.18 Detection of buried land mines.

mechanics, vibrational mode analysis and non-destructive testing of a variety of components in the automotive, aerospace and electronics industries. It uses laser light together with video detection, recording and processing to visualize static and dynamic displacements of object surfaces. Visualization is in the form of fringes on an image, where each fringe represents a displacement of half of the operational wavelength (which is of the order a fraction of a micron).

13.5.1 Operational Principle

The object surface under test is illuminated by an expanded coherent laser beam. Considering that the object surface is rough by optical standards, which is one of the requirements of speckle pattern interferometry, the optical image formed in the image plane of the CCD camera is a *subjective* speckle pattern. The detailed structure of the speckle pattern depends on the viewing system parameters such as the lens aperture (size of speckle) and position of the imaging system.

On the other hand, when laser light scattered off a rough surface falls on another surface, it forms an *objective* speckle pattern. Each point in the image can be considered to be illuminated by a finite area in the object surface. The amplitude, phase and intensity of the speckle pattern are all random and represent the microstructure of the corresponding area on the object surface.

In the next step, a reference light beam derived from the same laser source is superimposed on the CCD camera image. The image speckle pattern and the reference light field interfere and the resulting light field, which has random amplitude, phase and intensity, is therefore also a speckle pattern. This pattern is recorded by the CCD camera. When the object surface is displaced or deformed in response to some kind of loading, the distance between the object and image changes, thereby changing the phase of the image speckle pattern. This causes a change in the intensity of the combined light field. However, the intensity of the overall image emains unchanged if the phase change of the image light field is a multiple of 2π.

The new combined image of the deformed surface is also recorded. This new image is subtracted point by point from the first image. The resulting image is a fringe pattern, which reveals the displacement of

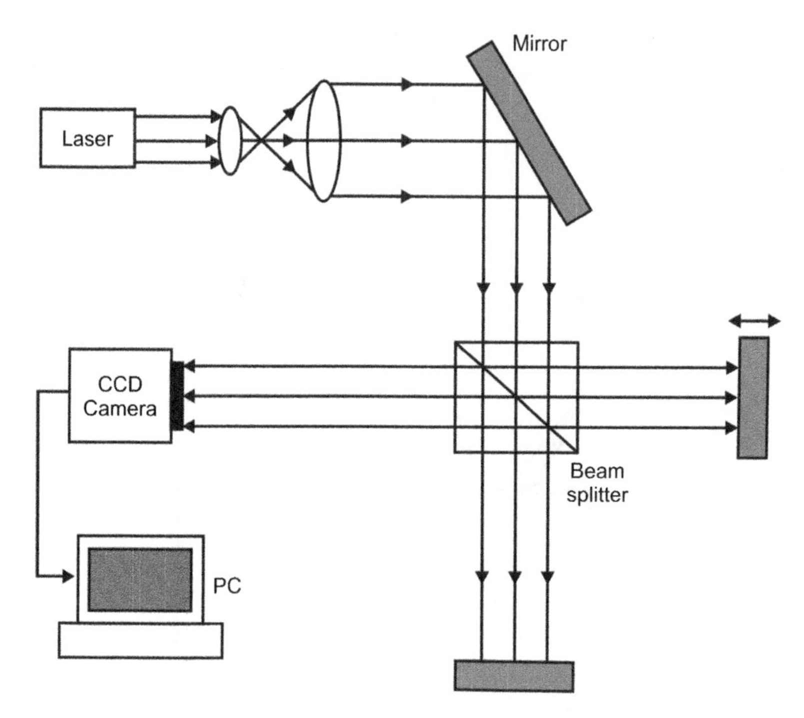

Figure 13.19 Basic electron speckle interferometry set-up.

the surface during loading as contour lines of deformation. Figure 13.19 illustrates an electron speckle pattern interferometry set-up.

The qualitative fringe patterns are noisy due to the presence of speckles and have low contrast. This is overcome by the process of phase shifting, which takes a series of speckle images for each surface state and calculates a quantitative phase map. The information in the phase map can be directly transformed into a displacement value.

13.5.2 Measurement Configurations

Electron speckle pattern interferometry can be employed in different configurations to suit the requirement of the intended measurement. These include configurations for out-of-plane displacement measurement, in-plane displacement measurement, in-plane displacement gradient measurement, 3D measurement and vibration analysis. Each of these is briefly described in the following sections.

13.5.2.1 Out-of-plane Displacement Measurement

Figure 13.20 shows the schematic diagram of the ESPI set-up used for measurement of the out-of-plane component of displacement. The concept here is the same as for the basic operational principle of ESPI. As shown in the diagram, the laser beam is split into two beams by a beam splitter. One of the beams (object beam) illuminates the measurement surface under test. The light is scattered back towards the CCD camera by the object surface. The second beam (reference beam) is combined with the object beam through the beam splitter and the combined beam is directed towards the camera sensor. The interference pattern (also called speckle pattern) formed is recorded by the CCD camera. After the object has become deformed, the new interference pattern is also recorded; the difference between the two images yields a measure of the deformation. Strictly speaking, the fringe pattern resulting from the difference between two speckle patterns only represents the purely out-of-plane displacement if the surface is illuminated

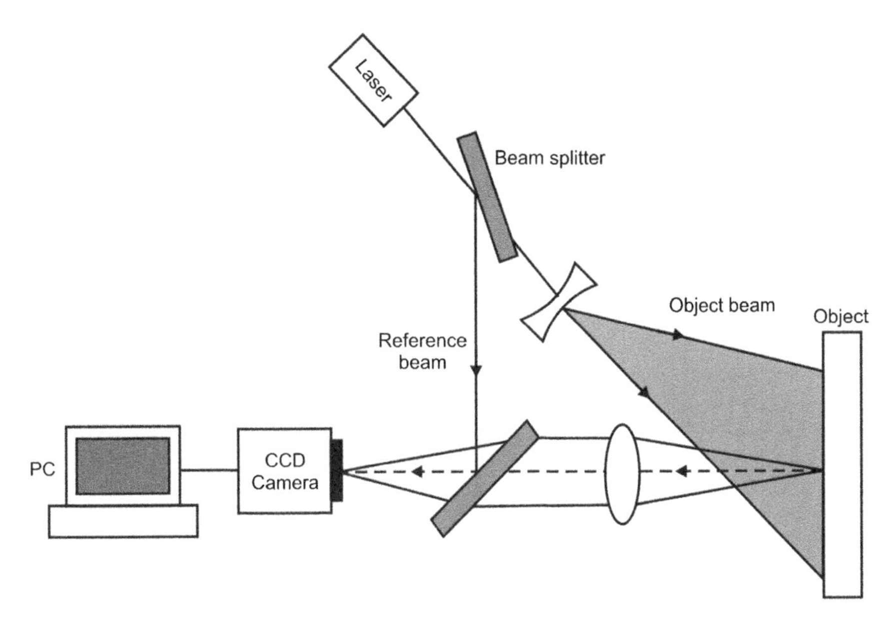

Figure 13.20 Schematic diagram of out-of-plane displacement measurement.

normally. However, since the dependence on in-plane movement is relatively small, the measurement error is negligible for small angular deviations of object illumination away from normal.

13.5.2.2 In-plane Displacement Measurement

For measurement of the in-plane component of displacement, the object surface is symmetrically irradiated from opposite sides of the object by two laser beams derived from a common laser output. It may be mentioned here that there can be two possible components of in-plane displacement measurement. In a given in-plane displacement measurement set-up, in-plane deformations formed by the two laser irradiation beams and the observation direction are measured. If the other orthogonal component of in-plane displacement is to be measured, a second pair of laser beams is required.

When the object surface which is displaced or deformed has a component in the direction normal to the viewing direction, the phase of one of the beams increases while that of the other decreases. As a result, the relative phase difference between the two beams changes. If the phase difference is a multiple of 2π radians, the speckle pattern remains unchanged. When the subtraction technique as described in the case of out-of-plane displacement measurement is used, fringes are obtained which represent in-plane displacement contours. Figure 13.21 shows the schematic diagram of in-plane displacement measurement.

13.5.2.3 In-plane Displacement Gradient Measurement

For measurement of the in-plane component of displacement gradient, the object is irradiated by the two laser beams derived from the same laser source as for in-plane displacement measurement. In the case of displacement gradient measurement however, the two laser beams are incident on the object surface from the same side but at different angles as shown in the schematic arrangement of Figure 13.22. When the object is displaced or deformed in its own plane, it can be shown that the relative phase difference between the two laser beams changes in proportion to the gradient of the in-plane displacement. Again, of the difference between the two is used to produce the fringe pattern representative of displacement gradient.

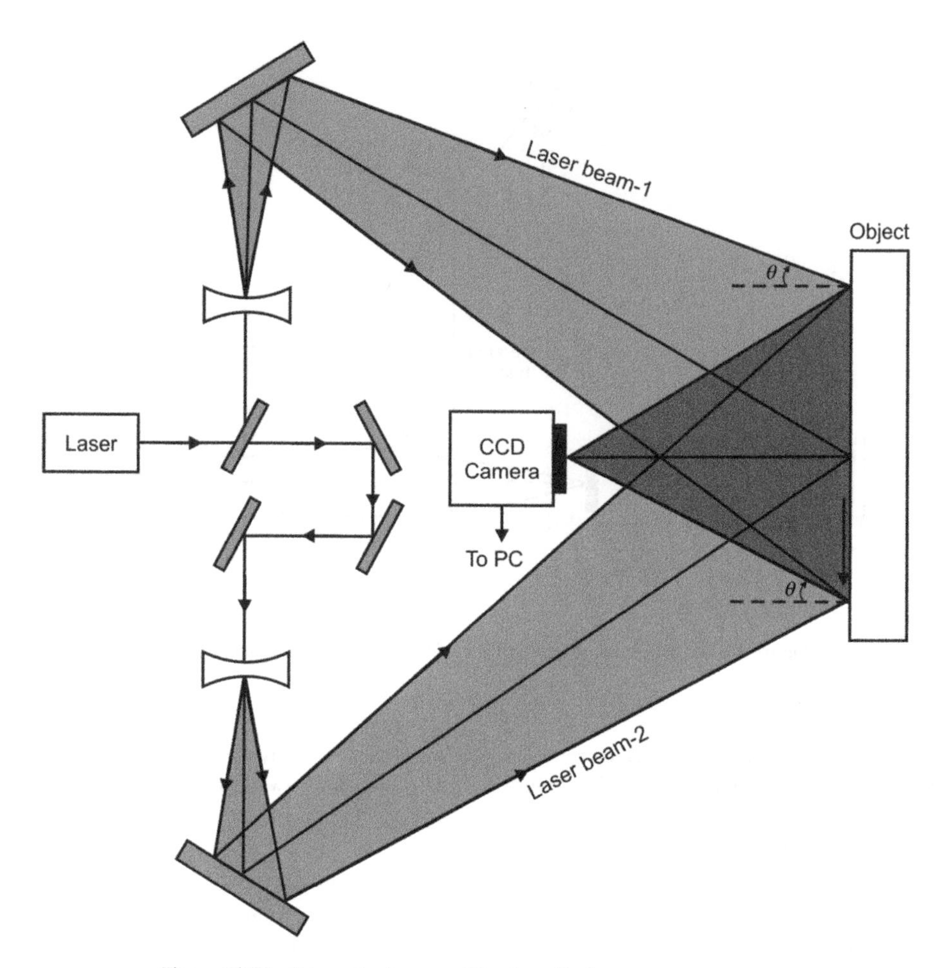

Figure 13.21 Schematic diagram of in-plane displacement measurement.

13.5.2.4 Three-dimensional (3D) Measurement

A 3D measurement of displacement allows measurement of the two in-plane and one out-of-plane components of displacement. It therefore requires a combination of the two techniques described in Sections 13.5.2.1 and 13.5.2.2 for in-plane and out-of-plane components of displacement. In order to record the two in-plane components of displacement, two orthogonal pairs of laser irradiation are required. In order to be able to measure the out-of-plane component of displacement, one of the radiations used for the in-plane component is coupled with an internal reference beam. The directions of the two laser radiations are usually switched automatically and the corresponding speckle patterns are recorded. These speckle patterns are then processed to extract the complete 3D information displacement or deformation.

A typical 3D ESPI camera comprises two internal laser sources and a CCD camera. Figure 13.23 depicts a 3D ESPI camera, Model: Q-300 from Dantec Dynamics, Denmark. The two horizontal and the two vertical ports provide laser illumination for the two in-plane measurements. Another port, as seen in front in the photograph, and a reference beam are used for out-of-plane measurement.

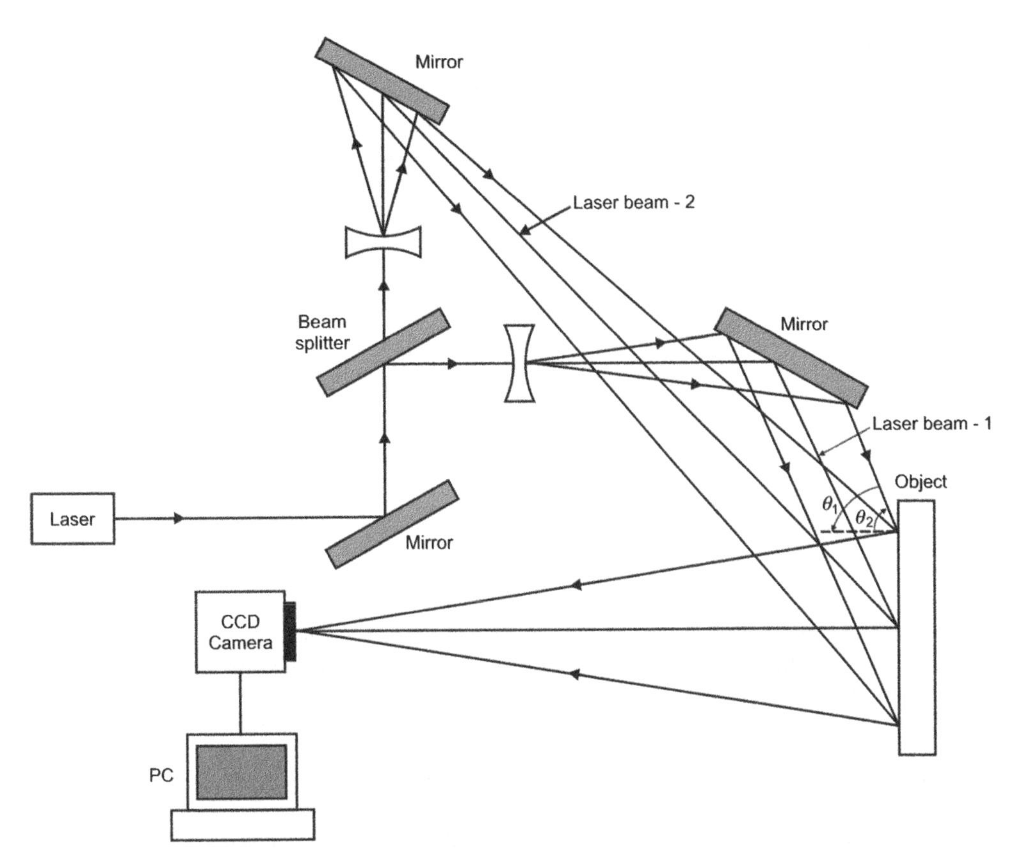

Figure 13.22 Schematic diagram of measurement of in-plane displacement gradient.

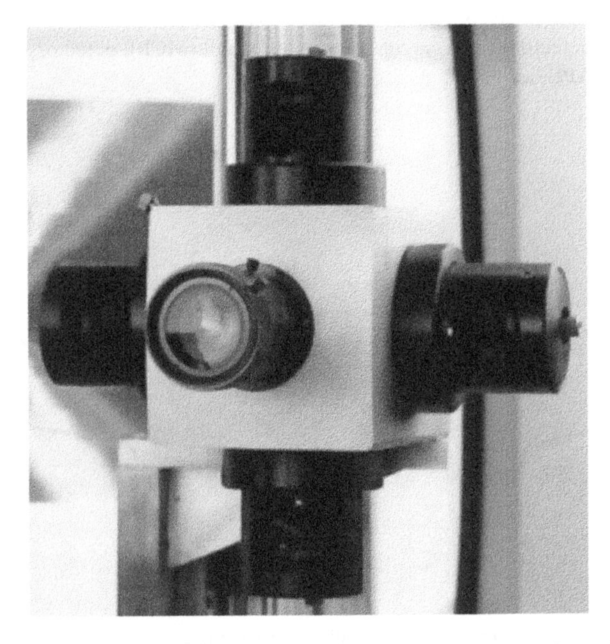

Figure 13.23 3-D ESPI camera, Model no. Q-300 (Courtesy of Dantec Dynamics A/S).

13.5.2.5 Vibration Measurement

Vibration mode analysis can be carried out using ESPI by either continuous wave laser radiation and shutter combination or a pulsed laser. An acousto-optic cell is generally used as a shutter for applied frequencies up to 40 kHz. Pulsed lasers are used for vibration frequencies greater than 40 kHz and for random vibrations. The laser is synchronized with the vibration source in order to provide illumination of the object surface under test at the maximum and minimum of the vibration modes of the material, allowing the ESPI system to quantitatively measure the displacements produced by the applied vibration.

13.6 Satellite Laser Ranging

Satellite laser ranging (SLR) measures the range of an Earth-orbiting satellite with the objective of determining the orbital parameters of satellites. By comparing these measured parameters to their predicted values, the temporal variation of the Earth's centre of mass can be accurately measured. A global network of ground stations measure the instantaneous time-of-flight of ultra-short laser pulses from the laser receiver to the satellite, equipped with special reflectors, back to the laser receiver. Instantaneous range information is calculated with a precision of a few millimetres, which can be used to generate orbital parameter data for the satellite. Figure 13.24 illustrates the technique.

The concept of satellite laser ranging is more than five decades old. It was initially experimented with in 1962 while looking for a solution to the inadequate ranging accuracy achievable at that time by means of optical and radar tracking of satellites; optical and radar tracking could provide tracking station coordinates to an accuracy of the order 100 m. Some of the geophysical processes such as Earth tides and plate tectonic motions affect tracking station coordinates to a level of a few centimetres over timescales ranging from less than one day to several years, however. Measurement accuracy of a few centimetres carried out for many years would therefore be required in order to obtain any experimental data capable of challenging/confirming any geodetic research.

Over the years, the concept of laser ranging to retro reflectors mounted on Earth-orbiting satellites has enhanced ranging accuracy to a level of a few millimetres, thereby significantly improving geodetic information. The improvement in accuracy has been assisted by the availability of ultra-fast pulsed lasers such as mode-locked Nd:YAG lasers and high-speed high-quantum-efficiency detectors such as microchannel plate (MCP) detectors and single-photon avalanche diodes (SAPD).

Figure 13.24 Satellite laser ranging concept.

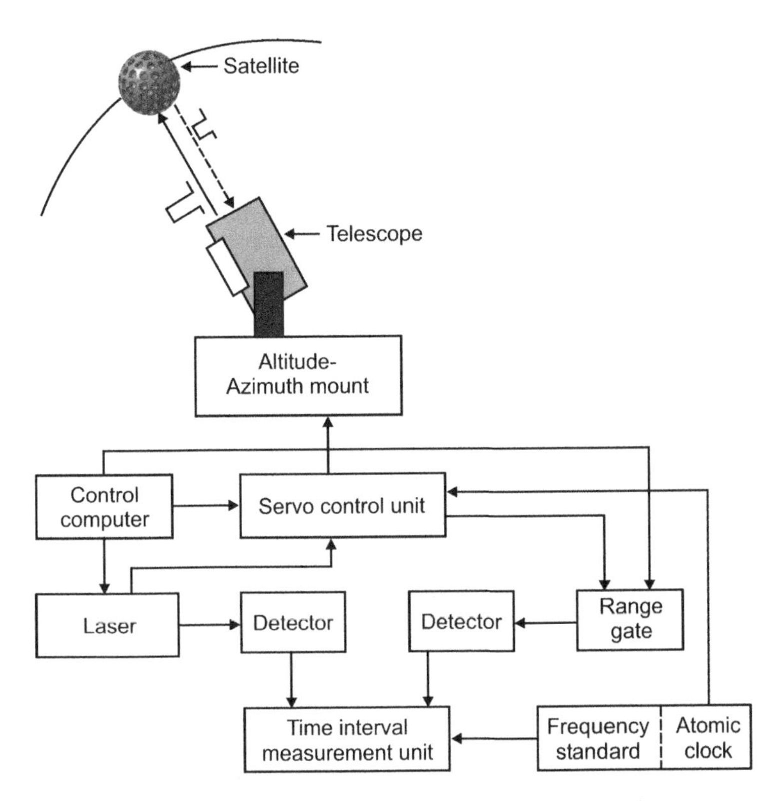

Figure 13.25 Block schematic arrangement of SLR station equipment.

13.6.1 Operational Principle

Figure 13.25 shows the block schematic arrangement of SLR station equipment. Ultra-short pulses from the laser are transmitted towards the orbiting satellite through the transmitting telescope, driven by a control system to point the laser beam towards the predicted position of the satellite. The laser beam is reflected from retro-reflectors mounted on the satellite to reach the detector system through the receiving telescope. In most cases, as shown in Figure 13.25, a common telescope is used for both transmit and receive purposes. The time instants of laser transmit and receive pulses as recorded by photodetectors are used to start and stop a time interval counter, thereby allowing computation of the round-trip time of the laser pulse. Multiplying half of the round-trip time by the speed of light gives the range to the satellite at the time instant of the laser transmit pulse. The measurement is time-stamped with an accurate epoch.

In order to be able to measure range with an accuracy of a few millimetres, two important factors need to be considered. These are the precision with which the epoch is recorded and the uncertainty due to transmit laser pulse width. Assuming the orbital velocity to be 5 km s^{-1}, which is typical of a geodetic satellite at a relatively lower height, the epoch would need to be recorded to an accuracy of better than 1 μs. This is achieved by most SLR stations by recording epochs with respect to the timescale broadcast by the global positioning system (GPS), which is precisely related to and closely aligned with the universal time clock (UTC).

The round-trip measurement time also needs to be measured with the required precision to obtain the desired level of ranging accuracy of a few millimetres. Quantitatively, 3 mm range inaccuracy corresponds to a round-trip time uncertainty of 20 ps, that is, a laser pulse width of 20 ps is needed for a measurement uncertainty of 3 mm in range. Other sources of error are from the internal optical and electronic path delays in the range measuring system. These can be accounted for by range calibration measurements carried out to a nearby reflective target board whose accurate distance from the invariant point of the range telescope has been independently surveyed.

13.6.2 Lasers for SLR

Choice of laser plays an important role in the measurement accuracy of round-trip time-of-flight and thereby the accuracy of measurement of range. Earlier experiments with satellite laser ranging during the 1960s used Q-switched laser pulses of pulse width of the order nanoseconds, pulse energy of few joules and a pulse repetition rate of one pulse every 3–5 seconds from a ruby laser emitting at 694.3 nm. The lower limit of the pulse width from the ruby laser at that time was limited to a few tens of nanoseconds by practical limitations on the rotation rate of the mirror Q-switch. With these lasers, the best achievable accuracy was of the order 0.5–1.0 m.

With the advent of non-mechanical electro-optic and acousto-optic Q-switches and also passive Q-switching using saturable absorbers, the ability to generate sub-nanosecond pulses improved accuracy of the range measurement to that of 10 cm or so.

A further quantum jump in measurement accuracy was achieved with pulse widths of the order of tens of picoseconds generated by mode-locked lasers. A mode-locked Nd:YAG laser with a Q-switched envelope is the preferred laser. In most cases, the laser pulse with highest peak power in the train of pulses in the Q-switched envelope is used. Peak power is generally of the order of gigawatts. Frequency-doubled output at 532 nm is generally used as opposed to infrared output at 1064 nm due to the relatively much higher speed and higher quantum efficiency of detectors in the visible spectrum.

13.6.3 SLR Telescopes and Stations

The telescope is another important part of the SLR station. The telescopes used for SLR measurements fall into two main categories: those using the same optical path for transmit and receive purposes and those using separate but co-mounted telescopes for transmit and receive operations. In some cases, two individually mounted and controlled telescopes have also been used.

Beam transport optics is used to transmit the laser beam from the laser output to the telescope input. The beam transport optics must ensure that laser beam lies along the optical axis of the telescope, irrespective of where the telescope is pointing in the sky.

Required tracking speeds depend upon the satellite altitude in addition to other factors. The time for which a satellite is visible to a given SLR station for carrying out measurements may vary from a few hours in the case of high-altitude satellites (e.g. ETALON and GLONASS) to few minutes for low-altitude satellites (e.g. STARLETTE and STELLA). Tracking speeds of up to $2^\circ\ s^{-1}$ may be required in order to track satellites at relatively lower heights while maintaining an absolute pointing accuracy of a few arc seconds.

Because of the altitude-azimuth mount used in modern telescopes, the azimuth velocity required to track satellites that pass near zenith can be very large. Telescopes are generally desired to slew at speeds of typically $20^\circ\ s^{-1}$ with higher acceleration capability. An alt-azimuth mount is a simple two-axis mount for supporting and rotating the telescope about two mutually perpendicular axes. Rotation about the vertical axis varies the azimuth of the pointing direction of the telescope, while rotation about the horizontal axis varies the altitude or angle of elevation of the pointing direction.

Currently, there are more than 30 active SLR stations connected in a worldwide network with most of them routinely achieving ranging accuracy of 1 cm RMS. The measurements carried out by these centres are regularly passed on to data centres for subsequent analysis by the scientific community. Some of the better-known satellite laser ranging stations include: Burnie TAFE in Australia; Concepcion TIGO in Chile; Herstmonceux in the UK; LURE in Hawaii; McDonald in the USA; MOBLAS-4, 5, 6, 7 and 8 in the USA, Australia, South Africa, USA and Tahiti, respectively; Mount Stromlo and Orroral in Australia; POTSDAM-2 and 3 and Wettzell in Germany; and Wuhan in China. Figure 13.26 is a photograph of the MOBLAS-7 SLR station, one of the most advanced SLR stations, located at the Goddard Geophysical and Astronomical Observatory (GGAO).

Figure 13.26 MOBLAS-7 SLR station (Courtesy of NASA).

13.6.4 SLR Applications

There are three categories of satellites that have been routinely tracked by these tracking stations for a range of scientific applications: geodetic, applications and navigational satellites.

Geodetic satellites are relatively small and inert satellites, spherical in shape with a high-density core and a surface uniformly covered with corner cube retro-reflectors (Figure 13.27). These retro-reflectors are optimized to reflect the wavelength of interest, which is 532 nm in most SLR systems. Common satellites in this category include STARLETTE (1975) with a diameter of 24 cm and an orbit height of 962 km; STELLA (1993) with diameter of 24 cm and orbit height of 810 km; AJISAI (1986) with a diameter of 215 cm and orbit height of 1500 km; LAGEOS-1 (1976) and LAGEOS-2 (1992) both having a diameter of 60 cm and an orbit height of 600 km; and ETALON-1 and 2 (1989) with a 129 cm diameter and orbit height of 19 000 km.

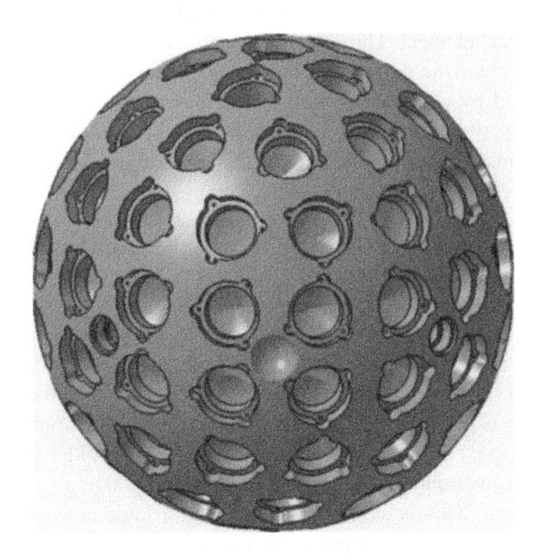

Figure 13.27 Typical geodetic satellite.

Observations from relatively low-altitude geodetic satellites are particularly suited to determining high-frequency terms in the expansion of the Earth's gravity field, used to study ocean tides and modelling variations in atmospheric density. Measurements from high-altitude geodetic satellites are extensively used for the determination of low-frequency terms in the models which represent Earth's gravity field and in the global terrestrial reference frame to study crustal motions.

In the category of applications satellites are the Earth observation and remote sensing satellites. SLR is currently the most accurate technique available for determining the geocentric position of an Earth satellite. By virtue of being the only space geodetic technique defining the Earth's centre of mass, SLR contributes to the definition of the International Terrestrial Reference Frame (ITRF). SLR measurements help to model the temporal and spatial variation of the Earth's gravitational field, contribute to the determination of ocean and Earth tides and are used to monitor tectonic plates and crustal deformation. SLR measurements are also used for monitoring Earth rotation and polar motion to relate it to the International Celestial Reference Frame (ICRF). Finally, SLR provides a unique capability for verification of the theory of general relativity.

In the case of navigational satellites, SLR measurements are used to provide an independent check on the accuracy of orbital information derived from radiometric tracking measurements.

13.7 Lasers in Astronomy

Lasers are used in astronomy as part of adaptive optics systems to improve the image quality in astronomical telescopes. One of the major factors that limit the image quality in ground-based astronomical telescopes is the image distortion resulting from the passage of light through Earth's atmosphere. Space-based telescopes are free from this problem. However, space-based telescopes cannot be as large as terrestrial telescopes and are very expensive to build, launch, operate and maintain.

13.7.1 Adaptive Optics

One effective method of overcoming the problem of image distortion from terrestrial telescopes is by using adaptive optics. An adaptive optical system involves measurement of wavefront distortion. A reference laser source is made to travel through the same atmospheric path, and is corrected or nullified by using an adaptive optical element such as a deformable mirror as one of the optical elements of the telescope. A deformable mirror is an adaptive optical element with a flexible surface profile that can be given any desired shape. This is usually done by driving the surface by piezo actuators. Once the image of the reference light source has undergone atmospheric distortion, corrected is obtained by controlling the shape of the flexible optical element. The object of interest is then viewed with the corrected element.

Figure 13.28 shows the schematic diagram of how laser-guide-star-assisted adaptive optics corrects wavefront distortions caused by atmospheric turbulence to improve the image quality. The adaptive optics system, comprising wavefront sensor, control system and deformable mirror, measures the distortion with light from the guide star and issues the appropriate control signals to the deformable mirror. Distortions in the light from the object of interest are eliminated after reflection from the deformable mirror. As an illustration, Figure 13.29 shows two images of the trapezium region in the Orion Nebula. The image of Figure 13.29a was acquired in 1999 without adaptive optics in place and Figure 13.29b was taken in 2006 using adaptive optics.

13.7.2 Laser Guide Star

The reference light source is one of the most critical elements of the adaptive optical system, as there aren't many high-brightness natural sources in the sky that can fulfil the brightness-level requirements. The reference source also needs to be close to the object of interest to be able to measure the phase distortions in the incoming wavefront. The solution lies in the creation of an artificial bright light source close to the object of interest. This is achieved by pointing a laser beam towards the sky. It is the backscattering of the laser beam in the upper atmosphere above the turbulent layers that acts as a reference source. This artificial bright light source is referred to as the *laser guide star*.

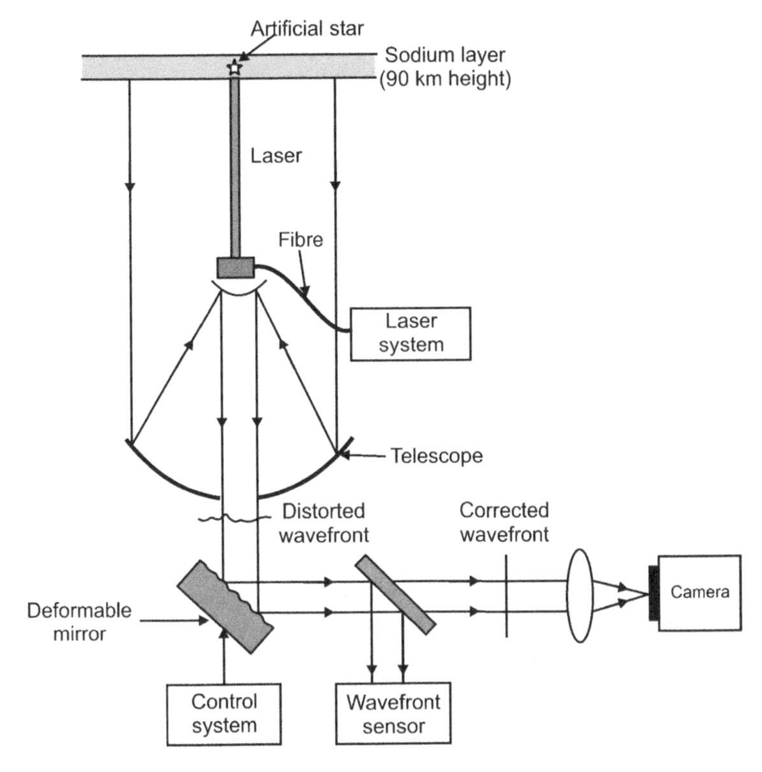

Figure 13.28 Schematic diagram of laser guide star optical system.

The laser guide star can be positioned anywhere in the sky, allowing a much larger fraction of the sky to be accessible for the adaptive optical system. The position of the artificial guide star may drift due to its passage through the atmosphere. This can however be corrected by monitoring the position of a nearby natural star and using a tip-tilt mirror. This star does not have to be as bright as the guide star.

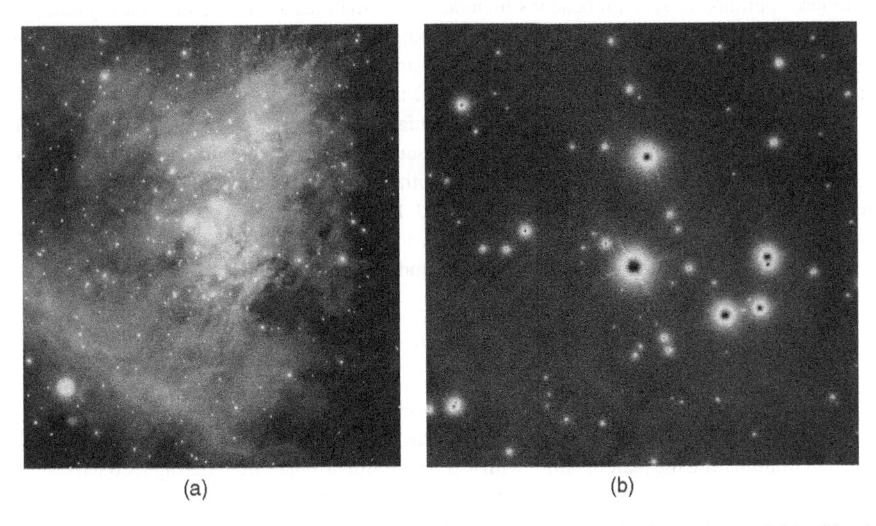

Figure 13.29 Image of trapezium region in the Orion Nebula: (a) without adaptive optics; and (b) with adaptive optics (Courtesy of the National Astronomical Observatory of Japan).

13.7.3 Laser Guide Star Mechanisms

There are two broad mechanisms by which the artificial light source is created using lasers. The first mechanism involves the process of Rayleigh scattering from air molecules such as nitrogen. The second mechanism makes use of sodium fluorescence caused by the sodium layer in the mesosphere. For maximum efficacy of the two mechanisms of producing laser guide stars, each of the two mechanisms must possess some specific properties.

13.7.3.1 Rayleigh Guide Star

The Rayleigh guide star is based on Rayleigh scattering in the lower atmosphere. In contrast to a sodium beacon that requires a narrow-line-width laser to produce resonant absorption, there is no such critical requirement for a Rayleigh guide star. The choice of laser wavelength is also not critical, except that it should be short to cause efficient Rayleigh scattering. Although the commonly used laser is the frequency-doubled solid-state laser, a copper vapour laser or an excimer laser can also be used. Pulsed lasers with nanosecond-wide laser pulses are preferred over continuous wave lasers. The pulsed format simplifies the non-linear frequency conversion in the laser source, and enables time-gated detection. Such laser sources are generally less complex and more powerful than those used with sodium beacon guide stars.

The disadvantage of the Rayleigh guide star is that the lower altitude of backscattered light compromises the quality of the wavefront correction. The William Herschel telescope located in La Palma, Canary Islands, Spain uses a Rayleigh guide star. The telescope is part of the Isaac Newton Group of telescopes, and is funded by research councils from the United Kingdom, the Netherlands and Spain. Figure 13.31 is a photograph of the 4.2 m Herschel telescope.

13.7.3.2 Sodium Beacon Guide Star

A sodium beacon guide star makes use of the sodium layer in the mesosphere at an altitude of 90 km above the surface of Earth. Sodium atoms are naturally present in the sodium layer. A sodium guide star is created by using a laser tuned to 589.2 nm, which is the resonant absorption wavelength of sodium atoms. This causes sodium atoms to absorb laser light and subsequently to emit fluorescence at the same wavelength.

The availability of a suitable laser at the desired wavelength and line width is a difficult proposition. Available laser options for sodium beacons include: (1) Raman laser based on a bulk crystal, pumped with a frequency-doubled Q-switched neodymium-based solid-state laser; (2) a frequency-doubled Raman fibre laser emitting at 1178 nm and pumped by an ytterbium-doped fibre laser; (3) sources based on sum frequency mixing of two laser sources; or (4) a pulsed dye laser.

The Lick Observatory of the University of California, the Palomar Observatory of Caltech and the Keck Observatory in Hawaii employ sodium beacon laser guide stars. Figure 13.30 is a photograph of the Shane telescope at the Lick observatory, equipped with a sodium beacon laser guide star.

Other large observatories that have either added laser-guide-star-aided adaptive optics to the facility or are considering including one are the Very Large Telescope (VLT) of the European Space Observatory (ESO) (Figure 13.32), Gemini North and the Multiple Mirror Telescope Observatory (MMTO).

13.8 Holography

Holography is the technique of forming a three-dimensional image of an object by recording the interference pattern formed by a split laser beam during the recording process on a photosensitive material, and illuminating the pattern either with a laser or with ordinary light during reconstruction. The concept of holography has a wide range of applications other than recording and reconstruction of 3D images, however. Holographic optical elements (HOE) are used in a myriad of technical devices as mirrors, lenses, gratings or combinations of these.

Figure 13.30 Shane telescope at (Courtesy of University of California Observatories 2010–2013).

Figure 13.31 William Herschel telescope (Photo courtesy of the Isaac Newton Group of Telescopes, La Palma).

Figure 13.32 Very large telescope of European space observatory (Reproduced by permission of ESO).

In the case of holographic interferometry, a recorded light field scattered from the object of interest is superimposed on the light field scattered from the object. If the object has undergone deformation, the phases of the two light fields will alter producing an interference pattern that can be interpreted to measure the deformation. One major application of holographic interferometry is in the measurement of microscopic displacements on the surface of an object and small changes in refractive index of transparent objects such as plasma and heat waves.

HOEs are likely to find application in a range of future photonic devices for optical computations, free-space interconnects and analogue and digital memory systems.

13.8.1 Basic Principle

Holography is a technique of recording the light field scattered off an object in such a manner that it can be reconstructed later when the object is no longer present. Holograms are produced by recording on photosensitive material the interference pattern of the light field scattered off the object and a part of the light beam acting as the reference light field. A hologram can be created by illuminating the recording medium by a part of the light beam falling directly and the other part reaching the recording medium, that is, the photosensitive material after scattering off the object.

The light source used to produce the interference pattern is a laser for obvious reasons: a laser is a coherent and monochromatic source of light, unlike sunlight or light from conventional sources which are incoherent and contain many different wavelengths. Holograms are generally recorded in darkness or in low-level light of a colour different from that of the laser light used to make the hologram in order to prevent interference from external light. The exposure time, which is a critical parameter, is controlled either by using a shutter or by electronically timing the laser.

Figure 13.33 shows the basic optical arrangement for recording of a hologram. The laser beam is split into two beams. One of the beams (object beam) is directed towards the object or scene after passing through some optical elements used to spread the beam, and is not shown in the figure. Some of this light energy scattered off the object falls onto the recording medium. The second beam (reference beam) is also spread through the use of optical elements, and is directed onto the recording medium without coming in contact with the object or scene. A common recording medium is a film of silver halide photographic emulsion, which has a much higher concentration of light-reactive grains compared to that of photographic film producing higher resolution. A thin layer of silver halide is deposited on a transparent substrate such as glass or plastic.

The two light fields interfere and the resulting interference pattern is imprinted on the recording medium. What is recorded is not the image of the object, but an interference pattern that can be

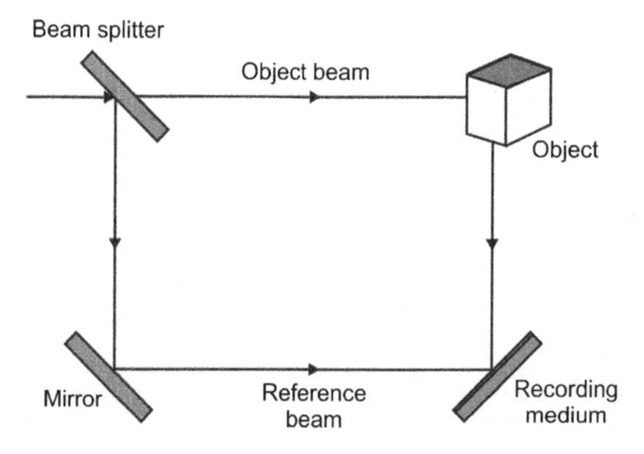

Figure 13.33 Optical arrangement for recording of hologram.

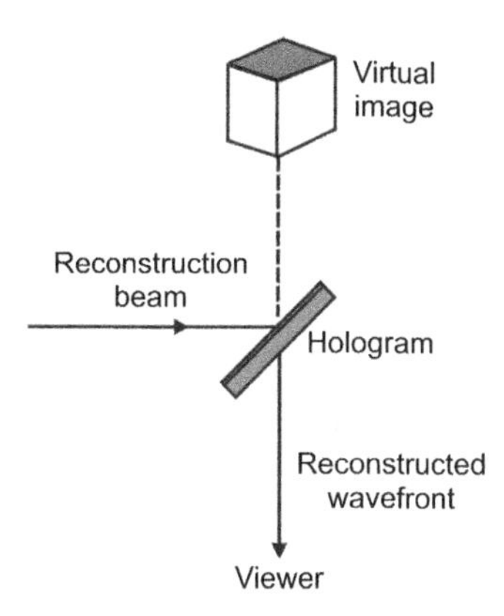

Figure 13.34 Optical arrangement for reconstruction of hologram.

considered as an encoded version of the object image. The interference pattern itself is apparently random, but it contains information on how light scattered from the object interfered with the original light source.

In order to reconstruct the object image, the encoded version of the object recorded in the hologram in the form of interference pattern only needs to be decoded. This is achieved by illuminating the hologram by the light beam identical to that used for recording the hologram. The light beam illuminating the hologram is diffracted by the hologram's surface pattern. The light field is identical to that originally produced by the laser beam scattered off the object. This can be seen as a virtual image. Figure 13.34 shows the optical arrangement for reconstruction of hologram.

A hologram records information about the object in the form of light as it scatters off the object in a range of different directions, rather than from only one direction as in a photograph. As a result, while each point in a photograph only represents light scattered from a single point in the object or scene, each point on a holographic recording includes information about light scattered from every point in the scene. A hologram therefore allows the object or scene to be viewed from different angles as if the object is still present. The ability to view the object from different angles in a hologram recording gives the perception of depth.

13.8.2 Types of Hologram

There are different ways of classifying holograms: *amplitude modulation* and *phase modulation* holograms; *transmission* and *reflection* holograms; and *thin* and *thick* holograms. There are many variations between transmission and reflection holograms, collectively known as *hybrid holograms*. Each of these types is briefly described in the following sections.

13.8.2.1 Amplitude- and Phase-modulated Holograms

In the case of an *amplitude-modulated hologram*, the magnitude of light diffracted from the hologram is proportional to the intensity of light used for recording. In a *phase-modulated hologram*, either the thickness or refractive index of the photosensitive material is varied in proportion to the intensity of the holographic pattern. A phase-modulated hologram reproduces the original wavefront when illuminated by a reference beam.

13.8.2.2 Transmission and Reflection Holograms

In the case of a *transmission hologram*, the object and reference beams are incident on the recording medium from the same side. During reconstruction, the typical transmission hologram is viewed with

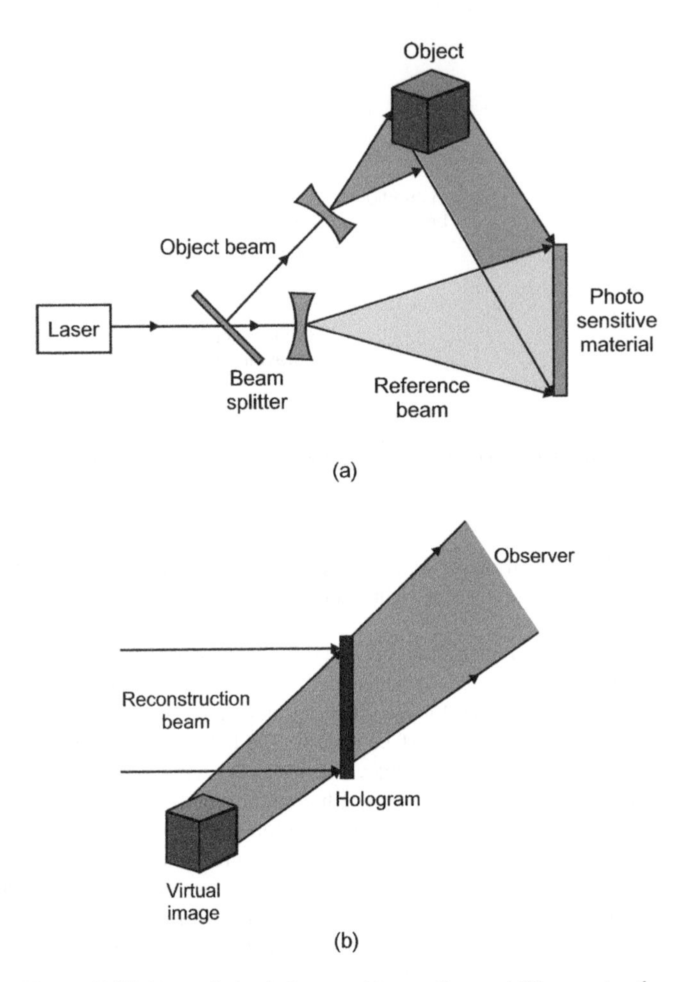

Figure 13.35 Transmission hologram: (a) recording; and (b) reconstruction.

laser light that is usually of the same type as that used for the recording. Figure 13.35a and b illustrate the experimental set-up for recording and reconstructing a typical transmission hologram, respectively.

In the case of a *reflection hologram*, the object and reference beams are incident on the photosensitive material from opposite sides. The hologram is viewed from the same side from which the reference beam is incident. The image therefore consists of light reflected by the hologram. Figure 13.36a and b show the experimental set-up for the recording and reconstruction of a typical reflection hologram, respectively. The reflection hologram allows a truly three-dimensional image to be seen near its surface and is the most common type shown in galleries. These holograms have been made and displayed in colour with their images optically indistinguishable from the original objects.

13.8.2.3 Thin and Thick Holograms

In the case of a *thin hologram*, the thickness of the photosensitive medium is much less than the spacing of the interference fringes which make up the holographic recording. In a *thick hologram* (also called a *volume hologram*), the thickness of the photosensitive medium is greater than the spacing of the interference pattern, which makes the recorded hologram a three-dimensional structure. The incident

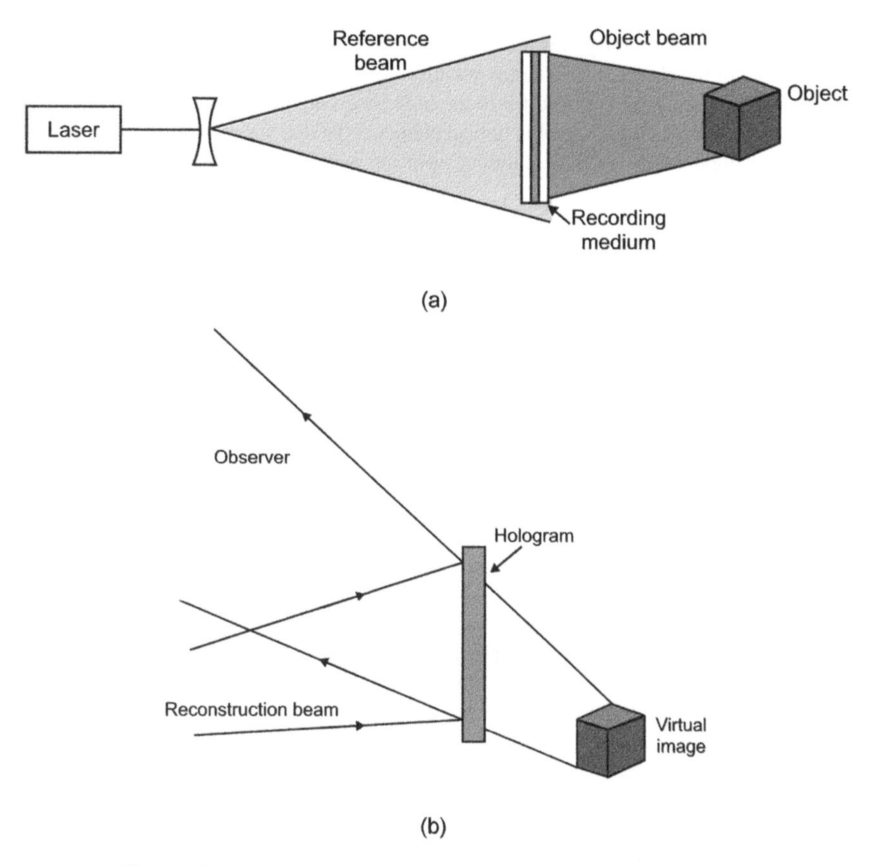

Figure 13.36 Reflection hologram: (a) recording; and (b) reconstruction.

light is diffracted only at a particular angle of illumination called the Bragg angle. If the angle of incidence is the same as the original reference beam angle and the hologram is illuminated with a light source having a broad spectrum of wavelengths, reconstruction occurs only at the wavelength of the original laser used. If the angle of illumination is different, reconstruction will also occur at a different wavelength causing the reconstructed image to appear in a different colour.

13.8.2.4 Other Commonly Encountered Holograms

Other commonly encountered holograms include *embossed holograms*, *integral holograms*, *multi-channel holograms* and *computer-generated holograms*.

Embossed holograms are the mass-produced holograms such as those used to build security into credit/debit cards. The original hologram is usually recorded on a photosensitive material and the replicas are made by pressing the 2D interference pattern onto thin foils. An *integral hologram* is a type of hologram that is automatically synthesized from a large collection of photographs, each taken from a slightly different position. *Multichannel holograms* allow completely different images to be observed in the same hologram with changes in the angle of the viewing light. This concept has enormous potential for massive computer memories.

Computer-generated holograms are digitally generated holographic interference patterns. The desired holographic image is digitally computed and then printed onto a mask or film and displayed by illuminating it with a suitable coherent source of light. Computer-generated holograms can be fully synthetic holograms without the need for the object to be a physical reality.

The whole process of computer-generated holograms involves three basic steps: computation of the scattered wavefront of the object; encoding the wavefront data; and modulating the interference pattern onto a coherent source of light to transport it to the user observing the hologram. Computer-generated holograms find extensive use as holographic optical elements (HOEs) for controlling laser light in many optical devices for applications such as scanning, splitting and focusing.

13.8.3 Applications

Due to the versatility of their underlying technology, holograms have been exploited in a wide range of applications in industry, scientific research, art, commerce and medicine. Applications range from the creation of pulsed holographic portraits and other works of art to adding security features to valuable documents; from authentication and decorative packaging of branded goods to medical diagnostics; from imaging to data storage; and from use of holographic optical elements for various functions to non-destructive testing to reveal structural stresses and strains without damaging the specimen.

A hologram is the established overt authentication feature incorporated into branded products and valuable documents to prevent counterfeiting, adulteration, substitution and parallel trading. They can also be integrated with other security technologies to provide a combination of overt and covert security with track-and-trace capabilities. Holograms are widely used on a range of security documents such as currency notes, fiscal stamps, cheques, bank cards and personal identification documents including ID cards, passports, and so on.

Holographic products are available in a wide variety of designs, materials and formats to interface with existing packaging design and production processes and are extensively used for packaging and promotion purposes to provide visual effects and brand differentiation.

HOEs are used to perform a variety of functions such as those performed by lenses, mirrors and gratings. Holography also allows the combining of several functions, not realizable with conventional optical elements. Some examples include bifocal contact lenses, supermarket barcode scanners and high-resolution spectrometers. A head-up display, a holographic windshield installed in a military aircraft, is another important application of HOEs. A holographic image of the cockpit control instrumentation appears to float in front of the windshield, thereby allowing the pilot to keep his eyes on the runway or the sky while handling the controls. This feature is also now available in some of the high-end automobiles.

Holography has applications in medical diagnostics. The hologram is embedded in a smart material capable of detecting certain molecules or metabolites. The chemical interaction between the added molecules in the holographic film and the molecules of interest manifests as a change in one of the properties of the holographic reflection in the form of change in either fringe periodicity or refractive index. This causes a change in the colour of the holographic reflection as is the case in a Bragg reflector.

Holographic interferometry allows the measurement of static and dynamic displacements of objects with optically rough surfaces to a precision of a fraction of a wavelength. Holographic interferometry is extensively used for non-destructive testing of engineering structures to measure stress, strain and vibration. The technique has also been used to visualize fluid flows.

Holographic memory is the next-generation data-storage device with the ability to store 1000 GB of data in a crystal approximately the size of a sugar cube; compare this to the 15.9-GB-capacity of a double-sided double-layer DVD. This tremendous increase in data storage capacity is due to the fact that holographic storage stores data throughout the volume of the medium, as opposed to only on the surface of the recording media in conventional magnetic and optical storage technology. Data storage capacity is further enhanced as it is capable of storing multiple data in the same area by utilizing light at different angles. Holographic data storage also offers much higher data transfer rates as it reads millions of bits in parallel; conventional magnetic and optical storage techniques record data 1 bit at a time in a linear fashion.

Holography technology is advancing very rapidly and many new applications will be added to the spectrum in the coming years. Holographic memory discussed in the preceding paragraph is also an emerging application. Two other applications that are likely to become established by 2020 include holographic television and optical computing. A prototype of holographic television has already been developed by the Massachusetts Institute of Technology (MIT). By 2020, computers may not only be using holographic displays, but may also be exploiting the principles of holography for computing using photons instead of electrons.

13.9 Summary

- As a source of electromagnetic radiation with exceptionally high frequency stability, the laser is the basis of length/distance metrology. Laser interferometry is at the core of length metrology, measurement of time-of-flight of laser pulses being the other commonly used technique. Optical frequency metrology allows high-precision measurements of optical frequencies. The atomic clock is an example.
- Lasers offer a non-contact methodology for precise measurement of distances and displacements. Most common laser-based methods of determining distances and displacements include interferometry, the time-of-flight method, the phase shift method, the triangulation method and the frequency modulation method.
- An *interferometer* is an optical device that makes use of interference phenomenon. Interferometry is particularly suited to measurement of changes in distance rather than absolute measurement of distance.
- A Michelson interferometer uses a single-beam splitter for separating and combining the beams. It is a very useful optical tool for the measurement of wavelength, distance, refractive index and coherence length of optical beams.
- A Twyman–Green interferometer is essentially an adaptation of Michelson interferometer which uses a monochromatic point source of light for characterizing optical elements such as lenses, prisms and flats.
- The Mach–Zehnder interferometer is an optical tool used to determine the relative phase shift between two collimated beams from a coherent light source.
- The Fabry–Pérot interferometer consists of two parallel optical flats with partially reflecting mirror surfaces facing each other. The monolithic version of the interferometer (etalon) consists of a transparent plate with two reflecting surfaces. A Fabry–Pérot interferometer differs from a Fabry–Pérot etalon in the fact that it allows the distance between the plates to be fine tuned to make transmission peaks correspond to the desired wavelength.
- The Sagnac interferometer is based on the Sagnac effect, which describes what happens when a ring laser comprising two counter-propagating laser beams is rotated around an axis perpendicular to its plane. The light beam travelling in the same direction as that of rotation experiences a longer path length, while the light beam travelling in the opposite direction experiences a shorter path length in one round trip. This produces a relative phase shift between the two counter-propagating beams, with the magnitude of the phase shift being proportional to the rotation rate.
- In the *time-of-flight* technique, the distance to an object is determined by measuring time-of-flight of a narrow laser pulse to travel to the targeted object and return after reflection.
- In the *phase shift* technique of range finding, the laser beam with sinusoidal power modulation is transmitted towards the target and the diffused or specular reflection from the target is received. The target distance is determined from the phase of the received laser beam.
- In the *triangulation method*, laser transmitter, laser receiver and the intended object form a triangle. The distance to the target is determined from known values of distance between the transmitter and the receiver and measured angles of lines joining transmitter and receiver, transmitter and object and receiver and object.
- In the *frequency modulation* method, the frequency of a narrow-line-width laser is modulated with a ramp or sinusoidal signal and then transmitted towards the target. The received signal corresponding to the reflected laser beam is mixed with the reference signal. The beat frequency produced as a result of homodyne detection is used to derive the distance information.
- The *optical frequency comb* technique has revolutionized optical frequency metrology. An optical frequency comb is simply an optical spectrum consisting of equidistant lines. It overcomes the major

limitations of the traditional frequency chain approach to generate frequency references, which allow frequency measurement around a single optical frequency only.
- An *optical clock* derives its output from an ultra-stable optical frequency standard and offers exceptionally high precision that is far superior to the best-available caesium atomic clocks. Timing precision of the order of 10^{-15} can be achieved.
- The line width of a laser is an indicator of the degree of its monochromaticity and is defined as the width of its optical spectrum, measured typically as FWHM. In the case of multimode lasers with line widths of the order of 10 GHz or more, the conventional diffraction-grating-based technique of optical spectrum analysis may be used. For single-frequency lasers with narrower line width, the self-heterodyne technique is often used. Line width in this case is measured as a beat frequency between the frequency-shifted laser output and a delayed version of it.
- An infrared thermometer in its most basic form consists of a focusing optic that focuses the received infrared energy onto a detector. It measures the temperature by detecting the infrared energy emitted by the object and converting it into an equivalent electrical signal which, after compensation for the ambient temperature, can be displayed in the units of temperature.
- Laser velocimetry is extensively used for measurement of one or more velocity components over one or more dimensions. Although the most widely exploited application of laser velocimetry continues to be in measurement of transparent and semi-transparent fluid flow, it has also been used for measurement of linear or vibratory motion of opaque reflecting surfaces. There exist many variations of laser veolocimetry, including laser Doppler and particle image velocimetry.
- Laser vibrometry, operationally similar to laser velocimetry, is a non-contact technique used for measurement of surface vibrations. It also relies on the Doppler effect to extract information on vibration amplitude and frequency from the Doppler shift undergone by the laser beam frequency due to vibratory motion of the surface of interest.
- Electronic speckle pattern interferometry (ESPI) is an optical interferometric technique enabling measurements of surface displacements or deformations on material surfaces caused by some kind of loading, which could be mechanical, thermal, vibrational or pressure. ESPI has been successfully used for the study of material properties, fatigue testing, fracture mechanics, vibrational mode analysis and non-destructive testing of a variety of components in automotive, aerospace and electronics industry.
- Satellite laser ranging (SLR) is a technique of measuring the range of an Earth-orbiting satellite using a laser with the objective of determining the orbital parameters of satellites and, from their variation from predicted values, accurately determining the temporal variation of Earth's centre of mass.
- Lasers are used in astronomy as part of an adaptive optics system to improve the image quality of terrestrial astronomical telescopes. A laser guide star serves the purpose of an artificial source of bright light at a location close to the astronomical object to be viewed. It is used to provide the reference source required by the adaptive optical system to counter or nullify the effect of atmospheric distortion.
- Holography is the technique of forming a 3D image of an object by recording on a photosensitive material the interference pattern formed by a split laser beam. This pattern is illuminated either with a laser or with ordinary light during the hologram reconstruction.

Review Questions

13.1. What is the role of an interferometer in optical metrology? Briefly describe the operational principle of the Michelson, Fabry–Pérot and Mach–Zehnder interferometers.

13.2. Compare the following distance-measuring techniques in terms of maximum range-measuring capability and range-accuracy specifications.

a. time-of-flight technique
b. triangulation technique
c. multiple phase shift technique
d. frequency modulation technique.

13.3. What is an optical frequency comb? Briefly describe how an optical frequency comb can be generated using ultra-short-pulse mode-locked lasers and how it can be used for the measurement of unknown frequency in a wide spectral band.

13.4. What is an optical clock? How does it compare with a caesium clock in terms of timing precision? With the help of a suitable block schematic arrangement, briefly describe the operational principle of an optical clock.

13.5. Define line width of a laser. How is line width related to monochromaticity? With the help of a suitable diagram, briefly explain the heterodyne principle of measurement of line width of the order of sub-kilohertz level.

13.6. What is laser velocimetry used for? With the help of a suitable optical set-up using backscatter detection, briefly explain the operational principle of laser Doppler velocimetry for measurement of fluid flow velocity.

13.7. What are the salient features of laser Doppler velocimetry? Where does laser Doppler velocimetry find application in flow research, industrial process control and medical diagnostics?

13.8. Compare and contrast Doppler velocimetry and particle image velocimetry. With the help of a suitable diagram, briefly explain the principle of operation of particle image velocimetry.

13.9. With the help of a suitable diagram, explain the principle of operation of electronic speckle pattern interferometry for studying material properties.

13.10. Briefly explain out-of-plane and in-plane displacement measurement set-ups using electronic speckle pattern interferometry.

13.11. What is satellite laser ranging? What is it used for? With the help of a block schematic arrangement, brief explain the operation of a satellite laser ranging station. Name two international satellite laser ranging facilities.

13.12. How does the use of adaptive optics help in improving the image quality of astronomical telescopes? What is the role of a laser guide star in building the adaptive optical set-up of an astronomical telescope?

13.13. With the help of a suitable diagram briefly explain the principle of operation of laser holography, highlighting both the recording and reconstruction procedures. Outline five important application areas of holography.

13.14. Write short notes on the following:
a. transmission and reflection holograms
b. laser vibrometry for detection of buried mines
c. infrared thermometer.

Self-evaluation Exercise

Multiple-choice Questions

13.1. An adaptation of a Michelson interferometer used for characterization of optical components such as lenses, flats and prisms is called a:
a. Fabry–Pérot interferometer
b. Twyman–Green interferometer
c. Mach–Zehnder interferometer
d. Sagnac interferometer.

13.2. For constructive or destructive interference to take place in a Michelson interferometer using a laser source, the path length difference needs to be
a. less than the coherence length of the laser
b. greater than the coherence length of the laser
c. precisely zero
d. none of these.

13.3. In a Fabry–Pérot etalon, the peak transmission wavelength can be fine tuned by varying
a. the distance between the two reflecting surfaces
b. the angle at which light travels in the space between the two reflecting surfaces
c. both (a) and (b)
d. none of these.

13.4. A ring laser gyroscope, used in inertial navigation systems, operates on the principle of a
a. Sagnac interferometer
b. Mach–Zehnder interferometer
c. Fabry–Pérot etalon
d. Michelson interferometer

13.5. Name the optical technique used for the measurement of distance:
a. time-of flight method
b. phase shift method
c. triangulation method
d. all of the above.

13.6. Of the caesium atomic clock and optical clock timing standards, which offers the higher timing precision?
a. caesium atomic clock
b. optical clock
c. a combination of the two
d. they offer the same timing precision.

13.7. Which of the following interferometers is an integral element of an optical clock timing standard?
a. Mach–Zehnder interferometer
b. Michelson interferometer
c. Sagnac interferometer
d. Fabry–Pérot interferometer.

13.8. Laser Doppler velocimetry is a
a. point technique based on interferometry
b. planar technique
c. imaging technique
d. none of these.

13.9. The preferred laser for satellite laser ranging experiments is the
a. mode-locked Nd:YAG laser
b. ruby laser
c. CW CO_2 laser
d. dye laser.

13.10. A laser pulse width of 50 ps in a satellite laser ranging experiment will result in a range measurement uncertainty of
a. 7.5 m
b. 7.5 cm
c. 7.5 mm
d. 15 mm.

13.11. The image quality of an astronomical telescope can be significantly enhanced by
a. light guide star adaptive optics
b. using a larger-aperture telescope
c. using better-quality optics
d. performing the experiment at the right time.

13.12. Hologram recording and reproduction are based on the properties of
a. interference
b. diffraction
c. interference and diffraction
d. none of these.

13.13. Holographic memory has high data storage density because
a. it stores data three-dimensionally
b. it can store multiple data on the same area by utilizing light at different angles
c. both (a) and (b)
d. none of these.

13.14. Peak wavelength emitted by a hot body is
a. inversely proportional to absolute temperature
b. directly proportional to absolute temperature
c. inversely proportional to the square of absolute temperature
d. directly proportional to the fourth power of absolute temperature.

13.15. When two coherent light beams interfere, the resultant intensity is equal to
a. the sum of the individual intensities
b. the difference between the individual intensities
c. the product of the individual intensities
d. none of these.

Answers

1. (b) 2. (a) 3. (c) 4. (a) 5. (d) 6. (b) 7. (a) 8. (a) 9. (a) 10. (c) 11. (a)
12. (c) 13. (c) 14. (a) 15. (d)

Bibliography

1. *Handbook of Laser Technology and Applications, Volume III*, 2003 by Colin E. Webb and Julian DC Jones, Institute of Physics Publishing.
2. *Optical Metrology*, 2002 by Kjell J. Gasvik, John Wiley & Sons.
3. *Handbook of Optical Metrology: Principles and Applications*, 2009 by Toru Yoshizawa, CRC Press.
4. *Optical Imaging and Metrology*, 2012 by Wolfgang Osten and Nadya Reingand, Wiley-VCH.
5. *Laser Velocimetry in Fluid Mechanics*, 2012 by Alain Boutier, Wiley-ISTE.
6. *Laser Guide Star Adaptive Optics for Astronomy*, 2000 by N. Ageorges and C. Dainty, Springer.

14

Military Applications: Laser Instrumentation

14.1 Introduction

From the early 1980s, lasers have penetrated almost every conceivable area of application from fundamental science and technology to industry, from medicine and healthcare to entertainment and from tactical battlefield to the strategic domain. The applications have grown at a very fast rate and not just in previously existing domains; laser devices have also found their place in many new areas. Due to the enormity of the subject and the interest it currently holds internationally, in terms of ever-increasing usage for a variety of deployment scenarios and the types of investments being made into their research and development, military applications of lasers and optoelectronics are covered in two chapters.

The focus in this present chapter is instrumentation and sensor military applications of lasers and related devices such as laser rangefinders, laser target designators, laser bathymetry, electro-optic guidance techniques (including detailed description of laser-guided munitions) and infrared-guided missiles. This chapter also focuses on important optronic sensors such as the ring laser and fibre-optic gyroscopes used for navigation. Chapter 15 focuses on the directed-energy applications of lasers and optoelectronics devices and systems.

We begin with an introduction to military applications of lasers, highlighting major functional disciplines including those already in use and those likely to be used in the not-too-distant future. We focus on the requirements of the application, various technological options, current state-of-the-art and future trends. The chapter concludes with a discussion on free space and underwater laser communication, with particular reference to their importance for military applications.

14.2 Military Applications of Lasers

While the expansion of non-military applications of lasers is primarily driven by the availability of a large number of wavelengths as well as ever-increasing power levels and diminishing price tags at which these wavelengths can be generated, military applications of lasers and related electro-optic devices have grown mainly because of technological maturity of the lasers that were introduced in the late 1960s and early 1970s. Technological advances in optics, optoelectronics and electronics leading to more rugged, reliable, compact and efficient laser devices are largely responsible for making them indispensable in modern warfare. Laser systems developed for existing military applications also continue to improve in terms of performance specifications and system engineering. Much work is being conducted, particularly in technologically advanced countries, to exploit the potential of lasers and optoelectronic devices in newer areas.

Lasers and Optoelectronics: Fundamentals, Devices and Applications, First Edition. Anil K. Maini.

Figure 14.1 Laser rangefinder on an M320 grenade launcher (Courtesy of Heckler & Koch GmbH).

The most common tactical battlefield applications of lasers and optoelectronic devices are as *laser rangefinders* and *laser target designators* for munitions guidance. A modern battlefield tank whose fire control system does not utilize the services of a laser rangefinder cannot be imagined; this is also true for other forms of armoured fighting vehicles. Short-range semiconductor diode-laser-based rangefinders are used on squad weapons such as assault rifles and light machine guns (Figure 14.1).

A laser target designator is an essential component of laser-guided munitions delivery systems, although there are other forms of electro-optic guidance such as that employed in the case of infrared-guided missiles. Laser-guided munitions, including bombs, projectiles and missiles, constitute an important class of precision-strike weapons and the laser target designator plays a key role in the overall delivery system. While a laser target designator is the key element of a laser-guided munitions delivery system, an optical-gyroscope-based inertial sensor such as ring laser gyroscopes (RLGs) and fibre-optic gyroscopes find extensive use in navigation systems of commercial airliners, ships, spacecraft, military aircraft and ballistic missiles.

There are other military applications that exploit the principles of laser rangefinding and target designation. *Laser-based proximity sensors*, *gap-measuring devices* and *obstacle-avoidance systems* are some examples where the laser rangefinding principle is put to use. *Laser tracking* is another example where a laser target designator/rangefinder mounted on a two-axis gimbal platform can be used to determine the 3D coordinates of a remote target and to track it.

Another application of the basic laser range-finding principle is in *laser bathymetry*. Laser bathymetry is a swath surveying technique that is complementary to conventional multibeam acoustic systems. Laser bathymetry is particularly attractive for surveying shallow coastal waters where the acoustic technique is not very effective due to limited swath width. With the arrival of airborne laser bathymetry systems that survey at aircraft speed and are effective in shallow and shoal-infested waters, laser bathymetry has become the ideal complement to acoustic swath mapping of coastal waters. Airborne laser bathymetry also offers a swath width that is independent of water depth.

Laser pointing is another common application. Small low-cost low-power semiconductor diode laser modules with provision of precise *x–y* movement are finding widespread use on squad weapons for target aiming and pointing, particularly during night-time operations. This increases weapon effectiveness by improving the single-shot hit probability and reducing collateral damage. Figure 14.2 shows the photograph of one such module, mounted on a riffle. These modules are generally integrated with the weapon through Picatinny or universal rails.

Figure 14.2 Laser aiming aid mounted on a rifle (Reproduced from public domain (http://en.wikipedia.org/wiki/File:INDIA5.jpg)).

Relatively newer types of laser systems that have found widespread recognition from the armed forces are the systems that can offer effective countermeasures against systems already in use. A typical electro-optic countermeasure (EOCM) system aims to provide protection to the military platform from laser-guided munitions attack. The EOCM system essentially comprises a laser warning system that provides information on the type and angle-of-arrival of laser threat and a suitable interface to a countermeasures system. The countermeasures system is either in the form of a smoke/aerosol screening system to block laser radiation or a high-energy laser system that can be used to neutralize the source of laser radiation, that is, laser target designator or seeker head of laser-guided munitions.

Lasers and optoelectronics devices and systems have also found widespread use in low-intensity conflict (LIC) operations. Many of the established military technologies and systems are being adapted for use in LIC scenarios. A low-intensity conflict is the most common form of warfare today and is likely to be so in the foreseeable future. Low-intensity conflict poses an alarming threat to national security and is an area of concern for the whole of the international community today. Its scope extends from emergency preparedness and response to domestic intelligence activities to riot and mob control; from combating illegal drug trafficking to protection of critical infrastructure; from handling counter-insurgency and anti-terrorist operations to detection of nuclear and biological agents; and from detection and identification of explosive agents to detection of concealed weapons.

Laser and optoelectronics technologies play an important role in handling low-intensity conflict situations. The key advantage of the use of laser technology in such applications mainly stems from its near-zero collateral damage, speed of light delivery and potential for building non-lethal weapons. Some of the well-established laser devices in LIC applications include *laser dazzlers* for close combat operations, mob/riot control and protection of critical infrastructures from aerial threats; *lidar sensors* for detection of chemical, biological and explosive agents; femtosecond lasers for imaging of concealed weapons; and lasers for sniper and gun fire location identification.

Use of *laser vibrometry* and *electron speckle interferometry* techniques for detection of buried mines and high-power lasers for the disposal of unexploded ordnances are emerging applications of laser technology for homeland security. Figure 14.3 is a photograph of a laser dazzler (Model CHP Laser DazzlerTM from M/s LE Systems Inc). This compact and handheld laser dazzler is configured around a 500 mW, 532 nm laser that is reportedly capable of causing vision impairment of the subject at a distance even in bright ambient conditions and is eye safe for specified non-ocular hazard distance (NOHD) and exposure time.

The last one decade or so has seen the emergence of some new potential areas of laser systems usage. The initial uses of lasers in defence as outlined above were in the form of devices such as laser pointers, laser rangefinders and target designators for enhancing the accuracy of the conventional weapon systems. Due to the unique nature of laser light, namely its coherence, monochromaticity, high degree of collimation and high intensity, it was also considered as a potent weapon system for directed-energy

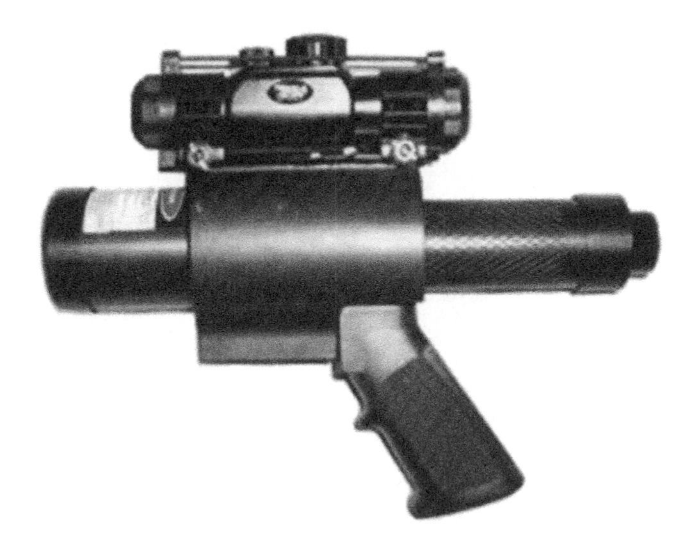

Figure 14.3 Compact high-power (CHP) Laser Dazzler (Courtesy of LE Systems, Inc).

applications. For these applications, the lethal laser energy is delivered to the target to cause some form of physical destruction or neutralize electro-optic sensors used on the target platforms. The main advantages with laser weapons are speed of light delivery, unlimited re-firing capability and reduced logistic support requirements. Such weapons can be grouped in two main categories based on the desired effect on the target: *non-lethal laser weapons* and *lethal laser weapons*. Lethal systems are commonly known by the name of *laser-based directed-energy weapons* (L-DEW). There are three major categories of directed-energy weapons: microwave-based DEW, laser-based DEW and particle beam weapons. A description of microwave-based DEW and particle beam weapons is beyond the scope of the present text.

Directed-energy laser weapons are systems designed for causing physical destruction of the target. The main use of these weapons is in providing air defence, where they aim to destroy the incoming munitions or aircraft before they can accomplish their mission. In such applications, the laser weapons are seen as the only solution to counter multiple simultaneous threats that would normally saturate the conventional air defence based on anti-aircraft guns and missiles.

The heart of laser-based DEW is a high-power laser, which has enough power in the case of CW lasers or sufficient pulse energy in the case of pulsed lasers, to inflict physical damage to the target. A target tracking and beam pointing subsystem is the other important component of the DEW system. These weapons are projected to play an increasingly dominant role in the future, rendering conventional high-tech weaponry obsolete. These weapons will begin to find use on land-based, ship-borne and aerial platforms for a range of surface-to-surface, surface-to-air and air-to-ground tactical applications. Tactical high-energy laser (THEL), a joint venture of the USA and Israel, is a tactical air defence system for use against rockets. Figure 14.4 is a photograph of the THEL system. The mobile version of THEL, also known as M-THEL, configured on three vehicular platforms has also been developed and successfully tested.

Another important area of military application of lasers is in secure point-to-point communication links. Free-space laser communication offers the inherent advantages of being hard to intercept, thus providing security. Free space here means communication in air, outer space, vacuum or any similar medium. It is comparatively easy to encrypt data travelling across a free-space laser communication link, further enhancing data security which is an important consideration for military communications. Other advantages include higher immunity to electromagnetic interference compared to microwaves, higher data and lower bit error rates, full duplex operation, protocol transparency and additional security provided by narrowness and directionality of laser beam. Laser communication also suffers from certain limitations however, which includes signal attenuation caused by beam dispersion, atmospheric

Figure 14.4 Tactical high-energy laser (THEL) (Public domain) Reproduced from public domain (http://en.wikipedia.org/wiki/File:THEL-ACTD.jpghttp://en.wikipedia.org/wiki/File:THEL-ACTD.jpghttp://en.wikipedia.org/wiki/File:THEL-ACTD.jpg).

absorption, scintillation and shadowing and pointing instability. All these factors lead to an increased bit error rate. These disadvantages are however outweighed by the significantly enhanced security provided by laser communications when considered in the requirements of military communication. Designers have found solutions to overcome these problems by employing multibeam or multipath architectures. Free-space communication is particularly attractive for communication in space such as satellite-to-satellite communication links, due to the absence of primary attenuation factors.

Laser communication is also attractive for underwater applications where operation in the blue-green wavelength region of the electromagnetic spectrum provides a high-speed communication link between submarines. Such a communication link overcomes data rate limitations of the existing acoustic communication systems due to multipath propagation, time variations of the channel, small available bandwidth and strong signal attenuation. Laser communication using the blue-green laser wavelength region has also become a viable system for providing communication between airborne or spaceborne platforms (such as aircraft and satellites) and submarines.

14.3 Laser-based Instrumentation

Laser-based instrumentation relates to those systems used for performing a measurement or sensor function. Common military applications of lasers covered in this section include laser devices and systems used to point or designate a target to enhance the aiming accuracy of the weapon, laser rangefinders used as standalone devices for determining the target range and also as an integral part of a fire control system, and related devices such as proximity sensors, a laser target designator used as a part of laser-guided munitions delivery system, laser-based tracking and imaging systems, laser bathymetry for swath mapping of large bodies of water, particularly the coastal waters, and lidar (light detection and ranging) sensors for detection and identification of chemical, biological and explosive agents.

14.3.1 Laser Aiming Modules

A very common application of a laser device when attached to a firearm is as an aiming device. Laser aiming modules are usually configured around low-power semiconductor diode lasers emitting in the visible or infrared. The CW power level is in the range 5–10 mW. Most laser aiming modules use red laser diodes emitting at either 635 nm or 650 nm. Green diode-pumped solid-state laser technology emitting at 532 nm is also presently in use. One of the limitations of using visible lasers for aiming and targeting is that it is visible to the naked eye, and thus inhibits a covert operation. Laser aiming modules configured around infrared diodes to produce an aim point on the target invisible to the naked human eye are now available. These aim points are detectable with night vision devices usually fitted to the firearm.

The aiming module is aligned to emit a laser beam parallel to the barrel. Due to an extremely low value of divergence of the emitted laser beam, it makes a very small spot even at long distances of up to hundreds of metres. As an illustration, full-angle divergence of 0.5 mrad would produce a spot diameter of 50 mm at 100 m distance to target. The user places the spot on the desired target and the barrel of the gun is aligned, not necessarily allowing for bullet drop, windage, distance between the direction of the beam and barrel axis and the target mobility during travel time of bullet.

Dual-wavelength laser aiming devices emitting at a visible wavelength (usually red) and a near-infrared wavelength are also commercially available. These modules are equipped with a mode select switch that allows the user to select either of the wavelengths at a time or both wavelengths simultaneously. Some devices are also equipped with a mechanism to provide adjustment for windage and elevation. One example of such an aiming device is Laser Aiming Module, Model: LAM-10 M from M/s Newcon Optik. Figure 14.5 is a photograph of a LAM-10 M laser aiming module. The module offers a visibility range of greater than 1000 m, both infrared (830–850 nm) and visible (650 nm) wavelengths of operation, beam divergence of 0.5 mrad and windage/elevation adjustment of $\pm$ 20 mrad. It offers two modes of operation at high output power for maximum distance and low power to maximize battery life. It may be mentioned here that most devices intended for a similar role have comparable technical specifications.

14.3.2 Laser Rangefinders

A *laser rangefinder* is used in a military application to determine distance to the intended target. It is used both as a standalone device as well as an integral part of a fire control system of main battle tanks and other armoured fighting vehicles. Laser rangefinders are available in a wide range of performance specifications in terms of operational range, range accuracy, size and weight to suit different application and platform requirements. Laser rangefinders are configured around different types of lasers including Nd:YAG, Nd:Glass, Er:Glass and semiconductor diode lasers. These are discussed later in Section 14.3.2.5. Commonly used techniques for determination of range include time-of-flight principle,

Figure 14.5 Dual wavelength laser-aiming module, Model LAM-10 M (Courtesy of Newcon Optik).

triangulation method and multiple-frequency phase-shift method. These are all discussed in detail in terms of their operational principle and salient features in the following sections.

14.3.2.1 Time-of-Flight Technique

In the *time-of-flight* technique, a narrow-pulse-width laser beam is transmitted towards the intended target. The target range is measured from the time taken by the laser pulse to travel to the target and back as shown in Figure 14.6. Target range or distance to target d is given by Equation 14.1:

$$d = c \times \Delta t/2 \tag{14.1}$$

where

d = target range (m)
c = speed of light $= 3 \times 10^8\,\mathrm{m\,s^{-1}}$
Δt = time interval between transmitted and received laser pulses

Range accuracy in this case depends on the receiver processing speed and rise and fall time of the laser pulses. Range here is measured as the time interval between the rising or falling edge of the transmitted pulse and the corresponding edge of the received pulse. Uncertainty due to finite values of rise or fall time causes range inaccuracy. The received pulse is processed in a high-speed counter and clock speed determines the processing speed. The time interval and hence range to target is measured in terms of the number of clock pulses counted by the time interval counter, started by the start pulse corresponding to the leading or trailing edge of transmitted laser pulse and stopped by the stop pulse corresponding to the leading or trailing edge of the relevant received pulse. In terms of number of clock cycles, distance to target is given by Equation 14.2:

$$d = \frac{cN}{2f_{\mathrm{clk}}} \tag{14.2}$$

where N is the number of clock pulses counted between start and stop signals and f_{clk} is the clock frequency. Range inaccuracy in this case is given by $\pm c/2f_{\mathrm{clk}}$.

The worst-case range inaccuracy equals one clock period. *Range resolution* is determined by the laser pulse width, which implies that the laser rangefinder based on time-of-flight principle cannot discriminate between two targets separated in radial range by a distance corresponding to the distance travelled by light in a time period equal to pulse width.

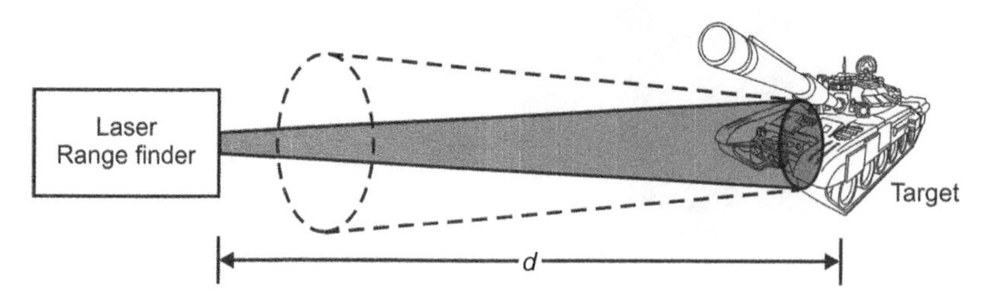

Figure 14.6 Time-of-flight principle of measuring target range.

Example 14.1

A laser rangerfinder operating on time-of-flight principle measures target range as 5.46 km. If the operating clock frequency is 50 MHz, determine the number of clock cycles the rangefinder counter would have recorded for this measurement. Also determine the range of distance in which the true range is likely to be.

Solution

- Measured target range, $d = \dfrac{cN}{2f_{\text{clk}}} = 3 \times 10^8 \times N/(2 \times 50 \times 10^6) = 3N$
- Therefore, $3N = 5.46 \times 10^3$ and the number of clock cycles $N = 1820$
- Also, worst-case range inaccuracy $= \pm c/2f_{\text{clk}} = \pm 3 \times 10^8/(2 \times 50 \times 10^6) = \pm 3$ m.
- Therefore, true range is 5.457–5.463 km.

14.3.2.2 Triangulation Technique

In the *triangulation* technique, range is determined by using simple laws of trigonometry. The principle of operation can be best explained with the help of Figure 14.7. In Figure 14.7, laser transmitter is located at point A, laser receiver is located at point B and point C indicates the target location. The distance-to-target d from the laser transmitter location can be computed from known values of length l and angles α and β using Equation 14.2:

$$d = l \times \frac{\sin \beta}{\sin (\alpha + \beta)} \tag{14.3}$$

The above equation is valid in flat or Euclidean geometry. The computed results become inaccurate if distances are appreciable compared to the curvature of the Earth. In that case, more complicated expressions derived using spherical trigonometry should be used.

Example 14.2

In a range measurement experiment using the triangulation method, the transmitter and receiver are separated by 10 m. The angles that the lines of sight between target and transmitter and target and receiver make with the line joining transmitter and receiver are measured as 85° and 90°, respectively. Determine the measured distance to the target.

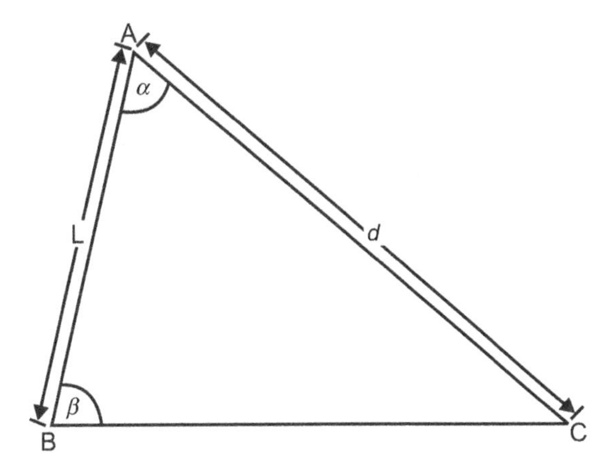

Figure 14.7 Triangulation method of range finding.

Solution

- The distance to target d can be computed from $d = l \times \dfrac{\sin \beta}{\sin (\alpha + \beta)}$
- $\alpha = 85°$, $\beta = 90°$ and $l = 10$ m.
- Therefore, $d = 10 \times \sin 90°/\sin (85° + 90°) = 10/\sin 175° = 10/0.087 = 115$ m.

14.3.2.3 Phase Shift Technique

In the *phase shift* technique of range finding, the laser beam with sinusoidal power modulation is transmitted towards the target and the diffused or specular reflection from the target is received. The phase of the received laser beam is measured and compared with that of the transmitted laser beam. The phase shift is 2π times the product of the time-of-flight and modulation frequency. This allows us to compute the time-of-flight and hence the distance to the target from known values of phase shift and modulation frequency. Higher-modulation frequencies can result in a higher spatial resolution. The phase shift method appears similar to the time-of-flight method, as the phase shift is proportional to the time-of-flight. However, time-of-flight conventionally refers to the technique where the time delay is measured more directly. In the case of the phase shift method, the range measurement is ambiguous as phase shift varies periodically with increasing distance. This ambiguity can be removed by measuring phase shift at two different frequencies.

14.3.2.4 FM-CW Range-finding Technique

The frequency-modulated continuous-wave (FM-CW) laser range-finding technique is similar to that followed in the case of its radar counterpart, that is, FM-CW radar. The frequency of a narrow-line-width laser is modulated with a ramp or sinusoidal signal, collimated and then transmitted towards to the target. The received signal corresponding to the reflected laser beam, specular or diffused, is mixed with the reference signal representing the transmitted laser beam. The beat frequency produced as a result of homodyne detection is used to derive the range information. Figure 14.8a shows the basic block schematic arrangement of a FM-CW laser rangefinder. Figure 14.8b shows the frequency variation in transmitted and received laser signals as a function of time when the transmitted laser is modulated by a ramp signal. Received signal is time-delayed from the transmitted signal as shown in the figure. The time delay and the corresponding beat frequency represent distance to target. The figure also shows beat frequency as a function of time. The beat frequency is measured in the portion where it is constant with respect to time and not in the transient region. The distance to target d is given by Equation 14.4:

$$d = \frac{f_{\mathrm{B}} c T_{\mathrm{R}}}{4\Delta f} \tag{14.4}$$

where f_{B} is the beat frequency, T_{R} is the ramp waveform time period and Δf is modulation frequency bandwidth.

Minimum measurable range in this technique corresponds to half the period of the beat frequency signal. Maximum measurable range equals half the distance travelled by light in half the time period of the ramp signal. Minimum and maximum ranges are given by Equations 14.6 and 14.7, respectively:

$$d_{\min} = \frac{c}{4\Delta f} \tag{14.5}$$

$$d_{\max} = \frac{cT_R}{4} \tag{14.6}$$

The measurement range and sensitivity in this case depends on the line width or coherence length of the laser. With a line width of a few kilohertz, it is possible to achieve a range of several hundred kilometres with an accuracy of better than 1.0 m. FM-CW laser rangefinders implemented with narrow-

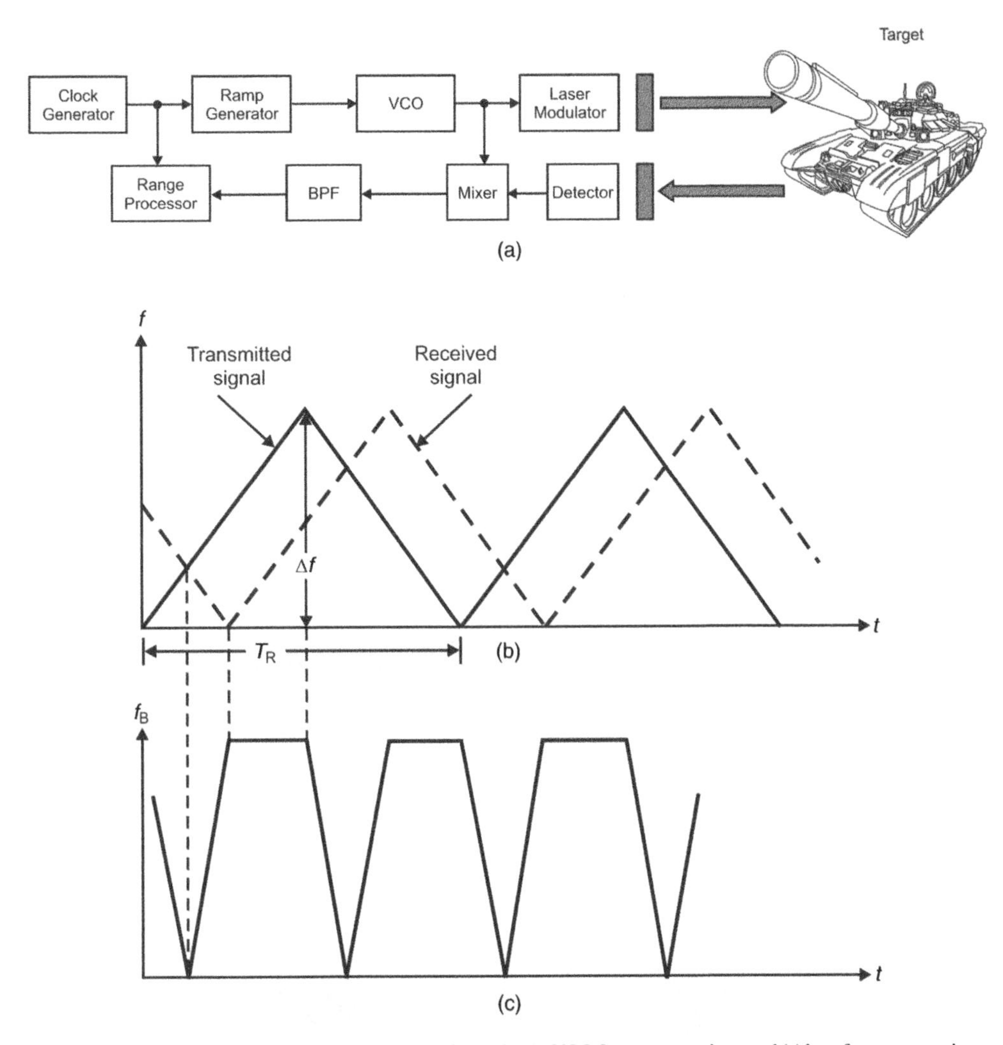

Figure 14.8 FM-CW laser rangefinder: (a) block schematic; (b) VCO frequency vs time; and (c) beat frequency vs time.

line-width fibre laser sources offer significant improvement in dynamic range, sensitivity, compactness, ruggedness and mounting flexibility.

Example 14.3

An FM-CW type of laser rangefinder uses a ramp-modulated narrow-line-width laser beam. The ramp frequency is 1 kHz and laser line width is 20 kHz. While measuring a designated target range with this rangefinder, beat frequency is measured to be 450 kHz. If the modulation bandwidth of the system is 3 MHz, determine target range. Also find the minimum and maximum range-measuring capability of the system.

Solution

- Target range is given by $d = \dfrac{f_B c T_R}{4\Delta f}$
- $T_R = 1/1 \times 10^3 = 1$ ms and $d = 450 \times 10^3 \times 3 \times 10^8 \times 1 \times 10^{-3}/(4 \times 3 \times 10^6) = 11.25$ km.

- Minimum and maximum range measuring capabilities of the system are:

$$d_{\min} = \frac{c}{4\Delta f} = 3 \times 10^8/(4 \times 3 \times 10^6) = 25\,\text{m}$$
$$d_{\max} = \frac{cT_R}{4} = 3 \times 10^8 \times 10^{-3}/4 = 75\,\text{km}$$

14.3.2.5 Lasers for Laser Rangefinders

Different types of laser sources have been exploited to build laser rangefinders for military applications. These include different types of solid-state lasers, semiconductor diode lasers, fibre lasers and CO_2 lasers.

Laser rangefinders used on land-based platforms, either in a standalone mode or as a part of integrated fire control system, are usually low-repetition-rate systems producing laser pulses in the range 5–30 ppm (pulses per minute) and are configured around *Q-switched Nd:YAG or Nd:Glass lasers.* These are capable of ranging up to a target range of 25 km with accuracy better than ±5 m and are available as compact, handheld devices similar in appearance to a pair of binoculars. Figure 14.9 is photograph of one such laser rangefinder, Model LH-30 manufactured by Bharat Electronics. This handheld Nd:YAG laser rangefinder has a maximum operational range of 20 km with range accuracy specification of ±5 m. Other features include beam divergence of 1.0 mrad, pulse repetition rate options of 10 and 30 ppm, pulse energy of 6–12 mJ, built-in magnification of 6 ×, RS-422A serial interface and remote triggering and bite readout.

Another class of rangefinders that combine the functions of target designation and range finding or that form the subsystem of a tracking system have relatively much higher repetition rates in the range of 10–50 pps (pulses per second). The inter-pulse period in this case depends upon the speed of the target to be tracked. These are invariably Q-switched Nd:YAG lasers, primarily due to the significantly superior

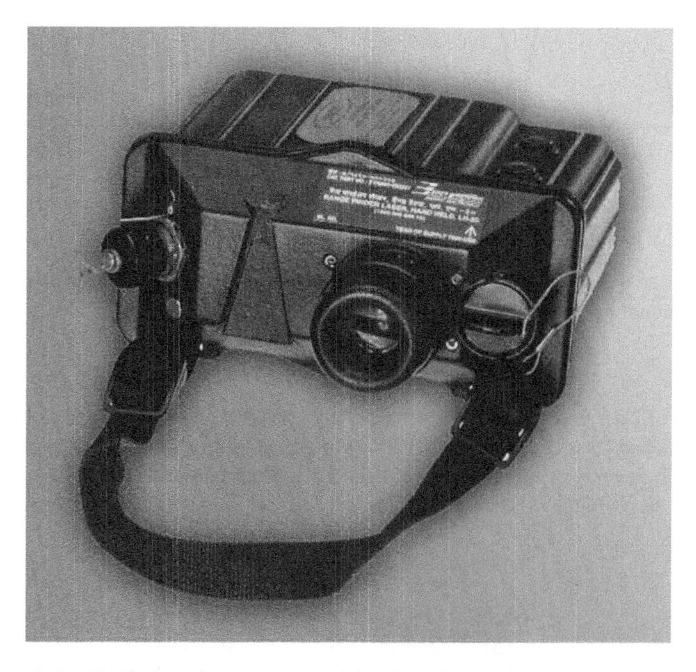

Figure 14.9 Handheld Nd:YAG laser rangefinder, Model LH-30 (Reproduced by permission of Bharat Electronics Ltd.).

thermal conductivity specification of YAG as compared to glass as a host material. These lasers are not only becoming smaller in size, but are now available with many additional features to give them in-built countermeasure capability against similar systems deployed by the adversary in an electro-optic countermeasure role.

Nd:YAG and Nd:Glass rangefinders emit at 1064 nm, which has a serious eye hazard. Personnel using these devices are at risk of being exposed to this hazardous radiation, and need to protect their eyes while using them to avoid scattered radiation falling onto their eyes and causing serious damage. Eye protection comes in the form of safety goggles, which attenuate the laser radiation to an absolutely safe level. This has led to the development of *eye-safe laser rangefinders* and target designators. The low-repetition-rate class of rangefinders discussed earlier is being gradually replaced by eye-safe versions.

One of the techniques is to use a different active medium that emits an eye-safe wavelength. Er: Glass is one such active medium and Er:Glass lasers emit at 1540 nm, a wavelength safe for human eyes. Figure 14.10 is a photograph of one such eye-safe laser rangefinder, having an operational range of 25 km and a range accuracy of ± 5 m.

The high-repetition-rate variety of eye-safe laser rangefinders is not possible with Er:Glass lasers due to the poor thermal conductivity characteristics of the glass host. Several non-linear techniques have been successfully used in conjunction with Nd:YAG lasers to generate eye-safe wavelengths. Raman-shifted YAG and OPO-based YAG have emerged as very strong contenders for building high-repetition-rate laser sources.

Another laser type exploited as a rangefinder is the *RF-excited waveguide CO_2 laser*. Advances in waveguide geometry and RF components have resulted in development of compact laser sources emitting at 10.6 μm. Rangefinders built using these lasers perform much better in adverse climatic conditions as compared to their counterparts operating at 1064 nm.

14.3.2.6 Applications and Related Devices

Laser rangefinders find extensive use as standalone devices for the purpose of observation and situational awareness of adversary movement of personnel and military assets. Most armoured fighting platforms are equipped with a laser rangefinder. While the basic rangefinder can be used to find target range, when combined with a digital magnetic compass and inclinometer it can also be used to determine target coordinates. The other important application of a laser rangefinder is in integrated fire control system of armoured fighting platforms. When interfaced with a fire control computer, target accuracy is significantly enhanced. Modern fire control systems of all main battle tanks today are invariably laser-rangefinder assisted. When interfaced with night vision, thermal and daytime optical aids, laser

Figure 14.10 Erbium-Glass laser rangefinder, Model LRB-25 000 (Courtesy of Newcon Optik).

rangefinders represent a useful and effective battlefield asset for observation, surveillance and situational awareness.

A high-repetition-rate laser rangefinder in a target designator is used for munitions guidance. They are also at the heart of laser trackers and 3D scanners. Other prominent systems with a laser rangefinder at their core include laser proximity sensors, laser bathymetric sensors, lidar and 3D imaging seekers.

14.3.3 Laser Target Designators

In military parlance, laser designation of targets is carried out for the guidance of laser-guided bombs, missiles and projectiles. The Paveway series of bombs and Lockheed-Martin's Hellfire missiles, both launched from aerial platforms, and canon-launched Copperhead projectiles are some examples of laser-guided munitions that require precise laser designation of the intended target.

A laser target designator for munitions guidance application is a high-repetition-rate Q-switched laser source whose pulse repetition frequency is coded in order to allow only a specific target designator and laser seeker combination to work in unison for the intended mission. This also allows multiple targets to be hit at the same time when illuminated by different target designators, in addition to giving the delivery system a countermeasure capability.

Modern laser target designators use Q-switched Nd:YAG laser source and operate at 1064 nm. All target designators have a range-finding channel either at the same wavelength of 1064 nm or at an eye-safe wavelength of 1540 nm. It may be mentioned here that the range-finding channel of almost all modern laser target designators uses an eye-safe wavelength of 1540 nm. The system is equipped with a suitable sighting system and a digital magnetic compass. It is mounted on a suitable angular platform. There is also trend towards building eye-safe laser target designators in the coming years; this would however necessitate a change of laser seeker heads, which have a spectral response to cover 1540 nm. This implies that the seekers would need to employ indium gallium arsenide (InGaAs) based sensors rather than silicon-based sensors. Laser guidance principles are discussed in detail in Section 14.4.2.

Laser target designators for handheld applications as well as for land-based and aerial platforms are available. Typical performance specifications of laser target designators for munitions guidance applications include pulse energy in the range of 50–120 mJ, pulse width in the range of 5–50 ns, pulse repetition frequency 5–20 Hz and laser beam divergence of 0.1–0.5 mrad. One such portable laser target designator is Model AN/PEQ-17 from M/s Elbit Systems, described earlier in Chapter 3 on solid-state lasers and depicted here in Figure 14.11.

Airborne laser target designators often form a part of a targeting pod. Targeting pods are basically target designation tools used for identification of ground targets and guiding munitions to their intended targets. Targeting pods are equipped with other electro-optic systems such as a laser spot tracker to locate the laser pulses reflected from the designated target or/and laser rangefinder to determine target range. In some cases, same sensor performs the functions of laser spot tracking and range finding. These systems are known as laser ranger and marked target seekers (LRMTS). Some targeting pods are equipped with a laser target designator to designate its own targets or for other friendly units.

Raphael's Litening targeting pod is an example, depicted by Figure 14.12. A Litening targeting pod contains a high-resolution forward-looking infrared (FLIR) sensor to provide an infrared image of the target with a wide field-of-view search capability and a narrow field-of-view acquisition and targeting capability, a CCD camera to produce target imagery in the visible portion of the electromagnetic spectrum, a navigation sensor on gimbal for automatic bore-sighting capability, a laser target designator for precise delivery of laser-guided munitions and a rangefinder.

14.3.4 Laser Proximity Sensors

A laser proximity sensor is a type of laser rangefinder designed to accurately measure relatively smaller distances to a target compared to those encountered in conventional laser rangefinders used for observation and surveillance and fire control. While a conventional battlefield laser rangefinder measures distance to target and the range information is displayed to the observer, in the case of a proximity sensor the processing circuitry generates a command signal when the distance to target is equal to the preset

Figure 14.11 Portable laser target designator Model AN/PEQ-17 (Courtesy of Elbit Systems of America).

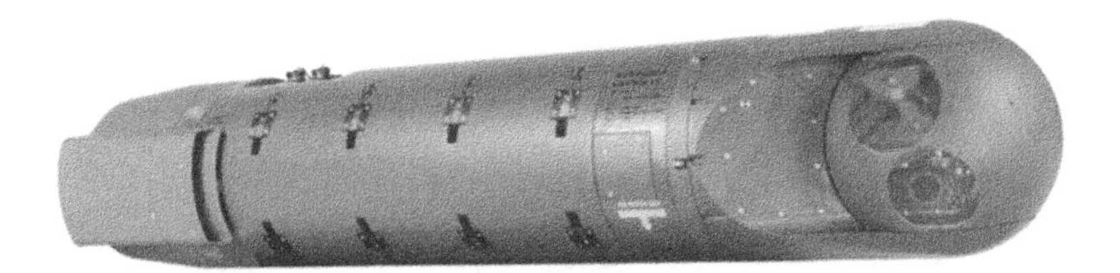

Figure 14.12 Litening targeting pod (Courtesy of Northrop-Grumman Corp).

distance value within a certain specified tolerance. The command signal in turn could be used to perform a variety of control functions. In military parlance, the most common application of a proximity sensor is in a laser proximity fuse where the command signal is used to initiate detonation of warheads of large artillery shells, aviation bombs and guided missiles.

Laser-based methods of determining target range are discussed in Section 14.3.2 on laser rangefinders. Theoretically, all the previously described range-finding techniques (time-of-flight, triangulation, phase shift and FM-CW techniques) can be used to build laser proximity sensors or laser proximity fuses. The optimum method for a given application however depends on the maximum range and range accuracy requirement of the intended application. The triangulation technique is much better suited to the design of laser proximity sensors or fuses, as it is capable of measuring shorter distances up to few metres with high accuracy. The range accuracy however falls off rapidly with increasing distance.

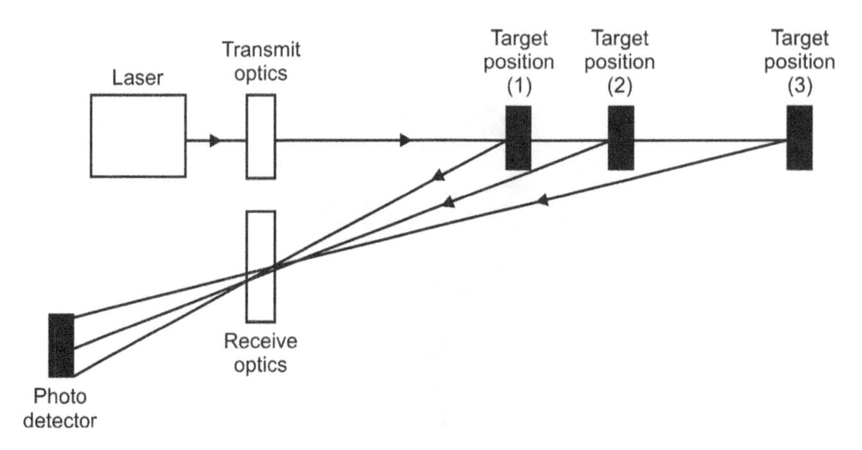

Figure 14.13 Laser proximity sensor principle of operation.

Figure 14.13 shows the basic principle of operation of a laser proximity sensor. The sensor has a transmitting channel usually configured around a semiconductor diode laser or a passively Q-switched diode-pumped solid-state microchip laser along with associated transmit optics and a receiving channel comprising receiving optics, PIN or APD sensor and range processing circuitry.

As shown in the figure, the transmitted laser beam reflected from the target located at three different distances produces the beam images that are displaced across the active area of the photosensor. The receiver is designed to produce the focused laser beam spot at the centre of the active area for a preset value of proximity distance. There will be a distance both longer and shorter than the desired distance, where the focused beam spot falls just outside the active area. This forms the basis of the operation of laser proximity sensor in general and laser proximity fuse in particular.

The performance of a laser proximity sensor can be enhanced by using an axially symmetric arrangement of multiple aperture photosensor as shown in Figure 14.14. The design offers better ballistics due to centre of gravity being located on the longitudinal axis of the ammunition round and higher signal-to-noise ratio due to averaging of multiple return signals.

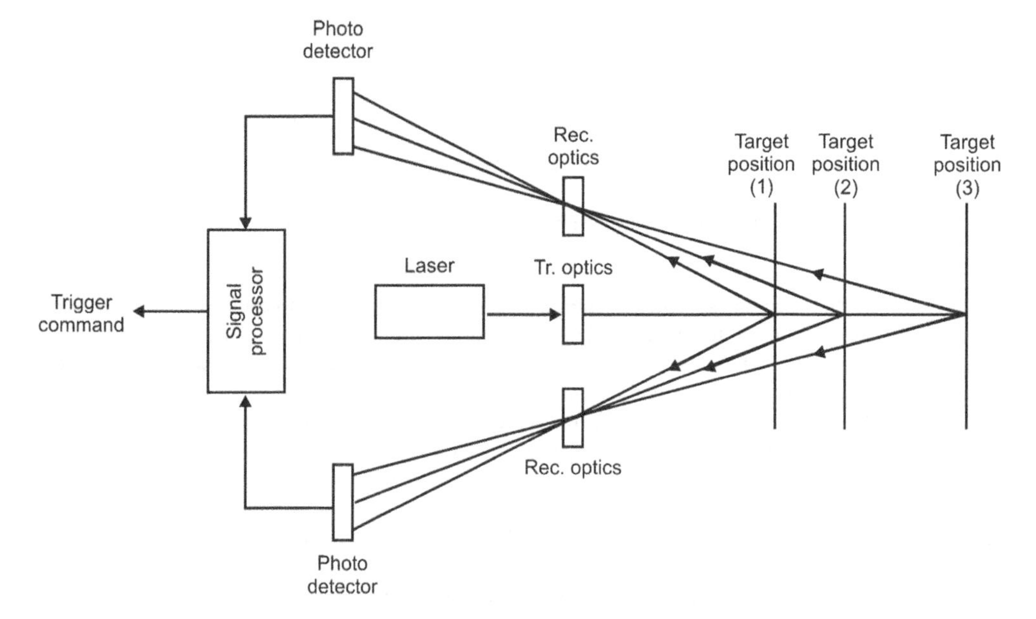

Figure 14.14 Laser triangulation proximity sensor in axially symmetric configuration.

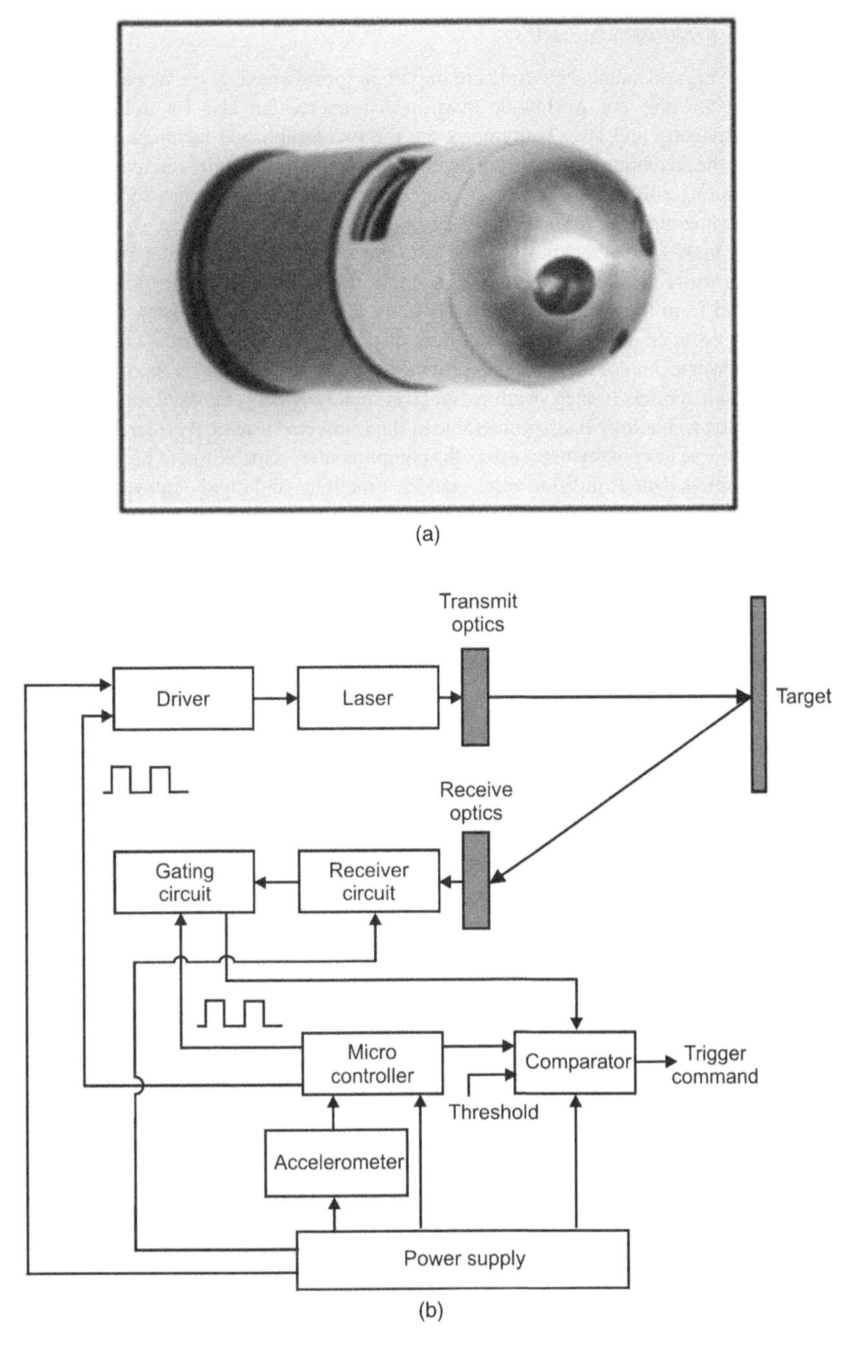

Figure 14.15 Laser proximity sensor type SOProF (14.15 (a) Courtesy of Physical Optics Corporation).

Figure 14.15a is a photograph of a laser proximity fuse assembled in M433 40 mm round and manufactured by M/s Physical Optics Corporation. Figure 14.15b shows the basic block schematic arrangement of the proximity sensor shown in Figure 14.15a. Desired proximity distance can be selected by choosing an appropriate threshold.

14.3.5 Laser Bathymetry Sensors

Hydrographic surveying and nautical charting are important for safe navigation in coastal waters and are therefore important not only for worldwide maritime commerce but also for military applications. Acoustic-based bathymetry and laser bathymetry are the two established techniques for the purpose. Laser bathymetry is the technique of measuring depths of water bodies, usually relatively shallow coastal waters, using a scanning pulsed laser beam. Measurement is usually done from an airborne platform, which gives it the name airborne laser bathymetry (ALB).

Laser bathymetry makes use of the transmissive and reflective properties of water and of the bottom of the water body to measure the depth. The measurement of depth relies on the differential timing of the laser pulses reflected from the surface of the water body and those reflected from the bottom.

Laser bathymetry from an airborne platform offers distinct advantages over the conventional vessel-mounted acoustic sensor technology, as the latter provides a swath coverage that is approximately twice the water depth. As a result, it offers limited swath coverage in shallow waters of coastal zones, as illustrated in Figure 14.16. Acoustic technology is also vulnerable in shoal-infected waters; it is therefore more suitable for deep waters. Airborne laser bathymetry offers the complimentary capabilities of high coverage rates in shallow waters, seamless data acquisition across land/sea interface and rapid deployment.

When the laser beam from the airborne platform hits the water column, a fraction of the incident laser energy is reflected off the surface and the remaining energy is mostly transmitted through the column. The laser beam transmitting through the water column suffers attenuation due to absorption, scattering and refraction. This limits the maximum measurable water depth. Maximum measurable depth by laser bathymetry depends on water clarity, and is generally three times the Sechhi depth. (Sechhi depth is an old method of quantifying water clarity and is equal to the depth at which a standard black and white disc is no longer visible to the naked eye.)

Figure 14.17 shows the operational concept of a typical airborne laser bathymetry system. The laser transmitter sends out two collinear pulsed laser beams simultaneously, one at a near-infrared wavelength (usually 1064 nm) and the other at 532 nm. The two pulsed laser beams strike the surface of water column at the same point and at the same time. The near-infrared beam is specularly reflected from the water surface. A part of it enters the receiver, which measures its intensity and time instant of arrival with reference to the time instant of the transmitted pulse train. The small fraction that is transmitted is absorbed within a few centimetres of the water column. In the case of green laser at 532 nm, most of it

Figure 14.16 Swath coverage in airborne laser bathymetry and acoustic sensor technology.

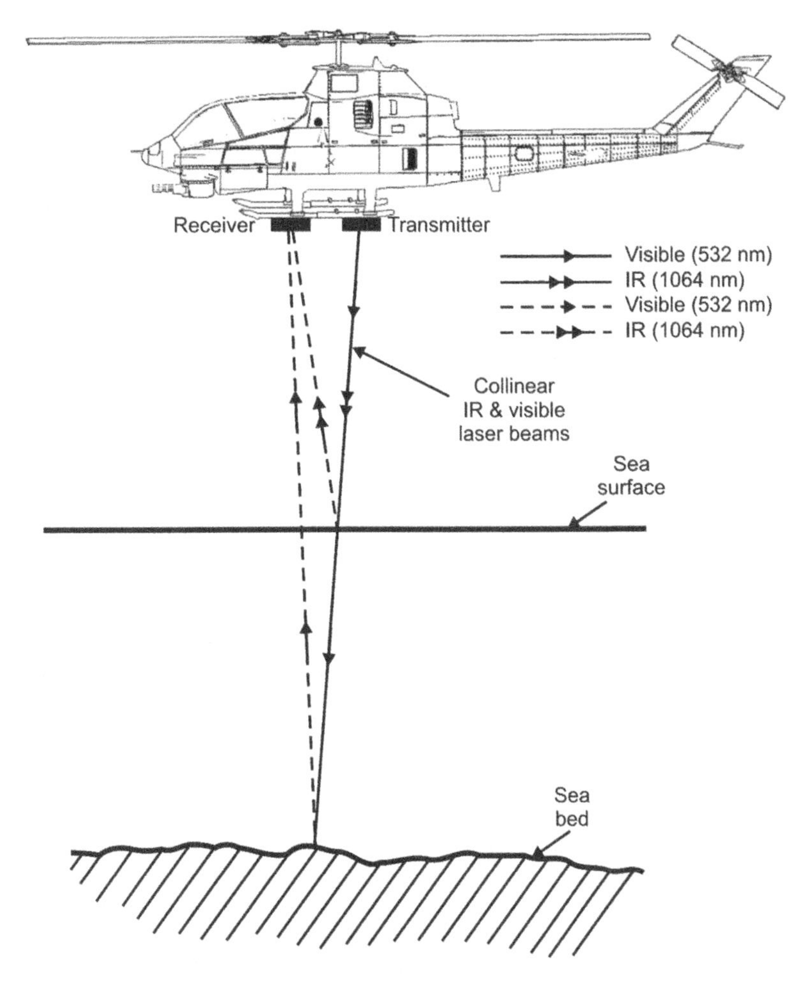

Figure 14.17 Operational concept of airborne laser bathymetry.

penetrates the water column and only a small fraction is specularly reflected. The fraction of green laser reaching the water column bed undergoes diffused reflection. After diffused reflection, it travels upwards through the water column and only a small fraction of it reaches the receiver within its field of view. The receiver measures the intensity and time instant of arrival again with reference to the transmitted pulse train. The water depth is computed from the difference in the round-trip propagation time of the surface reflected laser beam at 1064 nm and the one at 532 nm reflected off the bottom of the water body.

The whole process of transmission, reception and signal processing is far more complex than it appears from the basic description presented here. For example, in a practical system, specular reflection of the laser beam from the water surface is usual when the conditions are windy, as these conditions lead to formation of small capillary waves that act like mirror facets. In calm conditions, the surface acts like one large mirror, thus directing the reflection away from the sensor. It may be mentioned here that the transmitted laser beam makes an angle of about 20° with respect to the vertical. In calm conditions, the system depends on Raman scattering and fluorescence signals from the water volume just beneath the surface and uses an off-wavelength receiver to generate the received signal. The green laser beam hitting the surface is also refracted by about 15° depending on the surface wave structure at the time of impact. The sensor also needs to take into account the refraction effect while computing the water depth.

Armed forces around the world have equipped themselves with airborne laser bathymetry systems over the years for a range of applications. Some of these include acquisition of CZMIL system by US Navy in 2012 and US Army Corps of Engineers (USACE) in 2011, SHOALS (scanning hydrographic operational airborne laser survey) 3000 by Military Survey Department of UAE in 2010, SHOALS 3000 TH by US Navy in 2005, SHOALS 1000T by US Army Corps in 2003, SHOALS 1000 by Japan Coast Guard in 2003, Hawkeye by Swedish Navy in 1995, SHOALS 200 by US Army Corps of Engineers in 1994, ALARMS by DARPA in 1988 and FLASH by Swedish Defence Research Institute in 1988.

The SHOALS system from Optech Incorporated, Canada is a commonly used airborne laser bathymetry system. Different variants of the system in SHOALS-200, SHOALS-1000 and SHOALS-3000 have found themselves on the inventory of armed forces over the years as outlined above. The system meets International Hydrographic Organisation (ISO) Order 2 requirements for accuracy. The system offers depth measurement accuracy of better than 20 cm and horizontal positioning accuracy of better than 1.5 m. With its special shoreline depths processing mode, SHOALS can provide continuous topographic and bathymetric mapping through the shoreline from water onto land. The maximum depth measuring capability of the system is 40–50 m in clear ocean waters, 20–40 m in coastal waters and less than 20 m in more turbid inland waters. Other factors that limit the system depth measuring capability and accuracy include high surface waves, heavy fog and precipitation, sun glint, heavy bottom vegetation and fluid mud.

14.3.6 Laser Radar (Ladar) Sensors

Laser radar, also called ladar (laser detection and ranging), uses a laser beam instead of microwaves, that is, the transmitted electromagnetic energy lies in the optical spectrum in laser radars. The frequencies associated with laser radars are very high ranging from 30–300 THz and the corresponding wavelengths from 10–1.0 μm. The higher operating frequency means higher operating bandwidth, greater time or range resolution and enhanced angular resolution. Another advantage of laser radars compared to microwave radars is their immunity to jamming. The higher frequencies associated with laser radars permit detection of smaller objects. This is made possible by the fact that laser radar output wavelengths are much smaller than the smallest-sized practical objects. In other words, the laser radar cross-section of a given object would be much larger than the microwave radar cross-section of the same. Rain droplets and airborne aerosols have a significantly larger laser radar cross-section, allowing their range and velocity measurement which is very important for many meteorological applications. The higher resolution of laser radar allows recognition and identification of certain unique target features such as target shape, size, velocity, spin and vibration, which forms the basis of their use for target imaging and tracking applications.

Although there are numerous advantages that laser radars offer over microwave radars, laser radars are severely affected by adverse weather conditions. The narrow beam width of laser radars is also not conducive to surveillance applications. For surveillance applications, the laser radar needs to operate at very high repetition rates so that large volumes can be interrogated within the prescribed time. Alternatively, multiple simultaneous beams can be used.

Laser rangefinders, discussed in Section 14.3.2, are also a type of laser radar. A conventional laser rangefinder uses incoherent or direct detection, but the term laser radar is usually associated with systems that use coherent detection. Figure 14.18 shows the block schematic arrangement of the coherent laser radar. The laser beam is transmitted towards the target, and a fraction of the transmitted power/energy reflected from the target is collected by the receiver. The block diagram shown in Figure 14.18 is that of a monostatic system in which transmitter and receiver share common optics, made possible by the use of a transmit-to-receive switch. In a bistatic arrangement, transmitting and receiving optics are separate; the received laser beam is coherently detected in an optical mixer.

In the case of homodyne detection, a sample of the transmitted laser power is used as a local oscillator. In the case of heterodyne detection, another phase-locked-to-transmit laser is used as local oscillator. Heterodyne detection is used when transmitter and receiver are not collocated. The output of the optical mixer is imaged onto the photosensor. The electrical signal generated by the photosensor module is

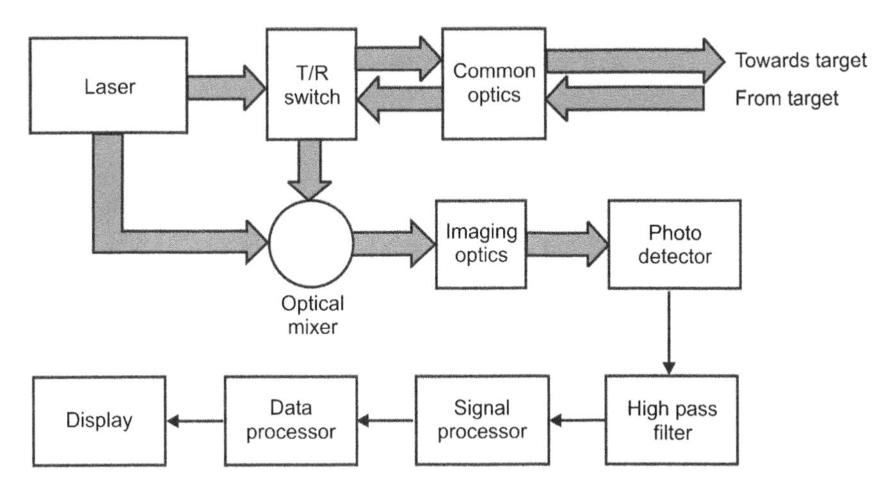

Figure 14.18 Block schematic of coherent laser radar.

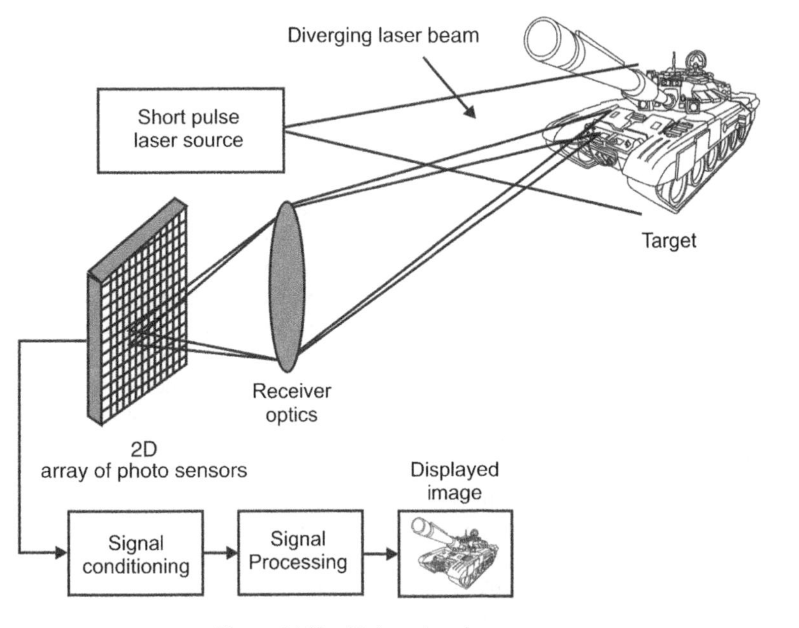

Figure 14.19 3D laser imaging concept.

processed to extract the desired information about the target. The photosensor module is a single sensor in the case of a scanning transmitted beam and a 2D array of sensors for the flash-type laser imaging radar.

Figure 14.19 illustrates the concept of a flash-type 3D laser imaging radar. In this case, a diverging pulsed laser beam illuminates the entire scene of interest. The transmitted pulse time is referenced by an auxiliary photosensor. The backscattered light is imaged onto a 2D array of photosensing elements called pixels. While a conventional camera measures the light intensity of the backscattered light pulse, in this case different sensor elements in the array measure time-of-flight. The time-of-flight is proportional to the distance between the point on the target from where the laser beam is reflected and the sensor element. The sensor array therefore produces a 3D image of the target (angle-angle-range). An alternative approach to generating a target image is to use a scanning laser beam and a single sensor.

Figure 14.20 Multimode enhanced ladar seeker (Courtesy of Lockheed Martin).

A common application of the ladar concept is in ladar seekers, used mainly in conjunction with other guidance systems on strategic payloads for intended target discrimination from advanced decoys and aim point selection. It is also well suited for combat identification, navigation of autonomous vehicles and topography. Ladar is also suitable for finding targets hidden by camouflage nets and foliage. A ladar seeker can detect and identify specific features of the target with very high definition up to a resolution of a few centimetres from a distance of a few kilometres. An automatic target acquisition algorithm processes the images to identify and acquire targets based on 3D templates stored in the weapon's memory before the mission. Ladar sensors are usually employed on loitering systems that look at the target from different angles, verify the target's identity and select the best attack position for desired results.

Figure 14.20 is a photograph of an advanced multimode ladar seeker from Lockheed Martin. The seeker can operate as a standalone semi-active laser (SAL) and in ladar modes as well as simultaneous SAL and ladar modes for target identification, acquisition and tracking. The seeker is designed to conduct a wide area search and identify actual or potential targets, including those obscured by camouflage or foliage. Such ladar seekers have been successfully tested on loitering attack missile (LAM) missions under DARPA's NLOS-LS (non-line-of-sight launch system) and USAF's LOCAAS (low-cost autonomous attack system) programs. With its multimode ladar seeker, the LAM searches a large area and relays the location of various targets back to the command centre, where these targets are engaged by direct attack or by other assets. In case of a priority target, LAM can be commanded to break off its search mission and attack the target.

14.3.7 Forward-looking Infrared (FLIR) Sensors

A forward-looking infrared (FLIR) sensor makes use of the thermal radiation emitted by the target or scene of interest to generate its image. Essentially, it comprises a front-end optical system, a 2D array of infrared detectors and image processing circuitry to produce output in the desired format. Infrared energy coming from the target or scene of interest is focused by the front-end optics onto the infrared detector subsystem. The electrical signal produced by the detector is sent to the processing electronics, which translates the data coming from the detector into an image. The image is displayed on a standard video monitor or LCD screen or anywhere on a network-enabled computer.

The term 'forward–looking' is used here to distinguish it from imaging and tracking systems known as 'pushbroom' systems that look sideways in a direction perpendicular to the direction of travel of the aerial platform they are mounted on. While a FLIR system uses a 2D detector array and is capable of producing target image in real time, a pushbroom system usually employs a single array of detectors and uses the motion of vehicular platform (aircraft or satellite) to generate a 2D image. Figure 14.21 shows a comparison of the FLIR and pushbroom concepts.

FLIR systems should not be confused with night vision devices; the latter devices operate in the visible and near-infrared regions of the electromagnetic spectrum extending from 0.4 μm to 1.0 μm. FLIR systems operate in the mid-infrared (3.0–5.0 μm) and far- or long-infrared (8.0–12.0 μm) bands. These bands allow them to detect heat sources such as heat from the engine parts of target platforms and from a human body from a distance of the order of several kilometres. Systems operating in the 3.0–5.0 μm band offer superior range performance as longer-distance imaging is more difficult in the 8.0–12.0 μm band due to absorption, scattering and refraction losses caused by air and water vapour. While more sensitive long-wave FLIR need a cryogenically cooled detector array, moderately sensitive systems do not use cryogenic cooling. Many uncooled long-wave FLIR systems are commercially available.

On the other hand, FLIR systems operating in the 3.0–5.0 μm band suffer much less attenuation due to water vapour but generally need a more expensive detector array and cryogenic cooling. Advanced fusion technologies are often used to blend an image produced by a visible spectrum sensor with an infrared spectrum image to produce better results than that from a single spectrum image.

There have been different generations of FLIR sensors. Each successive generation has incorporated not only a major change in the type of detector but also a major change in optical systems used to image the target onto the detector. First-generation FLIRs were scanning type LWIR (long-wave infrared) systems. The sensitivity of the first-generation FLIR systems was limited by the background radiation, which was overcome in second-generation FLIRs by using modified front-end optics that reduced unwanted flux. This however resulted in a fixed *f*-number for all fields of view. Second-generation FLIR systems operate either in MWIR (mid-wave infrared) or LWIR bands. Typically, MWIR FLIRs are staring systems and LWIR are scanning systems. While MWIR has distinct resolution advantage over LWIR because of its shorter wavelength, LWIR has the sensitivity advantage due to the increased number of photons at terrestrial temperatures. Third-generation FLIRs use a dual-band detector array and dual/variable *f*-number optical system.

FLIRs are used on land-based (armoured fighting vehicles), naval and aerial (aircraft, helicopters, missiles) platforms for surveillance, target acquisition and tracking applications. Some of the common non-military applications of FLIR systems include surveillance of living things, search-and-rescue operations during fire fighting, detection of gas leaks, monitoring of volcanoes and the detection of heat in faulty electrical joints.

In military applications, FLIR imaging offers several distinct advantages. The first advantage is its immunity to detection by the adversary as it is a passive sensor that does not emit any radiation for generation of an image. Secondly, it is extremely hard to camouflage the target from the sensor as the FLIR senses heat. Thirdly, FLIR can see through smoke, fog, haze and other atmospheric obscurants better than sensors operating in the visible spectrum. One of the limitations of FLIR sensors is that it is hard for it to discriminate friend from foe. Friendly forces can use heat beacons to overcome this problem.

Figure 14.22 is a photograph of a FLIR sensor designed for border and coastal surveillance applications. This HRC-series FLIR (HRC-E) uses a 640 × 480 pixel cooled microbolometer detector array. The sensor offers continuous zoom from 25° and 2° and a long range detection capability. The sensor is capable of detecting a man-sized target from greater than 15 km.

State-of-the-art sensors intended for surveillance, target detection and tracking applications employ multisensor configurations, often combining a visible spectrum sensor with an infrared sensor. Figure 14.23 is a photograph of one such sensor that combines a thermal imaging camera with

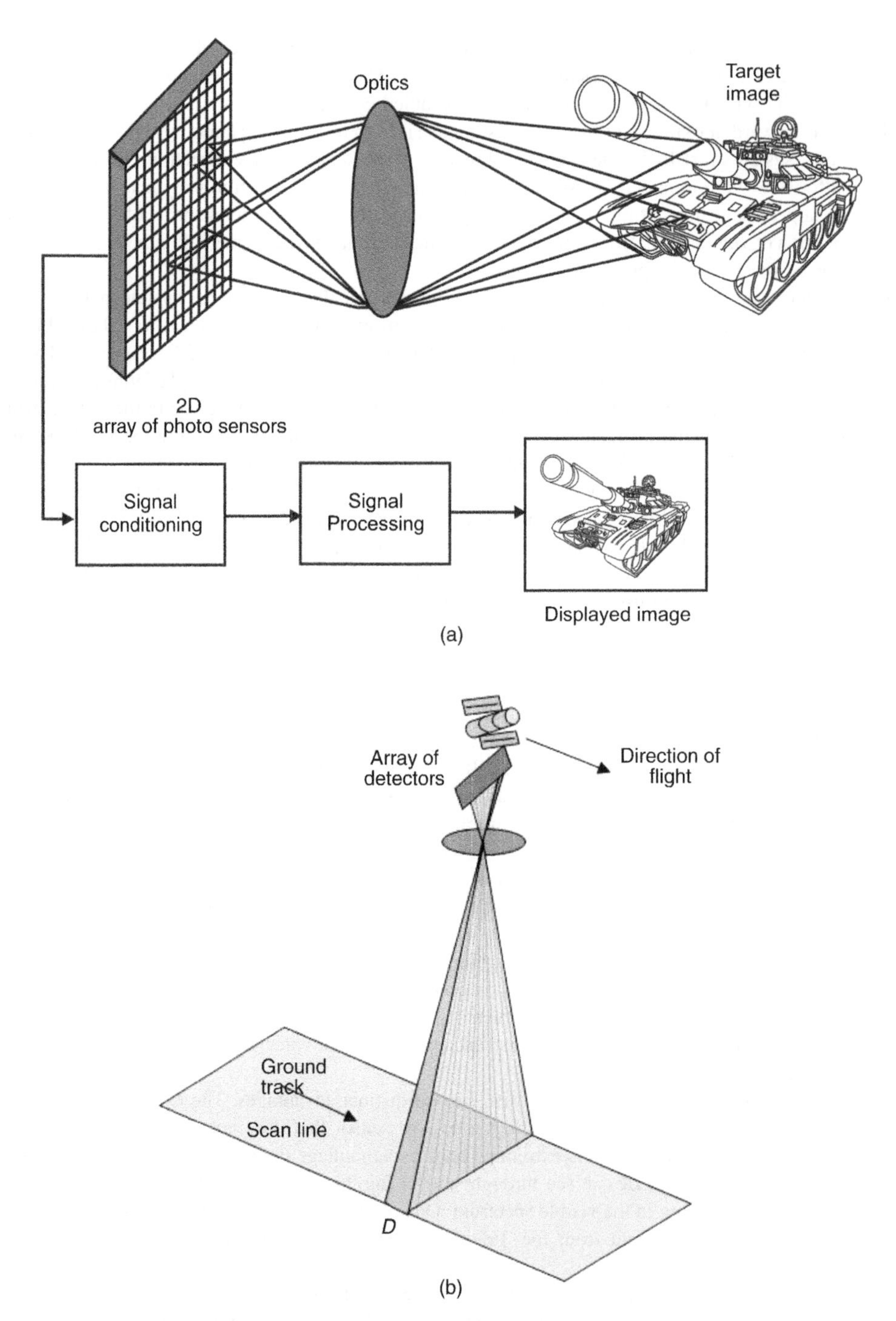

Figure 14.21 (a) FLIR concept; and (b) pushbroom concept.

Figure 14.22 HRC-E FLIR (Courtesy of FLIR Systems, Inc.).

Figure 14.23 Multisensor FLIR, type M-series (Courtesy of FLIR Systems, Inc.).

a low-light CCD camera. The FLIR camera here is a 640×480 pixel uncooled vanadium oxide microbolometer focal plane array operating in 7.5–13.5 μm band. FLIR offers a thermal sensitivity of better than 50 mK.

14.4 Guided Munitions

In this section we describe different techniques of munitions guidance with particular emphasis on electro-optically guided munitions including surface-to-air and air-to-air infrared-guided missiles and laser-guided munitions such as aerially delivered bombs and laser-guided missiles and projectiles. The discussion begins with a brief description of the basic guidance mechanisms in use, including both radar guidance as well as laser guidance, which is then followed by a detailed discussion on laser-guided and infrared-guided munitions.

14.4.1 Guidance Techniques

Different guidance techniques discussed in this section include beam rider, command, homing and navigation guidance, which are all described in the following sections.

14.4.1.1 Beam Rider Guidance

The concept of beam riding guidance of munitions is based on a radar beam or a laser beam constantly pointed towards the target throughout the flight time of the munitions. A launching station, possibly mounted on a vehicle, first directs a narrow radar or laser beam at the target, which could be a tank or an aircraft. The missile is then launched and, at some point after launch, it flies into the radar or laser beam. From this stage onwards, the missile attempts to keep itself inside the beam, while the aiming station keeps the beam always pointed at the target. The missile, controlled by a computer inside it, 'rides' the beam to the target. The missile's guidance sensors located at the rear of the missile receive information about the position of the missile within the beam. The missile interprets this information and generates its own correction signals. These correction signals are used to send command signals to the control surfaces of the weapon to keep the missile in the centre of the beam. The launch station keeps the beam pointed at the target throughout the engagement period and the missile rides the beam to the intended target. Both radar and laser beam rider guidance have been successfully employed for surface-to-surface, surface-to-air and air-to-ground weapons.

Figure 14.24 illustrates the concept of beam riding for a surface-to-air weapon using a radar beam. A laser beam could also have been used with similar performance. Figure 14.25 shows a laser beam rider missile launched from a helicopter against a tank target. As the beam moves further away from the launcher and towards the target, it spreads out and it becomes difficult to keep the beam in the centre of the target; this is why the beam rider concept is only effective for short to medium operational ranges.

Laser beam riding guidance became attractive for short-range anti-air and anti-tank missiles in the 1980s and 1990s with the introduction of low-cost and highly portable laser designators. In contrast to radar beam rider guidance, a laser beam can be made much narrower than a radar beam without increasing the size of the transmitter. Laser beam riding also allows the designer to encode additional information in the beam using digital means. Laser beam rider missiles are inherently more accurate. Also, the laser beam being used to guide the weapon is narrow, making it more difficult to be noticed by the target's warning sensor. Some common examples of laser beam riders include Star Streak, the RBS-70 and 9M119 Svir.

RBS-70 is a man portable air defence system (MANPADS) laser beam rider and is the product of SAAB Bofors Dynamics, Sweden. RBS-70 is a laser beam rider steered by a user-operated laser beam. The weapon is not susceptible to any deception by countermeasures employed by the target aircraft in the form of chaff or flares. Major specifications of the RBS-70 include an operational range of 0–6 km, target engagement height of 0–5 km and a speed up to 1.6 Mach. The RBS-70 New Generation (RBS-70NG; Figure 14.26) includes an improved sighting system capable of night vision. RBS-70 Mk-2 upgrade is called Bolide missile. It is faster with a speed of 2 Mach and has a range of 8 km.

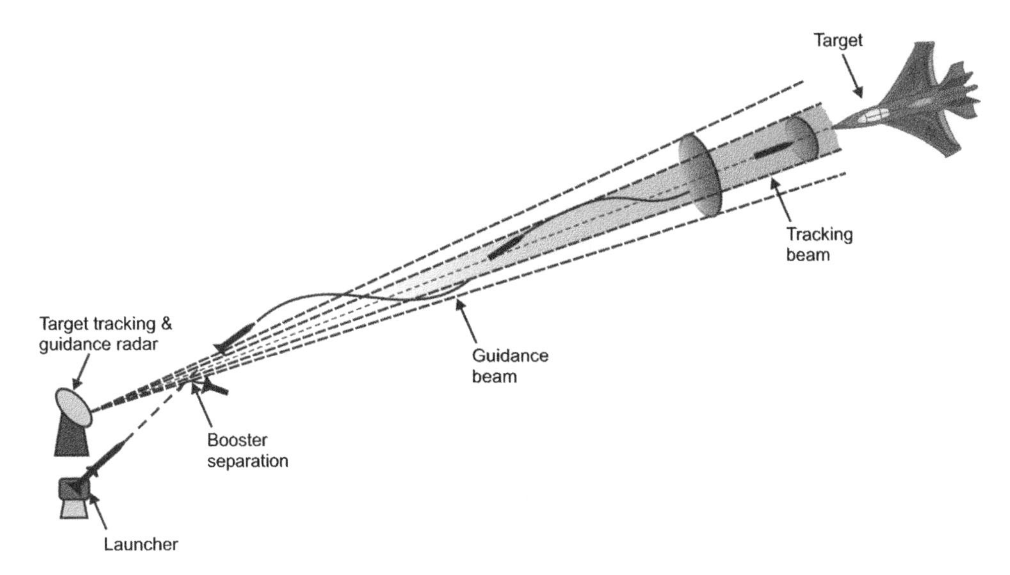

Figure 14.24 Beam rider guidance concept.

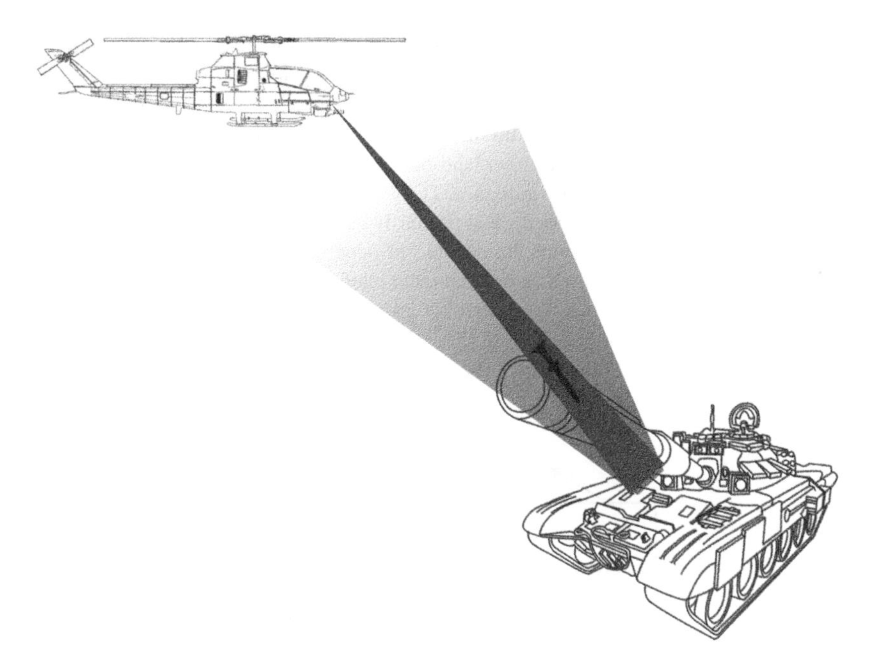

Figure 14.25 Laser beam rider concept.

14.4.1.2 Command Guidance

In the case of command guidance, the missile is commanded on an intercept course with the target. Conventionally, this is achieved by using two separate radars to continuously track the target and the missile. Tracking data from these radars is fed to a computer that computes the trajectories of the two vehicles. The computer in turn sends appropriate command signals over a radio link to the missile.

Figure 14.26 RBS-70 NG laser beam rider MANPADS (Courtesy of SAAB Sweden).

A sensor onboard the missile decodes the commands and operates the control surfaces of the missile to adjust its course in order to intercept the target in flight. Figure 14.27a shows the block schematic representation of command guidance and Figure 14.27b depicts the deployment scenario.

Wire-guided missiles are an example of command guidance. Command signals are sent to the missile through a conventional wire or a fibre-optic cable that actually reels out from the rear of the missile up to the launch platform. The missile trajectory in this case is controlled with the help of command signals transmitted via a wired link rather than a radio link. Wire-guided missiles are commonly used for short-range anti-tank operations launched from either land-based platforms or helicopters. In many cases, even torpedoes fired from submarines use wire guidance. TOW is a popular example of a wire-guided missile. Manufactured by Hughes Aircraft Company, it is primarily used in anti-tank warfare and is a command to line-of-sight weapon. Current versions are capable of penetrating 75 cm of armour at a maximum range of greater than 3 km. It can be fired from a vehicular platform, a helicopter or by infantrymen using a tripod stand. Images of a TOW missile system are provided by Figure 14.28.

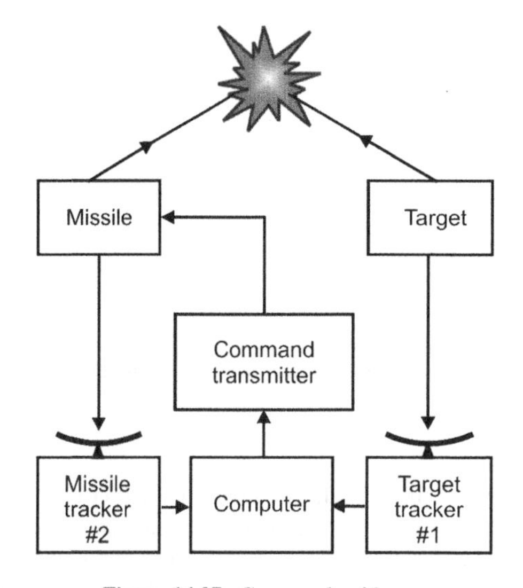

Figure 14.27 Command guidance.

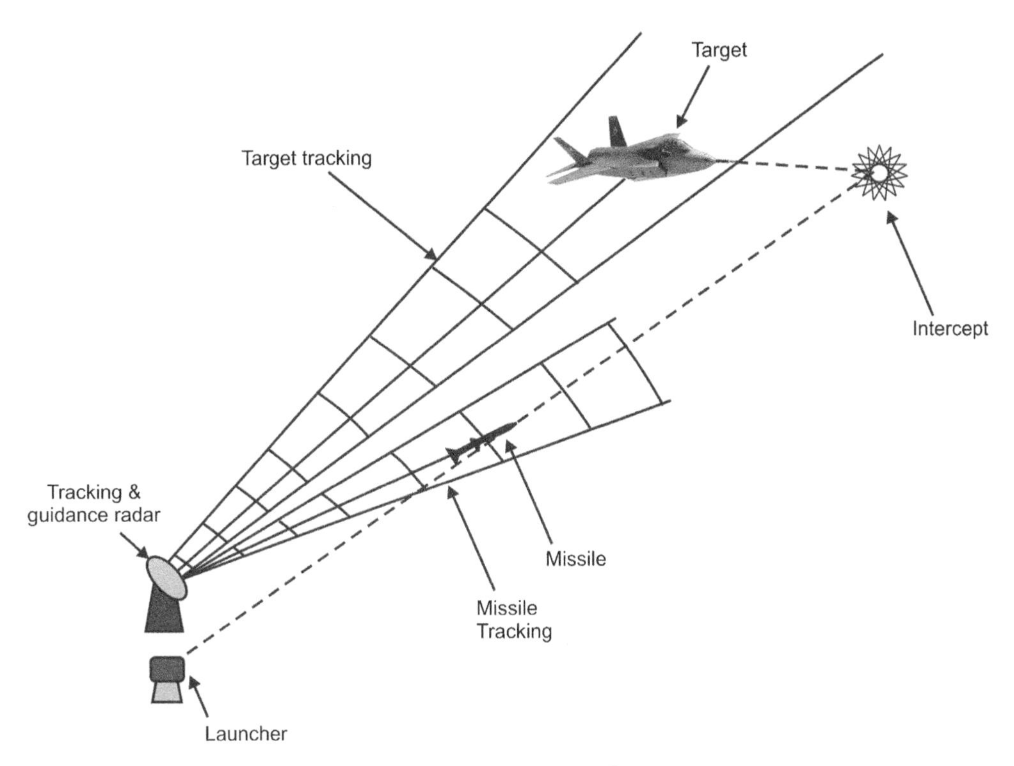

Figure 14.27 (*Continued*)

Figure 14.28 TOW missile system (Reproduced from public domain (http://en.wikipedia.org/wiki/File: Hires_090509-A-4842R-001a.jpg)).

Figure 14.29 Hellfire missile (Courtesy of Lockheed Martin).

Command guidance can be classified as command line-of-sight (CLOS) and command off line-of-sight (COLOS) guidance. CLOS systems are further subdivided into four groups: (1) manual command to line-of-sight (MCLOS) where target tracking, missile tracking and control functions are all performed manually; (2) semi-manual command to line-of-sight (SMCLOS) where target tracking is automatic but missile tracking and control functions are performed manually; (3) semi-automatic command to line-of-sight (SACLOS), where the target tracking is manual and the missile tracking and control functions are automatic; and (4) automatic command to line-of-sight (ACLOS) where all three functions are automatic.

SACLOS is the most common form of guidance in use against ground targets such as bunkers and tanks. Hellfire from Lockheed Martin is helicopter-launched fire-and-forget anti-armour air-to-ground weapon of the SACLOS category. Generations (1)–(3) of the weapon use a laser seeker while generation (4) uses radar seeker. Figure 14.29 is a photograph of a Hellfire missile.

Unlike the CLOS system, the COLOS system does not depend on angular coordinates of the missile and the target. The guidance system ensures missile interception of the target by locating both the missile and target in space for which the distance coordinate is needed. This is only possible if both missile tracker and target tracker are active. It may also be mentioned here that in the case of the COLOS system, the missile and target tracker can be oriented in different directions.

14.4.1.3 Homing Guidance

Homing guidance is the most common form of guidance methodology used in surface-to-air- and air-to-air-guided weapons. Homing guidance is further subdivided into four groups: semi-active homing, active homing, passive homing and track-via-missile homing, also known as retransmission homing.

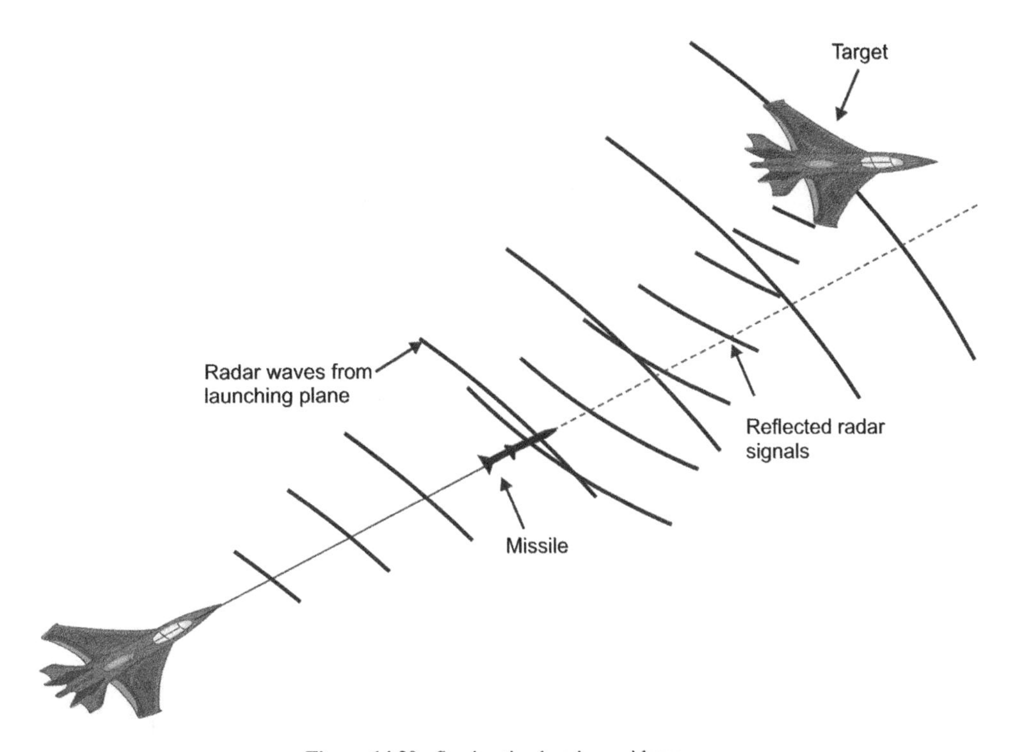

Figure 14.30 Semi-active homing guidance.

In the case of *semi-active* homing guidance, the target is illuminated by an external source which could be radar or laser. The electromagnetic energy reflected by the target is intercepted by the seeker head of the guided weapon. An onboard-computer processes the intercepted signal and determines the target's relative trajectory. It sends appropriate command signals to the control surfaces of the weapon to make it intercept the target. Figure 14.30 illustrates the concept of semi-active homing guidance in the case of an air-to-air missile. Semi-active homing is similar to command guidance except for the fact that, in the case of former, the command computer is onboard the weapon. The type of seeker head, whether it is a radar seeker or a laser seeker, depends on the type of external source designating the target. Both radar- as well as laser-guided semi-active homing weapons are in use. The Sparrow air-to-air missile and laser-guided weapons of the Paveway series are examples of semi-active homing guidance. Laser-guided munitions and infrared-guided missiles are described in detail in Sections 14.4.2.1 and 14.4.2.2, respectively.

In the case of *active* homing guidance, the source of target designation is also onboard the weapon with the result that this methodology does not require an external source. These features put it in the category of fire-and-forget missiles as the launch platform does not need to continue to illuminate the target after the missile has been launched. Active homing guidance weapons are usually radar guided. Also, in the case of active homing guidance, the transmitted and reflected waves are at the same angle with respect to the line-of-sight between the target and the missile. This is different from semi-active homing mechanism in which transmitted and reflected waves are at an angle. Fir this reason, semi-active and active homing guidance systems are sometimes called bistatic and monostatic systems, respectively. The RBS-15 anti-ship missile from Saab Bofors Dynamics, MBDA Exocet anti-ship and MBDA MICA (Figure 14.31) surface-to-air and air-to-air missiles from MBDA, AS-34 Kormoran anti-ship missile from EADS and Indian DRDO-Astra

Figure 14.31 MBDA MICA air-to-air missile (Creative Commons Attribution-Share Alike 3.0 Unported).

BVRAAM air-to-air missile are some examples of missiles that use active radar homing in the terminal phase.

Passive homing guidance makes use of some form of energy emitted by the target. This energy is intercepted by the missile seeker, which is processed to extract guidance information to guide the missile to the target. This energy could be in the form of heat energy generated by the target, which is made use of by the seeker in an infrared-guided missile. Infrared-guided missiles constitute an important category of electro-optically guided precision strike weapons, and are discussed in detail in Section 14.4.2.2. Anti-radiation missiles such as AGM-88 HARM air-to-ground missiles track the radio frequency energy emitted by the ground-based radar stations to generate guidance signals. Passive torpedoes make use of sound waves generated by engines of the ships or sonars to attack their targets. Missiles such as the AGM-65 Maverick are equipped with electro-optic sensors that rely on visual images to guide the weapon to the target.

In the case of *retransmission* homing guidance, the target is illuminated by an external radar. The energy reflected by the target is intercepted by the missile sensor. In this case, the missile does not have an onboard computer to process the sensor signal and generate guidance command. Instead, the sensor signal is transmitted back to the launch platform for processing. The command signals generated at the launch platform are retransmitted back to the missile for use by the missile's control surfaces to guide the missile to the target. The advantage with this guidance technique, also called 'track-via-missile', is that the expensive tracking and processing hardware is not destroyed along with the missile. The disadvantage is that it requires a high-speed communication link between the missile and the launch station. MIM-104 Patriot surface-to-air missile system of Raytheon Company, USA is an example of a track-via-missile homing guidance system (Figure 14.32).

14.4.1.4 Navigation Guidance

The term *guidance* not only refers to the determination of the desired path of travel, also called the trajectory from the vehicle's current location to a intended target, but also refers to the desired changes in velocity, rotation and acceleration required to follow the desired path. The term *navigation* refers to the determination, at a given time, of the vehicle's present state vector defined by location and velocity and also its attitude. The term *control* refers to the manipulation of the forces, by way of steering controls and thrusters, needed to track guidance commands while maintaining vehicle

Figure 14.32 MIM-104 Patriot surface-to-air missile system (Creative Commons Attribution-Share Alike 2.5 Generic).

stability. It is the combination of these three functions that comprises *navigation guidance*. *Navigation guidance* can also be classified as inertial navigation, ranging navigation, celestial navigation and geophysical navigation.

In the case of *inertial* navigation guidance, the vehicle uses onboard sensors to determine its motion and acceleration with the help of gyroscopes and accelerometers. The gyroscope is used to measure angular rotation and an accelerometer is used to measure the linear motion. The gyroscope and accelerometer are combined into a single unit along with a control mechanism, referred to as an inertial measurement unit (IMU) or inertial navigation sensor (INS). An INS system basically works by telling the vehicle where it is at the time of launch and the vehicle's computer uses the signals from the IMU to ensure that the vehicle travels along the programmed path. INSs are widely used on a range of aerospace vehicles including commercial airliners, military aircraft, missiles and spacecraft. The long-range all-weather subsonic cruise missile Tomahawk and medium-range all-weather beyond visual range air-to-air missile AMRAAM are examples that use inertial navigation for mid-course guidance.

While inertial navigation guidance technique makes use of onboard sensors, *ranging* navigation depends on external signals for guidance, which are usually provided by radio beacons. Based on the direction and strength of the signals received by the aircraft, it navigates its way along the desired trajectory. Ranging navigation guidance has been largely rendered obsolete with the arrival of the global positioning system (GPS). GPS-based navigation has largely replaced radio beacons in both military and civilian applications. GPS is a key enabling technology for existing and future military precision navigation applications. The joint direct attack munitions (JDAMs) series of guided bombs makes use of integrated INS and GPS guidance techniques to determine where they are with respect to the locations of their targets. An INS–GPS combination gives the precision-guided weapon a kind of all-weather capability and largely overcomes the vulnerability to adverse ground and weather conditions of weapons employing laser and imaging infrared seekers. In fact, state-of-the-art precision strike weapons use a combination of guidance technologies including inertial navigation, global position sensing and laser/infrared seeking to achieve higher performance levels.

Celestial navigation, one of the oldest navigation techniques, uses the positions of the stars to determine location, especially latitude, on the surface of the Earth. This form of navigation guidance requires good visibility of the stars, which makes it particularly useful at night or at very high altitudes.

In celestial navigation, the missile compares the positions of the stars to an image stored in its memory to determine its flight path. Submarine-launched ballistic missile (SLBM) Poseidon of Lockheed Martin, carrying multiple independent re-entry technology and with an operational range in excess of 4500 km, is an example of a ballistic missile using celestial navigation.

Geophysical navigation guidance depends on measurements made on the surface of the Earth for operation. It uses compasses and magnetometers to measure the Earth's magnetic field as well as gravitometers to measure the Earth's gravitational field. This technique has not found much application in missile guidance.

Another guidance technique makes use of *terrain contour matching*. It uses a radar altimeter to measure height above the ground. By comparing the contours of the terrain against data stored onboard the missile, the missile's autopilot navigates its way to the destined location. Terrain contour matching (TERCOM) is a navigation system used primarily by cruise missiles.

A related technique to terrain matching is *digital scene matching*. It is far more accurate than the terrain matching technique. For guidance, this technique compares the image seen below the weapon to satellite or aerial images stored in the missile computer. If the scenes do not match, the computer sends commands to control surfaces to adjust the missile's course until the images match to a certain acceptable level. The Tomahawk guidance systems use a combination of INS, GPS, TERCOM and digital scene matching techniques.

14.4.1.5 Ring Laser Gyroscope

A ring laser gyroscope (RLG) is an important optronic rate sensor and is at the heart of any inertial navigation system. There are different types of gyroscopic sensors such as spinning wheel mechanical gyro, dynamically tuned gyro, ring laser gyro and fibre-optic gyro. An RLG offers superior performance compared to a conventional spinning type mechanical gyro, as there are no moving parts and therefore no inherent drift terms due to absence of friction. Unlike a mechanical gyro, the RLG does not resist changes to its orientation. A combination of three orthogonally placed RLGs provides a rate sensor with three degrees of freedom.

Having only briefly described different types of guidance techniques in Sections 14.4.1.1–14.4.1.4, we discuss in detail the operational principle, salient features and performance parameters of a ring laser gyroscope in this section.

A ring laser gyroscope is primarily a rate sensor, which can be used to measure rotation by integrating the rate information. It essentially consists of two counter-propagating laser beams over the same path. It operates on the principle of the Sagnac effect, according to which the null points of the internal standing wave pattern produced by the counter-propagating laser modes shift in response to angular rotation. The shift in the nulls of the standing wave pattern manifests itself in the form of a moving interference pattern, observed by combining the two laser modes.

The basic concept of a ring laser gyroscope can be explained as follows. The frequency of oscillation in a linear laser is such that that the laser cavity consists of an integral number of wavelengths. Linear laser cavity is constituted by the double pass of the distance between the two mirrors. Since it is imperative that the beam replicates itself for successive passes over the cavity length, there will always be nodes at the two mirrors and the two oppositely propagating laser beams comprise a standing wave. In the case of a ring laser, the two oppositely directed laser beams can be considered as travelling waves and there is no such constraint of having a node at the mirror. The two beams in this case can be independent of each other and oscillate at a different amplitude and frequency. The oscillation frequency of each is determined by the optical path length (not the geometrical path length). Any mechanism that produces an optical path difference – rotation in this case – results in different frequencies of oscillation for the two laser beams. If the rotation is in the clockwise direction, the clockwise laser beam will see a larger optical path length than that encountered by the anticlockwise beam. For rotation in the anticlockwise direction, it would be the opposite. The frequency difference is proportional to the rotation rate of the cavity.

By measuring the frequency difference, the rotation rate of the cavity and the platform it is strapped onto can be measured. According to the Sagnac equation, the frequency difference $\Delta\nu$ is given by Equation 14.7:

$$\Delta\nu = \frac{4A\Omega}{\lambda P} \tag{14.7}$$

where

A = encircled oriented surface area of the ring laser
P = perimeter
λ = wavelength
Ω = rotation rate

Figure 14.33a depicts square ring laser geometry, which is one form of geometry in use. In the case of square ring laser geometry, four mirrors form a closed light path. A small part of the ring laser path houses the excited gain medium. The gain medium along with mirrors is responsible for producing the two counter-propagating laser beams. The difference frequency produced as result of rotation is obtained by interference of the two beams. The two beams are usually combined with the help of a corner prism as shown in the diagram. Figure 14.33b highlights the constructional features of the square type ring laser geometry. Another commonly used ring laser cavity structure is triangular geometry, as shown in Figure 14.34. This type of geometry is popular with ring laser gyroscopes from Honeywell.

As is evident from the Sagnac equation, constancy of A and P is very important to achieve the kind of angular resolution capable of present-day ring laser gyroscopes. It is due to this reason that a ring laser cavity block is made from material of an exceptionally high thermal stability. ZERODUR, the registered trade name of a near-zero thermal expansion transparent glass ceramic made by Schott, Germany, is one such material. The material has a thermal expansion coefficient of $\pm 10^{-7}\,°C^{-1}$ over a temperature range 0–50 °C.

RLGs suffer from an effect referred to as lock-in at very low rotation rates. When the rotation rate is very low, the frequencies of the counter-propagating laser beams become almost identical, which leads to the two beams being injection-locked to a common frequency. As a result of this, it does not respond to

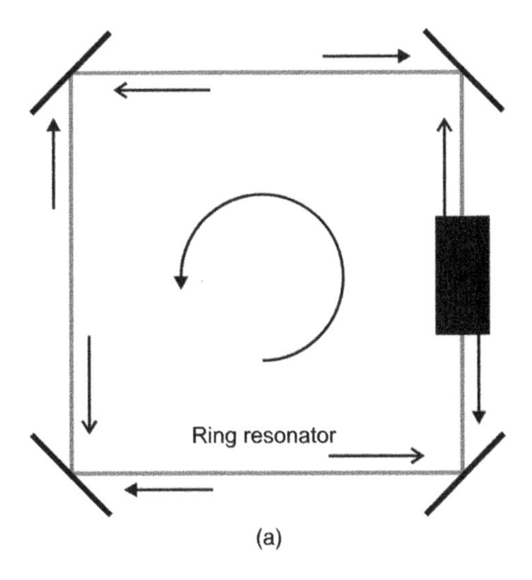

Figure 14.33 Square ring laser geometry.

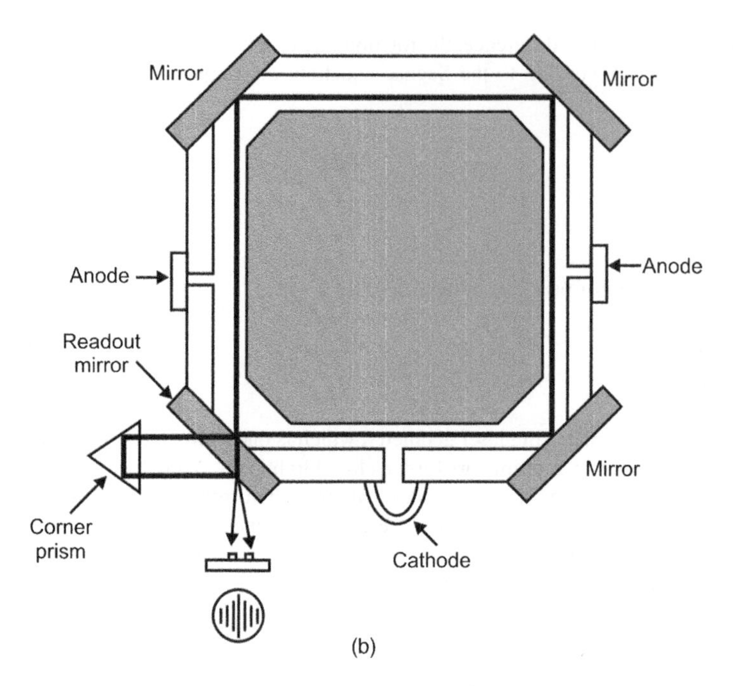

Figure 14.33 (*Continued*)

rotation. This is illustrated in the transfer characteristics of Figure 14.35. This problem is largely overcome by using forced dithering in which the ring laser cavity is rotated clockwise and anticlockwise about its axis using a mechanical spring driven at its resonance frequency. Typically, a peak dither rate of 1 Arcsec/sec is used and dither frequency is in the range of 400–500 Hz.

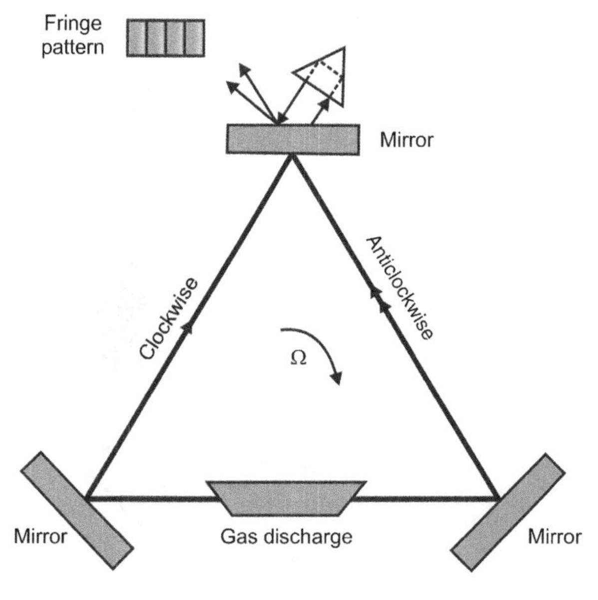

Figure 14.34 Triangular ring laser gyroscope.

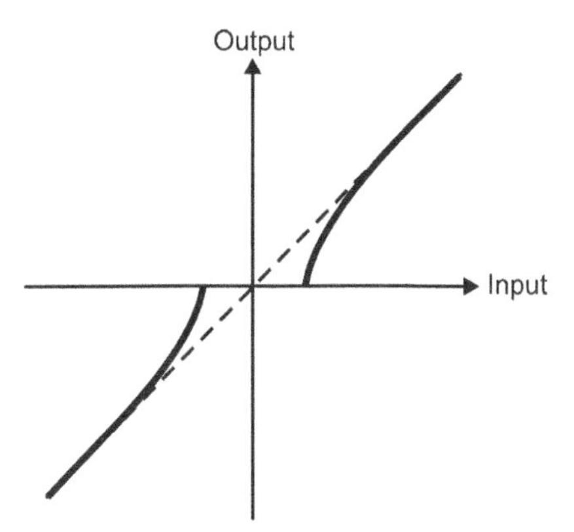

Figure 14.35 Lock-in effect in RLG.

Important parameters that describe the performance of a ring laser gyroscope include bias error, drift and instability, scale factor stability and angular random walk. *Bias error* is the output produced by the gyroscope when it is not experiencing any rotation. Bias error is defined as the voltage or percentage of full-scale output representative of the rotational velocity in $^\circ\,s^{-1}$ as measured by the gyroscope in the absence of any rotation. The bias error is primarily caused by a number of factors including calibration errors, bias drift, bias variation with temperature and effect of shock or 'g' level.

Bias drift refers to the variation of bias error over time, assuming all other factors remain constant. This is a warm-up effect usually caused by self heating of electrical and mechanical components of the gyroscope. This effect is more prevalent only during the warm-up time of first few seconds and is non-existent after that. It should not be confused with the more colloquial term drift rate.

Bias instability is the fundamental parameter that defines how good or bad the gyroscope is. It is defined as the minimum point on the Allan Variance curve measured in $\deg h^{-1}$. (The Allan Variance technique is a very powerful mathematical method which allows a user to assess the real performance of any gyroscope in relation to the dynamics of the application, taking into account the effects of bias, noise, drift and long-term sensor instability.) It represents the best bias stability achievable in a given gyroscope with bias averaging taken at the interval defined by the Allan Variance minimum. As an example, bias stability error in the case of a Honeywell gyroscope type GG 1320 is specified to be $\leq 0.002\,\deg h^{-1}$.

Bias instability is always measured by averaging successive data samples. The challenge lies in selecting a suitable time over which to average sampled data. Averages taken over short time intervals will be dominated by noise while those over a longer period are dominated by longer-term drift. The technique involves selecting a range of time intervals, typically 0.01–500 s, over which to average data. The standard deviation thus obtained from one averaged time period to the next is calculated and plotted against the averaging interval in log-log form. The resulting graph has a characteristic 'bathtub' shape. It is possible to compute the key defining characteristics of the gyroscope, namely angular random walk, bias instability and rate random walk, from this graph.

Scale factor and *scale factor stability* represent the transfer or output versus input characteristics of the gyroscope. It measures the signal produced in terms of voltage or bits for a given rotation rate. It is a measure of the slope of the best straight line through the points on a graph representing the gyroscope output against input rotation rate measured over the specified dynamic range of the gyroscope. As an example, for an analogue gyroscope it is measured in $V\,\deg^{-1}\,s$. Scale factor stability figure for a Honeywell gyroscope type GG 1320 is stated to be ≤ 10 ppm and maximum rate measuring capability of $\pm 500\,\deg s^{-1}$.

Angular random walk (ARW) is a measure of gyroscope noise and is measured in $\deg h^{-0.5}$ or $\deg s^{-0.5}$. It can be thought of as the standard deviation due to noise while integrating the output of a stationary gyroscope over a period of time. It is inversely proportional to the square-root of integration

Figure 14.36 MiniRLG2-based INS (Courtesy of Seatronics).

time. As an illustration, if a gyroscope had an ARW of 1 deg $s^{-0.5}$, then the result of angular position measurement in the ideal case would be zero. The longer the integration time, the larger the spread of the results from zero. The spread will follow the law of being inversely proportional to the square-root of integration time. In the case of a Honeywell gyroscope (and this is also true for typical inertial grade gyroscopes), the random walk noise is specified to be $\leq$0.0018 deg $h^{-0.5.}$

As outlined at the beginning of this section, the INS would have three orthogonally placed RLGs to provide measurement capability of rotation in three degrees of freedom. Along with accelerometers, the RLGs constitute the INS. Figure 14.36 shows one such RLG-based INS package. The system is a high-grade inertial measurement unit that meets the accuracy requirements of attitude, orientation, position and navigation. The system is characterized by a bias stability of 0.1 deg h^{-1}, scale factor stability of 75 ppm and angular random walk specification of 0.028 deg $h^{-0.5}$.

Example 14.4

A ring laser gyroscope uses a He-Ne ring laser operating at 0.633 μm. The ring laser employs an equilateral triangular laser cavity with each arm equal to 10 cm. Determine the beat frequency for a rotation rate 0.1 deg h^{-1}.

Solution

- According to the fundamental ring laser gyroscope equation, $\Delta\nu = \dfrac{4A\Omega}{\lambda P}$
- $P = 10 \times 3 = 30$ cm
- $\lambda = 0.633\ \mu m = 0.633 \times 10^{-4}$ cm
- $A = \sqrt{3} \times 10^2/4 = 173.2/4 = 43.3\ cm^2$
- Therefore, $\Delta\nu$ (in Hz) $= (4 \times 43.3/0.633 \times 10^{-4} \times 30) \times \Omega$ (in deg s^{-1}) $= 25.33 \times \Omega$ (in deg h^{-1}).
- For $\Omega = 0.1$ deg h^{-1}, $\Delta\nu = 25.33 \times 0.1 = 2.533$ Hz.

Example 14.5

Determine the percentage improvement in scale factor of two triangular ring laser gyroscopes with optical path lengths of 30 cm (each arm of the ring laser having a path length of 10 cm) and 45 cm (each arm of ring laser having an optical path length of 15 cm), assuming other parameters to be identical.

Solution

- Scale factor is proportional to the ratio of the area encircled by the optical path length to the optical path length. Assuming other parameters to be the same, this ratio is proportional to $\sqrt{3}/12 \times L$ where L is the optical path length.
- Therefore, improvement in scale factor $= L_2/L_1 = 45/30 = 1.5$
- That is, a scale factor of a 45 cm gyroscope is 1.5 times that of a 30 cm gyroscope.
- Percentage improvement is therefore 50%.

14.4.1.6 Fibre-optic Gyroscope

Unlike a classical mechanical gyroscope, which is based on the principle of conservation of momentum, but very much like the ring laser gyroscope described in Section 14.4.1.5, a fibre-optic gyroscope (FOG) has virtually no moving parts and no inertial resistance to movement. The operational principle of a fibre-optic gyroscope is also based on the Sagnac effect. It consists of a long coil of optical fibre, typically a few kilometres in length, and makes use of the interference of light to detect mechanical rotation. FOG provides extremely precise rotational rate information due to its lack of cross-axis sensitivity to vibration, acceleration and shock. Furthermore, it does not require any starting calibration and it consumes relatively little power. Fibre-optic gyroscopes can be very compact, as optical fibre lengths as long as a few kilometres can be wrapped around a small-diameter coil. The development of fibre-optic gyroscopes received a large boost and became a reality in the early 1970s due to technological advances in semiconductor diode lasers and low-loss single-mode optical fibre, mainly triggered by the demands of the telecommunications industry.

Figure 14.37 depicts a fibre-optic gyroscope. Two split laser beams from a single laser are injected into the same fibre but in opposite directions. Due to the Sagnac effect, the beam travelling against the rotation experiences a slightly shorter path length delay than the other beam. The resulting differential phase shift is measured through interferometry, translating one component of the angular velocity into a shift of the interference pattern which is measured photometrically. Beam-splitting optics launches light from a laser diode into two waves propagating in the clockwise and anticlockwise directions through a coil consisting of many turns of optical fibre. The strength of the interference signal depends on the effective area of the closed optical path according to the Sagnac effect. This is not simply the geometric area of the loop, but is enhanced by the number of turns in the coil. The FOG was first proposed by Vali and Shorthill in 1976. Development of both the passive interferometer type of FOG (IFOG) and a newer concept, the passive ring resonator FOG (RFOG), is proceeding in many companies and establishments worldwide.

A fibre-optic gyroscope is perhaps the most reliable alternative to the mechanical gyroscope. Because of their intrinsic reliability, fibre-optic gyroscopes are used for high-performance space applications. Compared to a ring laser gyroscope, it typically offers a higher resolution but has relatively poorer drift and scale factor performance specifications. Typically, its sensitivity figure of 0.1 deg h^{-1} it nowhere near the demonstrated RLG sensitivity of better than 0.001 deg h^{-1}. Unlike ring laser gyroscopes where zero beat frequency always means zero angular velocity, for a fibre-optic gyroscope the indication which corresponds to zero angular velocity needs to be determined.

The success of the fibre-optic gyroscope lies in its desirable features such as lightweight, small size, limited power consumption, projected long lifetime and low cost. This makes fibre-optic gyroscopes particularly attractive in applications such as automotive construction and robotics, which call for less-demanding performance levels in terms of sensitivity and drift specifications and where small size and low cost are important considerations.

Fibre-optic gyroscopes are used in a wide variety of applications, including inertial navigation, navigation of remotely operated vehicles (ROVs) and autonomous underwater vehicles (AUV),

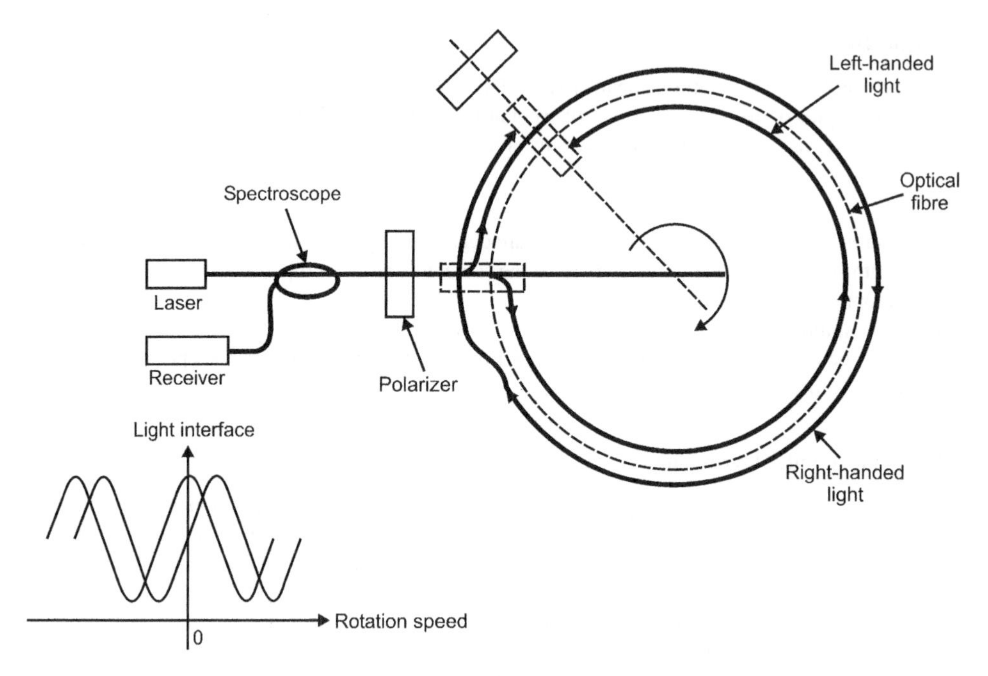

Figure 14.37 Principle of operation of fibre-optic gyroscope.

surveying, the automotive industry and robotics, EO/FLIR/radar stabilization, line-of-sight tracking and precision pointing, gunfire control systems, platform stabilization, gyroscope compassing and target acquisition systems, smart munitions and earth observation, scientific and telecommunication satellites.

Photographs of two such fibre-optic gyroscopes are provided in Figure 14.38. Figure 14.38a is Type EMP-1.2 K from Emcore Photonics Systems, which provides precise navigation with 1 mile/hr accuracy without GPS and fast gyroscope compassing to 1 mrad. Major performance specifications of the device include short- and long-term drift stability specifications of 0.005 deg h^{-1} and 0.01 deg h^{-1} respectively, scale factor stability and linearity specifications of 50 ppm and 25 ppm respectively and random walk specification of 0.0015 deg $h^{-0.5}$. Figure 14.38b shows a three-axis fibre-optic gyroscope assembly from AL Cielo. The sensors can be used in diverse navigation and control applications, such as for the stabilization of missiles, airborne sensor systems or land vehicle systems. State-of-the-art fibre-optic gyroscopes compare very closely with the specifications of ring laser gyroscopes.

14.4.2 Electro-optically Guided Precision Strike Munitions

This section provides an overview of electro-optically guided precision strike munitions, which includes laser-guided munitions and infrared-guided missiles, in terms of their operational basics, important performance parameters and need for periodic testing of operationally vital parameters. This is followed by a detailed discussion on laser-guided munitions.

Due to their precision strike capability, laser-guided munitions (bombs, projectiles and missiles) and IR-guided surface-to-air and air-to-air missiles are the most widely exploited electro-optically precision strike munitions on various land- and sea-based and airborne military platforms such as main battle tanks, armoured fighting vehicles, ships, fighter aircraft and attack helicopters. This class of guided weapons has proved their lethality and efficacy beyond any doubt during recent conflicts. These sophisticated weapons have huge price tags attached to them and are also of great tactical importance in the contemporary battlefield scenario.

All electro-optically guided precision strike munitions use a type of optoelectronic position sensor subsystem known as a seeker unit that determines the position of the weapon with respect to the target in

Figure 14.38 Fibre-optic gyroscopes: (a) Type EMP-1.2K (Courtesy of Emcore Photonics Systems); and (b) 3-axis FOG-based IMU (Courtesy of AL CIELO).

real time and feeds this information to a servo control subsystem. The servo control subsystem ensures that the weapon maintains its orientation in the desired direction of the target in order to ultimately hit it precisely. There are two broad categories of electro-optically guided precision strike munitions: laser-guided munitions and IR-guided missiles. Figure 14.39a and b show the constructional features of laser-guided and infrared-guided munitions, respectively. This classification is based on the type of

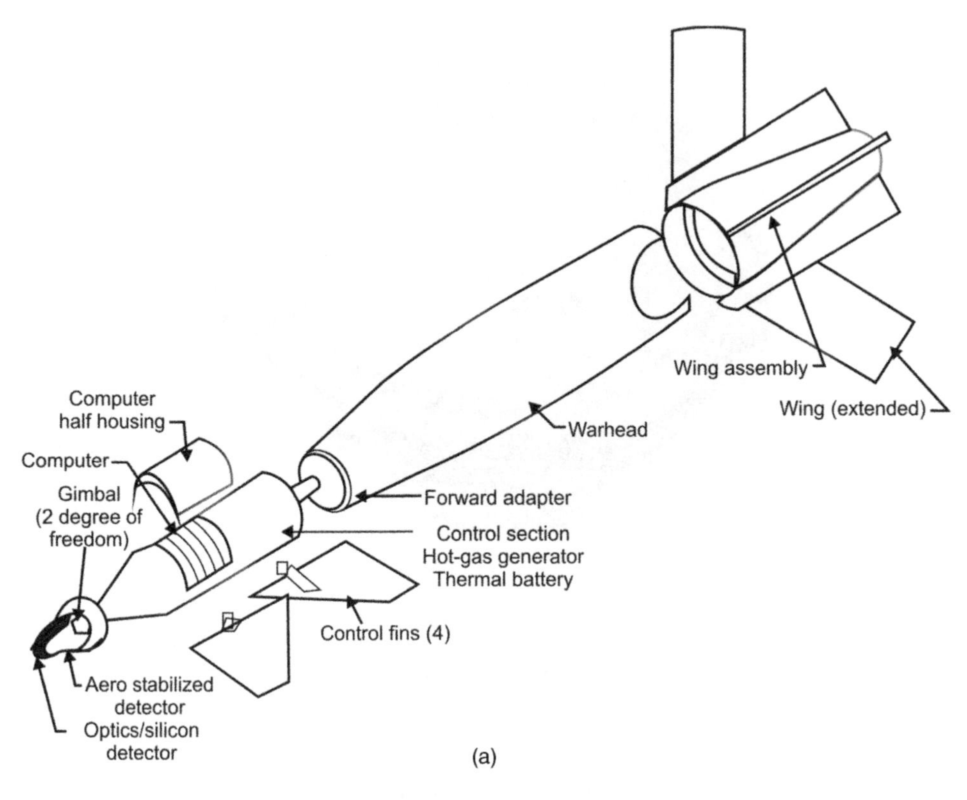

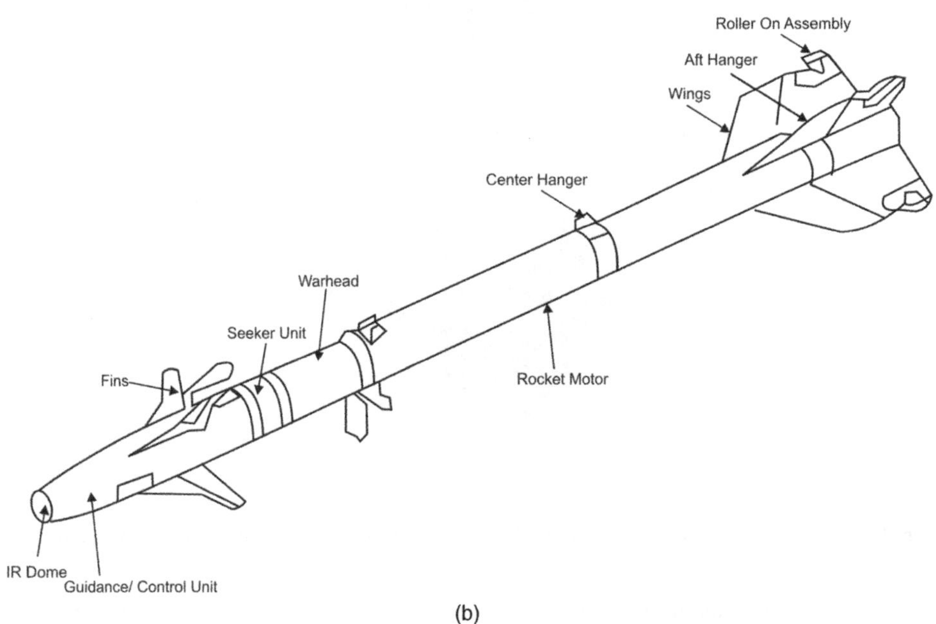

Figure 14.39 Laser-guided and infrared-guided munitions construction.

seeker unit and the associated principle of operation involved in the functioning of the two types. The seeker units used in the two types of guided munitions are known as laser seekers and IR seekers, respectively. IR seekers can be further classified into non-imaging and imaging.

14.4.2.1 Laser-guided Munitions: Operational Basics

In laser-guided munitions delivery operations, the target is illuminated (target designation) by a pulsed solid-state laser producing high peak power pulses with a known PRF. Peak power, pulse width and PRF are typically in the range 5–8 MW, 10–20 ns and 5–20 Hz respectively. The laser seeker head in the weapon makes use of laser radiation scattered from the target to generate information on the angular error, which in turn is used to generate command signals needed to guide the weapon to the source of scatter, which is the target (Figure 14.40).

Before the weapon locks on to the radiation scattered from the target, it ensures that the radiation is the intended one. The laser target designator and the laser seeker used in the guided weapon delivery mission use the same PRF code; the PRF code compatibility check forms the basis of identification of the desired radiation and is therefore essential to the function and mission success of the weapon. The PRF code is generally chosen to an accuracy of ± 1 to ± 2 μs in the time interval between two successive laser pulses in a nominal value that is usually in the range of 50–200 ms. The seeker head front-end uses a type of optoelectronic position sensor that determines the orientation of the weapon with respect to the target. Before it does that however, it deciphers the PRF of the received radiation; it is only processed further to extract information on the angular position of the weapon with respect to the target if the PRF matches the chosen PRF value within a specified tolerance.

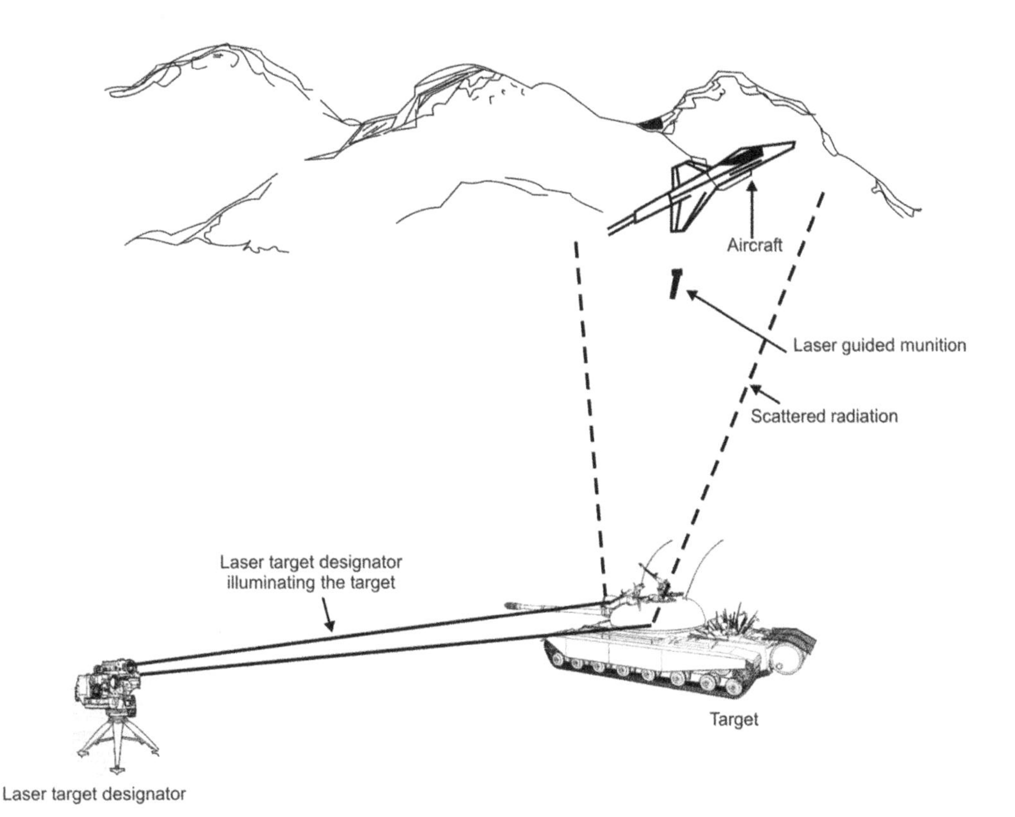

Figure 14.40 Laser-guided munitions delivery with ground designation and aerial delivery.

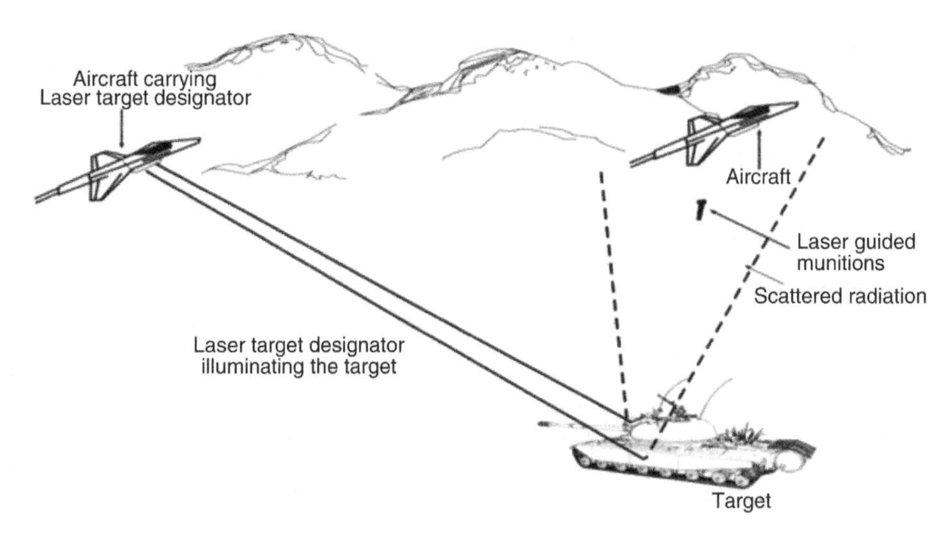

Figure 14.41 Laser-guided munitions delivery with target designated from another aircraft.

There are various possible laser-guided munitions deployment scenarios. The scenario illustrated in Figure 14.40 uses a ground-based laser target designator and aerially delivered bomb. Other possible scenarios include the case where the laser target designator is located on a different aircraft from that carrying the laser-guided bomb; such a deployment scenario is depicted in Figure 14.41. Alternatively, the target may be designated from a ground-based designator and the guided munitions are launched from a ground-based platform (e.g. cannon-launched laser-guided projectiles); this is depicted in Figure 14.42. Finally, both the laser target designation and laser-guided munitions delivery can be executed from the same airborne platform as shown in Figure 14.43.

The optoelectronic sensor employed for the purpose is usually a quadrant photosensor. A 2D array of photosensors is also used in some cases. Figure 14.44 describes the principle of operation of a quadrant photosensor when used for a position-sensing application. The quadrant photosensor is placed before the focal plane of the front-end optics. The focal spot is symmetrical about the centre of the quadrant

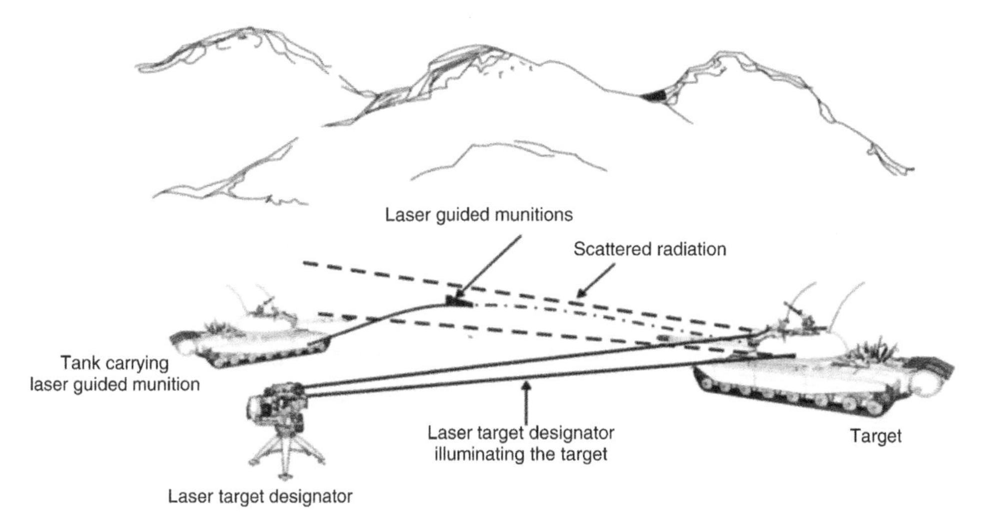

Figure 14.42 Laser-guided munitions delivery concept: laser-target designator operation and laser-guided munitions delivery from ground-based platforms.

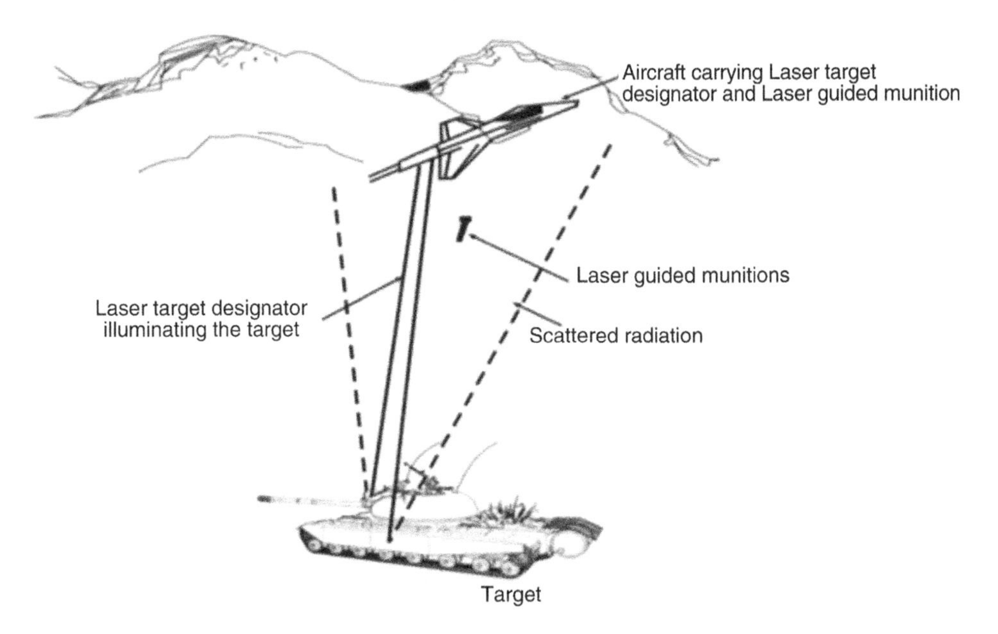

Figure 14.43 Laser-guided munitions delivery concept: laser-target designator operation and laser-guided munitions delivery from the same airborne platform.

photosensor when the perpendicular-to-focal plane of the detector is collinear with the axis of the received laser radiation scattered from the intended target, as shown in Figure 14.45a . This is the case when the weapon is pointing precisely towards the target. If the laser radiation is impinging on the laser seeker cross-section at an angle, which will be the case when the weapon is not pointing towards the intended target, the centre of the focused laser spot will shift, depending upon angular error in azimuth and elevation as shown in Figure 14.45b–e.

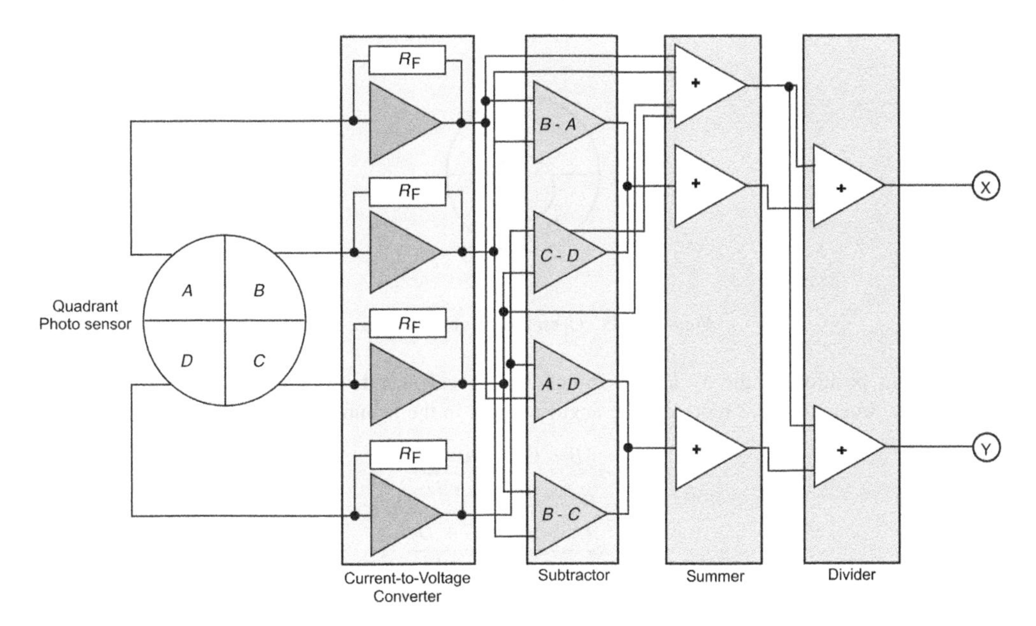

Figure 14.44 Principle of quadrant photosensor.

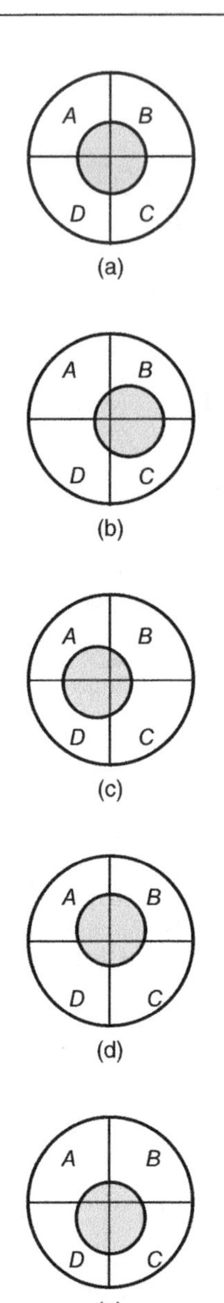

Figure 14.45 Operation of quadrant sensor.

The beam position in the x- and y- directions are calculated using Equations 14.6 and 14.7 respectively, where X and Y represent the angular errors in the azimuth and elevation directions:

$$X = \frac{(B+C)-(A+D)}{A+B+C+D} \tag{14.8}$$

$$Y = \frac{(A+B)-(C+D)}{A+B+C+D} \tag{14.9}$$

where A, B, C and D are the electrical voltages corresponding to laser power falling on the four quadrants.

In the case of precisely pointing towards the target, all four power levels are equal and therefore $X = 0$ and $Y = 0$. The sum $A + B + C + D$ represents total power; division by total power ensures that the calculated position error is independent of laser intensity variations.

The output analogue signals proportional to the magnitude of laser power falling on the four quadrants are digitized and then processed to compute X and Y. Error signals X and Y are then used to guide the weapon towards the desired position of null with the canards driven by a servo control system. The maximum value of proportional field of view offered by the laser seeker of this type in the x- and y-directions is proportional to $\pm R$, where R is the radius of the focused laser spot. A larger spot size gives a larger field-of-view but a lower angular resolution. The radius of the focused spot can at the most be equal to half of the radius of the quadrant active area.

In the case of a 2D-array-based sensor, each active element in the array is identified by a unique azimuth and elevation angle. The focused spot at any time covers more than one active element and the angular error in azimuth and elevation is computed from Equations 14.10 and 14.11. Figure 14.46 explains the concept.

$$X = n\theta + \frac{(B + C) - (A + D)}{A + B + C + D} \quad (14.10)$$

$$Y = m\theta + \frac{(A + B) - (C + D)}{A + B + C + D} \quad (14.11)$$

where θ is the proportional field-of-view of each miniature quadrant and n and m are constants depending upon the illuminating quadrant.

The major parameters of interest of laser-guided munitions include: (1) sensitivity; (2) field-of-view; (3) PRF code compatibility; and (4) response linearity. In addition to these, immunity to false codes and response to desired code in the presence of false code are also important in assessing the performance efficacy of laser-guided munitions.

Sensitivity is the minimum value of laser power density impinging on the seeker cross-section in a plane orthogonal to the optical axis of the seeker head to which it can respond satisfactorily. It is a characteristic of the seeker front-end optics and photosensor. The sensitivity of the seeker determines the

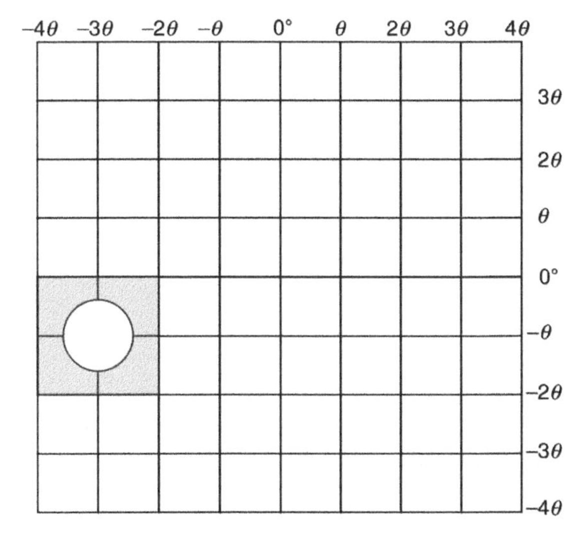

Figure 14.46 Operating principle of 2D-array-based position sensor.

maximum guidance range for known values of laser target designator parameters, target reflectivity, height of laser target designator (in the case of airborne laser designator) and laser seeker above sea surface and visibility condition.

Field-of-view determines the probability of the weapon finding itself within the laser basket at the maximum guidance range. Laser-guided munitions using a seeker head with a larger field-of-view would have a higher probability of finding themselves in the laser basket and subsequently hit the intended target.

PRF code compatibility is the primary requirement for the weapon to function and refers to the PRF code of the received laser radiation matching (within a certain tolerance) the PRF code programmed in the guidance unit before the start of mission. It is usually expressed as the time interval between two successive laser pulses in milliseconds up to a third decimal place. The two codes are considered compatible if the difference in time periods of the two codes is less than a certain specified value.

Response linearity predominantly determines the circular error probability (CEP).

Immunity to false PRF codes and the capability to stay locked to the desired code in the presence of false PRF codes enhances the probability of a target hit. The former test is performed by irradiating the seeker head with a PRF code different from the programmed PRF code, and the latter by irradiating the seeker head simultaneously with radiations of correct and false PRF codes.

14.4.2.2 IR-guided Missiles: Operational Basics

IR-guided missiles are of two types: those employing non-imaging and imaging seeker heads. The majority of the present-day surface-to-air and air-to-air IR-guided missiles are of the non-imaging type. Non-imaging IR-guided missiles make use of the IR emission corresponding to the thermal signatures of the exhaust and the mainframe of the target aircraft to home on it. Emission in the 3–5 μm and 8–12 μm bands is characteristic of the electromagnetic emission from a jet exhaust and mainframe of the aircraft. Figure 14.47 depicts the spectral profile of IR emission from a typical target aircraft. The spectral content of IR emission as received by the IR seeker head is the superposition of the spectral emission of the aircraft on the transmission characteristics of the atmosphere (Figure 14.48) . Figure 14.49 shows a

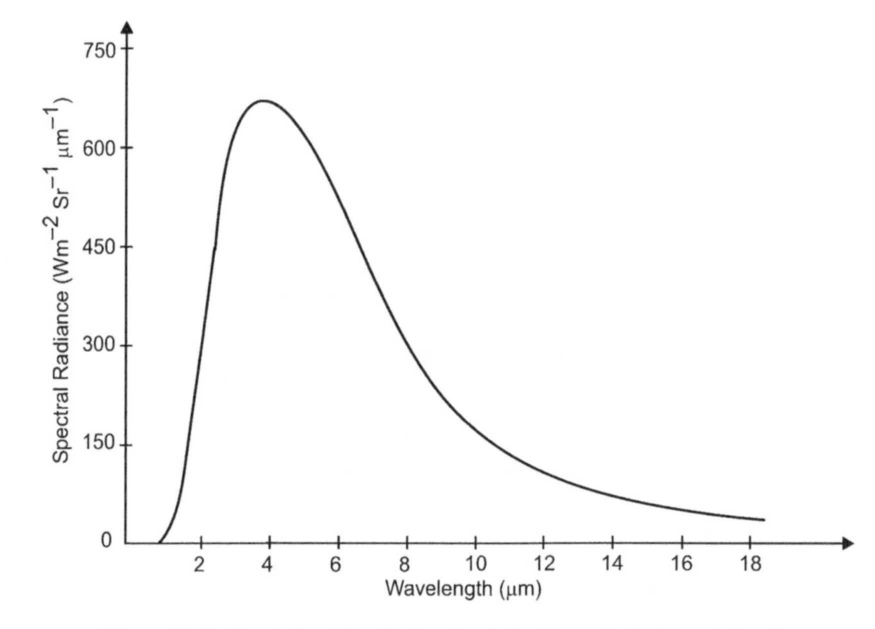

Figure 14.47 Spectral profile of infrared emission from a typical target aircraft.

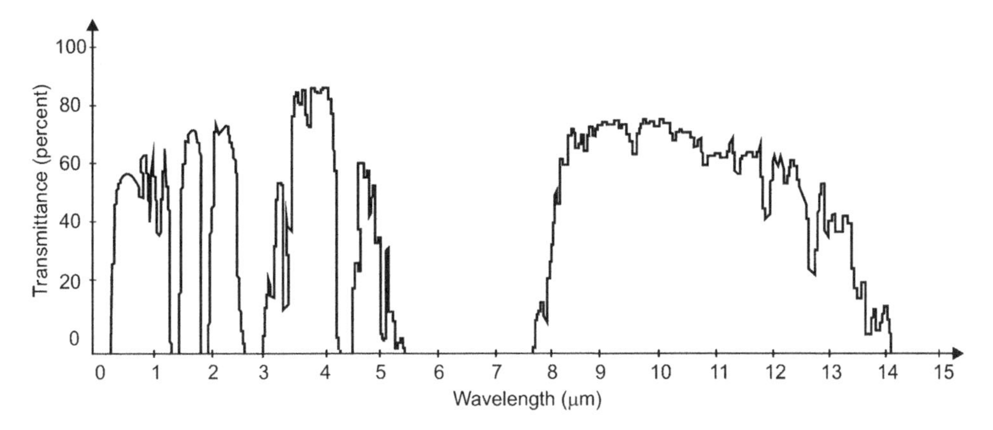

Figure 14.48 Transmission characteristics of atmosphere.

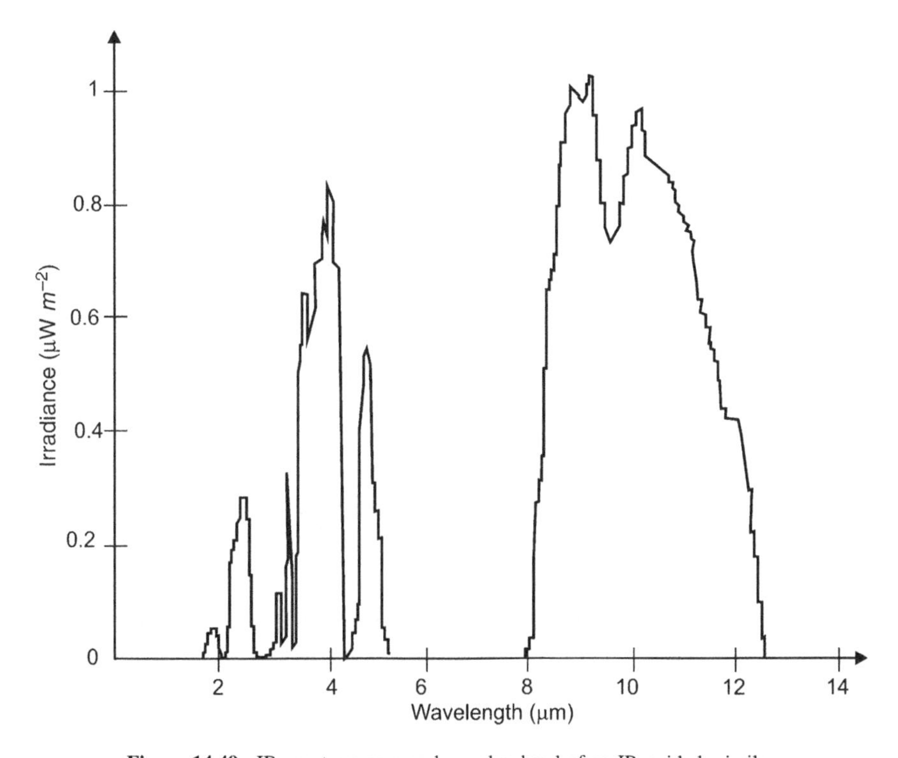

Figure 14.49 IR spectrum as seen by seeker head of an IR-guided missile.

typical IR spectrum that would be seen by the IR seeker head. This wavelength signature is judiciously used in the guidance of air-to-air and surface-to-air IR-guided missiles.

The IR-guided missiles developed in the 1970s and 1980s used single-colour IR seekers employing the 3–5 μm band. The MAGIC series of IR-guided air-to-air missiles from France and R-73 series of IR-guided air-to-air missiles from Russia are examples. State-of-the-art IR-guided missiles use two-colour seekers that employ both the 3–5 μm and 8–12 μm bands to offer improved false alarm rejection and immunity to deception by flares. The Python from Israel and RVVAE from Russia are examples of

IR-guided missiles using two-colour seeker heads. Another emerging trend is the use of imaging IR seekers even in surface-to-air and air-to-air IR-guided missiles, although their use is primarily in short-range anti-tank IR-guided missiles.

Both surface-to-air and air-to-air IR-guided missiles receive the IR signatures of the target in the presence of background radiation from the sky and the IR signatures of flares, if any, deployed by the target aircraft platform. The seeker head should be able to discriminate between the IR signatures of the background and flares from those of the target.

Important parameters of IR-guided missiles include spectral matching of received IR signatures with those of the target as known to the seeker, the response of the seeker head to target signatures in the presence of static IR background noise, immunity to deception by flares and field-of-view.

14.5 Laser Communication

Communication technology has experienced continual development to higher and higher carrier frequencies, starting from a few hundred kilohertz during the time of Marconi to several hundred terahertz since lasers were employed in fibre systems. The main driving force was that the usable bandwidth and the consequently the transmission capacity increased in direct proportion to the carrier frequency. Another asset comes into play in free-space point-to-point links. The minimum divergence obtainable with a freely propagating beam of electromagnetic waves is proportional to the wavelength. The jump from microwaves to light waves therefore means a reduction in beam width by orders of magnitude, even if we used transmit antennas of a much smaller diameter. The reduced beam width not only implies increased intensity at the receiver site but also reduced crosstalk between closely operating links and less chance for eavesdropping.

Space communication, as employed in satellite-to-satellite links, is traditionally performed using microwaves. For more than 25 years however, laser systems have been investigated as an alternative. It is hoped that mass, power consumption and size of an optical transceiver module will be smaller than that of a microwave transceiver. Fuel consumption for satellite attitude control when quickly re-directing antennas should also be less for optical antennas. On the other hand, a new set of problems would need to be addressed in connection with the extreme requirements for pointing, acquiring and tracking the narrow-width laser beams.

14.5.1 Advantages and Limitations

Optical communication is a technology that uses light propagating in the communication medium to transmit data for telecommunications or computer networking. The communication media can be free space (which means air), outer space, vacuum, water or an optical transmission line such as optical fibre cable.

The key advantages of using optical communications, whether fibre optic, free space or underwater, include high achievable data transmission rates, low bit error rates, immunity to electromagnetic interference, full duplex operation, higher communication security and no necessity for the Fresnel zone. The light beam can also be very narrow, making it hard to intercept.

Key disadvantages include beam dispersion, particularly in free-space and underwater communication applications, signal attenuation due to atmospheric absorption and adverse weather conditions and scintillation and signal swamping when the sun goes exactly behind the transmitter. These factors cause an attenuated receiver signal and lead to higher bit error ratio (BER). To overcome these issues, designers have found some solutions such as multibeam or multipath architectures, which use more than one sender and more than one receiver. Some state-of-the-art devices also have a larger fade margin (extra power, reserved for rain, smog and fog). To keep an eye-safe environment, good free-space optical communication systems have a limited laser power density and support laser classes 1 or 1 M. Attenuation due to atmospheric conditions, which are exponential in nature, limits the practical range of free-space optical (FSO) communication devices to several kilometres. In the following sections we describe free-space and fibre-optic communication.

14.5.2 *Free-space Communication*

Free-space optical communication is an optical communication technology that makes use of light as the carrier to transmit intelligence. Free space here means air, outer space, vacuum or something similar. The technology is useful where physical connections are impractical due to high costs or other considerations. The other advantages of optical communication, outlined in the previous paragraph, are all relevant. Figure 14.50 shows the basic block schematic arrangement of a free-space optical communication link.

Practical free-space point-to-point optical links are usually implemented by using infrared laser light, although low-data-rate communication over short distances is possible using LEDs. The maximum range for terrestrial links is of the order 2–3 km, but the stability and quality of the link is highly dependent on atmospheric factors such as rain, fog, dust and heat. Data transmission rates approaching 1 Gbps have been demonstrated in the case of free-space optical communication, but not for terrestrial links.

Preferred wavelengths for free-space optical communication are 850 nm and 1550 nm. Operation in the 3–5 μm and 8–14 μm bands have also been used due to the excellent atmospheric transmission characteristics in these bands. The selection of optimum wavelength for free-space communication depends upon many factors which include required transmission distance, eye safety considerations, availability of components and cost. Recent studies have revealed that operation in the 3–5 μm and 8–14 μm bands does not offer a significant advantage compared to the 850 nm and 1550 nm bands to counteract the scattering losses. The availability of sources and detectors is also limited in the mid-infrared and far-infrared bands. Another advantage with 1550 nm comes along with its eye safety feature. Regulatory agencies allow approximately 100 times higher laser power level for 1550 nm compared to 850 nm. In general, the choice of a specific wavelength is not so important as long as it is not strongly absorbed in the atmosphere.

In outer space, the communication range of free-space optical communication is currently of the order of several thousand kilometres. This can be extended to cover interplanetary distances of millions of kilometres, using optical telescopes as beam expanders. The use of optical communication technology

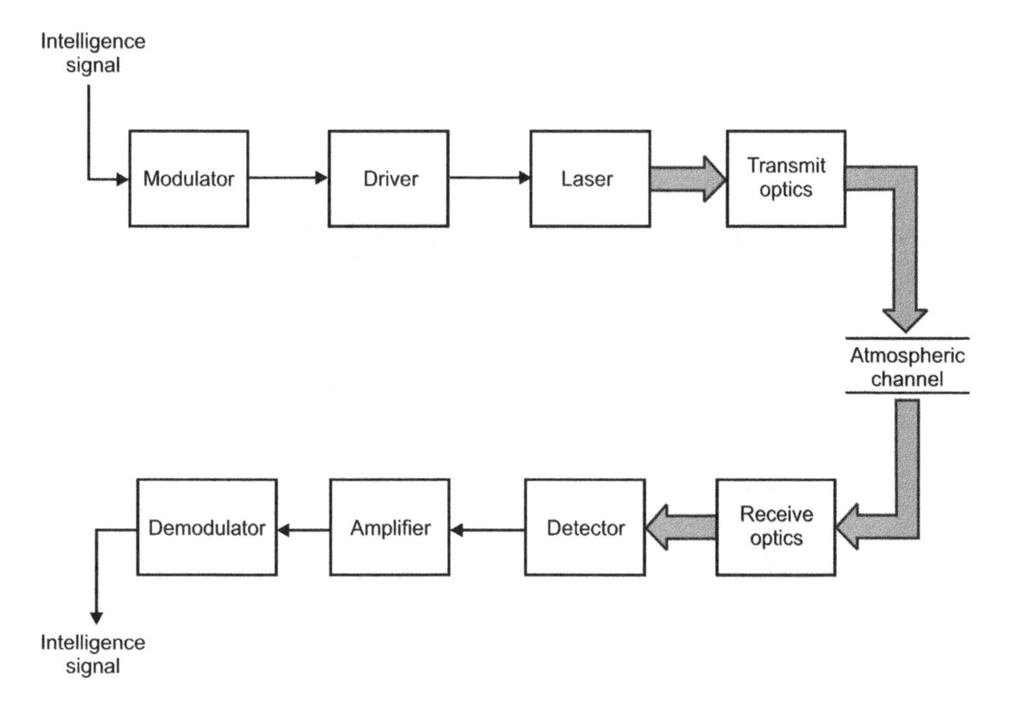

Figure 14.50 Block schematic arrangement of free-space optical communication link.

involving detection and emission of laser light by space probes has been investigated several times in the past. A two-way distance record for communication was set by the Mercury laser altimeter instrument aboard the Messenger (Mercury surface, space environment, geochemistry and ranging) spacecraft. This infrared diode-pumped neodymium laser, designed as a laser altimeter for a Mercury orbit mission and known as the Mercury laser altimeter (MLA), set a record of two-way communication across a distance of 24 million km (i.e. 24 Gm) as the craft neared Earth on a fly-by in May 2005.

The space-based free-space optical communication concept has been around for many years and significant advances have recently been made for the concept to fructify in both civilian and government-funded non-classified and classified applications. The primary market of free-space optical communication today is that of inter-satellite links (ISLs). There is scope for providing a space–Earth optical communication link despite the discouraging atmosphere-related issues. For example, research to develop a satellite–submarine optical communication link has been funded, which mainly interests military strategists. In such a link, the problems encountered are not only from atmospheric issues but also from the propagation of the laser through a turbulent ocean.

Returning to the primary application of free-space optical communication links in today's world, that is, inter-satellite links, it may be mentioned here that inter-satellite communications is mainly used for networking of a satellite constellation. The involved data rates could vary from hundreds of megabits per seconds to several gigabits per second. These inter-satellite links are in use for all types of satellite orbits including low Earth orbits (LEOs), medium Earth orbits (MEOs), geosynchronous Earth orbits (GEOs) and even highly elliptical orbits (HEOs). Although there are satellite constellations currently operational that employ RF inter-satellite links, examples being the Iridium satellite system and NASA's TDRSS (tracking and data relay satellite system), the future definitely belongs to optical ISLs. This is supported by the fact that most of the commercial satellite constellations being announced now will be using optical ISLs. The SILEX optical communication system is another example; its payload was embarked on the European Space Agency's ARTEMIS (advanced relay and technology mission satellite) spacecraft and also on the French Earth observation satellite SPOT-4. It uses GaAlAs laser diodes as the source and is used to transmit data at 50 Mbps from a low Earth orbit to a geostationary orbit.

TSAT (transformational satellite system) is another example of a contemporary satellite constellation (Figure 14.51) employing laser inter-satellite links. The system is designed to provide a protected, secure Internet-like communication system that integrates space, air, ground and sea networks. The TSAT program is composed of three segments and a systems engineering and integration function. The *space segment* will consist of five satellites in geosynchronous orbits interconnected by high-data-rate laser

Figure 14.51 TSAT satellite constellation.

cross-links. TSAT will use Internet-like technology to connect war fighters all over the world in a global information network with unprecedented carrying capacity, accessibility, reliability and immunity to jamming, eavesdropping and nuclear effects. It is the backbone of 21st century net-centric warfare and is projected to revolutionize military communications.

14.5.3 Fibre-optic Communication

Some of the key limitations of free-space optical communication encountered due to atmospheric propagation issues are overcome in the case of fibre-optic communication. In practice, laser-based communication is dominated today by fibre-optic transmission. Earlier, the lifetimes of semiconductor diode lasers and the fibre losses were too high to make laser-based fibre-optic communication an attractive alternative to other forms of communication. With advances in both semiconductor diode laser and fibre technologies, these shortcomings have been overcome. State-of-the-art semiconductor diode lasers have lifetimes of greater than 10^7 hours and fibre loss is as small as a small fraction of 1 dB km^{-1}. Today, fibre-optic communication links are a reality for both intra-city and trunk telephone lines, video data links and computer-to-computer communications.

A semiconductor diode laser is the natural choice for fibre-optic communication as they have a suitably small size and configuration for efficient coupling into the small diameter core of an optical fibre cable. Semiconductor diode lasers operating at CW power levels of a few milliwatts are suitable for fibre-optic communication. As described in Chapter 5, these lasers can be easily modulated by drive current modulation up to frequencies in the gigahertz range.

Laser wavelengths in use for fibre-optic communication are 0.85 μm, 1.3 μm and 1.55 μm. The first practical fibre-optic communication systems employed the wavelength 0.85 μm as it matched the available AlGaAs lasers. With advances in fibre technology and the opening up of lower-loss windows first at 1.3 μm and then at 1.55 μm, these became the preferred wavelengths for long-distance high-performance systems. Corresponding semiconductor diode laser and detector types for these operational wavelengths are AlGaAs and silicon for 0.85 μm, InGaAsP and InGaAs or germanium for 1.3 μm and InGaAsP and InGaAs for 1.55 μm. Typical fibre losses at these wavelengths are 2 dB km^{-1} at 0.85 μm, 0.5 dB km^{-1} at 1.3 μm and 0.2 dB km^{-1} at 1.55 μm.

Fibres with bandwidth-distance products approaching 3000 MHz km are available. Semiconductor diode lasers with 10 mW power level, modulation rate approaching 10 GHz and lifetime of 10^7 hours are also available. LEDs providing 0.1 mW into the fibre with a modulation rate of 200 MHz and lifetime of the order of 10^6–10^7 are also available. PIN-type photodiodes with responsivity and NEP figures of 0.5 A W^{-1} and 10^{-12} W $Hz^{-0.5}$, respectively, are also used. More sensitive APDs with responsivity and NEP figures of 80–100 A W^{-1} and 10^{-14} W $Hz^{-0.5}$ are commercially available for the purpose. Couplers and splices with insertion loss in the range of 0.1–0.5 dB are available for use. Splices are usually added to the link to allow fibre cable repair in case of need.

Figure 14.52 shows a typical fibre-optic communication link. The laser is pulse-code-modulated with intelligence to be transmitted through drive current modulation and is coupled into the fibre. On the receiver side, laser light is detected and the intelligence signal is recovered. Optical amplifiers are used to reinforce the signal strength every few kilometres of fibre cable length to counter signal degradation caused by various loss mechanisms such as absorption, scattering and modal and chromatic dispersion. To summarize, the process of fibre-optic communication involves generating the optical signal modulated with intelligence to be transmitted using a transmitter, relaying the signal along the fibre ensuring that the signal does not become too distorted or weak, receiving the optical signal, and converting it back into an electrical signal representing the original intelligence signal.

We have seen four generations of fibre-optic communication and are currently in the fifth generation. The *first-generation* system operated at 0.85 μm wavelength, at a bit rate of 45 Mbps, with repeater spacing of up to 10 km. In April 1977, General Telephone and Electronics sent the first live telephone traffic using fibre optics at a 6 Mbps data rate in Long Beach, California. The *second generation* of fibre-optic communication operated at 1.3 μm, and used InGaAsP semiconductor lasers. These fibre-optic systems were initially limited by dispersion of multimode fibres. The advent of single-mode fibres in

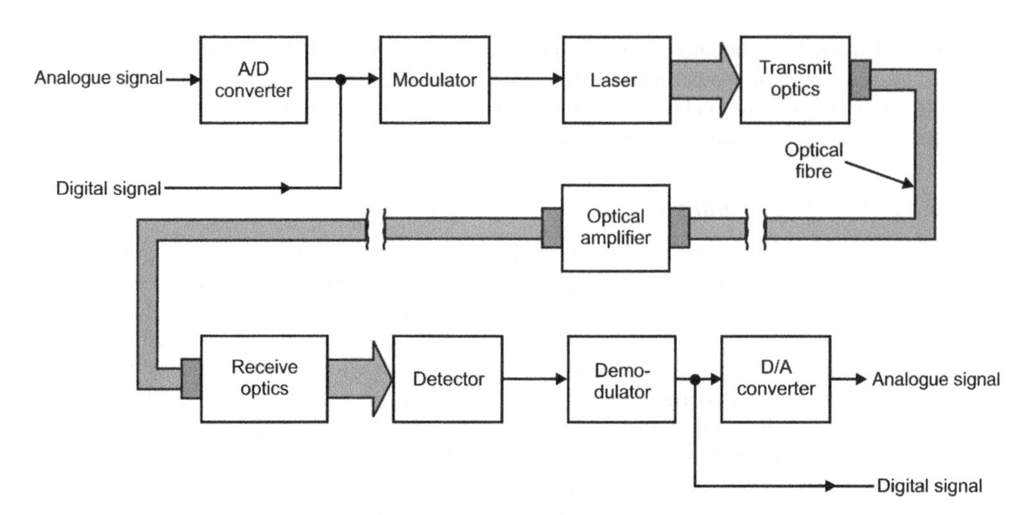

Figure 14.52 Basic block schematic of a fibre-optic communication link.

1981 significantly improved system performance. *Third-generation* fibre-optic systems operated at 1.55 μm. The difficulty faced earlier in terms of pulse spreading at 1.55 μm was largely overcome by using dispersion-shifted fibres designed to have minimal dispersion at 1.55 μm or by limiting the laser spectrum to a single longitudinal mode. The features of third-generation systems allowed commercial fibre-optic systems to operate at 2.5 Gbps with repeater spacing in excess of 100 km. The *fourth generation* of fibre-optic communication systems used optical amplification to reduce the need for repeaters. It also explored the use of wavelength-division multiplexing to increase data capacity. These features brought revolutionary improvements to the performance of fibre-optic systems. In fact, since the beginning of 1992, the data rate has doubled every six months until it reached a figure of 10 Tbps in 2000. In 2006 a bit-rate of 14 Tbps was reached over a single 160 km line using optical amplifiers.

In the *fifth generation* of fibre-optic communications systems, the focus is on extending the wavelength range over which a WDM system can operate. The conventional wavelength window, known as the C-band, covers the wavelength range 1.53–1.57 μm. The dry fibre has promised an extension of that range to 1.30–1.65 μm. Other features of fifth-generation fibre-optic systems include the concept of optical solitons, which involves the use of pulses of a specific shape that helps them preserve their shape by counteracting the effects of dispersion with the non-linear effects of the fibre.

Example 14.6

A laser power at 1550 nm of 1.0 mW coupled into a 30 km fibre length produces a power of 1.0 μW at the other end of the fibre. Determine fibre loss in dB km^{-1}. If this fibre is used in a fibre-optic communication link and the detector used is able to satisfactorily detect a 40 dBm signal, determine how long the fibre-optic link can be without a repeater?

Solution

- Power loss ratio in the fibre = $1/1.0 \times 10^{-3} = 1000 = 10 \log 1000$ dB = 30 dB
- Length of fibre = 30 km
- Therefore, fibre loss = 30/30 = 1.0 dB km^{-1}
- Minimum detectable signal power = 40 dBm
- Input laser power = 1.0 mW = 0 dBm
- Therefore, acceptable fibre loss = 40 dB
- Fibre loss = 1.0 dB km^{-1}
- Therefore, fibre length without repeater = 40/1 = 40 km.

14.6 Summary

- The most common tactical battlefield applications of lasers and optoelectronic devices are as laser rangefinders and laser target designators for munitions guidance.
- Military applications of lasers and related electro-optic devices are due to technological advances in optics, optoelectronics and electronics. Laser devices are indispensable in modern warfare due to their rugged, reliable, compact and efficient nature.
- Laser-based proximity sensors, gap-measuring devices and obstacle avoidance systems, laser bathymetry and laser tracking are some examples of use of the principle of laser range finding.
- Lasers and optoelectronics devices and systems are finding widespread use in low-intensity conflict (LIC) operations, due to their properties of near-zero collateral damage, speed of light delivery and potential for building non-lethal weapons.
- Some of the well-established laser devices in LIC applications include laser dazzlers for close combat operations, mob/riot control and protection of critical infrastructures from aerial threats; lidar sensors for detection of chemical, biological and explosive agents; femtosecond lasers for imaging of concealed weapons; and lasers for sniper and gun fire location identification.
- A very common application of a laser device when attached to a firearm is as an aiming device. Laser aiming modules are usually configured around low-power semiconductor diode lasers emitting in the visible or infrared.
- A laser rangefinder is used in military applications to determine distance to the intended target. It is used both as a standalone device as well as an integral part of a fire control system of main battle tanks and other armoured fighting vehicles. Commonly used techniques for determination of range include time-of-flight principle, triangulation method and multiple frequency phase shift method.
- A laser target designator for munitions guidance applications is a high-repetition-rate Q-switched laser source whose pulse repetition frequency is coded in order to allow only a specific target designator and laser seeker combination to work in unison for the intended mission. Laser target designators for handheld applications as well as for land-based and aerial platforms are also available.
- A laser proximity sensor is a type of laser rangefinder designed to accurately measure relatively smaller distances to targets, for applications such as observation and surveillance and fire control.
- Laser bathymetry is a technique of measuring depths of water bodies, usually of relatively shallow coastal waters, using a scanning pulsed laser beam. Measurement is usually performed from an airborne platform, which gives it the name airborne laser bathymetry (ALB).
- Laser bathymetry makes use of the transmissive and reflective properties of water and the bottom of the water body to measure the depth.
- Laser radar, also called ladar, uses a laser beam instead of microwaves. The higher operating frequency means higher operating bandwidth, greater time or range resolution, enhanced angular resolution and immunity to jamming. Higher frequencies associated with laser radars permit detection of smaller objects. This is made possible by the fact that laser radar output wavelengths are much smaller than the small-sized practical objects we can think of.
- A common application of ladar concept is in ladar seekers, used mainly in conjunction with other guidance systems on strategic payloads for intended target discrimination from advanced decoys and aim point selection.
- A forward-looking infrared (FLIR) sensor makes use of thermal radiation emitted by the target or the scene of interest to generate its image. It essentially comprises a front-end optical system, a 2D array of infrared detectors and image processing circuitry to produce the output in desired format.
- Different guidance techniques discussed in this section include beam rider guidance, command guidance, homing guidance and navigation guidance.
- The concept of beam riding guidance of munitions is based on radar or a laser beam constantly pointed towards the target throughout the flight time of the munitions. The missile is launched and flies into the radar or laser beam. From this stage onwards, the missile attempts to keep itself inside the beam, while the aiming station always keeps the beam pointed at the target.

- In the case of command guidance, the missile is commanded on an intercept course with the target. This is achieved by using two separate radars to continuously track the target and the missile. Tracking data from these radars is fed to a computer that computes the trajectories of the two vehicles. The computer in turn sends appropriate command signals over a radio link to the missile. A sensor onboard the missile decodes the commands and operates the control surfaces of the missile to adjust its course in order to intercept the target in flight.
- Homing guidance is divided into semi-active, active, passive and track-via-missile homing, also known as retransmission homing. In the case of semi-active homing guidance, the target is illuminated by an external source which could be radar or laser. For active homing guidance, the source of target designation is also onboard the weapon with the result that this methodology does not require an external source. Passive homing guidance makes use of some form of energy emitted by the target which is intercepted by the missile seeker. In retransmission homing guidance, the target is illuminated by an external radar and the reflected energy is intercepted by the missile sensor.
- Navigation guidance is divided into inertial, ranging, celestial and geophysical navigation. In the case of inertial navigation guidance, the vehicle uses onboard sensors to determine its motion and acceleration with the help of gyroscopes and accelerometers. Ranging navigation depends on external signals for guidance, which are usually provided by radio beacons. Celestial navigation, one of the oldest types of navigation, uses the positions of the stars to determine location, especially latitude, on the surface of the Earth. Geophysical navigation guidance depends on the measurements made by compasses and magnetometers of the Earth's magnetic field as well as gravitometers to measure the Earth's gravitational field. This technique has not found much application in missile guidance.
- A ring laser gyroscope is primarily a rate sensor, which can be used to measure rotation by integrating the rate information. It consists of two counter-propagating laser beams over the same path. It operates on the principle of the Sagnac effect, which describes how the null points of the internal standing wave pattern produced by the counter-propagating laser modes shift in response to angular rotation. Important parameters that describe a ring laser gyroscope's performance include bias error, bias drift and instability, scale factor non-linearity and angular random walk.
- The operational principle of a fibre-optic gyroscope is also based on Sagnac effect. It consists of a long coil of optical fibre (typically a few kilometres in length) and makes use of the interference of light to detect mechanical rotation.
- Optical communication technology uses light propagating in free space to transmit data for telecommunications or computer networking. Free-space optical communication and fibre-optic communication are widely used modes of optical communication. The key advantages of optical communication, whether fibre optic, free space or underwater, include high achievable data transmission rates, low bit error rates, immunity to electromagnetic interference, full duplex operation, higher communication security and no necessity for the Fresnel zone. Key disadvantages include beam dispersion, particularly in free space and underwater communication applications, signal attenuation due to atmospheric absorption and adverse weather conditions, scintillation and signal swamping when the sun goes exactly behind the transmitter.

Review Questions

14.1. List common military applications of lasers and briefly describe them in terms of key involved technologies, advantages and limitations if any.

14.2. Name different types of laser rangefinders. Briefly describe each of them.

14.3. What is the primary function of a laser target designator in a military application? What type of laser is generally used to build a laser target designator used for munitions guidance? Name major performance specifications of such a system along with typical values and briefly describe the significance of each parameter.

14.4. Describe the primary function and operational principle of the following laser based technologies/systems:
a. laser proximity fuse
b. laser bathymetry
c. LADAR seeker.

14.5. What is a FLIR system? What are the key advantages and limitations of a FLIR system? How does it compare with a pushbroom scanning system?

14.6. Name different guidance techniques used for munitions guidance. Briefly describe each of them with particular reference to their suitability for a given application.

14.7. With the help of a schematic diagram, describe the principle of operation of a semi-active laser guidance technique. Name major performance parameters that decide the guidance range and briefly describe their significance.

14.8. Draw the spectral emission profiles of a typical target aircraft and that seen by an infrared-guided missile targeting it. Describe the principle of operation of an infrared-guided missile.

14.9. What do you understand by single-colour and dual-colour infrared-guided missiles? Compare the performance of the two types with particular reference to countermeasures used by target aircraft.

14.10. Name different types of laser-based navigation sensors. Briefly describe the operational principle of a ring laser gyroscope sensor. What are the capabilities of state-of-the-art RLG sensors for inertial navigation sensors?

14.11. What are the advantages and limitations of free-space optical communication and microwave communication? With the help of a block schematic diagram, briefly describe operation of a free-space laser communication link.

14.12. Name the key components of a fibre-optic communication link. Briefly describe the present-day status of the performance level of these components and their technologies.

Problems

14.1. A laser rangerfinder operating on time-of-flight principle measures target range as 4.8 km. If the operating clock frequency is 30 MHz, determine the number of clock cycles which the rangefinder counter will record for this measurement. Also determine the range of distance in which the true range is likely to be.
[960 cycles, 4795–4805 m]

14.2. In a range measurement experiment using the triangulation method, the angles that the lines of sight between target and transmitter and target and receiver make with the line joining transmitter and receiver are measured as 60° and 85°, respectively. Determine the separation between transmitter and receiver if the measured distance to the target is 34.75 m.
[20 m]

14.3. An FM-CW type of laser rangefinder uses ramp frequency of 2.5 kHz. While measuring a designated target range with this rangefinder, beat frequency is measured to be 500 kHz. If the modulation bandwidth of the system is 3.75 MHz, find the minimum and maximum range-measuring capability of the system.
[20 m, 30 km]

14.4. Compare the scale factor of two ring laser gyroscopes, both using a He-Ne ring laser operating at 0.633 μm. The first gyroscope, RLG1, uses an equilateral triangular laser cavity with each arm of 10 cm optical path length. The second gyroscope, RLG2, uses a square ring laser geometry with

each arm having the same optical path length of 10 cm. Assume all other parameters for the two gyroscopes to be the same.
[RLG2 has a scale factor 3% higher than the scale factor of RLG1]

14.5. When coupled into a 60 km fibre length, 100 μW of laser power at 1550 nm produces a 0.1 μW signal at the other end. Determine fibre loss in dB km^{-1}. If this fibre is used in a fibre-optic communication link and the detector is able to satisfactorily detect a 50 dBm signal, what length can the fibre-optic link be without a repeater?
[0.5 dB km^{-1}, 80 km]

Self-evaluation Exercise

Multiple-choice Questions

14.1. A time-of-flight laser rangefinder uses 20 ns laser pulses at a repetition rate of 5.0 pps. The radial range resolution capability of this rangefinder is
a. 3 m
b. 6 m
c. 1.5 m
d. 12 m.

14.2. If the above rangefinder measured time between the leading edges of the start pulse and stop pulse as 25 μs, the measured target range is
a. 7.5 km
b. 3.75 km
c. 15 km
d. none of these.

14.3. Laser rangefinders based on the triangulation technique are better suited to measurement of
a. large target ranges with higher accuracy
b. large target ranges with relatively poorer accuracy
c. relatively small target ranges with higher accuracy
d. none of these.

14.4. It is desired to measure target ranges up to a maximum distance of 150 km with accuracy of the order of 1.0 m. The preferred laser rangefinder technology is
a. FM-CW laser rangefinder
b. time-of-flight laser rangefinder
c. triangulation laser rangefinder
d. none of these.

14.5. An acoustic bathymetry system is used to measure depths not exceeding 30 m. Maximum swath width in this case is
a. 30 m
b. 60 m
c. 90 m
d. 120 m.

14.6. The scale factor of a ring laser gyroscope for a given laser wavelength with area encircled by optical path length A and optical path length L is proportional to
a. L/A
b. L^2/A
c. $1/L^2$
d. A/L.

14.7. The operation of a ring laser gyroscope is based on the
a. Sagnac effect
b. Faraday effect
c. Fresnel effect
d. none of these.

14.8. The guidance technique that uses separate land-based radars for tracking the target and the missile is known as
a. command guidance
b. homing guidance
c. inertial navigation guidance
d. GPS guidance.

14.9. Which of the following infrared-guidance technologies performs better in the presence of aircraft-launched countermeasures such as chaff and flares?
a. single-colour infrared-guided missile
b. dual-colour infrared-guided missile
c. laser-guided missiles
d. None of these.

14.10. If you were designing a 1000 km long inter-satellite communication link with data rate approaching Gbps, which of the following technologies would you opt for?
a. Free-space laser communication link
b. microwave communication link
c. HF communication link
d. none of these.

14.11. The minimum rotation rate measurable by a state-of-the-art RLG designed for inertial navigation sensors of long-range strategic missiles is of the order of
a. $1\,\mathrm{deg\,h^{-1}}$
b. $10\,\mathrm{deg\,h^{-1}}$
c. $0.1\,\mathrm{deg\,h^{-1}}$
d. $0.0001\,\mathrm{deg\,h^{-1}}$.

14.12. The major advantages of an optical communication link include
a. high data transmission rate, immunity to jamming, low bit error rate
b. high data transmission rate, long operational range, immunity to jamming
c. long operational range, low bit error rate and communication security
d. none of these.

Answers

1. (a) 2. (b) 3. (c) 4. (a) 5. (b) 6. (d) 7. (a) 8. (a) 9. (b) 10. (a) 11. (d) 12. (a)

Bibliography

1. *Understanding Lasers: An Entry Level Guide*, 2008 by Jeff Hecht, IEEE Press.
2. *Solid State Lasers and Applications*, 2012 by Alphan Sennaroglu, CRC Press.
3. *Military Laser Technology for Defence*, 2012 by Alastair D. McAulay, Wiley-Interscience.
4. *Electro-Optics Handbook*, 2000 by Ronald Waynant and Marwood Ediger, McGraw-Hill, Inc..
5. *Fiber Optic Communications*, 2004 by Harold Kolimbiris, Prentice-Hall.
6. *Optical Fiber Communications*, 2010 by Gerd Keiser, McGraw-Hill, Inc..
7. *Laser Communications in Space*, 1995 by Stephen G. Lambert and William L. Casey, Artech House.
8. *Advances in Gyroscope Technologies*, 2010 by Mario Armenise, Caterina Ciminelli, Francesco Dell'olio and Vittorio M. N. Passaro, Springer.
9. *Fiber Optic Sensors*, 2008 by Shizhuo Yin, Paul B. Ruffin and Francis T. S. Yu, CRC Press.

15

Military Applications: Directed-energy Laser Systems

15.1 Introduction

Chapter 14 focused on the relatively low-power/energy military applications of lasers, covering instrumentation and sensors such as laser rangefinders, laser target designators, electro-optical guidance techniques, optronic sensors used for navigation and free-space and fibre-optic laser communication. This chapter describes optical and electro-optical devices for the detection of chemical, biological and explosive agents, and provides information on directed-energy applications of lasers and optoelectronics devices and systems. These include the use of laser technology for low-intensity conflict applications, electro-optic and infrared countermeasures and directed-energy laser weapons.

15.2 Laser Technology for Low-intensity Conflict (LIC) Applications

The present-day warfare scenario is that of low-intensity conflict (LIC), and is likely to remain so in the foreseeable future. Data suggest that more than 75% of the armed conflicts since World War II have been of the low-intensity variety. *Low-intensity conflict operation* is a military term used for deployment and use of troops and/or assets in situations other than conventional war. Compared to a war, armed forces engaged in low-intensity conflict operate with fewer soldiers, a reduced range of tactical equipment and limited scope to operate in a military manner. The use of artillery is avoided in the case of conflicts in urban territories, and use of air power is often restricted to surveillance and transportation of personnel and equipment.

Low-intensity conflicts pose an alarming threat to national security and are an area of concern for the whole of the international community today. LIC extends from combating illegal drug trafficking to protection of critical infrastructures; from handling counter-insurgency and anti-terrorist operations to detection of chemical and biological warfare agents; and from detection and identification of explosive agents (including improvised explosive devices frequently used by terrorist and anti-national elements) to detection of concealed weapons. In an urban terrorism scenario, even sanitisation of strategic areas from an attack by sharp shooters or constant observation and surveillance of suspect elements has become very important.

15.2.1 Importance of Laser Technology in LIC Applications

Laser and optoelectronics technologies play an important role in handling low-intensity conflict situations. The key advantage of the use of laser technology in such applications mainly stems from

Lasers and Optoelectronics: Fundamentals, Devices and Applications, First Edition. Anil K. Maini.

its near-zero collateral damage, speed of light delivery and potential for building non-lethal weapons. Some of the well-established laser devices in LIC applications include laser dazzlers for close combat operations, mob/riot control and protection of critical infrastructures from aerial threats; lidar sensors for detection of chemical, biological and explosive agents; femtosecond lasers for imaging of concealed weapons; and lasers for sniper and gun fire location identification.

Use of laser vibrometry and electron speckle interferometry techniques for detection of buried mines and high-power lasers for disposal of unexploded ordnances are emerging applications of laser technology for homeland security. In the subsequent sections are briefly described both established and emerging applications of laser and optoelectronics technologies for use in homeland security in terms of salient features, potential usage and international developments.

15.2.2 Detection of Chemical and Biological Warfare Agents

The importance of efficient and reliable technologies for stand-off detection and identification of chemical, biological and explosive agents has become increasingly important today. Chemical and biological weapons are relatively inexpensive to produce and are capable of unleashing a devastating effect as a terrorist weapon. The international community has already witnessed the devastating effects these weapons are capable of causing on more than one occasion. Millions have been affected by the destruction caused by these weapons of mass destruction in the past. The sarin attack by the Aum Shinrikyo cult in 1995 in a Tokyo subway, hydrogen cyanide, mustard gas attack by Iraq in the Anfal campaign against the Kurds (most notably in Halabja massacre in 1988) and the international dispersal of anthrax spores in the US in 2001 are some examples of chemical and biological weapons attack. Use of improvised explosive devices by terrorists and anti-national elements killing soldiers and civilians alike has become a routine affair. These incidents have highlighted the need for stand-off detection and identification of chemical, biological and explosive agents. In the following sections we discuss some laser-based methodologies/techniques that are either in use or being explored for the purpose.

15.2.2.1 Detection of Chemical Warfare Agents

Due to their acute toxicity, rapidity of action, indetectability by human senses and economic viability, nerve agents such as organophosphonate compounds and blister agents such as mustard compounds are two potentially very harmful classes of chemical warfare agents. Nerve agents are the organophosphorus esters, which initially stimulate and then paralyze certain nerve transmissions by interfering with the cholinesterase enzyme. Common chemical warfare agents belonging to this category include tabun, sarin, soman, VX, dichlorvos and malathion. Victims of chemical warfare agent attack are depicted in Figure 15.1.

Lidar-based detection is the most practically realizable technique for stand-off detection of chemical and biological warfare agents. Commonly exploited physical phenomena forming the basis of chemical and biological agent detection include elastic backscattering, laser-induced fluorescence and differential absorption. They are superior to point detection systems such as infrared spectroscopy and Raman spectroscopy because of their capability to range and discriminate the chemical and biological molecules in real time. *Ultraviolet laser-induced fluorescence* is used for the detection of biological molecules since most fluoresce when excited by ultraviolet radiation.

Differential absorption lidar (DIAL) is the most frequently used technique for the detection of chemical warfare agents along with detection of toxic gases and pollutants. It uses two wavelengths: one corresponds to the wavelength of peak absorption of the targeted molecule ('on' wavelength) and the other corresponds to the weak absorption of the targeted molecule ('off' wavelength). The ratio of the two received backscattered signals is a measure of the concentration of the targeted chemical warfare agent. Figure 15.2 shows the block schematic arrangement of DIAL system. Two laser pulses are transmitted towards the targeted area of interest in the atmosphere that is suspected to be contaminated with a potentially harmful chemical warfare agent.

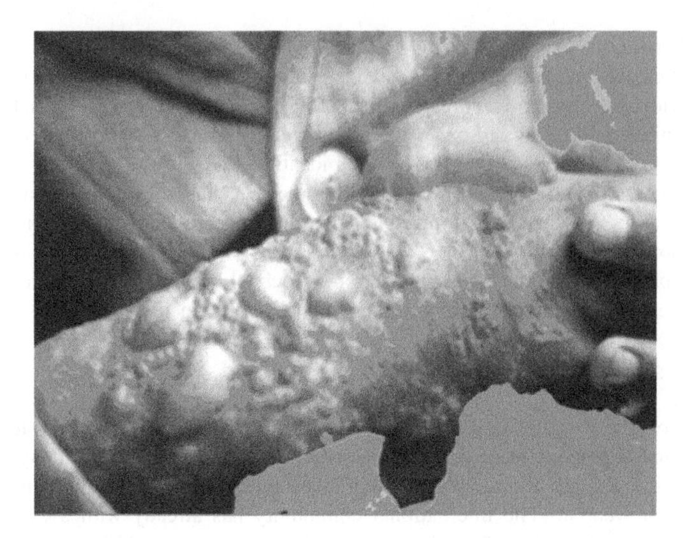

Figure 15.1 Effect of chemical weapon attack.

The choice of 'on' and 'off' wavelengths is unique for a given chemical species and is determined by using a tunable laser that scans the area of interest by transmitting pairs of these wavelengths. The on-wavelength λ_{on} encounters maximum absorption with the result that the corresponding backscattered signal is relatively weak compared to the λ_{off}, which encounters weak absorption and hence a relatively stronger backscattered signal. This forms the basis of knowing atmospheric constituents at that time. The ratio of the two backscattered signals yields the concentration of the molecule of interest.

Many of the chemical warfare agents experience significant absorption in the 3–4 μm and 9–11 μm bands. CO_2 lasers emitting in the 9–11 μm band have been commonly used for the detection of the majority of chemical agents. Some DIAL systems do make use of both these bands and therefore employ both tunable mid-infrared as well as CO_2 laser sources. The schematic arrangement of DIAL system shown in Figure 15.2 only makes use of a CO_2 laser, however. After suitable collimation, the transmit laser beam is directed towards the atmosphere in the desired direction with the help of a scanning gimbal mirror. The backscattered signal is received by the receiving telescope and is further focused onto the detector subsystem. An interference filter is used to block the undesired radiation and allow only the radiation of wavelength of interest to pass through it. The detected signal is then digitized and fed to the data processor to extract the desired information on the type and concentration of the chemical species.

15.2.2.2 Detection of Biological Agents

Stand-off detection of biological warfare agents and their discrimination from background aerosols is extremely challenging, as the distinction between innocuous ambient bacteria and virulent microbes amounts to subtle differences in their molecular make-up. As these subtle differences involve a very small percentage of molecules, they produce a very small change in their optical signatures. This makes it extremely difficult to reliably detect and discriminate these potentially dangerous molecules. The processes of detection and discrimination are further complicated by the fact that there is a constant variation in the growth media and the contaminants associated with the processing of biological warfare agents affect their optical signatures. Figure 15.3 shows the effects of a biological warfare agent attack.

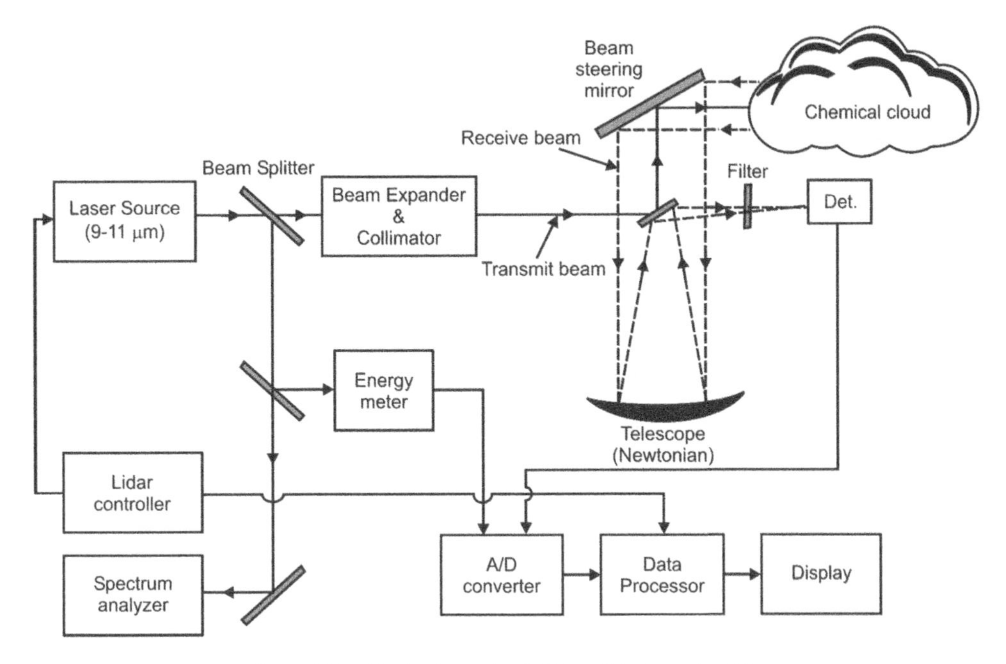

Figure 15.2 Block schematic of differential absorption lidar.

The stand-off detection of biological warfare agents is based on the concept of the laser-induced fluorescence (LIF) effect. Laser-induced fluorescence is the emission from atoms or molecules after they have been excited to higher energy levels by excitation by another laser. The emission takes place at a wavelength that is higher than the wavelength of the exciting laser light. The biological warfare agent molecules mainly constitute aromatic amino acids and coenzymes. Aromatic amino acids such as tryptophan, tyrosine and phenylanine absorb laser radiation at 280–290 nm and fluoresce in the 300–400 nm bands. It is therefore possible to detect biological warfare agents by using UV laser

Figure 15.3 Effects of biological warfare attack.

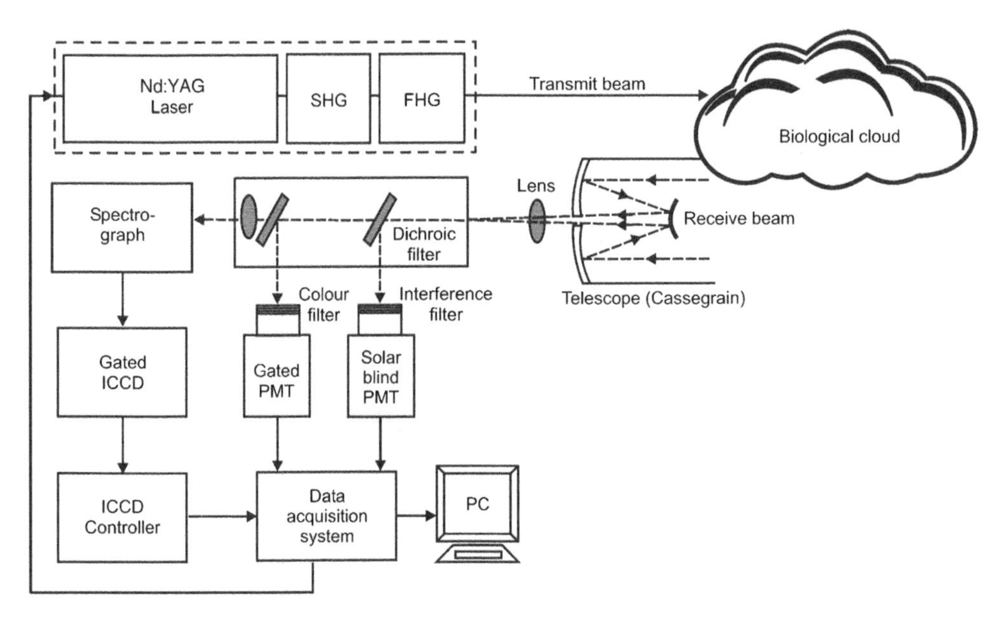

Figure 15.4 Block schematic arrangement of UV-LIF lidar system for detection of biological agents.

excitation at a suitable wavelength. Discrimination of biological agents can only be achieved from the LIF signal because the fluorescence cross-section for particles in the range 1–10 μm are sufficiently large to make single particle interrogation feasible.

Figure 15.4 shows the block schematic arrangement of a typical monostatic UV-LIF lidar system. The principle components of this lidar system include a frequency-tripled or -quadrupled Nd:YAG laser emitting at 355 nm and 266 nm respectively, a telescope used for both transmission and reception, a photomultiplier tube (PMT) used for recording the backscattered signal and the bio-fluorescence signal, a spectrograph with a gated intensified charged-couple device (ICCD) array for recording the dispersed fluorescence spectra and the lidar controller.

The fourth harmonic at 266 nm is transmitted towards the biological cloud. The telescope receives the backscattered and the bio-fluorescence signal, which is then fed to the PMT channel. The backscattered signal is received by the gated PMT channel and is present whenever there is a cloud along the beam path. This channel is used to measure the distance to the cloud. A solar-blind PMT channel is used to receive the fluorescence signal and is activated only when there is a suspicious cloud. A spectrograph with a gated ICCD is used to identify the nature of the bio-molecule that is responsible for the fluorescence. The bandwidth of the receiver channel is usually kept smaller in order to prevent unwanted background radiation entering the channel.

15.2.3 Detection of Explosive Agents

Stand-off detection of explosive agents using lasers and based on the trace detection method is one of the most widely researched technologies internationally. For the technology to mature to an extent where it can be transformed into a product usable in the kind of environment and field conditions usually encountered in homeland security applications is a great technological challenge.

One of the major problems is the decrease in the intensity of the backscattered light signal due to wavelength-dependent absorption and scattering losses and the intensity decreasing inversely with distance squared. The problem is compounded by the fact that the trace levels associated with common explosive agents are extremely low, in the range of a fraction of ppb (parts per billion) to a few ppm (parts per million) for common explosives.

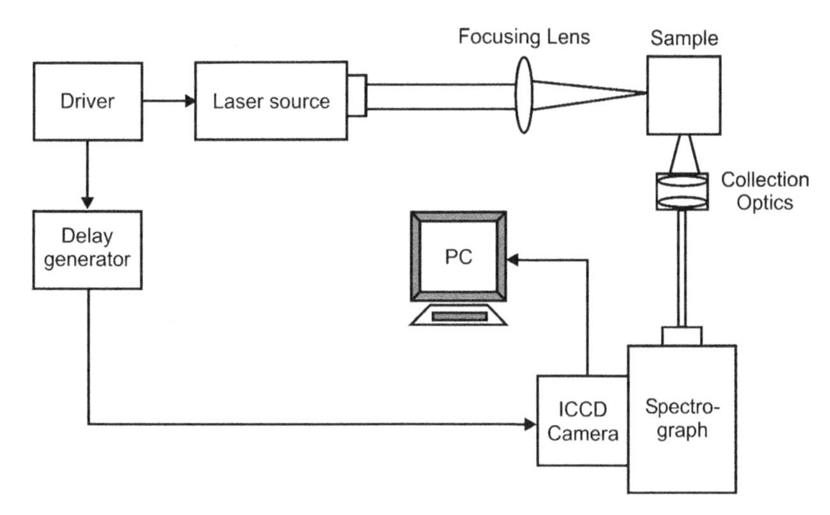

Figure 15.5 Laser-induced breakdown spectroscopy set-up.

The second major problem relates to the unique identification of the targeted explosive agent against a background of interferents. Many chemical agents have atomic compositions including sulphur, phosphorus, fluorine and chlorine in addition to nitrogen, oxygen, hydrogen and carbon present in organic molecules. The detection methodology therefore needs to be highly sensitive and selective. Laser-based spectrometric methods have the potential of being fast, sensitive and selective with the ability to detect and identify a wide range of explosive agents and to be upgraded to handle new threats.

Atmospheric transmission at the wavelengths concerned is an important factor while assessing suitability of a given stand-off detection methodology. Commonly used technologies are laser-induced breakdown spectroscopy (LIBS), Raman spectroscopy and its variants, laser-induced fluorescence (LIF) spectroscopy and IR spectroscopy. These are all trace detection methods; bulk detection methods such as millimetre wave imaging and terahertz spectroscopy are also available.

Laser-induced breakdown spectroscopy focuses a high-energy laser beam on the trace sample to break down a small part of it into a plasma of excited ions and atoms. The plasma emits light that is characteristic of emissions from ionic, atomic and small molecular species. These light emissions are detected by a spectrometer to identify the elemental composition. Figure 15.5 shows a typical laser-induced breakdown spectroscopy set-up. One of the challenges in the use of LIBS is to assess its efficacy to detect and identify explosive species in a real environment that is replete with many interfering substances. One method of obtaining the desired selectivity is to use a double-pulse LIBS; the first pulse is used to create so-called laser-generated vacuum and the second pulse, transmitted a few microseconds later, generates the return signal. Double-pulse LIBS is also observed to improve sensitivity in addition to enhancing selectivity. Selectivity can be further improved by adding temporal resolution to the LIBS emission analysis.

The wavelength 1064 nm has been widely employed for LIBS systems. Due to serious eye hazards posed by 1064 nm, scientists have also tried 266 nm. The latter also allows the designer to build Raman capability into the system. It may be mentioned here that the wavelength 266 nm has 600 times higher maximum permissible exposure (MPE) limit compared to that of 1064 nm.

Raman spectroscopy (Figure 15.6) offers another method for the stand-off detection of explosive agents. It has been extensively used for many years as a standard analytical tool for identification of chemical agents in the laboratory environment. The basis of detection in this case is the shift in the wavelength caused by inelastic Raman scattering by the target molecule. The inelastic scattering of impinging photons, where some energy is lost to (or gained from) the target molecule, returns scattered light with a higher (or lower) wavelength depending upon whether energy was lost to (or gained from)

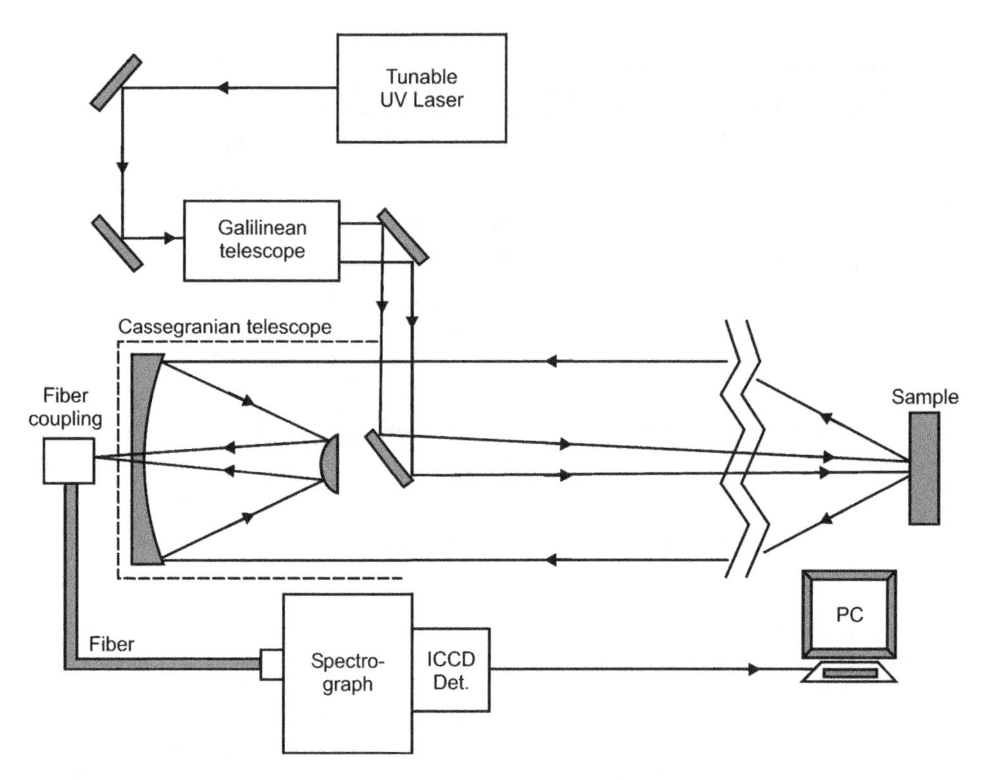

Figure 15.6 Raman spectroscopy set-up.

the target molecule. The difference is dictated by the energy of vibrational modes of the target molecule, and therefore constitutes the fingerprint or the basis of identification. Complex mixtures are identified using algorithms for pattern recognition.

A major drawback of the Raman technique is its extremely poor sensitivity caused by the fact that Raman scattering occurs for 1 in about 10^7 photons impinging on the sample. The weak return-signal intensity of Raman spectroscopy limits its use for trace detection, as it is also sensitive to ambient light and fluorescence from the sample itself or other chemicals in the vicinity. The fluorescence masks the Raman signal. These problems are overcome by use of resonant Raman spectroscopy. With a tunable laser, the wavelength can be chosen to match or nearly match a resonant absorption in the target molecule, leading to intensity enhancement of the order of 10^6. The problem of fluorescence masking the Raman signal can also be overcome by use of either infrared or ultraviolet radiation. Infrared radiation does not have sufficient energy to cause fluorescence and ultraviolet radiation will cause fluorescence in the visible, which is easily separated from the Raman signal.

Laser-induced fluorescence (LIF) is another important tool for similar applications. Although a very valuable tool in combustion diagnostics and for studying the decomposition of explosives, it has not been found to be very useful for the detection of explosive agents.

Almost all the laser-based stand-off detection methods have detection limits too high to make them suitable for the detection of explosives in field conditions, which require the ability to detect traces in the vapour phase or in the form of particles. Another problem is insufficient selectivity to identify the target explosive in the presence of interferents. None of the laser-based spectroscopic methodologies available today is ready for full functionality prototype manufacturing. There is much scope for further research to improve the performance of the more mature technologies in terms of both detection sensitivity and selectivity. The final solution probably lies in the integration of several technologies to derive the benefits from each.

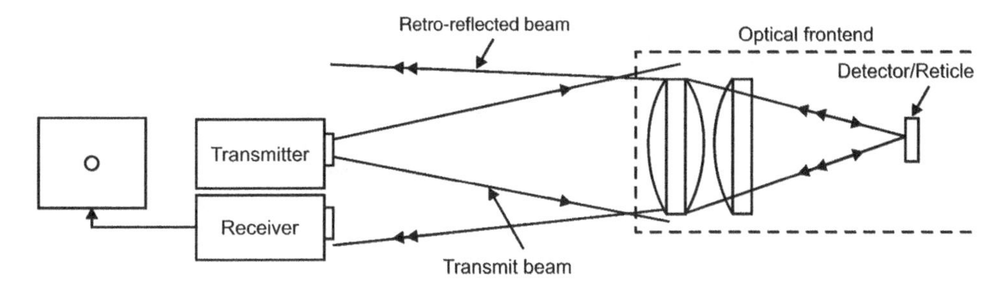

Figure 15.7 Cat's eye effect.

15.2.4 Detection of Optical and Electro-optic Devices

Another emerging application of laser technology is in the detection and identification of battlefield optical and optoelectronic sighting systems, which includes optical scopes, night-vision devices, thermal imagers, laser rangefinders and target designators. The device operates on the principle of the cat's eye effect as shown in Figure 15.7. The target optical device is illuminated by a laser beam and the optical system returns a fraction of it as backscattered energy, which is received by a sensitive receiver. With particular reference to homeland security, such a device could be very useful for the detection of optical scopes employed by snipers. Another security-related application could be surveillance of sensitive areas, particularly in urban territories. Figure 15.8 illustrates a typical deployment scenario of such a device in an urban environment when used for sanitization of the area.

One such international system is the Laser Sniper Detector from CILAS, France which is available in two variants: SLD-400 and SLD-500. Figure 15.9 is a photograph of the SLD-400, which can be fitted on a tripod or on a static vehicle. Its highly accurate threat detection and localization capability is compatible with the performance of the associated weapon or fire control system.

Figure 15.10 is a photograph of a similar system, the Optifinder-1200 from M/s Torrey Pines Inc. The Optifinder-1200 is an active system designed to detect the retro-reflected signal caused by the front-end optics of the target system. It uses a combination of a high-sensitivity detector and optical filter to offer a maximum operational range of 1200 m. It also makes use of time-of-flight calculation of the transmitted signal and the retro-reflected signal to compute the target range. The chosen detector offers day/night operational capability. The system allows penetration through bad weather and vegetation by range

Figure 15.8 Typical deployment scenario in an urban environment.

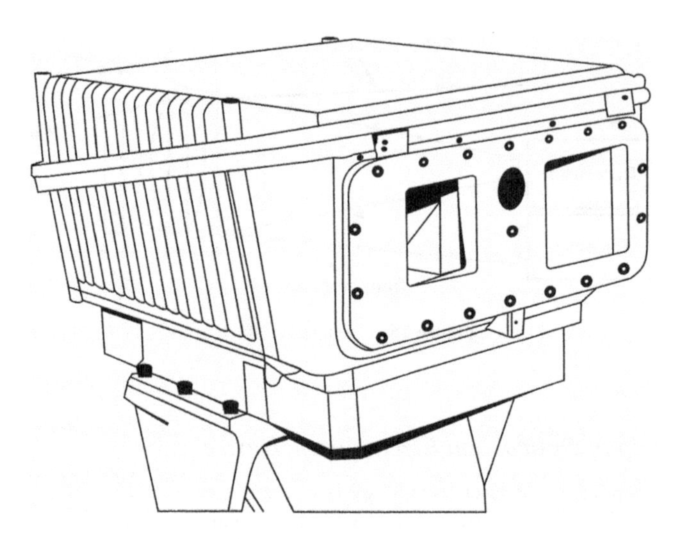

Figure 15.9 Laser sniper detector type SLD-400 (courtesy of Cilas).

gating. Three variants of the optifinder system are available with operational ranges of 350 m, 1200 m and 2000 m.

15.2.5 Disposal of Unexploded Ordnances

The disposal of unexploded ordnances including surface-laid mines, improvised explosive devices (IED), grenade shells, artillery/mortar rounds and cluster bombs from safe stand-off ranges using a high-power laser beam is an emerging application of directed-energy laser systems. Ordnance is disposed of by focusing a high-power laser beam on the ordnance casing, heating it until the temperature of the backplane of the casing exceeds ignition temperature of the explosive filler. The explosive filler ignites and begins to burn, independent of the type of fusing used by the target explosive. This leads to a low-level detonation or deflagration rather than full-power detonation. The advantages of the use of laser energy for ordnance disposal include large-magazine high-precision controllable effects with reduced collateral damage and safe and fast disposal from stand-off ranges.

One such international system is the ZEUS-HLONS (HUMMWV Laser Ordnance Neutralization System) of USA. Figure 15.11 is a photograph of the system. The concept of neutralization of live ordnances using laser energy was first demonstrated in the field in 1994 with development and field testing of a mobile ordnance disrupter system (MODS) that employed a 1.1 kW arc-lamp-driven solid-state laser mounted on an M113 A2 armoured personnel carrier. The ZEUS system initially employed a

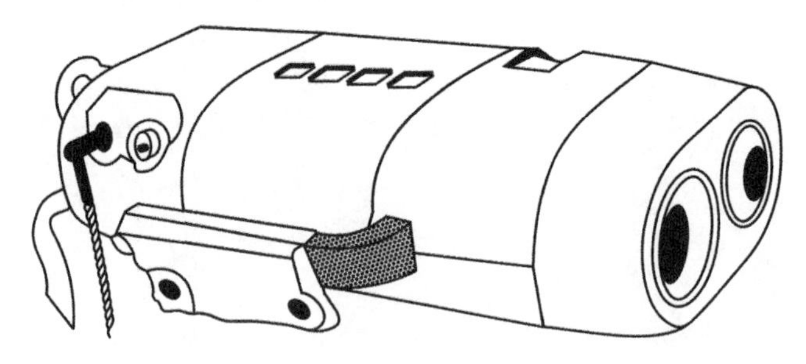

Figure 15.10 Optifinder-1200 sniper detection system (courtesy of LaseroptronixAB).

Figure 15.11 ZEUS laser ordnance neutralization system (Reproduced from public domain (http://en.wikipedia.org/wiki/File:Zeus_system.jpg)).

500 W laser; the latest version of the ZEUS system is equipped with a 2 kW fibre laser mounted on the roof of an adapted Humvee vehicle.

Another laser system designed for the disposal of unexploded ordnances is the Laser Avenger system from Boeing combat systems (Figure 15.12). Boeing successfully tested a 1 kW solid-state laser weapon mounted on a converted anti-aircraft vehicle in Redstone Arsenal in Huntsville Alabama by neutralizing multiple types of improvised explosive devices, including large-calibre artillery munitions and mortar rounds. The system was operated at safe distances from the targets under a variety of conditions including different angles and ranges.

Figure 15.12 Avenger laser ordnance disposal system (Courtesy of Boeing).

Figure 15.13 Laser dazzler in action (Courtesy of Freak Lasers).

15.2.6 Non-lethal Laser Dazzlers

Non-lethal weapons act as a force multiplier, enabling friendly forces to discourage, delay or prevent hostile action. They are particularly effective in situations where use of lethal force is not preferred, examples being limiting escalation and temporarily disabling facilities and equipment. Laser-based non-lethal weapons such as laser dazzlers can be used for counter-insurgency, anti-terrorism, crowd control and infrastructure protection applications. Laser dazzlers are emerging internationally as a new non-lethal alternative to lethal force for law enforcement, homeland security, border patrol, coastal protection, infrastructure protection and a host of other low-intensity conflict scenarios.

A laser dazzler emits a high-intensity laser beam in the visible band, usually in the blue-green region, to temporarily impair the vision of the adversary without causing any permanent or lasting injury to the subject's eyes. Figure 15.13 illustrates the use of a laser dazzler in a close-combat scenario and Figure 15.14 demonstrates the use of a vehicle-mounted laser dazzler system for unruly crowd/mob control applications. Such vehicles are usually equipped with remotely controlled weapon stations for a wide range of applications. The technology of the short-range and crowd-control laser dazzlers is similar, but the latter system employs a higher power to be able to produce the desired power density in a larger beam spot. Some means of scanning the laser beam spot may also be necessary.

Laser dazzlers are also being considered as a potential candidate for warning the crew of commercial airliners or military aircraft who violate (intentionally or unintentionally) a no-fly zone. Such systems in a networked configuration of multiple laser dazzler stations and radars could be effectively used for the protection of critical infrastructure or assets. In such a system, radar provides the initial cue to the rogue or suspect aircraft when it is more than 100 km away from the actual asset to be protected. The radar keeps a constant vigil on the suspect aircraft until it comes within the tracking range of the electro-optic tracker, which is usually an integral part of the laser dazzler station. The electro-optic tracker station takes the cue from the radar and tracks the target with the much higher accuracy needed for the dazzling action. The blue-green region is the chosen wavelength band as the response of the human eye is highest in this band, as shown in Figure 15.15. The most commonly employed wavelength for this purpose is 532 nm, usually generated by using either laser diodes or frequency-doubled Nd:YAG laser modules.

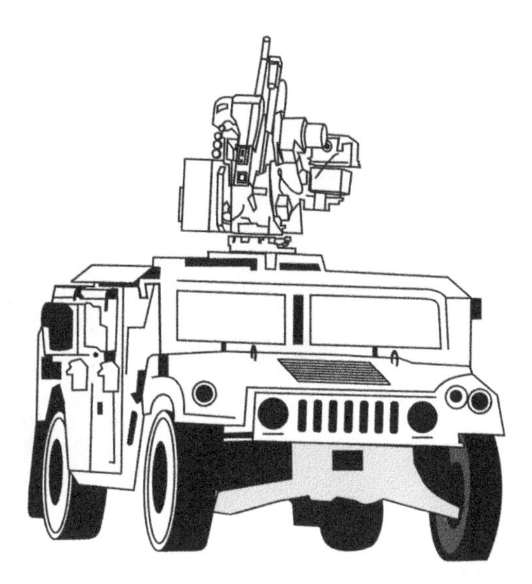

Figure 15.14 KLEG laser dazzler on CROWS-II platform (courtesy of US Army).

A laser dazzler is a non-lethal weapon specifically designed for applications where subject vision impairment is to be achieved at a distance ranging from a few tens of metres to several kilometres in bright ambient conditions.

The choice of performance parameters such as laser power, spot size at the target and laser power density are driven by the nature of deployment. The beam shaping and directing optics are designed to achieve the desired value of nominal ocular hazard distance (NOHD) and a laser power density that does not exceed the maximum permissible exposure (MPE) figure dictated by ANSI (American National Standards Institute) standards for eye safety.

The MPE is dependent on wavelength and exposure time. At 532 nm, maximum permissible power density is 2.5 mW cm^{-2} and 1.0 mW cm^{-2} for exposure times of 0.25 s and 10 s, respectively. These devices usually produce a randomly pulsed output in the range of 10–20 Hz, riding a DC level for better

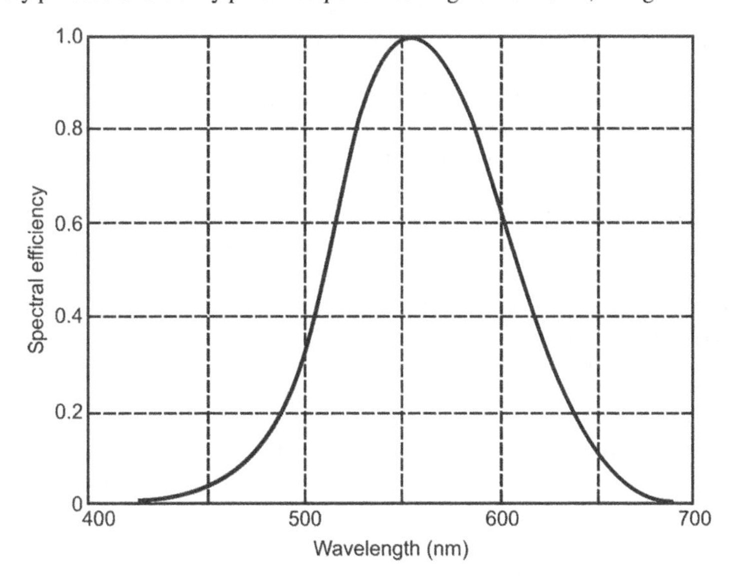

Figure 15.15 Response of human eye to different wavelengths.

Figure 15.16 Glare GBD-IIIC (Courtesy of BE Meyers Co. Inc).

overall effect. DC level is usually kept at 30–50% of the peak intensity level. Night-time maximum operational range is typically 3–4 times the maximum daytime operational range.

A large number of companies are offering short- to medium-range laser dazzler systems for various application scenarios. One such short-range laser dazzler is the Compact High Power (CHP) laser dazzler from LE systems, USA. The dazzler emits a 500 mW flashing green dazzling laser beam and is capable of engaging targets at distances of up to 200 m.

The GLARE MOUT is a non-lethal visual disruption laser with an effective range of 0.15–2 km. The device is ideally suited for small arms integration as well as mobile crew-served applications. When a suspect individual approaches a restricted or controlled area, such a device can be used to give warning to that person before shots have to be fired. Its shockingly bright green beam sends an irrefutable, multilingual, cross-cultural message that cannot be ignored. Reports confirm that the GLARE MOUT has saved numerous lives of both soldiers and non-combatants in Iraq and Afghanistan.

The GLARE GBD-IIIC (Figure 15.16) is a long-range variant of the GLARE MOUT, and is a visual deterrent laser device for hail and warning applications. The GLARE GBD-IIIC is an effective dazzler ideally suited to ship-to-ship signalling or airborne over-watch. With twice the power and a more concentrated beam than the GLARE MOUT, the GLARE GBD-IIIC can effectively hail and warn as far away as 4 km. Like the GLARE MOUT, the GLARE GBD-IIIC produces an overtly bright green diffused laser spot to impair the visual ability of a suspect.

The GLARE LA-9/P (Figure 15.17) is another long-range visual deterrent laser device for hail and warning applications and is intended to be effective to a range of 0.3–4 km for ship-to-ship signalling or airborne over-watch applications. The GLARE LA-9/P has an additional feature of automatically shutting off the device if the subject target is within the nominal ocular hazard (NOHD) zone, shutting off the glaring laser output if so to prevent any unintended eye injury.

The Sabre 203 grenade shell laser intruder countermeasure system (Figure 15.18) is a type of laser dazzler that uses a 250 mW red laser diode mounted in a hard plastic capsule in the shape of a standard 40 mm grenade. It is suitable for being loaded into a M203 grenade launcher. It has an effective range of 300 m. It is controlled via a box snapped under the launcher, with the batteries and firing switch housed in this box. In emergency it can be quickly ejected and replaced with a grenade. Sabre 203 dazzlers were used in Somalia in 1995 during operation United Shield.

The Chinese JD-3 laser dazzler and ZM-87 portable laser disruptor are also established systems. The JD-3 laser dazzler is reported to be mounted on the Chinese Type 98 main battle tank and is coupled with a laser radiation detector. It automatically aims for the enemy's illuminating laser designator, attempting

Figure 15.17 Glare LA-9/P laser dazzler (Courtery of BE Meyers Co. Inc).

to overwhelm its optical systems or blind the operator. The ZM-87 portable laser disturber is an electro-optic countermeasure laser device. It can blind enemy troops up to a range of 2–3 km and temporarily blind them at up to 10 km range. ZM-87 has reported been widely deployed, including in use on naval vessels.

The Photonic Disruptor (Figure 15.18), classified as a threat-assessment laser illuminator (TALI), is another non-lethal high-power green laser developed by Wicked Lasers, USA in cooperation with Xtreme Alternative Defence Systems. This tactical laser is equipped with a versatile focus-adjustable collimating lens to compensate for range and power intensity when used to either incapacitate an attacker in close range or safely identify threats from a distance.

The Dazer Laser from Laser Energetics Inc, USA is another very popular device. It comes in two variants: the Guardian (Figure 15.19) has a range of 1–300 m (model dependent) and the Defender has a range of 1–2400 m (model dependent). Both variants temporarily impair the vision of the target adversary and succeed in eliminating the threat's ability to see, engage or effectively target the user. This provides the user with a significant advantage over the threat at longer and safer stand-off distances. Both variants of the Dazer are designed to be eye safe at all ranges beyond 1 m, and meet the current American National Standard Institute (ANSI) safety standard Z136.1.

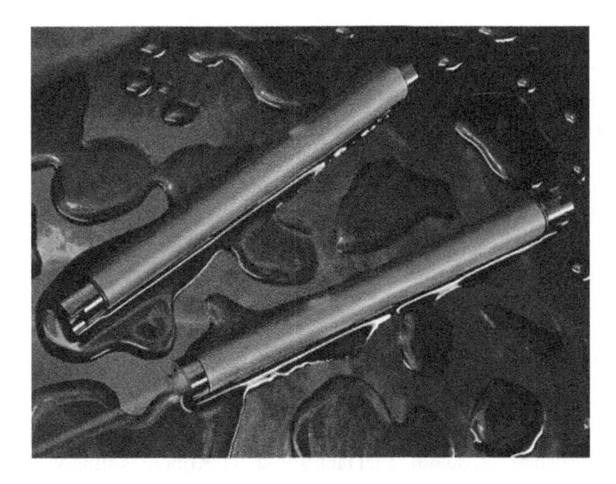

Figure 15.18 Sabre-203 laser dazzler system Photonic Disruptor (Reproduced by permission of Wicked Lasers).

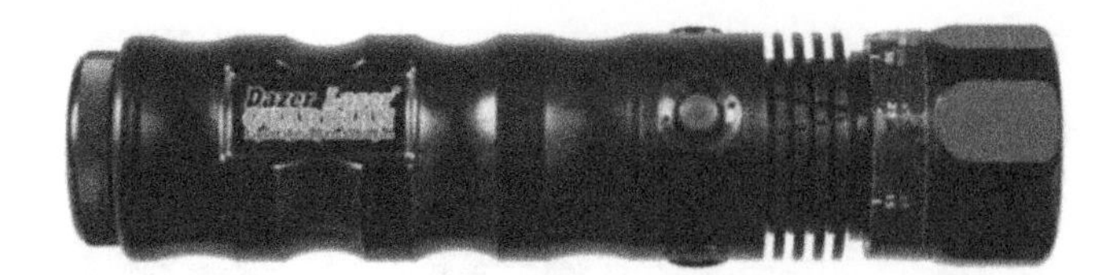

Figure 15.19 Dazer laser Guardian (Courtesy of Laser Energetics).

Laser dazzlers have shown much promise and potential for a variety of low-intensity conflict applications including close-quarter battle scenarios; border patrol and coastal surveillance; controlling unruly and violent crowds and mobs; countering asymmetric threats particularly encountered in the naval environment; and protection of critical infrastructure from terrorist attacks. Different variants of these devices are being developed to suit different application requirements. While handheld portable devices are offered to meet the short-range requirements of paramilitary and law enforcement agencies, tripod- and vehicle-mounted versions equipped with support electro-optic sighting, target acquisition and tracking devices are also being developed to meet the requirements of emerging scenarios.

15.3 Electro-optic Countermeasures

One of the relatively newer types of laser systems that have found recognition and wide acceptance by the armed forces in the last couple of decades are those that can offer effective countermeasures against the laser systems already in use. The relevance of electro-optic countermeasures (EOCM) equipment stems from the fact that, whenever a particular defence technology has matured to a high level leading to its widespread usage, investment is triggered in research and development efforts in the corresponding countermeasure technologies and systems. This has already occurred in the case of radar and other related military systems operating in the RF spectrum of electromagnetic radiation and in the case of lasers and related devices operating in the optical spectrum.

15.3.1 Need and Relevance

There is wide use of electro-optic devices such as sighting and observation devices and laser systems such as those used for range finding and target designation applications. Laser technology is quite mature today, which has led to introduction of laser devices and systems for many new applications not previously considered. Developed countries have expended much effort on the deployment of EOCM systems.

The need to develop and deploy the EOCM class of laser systems that can offer effective countermeasures against similar systems deployed by the armed forces of the adversary is more pertinent and relevant today. No military platform, be it land based, aerial or shipborne, is free from the risk of exposure to laser radiation. The activities of these platforms are under constant surveillance by various kinds of electro-optic devices and optronic sensors. Rendering these devices and sensors ineffective in the battlefield therefore makes a huge difference to the battlefield competence of a nation. The deployment of EOCM devices and systems designed to incapacitate or neutralize the more conventional laser devices and systems acts as a force multiplier, as a platform incapacitated in the enemy camp is a platform added to your own. The survivability quotient of armed forces equipped with such a capability is significantly enhanced.

15.3.2 Passive and Active Countermeasures

Electro-optic countermeasures are broadly classified as passive and active countermeasures. *Passive countermeasures* for platform protection include use of armour, camouflage, fortification and other protection technologies such as self-sealing fuel tank and so on. Armour is simply the protective covering used on the platform to prevent damage caused to it through the use of direct-contact weapons or projectiles. The use of reactive armour is a common example. Camouflage is a form of visual deception. With reference to electro-optic countermeasures, it is a methodology that helps military platforms remain unnoticed by the adversary and is achieved by blending with the environment or by resembling something else. Fortifications are military constructions designed to provide protection to specific equipment, platforms and military bases. Self-sealing fuel tank technology prevents the fuel tanks of aircraft from leaking and becoming ignited when attacked.

Active countermeasures comprise soft-kill and hard-kill countermeasures. Soft-kill countermeasures change the electromagnetic, acoustic or other forms of signatures of the platform to be protected. This in turn adversely affects the tracking or sensing capability of the incoming threat. With reference to electro-optic countermeasures, the incoming threat could be a laser-guided bomb or projectile or an infrared-guided missile. In the case of providing protection to land-based platforms from laser-assisted threats, one common method of achieving this objective is through the use of smoke or aerosol screen to block the radiation from the laser target designator, depriving the laser seeker in the guided munitions the required guidance information. In the case of active countermeasures, means are adopted to counter-attack the incoming threat. This may be achieved by neutralizing one of the elements responsible for the functioning of the guided threat or by physically destroying the incoming threat by launching a counter projectile in that direction. These are also known as active protection systems.

The hard-kill neutralizing capability of EOCMs can be illustrated by the case of laser-guided munitions. A laser sensor on the platform to be protected senses the existence, PRF code and direction of arrival of laser radiation from the laser target designator and then either generates a smoke/aerosol screen by launching a salvo of grenades or sends another high-power/energy laser beam in the same direction, thereby incapacitating the laser target designator. Another technique is to use the high-energy laser radiation to illuminate a dummy target 100–200 m away from the platform to be protected, thereby forcing the incoming laser-guided threat to land on the dummy target. This is a very attractive proposition for the protection of critical and high-value military assets such as aircraft shelters and ammunition depots during war. The use of infrared flares by target aircraft to protect it from the attack of infrared-guided surface-to-air or air-to-air infrared-guided missiles is yet another example.

The other type of hard-kill active protection system, launching a counter projectile towards the incoming threat, is generally used in the terminal phase of the incoming threat. This leads to the physical destruction of the threat by means of either blast and/or fragment action. The consequences of blast and/ or fragment action could be in the form of destabilization of kinetic energy penetrator, premature initiation of charge or destruction of airframe. The Arena by Russia's Kolomna-based engineering design bureau (KBM) and Trophy from Rafael, Israel are well-known active protection systems of this type, designed for the protection of land-based armoured fighting vehicles.

15.3.3 Types of EOCM Equipment

There are two broad categories of EOCM equipment: the EOCM class of lasers such as anti-sensor lasers and laser dazzlers, and the support devices such as laser threat detection systems.

From the viewpoint of countermeasures terminology, a laser dazzler may be viewed as a type of anti-sensor laser where the target sensor is the human eye. Laser dazzlers were discussed in detail in Section 15.2.6. In the category of anti-sensor systems, we have systems capable of causing either a temporary disability of electro-optic devices and optronic sensors deployed by the adversary, or those capable of inflicting a permanent damage. In both cases, the target is the front-end optics and optronic sensors. These are sometimes referred to as soft-kill systems, as they have no capability to inflict physical or structural damage to the platform carrying weapons. Hard-kill EOCM systems that are capable of inflicting physical damage to the front-end optics of any electro-optic system are usually vehicle mounted and are much larger in size and weight than their soft-kill counterparts. The pulse energy level in such lasers is of the order of several kilojoules, compared to a few joules in the case of soft-kill systems.

A *laser warning sensor system* is an indispensable component of any EOCM system, whether designed for soft-kill or hard-kill. It provides information on the type and direction-of-arrival of laser

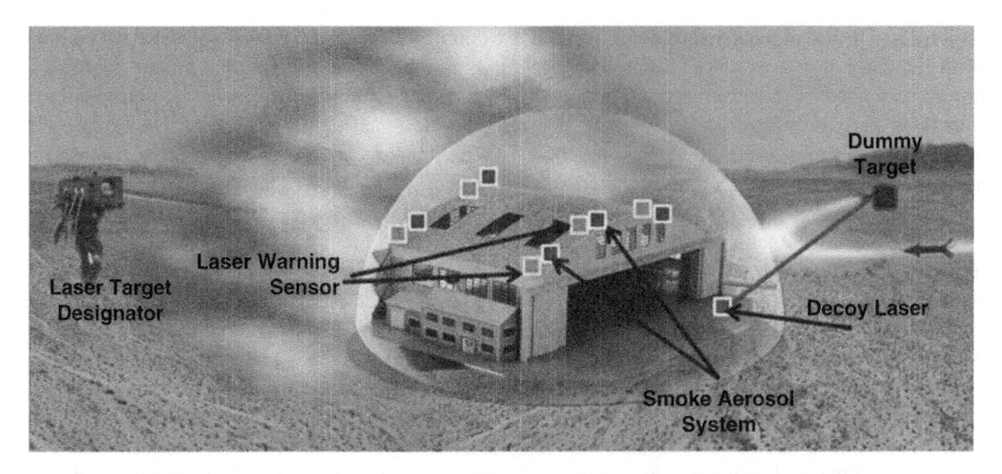

Figure 15.20 EOCM system as a decoy for the protection of high-value assets.

threats emanating from the adversary camp. A laser warning system provides valuable information that enables the platform crew to make an optimum choice regarding the type and *modus operandi* of the countermeasure system to be used. When used in conjunction with a smoke/aerosol screening system, the information provided by the sensor is used to trigger and deploy an optimal aerosol/smoke screen in the direction of the threat and block the incoming radiation. This allows the crew of the platform to take evasive action during those crucial 50–60 s for which smoke screen remains effective. This type of defensive countermeasure action is particularly effective against laser-guided munitions attack.

In the case of an active electro-optic countermeasures system equipped with high-energy lasers, the information on the incoming laser threat is fed to the servo control system which either reorients the high-energy laser precisely in the direction of the laser threat and neutralizes the source of radiation, or illuminates a dummy target to divert the incoming threat from its intended course. Figure 15.20 illustrates a typical deployment scenario of an EOCM used as a decoy system for protection of high-value assets such as aircraft shelters from laser-assisted threats.

Laser warning systems are available internationally covering a spectral response range from the visible to the far-infrared, encompassing the entire range of battlefield solid-state lasers typically operating at 1064 nm and 1540 nm, mid-infrared operating in the 3–5 μm band and far-infrared lasers operating in the 8–14 μm band. (Infrared countermeasures are discussed in the following section.) Operational ranges of 8–10 km and angle-of-arrival measurement resolution up to a fraction of a degree are available in commercial systems.

Laser warning sensors are used both as standalone devices as well as an integral part of larger EOCM systems. An example of a standalone system is the helmet-mounted laser warning system that uses multiple sensors to give full 360° coverage. One such system that uses four sensors has been patented in USA. Laser threat warning devices with operational ranges up to 10 km covering a spectral range from 400 to 1800 nm are being introduced. There are systems with direction detection capability in the form of predetermined sectors of 40–50° and also with very precision direction sensing accuracy approaching better than a degree.

Some common laser warning sensor systems are AVITRONICS LWS-300 and Goodrich's AN/AVR-2 series systems including AN/AVR-2, AN/AVR-2A and AN/AVR-2B systems. LWS-300 has a direction sensing resolution of 15° and an operational band of 500–1800 nm and Goodrich's AN/AVR-2A is a passive laser warning system specifically designed for the protection of aircraft from laser designators, range finders and beam riding missiles. The detection system provides full 360° azimuth and ±45° elevation coverage.

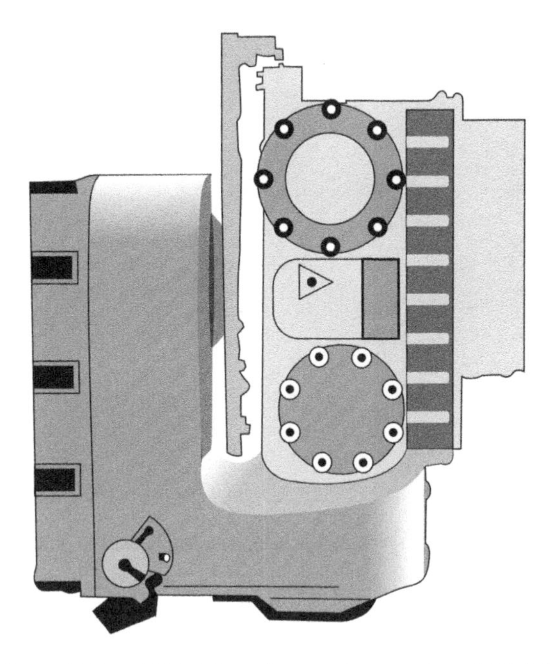

Figure 15.21 DHY-322 laser decoy system (courtesy of CILAS).

In the category of integrated EOCM equipment, there is a whole range of both application- and platform-specific systems. It is not feasible to discuss every system here; instead, an overview of some of the better-known EOCM systems is presented. Laser decoy systems Type-405 from GEC-Marconi and DHY-322 (Figure 15.21) from CILAS are two anti-sensor systems designed and built to divert the laser-guided munitions from their intended targets. The DHY-322 comprises a laser detection system and a high-energy laser. The high-energy laser is mounted on a controlled turret. The laser detection system detects the origin of the hostile laser radiation, deciphers its PRF code and then synchronizes the high-energy laser to operate at the decoded PRF code. The high-energy laser radiation is then used to illuminate a dummy spot where the incoming projectile cannot do any significant damage. DHY-322 is specifically designed to protect ships, strategic sites and other high-value assets.

The STINGRAY laser detection and countermeasures system from Lockheed Martin is another EOCM system designed to protect the frontline forces by accurately locating the optical and electro-optical fire control systems of the enemy. The system is particularly designed for and is popular with armoured fighting platforms. Another state-of-the-art EOCM system similar in capabilities and application potential to the STINGRAY system is the Russian SHTORA-1. SHTORA-1 has been designed to protect the armoured fighting platforms from laser-guided munitions, laser beam riders and anti-tank guided missiles. It is equipped with an array of laser warning sensors, including coarse sensors each having 135° azimuth coverage and fine sensors each with 45° azimuth coverage. All sensors provide elevation coverage of $-5°$ to $+25°$. The system is also equipped with anti-laser anti-thermal smoke grenades and active electro-optic jammers operating in the 0.7–2.7 μm band.

The other category of countermeasures system is the hard-kill systems, of which Russia's ARENA and Israel's TROPHY are examples. ARENA is an active protection system (APS) designed to protect armoured fighting vehicles from anti-tank weapons including anti-tank guided missiles and missiles with top-attack warheads. It uses a Doppler radar to detect incoming warheads and then fires a defensive rocket in the direction of the incoming threat, detonating near the approaching warhead and destroying it before it hits the platform. TROPHY (Figure 15.22) is also an active protection system designed to supplement the armour of both light- and heavy-armoured fighting platforms. It intercepts the incoming kinetic energy threats and destroys them by a shotgun-like blast.

Figure 15.22 TROPHY active protection system (Courtesy of Rafael).

15.3.4 Infrared Countermeasures

Infrared countermeasures (IRCM) are employed to protect aircraft from surface-to-air and air-to-air heat-seeking missiles. The IRCM systems achieve this by confusing the infrared-guidance system of the missiles and forcing them to deviate from their intended trajectory. The basic principle of operation of infrared-guided missiles was described in Chapter 14 (Section 14.4.2.2). The infrared missile seeker head uses either a spinning reticle and stationary optics called a spin scan or a stationary reticle and rotating optics called a conical scan. In both cases, a modulated signal is generated which allows the tracking logic to determine where the source of infrared energy is with respect to the direction of flight of the guided missile.

Infrared flares deployed by the aircraft are a common form of infrared countermeasures. Flares create infrared targets with a much stronger infrared signature than that from the engine of the aircraft and other parts. The flares mimic the target and force the missile to make an incorrect steering decision, which leads to the missile breaking off a target lock-on. This further causes the missile to deviate from its intended trajectory. State-of-the-art missiles are designed such that they are not deceived by flares and are be able to discriminate between real target and flares. This has led to the development of active infrared countermeasures.

Active IRCM systems use a modulated source of infrared radiation with a higher intensity than the emission from the target. If this modulated radiation is seen by a missile seeker and if the modulation scheme matches that used by the seeker of the target missile, the tracking logic of the missile confuses it as coming from the target aircraft and generates a false tracking command. As a result of this, the missile begins to deviate from the target, forcing the infrared seeker to go out of lock. Once the lock is broken, it is very hard for the missile to regain locking condition. The aircraft to be protected could then use infrared flares to force the incoming infrared-guided missile to lock onto the flare. The efficacy of IRCM system is determined by the ratio of jamming signal intensity J to the target signal intensity S and also on how close the modulation frequencies are to the actual missile frequencies. The required J/S for spin scan missiles is relatively much lower than for newer missiles.

Standard IRCM systems use incoherent and non-directional sources of infrared energy, usually configured around infrared-emitting arc lamps, which suffer from the disadvantage that the probability of the jamming signal penetrating the seeker head is low. Further, if the jamming signal is not effective against a particular seeker system, the IRCM acts as a strong source of infrared energy, thereby enhancing the ability of the missile to track the aircraft.

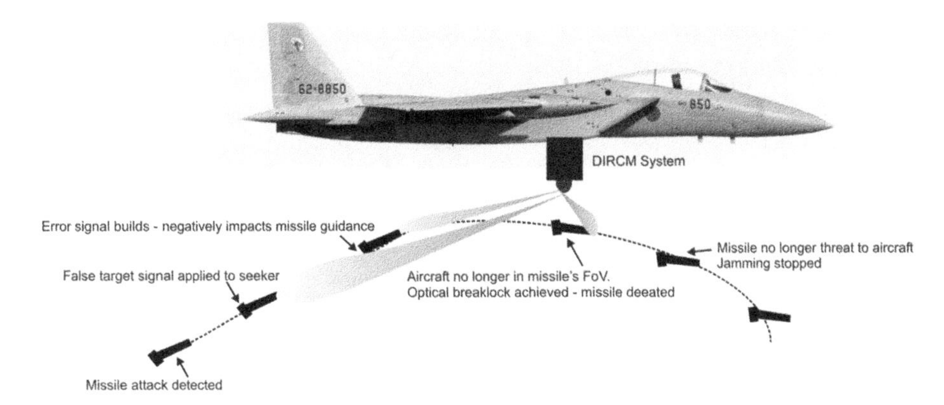

Figure 15.23 Concept of DIRCM system.

This shortcoming is overcome by directional infrared countermeasures (DIRCM), which integrate the infrared source of jamming signal with a missile approach warning system. The infrared jammers are mounted on a movable turret and operate in conjunction with the missile approach warning system. When an infrared-guided missile such as MANPADS is fired, it radiates electromagnetic energy in different wavelength bands. Most DIRCM systems use ultraviolet sensors to detect the approach of the missile, but detection in the infrared bands in addition to ultraviolet band has also been exploited. The detection system provides a cue to the tracking system, which uses information from the detection system to track the incoming missile. The infrared jammers direct the infrared jamming energy at the seeker. The modulation scheme can be cycled to try to defeat a variety of seekers. The efficacy of a DIRCM depends on how comprehensively the approaching threat can be analyzed and how accurately it can be tracked. Laser safety is another issue of concern if higher power needs to be employed in order to defeat advanced missiles. Figure 15.23 illustrates the concept of a DIRCM system.

Some common examples of DIRCM systems include multispectral infrared countermeasures (MUSIC) of Israel, BAE Systems' AN/ALQ-212 advanced-threat infrared countermeasures (ATIRCM) and Northrop Grumman's GUARDIAN and common infrared countermeasures (CIRCM) programs. ATIRCM is part of a directional infrared countermeasures suite and is fielded on US Army CH-47 Chinook helicopters. The suite provides protection against all infrared threat bands and, with its one or more infrared jamming heads, is capable of handling multiple missile attacks. MUSIC also uses active lasers instead of flares to provide protection to the aircraft against MANPADs. Northrop Grumman's GUARDIAN is a DIRCM system that uses ultraviolet sensors for the detection of an incoming missile and a turret-mounted infrared jammer to direct infrared energy at the missile seeker head when cued by the detection system. High-intensity arc lamps are used as a source of infrared energy in earlier versions, while newer versions use laser-based jammers. Figure 15.24 depicts a GUARDIAN DIRCM pod fitted underneath the aircraft fuselage.

The CIRCM program (Figure 15.25) envisages use of fibre coupling for the transport of laser energy to the jam head. The fibre-coupled approach enables the remote location of larger and more complex laser components that would be very difficult to mount on the jam head. CIRCMs jam-head interface supports both direct-coupled as well as fibre-coupled architectures.

15.4 Directed-energy Laser Weapons

One of the major areas of global interest for scientists and engineers today is that of directed-energy weapons (DEW). Three major categories of directed-energy weapons are microwave-based DEWs, laser-based DEWs and particle beam weapons, where the focus in this section is on laser-based DEW. A directed-energy weapon system primarily uses directed energy in the form of a concentrated beam of electromagnetic energy or atomic or subatomic particles in the targeted direction to cause intended

Figure 15.24 GUARDIAN DIRCM pod (Courtesy of Northrop Grumman).

damage to the enemy's equipment, facilities and personnel. The intended damage could be lethal or non-lethal.

A *microwave-based DEW* system is designed to produce the equivalent of electromagnetic interference to damage the enemy's electronics systems. Due to concerns regarding unintended side-effects on the host platform, it is usually preferred to put such weapons only on unmanned combat air vehicles (UCAV). The use of high-power microwaves as a weapon to attack underground and deeply buried targets that are resistant to high explosives is also under consideration. A *particle beam weapon* uses a high-energy beam of atomic or subatomic particles to inflict the intended damage on the target by disrupting its atomic and/or molecular structure.

At the core of *laser-based DEW* is a high-power laser that has enough power in the case of a CW laser or sufficient pulse energy in the case of pulsed laser to inflict physical damage on the target. Although the lasers intended for already-established applications such as those discussed in Chapter 14 will continue

Figure 15.25 CIRCM system (Courtesy of Northrop Grumman).

to improve as newer technologies evolve and develop, it is the use of lasers as weapons that is going to rewrite the military balance in the next 15–20 years.

The introduction of laser-based directed-energy weapon systems is set to dramatically alter the war-fighting capabilities of nations by enabling the execution of missions that would be extremely complex if not impossible to realize with conventional kinetic-energy weapons. These include ground-based laser systems for disabling low Earth orbit satellites and destroying missiles, airborne laser systems for destroying ballistic missiles and space-based laser systems for neutralizing of theatre and inter-continental ballistic missiles. A large number of experiments with laser-based directed-energy weapons to demonstrate these or similar capabilities have been carried out in different parts of the world. The realizability of these weapons has been established beyond doubt and these weapons have been projected by strategists as the weapons of the 21st century. Directed-energy weapons are described at length in this section in terms of salient features, deployment potential, the constituents of a DEW system, different types, international status and emerging trends.

15.4.1 Operational Advantages and Limitations

The primary advantages of laser-based DEWs include speed-of-light delivery, near-zero collateral damage, multiple target engagement and rapid re-targeting capability, immunity to electromagnetic interference and no influence of gravity. Deep magazine and low cost per shot are the other advantages.

Laser weapons engage targets at the speed of light with essentially no time of flight as compared to conventional kinetic energy weapons that require a finite travel time. As an example, one of the world's fastest cruise missiles BrahMos with a supersonic speed of 2.8–3 Mach would take about 5 min to reach its target located at its maximum operational range of 300 km. On the other hand, when targeted by a laser DEW, the same target would be hit in a millisecond.

Further, while the missile becomes destroyed during the mission, the laser weapon is re-usable. This feature makes them particularly suitable for engaging fast-moving targets. The rapid re-targeting feature of laser-based DEWs is attributed to their being powered by rechargeable chemical or electrical energy stores; their multiple target engagement simply involves the re-pointing and re-focusing of the beam directing optical system. The processes of generation and the transfer of lethal laser power to the target are purely in the optical spectrum; they are therefore immune to any electromagnetic interference and jamming. Laser pointing is practically without any inertia and a light bullet has no mass so is not influenced by gravity. As a result, no mid-course correction is required.

Although the initial costs involved in development and fabrication of laser DEWs are relatively much higher than for projectile weapon systems with similar applications, the operational costs in the case of former are practically negligible as compared to those in the case of latter. This is due to the fact that while conventional weapons are one-shot weapons, laser DEWs have unlimited magazine. The total number of shots a laser can fire is only limited either by the amount of chemical fuel for chemical lasers or electrical power for solid-state and fibre lasers. Projectile weapon systems, guided missiles in particular, expend a lot of expensive hardware such as rocket motors, guidance systems, avionics, seekers and airframes every time they are fired. In the case of laser weapons, the cost of each laser firing is essentially the cost of the chemical fuel or the electrical power consumed. As an example, a fourth-generation shoulder-fired surface-to-air missile MANPADS of the type FIM-92 Stinger series costs about US$ 40 000. This missile can be used against an aircraft on a single mission. A similar mission from a land-based laser DEW system can be carried out with a 50–100 kW solid-state laser by firing the laser beam for a dwell time of about 5 s. Going by present-day technology level, this laser system would draw about 400–500 kW of electrical power for a period of 5 s, which is same as the electrical power consumed by a 100 W bulb in 7 hours.

Laser-based DEWs also have some limitations. Some of these include their line-of-sight dependence, requirement of finite dwell time, problems due to atmospheric attenuation and turbulence and ineffectiveness against hardened structures. Unlike projectile weapons that instantly destroy the target upon impact, laser weapons require a minimum dwell time of the order 3–5 s to deposit sufficient energy for target destruction.

Laser weapons require direct line-of-sight to engage a target. Their effectiveness is reduced or neutralized by the presence of any object or structure in front of the target that cannot be burnt through. The effectiveness of the laser weapon is adversely affected by the atmospheric conditions. The laser beam suffers attenuation due to absorption and scattering by airborne particles and gas molecules, deterioration of beam quality in the form of deformation of the laser beam wave front and an increase in the laser beam spot size at the target. Laser weapons are not very effective against hardened structures; equipment such as antennas, sensors and external fuel stores mounted on these structures can however be targeted effectively.

15.4.2 Operational Scenario

The operational scenario of directed-energy laser weapons is broadly categorized as short- and medium-range tactical missions and long-range strategic missions. Some of the important application areas of tactical class laser weapons include stand-off neutralization of ordnances such as mines; unexploded ordnances and improvised explosive devices (IEDs); ground-based defence against rockets, artillery and mortars (RAM); ground-based capability to destroy unmanned aerial vehicles (UAVs) of the adversary; airborne defence of aircraft against man portable air defence systems (MANPADS) such as shoulder-fired surface-to-air missiles; and ship defence against manoeuvreing cruise missiles and tactical ballistic missiles.

Stand-off neutralization of ordnances requires laser power in the range 1–2 kW for operational ranges up to 300 m. Solid-state or fibre laser sources operating around 1.0 μm are used for the purpose. AVENGER and ZEUS of the US and Israeli THOR are laser ordnance systems with comparable specifications. Anti-UAV operations up to a range of 8–10 km require about 100 kW of laser power. Operational ranges of 5–6 km are possible with 50 kW laser systems. Again the preferred lasers are solid-state and fibre lasers. COIL may also be used. For applications such as air defence against RAM targets, rocket-propelled grenades (RPG), battlefield missiles and laser-guided munitions, typically 100 kW of laser power is needed for operational ranges of 5–10 km. The photograph in Figure 15.26 illustrates the concept of a laser weapon for air-defence application.

Long-range *strategic applications* of laser-based DEW systems mainly include ballistic missile defence, space control such as space-based lasers and anti-satellite applications. In all these applications, operational ranges are generally in the range of hundreds to thousands of kilometres and required power levels of the order 1–20 MW depending upon the actual mission. Space control applications such as anti-satellite applications require relatively much higher power than that needed for ballistic missile defence.

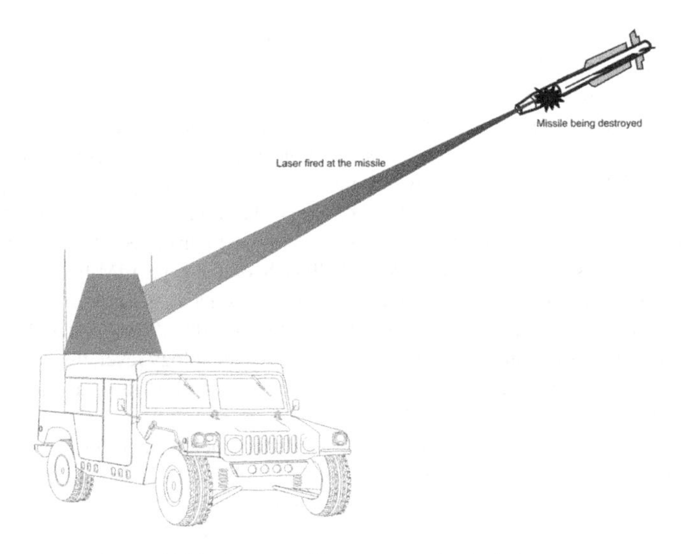

Figure 15.26 Concept of laser weapon in air defence role.

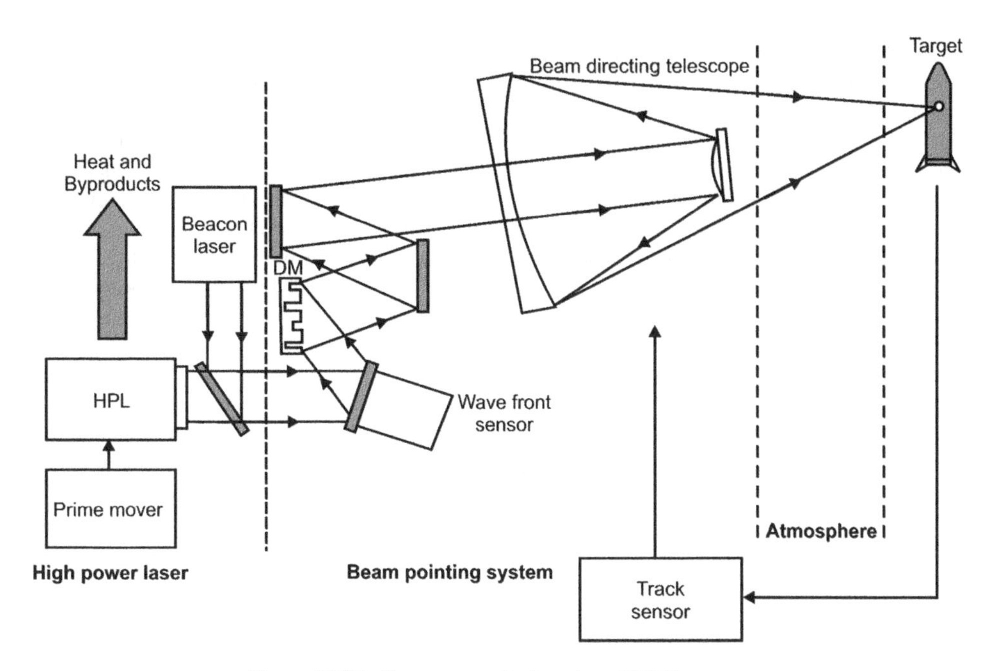

Figure 15.27 Components of a laser-based DEW system.

15.4.3 Components of Directed-energy Weapon Systems

Unlike conventional military applications of laser systems such as laser rangefinders and target designators which primarily comprise a laser source of desired specification, laser-based directed-energy weapons are much more than a high-power laser source. Figure 15.27 shows the block schematic arrangement representative of the different components of a laser DEW system and their interconnections. The subsystems other than the high-power laser source are required for the purpose of directing the laser beam to the intended point on the target, keeping it there for the desired dwell time and producing the desired value of fluence on the target. Laser-based DEW essentially comprises two major subsystems: the high-power laser source and the beam control system.

A *high-power laser source* generates a laser beam of the desired power level, beam quality and time duration to be able to inflict the intended damage on the enemy target. Potential candidates for building high-power laser sources include gas dynamic CO_2 lasers, chemical lasers including HF/DF and chemical oxy-iodine lasers, solid-state lasers and fibre lasers. The ultimate quality desired of a potential high-power laser source is its ability to produce as high a power density at the target for a given aperture size as possible. Power density achievable for a given source and beam director combination is inversely proportional to the square of wavelength and the beam quality parameter for given beam director aperture size and operational range. Lower wavelength and a smaller value of beam quality parameter therefore signifies greater suitability as a high-power laser source. A smaller value of beam quality parameter indicates higher beam quality, with unity being the ideal value. A shorter wavelength also has better target coupling efficiency. In addition, power density at the target also depends upon atmospheric losses. Other important factors include power scalability, operator safety and ease of mounting.

The primary function of the *beam control system* is to precisely point and focus the laser energy at the designated point on the target and keep it there for sufficient duration to cause intended damage to the target. In addition to target tracking and beam pointing and focusing, the beam control system has many other subsystems to perform different tasks which include beam shaping and beam stabilization and wave front control for atmospheric correction.

In addition to the laser source and beam pointing system, the propagation of the high-power laser beam through the atmosphere and its coupling with the target, also called target lethality, are the other important constituents determining the efficacy of laser weapon.

The laser beam has to propagate over long atmospheric paths, and the resulting propagation effects control the overall effective range or lethality range of the weapon system. Propagation effects include diffraction, absorption and turbulence. The *diffraction* effect tends to spread out the laser energy as the laser beam propagates; the amount of spreading is proportional to the wavelength to beam aperture diameter ratio. This limits the smallest spot diameter to which the laser beam can be focused.

Scattering is due to atmospheric constituents and results in a loss of laser energy from the beam path in other directions. Longer wavelengths have relatively lower scattering losses. *Absorption* by atmospheric constituents also causes loss of energy, consequently decreasing the laser power reaching the target end. There is a very strong dependence of absorption parameter on location, time of year and weather pattern.

Turbulence causes the beam to spread, resulting in an increase in laser spot size and reduction in power density at the target end. Turbulence is measured by the C_n^2 parameter, which is a strong function of altitude and wind speed and other atmospheric parameters. Thermal blooming is another parameter that causes laser beam aberration. This arises out of heating effects produced by absorption leading to a refractive index gradient in the laser beam path. This further leads to the formation of a negative lens and causes the laser beam to spread. Thermal blooming effects can be ignored for rapidly moving target such as airborne targets.

15.4.4 International Status

A large number of directed-energy laser weapon systems are presently reported to be under development and up-gradation. Some are experimental, some are technology demonstrators and others are being upgraded and ruggedized to become realistic battlefield weapon systems in the near future. Some of the more-talked-about systems are briefly described in the following.

The short-range high-energy laser (HEL) mounted on a tracked vehicle for use as an air defence system against low-flying high-performance aircraft, missiles and attack helicopters is a laser-based DEW system jointly developed by the two German companies Diehl and LFK. The system is configured around a gas dynamic CO_2 laser emitting at 10.6 μm and has associated target acquisition and tracking sensors. The beam director uses a 1 m diameter focusing mirror mounted on an extendible arm for delivery of a high-power laser beam to the target. Target damage is achieved by focusing the laser radiation into a small diameter spot, producing very high energy density and causing the target material to become successively heated, melted and vaporized. The system, which has been reported to be successfully tested, has an operational range of 8 km.

TRW's general area defence integrated anti-missile (GARDIAN) laser system is also a short-range complement to surface-to-air missile defence designed to engage discrete ballistic threats at longer ranges. The system generates a 400 kW laser beam, which is delivered through a 0.7 m beam pointer/tracker. Depending upon weather conditions, the laser can destroy targets at altitudes from ground-hugging heights up to 15 km. The system has a target acquisition and tracking system and, once locked, the laser stays locked to the target until it is destroyed. The system can carry fuel for firing 60 shots.

One of the very early directed-energy laser systems, a technology demonstrator, is TRW's mid-infrared advanced chemical laser (MIRACL). It is a 2.2 MW CW deuterium fluoride laser with a maximum lasing duration of 70 s. It uses a 1.8 m Sealite pointing and tracking device (Figure 15.28). The system was reported have undergone trials against different types of targets at the White Sands Missile Range in New Mexico. In one of the tests conducted in 1996, a small fraction of laser power was used to destroy a 122 mm short-range artillery rocket in flight. The laser beam was locked to the target for 15 s. The laser is also reported to have tested against sea-skimming missiles.

TRW have also developed the high-energy laser weapon system (HELWEPS), which is also a chemical laser using ethylene, hydrogen and fluorinated nitrogen as the active medium. The system is based on their earlier experience gained from building MIRACL, and has an integral electro-optic tracker.

Figure 15.28 Sealite beam director of MIRACL system (http://en.wikipedia.org/wiki/File:Slbd_front.jpg).

The chemical oxy-iodine laser (COIL) is in serious contention for use as a laser weapon, particularly for operation from an aerial platform. The famous airborne laser (ABL) uses COIL and is configured on a Boeing 747–400 freighter aircraft as shown in Figure 15.29. It is capable of destroying a ballistic missile in boost phase. In operation, the aircraft patrols the friendly air space. If an enemy missile launch is detected by a variety of sensors, this information is relayed to the aircraft configured as the high-power laser system. The nose of the aircraft is fitted with a 1.6 m beam director that focuses the high-power laser radiation from a megawatt-class COIL onto the missile once it rises above the cloud cover. The intercept range of ABL has been put at 200 km from its stand-off position. A performance evaluation of ABL began in July 2002. In February 2010, the system was successfully used to destroy a ballistic missile. The

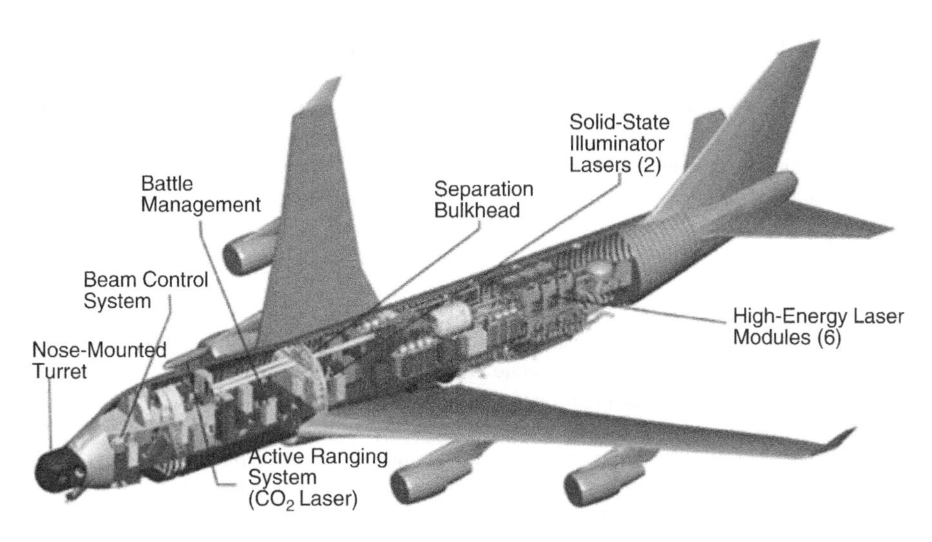

Figure 15.29 Airborne laser.

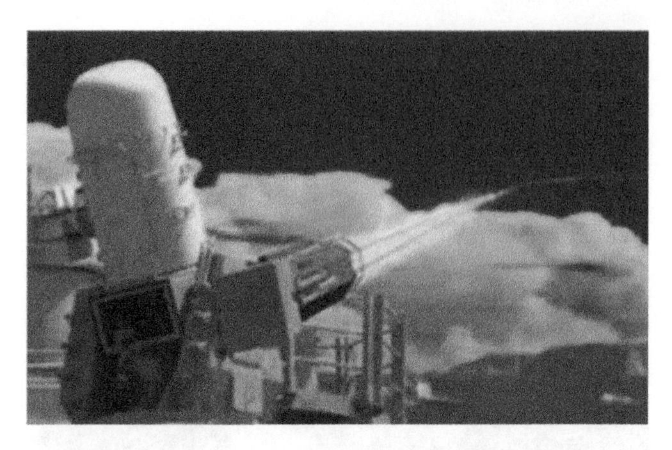

Figure 15.30 Laser Phalanx (Courtesy of Raytheon Company).

system was subsequently considered to be operationally unviable and the ABL program has been reportedly shelved since December 2011.

The advanced tactical laser (ATL) uses an 80 kW COIL and is mounted on a Boeing C-130H Hercules aircraft. It is intended to be used for covert activities such as setting fires to vehicles, disabling communication antennas, satellite and radar dishes and breaking electrical power lines. The ATL offers the mobility of a small aircraft, the high-resolution imagery for target identification and the ability to localize damage to a small area of less than a foot in diameter from a range of 8–10 km.

Another well-known laser-based DEW system is the tactical high-energy laser (THEL) and its mobile version M-THEL. The system uses a deuterium fluoride (DF) laser operating at 3.8 μm. It is a tactical air defence system against short-range rockets. The THEL demonstrator was successfully tested repeatedly between 2000 and 2004, destroying a number of 122 mm and 160 mm Katyusha rockets, multiple artillery shells and mortar rounds, including a salvo attack by mortar.

A space-based laser (SBL) is the ultimate objective of the US directed-energy laser program. A space-based laser, designed to intercept ICBMs and other strategic and tactical missiles, is proposed to be configured around a 20 MW hydrogen fluoride (HF) laser operating at 2.7 μm. The system will be deployed in a 800–1300 km orbit and will have an expected target engagement range of 4000–12 000 km.

Several directed-energy laser systems based on solid-state and fibre lasers are being developed and tested for tactical mission needs ranging from ordnance neutralization to anti-missile and anti-RAM applications. Raytheon has developed and successfully tested a directed-energy laser system called Laser Phalanx (Figure 15.30), employing a 20 kW industrial fibre laser. The system has been successfully demonstrated against a static mortar from a distance of 0.5 km. Raytheon has also successfully tested a ship-mounted solid-state laser weapon called Laser Weapon System (LaWS) in the shooting of four drones.

15.5 Summary

- Armed forces engaged in a low-intensity conflict operate with fewer soldiers, a reduced range of tactical equipment and limited scope to operate in military manner. Use of artillery is also avoided in the case of urban conflicts and use of air power is often restricted to surveillance and transportation of personnel and equipment.
- The key advantage of the use of laser technology in low-intensity conflicts mainly stems from its near-zero collateral damage, speed of light delivery and potential for building non-lethal weapons.
- Some of the well-established laser devices in LIC applications include: laser dazzlers for close combat operations, mob/riot control and protection of critical infrastructures from aerial threats; lidar sensors

for detection of chemical, biological and explosive agents; femtosecond lasers for imaging of concealed weapons; and lasers for sniper and gun fire location identification. The use of laser vibrometry and electron speckle interferometry techniques for the detection of buried mines and high-power lasers for disposal of unexploded ordnances are emerging applications of laser technology for homeland security.

- Due to their acute toxicity, rapidity of action, indetectability by human senses and economic viability, nerve agents (e.g. organophosphonate compounds) and blister agents (e.g. mustard compounds) are the two potentially very harmful classes of chemical warfare agents.
- Lidar-based detection is the most practically realizable technique for the stand-off detection of chemical and biological warfare agents. Commonly exploited physical phenomena forming the basis of chemical and biological agent detection include elastic backscattering, laser-induced fluorescence and differential absorption.
- Ultraviolet laser-induced fluorescence (UV-LIF) is used for the detection of biological molecules since most fluoresce when excited by ultraviolet radiation.
- The differential absorption technique known as DIAL uses two wavelengths: one corresponds to the wavelength of peak absorption of the targeted molecule and the other corresponds to the weak absorption of the targeted molecule. The ratio of the two received backscattered signals is a measure of the concentration of the targeted chemical warfare agent.
- The stand-off detection of biological warfare agents is based on the concept of laser-induced fluorescence (LIF), which is the emission from atoms or molecules after they have been excited to higher energy levels by another laser. The emission takes place at a wavelength that is higher than the wavelength of laser light exciting it.
- Commonly used technologies for stand-off detection of explosives include laser-induced breakdown spectroscopy (LIBS), Raman spectroscopy and its variants, LIF spectroscopy and IR spectroscopy. These are all trace detection methods; bulk detection methods such as millimetre wave imaging and terahertz spectroscopy are also available.
- LIBS focuses a high-energy laser beam on the trace sample to break down a small part of the sample into a plasma of excited ions and atoms. The plasma emits light that is characteristic of emissions from ionic, atomic and small molecular species. These light emissions are detected by a spectrometer to identify the elemental composition.
- The basis of detection in the case of Raman spectroscopy is the shift in the wavelength caused by inelastic Raman scattering by the target molecule. The inelastic scattering of impinging photons where some energy is lost to (or gained from) the target molecule returns scattered light with a higher (or lower) wavelength. The difference is dictated by the energy of vibrational modes of the target molecule and therefore constitutes the fingerprint or the basis of identification. Complex mixtures are identified using algorithms for pattern recognition.
- Almost all the laser-based stand-off detection methods have detection limits too high for them to be suitable for the detection of explosives in field conditions, which requires the capability to detect traces in the vapour phase or in the form of particles. Another problem is insufficient selectivity to identify a target explosive in the presence of interferents.
- The electro-optical target locator's functioning is based on the cat's eye effect. The target optical device is illuminated by a laser beam, returns a fraction of the beam as a retro-reflected beam and is received by a sensitive receiver.
- Moderately high-power lasers are a very attractive choice for neutralization of unexploded ordnances. The advantages their use include large magazine, high precision, controllable effects with reduced collateral damage and ensured and fast disposal from safe stand-off ranges. Unexploded ordnance is disposed of by focusing a high-power laser beam on the ordnance casing, thereby heating it until the temperature of backplane of the casing exceeds the ignition temperature of explosive filler, leading to a low-level detonation. AVENGER, ZEUS and THOR are well-known laser ordnance neutralization systems.
- Laser dazzlers are emerging internationally as a new non-lethal alternative for law enforcement, homeland security, border patrol, coastal protection, infrastructure protection and host of other low-

intensity conflict scenarios. A laser dazzler emits a high-intensity laser beam in the visible band, usually in the blue-green region, to temporarily impair the vision of the adversary without causing any permanent or lasting injury to the subject's eyes.

- The choice of laser dazzler performance parameters such as laser power, spot size at the target and laser power density are dictated by the nature of deployment. The beam shaping and directing optics is so designed to achieve the desired value of the nominal ocular hazard distance (NOHD) and a laser power density that does not exceed the maximum permissible exposure (MPE) figure dictated by ANSI for eye safety.
- Some of the well-known portable laser dazzler systems include the Compact High Power (CHP) laser dazzler, GLARE MOUT, GLARE GBD-IIIC, GLARE LA-9/P, Sabre 203, JD-3 laser dazzler, ZM-87 portable laser disruptor, Photonic Disruptor (classified as TALI) and Dazer Laser.
- Electro-optic countermeasures are broadly classified as passive and active countermeasures. Passive countermeasures for platform protection include use of armour, camouflage, fortification and other protection technologies such as a self-sealing fuel tank. In the case of active countermeasures, means are adopted to counterattack the incoming threat. This may be achieved by neutralizing one of the elements responsible for the functioning of the guided threat or by physically destroying the incoming threat by launching a counter projectile in that direction.
- Active countermeasures comprise soft-kill and hard-kill countermeasures. Soft-kill countermeasures change the electromagnetic, acoustic or other forms of signatures of the platform to be protected. Hard-kill countermeasures physically destroy the incoming threat.
- A laser warning sensor system is an indispensable component of any EOCM system, whether designed for soft-kill or hard-kill. It provides information on the type and direction of arrival of laser threats. A laser warning system provides valuable information that enables the platform crew to decide on the countermeasure system to be used. Some common laser warning sensor systems are AVITRONICS LWS-300 and Goodrich's AN/AVR-2 series systems.
- Laser decoy systems Type-405 from GEC-Marconi and DHY-322 from CILAS are two anti-sensor systems designed and built to divert the laser-guided munitions from their intended targets.
- STINGRAY laser detection and countermeasures system from Lockheed Martin is an EOCM system designed to protect the frontline forces by accurately locating the enemy's optical and electro-optical fire control systems. Russian SHTORA-1 has been designed to protect the armoured fighting platforms from laser-guided munitions, laser beam riders and anti-tank guided missiles. Russian ARENA and Israel's TROPHY are common examples of active protection systems.
- Infrared countermeasures (IRCM) are employed to protect aircraft from surface-to-air and air-to-air heat-seeking missiles. The IRCM systems achieve this by confusing the infrared guidance system of the missiles and forcing them to deviate from their intended trajectory. Active IRCM systems use a modulated source of infrared radiation with a higher intensity than the emission from the target. If this modulated radiation matches that used by the seeker of the target missile, the tracking logic of the missile generates a false tracking command, forcing the infrared seeker to go out of lock. Once the lock is broken, it is very hard for the missile to regain locking condition.
- In the case of a directional infrared countermeasures (DIRCM) system, infrared jammers are mounted on a movable turret and operate in conjunction with a missile approach warning system. Most DIRCM systems use ultraviolet sensors to detect the approach of the missile. Detection in the infrared bands have also been exploited. The detection system provides a cue to the tracking system, which uses information from the detection system to track the incoming missile. The infrared jammers direct the infrared jamming energy at the seeker. The modulation scheme can be cycled to try to defeat a variety of seekers. Some common examples of DIRCM systems include MUSIC of Israel, BAE Systems' AN/ALQ-212 ATIRCM and Northrop Grumman's GUARDIAN and CIRCM programs.
- The three major categories of directed-energy weapons are: microwave-based DEWs, laser-based DEWs and particle beam weapons. A directed-energy weapon system primarily uses directed energy in the form of a concentrated beam of electromagnetic energy or atomic or subatomic particles in the targeted direction to cause intended damage to the enemy's equipment, facilities and personnel. The intended damage could be lethal or non-lethal.

- The primary advantages of laser-based DEWs include speed-of-light delivery, near-zero collateral damage, multiple target engagement and rapid re-targeting capability, immunity to electromagnetic interference and no influence of gravity. Deep magazine and low cost per shot are the other advantages.
- Laser-based DEWs suffer the limitations of line-of-sight dependence, requirement of finite dwell time, problems due to atmospheric attenuation and turbulence and ineffectiveness against hardened structures. Unlike projectile weapons that instantly destroy the target upon impact, laser weapons require a minimum dwell time of the order 3–5 s to deposit sufficient energy for target destruction.
- The operational scenario of directed-energy laser weapons is broadly categorized as short- and medium-range tactical missions and long-range strategic missions. Some of the important application areas of tactical class laser weapons include stand-off neutralization of ordnances such as mines unexploded ordnances and IEDs; ground-based defence against RAM; ground-based capability to destroy UAV of the adversary; airborne defence of aircraft against MANPADS such as shoulder-fired surface-to-air missiles; ship defence against manoeuvring cruise missiles; and tactical ballistic missiles.
- Ballistic missile defence and anti-satellite applications are the common long-range strategic applications of laser weapons. Some of the well-known laser weapon technology demonstrators include THEL, M-THEL, ATL, ABL, MIRACL, LaWS, PHALANX and SBL.

Review Questions

15.1. What is a differential absorption lidar (DIAL)? With the help of a simplified block schematic arrangement, briefly describe how a DIAL system can be used for the detection and identification of chemical warfare agents.

15.2. Which lidar technique is suitable for the detection of biological warfare agents? Briefly describe the principle of operation of such a biological lidar system.

15.3. Name the different laser-based spectroscopy techniques suitable for stand-off detection of explosive agents. Compare these techniques in terms of their sensitivity and selectivity characteristics and also their ability to detect the explosive agents in the presence of interferents.

15.4. Briefly describe the following in the case of laser dazzlers:
  a. non-ocular hazard distance
  b. ANSI standards
  c. choice of suitable laser wavelength.

15.5. What is the cat's eye effect with respect to optical systems? How can this effect be exploited to detect the presence of optical and electro-optical devices? What is the significance of such a device in a homeland security scenario?

15.6. What is the basic operational principle behind the use of a high-power laser radiation to safely neutralize unexploded ordnances? Which laser wavelength is suitable for this purpose and why? Name any one laser-based ordnance neutralization system along with its source of origin and salient features.

15.7. Differentiate between
  a. passive and active electro-optic countermeasures
  b. hard-kill and soft-kill countermeasures systems
  c. infrared countermeasures and directed countermeasures

15.8. Briefly describe the following countermeasures systems
  a. ARENA active protection system
  b. DHY-322 decoy system
  c. GUARDIAN DIRCM.

15.9. Name the different categories of directed-energy weapon systems. What are the key advantages when compared to conventional kinetic-energy weapons? Outline some of the key limitations of laser-based directed-energy weapon systems.

15.10. What are the key elements of a laser-based DEW system? Briefly describe the role of different elements along with their significance in the overall performance of the weapon system.

15.11. Briefly describe at least three tactical and two strategic applications where laser-based DEW can be effectively used.

15.12. Name different potential high-power laser sources. Compare them in terms of their suitability for tactical and strategic applications. Name the high-power laser sources used in MIRACL, THEL, ABL and Laser PHALANX DEW systems.

Self-evaluation Exercise

Multiple-choice Questions

15.1. The lidar technique most suitable for the detection of chemical warfare agents is
a. differential absorption lidar
b. UV-LIF lidar
c. Raman spectroscopy
d. none of these.

15.2. The lidar technique most suitable for the detection of biological warfare agents is
a. differential absorption lidar
b. UV-LIF lidar
c. resonant Raman spectroscopy
d. photo-acoustic spectroscopy.

15.3. Which of the following spectroscopy techniques is likely to give the highest sensitivity?
a. Raman
b. resonant Raman
c. LIBS
d. LIF.

15.4. The most suitable laser wavelength for building laser dazzlers is
a. 1064 nm
b. 1540 nm
c. 650 nm
d. 532 nm.

15.5. The two important laser dazzler parameters from an operational viewpoint are
a. laser power and spot size
b. laser beam divergence and beam quality
c. laser wavelength and beam quality
d. MPE and NOHD.

15.6. Which of the following wavelengths is more suitable for laser-based ordnance disposal applications?
a. 975 nm
b. 2.7 μm
c. 1050–1070 nm
d. 8–12 μm.

15.7. Which of the following systems is a laser dazzler?
a. AVENGER
b. THOR
c. GLARE-MOUT
d. MIRACL

15.8. Which of the following is a laser-based DEW system?
a. DAZER
b. MIRACL
c. THOR
d. ZEUS.

15.9. Name the system designed for neutralization of ordnances from the following:
a. THOR
b. ABL
c. ATL
d. PHALANX

15.10. A tank is fitted with a laser countermeasures system comprising a laser warning sensor and a smoke/aerosol generation systems. This type of countermeasure system is classified as
a. an active countermeasures system
b. a passive countermeasures system
c. a hard-kill countermeasures system
d. none of these.

15.11. A DIRCM system is intended to protect the aircraft against
a. a laser-guided missile attack
b. MANPADS
c. an anti-aircraft gun attack
d. a laser-based DEW attack.

15.12. A DIRCM system comprises the following components:
a. set of sensors covering UV and IR bands
b. set of laser sources covering near-, mid- and far-infrared bands
c. set of UV and IR sensors interfaced with a mid-infrared jamming laser source
d. none of these.

15.13. The advantages of laser-based DEW systems include
a. speed of light delivery
b. deep magazine and low cost per shot
c. no gravity influence
d. all of the above.

15.14. The airborne laser program uses which type of laser source?
a. solid-state laser
b. fibre laser
c. gas dynamic laser
d. chemical oxy-iodine laser.

15.15. In which of the following laser DEW systems is a sealite beam director used?
a. ABL
b. M-THEL
c. MIRACL
d. LaWS.

Answers

1. (a) 2. (b) 3. (c) 4. (d) 5. (d) 6. (c) 7. (c) 8. (b) 9. (a) 10. (a) 11. (b) 12. (c) 13. (d) 14. (d) 15. (c)

Bibliography

1. *Understanding Lasers: An Entry Level Guide*, 2008 by Jeff Hecht, IEEE Press.
2. *Military Laser Technology for Defence*, 2011 by Alastair D. McAulay, Wiley-Interscience.
3. *Electro-Optics Handbook*, 2000 by Ronald Waynant and Marwood Ediger, McGraw-Hill, Inc..
4. *High Power Laser Handbook*, 2011 by Hagop Injeyan and Gregory Goodno, McGraw-Hill, Inc..
5. *Directed Energy Weapon Technologies*, 2012 by Bahman Zohuri, CRC Press.
6. *An Introduction to Laser Weapon Systems*, 2009 by Glen P. Perram, Directed Energy Laser Society.
7. *The Infrared and Electro-Optic Systems Handbook Volume-7*, 1993 by Joseph, S. Accetta, SPIE International Society for Optical Engineering.
8. *Space Weapons Earth Wars*, 2002 by Bob Preston, Rand Corporation.
9. *Strategic Technologies for the Military*, 2009 by Ajey Lele, Sage Publications Pvt. Ltd.
10. *Effects of Directed Energy Weapons*, 2012 by Philip E. Nielsen, CreateSpace.

Appendix A

Laser Safety

A.1 Laser Damage

Due to their low beam divergence and resultant high intensity even after having travelled long distances, lasers can be potentially dangerous to the eyes even at moderate power level and to the skin at relatively higher power level. Lasers may cause permanent damage to the eye at moderate power levels of only a few milliwatts due to the focusing action of the eye's lens that concentrates the laser energy into an extremely small spot on the retina. The damage to the photoreceptor cells of the retina is caused by the transient rise in temperature at the focal spot.

Damage can result from both thermal and photochemical effects. Thermal damage occurs when tissues are heated to the point of denaturation of proteins. In the case of photochemical damage, laser light triggers chemical reactions in the tissues. Different laser wavelengths produce different pathological effects. Photochemical damage mainly occurs with ultra-short-wavelength lasers emitting in the ultraviolet and blue regions of the electromagnetic spectrum, while visible and infrared wavelengths are harmful due to thermal damage. Table A.1 lists different wavelength ranges and the corresponding pathological effects of them.

A.2 Maximum Permissible Exposure

Maximum permissible exposure (MPE) is the highest power or energy density of the light source measured in $W\,cm^{-2}$ or $J\,cm^{-2}$, respectively, that is considered safe and has negligible probability of causing any damage to the eyes. MPE is usually taken as 10% of the power or energy density that has 50% probability of causing damage under worst-case conditions.

Table A.1 Pathological effects of different wavelength ranges

Wavelength range	Pathological effect
180–315 nm	Photokeratitis: inflammation of cornea, equivalent to sun burn
315–400 nm	Photochemical cataract
400–780 nm	Photochemical damage to retina, retinal burn
780–1400 nm	Cataract, retinal burn
1.4–3.0 μm	Aqueous flare, cataract, corneal burn
3.0–1000 μm	Corneal burn

Lasers and Optoelectronics: Fundamentals, Devices and Applications, First Edition. Anil K. Maini.

Table A.2 Laser safety standard: old system

Safety class	Description
I	A class I laser is safe and there is no possibility of eye damage. This can be either due to low output power with no risk of eye damage after hours of exposure, or due to the laser being contained inside an enclosure such as in the case of a compact disk player or laser printer.
II	This class refers only to lasers emitting in the visible spectrum with output power up to 1.0 mW. The lasers are safe due to the blink action of the eye unless deliberately staring into the beam for an extended period of time. Most laser pointers belong to this category.
IIa	For continuous exposure for a period of >1000 s, lasers at the low power end of the class II category may produce retinal burn.
IIIa	Lasers with power level >1.0 mW and <5.0 mW and power density <2.5 mW cm^{-2} belong to class IIIa. These lasers are dangerous when used in combination with optical instruments and also dangerous to the naked eye for direct viewing for more than 2.0 min.
IIIb	Lasers with power level 5.0–500 mW belong to Class IIIb. Direct viewing of these lasers may cause damage to the eye. A diffuse reflection is generally not hazardous but both direct viewing and specular reflection are equally dangerous. Class IIIb lasers, which are towards the high power end, may present a fire hazard or cause skin burn.
IV	Lasers belonging to class IV have power levels >500 mW. Class IV lasers can cause permanent eye damage or skin burn even when used without optical instrumentation. Diffuse reflection from these lasers can also be hazardous to eyes or skin within the nominal ocular hazard zone. Many industrial, medical, scientific and military lasers are of the class IV category.

Table A.3 Laser safety standard: revised system

Safety class	Description
1	The accessible laser radiation is not dangerous under reasonable conditions of use. A class 1 laser is safe under all conditions of normal use. This implies that while viewing a class 1 laser with the naked eye or with typical magnifying optics such as a telescope or a microscope; the MPE limit cannot be exceeded.
1M	The accessible laser radiation is not hazardous, provided that no optical instruments are used which may, for example, focus the radiation. A class 1M laser is safe for all conditions except when passed through magnifying optics such as telescopes and microscopes. The classification is applicable to lasers with power level greater than the limit specified for class 1 lasers provided the laser energy entering the pupil of the eye does not exceed the limits of class 1 lasers due to the large divergence of the laser.
2	The accessible laser radiation is limited to the visible spectral range (400–700 nm) and to 1 mW accessible power. Due to the blink reflex action of the eye, it is not considered dangerous for limited exposure up to 0.25 s.
2M	A class 2M laser is also safe because of the reflex action of the eye, with the additional restriction that it is not viewed through optical instruments. As for class 1M lasers, the classification is also applicable to laser beams with power level >1 mW provided the beam divergence is large enough to prevent laser energy passing through the pupil of the eye to exceed the limits of class 2 lasers.
3R	The accessible radiation may be dangerous for the eye, but can have at most 5 times the permissible optical power of class 2 lasers emitting in visible spectrum and class 1 for other wavelengths. The MPE can be exceeded with class 2M lasers with a low risk of injury.
3B	The accessible radiation may be dangerous for the eye and, under particular conditions, also for the skin. Diffuse radiation scattered from a diffuse target is normally harmless. The accessible emission limit is 500 mW for CW visible lasers emitting in visible spectrum and 30 mW for pulsed lasers. In the case of direct viewing of class 3B lasers, protective eye wear is required.
4	The accessible radiation of a class 4 laser is very dangerous for the eye and for the skin. Light from diffuse reflections and indirect viewing may be hazardous for the eye. Class 4 lasers must be equipped with a key switch and have an in-built safety interlock. Most industrial, medical, scientific and military lasers belong to this category.

A.3 Nominal Hazard Zone

Determining the required safety measures based on safety classification of the laser may not always be correct as it does not take into consideration the effect of beam divergence. A strongly focused laser beam can be so divergent that within a moderate distance after the focus, the intensity falls below the

Table A.4 MPE for ocular exposure (intra-beam viewing)

Wavelength (μm)	Exposure duration, t (s)	MPE ($J\,cm^{-2}$)	MPE ($W\,cm^{-2}$)
0.180–0.302	10^{-9}–3×10^{4}	3×10^{-3}	—
0.303	10^{-9}–3×10^{4}	4×10^{-3}	—
0.304	10^{-9}–3×10^{4}	6×10^{-3}	—
0.305	10^{-9}–3×10^{4}	10×10^{-3}	—
0.306	10^{-9}–3×10^{4}	16×10^{-3}	—
0.307	10^{-9}–3×10^{4}	25×10^{-3}	—
0.308	10^{-9}–3×10^{4}	40×10^{-3}	—
0.309	10^{-9}–3×10^{4}	63×10^{-3}	—
0.310	10^{-9}–3×10^{4}	0.10	—
0.311	10^{-9}–3×10^{4}	0.16	—
0.312	10^{-9}–3×10^{4}	0.25	—
0.313	10^{-9}–3×10^{4}	0.40	—
0.314	10^{-9}–3×10^{4}	0.63	—
0.315–0.400	10^{-9}–10	$0.56 \times t^{0.25}$	—
0.315–0.400	10–3×10^{4}	1.0	—
0.400–0.700	10^{-9}–18×10^{-6}	0.5×10^{-6}	—
0.400–0.700	18×10^{-6}–10	$1.8 \times t^{0.75} \times 10^{-3}$	—
0.400–0.450	10–100	1.0×10^{-2}	—
0.450–0.500	10–T_1	—	1×10^{-3}
0.450–0.500	T_1–100	$C_B \times 10^{-2}$	—
0.400–0.500	100–3×10^{4}	—	$C_B \times 10^{-4}$
0.500–0.700	10–3×10^{4}	—	1.0×10^{-3}
0.700–1.050	10^{-9}–18×10^{-6}	$5.0 \times C_A \times 10^{-7}$	—
0.700–1.050	18×10^{-6}–10	$1.8 \times C_A \times t^{0.75} \times 10^{-3}$	—
0.700–1.050	10–3×10^{4}	—	$C_A \times 10^{-3}$
1.050–1.400	10^{-9}–50×10^{-6}	$5.0 \times C_C \times 10^{-6}$	—
1.050–1.400	50×10^{-6}–10	$9.0 \times C_C \times t^{0.75} \times 10^{-3}$	—
1.050–1.400	10–3×10^{4}	—	$5.0 \times C_C \times 10^{-3}$
1.400–1.500	10^{-9}–10^{-3}	0.1	—
1.400–1.500	10^{-3}–10	$0.56 \times t^{0.25}$	—
1.400–1.500	10–3×10^{4}	—	0.1
1.500–1.800	10^{-9}–10	1.0	—
1.500–1.800	10–3×10^{4}	—	0.1
1.800–2.600	10^{-9}–10	0.1	—
1.800–2.600	10^{-3}–10^{-3}	$0.56 \times t^{0.25}$	—
1.800–2.600	10–3×10^{4}	—	0.1
2.6–1000	10^{-9}–10^{-7}	1.0×10^{-2}	—
2.6–1000	10^{-7}–10	$0.56 \times t^{0.25}$	—
2.6–1000	10–3×10^{4}	—	0.1

Notes:

1. To calculate MPE for $\lambda = 0.180$–$0.400\,\mu m$, use the $J\,cm^{-2}$ or $W\,cm^{-2}$ value or $0.56 \times t^{0.25}$ (whichever is lowest).
2. For multiple pulses for $\lambda = 0.400$–$1.400\,\mu m$, apply the correction factor C_P given in Table A.6.
3. For MPE for diffuse reflections for $\lambda = 0.400$–$1.400\,\mu m$, multiply the corresponding MPEs by correction factor C_E given in Table A.6.
4. For correction factors C_A, C_B and T_1, see Table A.6.

MPE level for the eye. Nominal hazard zone is defined as the zone within which safe exposure level may be exceeded.

A.4 Safety Classification

Based on how hazardous a given laser can be to the designers, operators and users of the device, it is assigned a safety class. The classification is based on the concept of accessible emission limit (AEL), which is defined for each of the laser classes. There are two classification systems: the American system which was in use until 2002 and the revised system which is part of the IEC 60825 standard and has been included in the ANSI Z136.1 laser safety standard of the US since 2007. The former designates safety class using Roman numerals in the US and Arabic numerals in the EU; the revised system uses Arabic numerals in all jurisdictions. In both old and revised classifications, lasers are classified into four main classes and a few subclasses in terms of maximum output power for different wavelength ranges. Salient features of laser safety classifications in the old and revised systems are listed in Tables A.2 and A.3 respectively.

Table A.5 MPE for skin exposure

Wavelength (μm)	Exposure duration, t (s)	MPE (J cm^{-2})	MPE (W cm^{-2})
0.180–0.302	10^{-9}–3×10^4	3×10^{-3}	—
0.303	10^{-9}–3×10^4	4×10^{-3}	—
0.304	10^{-9}–3×10^4	6×10^{-3}	—
0.305	10^{-9}–3×10^4	10×10^{-3}	—
0.306	10^{-9}–3×10^4	16×10^{-3}	—
0.307	10^{-9}–3×10^4	25×10^{-3}	—
0.308	10^{-9}–3×10^4	40×10^{-3}	—
0.309	10^{-9}–3×10^4	63×10^{-3}	—
0.310	10^{-9}–3×10^4	0.10	—
0.311	10^{-9}–3×10^4	0.16	—
0.312	10^{-9}–3×10^4	0.25	—
0.313	10^{-9}–3×10^4	0.40	—
0.314	10^{-9}–3×10^4	0.63	—
0.315–0.400	10^{-9}–10	$0.56\times t^{0.25}$	—
0.315–0.400	10–10^3	1.0	—
0.315–0.400	10^3–3×10^4	—	1×10^{-3}
0.400–1.400	10^{-9}–10^{-7}	$2C_A\times10^{-2}$	—
0.400–1.400	10^{-7}–10	$1.1C_A\times t^{0.25}$	—
0.400–1.400	10–3×10^4	—	$0.2\times C_A$
1.400–1.500	10^{-9}–10^{-3}	0.1	—
1.400–1.500	10^{-3}–10	$0.56\times t^{0.25}$	—
1.400–1.500	10–3×10^4	—	0.1
1.500–1.800	10^{-9}–10	1.0	—
1.500–1.800	10–3×10^4	—	0.1
1.800–2.600	10^{-9}–10^{-3}	0.1	—
1.800–2.600	10^{-3}–10	$0.56\times t^{0.25}$	—
1.800–2.600	10–3×10^4	—	0.1
2.6–1000	10^{-9}–10^{-7}	1.0×10^{-2}	—
2.6–1000	10^{-7}–10	$0.56\times t^{0.25}$	—
2.6–1000	10–3×10^4	—	0.1

Notes:

1. To calculate MPE for $\lambda=0.180$–0.400 μm, use the J cm^{-2} or W cm^{-2} value or $0.56\times t^{0.25}$ (whichever is lower).
2. For correction factor C_A, see Table A.6.

Table A.6 Parameters and correction factors

Parameters/correction factors	Wavelength (μm)
$T_1 = 10 \times 10^{20(\lambda - 0.450)^a}$	0.450–0.500
$T_2 = 10 \times 10^{(\alpha - 0.1.5)/98.5^b}$	0.400–1.400
$C_B = 1.0$	0.400–0.450
$C_B = 10^{20(\lambda - 0.450)}$	0.450–0.600
$C_A = 1.0$	0.400–0.700
$C_A = 10^{2(\lambda - 0.700)}$	0.700–1.050
$C_A = 5.0$	1.050–1.400
$C_P = n^{-0.25c}$	0.180–1000
$C_E = 1.0, \alpha < \alpha_{min}$	0.400–1.400
$C_E = \alpha/\alpha_{min}, \alpha_{min} \leq \alpha \leq \alpha_{max}$	0.400–1.400
$C_E = \alpha^2/(\alpha_{min}\alpha_{max}), \alpha > \alpha_{max}$	0.400–1.400
$C_C = 1.0$	1.050–1.150
$C_C = 10^{18(\lambda - 1.150)}$	1.150–1.200
$C_C = 8$	1.200–1.400

[a] $T_1 = 10$ s for $\lambda = 0.450$ μm and $T_1 = 100$ s for $\lambda = 0.500$ μm.
[b] $T_2 = 10$ s for $\alpha < 1.5$ mrad and $T_2 = 100$ s for $\alpha > 100$ mrad.
[c] Refer to ANSI Z136.1-2000, Section 8.2.3 for discussion of C_P and Section 8.2.3.2 for discussion on pulse repetition frequencies below 55 kHz (0.4–1.050 μm) and below 20 kHz (1.050–1.400 μm).

Notes:

1. For $\lambda = 0.400$–1.400 μm, $\alpha_{min} = 1.5$ mrad and $\alpha_{max} = 100$ mrad.
2. Wavelengths must be expressed in μm and angles in mrad for calculations.
3. Wavelength region λ_1 to λ_2 means $\lambda_1 \leq \lambda < \lambda_2$.

A.5 Calculation of MPE for Eye and Skin Exposure

Tables A.4–A.6 list MPE values for eye and skin exposure for different values of wavelength and exposure times. The information is derived from ANSI Z136.1 standard.

Index

3-D measurement (ESPI), 487
ABL (Air borne laser), 124
Absorption, 4
  probability of absorption, 10
Absorption length (Materials processing), 382
Alexandrite laser, 88–9
  energy level diagram, 89
  properties, 89
All gas phase iodine laser (AGIL), 124
ALPHA, 122
Amplitude modulated hologram, 499
ANSI Z 136.1 standard, 599–603
Arc lamps, 233–7
  electrical characteristics, 234–5
  modulated CW operation, 237
  power supply, 235–7
  quasi-CW operation, 236
ArF (Argon fluoride) laser, 120
Argon-ion laser, 118–20
  construction, 118
  energy level diagram, 118–19
  Innova-90C series Argon-ion laser, 120
Average power, 43, 49

Beam profiler, 57
Beam propagation analyzer, 58
  beam master profiler, 58
  mode master PC beam propagation analyzer, 59
Beam rider guidance, 532–3
Biological Warfare (BW) agent detection, 568–70
  laser induced fluorescence detection, 569–70
Boiling. *see* Vaporization cutting
Burning, 388

Capacitor charging power supply, 216–22
Carbon dioxide gas dynamic laser, 125
Carbon dioxide laser, 111–15
  energy level diagram, 111–12
  flowing gas CO_2 laser, 113
  gain curve, 112–13
  pressure broadening, 113
  Synrad 48-series sealed-off CO_2 laser, 113
  Synrad v30 RF-excited waveguide CO2 laser, 115
  TEA CO2 laser, 114–15
  transversely excited CO_2 laser, 113
  vibrational modes, 111
Cat's eye effect, 573
Cavity dumped output, 72
Characteristic parameters (photosensors), 318–24
  detectivity, 321
  D-star, 321
  noise, 323–4
  noise equivalent power, 321
  quantum efficiency, 321–2
  response time, 322–3
  responsivity, 318–20
Chemical degradation cutting, 388
Chemical lasers, 121–4
  all gas phase iodine laser (AGIL), 124
  ABL, 124
  ALPHA, 122
  chemical oxy-iodine laser (COIL), 123–4
  deuterium fluoride laser, 121–2
  hydrogen fluoride laser, 121–2
  MIRACL, 122
  THEL, 122
Chemical oxy-iodine laser (COIL), 123–4

Characteristic parameters-semiconductor lasers, 148–52
 beam divergence, 149–50
 beam polarization, 152
 line width, 151–2
 slope efficiency, 148–9
 threshold current, 148
Chemical warfare (CW) agent detection, 567–8
 blister agents, 567
 differential absorption lidar (DIAL), 567
 nerve agents, 567
CO_2 (carbon dioxide) laser power supplies, 257–60
 DC excited, 257–9
 Marx bank, 258
 RF excited, 259–60
Coherence, 36–9
 length, 37
 spatial, 38–9
 temporal coherence, 36–8
 time, 37
Coherence length, 37
Coherence time, 37
Colour centre lasers, 90–91
Command guidance, 533–7
 automatic command to line-of-sight (ACLOS), 536
 command to line-of-sight (CLOS), 535
 command off line-of-sight (COLOS), 535
 concept, 533–4
 manual command to line-of-sight (MCLOS), 535
 semi-automatic command to line-of-sight (SACLOS), 536
 semi-manual command to line-of-sight (SMCLOS), 535
 wire guided missile, 534
Computer generated hologram, 501
Concentric resonator, 22
Confocal imaging, 438
Confocal resonator, 22
Constant current sources, 186–91
 bipolar transistor based, 187–9
 current mirror, 190–191
 FET based, 186
 opamp based, 189
 three-terminal regulator based, 189
 Widlar source, 191
 Wilson source, 191
Copper vapour laser, 116–18
 energy level diagram, 117
Current mirror, 190–191
 basic current mirror, 190
 Widlar source, 191
 Wilson source, 191
Current-to-voltage converter, 197–9
CW output, 69
CW power, 42, 49–237
CW solid state laser electronics, 233–7
 arc lamp, 233–5
 arc lamp power supply, 235–7

Dentistry, 453–5
 considerations, 453–4
 lasers for, 454–5
Dermatology, 449–53
 hair removal, 452
 lasers for, 453
 pigmented lesions and tattoos, 451–2
 portwine stains, 450–451
Deuterium fluoride laser, 121–2
Differential absorption lidar (DIAL), 567
Diffused optical spectroscopy, 434
Diffusion length (materials processing), 382
Directionality, 39–40
Directed energy laser weapons, 585–92
 ABL (airborne laser), 591
 advantages, 587
 ATL (advanced tactical laser)
 components of, 589–90
 GARDIAN (general area defence integrated anti-missile laser), 590
 HELWEPS (high energy laser weapon system), 590
 international status, 590–592
 laser phalanx, 592
 limitations, 588
 MIRACL (mid infrared advanced chemical laser), 590
 operational scenario, 589–90
 space based laser, 592
 THEL (tactical high energy laser), 592
Directionality, 39–40
Displays, 349–63
 cathode ray tube displays, 361–2
 characteristics, 350
 digital light processing technology, 363
 electronic ink displays, 363
 field emission displays, 363
 light emitting diodes, 351–6
 liquid crystal displays, 356–61
 organic light emitting diodes, 362–3
 plasma display panels, 363
 types, 350–351

Distributed feedback laser, 138–9
Divergence, 46, 50
Doppler broadening, 35
Duty cycle, 44
Dye laser, 125–7
  active medium, 126
  energy level diagram, 126
  pumping mechanism, 126–7
  wavelength selection, 127

Einstein's coefficients, 6
Elastic scattering, 424
Elastic scattering spectroscopy, 434–5
Electronic speckle pattern interferometry (ESPI), 484–90, 510
  3-D measurement, 487
  configurations, 486–90
  displacement gradient measurement, 487
  in-plane displacement measurement, 487
  out-of-plane displacement measurement, 486–7
  principle, 485–6
  vibration measurement, 490
Electro-optically guided munitions, 546–56
  construction, 547–8
  infrared guided missiles, 554–6
  laser guided munitions, 549–54
Electro-optic countermeasures (EOCM), 580–583
  active countermeasures, 581
  equipment, 581–3
  hard kill, 581
  laser warning sensor, 581–3
  passive countermeasures, 580
  relevance, 580
  soft kill, 581
Embossed hologram, 501
Energy level diagram, 12–16
Erbium-glass laser, 85–8
  energy level diagram, 87
  LRB-25000 Er-Glass laser range finder, 86–7
  properties, 87
Erbium-YAG laser, 85–6
  energy level diagram, 86
  properties, 86
Excimer lasers, 120–121
  ArF laser, 120
  energy level diagram, 120
  KrCl laser, 120
  KrF laser, 120
  XeCl laser, 120
XeF laser, 120
Excimer laser power supplies, 261–2
Excited state, 3
Explosive agent detection, 570–572
  laser induced breakdown spectroscopy (LIBS), 571
  laser induced fluorescence spectroscopy (LIF), 572
  Raman spectroscopy, 571–2
  selectivity, 571
  sensitivity, 570
External cavity semiconductor diode laser, 141–3
  Littman-Metcalf configuration, 143
  Littrow configuration, 142
External trigger. *see* Flash lamp triggering circuit

Fall time, 45
Flash lamps, 24–7
  helical, 24
  krypton, 25–6
  linear, 24
xenon, 25–6
Fibre lasers, 91–100
  basic fibre laser, 92–3
  energy level diagrams, 93
  fibre versus bulk, 91–3
  operational regimes, 95–6
  photonic crystal fibre lasers, 96–100
Fibre-optic communication, 559–60
Fibre-optic gyroscope, 545–6
FieldMAXII Top meter, 49
Filters, 164–5
  capacitor filter, 164–5
  inductor filter, 164–5
  L-C filter, 165
Flame cutting. *see* Oxidation cutting
Flash lamp triggering circuits, 231–3
  external trigger, 231–2
  over-voltage trigger, 231
  parallel trigger, 233
  series trigger, 232–3
Fluorescence spectroscopy, 438–41
Flyback converters, 174–7
  externally driven type, 175–7
  off-line, 176–7
  self-oscillating type, 174–5
Flying optics configuration. *see* Laser cutting
Forming (Materials processing), 382
Forward converter, 178
Forward looking infrared (FLIR) sensor, 528–32
  FLIR applications, 529
  FLIR versus push broom, 529
Four-level laser system, 14

Free electron laser, 127–9
Free running output, 69
Free space communication, 557–9
Frequency metrology, 472–4
Frequency stabilization of gas lasers, 262–7
dither stabilization, 264–5
optogalvanic stabilization, 265–6
stabilization on saturation absorption dip, 266–7
stark-cell stabilization, 265
Fusion cutting, 387

Gain coefficient, 16
Gain guided diode lasers, 152
Gain of laser medium, 16
amplification factor, 16
gain coefficient, 16
Gain saturation, 16
Gas discharge characteristics, 242
Gas lasers, 105–29
active media, 105–6
chemical lasers, 121–5
CO_2 lasers, 111–15
excimer lasers, 120–121
helium-neon lasers, 107–11
inter-level transitions, 106
metal vapour lasers, 115–18
properties, 106
pumping mechanisms, 106–7
rare gas ion lasers, 118–20
Gas laser electronics, 242–67
frequency stabilization, 262–7
gas discharge characteristics, 242
gas laser power supplies, 242–63
Gas laser power supplies, 242–63
CO_2 laser power supplies, 257–60
excimer laser power supplies, 261–2
helium-neon laser power supplies, 244–57
ion laser power supplies, 262–3
metal vapour laser power supplies, 260–261
Gaussian distribution, 17
Gold vapour laser, 116–18
Ground state, 3
Guidance techniques, 532–46
beam rider guidance, 532–3
command guidance, 533–7
homing guidance, 537–8
navigation guidance, 538–40
Guided munitions, 532–56
electro-optically guided munitions, 546–56
guidance techniques, 532–46
Ground state, 3

Hard kill EOCM systems, 581, 583
ARENA system, 583
TROPHY system, 583
Helium-cadmium laser, 115–16
74-serieshelium-cadmium laser, 116
energy level diagram, 116
Heterojunction semiconductor laser, 134
Helium-neon laser, 107–11
construction, 109–11
energy level diagram, 107
gain curve, 108
P-122 helium-neon laser tube, 109
Helium-neon laser power supplies, 244–57
101T-series helium-neon laser power supplies, 211
1200-series helium-neon laser power supplies, 211
ballast resistance, 257
design, 247–57
ring laser gyro power supply, 254–5
Hemifocal resonator, 22
Hemispherical resonator, 22
High voltage triggering circuits, 200–202
Homing guidance, 537–8
active homing guidance, 537
passive homing guidance, 538
retransmission homing guidance, 538
semi-active homing guidance, 537
Homojunction semiconductor laser, 136
Host material, 67–8
Hydrogen fluoride laser, 121–2

Infrared countermeasures, 584–6
ATIRCM, 585
CIRCM, 585
directional infrared countermeasures, 585
GUARDIAN DIRCM, 585
infrared flares, 584
infrared jammer, 585
missile approach warning sensor, 585
MUSIC DIRCM, 585
Index guided diode lasers, 152
Inelastic scattering, 424
Infrared guided missiles, 554–6
Infrared thermometer, 476–8
In-plane displacement gradient measurement, 487
In-plane displacement measurement, 487
Integral hologram, 501
Interferometers, 466–72
Fabry-Perot interferometer, 469–71
Mach-Zehnder interferometer, 468–9

Michelson interferometer, 467–8
Sagnac interferometer, 471
Twyman-Green interferometer, 468
Inter-satellite links (ISL), 558
Inversion threshold. *see* Lasing threshold
Ion laser power supplies, 262–3
Irradiance, 45, 318

Joining (materials processing), 382
Junction photosensors, 329–45
photodiode, 329–40
photo FET, 343
photo-SCR, 343
phototransistors, 340–343
photo-triac, 343

Key-3 laser dentistry system, 454
KrCl laser, 120
KrF laser, 120
Krypton-ion laser, 118–20

Ladar sensor, 526–28
3-D laser imaging, 527–28
coherent radar, 526
Laser aiming, 512–13
Laser angioplasty, 456
Laser astronomy, 494–6
adaptive optics, 494
laser guide star, 494–6
Laser based ordnance disposal, 574–5
Avenger system, 575
ZEUS-HLONS system, 574–5
Laser bathymetry, 509, 524–6
airborne laser bathymetry, 524–6
principle, 524
SHOALS system, 526
Laser beam profiler, 57–8
beam view analyzer, 57
Laser characteristics, 34–40
coherence, 36–9
directionality, 39–40
monochromaticity, 34–6
Laser communication, 556–60
advantages, 556
fibre-optic communication, 559–60
free space communication, 557–9
inter-satellite links (ISL), 558
limitations, 556
Laser cutting, 381, 385–90
basic principle, 385–6
machine configurations, 388–90
processes, 387–8
versus plasma cutting, 387
Laser cutting processes, 387–8
basic principle, 385–6
boiling, 388
burning, 388
chemical degradation, 388
fusion cutting, 387
laser cutting versus plasma cutting, 387
machine configurations, 388–90
melt shearing, 387
oxidation cutting, 388
scribing, 388
thermal stress cracking, 388
vaporization cutting, 387
Laser dazzler, 578–9
GLARE MOUT laser dazzler, 578
GUARDIAN laser dazzler, 579
JD-3 laser dazzler, 579
KLEG laser dazzler, 576
Sabre-203 laser dazzler, 578
TALI laser dazzler, 579
ZM-87 laser dazzler, 579
Laser decoy system, 583
DHY-322 system, 583
type-405 GEC Marconi system, 583
Laser DEW, 586
Laser diagnostics, 430–442
basic principle, 431
comparison with other techniques, 431–2
confocal imaging, 438
diffuse optical spectroscopy, 434
elastic scattering spectroscopy, 434–5
fluorescence spectroscopy, 438–41
in-vitro diagnostics, 430
in-vivo diagnostics, 430, 432–42
optical coherence tomography, 435–7
Raman spectroscopy, 441–2
white light imaging, 432–3
Laser diode driver circuits, 278–91
automatic power control, 286–9
constant current source, 278–9
feedback control, 279–81
LDX-3200 series laser diode driver, 210
modulation input, 282
protection features, 284–6
quasi-CW laser diode driver, 289–91
temperature control, 291–307
Laser diode electronics, 271–307
driver circuits, 278–91
laser diode protection, 271–6
operational modes, 276–8
temperature control, 291–307

Laser diode operational modes, 276–8
  constant current mode, 276
  constant power mode, 277–8
Laser diode protection, 271–6
  current limit, 272
  drive and control, 272–4
  electrostatic discharge, 275
  interconnection and cabling, 274–5
  life time versus operating temperature, 273
  over voltage protection, 273
  power line transients, 273
  slow start, 272
  transient suppression, 275
Laser diode temperature control, 291–307
  drive and control circuits, 301–307
  LDT-5900 series laser diode temperature controller, 210
  thermoelectric cooling, 292–301
Laser Doppler velocimetry, 478–80
  advantages, 480
  applications, 480
  principle, 478–80
Laser drilling, 393–6
  advantages, 396
  basic principle, 393
  drilling lasers, 395–6
  processes, 394–5
Laser drilling processes, 394–5
  drilling on-the-fly, 395
  helical trepanning, 395
  imaged drilling, 395
  parallel percussion drilling, 395
  percussion drilling, 394
  single pulse drilling, 394
  trepanning, 395
Laser electronics, 208–13
  gas lasers, 211–12 and 242–67
  semiconductor diode lasers, 209–11, 271–308
  sensors, 213
  solid state lasers, 208, 213–38
  test and evaluation, 212–13
Laser engraving. *see* Laser marking
Laser gap measuring device, 509
Laser guided munitions, 549–54
  concept, 549–53
  field-of-view, 554
  GBU 24 Paveway-III laser guided munitions kit, 213
  parameters, 553–4
  PRF code compatibility, 554
  response linearity, 554
  sensitivity, 553
Laser guide star, 494–6
  Rayleigh guide star, 496
  Sodium beacon guide star, 496
Laser heat treatment processing, 385
Laser marking, 396–400
  advantages, 399
  basic principle, 396
  marking lasers, 398–9
  processes, 397–8
Laser marking processes, 397–8
  laser ablation, 398
  laser bonding, 397
  laser coating and marking, 398
  laser engraving, 397
  laser etching, 398
Laser micromachining, 401–7
  micromachining lasers, 403
  operations, 402
  techniques, 404–7
Laser micromachining operations, 402
  laser micro cutting, 402
  laser micro engraving, 402
  laser micro milling, 402
  laser micro scribing, 402
Laser micromachining techniques, 404–7
  direct writing, 404
  mask projection, 405
Laser obstacle avoidance system, 509
Laser parameters, 42–8
  average power, 43, 49
  CW power, 42, 49
  divergence, 46
  duty cycle, 44
  fall time, 45
  irradiance, 45
  M^2-value, 48
  peak power, 42, 49
  pulse energy, 43, 49
  pulse width, 44
  radiance, 45
  repetition rate, 43, 49
  rise time, 45
  spot size, 47
  wall plug efficiency, 48
  wavelength, 41
Laser pointing, 509–10
Laser printing, 414–17. *see also* Laser aiming
  anatomy, 415–16
  choice criteria, 416–17
  processes, 415
  versus ink jet printers, 417
Laser proximity sensor, 509, 520–523

axially symmetric configuration, 520
principle, 520
Laser radar sensor. *see* Ladar sensor
Laser range finder, 509, 513–20
applications, 518–19
FM-CW range finder, 519–20
lasers for range finding, 518–19
phase shift technique, 516
time-of-flight range finder, 514
triangulation technique, 515
Laser resonator, 17
concentric resonator, 22
confocal resonator, 22
hemifocal resonator, 22
hemispherical resonator, 22
plane parallel resonator, 21
stable resonator, 21
unstable resonator, 21
Laser safety, 599–603
ANSI Z136.1 standard, 599–603
laser damage, 599
maximum permissible exposure (MPE), 599
nominal hazard zone, 601
safety classification, 602
Laser spectrum analyzer, 56
Laser target designator, 509
Laser therapeutics, 442–3
Laser tracking, 509
Laser velocimetry, 478–82
laser Doppler velocimetry, 478–80
particle image velocimetry, 480–482
Laser vibrometer types, 484
3-D vibrometer, 484
continuous scan vibrometer, 484
differential vibrometer, 484
multi beam vibrometer, 484
rotational vibrometer, 484
self mixing vibrometer, 484
single point vibrometer, 484
Laser vibrometry, 482–4, 510
applications, 484
principle, 482–4
types, 484
Laser welding, 390–392
advantages, 392
processes, 390
welding lasers, 390–391
Laser welding processes, 390
deep penetration welding, 390
heat conduction welding, 390
Lasing threshold, 11
Lead salt laser, 147
Length metrology, 471–2
frequency modulation method, 472
interferometry method, 472
phase shift method, 472
time-of-flight method, 471
triangulation method, 472
Lidar sensors, 510
Light emitting diodes, 351–6
characteristic curves, 352–3
drive circuits, 354–5
parameters, 354
Light-tissue interaction, 422–30
diagnostic applications, 423–6
therapeutic applications, 427–30
Light-tissue interaction for diagnostics, 423–6
fluence rate distribution, 426
interaction mechanisms, 423–5
optical properties of tissues, 425–6
Light-tissue interaction for therapeutics, 427–30
photochemical effects, 427–8
photomechanical effects, 429–30
photothermal effects, 428–9
Linear power supplies, 161–73
constituents of, 161–2
filters, 164–6
linear regulators, 166–73
linear versus switched mode supplies, 173–4
rectifier circuits, 162–3
Linear regulators, 166–73
emitter-follower regulator, 166–7
linear IC regulators, 171–3
overload protection, 169
series-pass regulator, 167–70
shunt regulator, 170–171
three-terminal IC voltage regulators, 172–3
Line broadening, 34–5
Doppler broadening, 35
collisional broadening, 35
Line width measurement, 474–6
frequency comb, 55
recirculating fibre loop, 476
self-heterodyne technique, 475
Liquid crystal displays, 356–61
construction, 356–7
drive circuits, 357–8
response time, 358
types, 358–61
Longitudinal modes, 18–19
Low intensity conflict (LIC), 566
Low intensity conflict (LIC) applications, 566–80
BW agent detection, 568–70
CW agent detection, 567–8

Low intensity conflict (LIC) (*Continued*)
explosive agent detection, 570–572
laser dazzling applications, 576–80
optical target detection, 573–4
ordnance disposal, 574–5
Luminance, 318
LV-S01 laser vibrometer, 483

Machining (materials processing), 382
Materials processing applications, 381–414
cutting, 381, 385–90
drilling, 393–6
engraving, 396–400
marking, 396–400
micromachining, 401–7
photolithography, 407–11
rapid manufacturing, 411–14
welding, 390–392
Maximum permissible exposure (MPE), 599
Melt shearing cutting, 387
Metal vapour laser, 115–18
copper vapour laser, 116–18
gold vapour laser, 116–18
helium-cadmium laser, 115–16
Metal vapour laser power supplies, 261–2
Metastable state, 11
Michelson interferometer, 37
Microwave DEW, 586
Mie scattering, 424
MIRACL, 122
MOBLAS-7 SLR station, 493
Mode locked output, 72–5
Mode master beam propagation analyzer, 53
Monochromaticity, 34
Multi channel hologram, 501

Navigation guidance, 538–40
celestial navigation guidance, 539–40
digital scene matching, 540
geophysical navigation guidance, 540
inertial measurement unit, 539
inertial navigation guidance, 539
terrain contour matching, 540
Nd-glass laser, 84–5
energy level diagram, 84
properties, 85
Nd-GSGG laser, 82–3
Nd-YAG laser, 78–9
AN/PEQ-17 laser target designator, 79
energy level diagram, 78
LCY-series CW Nd-YAG laser, 79
LDY-series Q-switched Nd-YAG laser, 79
properties, 79
Nd-YLF laser, 79
energy level diagram, 81
LDY 300-series Nd-YLF laser, 81
properties, 82
Nd-YVO4 laser, 82
properties, 83
PULSELAS-P-1064–150-E Nd-YVO4 laser, 82
Nominal hazard zone, 601

Ophthalmology, 443–9
cataract surgery, 448–9
glaucoma treatment, 447–8
LASEK, 445
LASIK, 445
photocoagulation, 443–4
proliferative diabetic retinopathy, 449
refractive surgery, 444–7
retinopexy, 449
OPO shifted Nd-YAG laser range finder, 87–8
G-TOR laser range finder, 88
Optically allowed transition, 4
absorption, 4
spontaneous emission, 4
stimulated emission, 4
Optically pumped semiconductor lasers, 143–4
Optical metrology, 466–78
infrared thermometer, 476–8
interferometers, 466–71
length metrology, 471–2
line width measurement, 474–6
temperature measurement, 476–8
time and frequency metrology, 472–4
Optical coherence tomography, 435–8
Optical target detection, 573–4
cat's eye effect, 573
Optifinder-1200 sniper detector, 573
SLD-100 sniper detector, 573
Optifinder-1200 sniper detector, 573
Optocouplers, 363–70
application circuits, 366–70
characteristic parameters, 364
OS 523E/524E series infrared thermometer, 477
Out-of-plane displacement measurement, 486
Over voltage trigger. *see* Flash lamp triggering circuits
Oxidation cutting, 388

Parallel triggering. *see* Flash lamp triggering circuits
Particle image velocimetry, 480–482

Peak inverse voltage (Rectifiers), 162
Peak detector, 199–200
Peak power, 42, 49
Phase modulated hologram, 499
Photo chemical effects, 427–8
Photoconductors, 324–9
  application circuits, 326–9
  fundamentals, 324–6
Photodiode
  application circuits, 334–6
  avalanche photodiode, 331
  equivalent circuit, 331–2
  I-V characteristics, 333–4
  PIN photodiode, 331
  PN photodiode, 330
  Schottky photodiode, 331
  solar cell, 336–9
  types, 330
Photodynamic therapy, 456–9
  applications, 458–9
  considerations, 458
Photoelectric sensors, 315–47
  junction type photosensors, 329–45
  photoconductors, 324–9
  photoemissive sensors, 345–7
Photoemissive sensors, 345–7
  image intensifiers, 346–7
  micro channel plate (MCP), 346
  photomultiplier tubes, 345
  vacuum photodiodes, 345
Photolithography, 407–11
  photolithography lasers, 411
  processes, 408–11
Photolithography processes, 408–11
  barrier layer formation, 408
  hard baking, 410
  mask alignment, 408
  photoresist application, 408
  soft baking, 408
  UV exposure, 408
  wafer cleaning, 408
Photo mechanical effects, 429–30
Photometry, 316–18
  illuminance, 318
  luminance, 318
  photometric flux, 316
  photometric intensity, 316
Photonic crystal fibre lasers, 96–100
  applications, 98–100
  guiding mechanism, 97
  sub-classes, 97–8
Photosensors, 315–49
  characteristic parameters, 318–24
  photoelectric sensors, 315–47
  thermal sensors, 347–9
Photo thermal effects, 428
Pigmented lesions, 451
Population inversion, 3, 10–11
Portwine stains, 450–451
Pseudo simmer circuit, 224–5
Pulse energy, 43, 49
Pulse width, 44
Pulsed solid state laser electronics, 214–33
  capacitor charging power supply, 214
  flash lamp triggering circuit, 216
  pseudo simmer, 216
  pulse forming network, 216
  Q-switch driver
  simmer circuit, 216
Pulse forming network, 225–30
  critically damped condition, 226
  design, 228–30
  multiple mesh pulse forming network, 229
  over damped condition, 226
  under damped condition, 226
Pumping mechanisms, 23–9
  chemical reaction pumping, 29
  electrical pumping, 28–9
  electron beam pumping, 29
  optical pumping, 24–8
Push-pull converters, 178–81
  externally driven, 179
  full- bridge type, 181
  half-bridge type, 181
  self-oscillating type, 178–80

Q-300 3-D ESPI camera, 489
Q-switched output, 69–71
Quantum cascade laser, 145–7
Quantum dot laser, 138
Quantum mechanics (*Laser basics*), 3–4
Quantum well diode laser, 136–8
Quantum wire laser, 138

Radiance, 45
Radiometry, 316–18
  radiance, 318
  radiant incidence, 318
  radiant sterance, 318
  radiometric flux, 316
  radiometric intensity, 316
Raman spectroscopy, 441–2
Range finder and designator electronics, 237–8

Rapid manufacturing, 411–14
  advantages, 414
  lasers for, 413
  technologies, 412–13
  types, 412
Rapid manufacturing technologies, 412–13
  3-D printing, 413
  laminated object manufacturing, 413
  selective laser sintering, 413
  shape deposition manufacturing, 413
  stereo lithography, 413
Rapid manufacturing types, 412
  additive manufacturing, 412
  subtractive manufacturing, 412
Rare gas ion lasers, 118–20
  argon-ion laser, 118–20
  krypton-ion laser, 118–20
Ratio of rectification (Rectifiers), 162
Rayleigh guide star, 496
Rayleigh scattering, 424
Reactive cutting. *see* Oxidation cutting
Rectifier circuits, 162–3
  bridge rectifier, 162–3
  full wave rectifier, 162–3
  half wave rectifier, 162–3
specifications, 162
Reflection hologram, 500
Repetition rate, 43, 49
Ring laser gyroscope (RLG), 540–544
  angular random walk, 543
  bias drift, 543
  bias instability, 543
  lock-in effect, 541
  Sagnac effect, 540
  scale factor stability, 543
  scale factor, 543
Ripple factor (Rectifiers), 162
Ripple frequency (Rectifiers), 162
Rise time, 45
Ruby laser, 76–7
  energy level diagram, 76
  properties, 77
  sinon Q-switched laser, 77

Safety classification, 602
Satellite laser ranging, 490–494
  applications, 493–4
  lasers for, 492
  principle, 491–2
  stations, 492
  telescopes, 492
Scribing, 388
Semiconductor laser materials, 135
  direct band gap semiconductors, 135
  indirect band gap semiconductors, 135
  quaternary compounds, 135
  ternary compounds, 135
Semiconductor lasers, 132–53
  applications, 153
  characteristic parameters, 148–52
  emission bands, 136
  gain guided diode lasers, 152
  handling precautions, 152
  index guided diode lasers, 152
  I-V characteristics, 134
  operational basics, 132–5
  semiconductor laser materials, 135–6
  types, 136–47
Series trigger. *see* Flash lamp triggering circuits
Shane telescope, 496
Simmer circuit, 222–4
SLD-100 laser sniper detector, 5736
Sodium beacon guide star, 49
Soft kill EOCM systems, 581, 583
  SHTORA-1 system, 583
  STINGRAY system, 583
Solid state laser, 67–100
  Alexandrite laser, 88–9
  colour center laser, 90–91
  electronics, 208–38
  erbium-glass laser, 85–8
  erbium-YAG laser, 85
  host material, 67–8
  lasing species, 68
  Nd-glass laser, 84–5
  Nd-GSGG laser, 82–3
  Nd-YAG laser, 79
  Nd-YLF laser, 79
  Nd-YVO4 laser, 82
  operational modes, 68–75
  ruby laser, 76–7
  titanium-Sapphire laser, 90
  vibronic laser, 88–90
Solid state laser electronics, 208–38
  CW solid state lasers, 233–7
  pulsed solid state lasers, 214–33
  range finders and designators, 237–8
Spatial coherence, 38–9
Spectrum analyzer, 56
Spontaneous emission, 4
  probability of spontaneous emission, 10
Spot size, 47, 50

Stable resonator, 21
Stimulated emission, 4
probability of stimulated emission, 10
Surface engineering (*Materials processing*), 382
Switched mode power supplies, 173–86
connection in parallel, 184–6
connection in series, 184
flyback converters, 174–7
forward converters, 178
linear versus switched mode power supplies, 173–4
push-pull converters, 178–81
switching regulators, 181–4
Switching regulators, 181–4
boost regulator, 182–3
buck-boost regulator, 183
buck regulator, 182–3
three-terminal switching regulator, 183–4

Tattoos, 451
Temperature measurement, 476–8
Temporal coherence, 36–8
THEL, 122
Thermal sensors, 347–9
bolometer, 348
pyroelectric sensors, 348–9
thermocouple sensors, 347–8
thermopile sensors, 347–8
Thermal stress cracking, 388
Thermal therapy, 459–60
cancer treatment, 460
haemorrhages of peptic ulcers, 459–60
Thermoelectric cooler drive and control circuits, 301–7
error amplifier, 303
error signal processor, 303–6
output stage, 306–7
temperature sensing, 301–3
Thermoelectric cooling
drive and control circuits, 301–7
fundamentals, 292–4
heat sink selection, 299
TE cooler characteristics, 295–7
TE cooler selection, 297–9
Thick hologram, 500–501
Thin hologram, 500–501
Three-level laser system, 12
Time and frequency metrology, 472–4
optical clock, 474
optical frequency comb, 473
Timer circuits, 191–7
astable multivibrator, 194–5
digital IC based, 191–3
linear IC based, 193–7
monostable multivibrator, 195–7
retriggerable monostable multivibrator, 193
Titanium Sapphire lasers, 90
energy level diagram, 90
MBR-110 tunable CW Ti-Sapphire laser, 90
properties, 90
Transmission hologram, 499–500
Transverse modes, 19–20
Two-level laser system, 11

Unstable resonator, 21

Vaporization cutting, 388
Vascular surgery, 455–6
laser angioplasty, 456
Vertical cavity surface emitting laser (VCSEL), 140
Vertical external cavity surface emitting laser (VECSEL), 140–141
Vibronic lasers, 88–90
Alexandrite laser, 88–9
titanium sapphire laser, 89–90

Wall plug efficiency, 48
Waterlase laser dentistry system, 454
Wavelength meter, 56
WaveMaster wave length meter, 56
White light imaging, 432–3
William Herschel telescope, 496

XeCl laser, 120
XeF laser, 120
X-ray laser, 129

Very large telescope (European Space Observatory), 496

Young's double slit experiment, 38

Thick lens, 290
Thin lenses, 290